The Biosph

DATE DUE

PROTECTING OUR GLOBAL ENVIRONMENT

4th edition

Donald G. Kaufman
Miami University

Cecilia M. Franz
Miami University

 **KENDALL/HUNT PUBLISHING COMPANY**
4050 Westmark Drive   Dubuque, Iowa 52002

The Biosphere
Protecting Our Global Environment
4th edition

Donald G. Kaufman
Miami University

Cecilia M. Franz
Miami University

Kendall/Hunt Publishing Company

COVER PHOTO: Water droplet on jewelweed

DESIGN AND LAYOUT: Carole Katz
COVER PHOTO: Stephen G.Maka
PRINTER AND BINDER: Kendall/Hunt Publishing Company

Printed with non-toxic soy-based inks on Utopia 2 recycled paper, containing 10% post-consumer waste.

Biosphere 2000: Protecting Our Global Environment, third edition

Copyright © 2000, 2005 by Donald G. Kaufman and Cecilia M. Franz
ISBN 0-7575-1908-3

Library of Congress Catalog Card Number: 2005931170

Printed in the United States of America.
10 9 8 7 6 5 4 3 2 1

"A society whose youth believe only in the Now is deceiving itself. It denies man's basic and oldest characteristic, that he is a creation of memory, a bridge into the future, a time binder."

Loren Eiseley

Dedication

As with previous editions, the fourth edition of *The Biosphere* is dedicated to all of the students who helped to create it and who believed in it — and in the power of youth and love — to make a difference in the world. It is especially dedicated to the memory of Laura Luedeke, who was among the group of students who originally conceived of the book. Long before the book was completed, indeed, while it was still taking shape in our minds, she envisioned it, believed in it, and strove to realize it. Our fondest wish is that her belief and dedication — like the belief and dedication of all those very special students — may inspire others to action.

This labor of love continues....

D.G.K. and C.M.F.

About the Authors

Donald G. Kaufman, Professor Emeritus of Zoology at Miami University, is the Director of the university's Center for Environmental Education and Natural History and Director of the center's Hefner Zoology Museum. During his long tenure at Miami, Don earned several of the university's most prestigious honors, including the Outstanding Teacher of the Year, the E. Phillips Knox Teaching Award, and the Distinguished Educator, College of Arts and Science.

Cecilia Franz is the Assistant Director of Miami University's Center for Environmental Education and Natural History and the Senior Project Director of the center's Hefner Zoology Museum. Cecilia is responsible for numerous initiatives, notably the development of the Imaginarium, the center's facility for children ages 3-8, which opened in June 2005.

The authors' efforts in the field of environmental education are not limited to *The Biosphere*. Each summer, they offer introductory-, advanced- and master-level workshops for elementary teachers, a three-tiered approach to environmental education they call "The GREEN Teachers Institute." In these workshops, environmental themes are used to teach science concepts; teachers and students alike respond enthusiastically to topics concerning nature and the environment. Over the past 15 years, they have received grants totaling well over $2 million to support the GREEN Teachers Institute from diverse funding sources, including the National Science Foundation, the U.S. Department of Energy, the U.S. Environmental Protection Agency, the Ohio Environmental Protection Agency, the Procter & Gamble Company, and Cinergy Corporation.

The authors' commitment to environmental education and conservation are reflected in their personal lives as well as in their professional work. Don nurtures his life-long interest in the natural world at the 260-acre farm that he and wife Patricia own. They manage the property as a nature reserve, with the land protected in perpetuity by conservation easements. In 2002, Don and Cecilia founded The Oxford Society, a non-profit entity dedicated to environmental education and conservation in both the United States and Costa Rica. The Oxford Society has a Costa Rican sister organization, known as Tasbayam, a native word meaning "the land that saw you born." The Oxford Society and Tasbayam cooperate on joint projects designed to protect and preserve Costa Rica's natural and cultural diversity. Don is president of The Oxford Society, and Cecilia is a board member. Don is also a board member of Tasbayam.

To learn more about the authors' work, please visit their web site, www.EnvironmentalEducationOhio.org. Don and Cecilia can be reached in c/o the Center for Environmental Education and Natural History, 104 Upham Hall, Miami University, Oxford, Ohio 45056 or by email. Don's email address is kaufmadg@muohio.edu; Cecilia's is bergcf@muohio.edu

Contents

Unit II
An Environmental Foundation: Ecological Principles and Applications   47

CHAPTER 3
Ecosystem Structure   48

CHAPTER 4
Ecosystem Function   70

CHAPTER 5
Ecosystem Development and Dynamic Equilibrium   87

CHAPTER 6
Ecosystem Degradation   107

Unit III
An Environmental Imperative: Balancing Population, Food, and Energy   123

CHAPTER 11

Energy: Alternative Sources   220

**Unit IV
An Environmental
Necessity:
Protecting Biospheric
Components   237**

CHAPTER 12

Air Resources   238

Unit V
An Environmental Pandora's Box: Managing the Materials and Products of Human Societies   401

CHAPTER 18
Mineral Resources   402

Living On Round River

"One of the marvels of early Wisconsin," wrote the ecologist Aldo Leopold, "was the Round River, a river that flowed into itself, and thus sped around and around in a never-ending circuit. Paul Bunyan discovered it, and the Bunyan saga tells how he floated many a log down its restless waters.

"No one has suspected Paul of speaking in parables, yet in this instance he did. Wisconsin not only had a round river, Wisconsin is one. The current is the stream of energy which flows out of the soil into plants, thence into animals, thence back into the soil in a never-ending circuit of life."

Implicit in Leopold's writing is the idea that the entire Earth is also a round river. And we — the humans who live on the river — are bound up with it, part and parcel of the environment that we affect and are affected by.

Themes

This book is an environmental primer, a foray into the workings and the wonders of the Earth and the problems that beset it. It is written for use in introductory environmental science college courses or in general ecology courses with an environmental emphasis, but it contains a message and information for all individuals and communities. Throughout the text, we emphasize three themes.

First, despite the scope and gravity of environmental problems, we find reason to hope. *The Biosphere* is not a death knell for a doomed planet, nor is it an apologia for past failures. Inarguably, humanity faces many complex environmental problems, but focusing on the difficulty of resolving problems can only paralyze us. Instead, we take a holistic approach: We examine the many environmental problems that beset us, the dimensions of those problems, their varied and interrelated causes, and — importantly — the efforts underway to tackle those problems. You'll notice that our text contains no unit or section on pollution per se. Rather, Chapter 6, *Ecosystem Degradation*, presents a general discussion of the topic, including the types and sources of pollution and their effects and associated problems. Specific information on pollution is then integrated into appropriate chapters, especially those on air, water, and soil. Moreover, while we acknowledge the difficulty of the present situation, we clearly demonstrate that there are solutions to environmental problems — what's lacking, often, is the will to implement those solutions. We also show readers how they can make a difference. Throughout the text, we offer lists of practical things — called "What You Can Do" — for individuals who want to help protect and preserve resources and natural systems. These suggestions enable students, teachers, classes, and communities to become active participants in the most important endeavor of the twenty-first century: the effort to preserve our global environment.

The second major emphasis of *The Biosphere* is a focus on human attitudes and beliefs and their consequent impact on our environment. Beginning with the Prologue and continuing throughout the text, we consistently challenge readers to look at the human and social aspects of environmental issues. The message is clear: To solve environmental problems, we have to address the attitudes, values, and beliefs — and work within the human and social systems — that give rise to them.

The old adage, "Think globally, act locally," underscores the third major emphasis of *The Biosphere*. Through hundreds of examples and photos from around the world, we encourage students to look at the global nature of environmental problems. At the same time, we maintain a strong emphasis on North America, providing statistics for Canada, Mexico, and the United States for major issues such as population growth and energy. Our goal is to help students better understand the environmental impact of their own societies and countries in the context of the larger global picture.

How This Book Began

In Fall 1983, we taught an environmental science class for honors students at Miami University, Oxford, Ohio. At the end of the semester, the students agreed to do additional research on natural systems and resources with the long-term goal of incorporating their research into a textbook for college students taking their first, and possibly only, environmental science course. Many texts are targeted for these students, but none had been developed with the assistance of undergraduate nonmajors. One of the first things we did was send a questionnaire to university professors nationwide who teach environmental courses. We asked them to suggest resources or natural systems that best illustrate environmental principles. Each student selected a topic from these suggestions, developed a comprehensive research plan, and, for the next three years, pursued that plan in extensive library and field investigations. For example, one student traveled to Boston to meet with a leading authority on whales; another ventured to the Florida Everglades.

These students, and others who later joined the project, focused on the environmental problems associated with each resource; they summarized and evaluated various management strategies and made suggestions for future management. After years of effort, they completed extensive cases studies, which form the basis for *Environmental Science in Action* (an online supplement that can be accessed by going to www.EnvironmentalEducationOhio.org and clicking on "Biosphere Project"). These case studies supplement many of the chapters.

It's impressive enough to realize that these students — 18 and 19 years old when the project began — took on this work simply because they thought it important and worthwhile. What's even more impressive is that they took the initiative to seek and secure the financial support needed to fund their research, including a grant from the George Gund Foundation of Cleveland, Ohio, and grants from Miami University's Fund for Excellence.

Our project had its detractors. A fellow professor maintained that undergraduates were incapable of the intense research and the commitment necessary to accomplish such an ambitious goal. Clearly, the students proved him wrong; more importantly, they proved to themselves that individuals working toward a common goal can achieve it, despite the difficulty. In honor of the students whose work and enthusiasm gave birth to this text, we established the Global Heritage Endowment (GHE), which supports the efforts of Miami University's Center for Environmental Education and Natural History. We donated all of our royalties from the first edition to the GHE, which is managed by the Miami University Foundation. At least 10 percent of our royalties from all subsequent editions also are earmarked for the endowment.

Organization

The text's Prologue asks readers to examine their worldview in order to understand the values, beliefs, and attitudes that shape our actions toward the environment, how we use resources, and how we respond to environmental problems.

The body of the text is divided into five major units. Unit I: *The Biosphere and Environmental Science*, consists of two chapters; Chapter 1, *The State of the Biosphere*, looks at the development of the biosphere and presents an overview of environmental problems; Chapter 2, *Social Factors*, looks at the human and social systems — religion and ethics, politics and government, economics, and environmental education — that shape our impact on the biosphere.

Unit II: *Ecological Principles and Applications*, looks more closely at the biosphere. It includes three comprehensive chapters devoted to ecological concepts: Chapter 3, *Ecosystem Structure*; Chapter 4, *Ecosystem Function*; and Chapter 5, *Ecosystem Development and Dynamic Equilibrium*. Chapter 6, *Ecosystem Degradation*, examines the negative impact of human activities on natural systems.

Units III through V are devoted to a study of resources that have common qualities or are linked in a significant way. Unit III: *An Environmental Imperative: Balancing Population, Food, and Energy* presents an in-depth look at these three critical topics. Chapter 7, *Human Population Dynamics* and Chapter 8, *Managing Human Population Growth* discuss some of the most personal and controversial topics in the field of environmental science; one of the most complex topics, and surely the most heart-wrenching, is the focus of Chapter 9, *Food Resources, Hunger, and Poverty*. The unit closes with two chapters that detail the many issues related to energy consumption and production — Chapter 10, *Energy: Fossil Fuels* and Chapter 11, *Energy: Alternative Sources*.

Unit IV: *An Environmental Necessity: Protecting Biospheric Components* includes six chapters that examine specific parts of the biosphere; they are Chapter 12, *Air Resources*; Chapter 13, *Water Resources*; Chapter 14, *Soil Resources*; Chapter 15, *Biological Resources*; Chapter

16, *Public Lands*; and Chapter 17, *Cultural Resources*. Clean air, clean water, and fertile soil help to nourish our bodies; other living things, wild lands, and cultural artifacts nourish our spirits.

The resources and products critical to industrial societies are the focus of Unit V: *An Environmental Pandora's Box: Managing the Materials and Products of Human Societies*. The unit includes Chapter 18, *Mineral Resources*; Chapter 19, *Nuclear Resources*; and Chapter 20, *Unrealized Resources: Solid and Hazardous Wastes*.

Special Features of this Fourth Edition

While we retained the pedagogical features of previous editions (please see pages xviii-xxi), the fourth edition of *The Biosphere* is significantly different from the first three. The first thing you'll notice is the name change; we dropped "2000," a deletion that puts the focus on the subtitle: *Protecting Our Global Environment*. That subtitle is key to our philosophy, for we believe that environmental science is an active, evolving, and goal-oriented discipline.

The second major change is that the text now has 20 chapters rather than 28. We produced a shorter edition primarily by combining related topics — public lands and wilderness are now discussed in one chapter, as are solid wastes and toxic and hazardous substances — so that the information they share in common is given just once.

The third major change is that we combined the discussion of religion and ethics, politics and government, economics and environmental education, and we moved it to the front of the text, as Chapter 2, *Social Factors*. Putting this discussion much earlier in the text gives it greater emphasis and underscores our belief that to solve environmental problems, we must first examine our attitudes, values, and beliefs.

As in previous editions, *The Biosphere* asks readers to examine their worldview *before* they begin to study ecological principles and environmental issues. It includes several chapters that are unique to environmental science texts, including Chapter 17, *Cultural Resources* and Chapter 20, *Unrealized Resources: Solid and Hazardous Wastes*. You'll find an expanded discussion of certain critical topics, notably the disparity in resource consumption patterns between rich and poor nations (see Chapter 1, *The State of the Biosphere*); the relationship between population growth and women's status (see Chapter 7, *Human Population Dynamics*); population policies in both fast-growth and slow- or negative-growth countries (see Chapter 8, *Managing Human Population Growth*); international efforts to eradicate extreme poverty and hunger (see Chapter 9, *Food Resources, Hunger, and Poverty*); U.S. reliance on imported oil and its implications for national security (Chapter 10, *Energy: Fossil Fuels*); and air pollution in less-developed countries (see Chapter 12, *Air Resources*).

We've added two new *Focus On* essays to the fourth edition: *A Culture of Consumption* (see Chapter 1, *The State of the Biosphere*) and *Kazi Ya Mwanamke, The Work of a Woman* (see Chapter 7, *Human Population Dynamics*). Found throughout the text, *Focus On* essays give readers a closer look at a specific resource, illustrate an ecological principle, or delve more deeply into a particular environmental issue; many are written by guests, affording readers different points of views.

Supplementary Materials

Supplementary materials accompany *The Biosphere*:

- For the interested reader, *Delving Deeper* is an online component that can be found at our web site; go to www.EnvironmentalEducationOhio.org and click on "Biosphere Project." *Delving Deeper* includes in-depth discussions of specific topics, essays, graphics, and photos. We also include links to other helpful web sites and organizations. Periodically, we'll post relevant statistics to update tables and charts included in the textbook. For example, we will provide updated population statistics annually.

- *Environmental Science in Action*, a second online resource, is an extensive series of case studies. Each is a comprehensive study of an ecosystem or issue.

- *Teacher's Resource Guide* and *Test Bank* by Lisa Breidenstein, Donald Kaufman, Aaron Inouye, Bobbie Oh, and Lisa Rosenberger. For each chapter, we provide an overview, lecture outline, student objectives, teaching tips, and a list of resources for further study. An electronic test bank is available for instructors.

- *Student's Resource Guide*, authored by Stefanie Brown, Donald Kaufman, Aaron Inouye, and Tony Nardini. For each chapter, we provide an overview and outline, learning objectives, key terms, suggested activities, and review questions.

- Electronic Images: A complete package of electronic, full-color images is available for adopters. These high-resolution scans can be used electronically or printed for hand-outs or transparencies.

Features of the Text

CHAPTER 7

Human Population Dynamics

Learning Objectives

When you finish reading this chapter, you will be able to:

1. Explain the basic arguments of each side in the Great Population Debate.

2. Define demography and discuss why it is important.

3. Explain the three ways in which populations are measured.

4. Identify and discuss the factors that affect growth rates.

5. Discuss the relationship between population growth and the environment.

6. Discuss the relationship between population growth and quality of life.

7. Discuss the relationship between population growth and the status of women.

We are not faced with a single global population problem but, rather, with about 180 separate national population problems. All population controls must be applied locally; local governments are the agents best prepared to choose local means. Means must fit local traditions. For one nation to attempt to impose its ethical principles on another is to violate national sovereignty and endanger international peace. The only legitimate demand that nations can make on one another is this: "Don't try to solve your population problem by exporting your excess people to us."

Garrett Hardin

Why…have so few ecologists seriously applied their knowledge, skills, and talents to the question of the limits to human population?…The phenomena of human population growth and its impacts are all too apparent; is the ecological community willing to ignore the most pressing social and scientific issue of all time?

H. Ronald Pulliam and Nick M. Haddad

The topic of human population growth usually prompts a heated debate. Many people are alarmed about rapid population growth in critical regions of the world, particularly the tropics, and the sheer numbers in the world's most highly populated countries, China and India. Others point out that people in stable or slow-growing developed countries typically consume far more resources and generate far more pollution than those [in develop]oped countries; they maintain that many environmental p[roblems would] be nonexistent if resources were more equitably distribut[ed.] And the debate that always ensues in these discussions is th[e same one] that rages among ecologists, economists, philosophers, a[nd others] — a debate that has been raging for centuries. (See Box [7-1, The Great] Population Debate.) In simple terms, the topic of the debate [is "Is] population growth a problem?" Chapters 7 and 8 are de[signed to help] you decide where you stand in the Great Population Deba[te. In Chapter] 7, we describe the human resource in detail; in Chapter 8, [we describe how] populations have been — and are being — managed.

Chapter-opening quotes from a wide variety of sources help illustrate different views and provide context.

"Learning Objectives" indicate important information and prompt thought prior to reading the chapter.

What Factors Affect Growth Rates?

Among the factors affecting growth rates are fertility, age distribution, and migration. The importance of any one factor in determining growth varies among countries. Fertility and age distribution are particularly important when considering future growth in the developing world, while migration is one reason why the United States is one of the fastest growing nations in the industrialized world.

Fertility

For many countries and regions of the world, crude birth and death rates are the only demographic data available. Crude rates are so called because they do not include information about age and sex. Consequently, they do not enable us to make accurate predictions about the future dynamics of the population. For example, men, children, and the aged are included in the calculation

for crude birth rate. But only women bear babies, and women of certain ages are more likely to bear children than others. Thus, measures of **fertility**, the actual bearing of offspring, are more accurate indicators of the potential for future population growth. The most important of these are the general fertility rate, age-specific fertility rates, and the total fertility rate.

The **general fertility rate** is the number of live births per 1,000 women of childbearing age per year. Ages 15 to 49 are considered by many demographers (including the Population Reference Bureau or PRB, the nonprofit educational organization cited throughout this text) to be the childbearing years. A more helpful indicator of potential future growth is the **age-specific fertility rate**, the number of live births per 1,000 women of a specific age group per year, such as girls aged 15 to 19. The **total fertility rate** (TFR) is the average number of children a woman will bear during her life, based on the current age-specific fertility rate, assuming that current birth rates remain constant throughout the woman's lifetime. The TFR can be thought of as the average family size.

Figure 7-4: A young mother holds her child in Cochabamba, Bolivia.

Main headings phrased as questions promote inquiry learning and help readers absorb and internalize information.

Bolded terms identify key words and phrases.

Real-life, current examples illustrate principles and give readers a global perspective.

Useful illustrations and images help *show* readers important information and concepts.

"Focus On" essays give readers a closer look at a specific resource, illustrate an ecological principle, or delve more deeply into a particular environmental issue; many are written by guests, affording readers different points of view.

In general, population profiles for LDCs have a pyramidal shape, a depiction of the rapid growth these nations are experiencing. In Nigeria and many other African nations, for example, about half of the population is under age 15; just two percent or so is age 65 or older (Figure 7-6). Throughout the developing world, high fertility, coupled with reduced mortality rates among infants and children, results in relatively high growth rates. As more people survive to their childbearing years, they produce an even larger next generation and a larger, younger population. In contrast, older age groups have far fewer members, a reflection of both higher mortality rates among infants and children in the past and the increased incidence of mortality that comes with aging.

The rapid growth of developing countries is largely a result of a decline in the death rate. Beginning after World War II (WWII), death rates fell throughout the developing world as a result of improved medical care and sanitation. Before long, birth rates far outpaced death rates in developing nations, a trend that continues in most LDCs today. Demographers generally attribute high fertility in the developing world to a complex mix of social factors, particularly a lack of access to sex education, family planning services, and safe and reliable means of birth control. Economic security — or rather, the lack thereof — also plays a

role. That's because, while death rates among infants and children in the developing world have fallen over the past 60 years, they remain high. Consequently, parents are compelled to have child after child because of the reality that many will die as infants or youngsters. In the absence of insurance, welfare programs and other measures to care for the aged, ill and injured, children become their parents' social security. In many parts of the world, sons are seen as particularly desirable because they are expected by custom to care for aged parents. In rural areas, children help their mothers with planting and harvest, childcare, and all household activities. In urban areas, they help support the family; children may work as peddlers (selling small trinkets and goods), scavengers (scouring dumps for cast-off items that can be used by the family or sold), or prostitutes. Bangkok, Thailand, for example, is notorious for its high number of child prostitutes.

The population profile for slow-growing MDCs like the United States and Canada is constrictive (Figure 7-6). A post-WWII baby boom means that the 43- to 60-age groups are larger than other age groups; the children of these "baby boomers" make for age cohorts that are large relative to the over-65 age groups, which comprise just 12 and 13 percent of the U.S. and Canadian populations, respectively. Population growth in these countries will stabilize and begin to decrease as

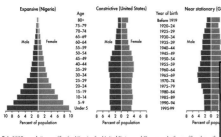

FIGURE 7-6: 2005 population profiles for Nigeria, the United States, and Germany. As the profiles show, the of Nigeria (and LDCs in general) is expansive. Africa, with the highest growth rate of any continent or regio youngest age structure and the greatest potential for future growth. Growth in the United States is constric Germany and some other developed countries, it is stable or declining.

134 AN ENVIRONMENTAL IMPERATIVE UNIT III

Kazi Ya Mwanamke
The Work of a Woman

Betsy A. Beymer

Kazi ya mwanamke is a kiswahili phrase meaning "the work of a woman." The following essay is a fictional, but nonetheless accurate, account of daily life for a woman in sub-Saharan Africa. It is a compilation of observations and notes made during my fieldwork in the West Usambara Mountains of Tanzania in the summer of 2003. The circumstances and challenges that it portray mirror those faced by rural women throughout the developing world.

Imagine, for a moment, that you are a young woman in sub-Saharan Africa. Somewhere in your 20s (no one in your village keeps records on births and deaths), you are responsible for a household that includes five children, ages 11, 9, 8, 3, and 1. Your husband works in a mine a long way from your village, returning home only six or eight times a year. With little or no education and living in a remote rural area, there is no such thing as a "career choice." You and your children depend for your livelihood on what you can grow and gather. You do not have the money to hire anyone to help you; rather, you must rely on your children. Sound difficult? Well, that's just the beginning. Imagine that you have no choice over the type of crops you grow for cash income, if there is any cash income to be made; those decisions are made by someone else. Nor do you have any choice about the number of children you will bear or the timing of their births; you do not know about contraceptives, and even if you did, you have no access to them. Moreover, the land that you work everyday — a small plot on a rocky stretch some distance from your home — is held in your husband's name only. You cannot own property. If your husband dies, there is a very real chance that the land will be taken away from you and your children.

Given these circumstances, what is your life like? Imagine.....

Although the sun has not yet risen, it's time for you to begin the day. So, at about 5 a.m., you rise quietly and leave your hut to collect water for the day's first meal and household chores. You bring along your youngest child, and he nurses as you walk. Your oldest child, an 11-year-old girl, is responsible for her siblings while you are gone. The water hole is more than a mile from your home, but you are so accustomed to the journey, you give the distance little thought. You fetch up to five gallons of water in an old plastic jug that you balance upon your head. Once home, you prepare tea for the rest of the household, then begin the daily ritual of washing the dirt floors with a wet rag in order to rid your house of insects. Soon after, you begin the day's laundry, which means another trip to the water hole. Without a washer and drier, washing clothes is a labor-intensive chore and takes you several hours. As you work, you talk with some of the women from your village who also have come to do their wash; their company makes the tedious task more enjoyable. After you finish, you return home to feed the livestock and tend to the crops. You perform all agricultural tasks — preparing the ground, plowing, planting the seeds, weeding, and harvesting — by hand. As you work, you again nurse your youngest son then listen to his cooing and babbling before he drops off to sleep.

While in the fields, you gather maize to prepare ughali, the staple of your diet. Returning home, you greet your four older children; your eldest son, nine, complains of hunger. The younger children play while their older sister helps you prepare the noon meal. Even food preparation is time-consuming and requires physical effort; you must shuck the maize, pick the kernels off the husks, and grind the kernels using a large, heavy pestle and mortar. You make enough ughali to last for several days; you eat meat only when you are able to

CHAPTER 7 HUMAN POPULATION DYNAMICS 143

Boxes highlight important, detailed examples of information within the text.

BOX 8–1 China's Program: The One-Child-Per-Family Policy

China offers the best-known example of a population control program, but the Chinese wouldn't call it that. Because China follows a strong Marxist ideology, the Chinese do not officially acknowledge population growth as a problem. Instead, they maintain that the program was designed to "maximize the health and well-being of each child born." According to official government policy, births are to be controlled because healthy, happy children are good for the state, and what is good for the state is good for all of the people (Figure 8-9). Even though China's delegation to the 1974 Bucharest population conference formally came out against population limitations, their country had already instituted one of the strongest family planning efforts that had ever been attempted.

Why did the Chinese feel it was necessary to initiate this program?

The first census taken of the Chinese population revealed that in 1953, China had a population of over 580 million people. By the late 1970s, that figure had reached over one billion. Fully 20 [percent of the world's] people were [on] seven percent [of the ar]able land. After [fa]mine (perhaps as [many as 30 milli]on people died), [and politi]cal turmoil, the [government bec]ame convinced [that to] adequately sup[port more th]an 650 to 700 [million people. T]he availability [of an adequate] food was espe[cially crucial in th]e determination

of carrying capacity. Because the population already exceeded the carrying capacity by more than 300 million people, a stringent family planning effort was needed to slow population growth and eventually reduce population size. The government estimated that, assuming growth rates continued to decline, it would take about a hundred years to reduce the population to 700 million. Because of population momentum, even if the growth rate fell to replacement level, China's population was predicted to peak between 1.2 and 1.5 billion people. China became the first nation to have as an official goal the end of population growth and the subsequent lowering of absolute numbers by a significant amount. Beginning in 1969, China achieved a transition from high to low birth rates. Let's examine how they did it.

In the late 1960s, the government instituted a family planning program that reached into every village across China. This program made birth control, including birth control pills, accessible to all. An extensive health care system was initiated that for the first time paid attention to China's rural poor. This important development would not have been possible without the major political change that took place in China. It was a change that stressed family planning as a way to increase the standard of living for all, gave women increased access to jobs and education, redistributed wealth to the masses, and provided for social security in the

form of old age pensions and food cooperatives.

The 1970s became known as the "later, longer, fewer" years. A mass educational effort, employing slogans, posters and radio and television programs, encouraged people to marry later, wait longer before starting a family, and have fewer children. The age for legal marriage for women was increased from as young as possible to the early twenties; contraceptives, abortions, and sterilizations were provided free of charge; and supplementary payments, longer maternity leaves, free education and health care for children, preference in housing, and retirement incomes were granted to couples who conformed to the goals of the program. Couples who did not conform lost benefits, paid heavy fines, and risked the loss of job promotions.

In 1979, China's leaders felt that population growth was not falling fast enough, and the emphasis switched to one child per family. Thus began the most restrictive family planning program ever attempted. The outside world deemed it extremely coercive. Intense pressure was brought to bear on women who became "unofficially" pregnant. Few women were able to resist this pressure; many had abortions or were sterilized. But even as the policy seemed to be having its intended effect — slower population growth — opposition to it increased, especially in rural areas, where children are needed to help with chores and food production. Critics claimed that the

[...]MPERATIVE **UNIT III**

WHAT YOU CAN DO ▶ Population Growth

- Cherish fewer children. Support relatives and friends who decide to have just one or two children or none. Avoid pressuring your children to bear children. And don't believe the stereotype that says single children and single adults are unhappy — it isn't true!

- Spread the love around. If you've got a strong parental urge, consider adopting children rather than having your own. Make enriching the lives of other people's children a part of your life.

- "Onlies" are OK. If you decide to have children, consider having only one or, at most, two. Each child born in the United States has an enormous impact on the environment due to our heavy consumption of water, energy, and goods.

- Mandate equal opportunity for women. Where women have better educational and economic opportunities, the birth rate has declined.

- Make contraceptives available globally. During the

next two decades, three billion young people will enter their reproductive years. Currently, only about half of fertile women have access to contraception. Encourage your congressional representatives to support expanded family planning programs in the United States and abroad.

- Work to eliminate the need for abortion. Outlawing abortion does not improve family planning or make for happy, wanted children; rather, it leads to dangerous illegal abortions, increased mortality rates for women, and unwanted children. Most people agree that eliminating the need for abortion is a worthier goal. Support family planning programs and efforts to make contraceptives widely available.

- Work to encourage an attitude of respect for human life — the unborn, the poor, the homeless, the parentless — so that everyone can enjoy a life of dignity.

- Limit development. Use your vote to promote land-use policies that preserve open space and farming, not only as a means of production, but as a way of life.

- Buy local produce or grow your own. When developing nations use scarce cropland to grow food for export, they deprive their populations of that land. Feed your family with foods produced from your area — even start your own garden.

- Sponsor a foster child in a developing country. For a small monthly fee, reputable organizations such as Plan USA link caring people in the United States with needy children and their families overseas. The programs help families and communities become self-sufficient. Write Plan USA, 155 Plan Way, Warwick, RI, 02886 or call 1-800-556-7918 toll-free (in RI, 401-738-5600).

Source: Adapted from a series on personal ecology by Monte Paulsen, editor and publisher of Casco Bay Weekly, in Portland, Maine.

Both men and women can be sterilized: Vasectomy is the cutting or tying of the tubes by which sperm leave the body, and tubal ligation is the tying of the oviducts by which eggs reach the uterus. Sterilization is becoming increasingly common in many LDCs. In China and India, for example, large numbers of women have undergone sterilization for contraceptive purposes.

Sterilization procedures, which are usually irreversible, are sometimes used in family planning programs that limit, rather than extend, the reproductive freedom of the individual. Among the countries and territories

where abuses have been reported, Puerto Rico has been singled out by women's groups as having the most abusive sterilization program. The government began promoting sterilization in the mid-1940s, and by 1965, Puerto Rico had the highest sterilization rate in the world. Fully one-third of all women who had ever been married were sterilized, 40 percent of them before the age of 25. Many women were not told that the procedure was irreversible, and some women were sterilized unknowingly, while under anesthesia for other procedures.

"What You Can Do" sections contain specific suggestions on ways that individuals can become involved in environmental issues, help to preserve natural systems, and protect resources.

Sources within the text and the bibliography aid further reading and research.

Discussion Questions

1. Describe the demographic transition. Which stage do you think the United States is in and why? Which stage do you think Nigeria is in and why? For developing nations that are attempting to free themselves from the demographic trap, what are the implications of the HIV/AIDS epidemic?

2. Explain the difference between pronatalist and antinatalist policies. What factors cause a country to adopt one or the other? Do you think the United States should adopt an official population control policy? Why or why not? If you answer yes, what policy should it adopt?

3. What are the goals of family planning? How is family planning different from policies and programs that aim for zero or negative population growth?

4. List some important features of China's "one child per family" policy. Do you think such a policy would work in India, the United States, Canada, and Mexico? Why or why not?

5. List at least five methods of birth control. Which ones are most effective in LDCs and in MDCs? Explain.

Summary

For most of its tenure on Earth, beginning some 40,000 or 50,000 years ago, *Homo sapiens* existed by gathering wild plants and hunting wild animals. Up until about 10,000 years ago, the Earth probably supported about five million people. With the rise of agriculture, populations began to grow more rapidly and the environmental impact of human activities increased. By the beginning of the nineteenth century, the human population had reached one billion. Advances in death control ushered in a period of rapid growth. Birth rates in the western world began to fall with the onset of the Industrial Revolution and subsequent rising standard of living, the introduction of safe and reliable means of birth control, and an increase in the cost of childrearing. However, the population continued to grow (though more slowly) because there were so many more people in total. Currently, growth rates for most industrial nations are either low, zero, or negative.

The population path followed by the industrial nations, the demographic transition, describes the movement of a nation from high growth to low growth. It consists of four stages. In Stage 1, birth and death rates are both high. In Stage 2, death rates fall, but birth rates remain high, and thus the population undergoes rapid growth. In Stage 3, birth rates begin to fall, and the growth rate declines until it eventually nears zero. In Stage 4, the growth rate is at or below zero.

Much of the growth in the human population in the past 50 years has occurred in less-developed countries (LDCs), as a result of falling death rates and constant or slightly rising birth and fertility rates. Rapid growth and the environmental deterioration it causes are fueling a downward spiral in the standard of living. Some developing nations are caught in a demographic trap, unable to break out of Stage 2 of the demographic transition. Their governments are in a state of demographic fatigue, worn down and financially strapped after decades of struggling to combat the consequences of rapid population growth. Some LDCs have even slipped back into Stage 1.

Any planned course of action taken by a government designed to influence its constituents' choices or decisions on fertility or migration can be considered a population policy. A pronatalist policy encourages fertility and a higher birth rate; an antinatalist policy discourages fertility and encourages a lower birth rate. Pronatalist or antinatalist policies are developed as a result of the way population changes are perceived. Family planning encompasses a wide variety of measures that enable parents to control the number of children they have and the spacing of their children's births. The goal of family planning is not to limit births, but to enable couples to have healthy children and to have the number of children they want.

China became the first nation to have as an official goal the end of population growth and the subsequent lowering of absolute numbers by a significant amount. Beginning in 1969, China achieved a transition from high to low birth rates by implementing the strongest measures ever attempted. These included free birth control, higher marriage ages for women, economic incentives to have fewer children, education, and media campaigns. The "one child per family" policy was effective but highly controversial. China began to relax its policies somewhat in the 1980s, due in part to an expanding economy and pressure from human rights activists.

Summaries identify and outline main concepts within each chapter.

"Discussion Questions" help students analyze and synthesize knowledge and provide a starting point for discussion.

War II, economic uses of land were emphasized over recreation... ness preservation. To counteract this emphasis, the Wil... tion Act, introduced in 1957 and passed in 1964, directed ... and Fish and Wildlife services to recommend areas for ... so enabled Congress to designate wilderness.

...ss Preservation Act of 1964 designated 9.1 million acres ...ares) of federal land as protected wilderness. These were ...tituents of the National Wilderness Preservation System, ...ts of 662 areas covering more than 150 million acres. ... facing the National Wilderness Preservation System is ...signate public lands as wilderness areas.

...nds are biologically significant for several reasons. First, ...n astounding diversity of living ecosystems. Second, these ...s perform a myriad of ecological functions, such as watershed protection. Finally, because the public lands encompass a wide array of ecosystems, they protect a diverse complement of living organisms. In addition to their biological wealth, the federal lands contain one-quarter of the nation's coal deposits, four-fifths of its huge oil-shale deposits, one-half of its uranium deposits, one-half of its naturally occurring steam and hot water pools, one-half of its estimated oil and gas reserves, and significant reserves of strategic minerals.

The federal lands are used for many purposes. Consumptive uses result in the depletion of a resource. Commercial consumptive uses include logging, mining, and oil development. The federal lands also offer a variety of recreational, scientific, and wilderness opportunities, many of which are nonconsumptive. However, even nonconsumptive uses can degrade an area if there are many users or if they are not careful about how they use the land.

Because the public domain is used for many purposes, a wide variety of problems threaten their beauty and integrity. The difficulty of managing federal lands is compounded by the fact that managing agencies must try to meet the interests of conflicting groups.

"Key Terms" provides a checklist of the chapter's important words and phrases. Definitions are given in the text and also appear in the glossary.

KEY TERMS

below-cost timber sales	consumptive use	public domain
chaining	multiple use	reserved water rights
clearcutting	non-consumptive use	utilitarianism

Acknowledgements

There's a Zen saying that goes, "Leap and the net will appear." Each edition of this textbook has been like that: We think there's no way we can pull it off, and yet once we decide to go ahead and try, the right people have appeared at just the right time to help. That's certainly true of this latest edition. *The Biosphere* has benefited from the talents and hard work of many people — students, professors, and environmental professionals. Any errors that may remain are, of course, our responsibility.

First of all, we wish to thank everyone at Kendall/Hunt who helped to guide the revision of *The Biosphere*. Special thanks to our project editor, Lynne Rogers; our permissions editor, Renae Heacock; our production editor, Charmayne McMurray; and our long-time editor and friend, Georgia Botsford. It's a pleasure to work with a company that is committed to producing a first-rate text in an environmentally sound manner.

We are grateful to Carole Katz, our design artist, for the aesthetically pleasing layout of *The Biosphere*. She has been with us since the second edition, and with each revision, she finds new ways to enhance the text and showcase its visual elements. We're fortunate to work with her: Carole *knows* — and is passionate about — this subject matter; her knowledge, combined with her technical expertise, results in a text that is both effective and inviting.

To quickly and accurately update this edition, we turned to family, friends, and colleagues for assistance with selected A number of friends from the Miami and Oxford community generously allowed us the use of their photographs; their contributions have made this a better and more interesting text. We wish to thank Scott Bagley, Lisa Breidenstein, Alan Cady, Jeff Davis, James Foley, David Gorchov, Kenneth Hanf, Paul Heideman, Susan Hoffman, Jason Irwin, Joseph Jacquot, Andy Jones, Alice Kahn, Carole Katz and Jonathan Levy, Patricia Kaufman, Joe Lamancusa, Orie Loucks, Jay

Mager, Ruth McCleod, Dave Russell, Stanley Toops, Louise Van Vliet, John Vankat, Mike Vanni, Tom Wissing, and Mike Wright.

A special thanks to the following three long-time Miami colleagues and friends, each of whom contributed numerous photographs to this effort: Dave Osborne, Dolph Greenberg, and Hardy Eshbaugh. Dave's images of the wildlife and people of Central America show his love and concern for that region. Dolph, who provided us with dozens of slides of natural areas throughout North America, is a talented photographer whose skill is exceeded only by his dedication to his students and to the Native American tribes with whom he works. Finally, we thank Hardy — a renowned botanist and birder and a world-traveler — not just for the many images he has lent us, but, more importantly, for his constant and continued support. He has been our patron, championing *The Biosphere* around the world, and for that, we are extremely grateful.

We want to thank Greg McNelly, with the Water Environment Foundation, for his help in updating Chapter 13, *Water Resources*. His help was invaluable!

Revising and producing a textbook and its ancillaries entails innumerable tasks and we would be unable to do it without the help of a talented and dedicated staff. Betsy Beymer managed the art program, finding compelling images and securing permissions for their use; she also contributed, in many valuable ways, to the content and organization of Chapter 7, *Human Population Dynamics*; Chapter 11, *Energy: Alternative Sources*; Chapter 13, *Water Resources*, Chapter 14, *Soil Resources*; Chapter 17, *Cultural Resources*, and Chapter 20, *Unrealized Resources: Solid and Hazardous Wastes*. She brought talent and good cheer, and we will miss her smiling face!

In taking on the daunting task of compiling and condensing related chapters (those that now comprise Chapter 2, *Social Factors*, Chapter 16, *Public Lands*, and Chapter 20, *Unrealized Resources: Solid and Hazardous Wastes*), Emily Crum was a model of efficiency, quickly and carefully distilling the material down to its essence. She also brought her considerable knowledge and experience to bear on the revisions of Chapter 12, *Air Resources*; Chapter 14, *Soil Resources*; Chapter 18, *Mineral Resources*; and Chapter 19, *Nuclear Resources*.

We have Kim Maher to thank for diligently organizing and updating the photo credits and preparing the detailed Table of Contents. She also did innumerable other tasks around the Hefner Zoology Museum, making it possible for the rest of us to focus on *The Biosphere*.

Mike Wright carefully and accurately updated the glossary and bibliography; he made valuable suggestions on several chapters, including Chapter 12, *Air Resources*; Chapter 13, *Water Resources*; and Chapter 14, *Soil Resources*. A talented naturalist and knowledgeable biologist, Mike helped to update and reorganize Chapter 15, *Biological Resources* and Chapter 16, *Pubic Lands*. We are grateful for his expertise.

Aaron Inouye was our jack-of-all-trades. A tireless and skilled researcher, he quickly tracked down needed information and data; he also read and proofread drafts, and his suggestions made the writing clearer and more concise. Aaron's help was particularly important in the revision of Chapter 15, *Biological Resources* and Chapter 16, *Pubic Lands*. He also did the initial layouts for many of the chapters, a task that helped expedite the revision process. And finally, he updated the supplementary materials (*Teacher's Resource Guide*, *Test Bank*, and *Student's Resource Guide*) and he helped to create *Delving Deeper*, one of our online resources.

Finally, we owe our deepest thanks to our colleague, Lisa Rosenberger, who oversaw the entire revision from start to finish. She developed the master schedule (and made certain that the rest of us met it!); supervised the student workers; revised Chapter 10, *Energy: Fossil Fuels* and Chapter 11, *Energy: Alternative Sources*; provided helpful comments and suggestions on many other chapters; proofread final copy; and helped Aaron create *Delving Deeper* and update the supplementary materials. Lisa's work ethic is exceptional; she put in countless extra hours in the evenings and on weekends in order to help us meet our deadlines. For that — and for other reasons to numerous too mention — we are thankful. We wish her all the best — and a smooth landing — with each leap she takes.

Donald G. Kaufman
Cecilia M. Franz

Reviewers

We would like to thank the following academic and technical reviewers:

Clark E. Adams, Texas A&M University

David Adams, North Carolina State University

Will Ambrose, Bates College

Doug Ammon, Clean Sites, Inc., Alexandria, Virginia

Valerie A. Anderson, California State University, Sacramento

Susan Arentsen, U.S. EPA, Cincinnati, Ohio

Lisa Bardwell, University of Michigan, Ann Arbor

David Berg, Miami University

Jeff Binkley, Roy F. Weston, Inc. (hazardous waste remediation), Okemos, Michigan

Neil Blackstone, Northern Illinois University

Dale J. Blahna, Northeastern Illinois University

Gary Blevins, Spokane Falls Community College

Anne Brataas, Pioneer Press, St. Paul, Minnesota

W.M. Brock, Great Basin National Park, Baker, Nevada

Moonyean Brower, Armstrong State College

Warren Buss, University of Northern Colorado

William Calder, University of Arizona

Edwin Toby Clark, Clean Sites, Inc.

LuAnne Clark, Lansing Community College

Paul Doscher, Society for the Protection of New Hampshire Trees, Concord, New Hampshire

Robert Dulli, National Geographic Society, Washington, D.C.

Steve Edwards, Miami Oxford Recycling Enterprise, Oxford, Ohio

Hardy Eshbaugh, Miami University

Eric Fitch, Director, Coastal Zone and Natural Resource Studies, University of West Florida

Lloyd Fitzpatrick, University of North Texas

Herman S. Forest, State University of New York, Geneseo

Judith Franklin, Oakland Community College

Dwain Freels, FFA Director, Oxford, Ohio

Gerald Gaffney, Southern Illinois University

Greg Githens, O.H. Materials Company (hazardous and toxic waste management), Findlay, Ohio

Wright Gwyn, Director of the Environmental Awareness and Recycling Program, Forest Park and Greenhills, Ohio

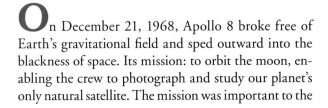

PROLOGUE

Discovering Our Worldview

Every form of refuge has its price. *

The Eagles

On December 21, 1968, Apollo 8 broke free of Earth's gravitational field and sped outward into the blackness of space. Its mission: to orbit the moon, enabling the crew to photograph and study our planet's only natural satellite. The mission was important to the lunar program of the United States because it would supply information about the lunar surface and pave the way for a future moon landing. But the craft's journey turned out to have a greater, unexpected significance. As Apollo 8 emerged from the dark side of the moon

FIGURE P-1: Photo of earthrise taken by Apollo 8 crew. The most important result of humankind's ventures into space is that we have a clearer, more accurate view of Earth. The physical reality of the planet — its significance in comparison to the universe — and the possibility that it is the only planet capable of supporting life as we know it have profoundly altered the way we think about Earth.

FIGURE P-2: Diverse worldviews. (a) The Senecas, a native American tribe, believed that the Earth formed when a woman dropped from the sky, fell to the sea, and was helped by a turtle who rose to the surface and allowed her to ride upon its back. (b) According to Egyptian mythology, Sky is a feminine god who arches over the male Earth. (c) Ge, or Gaia, is the ancient Greek Earth goddess, whose abundance nurtures life. (d) The god Shiva symbolizes the ancient Hindu belief in a cyclic universe. The halo of fire surrounding the god represents his dominion over the cycle of creation, destruction, and rebirth.

FIGURE P-3: How different worldviews shape attitudes and behavior toward nature and resources. (a) Michelangelo's painting on the ceiling of the Sistine Chapel depicts a fundamental Christian belief, that God created humankind in His image. God's touch imbues humanity with the soul that sets humans apart from all other creatures. (b) Developed by the philosopher Lao-tzu, the yin and yang represent complementary opposites — earth/sky, positive/negative, male/female, light/dark — that together form an unbroken whole and symbolize the harmonious balance of nature. (c) A leading scientific explanation for the origin of the universe is that it resulted when an incomprehensibly small, dense mass (or superatom) exploded, a concept popularly known as the big bang. The painting is by Adolph Gottlieb.

at the end of its fourth orbit, the crew — Frank Borman, James A. Lovell, Jr., and William A. Anders — rotated the craft. As they did so, they became the first human beings to witness an earthrise (Figure P-1). Anders, recalling the sight said, "We saw the beautiful orb of the planet coming up over this relatively stark, inhospitable lunar horizon and it brought back to me that indeed even though our flight was focused on the moon, it was really the Earth that was the most important to us." That realization resonated in the hearts of millions who saw the photographs and video footage the Apollo 8 crew sent back to a waiting Earth. Those images forever altered humankind's view of our home planet. For the first time, we gazed at the Earth from the vantage point of outer space, and what we saw was a strikingly beautiful oasis of life dwarfed by the black void through which it traveled.

The heavens have always been an object of wonder, mystery, and speculation. As human cultures arose, people developed different explanations of the origin of the Earth and the universe and of the human place in the cosmos. These accounts helped to form their worldview, a way of looking at reality. The earliest worldviews took voice in many diverse stories that explained the creation of both the world and human beings; they took shape in drawings, paintings, and sculpture (Figure P-2). Ultimately, worldviews take life in the beliefs, attitudes, and values of people. Worldviews are reflected in endeavors of the human spirit, like religion, philosophy, art, music, and literature (Figure P-3). What do worldviews have to do with environmental science? As it turns out, quite a bit. Because a worldview includes basic assumptions about the self, others, nature, space and time, and about the relationships between the self and everything else, it shapes our attitudes toward nature and influences how we use resources.

More than three decades have passed since the world watched its first earthrise. Successive space missions have sharpened our first images of Planet Earth. In 1972, astronauts took the first picture of the full Earth, a planet of deep blues, verdant greens and variegated browns, cloaked in swirling clouds of pure white. Pictures from the space shuttles and human-made satellites have further clarified our image of Earth. What we are learning is that this planet we call home is an unbelievably complex and possibly unique entity. We are learning, too, just how much humans have changed and continue to change the Earth. As our physical understanding of the Earth becomes more sophisticated, our worldview is also changing. Slowly — perhaps too slowly — we are realizing that the Earth is not a sophisticated mechanical "spaceship," an analogy popular during the exciting era of the nation's space program three decades ago, nor are we the drivers of this craft. Many people feel that we need a new way of looking at our planet, a worldview that accurately accounts for the complexity of Earth and the interdependence of its life forms.

Take some time to think about your own worldview. Jot down some notes in your text or a notebook. As you read, think about the consequences of your beliefs, attitudes, and values. What is the price of the refuge you take in Earth? What sort of Earth do you want to leave for your children, for all future generations, to inherit? What will it take to make that legacy possible?

This grand show is eternal. It is always sunrise somewhere; the dew is never all dried at once; a shower is forever falling; vapor is ever rising. Eternal sunrise, eternal sunset, eternal dawn and gloaming, on sea and continents and islands, each in its turn, as the round earth rolls.

John Muir

UNIT I

The Biosphere and Environmental Science

CHAPTER 1

The State of the Biosphere

Learning Objectives

When you finish reading this chapter, you will be able to:

1. Describe the biosphere and explain how scientists believe it evolved.

2. Summarize the current state of the biosphere.

3. Identify the three root causes of environmental problems.

4. Distinguish between linear and exponential growth and explain the significance of exponential human population growth.

5. Define science, ecology, and environmental science.

6. Briefly describe the process of scientific inquiry and explain its strengths and limitations.

7. Explain how knowledge from the social sciences, natural sciences, arts and humanities, communication, and education contributes to solving environmental problems.

Right now, at this moment in history and for the long haul into the next century and centuries beyond, no other issue is more relevant to the physical quality of life for the human species than the condition of our environment and status of our natural resources — air, water, soil, minerals, scenic beauty, wilderness, wildlife habitat, forests, rivers, lakes, and oceans. These resources determine the physical condition of our lives and dramatically influence the human condition....

Gaylord Nelson

Hope is a risk that must be run.

George Bernarnos

Consciously or unconsciously, we mark our lives by change. Some change is circular — the daily cycle of sunrise and sunset and the passing seasons that delineate the year — and its regularity brings a comfortable rhythm to our lives. Other change is linear — the birth of a child, a toddler's first unsteady steps, the first day of school, graduation, a marriage, a terminal illness, a death — and its occurrence reminds us that just as each day is unique, it is also fleeting. Thus, while change can be welcome, it is often unsettling because it means the loss of the familiar and the arrival of the unknown.

The unsettling nature of change inspired the ancient Chinese curse, "May you live in interesting times." Whether they view change as unsettling or inspirational, most people would agree that we live in interesting times, indeed. For evidence, we only have to look back at the last century.

In 1900, the horseless carriage was a novelty and human flight was the fanciful dream of a few; the letter and the newspaper were the most common forms of communication; and the sun never set on the British Empire. But by the dawn of the twenty-first century, jetliners regularly crossed the oceans in mere hours; people around the world communicated nearly instantaneously via the Internet; and independence was the rule, rather than the exception, for the many nations that once had been the colonies of Britain and other world powers.

Change — in travel, communications, political structures and alliances, and many other facets of human culture and society — has become the norm. But our impact is not limited to human institutions; increasingly, through sheer numbers and the power of our technologies, we are changing the very nature of the world. Consider these examples. Humans have transformed between one-third and one-half of the Earth's surface and use more than half of all accessible fresh water. Our activities fix more atmospheric nitrogen than all natural terrestrial sources combined. The atmospheric concentration of carbon dioxide has increased nearly 30 percent since the beginning of the Industrial Revolution, and most scientists believe that carbon dioxide and other "greenhouse gases" will raise the average global temperature by four to seven degrees Fahrenheit by 2100. Approximately one-quarter of the Earth's bird species have been driven to extinction, one-third of amphibian species face extinction in the next few years, and two-thirds of major marine fisheries are fully exploited, overexploited, or depleted.

Some people argue that these changes are just the natural result of a very successful species — ours. They point out that all species alter their surroundings. To a certain extent, that's true. Like all other species, humans always have altered the natural world to suit their needs. But what's different, and what's critically important, is the unprecedented rate and scale of the human-induced changes now underway. Moreover, the effects of those changes — on the Earth and its ability to support life as we know it — are not fully known.

We live in interesting times, of that there is no doubt. But this need not be a curse. In fact, it could be a blessing. Which it will be is largely up to us and to the actions we take now and in the decades to come. Our goal in writing this textbook is to help people become environmentally educated, so that we are equipped to live in this time of rapid change. That goal shaped the text's organization and content, with each chapter answering a number of major questions concerning a specific topic or topics. If you are interested, you can explore many of these topics in greater depth by visiting *The Biosphere: Delving Deeper,* an online supplement found on our website; to access the supplement, go to www.EnvironmentalEducationOhio.org and click on the "Biosphere Project" tab.

This first chapter presents an overview of the environmental problems that are explored in greater detail throughout the rest of the text and briefly discusses what we need to know in order to solve those problems. A good grasp of environmental problems, however, is impossible without an understanding of the biosphere and how life on Earth evolved.

What Is the Biosphere?

Viewed from a great distance, Earth presents a startling and beautiful contrast to the vast blackness of space. Teeming with living things and bejeweled with colors, it is home to millions of different species of organisms intricately woven into the complex tapestry we call life (Figure 1-1). That life exists on Earth is due to the **biosphere**, the thin layer of air (atmosphere), water (hydrosphere), and soil and rock (lithosphere) that surrounds the planet and contains the conditions to support life. The biosphere is our **global environment** — the sum of all living organisms, the relationships among organisms, and the relationships between organisms and their physical surroundings. The biosphere includes any place on, above, within, or below the Earth's surface where life can be found, such as areas around thermal vents in the ocean deeps, underground caverns, the deepest penetration of roots into soil, and the highest reaches of avian flight. Each of these sites (and millions more) constitutes a specific local **environment**, defined as a system of interdependent living and nonliving components and their interactions in a given area, over a given period. However vast it might seem, the biosphere is relatively small compared to the Earth's total mass. Further, this life zone is unevenly distributed across the Earth; some places are too hot, too cold, or too dry to support life.

The biosphere is made up of nonliving and living components. Nonliving physical surroundings comprise the abiotic component, or **abiota**; living organisms collectively comprise the biotic component, or **biota**. The millions of different species (including us, *Homo sapiens*) that depend upon the biosphere for their existence are at the same time an integral part of the biosphere. They influence the cycling of nutrients and water between the abiotic and biotic components and thus help to regulate the global environment. Because humans are part of the biota, the biosphere encompasses human systems such as farmlands and cities as well as natural systems such as marshes and forests.

To see how the abiotic and biotic components are intricately linked, let's look more closely at our atmosphere. It is composed of about 78 percent nitrogen, 21 percent oxygen, 0.03 percent carbon dioxide, and trace gases. Incoming solar radiation is reradiated from the Earth's surface back into space as heat energy. Carbon dioxide, along with other gases, traps some of the reradiated heat energy. This trapped heat warms the Earth, much as the glass of a greenhouse traps the reradiated energy from sunlight and thereby warms the interior of the structure. Earth's proportion of greenhouse gases enables life to flourish.

Life on Earth also helps to maintain the composition of the atmosphere. Water, nitrogen, and trace atmospheric elements such as sulfur and phosphorus all continually cycle through the biotic and abiotic components of the biosphere. For example, when they respire, both plants and animals take in oxygen and give off carbon dioxide. However, during photosynthesis, the process by which green plants capture and convert sunlight to chemical energy, carbon dioxide is taken in and oxygen is given off. Thus, the biota help to maintain the composition of the atmosphere at relatively constant levels.

What Are the Characteristics of Living Organisms?

Life is the predominant characteristic of the biosphere. In all its diversity — from the familiar to the exotic, the microscopic to the gigantic, the fiercely beautiful to the beautifully fierce — life is the hallmark of our planet. Earth may be the only planet in the galaxy, even in the universe, capable of supporting life as we know it. But this statement begs the question, how do we know life? Indeed, what is life? Scientists, philosophers, and theologians have been trying to answer that question for ages. So far, no one has come up with a definition that satisfies everyone. However, scientists have compiled a list of characteristics that describe living organisms.

First and foremost, all organisms live at the expense of their environment. From the simplest one-celled bacteria (which, in one of life's delightfully ironic twists, are far from simple) to the most complex mammals, all organisms must extract materials and energy from the biosphere in order to live. All living things have a cellular structure. The cell is the smallest unit of life that has the structures and chemical mechanisms needed to conduct the activities associated with living. All living organisms exhibit movement of some type, even if it is only at the cellular level. Movement is one of the most reliable indicators of life (not surprising when you remember that someone or something which is being hunted will often "play possum" to confuse its preda-

FIGURE 1-1: The fabulous diversity of life on Earth. Among Earth's estimated 30 million species are (a) Tunicate worms, (b) African elephants, (c) Quaking Aspen, (d) Agama lizard, (e) Acetabularia, (f) Day lilies and Liatris, (g) Heliconia (species unknown), and (h) Mushroom (species unknown).

tor). Living things also grow, gaining mass over time. A plant manufactures the organic molecules that form the raw material for growth. In contrast, an animal acquires the raw materials it needs by consuming plants, animals, or both. Living organisms also reproduce. Not all individuals of a given species may reproduce, but the species as a whole must be capable of reproduction or it would die out as its members died. Living things respond to stimuli. Response allows an organism to react to changes in the environment, such as the movement of the sun or a scent on the wind. Finally, living things evolve and adapt. Just as an individual must be able to respond to changes in its immediate environment, species evolve, or change over time, as they adapt to long-term changes in the environment. **Evolution** is the process of change with continuity in successive generations of organisms. Charles Darwin, the English naturalist who first proposed the theory of evolution by natural selection, called it "descent with modification."

Describing characteristics of living organisms is quite different from defining life. Keep in mind that none of the characteristics listed above alone defines life; they must be considered as a sort of tableau that distinguishes the living from the nonliving. Moreover, some of these characteristics do not apply to all organisms, and others apply to nonliving things as well.

How Did the Biosphere Develop?

To understand how the biosphere developed, we must first understand the milieu in which it developed, that is, we need some inkling of the origins of the universe and the Earth. Of course, no one can explain with absolute certainty the origins of the universe. The most widely accepted scientific theory, developed by theoretical physicists using powerful mathematical tools, holds that the universe came into being as a result of what is popularly known as the big bang. According to the **big bang theory**, the universe — all matter, energy, and space — arose from an infinitely dense, infinitely hot point called a singularity, roughly the size of a speck of dust. About 13 to 20 billion years ago, the singularity exploded and space began to expand. Only energy had existed in the singularity, where the temperature reached billions of degrees Celsius, but as the expanding universe began to cool, matter gradually formed. Hydrogen and helium formed during the first 500,000 years after time zero, the moment of explosion. In the young expanding universe, gravity caused the newly

formed matter to clump together in large masses, the precursors of galaxies. As matter rushed outward into the void, some of the galaxies fragmented, forming stars. Nuclear reactions within the cores of the stars gave off heat and light and created the heavier elements that comprise the planets and our bodies. About five to 10 billion years ago, our solar system developed from gas and dust on the outer edge of the galaxy we call the Milky Way.

The big bang theory may be difficult to imagine, but astronomic observations lend credence to the concept. In the 1920s, the American astronomer Edwin Hubble observed that the universe is expanding, an observation that would be expected if the universe had indeed arisen from some sort of explosion. Moreover, astronomers have noted the presence of cosmic background radiation — radiation that seems to originate equally from all directions in the cosmos. The characteristics of this type of radiation are consistent with what would be expected if the universe had arisen from a small, dense area that underwent a big bang.

The Earth formed about 4.6 billion years ago, condensing out of interstellar gas and dust. Hot, molten and volcanic, the planet was a lifeless sphere veiled by a thin layer of hot gases. Farther above lay a heavy cloud cover, cloaking the Earth in darkness. The only light came from cracks in the planet's molten surface, revealing the fiery heat below, and the almost continual lightning that split the dark from above. Water vapor that condensed and fell to the Earth's scorching surface was immediately vaporized and returned to the atmosphere. The desolate, surreal setting gave little evidence that upon this stage the great drama of life would soon begin. Yet life on Earth developed soon after the surface of the young planet cooled, somewhere between 3.5 to four billion years ago. And it began, scientists believe, as a result of chemical and physical processes on the planet's surface and in its atmosphere.

The early atmosphere is believed to have consisted largely of hydrogen, far different from the present atmosphere. Hydrogen tends to combine with many other elements, among them nitrogen, oxygen, and carbon. Consequently, the primitive atmosphere was likely characterized by ammonia (hydrogen and nitrogen), water (hydrogen and oxygen), and methane (hydrogen and carbon).

As the planet's surface cooled, water pooled in valleys and other low places. But the surface was still too warm for the water to remain liquid for long. It evaporated, cooled in the atmosphere, and fell to the Earth once more. The rains that pelted the planet's surface washed minerals and salts from the exposed rocks. The mineral-laden waters filled cracks and crevices; streams and rivers formed, eventually giving way to oceans as the waters filled the Earth's deepest valleys and canyons. The hydrogen-rich molecules that had formed in the atmosphere became dissolved in the primitive planet's seas. Ultraviolet radiation and lightning, coupled with the heat trapped within the Earth, supplied the energy that broke apart the hydrogen-rich molecules, enabling them to recombine into new and more complex molecules. Thus, the oceans became a sort of "primordial soup," a mixture of molecules and substances of increasing complexity. And it was from this primordial soup, scientists theorize, that life on Earth originated.

In the harsh environment of the young planet, evolution proceeded slowly. (Table 1-1 summarizes what are believed to have been the major events in evolution.) The factors that drive evolution are mutation and natural selection. A **mutation** is a random change within the genetic material of an individual that can be passed on to that individual's offspring. Most mutations are harmful or result in changes that are useless to the organism. Some mutations, however, cause the organism to differ in a way that allows it (or its offspring) to adapt to a change in environmental conditions. Because the individual is better suited to the environment than others in the population, it is more likely to survive and reproduce. This phenomenon is known as **natural selection**, the process that enables individuals with traits that better adapt them to a specific environment to survive and outnumber other, less well-suited individuals. To be advantageous, natural selection must improve an organism's chances for successful reproduction. The flip side of natural selection is that other organisms less suited to environmental conditions eventually are eliminated. Over time, natural selection may lead to **speciation**, the separation of populations of organisms, originally able to interbreed, into independent evolutionary units (or species) that can no longer interbreed because of accumulated genetic differences.

Once the evolutionary process took hold, the planet's watery surface teemed with a rich diversity of life forms. Organisms capable of capturing the sun's energy proliferated. Because these organisms gave off oxygen, they slowly changed the atmosphere to an oxygen-rich environment. This change, about 600 million years ago, signaled the start of the Cambrian period. Many species, particularly bacteria that had thrived in the low-oxygen environment, could not tolerate the increased oxygen atmosphere and died out.

But it is during the Cambrian period that speciation really took off, and a great many new marine organisms appeared. Approximately 425 million years ago, atmospheric oxygen reached its present level. About the same time, the land mass became increasingly drier as the great oceans began to shrink. The first land plants and animals evolved. Seed plants, bony fishes, and the first amphibians appeared soon after. Some 310 million years ago, reptiles and nonflowering seed plants known as gymnosperms (a group that includes conifers, cycads, and gingkoes) appeared. Vast forests and extensive swamps covered the land surfaces. Dinosaurs appeared about 220 million years ago, and before long (40 million years later), birds and mammals appeared. The end of the Cretaceous period, about 65 million years ago, marked a period of mass extinctions during which many species, including the great dinosaurs, disappeared. Some scientists believe that prehistoric glacial or meteoric activity altered environmental conditions and caused these mass extinctions. About 25 million years ago, mammals began to evolve more rapidly; the first humanlike primates appeared. *Homo sapiens* came on the scene only about three million years ago, scarcely more than a blink of the eye in geologic time.

To get a better understanding of just how young our species is, in geologic terms, let's compress the 15-billion-year history of the universe into a single 24-hour day. Starting at midnight, atoms form in the first four seconds. Stars and galaxies appear by about 5 a.m. Not until 6 p.m., a full 13 hours later, do our sun and solar system form. Life on earth appears about 8 p.m., but the first vertebrates (animals with backbones) do not appear until about 10:30 p.m. Dinosaurs roam the earth from approximately 11:35 to 11:56 p.m. Finally, 10 seconds before midnight, *Homo sapiens* appears. The Age of Exploration to our nuclear age (fifteenth century to the present) occurs in the last thousandth of a second.

It has been said that the only constant we can count on is change. Individuals change. Countries and nations change. And as our abbreviated discussion illustrates, the Earth's environment and its life forms have changed, and changed dramatically, over the millennia. In the last 450 years, a new agent of change has begun to exert its influence on the planet. Since the beginning of the Age of Exploration, the activities of a single species, *Homo sapiens*, have accelerated the rate of environmental change and species extinctions to levels not seen since the end of the Age of Dinosaurs. To understand how our species is altering the planet, let's take a close look at the state of the biosphere.

TABLE 1–1: Major Events in Evolution (mya = millions of years ago)

Eon	Era	Period	Epoch	millions of years ago	Life forms	Major events
PHANEROZOIC (Phaneros = "evident;" zoic = "life")	CENOZOIC	Quaternary	Recent, or Holocene		Spread of modern humans / Extinction of many large mammals and birds / *Homo erectus*	Eruption of volcanoes in the Cascades
			Pleistocene	1.6		Worldwide glaciation / Fluctuating cold to mild in the "Ice Age" / Uplift of the Sierra Nevada
		Tertiary — Neogene	Pliocene	5.3	Large carnivores / Earliest hominid fossils (3.4-3.8 mya)	Linking of North and South America / Beginning of the Cascade volcanic arc
			Miocene	23.7	Whales and apes / Large browsing mammals; monkey-like primates; flowering plants begin	Beginning of the Antarctic ice caps / Opening of the Red Sea / Rise of the Alps; Himalaya Mountains begin to form
		Tertiary — Paleogene	Oligocene	36.6	Primitive horse and camel; giant birds; formation of grasslands	Volcanic activity in Yellowstone region and Rockies / Ice begins to form at the poles
			Eocene	57.8	Early primates	
			Paleocene		Extinction of dinosaurs and many other species (65 mya)	Collision of India with Eurasia begins / Eruption of Deccan basalts
				66.4		
	MESOZOIC	Cretaceous		144		Formation of Rocky Mountains
		Jurassic		208	Placental mammals appear (90 mya) / Early flowering plants / Flying reptiles	
		Triassic		245	Early birds and mammals / First dinosaurs	Breakup of Pangaea begins / Opening of Atlantic Ocean
	PALEOZOIC	Permian		286	Coal-forming forests diminish	Supercontinent Pangaea intact / Culmination of mountain building in eastern NA (Appalachians); extensive glaciation of southern continents
		Carboniferous — Pennsylvanian		320	Coal-forming swamps abundant / Sharks abundant / Variety of insects	Warm conditions, little seasonal variations; most of NA under inland seas
		Carboniferous — Mississippian		360	First amphibians / First reptiles	
		Devonian		408	First forests (evergreens)	Mountain building in Europe (Urals, Carpathians)
		Silurian		438	Early land plants	
		Ordovician		505	Invertebrates dominant / First primitive fishes	Beginning of mountain building in eastern NA (rest of NA low and flat)
		Cambrian		570	Multicelled organisms diversify / Early shelled organisms	Extensive oceans cover most of NA
PRECAMBRIAN	Proterozoic ("early life")				First multicelled organisms	Formation of early supercontinent (~1.5 billion years ago)
				2500	Jellyfish fossil (~670 mya)	Abundant carbonate rocks being deposited; first iron ore deposits / Oldest known sedimentary rocks
	Archean ("ancient")				Early bacteria and algae	Primitive atmosphere begins to form (accumulation of free oxygen) / Earth begins to cool
				3800		Oldest known rocks on Earth (~3.96 billion years ago)
	Hadean ("beneath the Earth")				Origin of life?	Oldest moon rocks (~4 billion years ago) / Earth's crust being formed
				4600	FORMATION OF THE EARTH	

(Life forms column note — Age classifications along epoch/period divisions: Age of Mammals (Cenozoic), Age of Reptiles (Mesozoic), Age of Amphibians (Permian–Carboniferous), Age of Fishes (Devonian–Silurian), Age of Marine Invertebrates (Ordovician–Cambrian))

What Is the State of the Biosphere?

In 2006, most traditional first-year college students — those approximately 18 years of age — will begin their post-secondary educations unaware that their birth year, 1988, brought about a sea-change in American culture. That's because in 1988, *Time* Magazine broke with tradition; in place of its Man of the Year award, given to the person who most influenced events in the world that year, *Time* named Endangered Earth the Planet of the Year. This was a surprising move by a mainstream publication that had no affiliation with an environmental organization or cause, and it grabbed the public's attention.

The reasons behind the magazine's unusual action were obvious. In 1988, the United States experienced one of the worst heat waves ever recorded, with temperatures across the nation soaring into the 100s for days on end. At the same time, many regions of the nation suffered through a severe three-month drought. Stunted crops withered in the fields; the country's grain harvest was reduced by almost a third. The drought also affected natural systems. With awe and horror, the public watched as fires raged through rain-parched Yellowstone National Park and many western forests. Elsewhere, devastating hurricanes tore through the Caribbean, and a catastrophic flood ravaged Bangladesh.

In addition to the natural catastrophes of 1988, many human-caused environmental calamities generated public attention. Along the U.S. coastline, particularly in the East, sunbathers found beaches despoiled by garbage, raw sewage, and medical wastes. On the beaches of the Mediterranean and the North seas, Europeans encountered similar kinds of pollution. Scientists watched a growing hole in the ozone layer. The hole had appeared over the Antarctic each fall for the previous several years, and scientists had linked its appearance to chemicals produced and released to the atmosphere by humans. Ozone, O_3, a naturally occurring gas in the upper atmosphere, shields the planet from ultraviolet (UV) radiation, which is harmful to life. The destruction of the ozone layer, scientists warned, meant an increase in the amount of UV radiation that would reach the Earth, with possible adverse health effects.

The extraordinary events of 1988 spawned a global environmental awakening. Many people believed that the year's weather-related catastrophes — the heat wave, drought, flood, and hurricanes — were evidence of global climate change caused by human activities.

Whether or not they were can't be proven. Variations in temperature and rainfall can be caused by a number of natural phenomena, including ocean currents and increased solar radiation brought on by sunspots and solar flares. The 1991 eruption of the Philippines' Mt. Pinatubo, for instance, spewed so much volcanic debris into the upper atmosphere that it reduced the amount of light and UV radiation reaching the Earth, thus leading to slightly cooler global temperatures in 1992. Nevertheless, the public perception was that human actions were responsible for the year's environmental woes. That perception persists, bolstered by more recent events, including the record 2003 heat wave in Europe that killed an estimated 35,000 people; the 2004 monsoon-driven floods and mudslides in Southeast Asia that killed approximately 2,000 people and left millions more homeless; and the record 2004 Caribbean hurricane season, which pounded Florida (the state was hit by four major hurricanes in six weeks) and devastated Haiti (where more than 1,500 people were killed by a single hurricane, Jeanne). Such catastrophic events exact an enormous human, economic and environmental toll, and they appear to be part of a troublesome trend. According to the Worldwatch Institute, an independent and non-partisan research group, the frequency of weather-related disasters is rapidly increasing: In the United States, the number of such events has increased fivefold since the 1970s; worldwide, the number has quadrupled since the 1960s.

Now, in the early years of the twenty-first century, environmental issues are widely discussed and debated, not just in the halls of Congress and the United Nations, but also in living rooms and classrooms, in corporate headquarters, and at corner markets. By naming Endangered Earth the Planet of the Year, *Time* held up a mirror to the world, and what we saw was a small and fragile planet. The vision in that mirror was frightening and sobering, for it reflected the three root causes of environmental problems.

What Are the Three Root Causes of Environmental Problems?

There are three root causes of environmental problems: growth of the human population, abuse of resources and natural systems, and pollution. They exert the most disruptive pressures on the biosphere, altering environmental conditions in such a way that one or more members of the biota are adversely affected. As we will see, the three root causes are interrelated; consequently, environmental problems are complex and difficult to solve.

Population Growth

The human population reached 6.4 billion in 2004. Person number 6.4 billion (6.4B) is an unwitting participant in an already crowded race. By the end of 6.4B's teen years, another billion contestants may have joined the race. When 6.4B is 60 years old, another seven billion racers could swell the ranks of contestants!

What is this race and what are the racers trying to win? It is the global population growth race, and the contestants are trying to win some share of the world's space, food, water, and shelter. How much they are likely to win depends on where they were born, to whom, and the educational opportunities they enjoy. But the racers have no say in these matters, and they cannot decline the invitation to race. Birth automatically enters each one of us.

The race only can become progressively more intense because global population growth is exponential rather than linear. **Linear growth** occurs when a quantity increases by a fixed amount in a given time period (1, 2, 3, 4, 5, 6). **Exponential growth** occurs when a quantity increases by a fixed percentage of the whole in a given time period. With exponential growth, also known as geometric or compounding growth, doubling can occur (1, 2, 4, 8, 16, 32, 64). To illustrate the difference between linear and exponential growth, imagine that you are given a job that will last 30 days. You are offered a choice of one of two salaries. With option #1, you will be paid $1,000 on the first day with an increase of $100 on each succeeding day ($1,000; $1,100; $1,200 and so on). With option #2, you will be paid one cent on the first day, with the promise of an exponential increase on each succeeding day ($0.01; $0.02; $0.04 and so on). Which option would you choose? If you like, you can do the math yourself, but the totals are as follows: Option #1 will pay you $73,500 at the end of the month. Option #2 will pay you over $5 million on the last day alone!

Exponential doubling may be a good thing where your salary is concerned, but with regard to population growth, it presents serious problems. The increase in numbers (that is, additional births) depends upon reproduction by individuals in the population. When human populations were small, the doubling caused by exponential growth did not produce a dramatic increase in total population numbers. In fact, the human population did not reach one billion until about 1800, and 130 years passed before it doubled to two billion. But from 1930 to 1975, just 45 years, it doubled again, to four billion. Clearly, once the population reached a critical size, the doubling effect became significant. At current growth rates, the global population would double again in 54 years, rising to nearly 13 billion by 2058. As Figure 1-2 illustrates, the historical growth of the human population, when graphed, takes the shape of the letter J. That growth, which began so slowly, speeded up dramatically once we rounded the J-curve. It will be difficult to slow the pace of human population growth (and as we shall see in Chapters 7 and 8, virtually impossible within the next few generations), with the consequence that contestants in the race for the necessities of life face a continual and increasing struggle.

Numbers tell only part of the story. What does the global population mean in terms of how people live? According to the United Nations, of the approximately 6.4 billion people who now inhabit the Earth, only one-fifth have adequate food, housing, and safe drinking water. Four-fifths do not. Each day, malnutrition and related illnesses result in the deaths of at least 19,000 children worldwide. War, internal political strife, and natural disasters such as droughts and floods exacerbate the stress caused by growing populations and thus contribute to hunger. Most of the severely stressed are in the developing (also called the less-developed or less-industrialized) countries, but even in some developed (industrialized) nations, many people do not have enough food to eat and cannot afford a warm, dry place in which to live.

Resource Abuse

A **resource** is anything that serves a need; it is useful and available at a particular cost. All organisms consume resources (another way of saying that all living things live at the expense of their environment). They produce wastes, which then become resources for other organisms. Resources can be categorized as perpetual, renewable, or nonrenewable. A **perpetual resource**, such as solar energy, is one that originates from a source which is essentially inexhaustible. A **renewable resource** can be replaced (renewed) by the environment, and as long as it is not used up faster than it can be restored, the supply is not depleted. Forests and the famed blue crabs of Chesapeake Bay are renewable resources. In contrast, a **nonrenewable resource** exists in finite supply or is replaced by the environment so

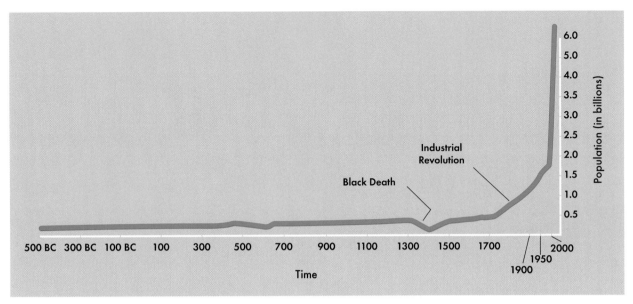

Figure 1-2: Growth in human population over time. The current rapid growth of the human population holds potentially catastrophic consequences for civilization and the biosphere.

slowly (in human terms) that, for all practical purposes, the supply might as well be finite. With use, the supply of a nonrenewable resource is depleted. Coal, oil, and other fossil fuels are nonrenewable resources.

All organisms exploit the environment to the best of their ability. Humans are one of the most successful of all species at doing so. Technology has enabled us to extract and use resources more easily and has allowed us to overcome or minimize the effects of such things as extreme climate, other predators, and certain diseases that might otherwise keep our population at lower levels. The technologies we have developed, the fact that we tend to satisfy wants rather than needs, and our sheer numbers have resulted in an environmental impact greater than that exerted by any other single species. To understand the extent of this impact, let's look at a concept scientists call net primary productivity (NPP). The **net primary productivity** is the total amount of solar energy fixed biologically through the process of photosynthesis minus the amount of energy that plants use for their own needs. Humans appropriate an estimated 40 percent of the potential NPP on land, chiefly for food but also for fiber, lumber, and fuel. What are the implications of this for other species?

Human demand for renewable resources appears to be accelerating faster than the biosphere can replenish them. To fully appreciate that statement, think of the environment as a savings and investment portfolio. When it comes to your own personal finances, your goal is to have enough money in savings and investments — your capital — so that you can live off of the

interest that your money earns. If you suffer a personal catastrophe, you can always dip into your savings, but you try to spend no more than you can comfortably afford using only your interest earnings. In that way, your capital is protected and continues to generate the income that sustains you. Unfortunately, as a species, we are living off the capital of our environment rather than the interest. Consider the following. According to the United Nations Food and Agriculture Organization, nine of 17 major ocean fisheries are in decline, chiefly due to overharvesting, pollution, and human population pressures on sensitive coastal areas. Land systems are also threatened. Some experts estimate that human activities have transformed 10 percent of the land surface from forests and rangelands into desert, and another 25 percent is at risk (Figure 1-3). Existing agricultural land lost through erosion alone is estimated at 15 to 17 million acres (six to seven million hectares) annually; an additional 3.7 million acres (1.5 million hectares) are lost to other factors related to human activities. The pace of tropical deforestation, which has slowed little since reaching an all-time high in the 1980s, is also alarming. Over 31 million acres (12 million hectares) of tropical forest are lost each year. The nonprofit World Resources Institute estimates that almost half of the Earth's original tropical mature forests have been cleared to accommodate other uses. Tropical rain forests house about half of all the species on Earth, and they help to regulate the global climate. Moreover, like all forests, they prevent erosion and thus protect streams and rivers from excess siltation.

FIGURE 1-3: Desertification attributable to overgrazing, Yatenga Province, Burkina Faso.

To measure humanity's demand on the Earth's renewable natural resources, scientists with the World Wildlife Fund, in partnership with the Global Footprint Network, have developed a tool called the Ecological Footprint. The graph in Figure 1-4 tracks humanity's Ecological Footprint from 1961 through 2001. The horizontal axis marks time in ten-year increments. The vertical axis charts the number of planets needed to support the human population; one planet equals the total biologically productive capacity of the Earth in any single year. In just 40 years, humanity's Ecological Footprint increased 2.5 times, and in 2001, it exceeded the Earth's biological capacity by 20 percent. The bottom line? We are depleting the Earth's natural capital, a trend that simply is not sustainable.

A debate rages regarding the actual cause of resource abuse and depletion. Some people contend that the growing global population is consuming resources faster than the Earth's living systems can replace them. Others argue that sufficient resources are available to support the world's people, if we can find the means to distribute the resources more evenly and to stop their wasteful use. They point out that about 20 percent of the world's people — those living in the developed world — consume the lion's share of the world's resources, including about 60 percent of fossil fuels, 80 percent of metals, and 75 percent of paper. The United States alone, with just five percent of the global population, uses 25 percent of the world's fossil fuels, 20 percent of metals, and 33 percent of paper.

The developed world as a whole generates 90 percent of the world's hazardous waste, with the United States accounting for 72 percent of the total. What are the environmental implications if countries with a standard of living barely above subsistence want a standard of living similar to that of the United States, Canada, Australia, western Europe, or Japan? *Focus On: A Culture of Consumption* (page 19) looks more closely at the issue of resource consumption and its implications for people and natural systems.

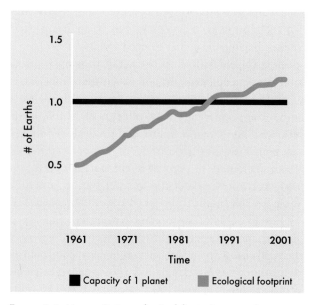

FIGURE 1-4: Humanity's ecological footprint over time.

Pollution

Most of us find it easy to identify and describe pollution. Newspapers, radio, and television draw our attention to dramatic pollution events, such as the catastrophic 1984 release of deadly methyl isocyanate from Union Carbide's pesticide production facility in Bhopal, India. Worldwide, there are countless constant reminders of the less dramatic, but more insidious, continual and increasing pollution of our environment — the recent, precipitous worldwide decline in amphibian species; the rapid warming of the Arctic, where temperatures are climbing at almost twice the global average; dead fish on stream banks; litter in national parks; decaying bridges and buildings; leaking landfills; and dying forests and lakes.

In the United States, many people experience the effects of pollution firsthand. Some are forced to evacuate their homes because train and truck accidents occurring nearby have caused the release of toxic contaminants. Others become ill after drinking contaminated water or breathing contaminated air. In certain areas of the country, residents no longer can swim at favorite beaches because of sewage contamination. Restrictions are placed on the consumption of fish, shellfish, and meat because of the presence of harmful chemicals, cancer-causing substances, and hormone residues (Figure 1-5). Nuclear contaminants released to the air and water from uranium-processing plants in areas such as Hanford, Washington, and Fernald, Ohio, spread fear and concern among the people living nearby. Exposure to these contaminants has been linked to fatal and debilitating conditions such as liver, kidney, and nerve damage; lung and bone marrow cancers; birth defects; stillbirths; spontaneous abortions; and reduced fertility.

Serious pollution problems do not occur only in developed, affluent countries. In much of the developing world, untreated human sewage contaminates local water sources; industrial boilers and furnaces spew choking soot and emissions into the air; and open garbage dumps pollute local water and air supplies. There are few regulations on polluting industries, since the processes that produce the pollution are associated with jobs and economic development; even fewer are enforced. Indeed, those countries that are most aggressively pursuing economic development, in the style of the industrialized world, are paying the highest price. A 1998 report by the World Health Organization (WHO), for example, found that China is home to seven of the world's 10 most polluted cities.

FIGURE 1-5: Don't eat the shellfish! A sign in Wells, Maine, near Laudholm Beach, warns people not to eat mussels and other shellfish due to contamination.

While pollution is often easy to identify, it is less easy to define. A **pollutant** is a substance or form of energy, such as heat, that adversely alters the physical, chemical, or biological quality of natural systems or that accumulates in the cells or tissues of living organisms in amounts that threaten their health or survival.

Pollution can come from natural or cultural sources. It is an inevitable consequence of the biological, chemical, and physical processes of the Earth. Volcanoes, fires caused by lightning, products from the decomposition of materials in swamps and soils, sulfurous gas seeps, and other natural processes all contribute pollutants to the Earth's soil, water, and air. Further, every living organism produces wastes.

Historically, as long as human numbers remained small, the environment could adequately process human wastes and discards. But as population numbers increased, those wastes and discards began to overwhelm natural systems. As societies became technologically sophisticated, humans began producing large numbers and volumes of new substances. It is difficult, and sometimes impossible, for the environment to absorb and process these substances.

What Must We Know to Solve Environmental Problems?

Human population growth, resource abuse, and pollution occur when human behaviors and activities, or the products of those behaviors and activities, are incompatible with natural systems. In other words, environmental problems stem from the interaction of natural and cultural systems. Obviously then, to solve environmental problems, it's essential that we understand both systems.

Understanding Natural Systems

To understand natural systems, we rely on **science**, the rigorous and systematic search for unbiased explanations of natural phenomena, from the world within the cell to the far reaches of the cosmos. The term "science" also refers to a body of knowledge about nature and the physical world that has been derived from observation, study, and experimentation. If you learn only one thing from this chapter, let is be this: The body of scientific knowledge is not static and complete. Rather, it continues to grow and deepen — and sometimes change — as our understanding of the world changes.

There are many different scientific disciplines, or natural sciences, including biology, chemistry, physics, and geology. These disciplines help us understand natural systems and the biological, chemical, and physical processes of our environment. Particularly helpful is the natural science of **ecology**, the study of the structure, function, and behavior of the natural systems that comprise the biosphere. Ecologists study the interrelationships among and between organisms, and between organisms and all other components, living and nonliving, of their environment. The term "ecology," first proposed by Ernst Haeckel in 1866, comes from the Greek root *oikos*, meaning "home," and *logos*, meaning "the study of." Literally, it refers to the study of the home.

As a discipline, ecology grew out of the scientific interest in the natural history of plants and animals and from the pioneering work of plant geographers. In the 1940s, many biologists urged that a course in ecology be offered for all majors in the biological sciences. The first widely used college textbook for ecology students, Eugene Odum's *Fundamentals of Ecology*, was published in 1953 and focused attention on understanding the dynamics of the biosphere through the study of natural systems such as lakes, forests, and tidal pools. Ecologists study nature holistically, that is, as a functioning system instead of as a collection of distinct, unrelated parts. They try to understand how the parts function together to form a wetland, prairie, stream, or coral reef. Ecological research, for example, can help us to understand how nutrients and energy are transferred from organism to organism, how climate is modified by forest destruction, and how one species affects the population dynamics of another (Figure 1-6).

FIGURE 1-6: Researchers catch larval fish to estimate the population size of yellow perch in Lake Mendota, Wisconsin.

A Culture of Consumption

One of the most important aspects of environmental educa-tion is to ask students to examine deeply held beliefs about our culture, society, and lifestyle. It may be uncomfort-able to think about the ways in which our attitudes and behaviors contribute to environmental problems, but it also can be enlightening and liberating. In the following essay, we ask you to consider the topic of consumption, specifically the degree of consumption prevalent in the United States. We also invite you to find out if you have the "consumption bug"—affluenza—by visiting the website www.pbs.org/kcts/affluenza/. There, you can find ways to "beat the bug," should you choose to do so.

In *The Price*, the American playwright Arthur Miller describes the national character in this way:

"Years ago a person, he was unhappy, didn't know what to do with himself — he'd go to church, start a revolution, something. Today you're unhappy? Can't figure it out? What is the salvation? Go shopping."

Think about it. In every facet of life, Americans appear obsessed with having the latest and greatest: from iPods to online gaming systems, from personal computers to plasma screen television sets, from trendy sportswear to designer shoes, from sport utility vehicles to Hummers. The palace of Kubla Khan has nothing on the twenty-first century American home, replete with a huge master bedroom suite, his-and-hers walk-in closets, surround-sound theater, exercise room, and a kitchen outfitted for a master chef. As a society, we equate "more" with "better," and as a result, we have raised the acquisition of material goods to an art form. Ours is a culture of consumption.

Some people argue that consumption is positive, for it raises one's standard of living. To a point, they're right. In fact, the world's poorest people *must* increase their consumption — of food, water, energy, and other basic necessities — just to survive. Even in developed nations, those with low incomes need to consume more

in order to break the cycle of poverty. But studies show that beyond a certain level, increasing consumption does not correlate with increasing happiness. In fact, the percentage of Americans who describe themselves as "very happy" peaked in 1957 and has remained stable or declined ever since.

Most of us enjoy a standard of living that ensures a comfortable life, and yet we never seem to have enough. For proof, just look around you. If you're like many Americans, there's *stuff* everywhere: in your cabinets, closets, basement, garage, even the trunk of your car. In 1950, the average American home was 900 square feet; by 2004, it had risen to 2,200 square feet, despite the fact that average household size had fallen.

In the mid-1990s, the Public Broadcasting Service aired *Affluenza*, a television program that examines the high social and environmental costs of unrestrained consumption. The program focused on our era's fasci-nation with materials goods. How did it get to be this way? Centuries ago, affluenza was confined to the very wealthy. The aristocracy of eighteenth-century Europe was able to live in the lap of luxury in part because of the wealth it drained from its colonies, particularly those in the Americas. By the late nineteenth century, however, three major developments jump-started our culture of consumption: the explosion of advertising, the advent of planned obsolescence, and the rise of consumer credit.

The culture of consumption is fueled by advertising — the art of making wants appear to be needs. The basic premise of the culture of consumption is that we can never have enough; advertising succeeds when it convinces us that we are not satisfied with what we already have. In 2002, global spending on advertising climbed to $446 billion, about nine times the amount spent in 1950. That's $446 *billion* with a *b*, all of it designed to make us want more *stuff*: vehicles, electron-ics, appliances, toys, clothing, personal-care products, and countless gadgets. More than half of the global advertising total is spent in U.S. markets.

Planned obsolescence is the development of products designed to wear out, break down, or become dated. Razor blades and the "model year" automobile are two early examples of planned obsolescence. In 1895, a traveling salesman named King Gillette invented the disposable razor, a product touted as a safe and convenient alternative to the straight-blade razor. Gillette knew that customers who opted for a disposable razor would be back again and again to buy replacements. The product was a hit; within 20 years, 70 million razors had been sold, and today, Gillette is a multinational company with annual sales of $10 billion.

Similarly, around 1925, General Motors Company introduced the "model year" car. The idea was simple: Make slight changes to the product each year — color, for example — to entice consumers looking for the most fashionable automobile. Having the latest model became a status symbol, and "keeping up with the Joneses" entered the national psyche.

Advertising and planned obsolescence fueled consumer desire, but it was the widespread use of credit that allowed the average citizen to act on impulse. By 1920, credit plans were touted as the best way to buy big-ticket items; with a down payment and a credit plan, you could instantly realize the American Dream. Credit cards, introduced in the 1940s, dramatically upped the ante. For the first time, consumers could purchase just about anything, anywhere, with plastic.

Fast forward to the twenty-first century. Consumption has become an end in itself, and we are encouraged to buy, buy, buy. This is particularly true in December, of course, but virtually every holiday has become a gift-giving occasion. Often, politicians lead the refrain, arguing that consumer spending keeps the economy chugging along. Company executives, who ultimately answer to their shareholders, agree with marketing analyst Victor Lebow (1950): "Our enormously productive economy...demands that we make consumption a way of life.... We need things consumed, burned up, worn out, replaced, and discarded at an ever-increasing rate." But unrestrained consumption has its costs, and sooner or later, the bill must be paid.

For the individual, unrestrained consumption jeopardizes one's financial security. As a percentage of personal income, U.S. consumer credit — the amount of goods and services purchased on credit — has reached a historic high. Simply put, we are buying more and more of the things we *want* on credit, amassing greater levels of debt than ever before. At the same time, we are saving less for the future; Americans save just three to five percent of their disposable income, far less than our counterparts in other developed nations. Mortgage debt as a percentage of personal income also has reached a historic high. And more Americans are using the equity in their homes to buy consumable items like vehicles and luxury vacations. Some home-equity lines of credit even allow you to buy groceries and gasoline. As one financial writer noted, that's like eating your house.

The culture of consumption takes a toll on one's health, as well. "Super-size me" has taken on a whole new meaning, as obesity continues to rise unabated. Approximately 65 percent of U.S. adults are either overweight (10 percent over ideal body weight) or obese (20 percent over ideal body weight). Thirty percent of the nation's children are either overweight or at risk of becoming so. Meanwhile, Americans work longer hours and take less vacation time in order to afford their consumption habit. It's little wonder that so many of us feel overworked and stressed out.

At the national level, our culture of consumption has dramatically increased the use of materials and energy, pollution, and wastes. According to some measures, in the years since the end of World War II, humans have consumed more physical resources than in all of history prior to that global conflict. The amount of waste generated in the United States each year would fill a convoy of garbage trucks long enough to circle the Earth six times. And there are serious economic consequences, as well. Massive debt and a trade imbalance weaken the dollar against foreign currencies. We import goods because things can be made more cheaply overseas.

The one thing we continue to export is our lifestyle. Resource use is rising sharply in some developing nations, notably China and India, as these countries become more prosperous. In *State of the World 2004*, the Worldwatch Institute warns of the consequences of a growing global consumer class. Members commonly own or have access to televisions, telephones and the Internet, as well as the ideas, images, and values these technologies transmit. Half of the global consumer class lives in the developed world, half in the developing world, but it is the latter where the potential for new members is greatest. Fully 80 percent of people in developed nations belong to the consumer class, compared to just 17 percent of those in developing nations. If the consumer class of developing nations grows, as it's likely to, what will be the impact on global water supplies, air quality, natural resources, wild places, and wildlife? What will be the impact on the world's poorest people, those who face a daily struggle for survival? As you'll see throughout this text, poverty is at the heart of many environmental issues, from hunger to deforestation to diminishing biological diversity. And finally, if the gap between rich and poor continues to widen, what will be the impact — environmentally, politically, and socially — on all of us?

Ecology and all other natural sciences share a common belief: Knowledge must be acquired in an objective manner through the **scientific method**, a process of observation, hypothesis development, and experimentation. The scientific method, or scientific inquiry, is a rigorous process; fundamentally, it consists of the posing of questions and the painstaking search for precise answers to those questions. Scientific inquiry is limited to questions that can be answered objectively through verifiable observation and experimentation — "how," "when," "where," and "what." Such questions enable us to probe the structure and function of nature. Based on observation and experimentation, scientists draw conclusions to explain a phenomenon or establish a causal link between two events.

Strengths of Scientific Inquiry. When scientists design their experiments and interpret data from their research, they should strive to be free of bias or coercion. Freedom from bias is perhaps the greatest strength of scientific inquiry, for it enables the scientist to obtain objective answers to the questions posed. Two hallmarks of scientific inquiry help to reduce bias and coercion: the controlled experiment and the complete reporting of methods and results to other scientists who may want to repeat the experiment in order to validate the results.

Controlled experiments are designed to compare two situations differing in only one variable. If the comparison yields different results, the difference is attributed to that variable. If the results of the experiment support the hypothesis and the same results are obtained when the experiment is repeatedly conducted under the same conditions, the hypothesis is accepted by the scientific community as an explanation for the original observation. If the experimental results do not support the hypothesis, it is disregarded or revised for further testing. A new look at the observation might lead to new questions and new hypothesis formation, and the process continues.

The more data collected to support the hypothesis, the more acceptable the hypothesis becomes. Typically, experiments and data collection take place over several years; major scientific research requires several decades or more of experimentation and validation of hypotheses.

If, over long periods of time, many verifiable facts support the hypothesis, it might then be considered a theory. Theories unify many related facts or observations. Theories that withstand repeated testing over time become principles or laws. For example, four commonly held biological principles are: (1) all organisms and living systems evolve; (2) all organisms are made of cells; (3) all life is a competition for the capture and use of energy; and (4) organisms share the common characteristics of metabolism, growth, movement, responsiveness (irritability), reproduction, and adaptability.

Because we refer to a statement as a principle or law doesn't mean that all observations support it or that all scientists agree about all aspects related to it. Nor does it mean that our understanding of it might not change as new information is discovered. The natural world is exciting precisely because there are so many interesting unanswered and, as yet, unposed questions to explore.

Limitations of Scientific Inquiry. Scientific inquiry is a powerful tool for acquiring knowledge, but it does have limitations. First, it isn't always possible to conduct controlled experiments. The system under observation might be too large or complex or the time frame may be too long. For example, scientists have observed that the carbon dioxide content of the atmosphere is increasing. They suspect the increase is due to rising fossil fuel combustion and the widespread clearing and burning of forests. Further, they suspect that increasing levels of carbon dioxide will amplify the warming effect of the atmosphere, perhaps raising the average temperature of the planet by several degrees. Since scientists do not have an unindustrialized, uncleared "spare" Earth to serve as a control group in experiments, they base their conclusions about the greenhouse effect on mathematical models. By constructing mathematical models of the atmosphere, scientists simulate the effects of increased levels of carbon dioxide on the Earth.

A second, related limitation of science is that it tends to be reductive. In order to isolate variables and determine their effects, scientists must often look at ever-smaller units in laboratory settings. But determining the effect of a single variable in isolation from other related variables does not afford an accurate picture of how nature works.

The way that scientists approach their work constitutes a third limitation of scientific inquiry. The objective, value-free approach that is a hallmark of good science means that science cannot determine when or if something is good or bad or beautiful or worth preserving; neither can it answer questions about the spiritual or supernatural. Those conclusions and answers are not subject to scientific proof. Instead, they arise out of value judgments. Consequently, science cannot take the place of other ways of knowing, such as philosophy, sociology, theology and religion, art, music, and literature.

Understanding Cultural Systems

As humans, our behaviors and activities are determined in large part by our attitudes, values, and beliefs. Collectively, attitudes, values, and beliefs comprise a person's, or a society's, **worldview** — a way of perceiving reality that includes basic assumptions about the self, others, nature, space, and time. Worldviews, in turn, are both shaped by and reflected in the social and behavioral sciences, arts and humanities, communication, and education. These disciplines help to explain the human dimensions of environmental problems, and they can help us to modify our worldviews and behaviors in order to alleviate or solve those problems.

The social and behavioral sciences, particularly political science and economics, provide us with the information we need to understand local, national, and world events and their implications. Our understanding of these events is greatly enhanced by a knowledge of geography, the descriptive science that deals with the surface of the Earth; its division into continents and countries; and the climate, plants, animals, natural resources, inhabitants, and industries of the various divisions. Through anthropology, history, psychology, sociology, theology and ethics, we study human origins and diverse cultures, beliefs, and values. These disciplines enable us to better understand other societies, civilizations, and worldviews.

The arts and humanities, which reflect societal beliefs and values, allow us to define, record, interpret, and teach each other what it means to be human. Through philosophy, literature, aesthetics, art, music and architecture, we seek to understand and express the value we place on resources, the natural world, and the relationships between natural and cultural systems. Moreover, we often learn about the beliefs and values of other cultures through the arts they inspire. The knowledge we gain enables us to share in the thoughts and experiences of the artist (and by extension, his or her society), and so it broadens our perspective.

Other fields that contribute to environmental studies are communication and education, which help us to acquire, organize, and transmit information. As shapers of public perception, communication and education are powerful agents of change.

The social and behavioral sciences, arts and humanities, communication, and education contribute in important ways to **environmental science**, which we define broadly as the study of the human impact on the environment. This broad definition underscores the fact that humans are part of nature; consequently, their cultural systems must be taken into account when studying the environment. Given the breadth of environmental science, it's not surprising that those who practice it hail from a wide range of academic fields, from anthropology to zoology — with a good many ecologists in between! That's a positive thing, because environmental scientists tackle a host of complex problems: the effect of acid precipitation on the growth and health of vegetation, the impact of agricultural pesticides on insect-foraging spiders in croplands, the consequences of human population shifts between regions and countries, and the true economic value of clean air, functioning wetlands, and healthy forests.

Summary

The biosphere is the thin layer of air, water, soil, and rock that surrounds the Earth and contains the conditions to support life. Nonliving physical environments comprise the abiotic component of the biosphere, and living organisms comprise the biotic component. A number of characteristics describe living organisms. All organisms live at the expense of their environment. All living things have a cellular structure. All living organisms exhibit movement of some type, even if it is only at the cellular level. Living things grow, gaining mass over time. They also reproduce; not all individuals of a given species may reproduce, but the species as a whole must be capable of reproduction or it would die out as its members died. Living things respond to stimuli. Finally, living things evolve and adapt.

According to the big bang theory, all matter, energy, and space arose from an infinitely dense, infinitely hot point called a singularity about 13 to 20 billion years ago. Approximately five to 10 billion years ago, our solar system developed from gas and dust on the outer edge of the galaxy we call the Milky Way. The Earth formed about 4.6 billion years ago, condensing out of interstellar gas and dust. Life on Earth developed soon after the surface of the young planet cooled, somewhere between 3.5 to four billion years ago. In the harsh environment of the young planet, evolution proceeded slowly, driven by mutation and natural selection. A mutation is a random change within the genetic material of an individual that can be passed on to that individual's offspring. While most mutations are harmful or result in changes that are useless to the organism, some mutations cause the organism to differ in a way that allows it (or its offspring) to adapt to a change in environmental conditions. Because the individual is better suited to the environment than others in the population, it is more likely to survive and reproduce. This phenomenon is known as natural selection, the process which enables individuals with traits that better adapt them to a specific environment to survive and outnumber other, less well-suited individuals.

Change has been a constant characteristic of the planet throughout its history. Recently (in geologic time), however, a new agent of change has altered the planet in unprecedented ways. Modern culture has enabled humans to exert an influence far greater than that of any other single species in the history of the planet. We have been able to modify our environment significantly through technology and social institutions. Our achievements, however, have blinded us to the intricacies of nature, for while we are able to modify certain aspects of our existence, we cannot control natural processes. The environment sets limits on all creatures. Many people believe that we have reached those limits; as evidence, they point to worldwide environmental degradation and problems such as global climate change.

The three root causes of environmental problems are population growth, abuse of resources and natural systems, and pollution. They are interrelated; two or three of these factors acting together underlie many environmental problems.

The global human population, which in 2004 was approximately 6.4 billion, is growing exponentially. Exponential growth occurs when a quantity (or population) increases by a fixed percentage of the whole in a given time period. It is also known as geometric or compounding growth, and can

Discussion Questions

1. Give at least three examples of the ways in which humans have become a force of significant environmental change.

2. How are the three root causes of environmental problems interrelated?

3. Predict the implications for the biosphere (including humans) if the human population doubles in your lifetime and resource use continues to increase.

4. Humans appropriate 40 percent of the Earth's net primary productivity. Explain the implications of this for all other species.

5. List three local and three national or global environmental problems. What role do population growth, resource abuse, and pollution play in each of these problems?

6. Refer to the problems you identified in Question 5. How might scientific research contribute to a better understanding of each of these problems?

7. Why is scientific inquiry such a powerful tool for acquiring knowledge?

8. Again refer to the problems you identified in Question 5. Which of these problems can be solved by scientists alone? Please explain. If they cannot be solved by scientists alone, what kind of knowledge (from the social and behavioral sciences, arts and humanities, communication, and education) is needed in order to develop effective solutions?

cause a sum or population to double. When graphed, exponential growth takes the shape of the letter J. When human populations were small, the doubling caused by exponential growth did not produce a dramatic increase in total population numbers. But if current growth rates remained steady, the global population would double again, to nearly 13 billion by 2058.

The growing human population will exert increasing pressure on natural systems, with the likely result being increased resource abuse. A resource is anything that serves a need; it is useful and available at a particular cost. A perpetual resource is one that originates from a source that is essentially inexhaustible. A renewable resource can be replaced (renewed) by the environment, and as long as it is not used up faster than it can be restored, the supply is not depleted. A nonrenewable resource exists in finite supply or is replaced by the environment so slowly (in human terms) that, for all practical purposes, the supply might as well be finite. As it is used, a nonrenewable resource is depleted. Our use and abuse of resources has a significant impact on other species. Net primary productivity is the total amount of solar energy fixed biologically through the process of photosynthesis minus the amount of energy that plants use for their own needs. Humans appropriate an estimated 40 percent of the potential net primary productivity on land, mostly for food, but also for fiber, lumber, and fuel. As a species, we are living off the capital of our environment, rather than the interest. There is widespread evidence to support this statement: depleted fisheries, degraded croplands and grazing lands, and rapid deforestation.

Population growth and resource abuse result in increased pollution. A pollutant is a substance that adversely affects the physical, chemical, or biological quality of the Earth's environment or that accumulates in the cells or tissues of living organisms in amounts that threaten the health or survival of those organisms. Pollution can come from natural or cultural sources. Examples of the former include volcanoes, fires caused by lightning, products from the decomposition of materials in swamps and soils, and sulfurous gas seeps. Examples of the latter include electric generating plants, automobiles, hazardous waste dumps, and litter.

Environmental problems stem from the interaction of natural and cultural systems. Therefore, to solve environmental problems, we must understand both systems. To understand natural systems, we rely on science, the rigorous and systematic search for unbiased explanations of natural phenomena. Particularly relevant is the natural science of ecology, the study of the structure, function, and behavior of the natural systems that comprise the biosphere. In their quest to understand nature, scientists employ the scientific method, a systematic process of observation, hypothesis development, and experimentation. Scientific inquiry is a rigorous process in which scientists pose questions that can be answered objectively through verifiable observation and experimentation — questions such as "how," "when," "where," and "what."

One question that science cannot answer is "why." That's because "why" questions are the result of human behaviors and activities, which are determined in large part by our worldview, a way of perceiving reality that includes basic assumptions about the self, others, nature, space, and time. A worldview is the collective product of a person's — or a society's — attitudes, values, and beliefs. The social and behavioral sciences, arts and humanities, communication, and education help to explain the human dimensions of environmental problems, and they can help us to modify our worldviews and behaviors in order to alleviate or solve those problems. In so doing, they contribute in important ways to environmental science, broadly defined as the study of the human impact on the environment.

KEY TERMS

abiota	exponential growth	perpetual resource
big bang theory	global environment	pollutant
biosphere	linear growth	renewable resource
biota	mutation	resource
ecology	natural selection	science
environment	net primary productivity	scientific method
environmental science	nonrenewable resource	speciation
evolution		worldview

CHAPTER 2

Social Factors

Learning Objectives

When you finish reading this chapter, you will be able to:

1. Explain what religion is and how it is tied to ethics and environmental interactions.

2. Explain what ethics is and how ethics affect a person's or society's interactions with the environment.

3. Describe the study of economics and how economic growth can be damaging to the environment.

4. Discuss the connection between ecological economics and sustainable development and explain the importance of these concepts in managing environmental issues.

5. Explain how government activities, including environmental policy and law, contribute to environmental problem solving and resource management.

(continued on next page)

If we cannot end our differences at least we can help make the world safe for diversity. For in the final analysis our most basic link is that we all inhabit this small planet. We all breathe the same air. We all cherish our children's future. And we are all mortal.

John F. Kennedy

Protecting the world's environment is part of the natural order, and those who damage it are showing contempt for the divine nature of created things.

Pope John Paul II

We have an assignment for you. Don't worry; it should take only a few minutes. Read this paragraph, then close your eyes and imagine the kind of world you want to live in 10 or 20 years from now. Not the larger world, but the world *you* personally inhabit: where you'll live, the kind of job you'll have, the activities and pastimes you'll enjoy. Conjure up a particular day. What do you see yourself doing? What's your state of mind — are you feeling happy or sad, confident or insecure? Are you hopeful or worried about your future? Go ahead: Take some time and start imagining.

All finished? Well, before you start thinking this was just a silly exercise, consider the following: The scenario you imagined is being shaped, for better or worse, by the decisions you made yesterday, those you're making today, and those that you'll make tomorrow. Inevitably, those decisions affect not just you, but the environment as well. And that's true for every one of us.

Take a simple example. How do you get to class or work? Do you walk, ride a bicycle, take a bus, or drive your own car? If you drive, what kind of gas mileage does your car get? Try another example. When you shop for groceries, do you ask for paper or plastic bags, or do you bring your own? Do you shop at a major grocery store, or do you make an effort to buy locally grown foods?

These may seem like little things, but multiply them by the current population of your city, state or country, and you'll see that they quickly add up. Moreover, with time, your decisions will have an even greater environmental impact. Think about it. When you're ready to take that step, what kind of permanent living quarters will you choose? Will you be satisfied with a small house or apartment that offers adequate space or do you have your heart set on a spacious home? If you want children, how many? At the polls on Election Day, will you support candidates who believe in strong government regulation of pollution sources or will you support those who favor a more hands-off approach in the name of economic growth?

Just as you help shape the world on both a personal and societal level by the decisions you make daily, logically, it follows that *your world* is being shaped by the decisions of many others: your family members, friends, neighbors, and fellow citizens — not to mention people around the world. So, while we asked you to concentrate on the world you personally inhabit, that's a trick question, because we all share this one, finite planet. Think about that last statement in light of the events of September 11, 2001. In your 10- or 20-year scenario, do you imagine yourself and your country more or less secure than now? Is it possible for those of us in the developed world to be more secure in the future if we continue to consume the lion's share of the world's resources? And even if it is possible, is it ethical?

All but the most naive individuals — or those determined to ignore the obvious — can see that humans are the greatest agents of change on the planet. The social, economic, and political decisions that we make all greatly affect the environment. Consequently, any study of environmental issues is incomplete without a basic understanding of the institutions and mechanisms of social change. In this chapter, then, we shine the spotlight on some of the social factors that play the most significant roles in shaping our world: religion, ethics, economics, government, and education.

(continued from previous page)

6. Discuss what is included in environmental education, how it can be taught, and how it affects environmental problem solving and resource management.

7. Discuss how religion and ethics, economics, government and politics, and environmental education can work together to contribute to environmental problem solving and resource management.

What Is Religion?

Religion is the expression of human belief in, and reverence for, a superhuman power, which may be recognized as a creator and governor of the universe, a supernatural realm, or an ultimate meaning. In many religions, rites and rituals are formal expressions of the followers' belief in a supreme being or a spiritual realm beyond physical existence.

Religions are characterized by a core of beliefs or teachings that answer basic questions about the universe, existence, the world, and humankind: Is there an ultimate spiritual presence in the universe? If so, what is the nature of the Divine? How was the world created? Who are we and why are we here? Two important questions concern humanity's place in the world: Are humans part of or apart from "nature"? Is it humanity's role to act as lord and master of creation or as its steward?

The majority of the world's peoples adhere to some particular religion (Table 2-1). The diversity of these religions reflects the diversity of human cultures. Despite this diversity, all major religions share two common core beliefs. First, each has some version of the golden rule: "Do unto others as you would have them do unto you." Second, every major religion is based on the sacredness, or divine nature, of the creation and instructs its followers to care for, or steward, the natural world.

How Do Religious Beliefs Affect the Environment?

Religion is a powerful force in many people's lives, and it can be a primary influence on the relationship between humans and nature. The values, beliefs, and attitudes inspired by religions may influence how the faithful respond to environmental concerns. For example, a religion that espouses the belief that God ultimately will solve all earthly problems can cause followers to delay taking action on a specific environmental problem or to take only moderate action. Religions also may restrict the options available to followers who are struggling with such issues as family size. Perhaps most important, religious beliefs are a powerful shaper of worldviews. To see how, let's take a closer look at the impact of Judeo-Christian thinking on the dominant worldview of western societies.

Prior to the dawn of Judaism, human societies practiced **pantheism**, the belief that multiple gods were responsible for the various forces and workings of nature. These gods inhabited and were active in the world — in the waters, the winds and weather, the Earth itself, and in the rich abundance of plants and animals (Figure 2-1). In contrast, Judaism, and later Christianity, are based on **monotheism**, the belief in a single God who created the universe but is separate from and outside of His creation. Some believers interpreted the Book of Genesis to mean that God intended humans to steward Earth and its creatures, but others saw it as a license for humans to dominate all of creation, using the environment and its resources as they saw fit. Another important Judeo-Christian belief predicted that God would bring a cataclysmic end to the Earth sometime in the future. One consequence of this belief is the view that the Earth is only a temporary way station on the soul's journey to the afterlife.

Not surprisingly, those Judeo-Christians who believed in human dominion over all creation and the temporary nature of the Earth generally devalued the natural world and focused instead on the spiritual realm. Theirs was — is — an **anthropocentric worldview**, a way of perceiving reality that places humans in

---

TABLE 2–1: **Estimated World Percentages of Adherents, Various Religious Beliefs, 2002**

Religion	Percentage of World Adherents
Christians	32.71
Roman Catholics	17.28
Protestants	5.61
Orthodox	3.49
Anglicans	1.31
Others	5.02
Muslims	19.67
Hindus	13.28
Buddhists	5.84
Sikhs	0.38
Jews	0.23
Others	13.05
Nonreligious	12.43
Atheists	2.41

Source: *CIA World Fact Book, est. 2002 values.*

Figure 2-1: Initiation ceremony, Angola. Dancers in this ceremony represent ancestors and animal spirits.

a preeminent position that is both above, and separate from, the rest of nature. Anthropocentrism *tends* to foster attitudes and behaviors that have a negative effect on the environment, and it *may* cause people to abuse natural resources, disregard the needs of other species for habitat and food, and consider pollution and resource depletion as simply the necessary and natural consequences of economic progress.

In contrast, a **biocentric worldview** is a way of perceiving reality that recognizes an inherent worth in all life and maintains that humans are no more or less valuable than all other parts of creation. Biocentrism encourages humans to live in harmony with nature; it *tends* to foster a sense of care and concern for the natural world. Native American religious beliefs see humankind as part of nature, as do many Eastern religions, particularly Taoism-Confucianism and Buddhism. (See *Focus On: Native American Beliefs*, page 30.)

Being an observant reader, you surely noticed that in the preceding paragraphs, we italicized the words "tends" and "may." That's because those statements are generalizations; they do not apply in all cases. Personal interpretations of religious teachings, and how carefully people follow their professed beliefs, play a significant role in shaping human behavior toward the environment. For example, even Buddhist countries such as Thailand struggle with serious environmental degradation.

Similarly, anthropocentrism doesn't necessarily lead to environmentally unsound behaviors. In recent years, the **stewardship worldview** — a way of perceiving reality that maintains humans have a responsibility to care for the Earth — has gained wider acceptance in the West. While still anthropocentric, the stewardship worldview fosters a sense of nurturing and care for creation. Some Jews and Christians argue that God never intended humans to abuse or exploit the natural world, its resources, and other species. They contend that humans are a link between God and nature, part of a system created by God and pronounced by God as good. Accordingly, the stewardship worldview is really a return to early Judeo-Christian values. As the theologian Richard A. Baer, Jr., points out, "though Biblical writers demythologized nature, they always treated it with deep respect as the glorious creation of a Supreme Architect."

What Are Ethics?

Ethics is the branch of philosophy concerned with standards of conduct and with distinguishing between right and wrong behavior. Often, but not always, an individual's ethic, or personal code of right and wrong behavior, is based on his or her **morals**, principles that help to distinguish between good and evil. Morals imply a faith component: faith in God or faith in an ultimate, universal meaning or truth.

Distinctions between right and wrong and good and evil are not always simple ones. Not surprisingly, then, there is no universal human ethic. What one person considers acceptable behavior may be considered unethical by another. For example, there are two radically different positions with respect to the proposal to drill for petroleum in Alaska's Arctic National Wildlife Refuge (ANWR). One side maintains that drilling is ethical and necessary because oil is vital to our nation's economy and national security. The opposing side maintains that it is neither ethical nor necessary; they contend that drilling in ANWR will degrade irreplaceable habitat for numerous species and generate significant pollution — unacceptable consequences, given the fact that the Refuge likely contains only a six-month supply of oil (at current U.S. rates of consumption), an amount that will do little to help either our economy or our national security.

How Do Ethics Affect the Environment?

Individually and collectively, our ethics influence the choices and decisions we make regarding our own

Native American Beliefs

My people, the Blackfeet Indians, have always had a sense of reverence for nature that made them want to move through the world carefully, leaving as little mark behind them as possible. My mother once told me: "A person should never walk so fast that the wind cannot blow away his footprints."

Jamake Highwater, Blackfeet

In Ani Yonwiyah, the language of my people, there is a word for land: *Eloheh*. This same word also means history, culture, and religion. This is because we Cherokees cannot separate our place on Earth from our lives on it, nor from our vision and our meaning as a people. From childhood we are taught that the animals and even the trees and plants that we share a place with are our brothers and sisters.

So when we speak of land, we are not speaking of property, territory or even a piece of ground upon which our houses sit and our crops are grown. We are speaking of something truly sacred.

Jimmie Durham, Cherokee

Sell the earth? Why not sell the air, the clouds, the great sea?

Tecumseh, Shawnee

Did you know that trees talk? Well, they do. They talk to each other, and they'll talk to you if you listen. Trouble is, white people don't listen. They never learned to listen to the Indians so I don't suppose they'll listen to other voices in nature. But I have learned a lot from trees: sometimes about the weather, sometimes about animals, sometimes about the Great Spirit.

**Walking Buffalo,
Stoney tribe, Canada**

We did not think of the great open plains, the beautiful rolling hills, and winding streams with tangled growth as "wild." Only to the white man was nature a "wilderness" and only to him was the land infested with "wild" animals and "savage" people. To us it was tame. Earth was bountiful and we were surrounded with the blessings of the Great Mystery.

**Luther Standing Bear,
Oglala band of Sioux**

Every part of this soil is sacred in the estimation of my people. Every hillside, every valley, every plain and grove has been hallowed by some sad or happy event in days long vanished. Even the rocks, which seem to be dumb and dead as they swelter in the sun along the silent shore, thrill with memories of stirring events connected with the lives of my people, and the very dust upon which you now stand responds more lovingly to (our) footsteps than to yours, because it is rich with the blood of our ancestors and our bare feet are conscious of the sympathetic touch....

And when the last Red Man shall have perished, and the memory of my tribe shall have become a myth among the White Men, these shores will swarm with the invisible dead of my tribe, and when your children's children think themselves alone in the field, the store, the shop, upon the highway, or in the silence of the pathless woods, they will not be alone. In all the Earth there is no place dedicated to solitude. At night when the streets of your cities and villages are silent and you think them deserted, they will throng with the returning hosts that once filled them and still love this beautiful land. The White Man will never be alone.

Let him be just and deal kindly with my people, for the dead are not powerless. Dead, did I say? There is no death, only a change of worlds.

Translation by Dr. Henry Smith of a speech attributed to Chief Sealth of the Duwamish tribe, 1853

welfare as well as the welfare of other people, other species, and the environment. Those choices and decisions have a profound impact on the health and wholeness of our world.

For some individuals, religious upbringing largely shapes their personal ethic. For others, secular values and concerns are the dominant forces. But for societies as a whole, ethics are generally the product of mixed influences. Nowhere is that more apparent than in the recent history (geologically speaking) of North America, where the frontier ethic has held sway for over 500 years.

The Frontier Ethic

In the fifteenth century, Europeans began arriving in the New World. They found a wild and beautiful land of seemingly unlimited natural resources, peopled by natives whose worldview differed dramatically from their own. Over the next four centuries, their descendants, and the millions who followed them, enjoyed the confluence of a large resource base, the technological means to exploit this base, and the social acceptance of maximum resource exploitation. In the process, they drastically altered the landscape of the American continents and significantly influenced the course of human history and nonhuman life (Figure 2-2). And as their heirs, whether genetically or culturally, we continue that legacy today.

What afforded them — affords us — such power? To a large extent, the answer is to be found in the ethic that dominates our society. The **frontier ethic** is a code of conduct based on the premises that resources are essentially unlimited, that exploration or human inventiveness will discover new resources to replace those resources that are depleted, and that technology

Figure 2-2: Conquest of the bison, symbol of the Great Plains. In 1800, an estimated 30 million bison roamed the Great Plains of North America; by 1889, that number had plummeted to 1,000. Hunters decimated herds for meat, hides, and sport, leaving the Plains littered with the remains of the carnage. Bone pickers filled wagons with skeletons, then sent them east, by rail, to plants like the Michigan Carbon Works in Detroit, where mountains of skulls were crushed into fertilizer.

will solve any problems arising from human exploitation of the environment. The frontier ethic encourages the aggressive exploitation of nature.

The frontier ethic arose from the dominant western worldview — humans as lord and master over all creation. Earlier, we learned that Judeo-Christian beliefs were a major factor in the development of that worldview. But other factors also played important roles, particularly science, democratic ideals, economic concepts, and the Industrial Revolution.

In seventeenth-century Europe, science began to eclipse other ways of knowing, rising to greater prominence in society. The English philosopher and essayist Francis Bacon believed that God's kingdom would be reestablished on Earth when humans, via science, achieved dominion over nature. According to Bacon, God had created a "clockwork" universe that operated by certain rules and patterns, and science could arrive at the ultimate truth by discerning these patterns. Scientifically derived knowledge translated into new technologies that enabled humans to exploit resources more fully.

The eighteenth century brought the Enlightenment, when profound changes in political, economic, and social orders further influenced the Western worldview. The revolutions that established democracies in North America and France affirmed the rights of individuals to determine their own destinies by making and enforcing laws, owning property, and taking the initiative to develop the resources associated with private property. Capitalism, an aggressive economic system based on the accumulation of wealth, gained wide acceptance. About this time, Europe shifted from a largely agrarian, or farming, society to an industrial one. As mechanical inventions such as seed sowers allowed farmers to increase crop yields, they were able to raise enough food to feed many more people. Consequently, fewer individuals were needed to farm the land. Many people found work in factories, as new industries emerged and flourished. The Industrial Revolution began in England at the end of the eighteenth century; its success depended, to a large extent, upon the availability of a large labor force.

The dominant Western worldview, and the frontier ethic that it spawned, had a tremendous impact on the environment in England, the United States, Canada, and parts of Europe. For nearly a century, industrialization dominated the Western world, bringing with it urbanization, accelerated resource use, and increased pollution. Steam engines drove the Industrial Revolution, and coal fueled the steam engines. More factories and more engines meant more coal; the burning of more coal meant increased air pollution, especially in and around urban centers. In the twentieth century, the development of the internal combustion engine, with its reliance on petroleum, exacerbated the trend of resource exploitation and consequent pollution. The declining quality of life also was due, in no small part, to mere numbers. With many people concentrated in urban areas, refuse was burned in open dumps, further fouling the air, and sewage and wastes were poured into rivers and streams.

Approximately 100 years after the dawn of the Industrial Revolution, some people began to voice concern about the wisdom of unrestrained resource exploitation and the impact of continued environmental degradation. George Perkins Marsh was one of the earliest. Around 1864, Marsh, a visionary scholar, diplomat and lawyer, warned of the destructive effects on the environment of the dominant cultural beliefs and practices. Late-nineteenth-century naturalist and writer John Muir echoed and expanded upon Marsh's warnings. Arguing that nonhuman life had intrinsic value, Muir was a vocal and influential advocate for the preservation of wild places. Gradually, the arguments voiced by Marsh, Muir and others began to resonate with the public, laying the groundwork for the development of environmentally sound alternatives to the frontier ethic.

Environmental Ethics

There are many different types or degrees of **environmental ethics**, but they share certain qualities. Robert Chan, author of *Footprints on the Planet: A Search for an Environmental Ethic*, contends that "the main ingredients of an environmental ethic are caring about the planet and all of its inhabitants, allowing unselfishness to control the immediate self-interest that harms others, and living each day so as to leave the lightest possible footprints on the planet." Two of the most prominent environmental ethics are the land ethic and the stewardship ethic.

Aldo Leopold, a twentieth-century forester and ecologist, argued for the evolution of a **land ethic**, a biocentric code of conduct based on the premise that nonhuman nature has intrinsic value (Figure 2-3). Unlike Muir, however, Leopold grounded his argument in a solid understanding of the young field of ecology, thus lending the weight of science to the case for conservation and preservation:

"An ethic, ecologically, is a limitation on freedom of action in the struggle for existence. An ethic, philosophically, is a differentiation of social from anti-social

FIGURE 2-3: Aldo Leopold, former U.S. Forest Service employee, professor of wildlife management at the University of Wisconsin, and one of the founders of The Wilderness Society.

conduct....All ethics so far evolved rest upon a single premise: the individual is a member of a community of interdependent parts. His instincts prompt him to compete for his place in that community, but his ethics prompt him also to cooperate....The land ethic simply enlarges the boundaries of the community to include soils, waters, plants, and animals, or collectively: the land."

Leopold realized that a land ethic would not prevent the abuse of resources and natural systems, but he maintained that it would affirm "their right to continued existence, and, at least in spots, their continued existence in a natural state." And he summed up his ethic in two simple sentences: "A thing is right when it tends to preserve the integrity, stability, and beauty of the biotic community. It is wrong when it tends otherwise."

Given our society's faith in science, the ideas developed by Leopold helped to crystallize the country's environmental movement in the 1960s. Those ideas also inspired others to develop ethical codes that are environmentally friendly. One of the most influential of those codes is the stewardship ethic. Where the land ethic is biocentric, the **stewardship ethic** is anthropocentric; it is a code of conduct based on the premise that humans are to act as stewards of nature, with the responsibility of caring for and nurturing our planet. This ethic flows from the stewardship worldview and is closely aligned with the basic beliefs of all major religions, which teach that stewardship is the proper relationship of humans to nature. Around the globe, concern for the Earth and the belief that humans are to care for God's creation are increasingly popular among many religious groups.

It's important to acknowledge that most people tailor their ethic to specific situations: They do not always behave as either a "frontiersman" or a "steward." Even those people with the best of intentions sometimes use resources wastefully. By recognizing this, we can begin to modify our behaviors and more consistently apply an environmental ethic. It matters little to the effectiveness of environmental efforts whether our personal ethic rests on a religious or secular foundation. What counts is how we apply that ethic, in our private lives and in our society.

What Is Economics?

As the world's human population increases, the demand for limited resources will continue to rise and the potential for conflict over access to resources will continue to grow. **Economics** — the discipline concerned with the production, distribution, and consumption of wealth and with the various related problems of labor, finance, and taxation — plays a large role in determining how the world's resources are controlled.

All human societies use **natural capital** (natural resources) and **human capital** (skill and labor) to produce **economic goods**, or **manufactured capital**, such as clothing and appliances. The natural capital (materials and energy) used to produce economic goods is known as **throughput** because these resources are taken from the environment, used by humans to produce goods, and eventually returned to the environment.

The system of production, distribution, and consumption of economic goods or services is an **economy**. There are three major types of economies: traditional, pure command, and pure market. In a **traditional economy**, people grow their own food and make the goods, such as clothing and spears for hunting, that they need to survive; traditional economic communities are essentially self-sufficient. In a **pure command economy**, the government makes all economic decisions, such as what goods to manufacture and how to distribute goods or services among the populace. In a

pure market economy, also known as pure capitalism, economic decisions are made by buyers and sellers in the marketplace and are based on the interactions of demand, supply, and price. In reality, every nation has a **mixed economic system**, one that combines elements of all three major economic systems (Figure 2-4).

Most countries, whatever economic system they adopt, aspire to modernize. They generally agree that the best way to modernize is through **economic growth**, an increase in the capacity of the economy to produce goods and services.

To measure economic growth, economists look at various indicators. One commonly used indicator is **gross national product** (GNP), which is the total national output of all goods and services valued at market prices in current dollars for a given year. The GNP includes all personal and government expenditure on goods and services, the value of net exports (exports less imports), and the value of private expenditure on investment. The **real GNP** is the GNP adjusted for any rise in the average price of final goods and services; it enables economists to compare growth from year to year. The **real GNP per capita**, the real GNP divided by the total population, gives some idea of how the average citizen is faring economically. If population growth outpaces growth in the real GNP, then the real GNP per capita will decline. On the other hand, if growth in the real GNP outpaces population growth, the real GNP per capita will rise. In this scenario, the economic situation of the average citizen is presumably improving. But the real GNP per capita can be misleading; the wealthiest segment of the population may prosper while the poorest segment realizes little if any economic improvement.

FIGURE 2-4: A vendor displays goods for sale in a Chinese market.

Measures of GNP are sometimes incorrectly interpreted as indicators of the quality of life. Bear in mind that the GNP is a limited *economic* measure. It tells us nothing about social conditions within a nation: the degree of freedom enjoyed by citizens, their health or well-being, the status of women and minority groups, or the literacy rate. Similarly, the GNP gives no indication of a nation's environmental conditions: the status of resource reserves, the degree of environmental degradation, or the health of natural systems. Neither does it account for the value of **ecosystem services** — the functions or processes of a natural ecosystem and its biota that provide benefits to humans. Examples of ecosystem services include the ability of wetlands to mitigate or prevent damaging floods, the ability of forests to cleanse air and water supplies, the ability of forests to moderate the climate, the ability of terrestrial systems to generate soil, and the beneficial activities of insect pollinators. Perhaps the most egregious limitation of the GNP, however, is that it actually *rises* because of expenses associated with environmental degradation, such as pollution control, remediation, and health care.

To compensate for what the GNP does not tell us, economists have developed a number of alternative indices. The **net national product** (NNP) is based on the GNP but factors in the depletion or destruction of natural resources. The **human development index**, proposed by the United Nations Development Programme, uses three indicators — life expectancy at birth, literacy rates, and real GNP per capita — to estimate the average quality of life in a country. The **index of sustainable economic welfare** (ISEW) adjusts the per capita GNP according to inequalities in income distribution, resource depletion, loss of wetlands, loss of farmland, and the cost of air and water pollution.

How Does Economic Activity Affect the Environment?

All economic activity affects the environment in some way. For the most part, these effects are negative: resources are consumed, energy is degraded, and wastes are produced. Therefore, with respect to economic activity, some degree of environmental impact is unavoidable. However, five factors render economic activity particularly damaging. These are external costs, the inadequacy of market valuation for ecological goods and services, the undervaluation of natural resources, an emphasis on economic growth, and population growth.

External Costs

The maker of a product does not pay all of the costs associated with its production. For instance, a farmer who grows corn sells it at a price that covers the direct costs, or **internal costs**, of its production. These include seed, fertilizer, pesticides, and labor. But growing corn may create additional costs for which the farmer does not pay. Soil eroded from his fields, for instance, may enter a nearby river, costing cities downstream more money to treat their drinking water supplies. **External costs** are harmful social or environmental effects of the production and consumption of an economic good that are not included in the market price of the good (Figure 2-5).

In addition to the external costs that are paid by other people or society at large, there are some external costs for which no one pays. For example, the soil that erodes from the farmer's fields may destroy fish spawning areas, decreasing fish populations. Although some people may be affected by a decline in fish populations, no one directly pays out money for the effects.

The failure to include external costs in cost-benefit analyses can result in development projects that cause significant damage to the environment and both human and nonhuman life. **Cost-benefit analysis** is a technique used to compare the estimated costs (losses) of a proposed project with the benefits to be gained. If the benefits outweigh the costs, the project is deemed worthwhile. Estimating benefits is relatively straightforward. For example, a developer who wants to start a new subdivision can predict that he will be able to build x number of homes and sell them at an average price of y. Costs are more difficult to estimate. There are the obvious internal costs of home and road construction, for which the developer must pay. But there are also significant external costs, such as soil erosion and runoff from the construction site, air pollution caused by increased automobile traffic, and diminished biological diversity.

The Inadequacy of Market Valuation for Ecosystem Goods and Services

Conventionally, economics defines the value of a commodity according to the price it brings in the free marketplace. It presupposes that the buyer knows what commodity she wants and understands its benefit to her. Using this reasoning, the value of a forest may be equated with the price of the lumber it yields, which is itself a product of the species present, their size, and quality. The market price *does not* include the value of the ecosystem services provided by the intact forest, the biological diversity it harbors, and its aesthetic and recreational aspects. Market valuation is inadequate for ecosystem goods and services, which are long-term in nature and are not "traded" in the marketplace. That inadequacy is compounded because most consumers have little or no understanding of ecosystem services.

FIGURE 2-5: Smokestacks in Kunming, China. As the country continues to industrialize, the external costs of increased energy production include worsened air and water quality.

The Undervaluation of Natural Resources

Undervaluing natural resources also contributes to environmental degradation because it discourages conservation and recycling. For example, low federal and state taxes on petroleum keep down the cost of this commodity in the United States; automobile owners pay little for gasoline relative to consumers in other industrialized nations. As a result, U.S. consumers have less incentive to limit car trips, form carpools, or use public transportation. In some cases, government subsidies make it cheaper for companies to use virgin resources (such as minerals and timber) than recycled materials. For example, the federal government pays to build logging roads in national forests; companies who buy rights to log in national forests are not required to pay for roads. As a result, the cost to companies is artificially low, discouraging them from increasing their reliance on recycled paper and pulpwood.

Emphasis on Economic Growth

Perhaps the most serious threat to environmental quality is the global emphasis on economic growth. In the United States and other industrialized nations, growth-mania has created a **culture of consumption** in which disposable products proliferate, products quickly become outdated and are replaced by newer models, and fashion dictates that we change clothing and home styles frequently. (See *Focus On: A Culture of Consumption* in Chapter 1.) Historically, when the economy was very small relative to the ecosystem, it was reasonable to believe that growth could continue unabated. Growth, in fact, was restricted only by human capital, while natural capital was seen as unlimited. In the present era, however, the economy is a global one and natural capital — not human capital — is the limiting factor to growth. Faced with myriad threats to natural resources and systems, we would be wise to remember the analogy offered by the writer Edward Abbey: Growth is the ideology of the cancer cell.

Population Growth

The exponential growth of the human population has dramatically increased the environmental impact of economic activity. It's obvious that, in the absence of conservation and efficiency measures, more people will require more materials and energy and create more waste. What may be less obvious is that rapid population growth can overwhelm economic growth. This is what happened, to a large extent, in many of the world's less-developed countries (LDCs) over the last 50 or 60 years. They borrowed heavily from international lending institutions and wealthier nations in an effort to modernize in the style of the industrialized world. The lenders were willing to finance large-scale development projects, such as the construction of major roads, dams, mining operations and irrigation systems, without regard for their social and environmental impacts — including deforestation, displaced peoples and communities, and pollution. By the 1980s, as these loans began to come due, LDCs found that they did not have enough money to pay even the interest on the loans, let alone the principal. As long as LDCs remain in heavy debt, they will be unable to institute the strong social and environmental programs needed to improve their growing populations' standards of living.

What Are Ecological Economics and Sustainability?

Ecological economics is an transdisciplinary field of study that addresses the relationships between ecosystems and economic systems in the broadest sense. Developed in response to the limitations of conventional economics, ecological economics integrates different disciplinary perspectives in order to focus on environmental and economic problems.

The goal of environmental economics is **sustainability**, defined as the relationship between dynamic human economic systems and larger dynamic, but normally slower-changing, ecological systems, in which human life can continue indefinitely, human individuals can flourish, and human cultures can develop. According to sustainability, the effects of human activities remain within limits, so as not to destroy the diversity, complexity, and function of the ecological life support system.

Achieving sustainability requires the development of a **steady-state economy** (SSE), one characterized by a constant level of human population and a constant level of artifacts, known as stock. The levels of both the population and stock allow for a good quality of life that is sustainable well into the future. With respect to population, birth rates are equal to death rates at relatively low levels, and life expectancy is high. With respect to stocks, matter and energy are used as efficiently as possible and production equals depreciation at low levels; as a result, economic goods are long lasting, minimizing resource depletion and pollution. In a SSE, the rate of throughput needed to maintain the population and stock is reduced to the lowest feasible level.

A SSE is characterized not by growth, but by **sustainable development**, defined as improving the quality of human life while living within the carrying capacity of supporting ecosystems. Sustainable development seeks to maximize human resource potential as well as the wealth provided by natural resources by managing all resources — natural, human, financial, and physical — so that they can be used to serve the common good. Development is only sustainable when it meets the needs of present generations without compromising the ability of future generations to meet their own needs.

Sustainable development presupposes **sustainable resource use**, that is, the use of renewable resources at rates that do not exceed their capacity for renewal (Figure 2-6). By definition, sustainable use does not apply to nonrenewable resources; because the supply of nonrenewable resources is finite, they cannot be used sustainably. At best, it is possible only to extend the life of nonrenewable resources through recycling, conservation, and substitution measures.

Benefits of Ecological Economics

Ecological economics has several advantages over traditional economics. First, it internalizes the external costs of an economic good, yielding a price that reflects the true value of the commodity and the costs of its production. True-value pricing results in higher market prices for goods but lowers hidden costs, such as health care and waste cleanup. Ecological economics also eliminates harmful subsidies, markedly improving the ways in which we use and manage natural resources. For example, in the United States, eliminating subsidies to mining, logging, and other resource extraction industries would conserve natural resources and help make reusing and recycling more competitive options. One of the most important tasks for ecological economists is to develop better measures of the value of ecological goods and services and to create accounting systems that more accurately reflect the cost of natural resource depletion and ecosystem degradation.

One important tool for ecological economists is environmental or **green taxes**, fees assessed to discourage the use of environmentally harmful practices or products or to extend the life of nonrenewable resources. Examples of green taxes (and the environmental objective they are designed to achieve) include fees on carbon emissions resulting from the combustion of coal, oil,

Figure 2-6: Sustainable resource use, Guatemala. This woman is part of an enterprising cooperative that grows medicinal herbs.

and natural gas (to help slow global warming); taxes on virgin materials (to encourage reuse and recycling); and levies on the overpumping of groundwater (to foster conservation and more efficient water use). Proponents of green taxes argue that they can help achieve multiple objectives. For example, a significant increase in the gasoline tax would encourage conservation of petroleum and provide a designated source of revenue for the support of less-polluting transportation alternatives, including mass transit, carpools, and biking paths.

What Are Government and Politics?

A **government** is an established system of administration through which a nation, state, or district is ruled. Governments produce the basic infrastructure, such as roads and sewer and water systems, businesses and citizens need to support their daily activities. Governments provide financial and technical support and pass laws and regulations to help individuals and businesses reach their social and economic goals. They enact additional laws to prevent individuals and businesses from engaging in activities that are threats to public health, safety, and welfare or that violate accepted norms of behavior. Citizens must comply with the laws of the land or risk some penalty. At the highest level, governments are responsible for national security, economic health, and environmental sustainability.

Politics encompasses the principles, policies, and programs of government; it is also the arena in which citizens debate those issues that affect their individual and collective health and welfare. Different political systems, such as democratic republics (United States) and communist states (People's Republic of China), are based on different principles and consequently adopt different policies and programs in order to govern their societies.

Even within a particular nation, however, all citizens do not agree on which policies and programs will ensure national and environmental security and economic health. Take the United States, for example. In the wake of September 11, 2001, national security is a major concern, if not *the* chief concern, for a majority of U.S. citizens. But how do we achieve national security? Some people favor unilateral, preemptive action against nations suspected of harboring or supporting terrorists. Others favor multilateral efforts aimed at putting widespread economic and public pressure on those nations. Still others argue that we cannot achieve real security until we understand and resolve the issues that compel terrorists to act.

Like national security, economic health and environmental sustainability are widely debated in the political arena. Many people argue convincingly that we cannot achieve national security while pursuing economic and environmental policies that are not sustainable. Nowhere is this more apparent than in terms of our national energy needs. Consider this: Our economy — our society — is *absolutely* dependent upon fossil fuels, and while we have abundant coal reserves, our domestic supplies of petroleum are limited and dwindling. Currently, we rely on imports to meet 50 percent of our annual petroleum needs. Clearly, if our goal is national security, we are on shaky ground when we import half of our oil supplies, particularly given that the world's largest oil reserves are in the troubled Middle East. So what energy policy should we pursue? Should we increase oil drilling in environmentally sensitive areas

in Alaska, *and if we do*, just how long will those finite supplies last? Or, do we increase conservation and efficiency measures to make our oil supplies last longer and invest in alternative fuels to take their place?

How Does Politics Affect the Environment?

Governments affect the environment through the policies and programs they adopt and through the laws they enact. Thus, politics has a tremendous effect on a nation's environment, and as the preceding paragraph illustrates, on its economic health and national security. Increasingly, however, the environmental effects of a government policy — whether it concerns water rights, biological diversity, or the transport of toxic wastes — extend well beyond that nation's borders.

Typically, it is not a lack of knowledge that prevents us from alleviating or eliminating environmental problems; rather, it is a lack of political will. Arguably, the most pressing environmental problem facing the global community today is the climate change associated with the release of greenhouse gases. The solution is relatively simple: Reduce emissions of greenhouse gases and protect the planet's remaining forests, which absorb carbon dioxide (the principal greenhouse gas) from the atmosphere. We can make the most significant reductions in greenhouse gases by curbing our use of fossil fuels.

In fact, 126 nations of the world have consented to do just that by signing and ratifying the 1997 Kyoto Protocol, an international attempt to set binding limits on greenhouse gas emissions. Unfortunately, the United States, which originally signed the Protocol under then-President William Jefferson Clinton, failed to ratify it, and in 2001, President George W. Bush withdrew from the agreement altogether. He cited concerns that the restrictions on emissions were too costly and that the Protocol unfairly excluded developing nations. Instead of action, President Bush has called for further scientific study of climate change and its environmental impact. Not surprisingly, that call does not sit well with the scientific community, which overwhelmingly agrees that human-induced climate change is now or soon will be occurring. Nor does it play well with the international community, which points out that the United States accounts for about one-quarter of annual global emissions of carbon dioxide.

Worldwide, citizens are calling on their governments to safeguard their environment and thereby enhance their economic and national security (Figure 2-7). In the U.S., where the people elect government officials, public demand can strongly influence policy decisions. The political will to take action on issues usually stems from sufficient public demand. When enough people make enough noise, politicians begin to find ways to satisfy their constituents' demands. Public support and strong lobbying efforts were key to the passage of many environmental laws, such as the Wilderness Act of 1964, the Federal Lands Policy and Management Act, and the Superfund Amendments Reauthorization Act. Public support may yet spur U.S. action on the issue of global climate change. Whether or not it does is up to us.

Before we leave this topic, we want to point out that concern for the environment is not the sole domain of either party. Arguably, our nation's two most environmentally aware and concerned presidents were Theodore Roosevelt and Jimmy Carter, one a Republican, the other a Democrat. Clearly, environmentalism is a philosophy that transcends party lines. That's as it should be. Environmentalism is conserving, and hence *conservative*, in the true sense of the word. Political decisions about environmental and natural resource issues should err on the side of caution — on the side of conservation and preservation.

What Is Environmental Law?

As we mentioned earlier, governments establish laws that require compliance under penalty. The system of **environmental law** is an organized way of using *all* of the laws in a nation's legal system to minimize, prevent, punish, or remedy the consequences of actions that damage or threaten the environment. Specific laws govern the activities of persons, corporations, government agencies, and other public and private groups in order to regulate the impact of activities on the environment and natural resources. Environmental protection resides in common and statutory law.

Common law is a body of written and unwritten rules based on **precedent**, a legal decision or case that may serve as an example, reason, or justification for a later decision. Cases involving common law are based on three legal concepts: nuisance, trespass, and negligence. **Nuisance**, the most common cause of action in the field of environmental law, is a class of wrongs that arise from the unreasonable, unwarrantable, or unlawful use of a person's own property that produces annoyance, inconvenience, or material injury to another. **Trespass** is the unwarranted or uninvited entry upon another's property by a person, the person's agent, or an object that he or she caused to be deposited there. **Negligence** is the failure to exercise the care that a "prudent person" usually takes, resulting in an action or inaction that causes personal or property damage.

Statutory law is the body of facts passed by a local legislature or Congress. Statutes in the area of environmental law govern activities that affect the environment and human health and the management of resources. Statutes generally state the broad intentions of their specific provisions, such as to protect health and the environment by reducing air pollution or to use resources wisely by mandating conservation measures and standards.

Enacting a statute is often a lengthy process marked by give and take. In the United States, at the federal level, both the House of Representatives and the Senate must approve a bill before it can be signed into law by the President. This process can take months of back and forth between the two houses of Congress. As mentioned, public involvement can help to shape political decisions, and individuals, citizen groups, and lobbyists can work to ensure that their interests and concerns are incorporated into a bill.

Figure 2-7: Logging protest, Ormoc City, Leyte Island, Philippines. Flooding caused by deforestation resulted in the deaths of approximately 8,000 people, most of them poor shantytown dwellers.

How Do Environmental Laws Affect Society?

Imagine that your Saturdays and Sundays differed little from any other day of the week; that the workday was 12 to 14 hours long; that you earned pitifully low wages in return for your work; and that child labor, even for the very young, was a given. It may be difficult to envision such a world, but less than a century ago, those conditions were the reality for many people in our society. In the same way, prior to the environmental movement, many people lived in areas where the air was thick with acrid smoke and other pollutants; where sewage and industrial effluent fouled waterways; and where developers and businesses operated with few, if any, restrictions on their activities.

The fact that this latter scenario — which was common less than 40 years ago — is foreign to you is a testament to the social impact of U.S. environmental laws. Just as the labor movement improved the quality of life for the nation's working classes, the environmental movement that began in the 1960s led directly to the passage of numerous laws that dramatically improved the health and welfare of millions of Americans. These laws concern disparate aspects of the environment — among them air, water, endangered species, forests, hazardous waste sites, and agricultural pesticides.

All environmental laws protect or improve the environment and/or human health in one of the following ways:

- By setting pollution level standards or limiting emissions of effluents for various classes of pollutants (for example, the Safe Drinking Water Act)
- By screening new substances to determine their safety before they are widely used (for example, the Toxic Substances Control Act, or TSCA)
- By evaluating the environmental impacts of an activity before it can be started (for example, the National Environmental Policy Act, or NEPA)
- By setting aside or protecting ecosystems, resources, or species from harm (for example, the Endangered Species Act and the Wilderness Preservation Act)
- By encouraging resource conservation (for example, the Food Security Act of 1985)
- By cleaning up existing environmentally unsound hazardous waste disposal sites (for example, the Comprehensive Environmental Response, Compensation and Liability Act, or CERCLA)
- By preventing the creation of new unauthorized hazardous sites (for example, the Resource Conservation and Recovery Act, or RCRA)

Generally, each environmental law specifies how it is to be enforced and who is to enforce it. In the U.S., the Environmental Protection Agency (EPA), perhaps the best known of the federal agencies charged with enforcing environmental policies, is the federal government's primary "environmental watchdog." Unfortunately, a variety of factors often prevent environmental laws from being enforced effectively. In some cases, the allocation of authority is unclear, especially when either state or federal authorities could be in charge. Environmental laws are also ineffective when the administration or Congress fails to provide the funds needed for enforcement or if a government agency has conflicting priorities.

Given the strong connection between government and economics, it is not surprising that economic analyses factor into environmental laws. Many environmental laws attempt to put a monetary price on the value of resources such as clean air, clean water, and productive soil. Many environmental laws also include provisions that require the economic analysis of various alternative solutions or economic penalties for failure to protect resources. For example, CERCLA established a fund to cover the costs of restoring natural resources that have been damaged by toxins. As we saw earlier in the chapter, however, economic cost is not equivalent to the true value of a resource, and thus, the tax imposed on the producer does not equal the value of the damaged resource or ultimate costs to society. Clearly, our laws have not been able to account for the real value of natural resources and the costs to society of restoring damaged resources and ecosystems.

What Is Education?

Education is the study of the processes through which students learn, developing knowledge, skills, and expertise in diverse subjects. Typically, education refers to learning that occurs in formal settings, but we all know that learning can and does take place anywhere. And learning about any subject — from religion and ethics to economics and politics — greatly affects a person's beliefs, values, behaviors, and decisions. In a sense, then, every subject you can name relates, in some way, to our environment — past, present, or future. However, the formal discipline of **environmental education** is concerned with the process of learning about the biosphere, its associated problems, and the human role in causing and solving those problems.

We believe that an environmental education should achieve five goals. First, it should help us to understand how nature works. Second, it should compel us to appreciate and respect the interdependence of all life and the living and nonliving components of the environment. Third, it should enable us to understand how environmental problems are caused. Fourth, it should demonstrate the absolute necessity of adopting an

FIGURE 2-8: Effective environmental education sparks our innate sense of wonder.

For young children, we believe that the emphasis in environmental education should be on the wonder of nature and its many interdependent parts. Educators who work with children pre-K through middle school should focus on the first two goals of environmental education — helping their students to discover how nature works and to appreciate and respect the interdependence of all life and the living and nonliving components of the environment (Figure 2-8).

By the time most students reach junior high or high school, they are mature enough to begin learning about environmental problems and the roles that various people play in finding solutions to those problems. Very likely, they are beginning to make many decisions that affect the environment: How will I get to and from school — by car, bus, bike, or footpower? Will I pack my lunch or eat from a vending machine? When it's time to drive, will I agitate for a fuel-efficient car or the latest, greatest gas-guzzler? And finally, if they choose to do so, junior and senior high school students are ready and able to act in ways that help prevent or solve environmental problems.

How Is Environmental Education Taught?

Environmental education is taught both formally, in the classroom, and informally, outside of the classroom. Formal environmental education can be taught as a discrete course, or it can be integrated across all courses within a curriculum. Each approach has advantages. Teaching a separate environmental course allows a more in-depth focus on environmental issues and does not require instructors to modify other courses. On the other hand, integrating environmental education across all courses helps students to more fully understand that all disciplines impact the environment. Integration also adds a contextual framework to environmental issues, so that students can apply environmental principles to all aspects of their lives.

Informal environmental education is designed to reach groups outside of the formal school system, including local elected officials, civic groups, industrial and commercial leaders, general adult audiences, and the media. Making these groups environmentally aware is important because their action or lack thereof influences local environmental quality.

Schoolchildren also can benefit from informal environmental education programs. Informal programs

interdisciplinary approach to solving those problems. And fifth, it should allow us to take action to prevent or solve environmental problems, should we choose to do so.

As a field, environmental education emerged during the environmental movement and has been evolving ever since. The early focus of environmental education was to educate all citizens, children and adults, in the skills needed to solve environmental problems and to motivate them to be actively involved in environmental issues. Today, environmental education increasingly seeks to awaken and explore people's values and to help us understand and appreciate our dependence on the natural world. Our technocratic society can separate, even isolate, us from the natural world; through environmental education, we can break that isolation. Environmental education seeks to nourish our inherent joy and awareness and to enhance our environmental literacy.

held at zoos, nature centers, museums, aquariums, and parks complement or build upon classroom lessons, thus reinforcing important ecological and environmental concepts (Figure 2-9).

Whether formal or informal, environmental education can shape our attitudes and actions toward the environment. Because the world is ever-changing, the process of environmental education must be lifelong. "The waves echo behind me," wrote Anne Morrow Lindbergh. "Patience — Faith — Openness is what the sea has to teach. Simplicity — Solitude — Intermittency....But there are other beaches to explore. There are more shells to find. This is only a beginning."

FIGURE 2-9: Children at Miami University's Hefner Zoology Museum (Oxford, Ohio) participate in an intriguing informal environmental education program.

Summary

Religion is the expression of human belief in, and reverence for, a superhuman power, which may be recognized as a creator and governor of the universe, a supernatural realm, or an ultimate meaning. Religions are characterized by a core set of beliefs or teachings that answer basic questions about the universe, existence, the world, and humankind.

Religious beliefs can be a primary influence on the relationship between humans and nature, and they are often a powerful shaper of worldviews. An important part of any worldview is a person's belief about the place of humans in the natural world. An anthropocentric worldview is a way of perceiving reality that places humans in a preeminent position above, and separate from, the rest of nature. A biocentric worldview is a way of perceiving reality that recognizes an inherent worth in all life and maintains that humans are no more or less valuable than all other parts of creation. Anthropocentrism tends to foster attitudes and behaviors that have a negative effect on the environment, and biocentrism tends to foster a sense of care and concern for the natural world.

Ethics is the branch of philosophy concerned with standards of conduct and moral judgment. An ethic is a code of morals that governs or shapes behavior. Morals are principles that help us distinguish between right and wrong. There is no universal human ethic. Individually and collectively, our ethics influence the choices and decisions we make regarding our own welfare as well as the welfare of other people, other species, and the environment.

The frontier ethic is based in large part upon anthropocentric interpretations of the Judeo-Christian worldview, which holds humans as having dominion over all living things. The frontier ethic encourages the aggressive exploitation of resources and natural systems. In contrast, environmental ethics foster a sense of caring about the natural world. The land ethic, developed by Aldo Leopold in the mid-twentieth century, is the best-known environmental ethic. It holds that the natural world — soil, water, air and all species, including humans — is a community of interdependent parts and that all parts have intrinsic value. Leopold grounded his argument in a solid understanding of ecology, thus lending the weight of science to the case for conservation and preservation. A second influential environmental ethic is the stewardship ethic; it is based upon the Judeo-Christian worldview that sees humans as stewards of nature.

The degree to which a person follows a particular ethic varies according to societal situations and personal interpretations of religious teachings. Most people tailor their ethic to specific situations. What counts is how we apply that ethic in our private lives and in society.

Economics is the discipline concerned with the production, distribution, and consumption of wealth and with the various related problems of labor, finance, and taxation. All societies use natural and human capital to produce economic goods, or manufactured capital, such as clothing and appliances. The natural capital used to produce economic goods is known as throughput.

The system of production, distribution, and consumption of economic goods or services is an economy. There are three major types of economies: traditional, pure command, and pure market. In reality, every nation has

Discussion Questions

1. What are ethics? What are morals? How are they related to religion?

2. What is the difference between a frontier ethic and a stewardship ethic? How and why have these two different ethics developed?

3. What is the link between ecology and economics? How does conventional economics fail to account for the value of ecological goods and services, the degradation of natural systems, and resource depletion?

4. Define and distinguish between a traditional economy, a command economy, and a market economy. In your opinion, which system is mostly likely to result in environmentally sound decisions? Explain.

5. How can politics contribute to environmental problems? How can politics contribute to solutions to these problems?

6. How do environmental laws reflect society's priorities for environmental management? Give at least three examples.

7. What is environmental education? How do contemporary definitions differ from older definitions?

a mixed economic system, which combines elements of all three major economic systems.

Most countries aspire to modernize or develop through economic growth. Economic growth is an increase in the capacity of the economy to produce goods and services and is viewed as the best way to modernize. The gross national product (GNP) is the total national output of all goods and services valued at market prices in current dollars for a given year. The real GNP is the GNP adjusted for any rise in the average price of final goods and services; it enables economists to compare growth from year to year. The real GNP per capita, the real GNP divided by the total population gives some idea of how the average citizen is faring economically. Measures of GNP are often good indicators of the economic quality of life, but they do not adequately reflect social or environmental conditions.

With respect to economic activity, some degree of environmental impact is unavoidable. Five factors render economic activity particularly damaging. These are external costs, the inadequacy of market valuation for ecological goods and services, the undervaluation of natural resources, an emphasis on economic growth, and population growth.

Ecological economics is a transdisciplinary field of study that addresses the relationships between ecosystems and economic systems in the broadest sense. It integrates many different disciplinary perspectives in order to focus on environmental and economic problems. The goal of environmental economics is sustainability, which is defined as the relationship between dynamic human economic systems and larger dynamic, but normally slower-changing, ecological systems, in which human life can continue indefinitely, human individuals can flourish, and human cultures can develop; the effects of human activities remain within limits, so as not to destroy the diversity, complexity, and function of the ecological life support system.

Achieving sustainability requires the development of a steady-state economy (SSE), one characterized by a constant level of human population and a constant level of artifacts, known as stock. A SSE is characterized not by growth, but by sustainable development, defined as improving the quality of human life while living within the carrying capacity of supporting ecosystems. Sustainable development seeks to maximize human resource potential as well as the wealth provided by natural resources by managing all resources so that they can be used to serve the common good.

A government is an established system of administration through which a nation, state, or district is governed. Governments produce basic infrastructure to support daily activities, provide financial and technical support to individuals and businesses, and pass laws and regulations to help citizens reach their goals and to prevent activities that are viewed as threats to public health and safety or that violate accepted norms of behavior. Politics encompasses the principles, policies, and programs of government. A primary goal of any government is to ensure economic growth for its people; thus, the political and economic systems of a country go hand-in-hand.

Governments create policies and laws in order to work towards economic growth and a higher quality of living for their people. These policies and laws, which all members under a government are obligated to follow, can be very effective means for influencing how a society treats the environment.

Lack of political will often prevents us from alleviating or eliminating many serious environmental problems. The political will to take action on issues usually stems from sufficient public demand. Politics is an important component of national and international environmental issues.

Governments establish laws that require compliance under penalty. Such laws are perhaps the most effective means through which governments can impact environmental and natural resources management.

The system of environmental law is an organized way of using all of the laws in a nation's legal system to minimize, prevent, punish, or remedy the consequences of actions that damage or threaten the environment. Environmental protection resides in common and statutory law. Common law is a body of written and unwritten rules based on precedent. Cases involving common law are based on three legal axioms: nuisance, trespass, and negligence. Statutory law is the body of facts passed by a local or national government.

Congress has enacted a number of important environmental and resource protection laws. These laws set pollution level standards; screen new substances for safety before they are widely used; evaluate the environmental impacts of activities before they are undertaken; protect various ecosystems, resources, or species; clean up environmentally unsound toxic waste disposal sites; and prevent new unauthorized toxic sites from forming.

Each environmental law specifies how it is to be enforced and who is to enforce it. A variety of factors often prevent environmental laws from being enforced effectively. In some cases, the allocation of authority is unclear, especially when either state or federal authorities could be in charge. The administration or Congress may fail to provide the funds needed for enforcement, or a government agency may have conflicting priorities.

Environmentally sound laws are important to environmental problem solving and management. The legal system comes into play at each step of the problem-solving process. Certain environmental laws illustrate and support the stewardship ethic, a biocentric worldview, knowledge of political systems, economic analysis, and public participation.

Environmental education seeks to awaken and explore people's values and to help us understand and appreciate our real relationship to nature. It attempts to educate all citizens about the environment and its problems and to motivate them to solve these problems.

Five elements are especially important to environmental education programs. They are an understanding of, and appreciation for, how nature works; an awareness of the interdependence of all life and of the living and nonliving; an understanding of how environmental problems are caused; an interdisciplinary approach; and action.

Formal environmental education can be taught as a discrete course, or it can be integrated across all courses within a curriculum. Informal environmental education seeks to reach groups outside of the formal school system, including local elected officials, civic groups, industrial and commercial leaders, general adult audiences, and the media. Zoos, nature centers, museums, aquariums, and parks are just some examples of the diverse forums through which environmental education programs are offered.

Environmental education emphasizes the role of values, morality, and ethics in shaping attitudes and actions affecting the environment. And, because environmental education is a lifelong process, it continually enriches our lives and serves to remind us of the importance of problem solving and resource management.

KEY TERMS

anthropocentric worldview

biocentric worldview

common law

cost-benefit analysis

culture of consumption

ecological economics

economic goods

economic growth

economics

economy

ecosystem services

education

environmental education

environmental ethic

environmental law

ethics

external costs

frontier ethic

government

green taxes

gross national product

human capital

human development index

index of sustainable economic welfare

internal costs

land ethic

manufactured capital

mixed economic system

monotheism

morals

natural capital

negligence

net national product

nuisance

pantheism

politics

precedent

pure command economy

pure market economy

real GNP

real GNP per capita

religion

statutory law

steady-state economy

stewardship ethic

stewardship worldview

sustainability

sustainable development

sustainable resource use

throughput

traditional economy

trespass

UNIT **II**

An Environmental Foundation

Ecological Principles
and Applications

CHAPTER 3

Ecosystem Structure

Learning Objectives

When you finish reading this chapter, you will be able to:

1. Identify the levels of ecological study and briefly define each.

2. Define ecosystem and list the two major components of any ecosystem.

3. Define producers and consumers and give examples of each.

4. Describe the roles played by producers, consumers, and decomposers in an ecosystem and the interactions among them.

5. Identify three important limiting factors and give an example of how each helps to regulate the structure of an ecosystem.

A thing is right when it tends to preserve the integrity, stability and beauty of the biotic community. It is wrong when it tends otherwise.

Aldo Leopold

We have what may seem to be a strange request. Think, for a moment, of the languages you studied during high school or college, such as Spanish, French, Russian, or Greek. Before you could hope to read and speak Spanish or Greek, you had to learn the letters and words that are the building blocks of language. Only then could you begin to learn the rules that govern how the letters and words are formed into sentences to communicate ideas. As your understanding grew, you were able to express your own ideas and to read complete passages written in a language that at one time had seemed so alien.

You are probably wondering what this has to do with environmental science. In a real sense, nature, too, has its own language. Ecosystem structure is an alphabet, a collection of letters and words that can be combined in many different ways. In this chapter, we will take a close look at the structural components — the alphabet and words — all ecosystems share.

Knowing letters and words alone does not allow one to speak a language, and so it is with nature. Simply knowing the structure of ecosystems will not allow us to understand how they function, change, and respond to human activities — in short, how to speak the language of nature. To do that, we must learn how the structural components of an ecosystem function as part of a greater whole. That topic is the focus of Chapter 4.

Language, of course, is a cultural artifact, something created by humans for humans. As the makers of language, we establish the rules — grammar — that enable us to communicate with one another. Natural systems, too, have rules that govern how they function and respond to change. Unfortunately, we do not always understand — and sometimes are not even aware of — those rules. Ecology, the study of the structure, function, and behavior of natural systems, is still in its infancy, but already it shows great promise in helping us to understand the rules that govern natural systems. Understanding these rules, in turn, may enable us to decipher the intricate, beautiful language of the biosphere.

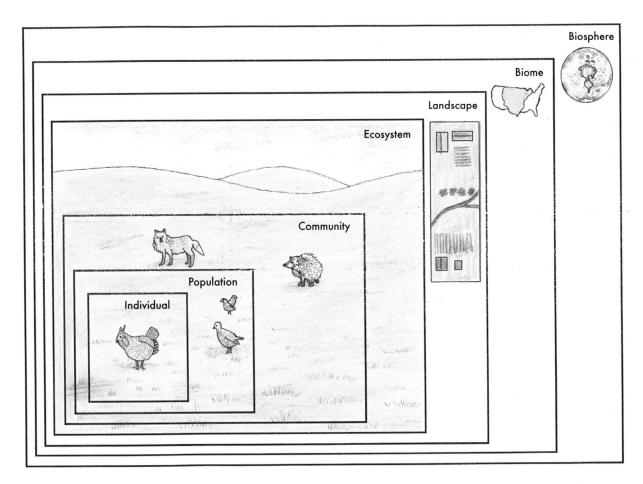

FIGURE 3-1: Levels of ecological study. The study of ecology spans the continuum from the individual organism (like this greater prairie chicken), to populations, communities, ecosystems, landscapes, biomes (like the temperate grassland shown here), and finally, the biosphere. Information at all these levels contributes to our understanding of the "ecos," our home.

What Are the Levels of Ecological Study?

In their quest to understand the structure and function of ecosystems, ecologists study natural systems at many levels: individual, population, community, ecosystem, landscape, biome, and biosphere (Figure 3-1). An individual is a single member of a species. A **species** includes all organisms of a particular kind that are capable of producing viable offspring (that is, individuals which can themselves produce offspring). For species that reproduce sexually, the individual members reproduce by mating, or interbreeding.

A **population** is a distinct group of individuals of a species that live, interbreed, and interact in the same geographic area, for example, largemouth bass in a particular pond or Canada goldenrod in a single field. Populations have measurable group characteristics such as birth and death rates or seed dispersal and germination rates. Both individual organisms and populations have a place within the physical environment commonly called **habitat**, the place where the organism or population lives. Forests, streams, and soils are just a few examples of habitats where organisms or populations are found.

A **community** includes all of the populations of organisms that live and interact in a given area at a given time. Plants, animals, and microorganisms are bound together by feeding relationships and other interactions, thus forming a complex whole. These interactions are discussed in Chapters 4 and 5.

An **ecosystem** is a self-sustaining, self-regulating community of organisms interacting with the physical environment within a defined geographic area (Figure 3-2). The term 'ecosystem' is derived from two Greek words: *oikos,* meaning home, and *sustéma,* meaning a composite whole. Rivers, lakes, and tidal pools are

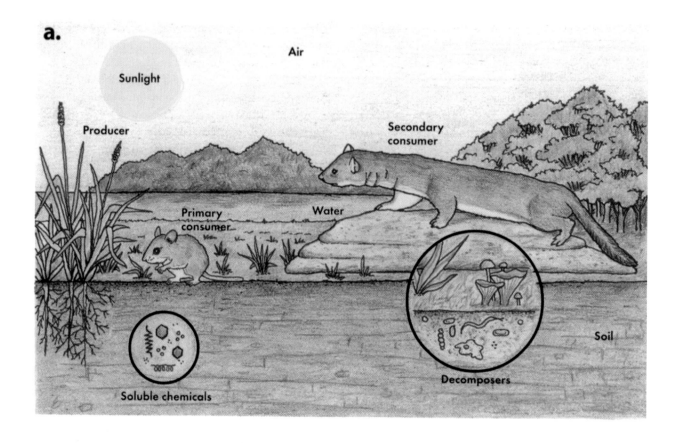

a.

Air

Sunlight

Producer

Secondary consumer

Primary consumer

Water

Soil

Soluble chemicals

Decomposers

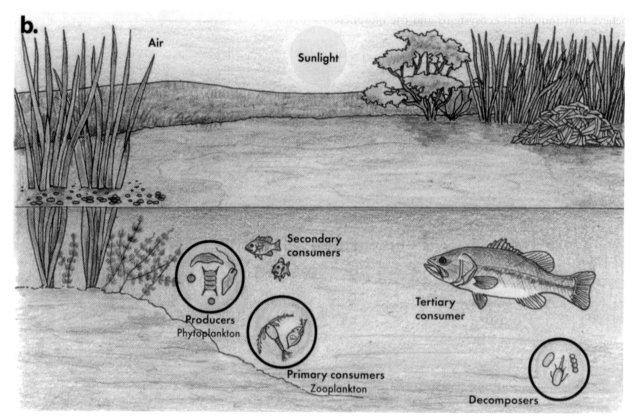

b.

Air

Sunlight

Secondary consumers

Tertiary consumer

Producers
Phytoplankton

Primary consumers
Zooplankton

Decomposers

FIGURE 3-2: Examples of terrestrial and aquatic ecosystems. Ecosystems vary in size and location, but whether it is a field (top) or pond (bottom), every ecosystem is a self-sustaining community of organisms interacting with the physical environment within a defined geographic space.

examples of aquatic ecosystems. Oak-hickory forests, alpine meadows, and high mountain deserts are examples of terrestrial ecosystems.

Ecosystems are not static. Even mature ecosystems like hardwood forests constantly undergo internal changes as a result of changes in their external environments. For example, lack of rainfall can make older trees more susceptible to disease. Diseased trees may die, allowing younger, healthier trees to dominate. Heavy winds can topple trees, opening up the forest floor to increased sunlight and temperature, which promote, for a time, the growth and success of species adapted to these changed conditions.

Sometimes, seemingly minor changes in one component of an ecosystem, such as spraying a field with an insecticide, can have far-reaching effects on other components of the system. The insecticide might destroy nontarget insects more successfully than it does the insect pest. For example, killing pollinators like honeybees can reduce pollination success in flowers and nectar production for the beehive. In addition, insecticides can kill predators that previously helped to control the target insect population.

Ecosystems usually can compensate for the stresses caused by external changes. But many ecologists believe that individual ecosystems and the biosphere have thresholds beyond which catastrophic, potentially permanent change occurs. For example, at the ecosystem level, thresholds may be exceeded when a lake is acidified, a stream is overloaded with sewage, or a forest is fragmented. Perhaps, if the external stress is eliminated, the system may recover over time. It is uncertain, however, if the system will return to its pre-stress state. The rapid loss of biological diversity is one example of a potentially disastrous biospheric change.

Many ecosystems taken together are referred to as a **landscape**. Each ecosystem in a landscape has its own distinct elements that, in combination, form the heterogeneous character of the whole. Landscape ecologists are interested in knowing how the ecosystems in a landscape are arranged and how this arrangement affects ecological processes.

A **biome** consists of landscapes grouped across large terrestrial areas of the Earth (Figure 3-3). Biomes are identified and classified according to their dominant vegetation type. Vegetation type is largely the product of climatic conditions, that is, long-term weather patterns, including temperature, precipitation, and the availability of light. The eastern United States, for example, is characterized by temperate deciduous forest, which is dominated by hardwood trees. Beech and maple might dominate in one area, oak and hickory in

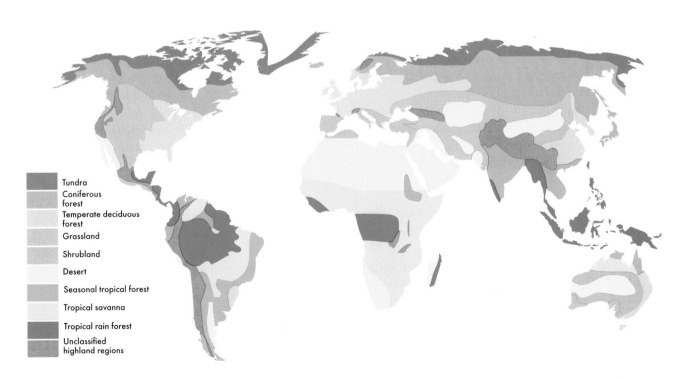

Tundra
Coniferous forest
Temperate deciduous forest
Grassland
Shrubland
Desert
Seasonal tropical forest
Tropical savanna
Tropical rain forest
Unclassified highland regions

Figure 3-3: World map showing major biomes of the Earth.

another and tulip poplar in a third, but the entire area, collectively, is one biome. *Focus On: Biomes* (page 54) takes a closer look at these large terrestrial systems.

Biomes and aquatic ecosystems (such as lakes, the open oceans, coral reefs, and tidal pools) collectively comprise the **biosphere**, the thin layer of air, water, and soil that surrounds the planet and contains the conditions to support life.

What Are the Components of an Ecosystem?

Important research is conducted at all levels of ecological study. In this text, however, we focus on the ecosystem; doing so provides a holistic understanding of natural system structure and function. Ecosystems are composed of nonliving, or abiotic, components and living, or biotic, components. The abiotic and biotic components interact to provide the materials and energy necessary for organisms to survive.

Abiotic Components

The abiotic components of ecosystems include energy, matter, and other factors such as temperature and water.

Energy. The ability to do work — to move matter from place to place or to change matter from one form to another — is known as **energy**. We use energy for many purposes: to build shelters and warm or cool them, to process and transport food, and to keep the cells of our bodies active and functioning properly.

Energy reaches the Earth in a continuous but unevenly distributed fashion as solar radiation. Solar radiation is about 45 percent light; the rest is ultraviolet radiation or heat. Less than one percent (0.1-0.3 percent) of the total energy reaching the Earth's atmosphere each day is actually captured through photosynthesis by living organisms; the rest is reflected by the Earth's cloud cover and never reaches the biosphere, or it is radiated by the Earth's surface back into space as heat. Huge amounts of energy, trapped when the planet formed, are stored deep below the Earth's surface. Stored energy makes its way to the surface through volcanoes, deep sea vents, hot springs, and geysers.

Energy cannot be recycled; when it is used, it is changed to another less concentrated form and eventually radiated into space as heat. Consequently, the Earth must be supplied with a constant flow of energy

in order to support life. The internal energy of the Earth accounts for only a small percentage of its energy; the vast majority is supplied by the sun. Thus, the Earth is an open system for energy, continuously receiving and using energy from the sun and radiating waste heat into space.

The first law of energy, or **first law of thermodynamics**, states that during a physical or chemical change, energy is neither created nor destroyed. However, it may be changed in form and it may be moved from place to place. Think for a moment about the simple act of eating, which provides us with both nutrients and energy. The ultimate origin of the energy present in all food is the sun. Solar energy is captured by green plants, which then convert it to energy in the chemical bonds of sugars. Thus, plants do not *create* energy; they *capture and convert* solar energy to chemical energy. Here's a second example: The combustion of gasoline in an automobile engine simply releases the energy stored within the chemical bonds of the gasoline. This chemical energy is neither created nor destroyed during the process but is instead transformed through a series of steps into mechanical energy, which then causes the car to move.

The second law of energy, or **second law of thermodynamics**, states that with each change in form, some energy is degraded to a less useful form and given off to the surroundings, usually as low-quality heat. In the process of doing work, high-quality (concentrated) energy is converted to low-quality (dispersed) energy. For example, the energy stored in the sugars we eat is converted to chemical energy; this, in turn, is converted to mechanical energy that moves our muscles and creates heat. Similarly, the high-quality energy available in the chemical bonds of gasoline is converted to mechanical energy (which powers the car) and heat. With each transfer of energy, heat is given off to the surroundings, eventually dissipating to the external environment, through our atmosphere to space, and throughout the universe.

As a consequence of the second law of thermodynamics, energy constantly flows from a high-quality, concentrated, and organized form to a low-quality, randomly-dispersed and disorganized form, a phenomenon called **entropy**. Increasing entropy means increasing disorder. In general, life slows down, but does not stop, the process of entropy. Living organisms temporarily concentrate energy in their tissues and thus, for a time, create a more ordered system. Perhaps the best example of this is the capture and storage of the sun's energy by green plants, algae, and some bacteria. Inevitably, however, these organisms are eaten or die and the energy stored within them is transferred through the ecosystem. As chemical bond energy eventually degrades to heat, the system tends toward entropy (Figure 3-4).

Matter. Anything that has mass and takes up space is **matter**. Although meteors and meteorites sometimes enter the Earth's atmosphere, adding matter to the biosphere, the Earth is essentially a closed system. Most of the matter that will be incorporated into objects in future generations is already present and has been present since the planet came into being.

An **element** is a substance that cannot be changed to a simpler substance by chemical means. Each element has been given a name and a letter symbol. Some familiar elements are oxygen (O), carbon (C), nitrogen (N), sulfur (S), and hydrogen (H). All matter is composed of elements.

There are 92 naturally occurring elements and 15 synthetic ones; each has special characteristics that make it unique from all others. What makes an element unique is its atomic structure. All elements are composed of atoms. Thus, atoms are the basic building blocks of all matter. An **atom** is the smallest unit of an element that retains the unique characteristics of that element; it is the smallest particle of an element that can participate in a chemical reaction.

Only a few naturally occurring elements exist as single atoms; these are known as the 'noble gases' — helium, neon, argon, krypton, xenon, and radon. Single atoms of most other elements cannot exist independently for more than an instant. Oxygen, hydrogen, nitrogen and chlorine, for example, exist in the gaseous state as two atoms joined to form a molecule. A **molecule** is the smallest particle of a substance that has the composition and chemical properties of that substance and is capable of independent existence. A molecule of the substance helium, then, consists of a single atom of helium; a molecule of oxygen consists of two atoms of oxygen; and a molecule of water consists of two atoms of hydrogen and one atom of oxygen. Chemical notation allows us to distinguish between atoms and molecules. For example, O, written by itself, represents one atom of oxygen; 2O represents two atoms; and O_2 represents two atoms of oxygen joined to form a molecule of oxygen.

When two or more elements chemically combine in definite proportions, they form a substance known as a **compound**. For instance, water is a compound in which hydrogen and oxygen combine in a ratio of 2:1; that is, each water molecule consists of one oxygen atom

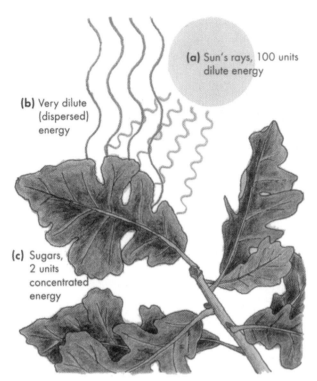

FIGURE 3-4: The two laws of thermodynamics. The conversion of solar energy (a) to chemical bond energy in food molecules (c) through the process of photosynthesis illustrates the first law of thermodynamics. The second law of thermodynamics maintains that (c) is always less than (a) because heat (b) is lost during the conversion.

joined to two hydrogen atoms. The chemical notation for water is H_2O. Glucose sugar is a compound formed of 6 carbon atoms, 12 hydrogen atoms, and 6 oxygen atoms; its chemical notation is $C_6H_{12}O_6$. Most matter exists as compounds held together by the forces of attraction in the chemical bonds between their constituent atoms. A compound can be separated into its constituent elements by chemical reactions but not by physical methods.

Organic compounds all contain atoms of carbon. These may be combined with other carbon atoms or with atoms of one or more other elements. Hydrocarbons such as methane (CH_4), chlorinated hydrocarbons such as DDT ($C_{14}H_9Cl_5$), and simple sugars such as glucose ($C_6H_{12}O_6$) are representative of the millions of organic compounds. All other compounds are inorganic.

The **law of the conservation of matter** states that during a physical or chemical change, matter is neither created nor destroyed. However, its form may be changed, and it can be moved from place to place. Ecosystems function within the law of the conservation of matter by using processes that constantly recycle matter.

Biomes

Climate, especially temperature and precipitation, is the chief determinant in the formation of biomes; other abiotic factors, including the availability of light and the topography, are also important. While there are many different methods of classifying the world's biomes and some disagreement about the exact number, we discuss nine. Biomes can be distinguished along a continuum of latitude, precipitation, and/or altitude. Beginning at the equator and moving northward, the biomes we discuss are: evergreen tropical rain forest, seasonal (or dry) tropical forest, tropical savanna, desert, chaparral, temperate grassland, temperate deciduous forest, coniferous forest, and tundra.

Evergreen Tropical Rain Forest

The most diverse of all biomes, evergreen tropical rain forests boast a spectacular array of plant and animal life (Figure 3-5). The lush vegetation is supported by abundant annual rainfall — 100 to 180 inches (250 to 450 centimeters) — evenly distributed throughout the year. The seasonal variation in temperature is slight, often less than the daily variation (day and night temperatures). Evergreen tropical rain forests occur in three main areas: the Amazon and Orinoco river basins of South America (the world's largest continuous rain forest) and the Central American isthmus; the Indonesian Archipelago; and Madagascar and the Congo, Niger, and Zambezi river basins of central and western Africa.

The natural functioning of tropical forests provides numerous ecosystem services. Rain forests are an important carbon sink, storing large amounts of carbon in trees and other vegetation. Because the clearing of forests accounts for about 25 percent of the carbon released to the atmosphere each year, deforestation has potentially serious implications for global climate change. The dense forest cover also protects the soil, preventing erosion of the thin tropical soils and siltation of local streams and rivers. Rain forests also play a vital role in moderating the global climate, as well as the regional hydrologic cycles where they occur. For instance, the Amazon rainforest generates about half of the

FIGURE 3-5: Tropical rain forest, Costa Rica.

rain that falls in the Amazon Basin by the recycling of water through the processes of evaporation and transpiration.

The biological diversity of the tropical rain forests is unequaled. In some Amazonian and Southeast Asian forests, a single hectare (about 2.5 acres) may be home to over 200 different tree species. A study conducted in Peru found over 300 tree species in a one-hectare plot! By comparison, 10 or so are commonly found in temperate forests. In general, the trees in tropical forests are almost entirely broadleaved, meaning that the leaves have not been modified into needles or spines. A few conifers, or cone-bearing trees, can be found, such as the klinki pines and hoop pines of New Guinea and Australia. Because of the plentiful rainfall and warm temperatures, the trees retain their leaves year-round and are thus classified as evergreen.

Typically, vegetation in the tropical rain forests is stratified. Broadleaved, evergreen trees 80 to 100 feet tall (25 to 30 meters) form an almost unbroken canopy. Emerging from this great canopy are a few very tall trees (about 200 feet or 60 meters) scattered throughout the forest, perhaps one or two emergent trees per acre. An understory, consisting of smaller tree species, younger trees and shrubs, may also be present, but most of the biodiversity is found in the canopy. Rain forest trees tend to be shallow rooted because the roots do not have to penetrate far to obtain moisture. Many have buttressed stems, swollen bases that help provide support. Common to the tropical rain forest are lianas, climbing plants with a rope-like stem, and epiphytes, plant species that live on the stems and branches of trees and do not touch the soil below. Epiphytes absorb water directly from the humid air or capture the abundant rainfall for later use. Orchids, bromeliads, mosses, and ferns grow on the tree branches, further enriching the plant community. The forest floor is usually dark, owing to the dense tree cover. Even so, it teems with life, chiefly insect scavengers, bacteria, and fungi. Organic material decomposes rapidly under the dark, warm, and very humid conditions. Nutrients are quickly recycled and returned to producers, and detritus does not accumulate to enrich and deepen the soil. Most of the nutrient load in the rain forest is stored in the biomass, and consequently the soils are shallow and nutrient-poor. Thus, if an area is deforested, the soils remain productive for only two to five years.

Like plant species, insect and bird species are found in great abundance and diversity in the tropical rain forest. For example, there are over 20,000 species of insects in a six-square-mile area on Barro Colorado Island, a patch of rain forest in the Panama Canal Zone. In contrast, there are only a few hundred insect species in all of France. One reason given for the multitude of bird species is that many of them are herbivores that feed on the plentiful fruits and seeds of the forest trees. These include the fruit-eating parakeets, toucans, hornbills, cotingas, trogons, and birds-of-paradise. Reptiles, amphibians, and small mammals are also numerous. Most of the animals of the tropical rain forest are arboreal, or tree-dwelling, adapted to life in distinct parts of the canopy or subcanopy. Examples include sloths, monkeys, and small cats. The Costa Rican rain forest harbors 14 ground-foraging mammal species, 59 subcanopy species, and 69 canopy species. Two-thirds of the mammals are arboreal.

Seasonal Tropical Forest

Seasonal tropical forests, sometimes called dry tropical forests, are found in tropical climates that have a pronounced dry season (Figure 3-6). According to the length and severity of the dry season, some or all of the trees lose their leaves. A sweep of color may brighten the landscape just before the leaves fall, or when new leaves begin to bud after the rains return. Asian monsoon forests, dominated by teak and sal trees, have the longest dry season; when the rains fall, they are torrential. Annual rainfall, though seasonal, is still quite high, fostering diverse communities. In fact, among terrestrial systems, seasonal tropical forests are second only to tropical rain forests in terms of species richness.

Tropical Savanna

Savannas are tropical grasslands that contain scattered trees or clumps of trees; they often border tropical rain forests (Figure 3-7). Annual rainfall is 40 to 60 inches (100 to 150 centimeters). Central and East Africa boast the largest tropical savannas, but they are also found in Australia and South America.

Savannas are subject to frequent and extensive fires during the dry season, which are thought to stop the spread of forests. Trees and grasses must be resistant

FIGURE 3-6: Seasonal tropical forest, Santa Rosa National Park, Costa Rica.

FIGURE 3-7: The tropical savanna biome, shown here in Serengeti National Park, Tanzania.

to drought and fire, and consequently the number of species is rather small, especially when compared with the adjacent equatorial forests. A single species of grass or tree may dominate over a large area. The vast African savanna is dotted with thorny acacias and other leguminous trees and shrubs, the large-trunked baobab, and palms.

The distinguishing characteristic of the African savanna may well be its populations of large herbivores, the hoofed mammals. Zebras, giraffes, and antelopes of many kinds (including wildebeests, eland, and impala) graze or browse on the vegetation, hunted by such predators as lions and cheetahs. Insects are plentiful during the wet season, when the area is visited by large numbers of nesting birds.

Desert

The word "desert" may conjure an image of a relentless sun beating down on endless sand dunes, but the world's deserts are actually quite varied (Figure 3-8). Some examples are the hot, very dry Sahara (the world's largest desert), the vast Australian desert, the Atacama in Chile, the Negev in southern Israel, and the cool high deserts of North America, which often see winter snows. All deserts share one characteristic: limited precipitation, less than 10 inches (25 centimeters) per year.

Naturally occurring deserts (unlike areas desertified due to human activity) are healthy ecosystems with communities uniquely adapted to life in a dry and sometimes harsh climate. The vegetation are xerophytes, plants adapted to extremely dry climates. Xerophytic plants have various modifications, such as small leaves, to reduce water loss by transpiration and enable them to live for long periods with little water. Cacti, for example, have leaves that are modified into spines, and they produce a thick, waxy cuticle over their stems, where photosynthesis occurs. However, they are not found in all deserts, as they occur only in the Americas. Similarly adapted species occur in the world's other deserts, but are in different plant families, such as the *Euphorbiaceae*. Saguaro, ocotillo, Joshua tree, creosote bush, sagebrush, agave, cholla, and paloverde are found in the deserts of southwestern North America. Colorful ephemeral annuals bloom soon after the infrequent rains and have a very short life cycle. For many species, a critical amount of rain must fall in order to trigger germination; if the rains fail or too little rain falls, the tough seed coats do not open. They remain dormant, perhaps for several seasons, until sufficient rain finally falls. This favorable adaptation allows them to survive long dry periods.

Desert animals include arthropods (especially insects, spiders, and centipedes), reptiles, birds, and small mammals. They too are adapted to desert conditions. The camel, a rare example of a large desert-dwelling mammal, can go for long periods without water because its body tissues can tolerate elevated body temperatures and a high degree of dehydration (conditions fatal to most animals, including humans). The camel does not store water in its hump(s), as is often believed; rather, it conserves water through a special heat exchanger in its nose. Many desert animals, including such predators as owls and rattlesnakes, are nocturnal, avoiding activity during the hot daylight hours. One such species is the desert kangaroo rat of southern California, which remains in its burrow throughout the day. The kangaroo

Figure 3-8: Desert sunset, Guadalupe Mountains National Park, Texas.

rat is unusual in that it never drinks. Rather, it obtains the water it needs through the moisture contained in its food and the metabolic water it produces through respiration. The animal's efficient kidney produces only small quantities of concentrated urine; most water loss occurs as it breathes. Other adaptive characteristics include oversized ears for cooling and a light-colored coat to reflect the intense sunlight.

Chaparral

The chaparral, or Mediterranean scrub forest, is exclusively coastal, found chiefly along the Pacific coast of North America (California and Mexico) and the coastal hills of Chile, the Mediterranean, southern Africa, and southern Australia (Figure 3-9). Limited winter rainfall of approximately 10 inches (25 centimeters) is followed by drought the rest of the year. However, cool, moist air from the adjacent marine environment moderates the climate of the chaparral, and the vegetation consists of trees or shrubs (or both), that have hard and thick evergreen leaves.

The life forms of the chaparral, like those of the desert, are adapted to drought conditions. In California, for example, mule deer and many birds inhabit the chaparral during the rainy season (November to May) then move north or to higher, cooler altitudes during the dry, hot summer. Resident species include small brush rabbits,

wood rats, chipmunks, lizards, wren-tits, and brown towhees. Many chaparral plants are adapted to fire as well as drought. Fire is a natural occurrence brought on by prolonged dry conditions, resinous plants, and a thick layer of dry, slowly decomposing litter on the forest floor. After periodic brush fires sweep through the chaparral, recovery is usually rapid, with shoots sprouting from charred stumps and fire-resistant seeds.

Temperate Grassland

Temperate grasslands occur in areas where average annual rainfall is between 10 and 40 inches (25 to 100 centimeters), depending on the temperature, seasonal distribution of the rainfall, and the water-holding capacity of the soil. Soil moisture is especially important because it limits decomposition and nutrient cycling.

Large grasslands are found in the interior of North America (the prairie, Figure 3-10), southern South America (the Argentine pampas), Eurasia (the steppes), South Africa (the veldt), and Australia. The seasonal nature of the rainfall, and the natural occurrence of periodic fires, discourage the growth of trees. Accordingly, grasses are the dominant vegetation. Which species of grasses are present varies according to local conditions, particularly precipitation. Grasslands may be dominated by annual grasses, perennial bunchgrasses, or perennial sod-forming grasses. Non-grass herbaceous plants may also be seasonally dominant.

This variation in dominant vegetation occurs within all grasslands, and is illustrated by the prairie of central North America. The precipitation increase from west to east results in a continuous change from short-grass to mid-grass to tall-grass prairie.

Grasses have adapted numerous mechanisms to cope with limited water. The roots of some species penetrate deeply into soil (as much as six feet) to reach the permanent water table. Other species have diffuse, spreading fibrous roots. Additionally, grasses may become dormant during prolonged dry spells, reviving with the onset of the

FIGURE 3-9: Chaparral biome, Santa Monica Mountains, California.

FIGURE 3-10: Native tallgrass prairie, Agassiz National Wildlife Refuge, Minnesota.

rains. Some species, such as big bluestem and buffalo grass, have rhizomes, horizontal underground stems with leaves and buds, that serve as a storage organ and means of vegetative propagation. The presence of rhizomes helps to form sod, a dense mat of roots and soil. Much of the biomass of grasses is below ground in the roots; in fact, the roots of healthy perennials may weigh several times as much as the part of the plant above the soil!

Thanks to their productivity, grasslands can support large populations of herbivores, chiefly hooved and burrowing types. In North America, hooved grazers include bison and antelope, while in South America, llamas and alpacas roam the pampas. The large herbivore of the Australian grasslands is the kangaroo. Burrowing mammals include ground squirrels and prairie dogs in North America. Birds and insects are also common grassland residents; the latter are found in especially high numbers in undisturbed areas.

Temperate Deciduous Forest

Much of eastern North America, eastern Asia, and central Europe belong to the temperate deciduous forest biome (Figure 3-11). As its name suggests, the dominant vegetation is deciduous trees, which shed their leaves

seasonally. Significant variation occurs within the biome, resulting in certain groups of dominant tree species characteristic of particular regions. Seven such regions are recognized in North America, with widespread dominants such as maples, basswoods, oaks, hickories, and beeches occurring in different abundances throughout the range of the deciduous forest.

With an average annual rainfall of 40 to 60 inches (100 to 150 centimeters), distributed evenly throughout the year, the temperate deciduous forest supports a variety of plant and animal life. The climate is characterized by distinct seasons, with the long growing season giving way to a shorter, but sometimes extreme, period of cold. The plants are characterized by seasonal flowering and fruiting. Depending on the length of the growing season, the forest undergoes a transformation as the leaves change their color from green to brilliant combinations of red, orange, and yellow. The huge amount of biomass falling to the ground as leaves is rapidly decayed, resulting in a rich organic layer and well-developed soils. Animal species have also adapted to the forest's seasons; many migrate (chiefly birds), become dormant, or hibernate. Other animals, such as the rabbit, deer and fox, simply go about their business, managing as best they can until spring returns once again.

FIGURE 3-11: Temperate deciduous forest biome, as seen in the Great Smoky Mountains National Park.

Coniferous Forest

Most of the world's coniferous forests occur in the Northern Hemisphere, but they are also found in smaller areas in South America, Australia, and New Zealand. Coniferous forests are typically dominated by cone-shaped evergreen trees with needlelike leaves. Three of the most common types of coniferous forests include the temperate rain forest, the boreal coniferous forest, and the montane coniferous forest.

The temperate rain forest is a relatively small, but biologically important, vegetation type. Widely scattered throughout the world, temperate rain forests are found on the northwest coast of North America, in southeastern Australia, in New Zealand, and in southern South America (Figure 3-12). Annual precipitation is high, 80 to 150 inches (200 to 380 centimeters). A coastal forest, it benefits from the moderating climatic effects of the ocean. Thus, winters are mild and summers are cool. Thanks to abundant moisture, all temperate rain forests are characterized by a profuse growth of mosses, liverworts, and ferns. A cloak of green covers every tree trunk and rock. Unlike the boreal coniferous forest, this thick understory is well developed wherever light filters through from above. The tree species of the temperate rain forest vary widely throughout the world. In North America, the temperate rain forest on the Olympic Peninsula consists chiefly of conifers such as western hemlock, western arbor vitae, grand fir, and Douglas fir. The southern reaches of the North American rain forest are dominated by the magnificent redwood, the tallest organism on Earth, while Sitka spruce dominates the northern range. The Northern spotted owl, the marbled murrelet, and the red-backed vole are just a few of the many animals that inhabit the North American temperate rain forest. In New Zealand, magnificent stands of southern beeches, kauri pines, and the yew-like podocarps or yellow-woods dominate. The temperate rain forest of Tasmania is home to myrtle, sassafras, and tree ferns.

The vast boreal coniferous forest, also called the taiga, forms a broad belt across both North America

FIGURE 3-12: Coastal temperate rain forest, Glacier Bay National Park, Alaska.

and Eurasia. There is no comparable forest in the Southern Hemisphere, which lacks a sufficient land mass at the equivalent latitude. The boreal forests experience long, severe winters and short growing seasons with low temperatures and limited precipitation. They are one of the great lumber-producing regions of the world. Unfortunately, in recent years, unrestricted timbering and deforestation have increased rapidly in some areas, including Siberia, site of one of the world's largest stretches of boreal forest.

Evergreen trees — spruce, fir, and pine — are the dominant vegetation of the boreal coniferous forest. Their needle-like or scale-like leaves are covered by a waxy cuticle that retards water loss. This adaptation is essential in the arid environment of the taiga, where groundwater is frozen and thus unavailable to plants for much of the year. The dense evergreens shade the ground, resulting in poorly developed shrub and herb layers. The soil is acidic and covered by a thick layer of partially-decomposed pine needles. Fire is a very

important factor in forest development, as most of the tree species have serotinous seed cones that require very high temperatures to release the seeds. Conifer seeds are a primary food source for squirrels and finches such as siskins and crossbills. In disturbed areas, developmental communities of birch, poplar, alder, and willow may be found. These broad-leaved communities are an important food source for many of the taiga's herbivorous inhabitants, such as moose, snowshoe hare, and grouse. Other residents of the coniferous forests include rabbits, porcupines, rodents, elk, deer, grizzly and black bears, wolves, lynx, and wolverines.

Although precipitation is low, the taiga appears quite wet. The topographic relief is largely the product of past glaciation, with extensive bogs, or muskegs. Long ago, such areas were large ponds; they have since developed into the spongy, quaking bogs we see today. They are colonized by plants that are acid-tolerant, including sphagnum moss, cranberry, pine, spruce, and tamarack. Decomposition in a bog is slow, in part because of the cool climate and in part because the oxygen supply quickly becomes depleted in the standing water. Consequently, bogs tend to be far less productive than surrounding forests. Deposits of peat, a brown, acidic material made up of the compressed remains of partially-decomposed plants, accumulate in bogs. Peat can be used to enrich soil or as a potting compost, or it can be dried and used as a fuel.

Montane coniferous forests are found in the mountains of western North America from northern Canada and Alaska down into Mexico, as well as along the higher ridges in the Appalachian mountains of the eastern United States. The coniferous forests of the Rocky Mountains, Sierra Nevadas, and Cascade Mountains are characterized by bands of vegetation corresponding to changing altitude. These bands are lower in the Sierras and Cascades due to the moderating effect of the Pacific Ocean. These mountain ranges are known for high precipitation levels on their west slopes and relatively small amounts on their east sides. This phenomenon is known as the rain shadow effect. As humid air masses move into the mountains and rise, the air cools, the moisture condenses, and the precipitation falls before reaching the eastern slopes.

In addition to altitude, other important environmental factors help to determine vegetation distribution in the montane coniferous forest. Topographic relief (slope exposure and steepness) can determine the extent of such species as subalpine fir, Engelmann spruce, lodgepole pine, Douglas fir, and ponderosa pine. Some of these species have adapted specialized growth forms to deal with the harsh conditions of the timberline, the upper elevational extent of tree growth. Like the boreal coniferous forest, montane coniferous forests have evolved with fire as a naturally reoccurring part of the environment.

Tundra

The tundra is the northernmost biome; it has no equivalent in the Southern Hemisphere, except in a few mountainous areas, where harsh conditions result in an alpine tundra (Figure 3-13). Precipitation is scarce; a scant six inches (15 centimeters) falls annually, much of it as snow. The distinguishing feature of the tundra is the permafrost, a layer of permanently frozen soil lying several inches below the surface.

Low temperatures, a short growing season, and strong winds are the major limiting factors to life. For much of the year, precipitation is also a problem, as the water of the tundra is frozen and unavailable to most life forms. During the long winter, the tundra is a windswept, frozen, and seemingly barren land. But during the summer growing season, which lasts about two months, surface waters and the top few inches of soil thaw. The permafrost prevents the water from percolating downward and ponds form, dotting the landscape. Although summers are short, summer days are very long, and at these times a vibrant, varied plant community blankets the tundra. Essentially, the "low tundra" of the Alaskan coastal plain is a wet Arctic grassland dominated by grasses, sedges, and dwarfed woody plants such as willows and birches. In many areas, it resembles a thick spongy mat of living and decaying vegetation. A far less profuse growth of lichens, mosses, and grasses is found in the "high tundra," particularly in areas with considerable relief. Most plants of the tundra are long-lived perennials with much of their biomass below ground. Many of the plant species have pigments such as anthocyanin, which help absorb sunlight. Tundra animals include caribou, musk oxen, polar bears, wolves, foxes, ptarmigans, bald eagles, snowy owls, snowshoe hares, Arctic ground squirrels, lemmings, and reindeer. In addition, swarms of mosquitoes and flies are legendary. Some species, such as caribou, migrate with the onset of winter, while others, including lemmings and ptarmigans, remain for the season.

FIGURE 3-13: The vast tundra, seen here in Denali National Park, Alaska, is dominated by lichens, grasses, and dwarfed trees.

Carbon (C), oxygen (O), hydrogen (H), nitrogen (N), phosphorus (P), potassium (K), calcium (Ca), magnesium (Mg), and sulfur (S) are **macronutrients**, chemicals needed by living organisms in large quantities for the construction of proteins, fats, and carbohydrates. These nine macronutrients are the major constituents of the complex organic compounds found in all living organisms. **Micronutrients** are substances needed in trace amounts, such as copper (Cu), zinc (Zn), selenium (Se), lithium (Li), iron (Fe), sodium (Na), cobalt (Co), boron (B), molybdenum (Mo), and chlorine (Cl). These and other micronutrients, along with macronutrients, are regulated by cycles so that they remain available in the physical environment. The chemicals and water that form the complex compounds found in living organisms continually cycle between the abiotic and the biotic components.

Abiotic Factors. In addition to energy and matter, a number of other factors are important to consider when studying the abiotic component of the ecosystem. They include temperature, precipitation, humidity, wind, light, shade, fire, salinity, and available space. These factors do not remain constant, but vary over space and time. They determine the native vegetation found in a particular area, and thus what biome and ecosystems will occur there (Figure 3-14). Together with the available energy and the type, amount and distribution of nutrients, these physical factors help to determine which organisms will comprise the biotic components of that system.

Biotic Components

The biotic components of ecosystems are grouped into two broad categories, producers and consumers, based on nutritional needs and manner of feeding.

Producers. **Producers**, or **autotrophs**, are self-nourishing organisms (auto, "self"; troph, "nourishment"). Given water, nutrients and a source of energy, they can produce the compounds necessary for their survival.

Most producers, including green plants, algae and cyanobacteria (blue-green algae), are **phototrophs**. They contain chlorophyll, a green pigment that absorbs light energy from the sun. Through the process of **photosynthesis** (photo, "light"; synthesis, "to put together"), phototrophs use the sun's light energy to convert carbon dioxide and water into complex chemical bonds forming simple carbohydrates such as glucose and fructose. Essentially, light energy is transferred to the carbon bonds that form carbohydrates. In the pro-

cess, oxygen is given off. Photosynthesis is roughly one to three percent efficient at converting light energy to chemical energy, that is, 100 units of light energy result in one to three units of chemical energy produced.

Phototrophs can then convert simple carbohydrates into more complex carbohydrates (starches and cellulose), lipids (fats and oils), and proteins. They use some of the energy to manufacture cell contents and to carry out life processes. Complex carbohydrates are stored in their tissues, to be used later to meet energy needs. This stored energy (in seeds, roots, or sap) enables plants to call on reserves during germination, after winter, or during prolonged periods of cloudy days.

A small percentage of the Earth's biota depends on the energy stored by **chemotrophs**, autotrophs that use the energy found in inorganic chemical compounds (rather than light energy from the sun) for their energy needs. Chemotrophs are represented primarily by species of bacteria that live in and around deep thermal vents in the oceans, at the mud-water interface in high mountain lakes during winter, or in wetlands. Through **chemosynthesis**, they convert the energy in the chemical bonds of hydrogen sulfide to make and store carbo-

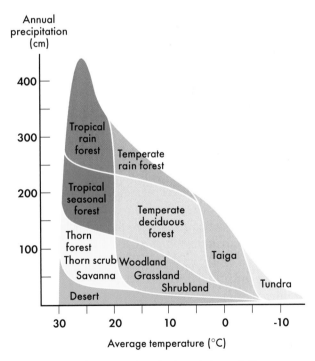

FIGURE 3-14: Biomes are largely the product of climate, specifically temperature and precipitation. Boundaries between biome types are approximate; other physical factors, including soil type, exposure to frequent fire and maritime versus continental climates, can determine which of the two or more biomes, both adapted to the climate of an area, will actually occur there.

Figure 3-15: Dung beetles, detritivores that feed on the organic wastes of other organisms, gather and transport dung in tandem.

hydrates and, in doing so, give off sulfur compounds into the water. Unlike photosynthesis, chemosynthesis takes place without sunlight or chlorophyll.

The importance of producers cannot be emphasized enough, for the energy that they capture and fix supports the Earth's biota. To put it another way, whereas producers capture new energy from the environment, all other organisms simply transfer portions of that energy from their food to themselves. In terrestrial systems, trees, grasses, herbs, and shrubs are the major producers; in aquatic systems, **phytoplankton**, microscopic floating plants and algae, function as the major producers.

Consumers. Consumers, or **heterotrophs** (hetero, "other, different"), eat by engulfing or predigesting the fluids, cells, tissues, or waste products of other organisms. Because consumers cannot make their own food, they rely on other organisms for their energy needs. They can be broadly categorized as macroconsumers or microconsumers.

Macroconsumers feed by ingesting or engulfing particles, parts, or entire bodies of other organisms, either living or dead. They include herbivores, carnivores, omnivores, scavengers, and detritivores. **Herbivores**, or **primary consumers**, (grasshoppers, mice and deer, for example) eat producers directly. Other consumers (such as meadowlarks, black rat snakes, and bobcats) feed indirectly on producers by eating herbivores.

Because they eat other animals, they are referred to as **carnivores**, or **secondary consumers**. Consumers that eat both plants and animals (like black bears, Norway rats, and humans) are **omnivores**. Carnivores that eat secondary consumers (such as hawks and large-mouth bass) are **tertiary consumers**.

Many heterotrophs consume dead organic material. Those that consume the entire dead organism are known as **scavengers**. Two familiar examples are vultures and hyenas. Consumers that ingest fragments of dead or decaying tissues or organic wastes are called **detritivores**, or **detritus feeders** (blue crabs, dung beetles, earthworms and shrimp, for example) (Figure 3-15).

Like detritivores, **microconsumers** feed on the tissues of dead organisms; they also consume the waste products of living organisms. They differ in that they digest materials outside of their cells and bodies, through the external activities of enzymes, and then absorb the predigested materials into their cells. Consequently, they are often referred to as **decomposers**. Decomposers live on or within their food source. The result of the activity of decomposers is what we call rot or decay. Eventually, decomposers reduce complex molecules to simple molecules and return them to the physical environment for reuse by producers. Microconsumers include some bacteria, some protozoans, and fungi (for example, yeasts and molds).

Decomposers play the major role in reducing complex organic matter to inorganic matter and returning nutrients and necessary chemicals to the physical environment in a form that can be used by producers. The importance of decomposers is often overlooked. We are only now beginning to get a clearer picture of how they perform their vital functions. Consider what it would be like if leaves were not decomposed after they fell from trees. How quickly nutrients in the soil would be depleted if decomposers did not continually recycle them after the deaths of producers and consumers! Decomposers form the vital link in the cycle that returns nutrients and chemicals to the soil, enabling material to proceed from death to life in ecosystems.

What Determines the Structure of Ecosystems?

No organism, population, or community is distributed evenly about the Earth. Instead, each occupies a particular environment or habitat. Some organisms are distributed throughout large areas, while others live in very specific habitats (Figure 3-16). Some organisms are found only in the leaf litter of deciduous forests, under rocks in fast moving streams, or in the minute cracks of rocks in the Antarctic. Some organisms are found on the forest floor, but not in the forest canopy; some survive at great ocean depths, but not in shallow waters. Some organisms are successful in the wetter, cooler conditions on the western slopes of the coastal ranges of western North America, while others are successful in the drier, warmer conditions of the eastern slopes.

The species, populations, and communities found in an ecosystem or biome are the result of **limiting factors**, abiotic and biotic regulators that determine the distribution and success of living organisms.

Abiotic Limiting Factors

Temperature, light, precipitation, and available phosphorus, oxygen, and carbon are examples of abiotic limiting factors. As we saw in Figure 3-14, a difference in average rainfall separates the major plant communities into forests, grasslands, or deserts. Typically, regions with more than 40 inches (100 centimeters) of precipitation per year are forests, 10 to 40 inches (25 to 100 centimeters) are grasslands, and less than 10 inches (25 centimeters) are deserts. Temperature is another important limiting factor. If an area usually gets 40 inches of rainfall per year and is hot, it will sustain a tropical savanna; if the area is temperate, it will sustain a deciduous forest (beeches, maples, oaks). Soil types within a forest are also a limiting factor. In a temperate climate with adequate rainfall, oaks and hickories are more successful on low-nutrient soils, while beeches and maples are more successful on high-nutrient soils. Abiotic factors form a complex set of interactions that limit or control the activities of organisms, populations, and communities. Ecologists do not understand fully all the ways in which these factors interact.

In an aquatic ecosystem, oxygen, sunlight, and nutrients (phosphorus and nitrogen) are the most significant abiotic limiting factors. Generally, the availability of phosphorus (as the compound phosphate) in lakes and streams limits the growth of aquatic plants and algae. Increasing the amount of phosphates increases the growth of plants and algae. Nutrient enrichment of a lake, stream, or estuary can set in motion a mix

FIGURE 3-16: Habitat and distribution of species. (a) Some species are widely distributed. The giant ragweed is found throughout North America from Mexico to Canada. (b) Other species, like the giant, blood-red tube worm, are found in very specific habitats, such as near thermal vents on the ocean floor.

of physical, chemical, and biological changes that collectively are known as **eutrophication**, the natural aging of a lake. The high input of nutrients may be the result of natural erosion or it may be related to human activities, as in the case of runoff from agricultural fields that have been treated with fertilizers. When human activities lead to nutrient enrichment of a body of water, ecologists say that the aquatic system has undergone cultural eutrophication.

The single largest marine system affected by cultural eutrophication is an area known as the "Dead Zone" in the Gulf of Mexico. Eutrophication in the Dead Zone has resulted in hypoxic conditions, which are characterized by low oxygen levels. According to a 2000 report by the National Science and Technology Council (NSTC) Committee on Environment and Natural Resources, the primary cause of the hypoxia is excess nitrogen that flows into the Gulf from the Mississippi-Atchafalaya River Basin. This river basin is the largest in the country, covering 41 percent of the continental United States and containing 47 percent of the nation's rural population and 52 percent of its farms. Waste, including nitrogen-based fertilizer, from this area drains directly into the Gulf of Mexico, resulting in spring and summer "blooms," or dense mats, of algae. The first Dead Zone was recorded in the early 1970s. Originally, the hypoxic events occurred every two to three years; however, they now occur annually. By the early 1990s, the zone covered approximately 3,670 square miles (9,500 square kilometers); by 2001, the area had doubled to 8,000 square miles (20,800 square kilometers), an area larger than the State of New Jersey. The zone occurs in one of the most important commercial and recreational fisheries in the country and could threaten the economy of Gulf States such as Texas and Louisiana. According to the NSTC report, efforts are underway to reduce nitrogen flows into the Gulf.

Cultural eutrophication is also a problem in many freshwater systems, including Lake Erie, part of the 2,000-mile-long Great Lakes system. Historically, the ecosystem teemed with diverse wildlife, including such fishes as blue pike, lake whitefish, lake sturgeon, and cisco. The young of these and other species fed on the mayfly, a once-plentiful insect that, in its larval stage, lives on the lake bottom. However, by the 1950s, dense algal blooms, particularly of *Cladophora*, became commonplace in the lake's western and central basins during periods of calm, warm weather in mid to late summer. The blooms were caused by a high influx of phosphates, which entered the lake in agricultural runoff, detergent-laden wastewater, and insufficiently treated sewage. When the algae died and sank to the bottom, the decomposition of such large amounts of organic matter depleted the water's oxygen supply, and the mayfly population crashed. As it did, the populations of fish species whose young depended on the larva also plummeted. By the 1960s, the blue pike, lake whitefish, lake sturgeon, and cisco were extirpated from Lake Erie. To learn more about the effect of cultural eutrophication on Lake Erie's biotic community, the progress that has been made in reducing nutrient loads to the lake, and the challenges that remain in the effort to restore and protect the ecosystem, read *Environmental Science in Action: Lake Erie* by going to www.EnvironmentalEducationOhio.org and clicking on "Biosphere Project."

Living organisms, populations, and communities have a range of tolerances for each of the abiotic limiting factors, a concept known as the **law of tolerances**. Tolerances range along a continuum from the maximum amount or degree that can sustain life to the optimum or best amount for sustaining life to the minimum amount that can sustain life (Figure 3-17). Any change that approaches or exceeds the limits of tolerance, either the maximum or the minimum, becomes a limiting factor. For example, laboratory trials have demonstrated that speckled trout prefer water temperatures of 57° to 66° F (14° to 19° C), although they can tolerate temperatures as high as 77° F (25° C) for short periods of time. This finding is corroborated by field studies, which show that speckled trout are not found in streams where the temperature exceeds 75° F (24° C) for an extended period of time.

Some organisms have wide ranges of tolerances and others have narrow ones. Aquatic insects such as mayflies, fish such as hogsuckers, and flatworms such as planaria all have narrow tolerances for oxygen that limit their success in aquatic habitats. When the concentration of dissolved oxygen in streams or lakes is reduced, these species disappear quickly and are replaced by species more tolerant of lower oxygen levels, such as mosquito larvae, rat-tailed maggot, and carp. A mayfly (or hogsucker or planaria) is an **indicator species**, a species that indicates, by either its presence or absence, certain environmental conditions (in this case, the amount of dissolved oxygen). If mayflies are abundant in a stream, the amount of dissolved oxygen must be relatively high; if they are absent, it is probably low. Scientists look for the presence of such organisms when determining the ecological health of both aquatic and terrestrial ecosystems.

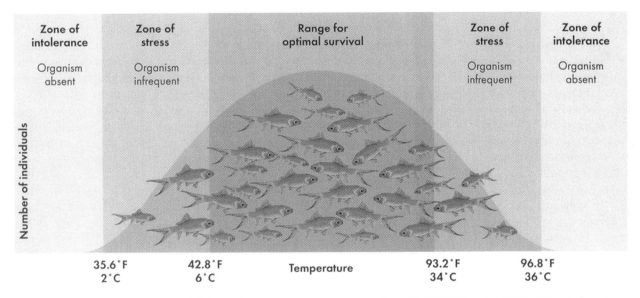

FIGURE 3-17: Law of tolerances. Goldfish can live at temperatures ranging from 35.6°F (2°C) to 96.8°F (36°C) but cannot tolerate temperatures above or below this range.

Limiting factors have important implications for humans. For example, the success of a crop is sometimes limited by micronutrients, such as iron for soybeans or molybdenum for clover, even if macronutrients (phosphorus, nitrogen, and potassium) are applied in large quantities for that particular crop. We manipulate limiting factors when we try to maintain optimum conditions for growth to ensure or increase our harvest from gardens, farms, and ponds.

Biotic Limiting Factors

Ecosystem structure is also affected by the interactions between species, such as competition for food or shelter. One such interaction occurs when **prey**, living organisms that serve as food for other organisms, are inhibited or eliminated by **predators**, organisms that obtain their food by eating other living organisms. For example, if the population of bass, a predator, in a pond increases, the population of prey species such as bluegills may decrease. The resultant drop in the bluegill population may then lead to an increase in the small animals (zooplankton) that the bluegills eat. The increase in zooplankton, in turn, may cause a decrease in the population of producers, or phytoplankton, in the pond.

Ecologists are beginning to understand the effects that organisms can have on community structure. Beavers are an excellent example of a biotic regulator. Through the construction of dams, beavers significantly alter the ecosystem structure of their habitats. Beaver dams alter stream channels and slow down the flow of water, creating and maintaining wetlands, thereby encouraging the growth of plants and animals that are successful in flooded areas and discouraging the growth of those successful in drier areas. By altering their habitats, beavers significantly influence the diversity of the community. For this reason, they are known as a **keystone species**, a species that has a significant role in community organization due to its impact on other species.

The rabbits of southern England are a second example of a keystone species. A drastic reduction in the rabbit population through disease allowed a thick growth of meadow grass. The heavy grass caused the local extinction of open-ground ants and, in turn, the extinction of the large blue butterfly, whose caterpillar fed on the ants. The loss of a keystone species, the rabbit, caused the successful growth of a plant species, the meadow grass, and the local extinction of two animal species, the ant and the butterfly.

Discussion Questions

1. A bumper sticker reads "Nature bats last." Explain what that saying means in ecological terms; be sure to consider its implications in terms of the structure of the biotic community.

2. Because they are internally ordered, organisms, populations, and living systems tend to slow down the disorder or randomness of the physical Earth, thus slowing, but not stopping, the process of entropy. The order found in an organism is paid for by creating disorder in its immediate environment. Humans, it seems, are an exception to this rule. Our societies have grown more complex and theoretically more ordered. Can you explain this seeming paradox?

3. According to the law of the conservation of matter, matter can neither be created nor destroyed. Many people argue that humans ignore this law. What might be some consequences of our failure to heed the law of the conservation of matter?

4. Differentiate between primary consumers, secondary consumers and tertiary consumers, and between microconsumers and detritivores. List an example of each of these that is found in your area.

(continued on next page)

Summary

Ecologists study the natural world at many levels: individual, population, species, community, ecosystem, landscape, biome, and biosphere. An individual is a single member of a species. A species includes all organisms that are capable of breeding to produce viable, fertile offspring. Individuals of a particular species that live in the same geographic area comprise populations; populations have measurable group characteristics such as birth rates, death rates, seed dispersal rates, and germination rates. The place where the individual organism or population lives is its habitat. All of the populations of organisms that live and interact with one another in a given area at a given time are collectively known as a community. A community and its interactions with the physical environment comprise an ecosystem. A landscape is a heterogeneous "patchwork" of many ecosystems taken together. A grouping of many landscapes is a biome, which is identified and classified by its dominant vegetation type. The union of all terrestrial and aquatic ecosystems — and the largest system of life-physical interactions on Earth — is called the biosphere.

Ecosystems are composed of biotic components (communities) and abiotic components (physical surroundings). Energy, matter, and physical factors like temperature and rainfall constitute abiotic components. According to the first law of energy, or first law of thermodynamics, energy can neither be created nor destroyed but may be changed in form and may be moved from place to place. The second law of energy, also known as the second law of thermodynamics, states that with each change in form, some energy is degraded to a less useful form and given off to the surroundings, usually as heat. Consequently, in the process of doing work, high-quality energy is converted to low-quality energy. Energy constantly flows from a high-quality, concentrated, and organized form to a low-quality, randomly-dispersed and disorganized form, a phenomenon called entropy.

Matter is anything that has mass and takes up space. Elements, substances that cannot be changed to simpler substances by chemical means, comprise all matter. An atom is the smallest unit of an element that retains the unique characteristics of that element. Molecules are formed when two or more atoms combine. Compounds are molecules composed of two or more different elements. Compounds containing atoms of carbon are known as organic compounds. According to the law of the conservation of matter, matter is neither created nor destroyed, but its form may be changed, and it can be moved from place to place.

Biotic components are composed of producers (autotrophs), consumers, and decomposers (heterotrophs). Autotrophs that convert the energy of the sun into chemical energy are called phototrophs. Autotrophs that use the energy found in inorganic chemical compounds in order to produce starches and sugars are known as chemotrophs. Consumers are categorized as macroconsumers or microconsumers. Macroconsumers include herbivores (primary consumers), which eat plant matter; carnivores (secondary consumers), which feed on animals; and omnivores, consumers that eat both plants and animals. Carnivores that feed on secondary consumers are called tertiary consumers. Some consumers feed on dead organisms:

Scavengers consume the entire organism; detritivores ingest fragments of dead or decaying tissues or organic wastes. Microconsumers, or decomposers, digest organic material outside of their bodies through the activities of enzymes; they then absorb the predigested material into their cells. Because they break down wastes and dead plant and animal matter, microconsumers return nutrients to the environment to be used once again.

The species, populations, and communities in an ecosystem or biome are the result of limiting factors, abiotic and biotic regulators that determine the distribution and success of living organisms. Abiotic factors include precipitation, temperature, and nutrient levels. According to the law of tolerances, living organisms, populations, and communities have a range of tolerances for each of the abiotic factors that operate in a specific ecosystem. An indicator species is a species that indicates, by its presence or absence, certain environmental conditions with respect to limiting factors (such as the amount of dissolved oxygen in a stream). Examples of biotic limiting factors include predator-prey interactions and keystone species. A keystone species is one that has a significant role in community organization due to its impact on other species.

(continued from previous page)

5. Can you think of ways that nonhuman members of the biotic community alter or affect the structure of the ecosystems found in your area?

6. Define limiting factors; give several examples of both abiotic and biotic limiting factors; and explain the relationship between abiotic limiting factors and the law of tolerances.

KEY TERMS

atom	first law of thermodynamics	omnivore
autotroph	habitat	organic compound
biome	herbivore	photosynthesis
biosphere	heterotroph	phototroph
carnivore	indicator species	phytoplankton
chemosynthesis	keystone species	population
chemotroph	landscape	predator
community	law of the conservation of matter	prey
compound		primary consumer
consumer	law of tolerances	producer
decomposer	limiting factor	second law of thermodynamics
detritus feeder	macroconsumer	
detritivore	macronutrient	scavenger
ecosystem	matter	secondary consumer
element	microconsumer	species
energy	micronutrient	tertiary consumer
entropy	molecule	
eutrophication		

CHAPTER 4

Ecosystem Function

> *We see, then, that chains of plants and animals are not merely "food chains," but chains of dependency for a maze of services and competitions, of piracies and cooperations.*
>
> **Aldo Leopold**

> *Only the sun gives without taking.*
>
> **Anonymous**

Learning Objectives

When you finish reading this chapter, you will be able to:

1. Define gross and net primary productivity.

2. Trace the flow of energy through the biotic community.

3. Describe how energy flow affects the structure of an ecosystem.

4. Explain, in general, how materials cycle in an ecosystem.

5. Explain how the flow of energy and the cycling of matter bind together the structural components of an ecosystem.

While the biotic and abiotic components form the structure of an ecosystem, it is process — energy flow and materials cycling — that links them together as a functional unit. All ecosystems — aquatic or terrestrial, frigid or tropical, desert or rain forest — are dependent upon the flow of energy and the cycling of materials through the community of living organisms (Figure 4-1). In this chapter, we will learn what primary productivity is, how energy flows through the biotic community and affects its structure, and how materials cycle in an ecosystem. We'll also learn about ecosystem services and the benefits they provide to humanity.

FIGURE 4-1: The sun powers all ecosystem processes.

What Is Primary Productivity?

Annually, about 34 percent of the energy input from the sun is reflected directly back to space by clouds, dust, water, and chemicals in the atmosphere (Figure 4-2). Approximately 1.1 to 1.5 million kilocalories per square meter per year (kcal/m²/yr) reach the Earth's surface. Of the 66 percent that is absorbed by the biosphere, 42 percent heats the land and warms the atmosphere, 23 percent helps regulate the water cycle through evaporation, and about 1 percent generates wind currents. Only 0.1 to 0.3 percent of the sunlight reaching the Earth is actually captured by phototrophs, yet that small fraction results in the huge amount — hundreds of billions of tons — of living matter or **biomass** that exists on the planet.

The total amount of energy fixed by autotrophs over a given period of time is called **gross primary productivity** (GPP). Usually, figures for GPP are given for a particular environment, such as a forest or estuary. After the producers' own energy needs are met through **respiration** (R), the release of energy from fuel molecules, the amount of energy available for storage is the **net primary productivity** (NPP): GPP - R = NPP. Measuring NPP can tell us how much energy is available for primary consumers. Respiration enables all living organisms, including autotrophs, to carry out life processes such as growth, reproduction, and movement. Only organisms with chlorophyll can photosynthesize, thereby capturing the energy needed for life, but all living things respire. Plants and animals require oxygen to release energy, a process known as **aerobic respiration**. Some microorganisms do not require oxygen to obtain energy from fuel molecules, a process known as **anaerobic respiration**.

It's important to realize that ecosystems vary widely in terms of net primary productivity. In specific ecosystems and for specific crops (like fish or timber), we

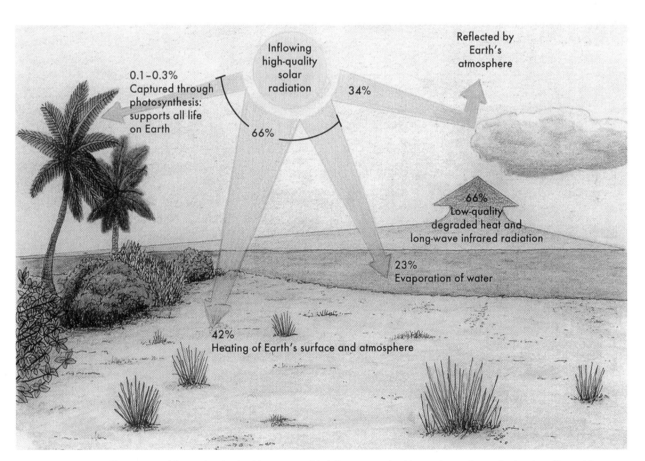

FIGURE 4-2: Energy flows to and from the Earth. About one-third (34 percent) of incoming solar radiation is reflected by the Earth's atmosphere back into space. Of the two-thirds (66 percent) that is not, 42 percent heats the Earth's surface and atmosphere, 23 percent causes the evaporation of water (from oceans, lakes, rivers, and plant leaves), and less than 1 percent is captured by phototrophs. Notice that outflowing radiation (the amount immediately reflected by the Earth's atmosphere plus the amount of degraded heat energy and long-wave infrared radiation) equals incoming radiation.

can determine the amount of food and fiber available for harvest that does not endanger the system's ability to sustain itself. Ecologists also try to determine how efficient our crops and animals are at providing energy and materials for our use. Small changes in efficiency can mean substantial increases in the production of food and fiber over an entire crop or herd.

One glance at Table 4-1 reveals some startling insights. The ecosystems that are most efficient at capturing energy are those most threatened by human encroachment. Look at the energy fixed by tropical forests, swamps and marshes, reefs, estuaries, and continental shelves. Can you understand why it is vitally important, in terms of energy flow, to maintain the integrity of certain ecosystems? What areas should we protect to maintain the integrity of ecosystem energy flow? What areas should we explore for new food sources? Can we improve the productivity of ecosystems we now exploit?

In the United States, one of the ecosystems most efficient at capturing energy, the Chesapeake Bay, is under siege. The bay is the nation's largest and most productive **estuary**, a semi-enclosed coastal body of water composed of fresh and saline (salt) water. The "Queen of Estuaries" is the largest producer of blue crabs in the world, and the value of its finfish and shellfish harvests is approximately $1 billion annually. In the United States, only two other areas outproduce the Chesapeake — the Atlantic and Pacific oceans. Unfortunately, the bay is much less productive than in earlier years. Commercial harvests of rockfish, shad, white perch, and oyster have fallen over the past several decades. The bay's declining productivity is directly linked to the explosion in its human population. Half of the Chesapeake basin's original forests and wetlands — which buffered the bay against pollutants and sediments — are gone, much of it replaced by suburban sprawl. Developed land comes at the expense of natural systems and farmland, and often exacerbates soil erosion and siltation. Population growth and urban development have brought increased municipal wastewater discharge and concentrated industry, thus increasing pollution in certain areas. After over 30 years of efforts to reverse the bay's decline, some progress has been made, but much work remains. To learn more about the Chesapeake Bay and the efforts underway to restore its health and biological productivity, read *Environmental Science in Action: The Chesapeake Bay* by going to www.EnvironmentalEducationOhio.org and clicking on "Biosphere Project."

TABLE 4–1: Primary Production of the Earth

The energy of sunlight reaching the Earth averages 700 cal/cm^2/day for all wavelengths outside the atmosphere, about 5.5 x 10^5 kCal/m^2/yr in the visible range of the Earth's surface. A fraction of this energy is absorbed by chlorophyll and used in photosynthesis and primary productivity, the amounts of which vary widely in different kinds of communities. The energy content of organic matter averages 4.26 kCal/g of dry matter in land plants, 4.9 kCal/g in open-ocean plankton, and 4.5 kCal/g in other algae.

Ecosystem type	Area (10^6 km^2)	Net primary productivity per unit area (dry g/m^2/yr) Normal Range	Mean	World net primary production (10^9 dry t/yr)
Tropical forest	24.5	1000-3500	2000	49.4
Temperate forest	12.0	600-2500	1250	14.9
Boreal forest	12.0	400-2000	800	9.6
Woodland and shrubland	8.5	250-1200	700	6.0
Savanna	15.0	200-2000	900	13.5
Temperate grassland	9.0	200-1500	600	5.4
Tundra and alpine	8.0	10-400	140	1.1
Desert and semidesert	42.0	0-250	40	1.7
Cultivated land	14.0	100-3500	650	9.1
Swamp and marsh	2.0	800-3500	2000	4.0
Lake and stream	2.0	100-1500	250	0.5
Total continental	149.0		773	115.0
Open ocean	332.0	2-400	125	41.5
Continental shelf, upwelling	27.0	200-1000	360	9.8
Algal beds, reefs, estuaries	2.0	500-4000	1800	3.7
Total marine	361.0		152	55.0
World total	510.0		333	170.0

Source: *Robert Whittaker, Communities and Ecosystems, 1970, and Whittaker and Likens, Human Ecology, Vol. 1, No. 4, 1973.*

How Does Energy Flow through a Community?

To understand the movement of energy and materials through an ecosystem, we must look at feeding relation-

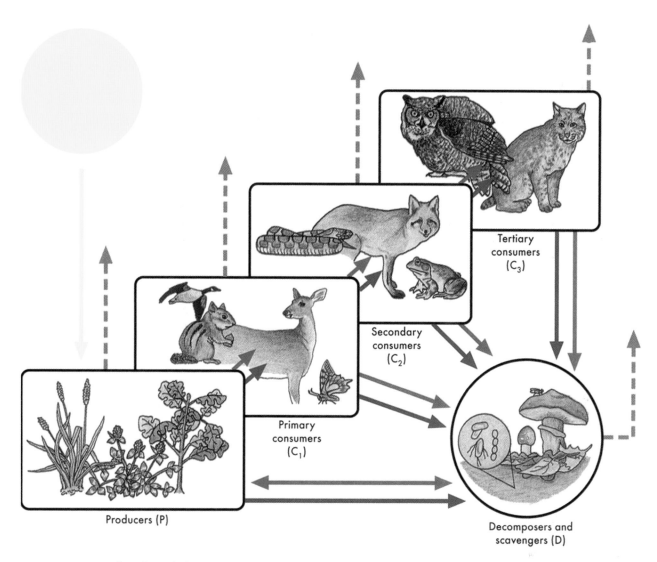

Figure 4-3: Energy flow through the components of an ecosystem. The energy of sunlight is captured by phototrophs through the process of photosynthesis. It then flows, as the energy in chemical bonds in food, through successive trophic levels (solid blue lines). Ultimately, the energy passes to decomposers. At each transfer, some energy is lost as heat, a product of the respiratory activities that support all organisms (dashed brown lines). The red lines represent the cycling of materials through the ecosystem. Note the two-way red arrow between decomposers and producers.

ships. In every ecosystem, successive levels of consumers depend upon the organisms at lower levels. These various levels — the producers and successive steps removed from the producers — are called **trophic levels**. Thus, the community of organisms forms many **food chains**. The first trophic level consists of the producers (P); the second level, primary consumers (C_1), or herbivores; the third level, secondary consumers (C_2), or carnivores and omnivores; and the fourth level, tertiary consumers (C_3) (Figure 4-3). For example, weedy plants (P) are eaten by grasshoppers (C_1), which in turn are eaten by blue jays (C_2), which in turn are eaten by Cooper's hawks (C_3) (Figure 4-4). In both aquatic and terrestrial ecosystems, decomposers (microconsumers, D) operate

at each trophic level, breaking down wastes from living organisms and the tissues of dead organisms. Unlike other food chains, detritus food chains begin with dead or decaying materials (Figure 4-5).

Food chains are simplified illustrations of the path that energy and materials follow in an ecosystem. But feeding relationships are more complicated than food chains imply. Interlocking chains, woven into complex associations called **food webs**, more accurately describe the feeding relationships among organisms in a community and the movements of energy and materials (Figure 4-6). Just like food chains, food webs start with producers of many kinds, consumed by many species of consumers at several trophic levels, and culminate with

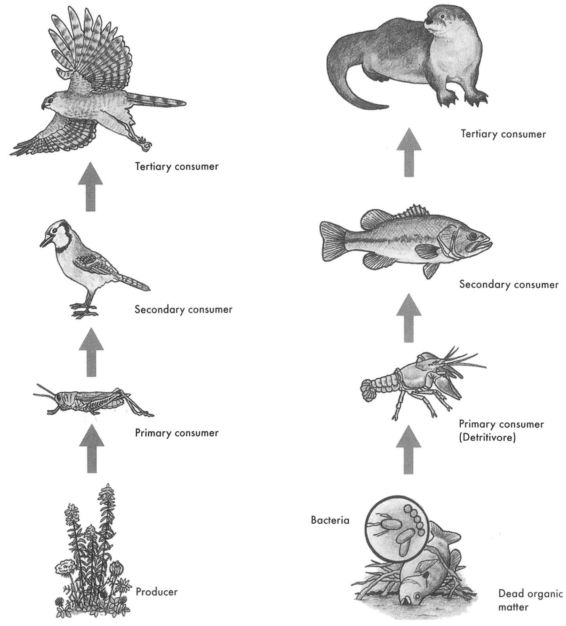

FIGURE 4-4: A typical food chain. Queen Anne's lace, yarrow, and clover capture and store the sun's energy, forming the basis of this food chain. Herbivorous grasshoppers (primary consumers) feed on the plants, and carnivorous blue jays (secondary consumers) feed on the grasshoppers. Carnivorous Cooper's hawks (tertiary consumers) feed on blue jays.

FIGURE 4-5: A typical detritus food chain. Detritus food chains are unique because they begin with dead matter. Bacteria decompose dead organic matter (both plant and animal) which is then eaten by detritus feeders, such as crayfish. Fish that feed upon the crayfish, such as bass, are in turn fed upon by top carnivores such as otters.

decomposers working at all levels. **Detritus food webs** are based on decomposing plant and animal material or animal waste products. They include several levels of consumers, perhaps as many as four in an aquatic habitat. It is important to remember, however, that the energy derived from the detritus originated with living producers.

How Does Energy Flow Affect Ecosystem Structure?

Energy, which flows through an ecosystem according to the laws of thermodynamics, affects the structure of the system because it determines trophic relationships.

Figure 4-6: A simplified food web for a wetland in the eastern United States. Arrows indicate the movement of energy and materials through the biota.

In general, each successive trophic level contains less energy, less biomass and fewer numbers of organisms, resulting in pyramidal relationships for energy, biomass, and numbers.

The **pyramid of energy** depicts the production, use, and transfer of energy from one trophic level to another (Figure 4-7). Energy transfers from one trophic level to another are not efficient and vary widely from species to species. As a general rule, 90 percent of the available energy is lost as heat when members of one trophic level are consumed by members of another. This is not a precise figure. In some cases the energy efficiency might be less than 10 percent, in others as high as 30 percent. On average, however, only 10 percent of the available energy is actually transferred at each step, a phenomenon known as the **10 percent rule**. The 10 percent rule explains why food chains are generally short and why pyramidal relationships exist.

The 10 percent rule also has significant implications for human populations. Humans are omnivorous,

capable of eating both plant and animal tissue. When they consume most of their food at the secondary consumer level, the transfer of energy is far less efficient than it is when they consume at the primary level. The trophic level at which people consume most of their calories, then, depends upon the availability of food. Meat production is an expensive and energy-inefficient process; for this reason, people who cannot afford a diet heavy in meat or meat products tend to eat foods derived from plants. Consumption at the primary consumer level allows them to "get more (energy and biomass) for their money." In many areas of Asia, by far the planet's most populous region, people obtain most of their calories at the primary consumer level, supplementing their diet with protein-rich seafood or livestock such as chickens and hogs. Theirs is an energy-efficient diet when compared to the meat-rich diet of the United States and other Western nations.

The **pyramid of biomass** depicts the total amount of living material at each trophic level (Figure 4-8).

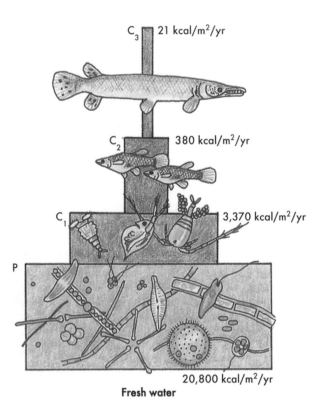

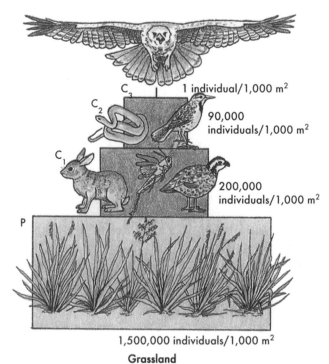

FIGURE 4-7: Pyramid of energy. A freshwater aquatic community at Silver Springs, Florida, illustrates the inefficiency of energy transfers among trophic levels. Energy flow is expressed as kilocalories per square meter per year.

FIGURE 4-9: Pyramid of numbers. The number of organisms at each trophic level of a grassland community results in a stepped pyramid.

FIGURE 4-8: Pyramid of biomass. Measurements of biomass (grams per square meter) in a Panamanian rain forest reveal a large difference between the producer and consumer levels: 40,000 grams per square meter supports a total consumer mass of just five grams per square meter. D represents the decomposer biomass, which we would expect to be large in a community where nutrients are recycled rapidly.

Measuring biomass at each level enables us to estimate the total amount of living matter in any community or ecosystem at any one time. Together, estimates of the rate of energy fixed and the amount of biomass produced enable us to better understand the limitations of our use of energy and materials (food and fiber) of any particular ecosystem.

The **pyramid of numbers** depicts the relative abundance of organisms at each trophic level (Figure 4-9). For example, herbivores at the first consumer level, such as grasshoppers, mice, daphnia (a type of microscopic animal) and antelopes, are usually found in large numbers, while carnivores at higher levels, such as wolves, lions, bobcats, owls and eagles, are found in smaller numbers. Carnivores at the top of the food chain are dependent upon huge numbers of organisms at the trophic levels below them. At the same time, they may regulate the populations of organisms at lower trophic levels. What would be the result if all top carnivores were eliminated from an ecosystem?

How Do Materials Cycle Through an Ecosystem?

Unlike energy, materials such as water, oxygen, carbon, phosphorus, and nitrogen are used over and over again.

Materials cycle through ecosystems by the workings of many processes. We call these processes **biogeochemical cycles** because they involve living organisms and geologic and chemical factors.

Materials cycle from the air, water, and soil through food webs and back to the air, water, and soil. Throughout the Earth's history, materials have continued to cycle in this way. Nutrients and gases are released into soil, water, and air through the respiration of living organisms and when microbes decompose once-living tissue to simpler molecules; the molecules are subsequently altered by chemical and physical changes until they are in a form that can once again be used by living organisms. Although natural "sidetracks" of materials do occur — the formation of coal beds, for example, or the loss of phosphorus to the deep sediments of the oceans — these take place very slowly, over millions of years. Moreover, processes such as the movement of tectonic plates, eruption of volcanoes, and the upwelling of deep ocean currents eventually recycle even these materials. Thus, the chemicals making up our bodies and the pages of this book might once have been part

of a tree fern in an ancient swamp, a dinosaur's bone, or a rock in the Blue Ridge Mountains. Perhaps the nutrients in the food or the oxygen in the air that once nourished Confucius or Socrates or Joan of Arc now nourish one of us. Perhaps some of these nutrients will be sidetracked for millions of years in one of humankind's many dumps or non-biodegradable synthetic substances!

Biogeochemical cycles are grouped into three categories: hydrologic, gaseous, and sedimentary.

Hydrologic Cycle

The **hydrologic cycle** includes all of the processes that account for the circulation of water through bodies of water, air, and land (Figure 4-10). Water enters the atmosphere through solar-driven vaporization from lakes, rivers, oceans (evaporation), and the leaves of plants (transpiration). It cools and condenses, forms clouds, and returns to the Earth as some form of precipitation (rain, snow, hail, etc.) to begin the cycle once more. The Earth's total water supply is not added to or sub-

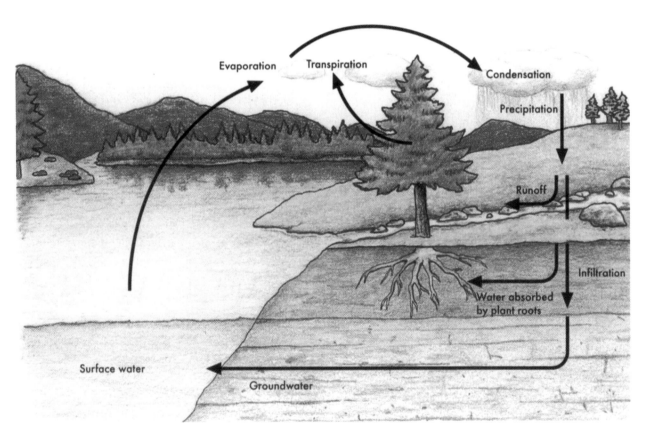

FIGURE 4-10: Hydrologic cycle. Water cycles continually from the hydrosphere through the atmosphere and lithosphere and back to the hydrosphere.

tracted from over time. The quality and availability of water at any one time or place, however, are dependent upon several factors: the uses to which humans put it, the condition and average temperature of the Earth's surface where it falls, and the continued ability of the soils, forests, wetlands, and lakes to absorb, clean, and help store it.

Gaseous Cycles

Gaseous cycles take place primarily in the atmosphere, as materials circulate from the air, through land and water, and back again. Two of the most significant gaseous cycles are the carbon and nitrogen cycles.

Carbon Cycle. Each year, phototrophs produce billions of tons of organic matter. These producers, primarily terrestrial plants and aquatic algae, remove carbon, in the form of carbon dioxide (CO_2), from the atmosphere

and hydrosphere to produce sugars and other complex organic molecules. In doing so, they also produce oxygen, which is released to the environment. Simultaneously, respiration by living organisms (including plants) converts organic matter back to carbon dioxide and water (Figure 4-11).

About 300 million years ago, a surplus of organic matter began to accumulate in the sediments of oceans, seas, and swamps faster than consumers or decomposers could use it. Mineral sediments then buried the organic material. Physical processes of great pressure and heat slowly converted this detritus into huge deposits of coal, petroleum and natural gas, materials we now consume in large quantities to meet our energy needs. Combustion of these fuels adds millions of tons of carbon dioxide to the atmosphere each year. Additional amounts are released to the atmosphere annually through the cutting, clearing, and burning of major portions of our forests, both temperate and tropical.

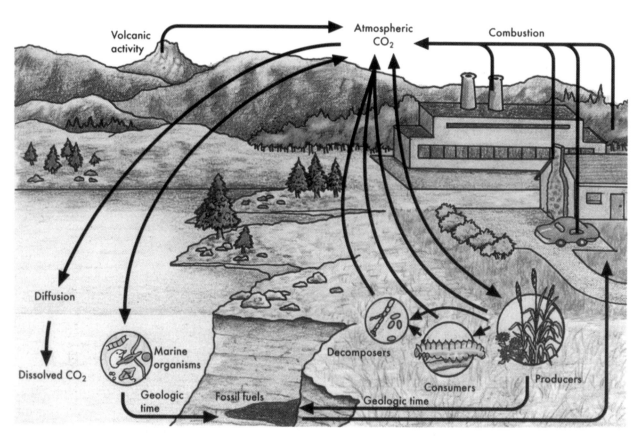

FIGURE 4-11: Carbon cycle. The cycling of carbon begins when carbon dioxide (CO_2) enters plants and algae during photosynthesis. Through respiration, phototrophs release some carbon back to the atmosphere (and to the soil) as carbon dioxide. Phototrophs give off oxygen as a product of photosynthesis, which is then used by organisms during respiration. When animals eat plants, respiration returns more carbon dioxide to the atmosphere. Volcanic activity and erosion from carbonate rock are other natural sources of carbon dioxide. The burning of fossil fuels is a significant cultural source of carbon dioxide.

Because green plants use carbon dioxide in the process of photosynthesis, forests act as carbon sinks, that is, they absorb and store significant amounts of carbon dioxide. The oceans also act as a carbon sink. Large amounts of CO_2 dissolve in them, where it may combine with calcium and magnesium to form precipitates of limestone or dolomite.

Over the last 60 million years, the Earth has experienced oscillations in the balance of oxygen and carbon dioxide in the atmosphere. During the last century, however, human activity has resulted in an increase of atmospheric carbon dioxide. The increase seems to be accelerating and could signal significant warming of the Earth's climate, since carbon dioxide (and other gases) help to retain heat in the atmosphere. Changes of just a few degrees in average temperature could melt a portion of the polar ice caps, raising ocean levels. Ocean levels would also rise as the warmer temperatures caused the waters to expand. Warmer oceans will not hold as much carbon dioxide, thus releasing more carbon dioxide to the atmosphere. Temperature changes might also affect the amount of precipitation. Both precipitation and temperature are major limiting factors in an ecosystem. Some scientists speculate that increases in carbon dioxide will be followed by a long period of significant temperature and precipitation changes that may accelerate unless or until phototrophs absorb enough carbon dioxide to reduce its level in the atmosphere.

Nitrogen Cycle. Nitrogen is the fourth most common element in living tissues (after oxygen, carbon, and hydrogen). A chief constituent of chlorophyll, nucleic acids and amino acids, the "building blocks" of proteins, nitrogen is essential to life. Proteins are used as structural components in cells and as constituents of the enzymes that catalyze cellular processes. Though nitrogen (N_2) is the most common gaseous constituent of the atmosphere (78 percent), plants and animals cannot use it directly. Plants can use nitrogen only when it is in the form of inorganic compounds, principally nitrate and ammonium; animals obtain the nitrogen they need by eating plant or animal tissues. Because plants cannot use gaseous nitrogen directly, and the conversion of nitrogen to ammonium or nitrate is slow and complicated, nitrogen is often a limiting factor for plants, especially terrestrial plants.

Atmospheric nitrogen is converted to a form usable by plants chiefly through **nitrogen fixation**, the process by which certain kinds of bacteria, either free-living in the soil or living in nodules (swellings) on plant roots, convert nitrogen to ammonia (NH_3) (Figure 4-12). Some of the ammonia then combines with H^+

present in soil water to become ammonium (NH_4^+), a form some plants can use. For example, the bacterium *Rhizobium* lives within the root nodules of legumes (such as peas, beans, alfalfa, clover, locust trees, and redbud trees). It fixes nitrogen, which is then used by the legume; the legume, in turn, provides the bacterium with a protective environment and an energy source for nitrogen fixation and other metabolic activities. It's important to realize that, although we think of plants and animals as complex and sophisticated, all are dependent on nitrogen-fixing bacteria and cyanobacteria (blue-green algae).

Once nitrogen has entered the biotic realm, bacteria and fungi decompose plant and animal tissues and animal wastes, reducing them to less complex forms and eventually converting proteins and amino acids to ammonia or ammonium. Several species of soil bacteria (such as *Nitrosomonas*) are able to oxidize ammonia or ammonium — that is, combine it with oxygen — to form nitrite (NO_2^-). Although nitrite is toxic to many plants, it usually does not accumulate in the soil. Instead, other bacteria (such as *Nitrobacter*) oxidize the nitrite to form nitrate (NO_3^-), the most common biologically usable form of nitrogen.

A very small amount of atmospheric nitrogen is converted to a usable form by lightning. Lightning transforms nitrogen to nitrates which then enter the soil in rain.

Nitrogen reenters the atmosphere through the action of denitrifying bacteria, which are found in waterlogged soils and nutrient-rich habitats such as swamps and marshes. These bacteria break down nitrates into nitrogen gas and nitrous oxide (N_2O), which reenter the atmosphere. Nitrogen also reenters the atmosphere from volcanoes, exposed nitrate deposits, and emissions from such cultural sources as factories, electric power plants, and automobiles.

There is growing scientific consensus that human activity has a substantial adverse effect on the nitrogen cycle. *Focus On: The Impact of Human Activities on the Nitrogen Cycle* (page 82) summarizes a 1997 report in which a panel of distinguished scientists discusses the findings of over 140 references in the primary scientific literature regarding the human impact on the global nitrogen cycle. The panel, composed of eight scientists chosen to represent a broad range of expertise, concludes that human activities have at least doubled the rate at which nitrogen enters the land-based nitrogen cycle. A follow-up report published in 2000 corroborates these findings and adds that the human conversion of atmospheric nitrogen into biologically usable nitrogen (fertilizer) now matches the natural

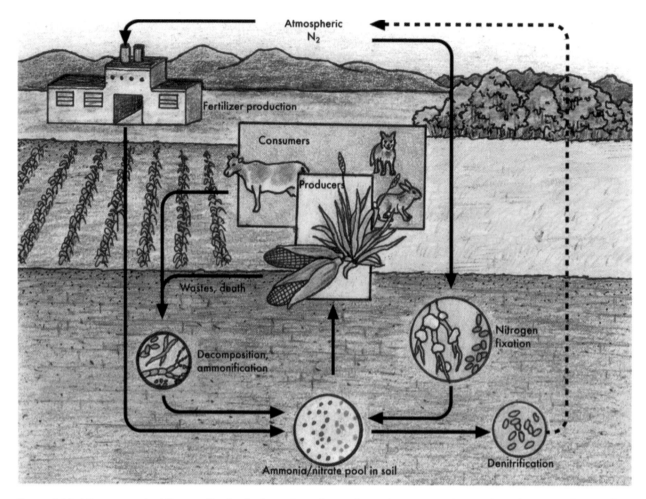

FIGURE 4-12: Nitrogen cycle. Nitrogen fixation by bacteria and cyanobacteria removes nitrogen from the atmosphere and converts it to a form that can be used by producers. The nitrogen then passes through successive trophic levels as proteins, nucleic acids, and other organic molecules. Nitrogen also is converted to a usable form through the decay of organic matter. The dashed line indicates the pathway by which nitrogen is returned to the atmosphere.

rate at which nitrogen is biologically fixed. This report also states that nutrient over-enrichment — specifically of nitrogen and phosphorus — in coastal ecosystems has triggered ecological changes resulting in decreased coastal biological diversity.

Sedimentary Cycles

Sedimentary cycles involve those materials that move primarily from the land to the oceans and back to the land again (although the cycles may include a gaseous phase). Phosphorus, sulfur and other nutrients, such as potassium, calcium and magnesium, follow essentially the same pathway.

Phosphorus Cycle. Phosphorus is a major component of genetic material (DNA and RNA), energy molecules (ATP), and cellular membranes. It is also the major structural component of the shells, bones, and teeth of animals.

Much of the phosphorus found in soil is unusable and tends to remain that way for long periods of time. Additionally, it cannot be used in its elemental form by organisms (Figure 4-13). Decomposers convert phosphorus into phosphate (PO_4), a form producers can use. Wind and water erode phosphate from phosphate-rich rocks. In terrestrial systems, the cycling of phosphate is usually very efficient. Even though small amounts are slowly lost to streams, rivers and eventually to the oceans, most phosphate, because it is in a readily usable form, continues to cycle in the ecosystem. The phosphate that does eventually make its way to the oceans remains there for long periods of time. Sea bird droppings, known as guano, deposited on the land, and ocean floor uplifting help counteract this loss, but these are slow processes.

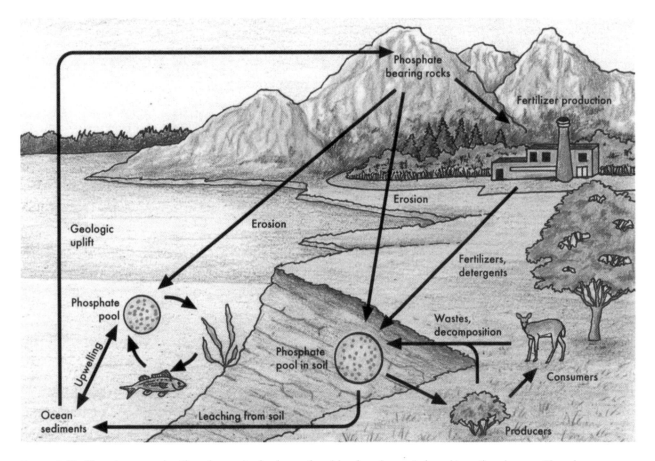

FIGURE 4-13: Phosphorus cycle. Phosphorus, in the form of usable phosphates, is found in soil and water. Phosphates pass to consumers through trophic levels. While decomposers make most of the phosphates available again, a portion remains locked for long periods of time in animal bones, teeth, and shells and in the deep sediments of the oceans.

The implications for interrupting phosphorus cycling are particularly serious because phosphates are an important limiting factor in aquatic habitats. When aquatic plants or algae get extra phosphate from soil runoff, phosphate detergents or sewage, a dense bloom often results and the water becomes choked with plant life. Lake Erie and Lake Washington (near Seattle), among others, have been adversely affected by increased phosphate loads.

Sulfur Cycle. Sulfur is a constituent of amino acids and proteins. It helps to activate enzymes, is important in energy metabolism, and becomes a part of some vitamins. Like phosphorus, sulfur cannot be absorbed by plants in its elemental form. Instead, it is absorbed as sulfates (SO_4), which are produced from the weathering of rocks and soils. Sulfur oxides released to the atmosphere contribute to the formation of sulfuric acid (H_2SO_4) and ammonium sulfide (NH_4S). Sulfur also enters the atmosphere as hydrogen sulfide (H_2S), which is produced by volcanoes, hot springs, and decomposers.

Cultural sources such as power plants and industrial activities now account for the greatest release of sulfur oxides into the atmosphere.

What Are Ecosystem Services?

The flow of energy and the cycling of materials through the biosphere are examples of **ecosystem services**, functions or processes of a natural ecosystem that provide benefits to human societies. Energy flow and materials cycles help sustain biotic communities, making possible the diversity of living organisms found on Earth.

Two reports published in 1997 highlight the role and importance of ecosystem services and their tremendous value to humanity. In "Ecosystem Services: Benefits Supplied to Human Societies by Natural Ecosystems," a panel of distinguished scientists provides a comprehensive overview of ecosystem services, the threats to

The Impact of Human Activities on the Nitrogen Cycle

The availability of nitrogen — an element essential for life — plays a vital role in the structure and function of ecosystems. In many terrestrial and aquatic ecosystems (including managed systems such as croplands and unmanaged systems such as wilderness areas), the supply of nitrogen helps to control the nature and diversity of plant life, the population dynamics of herbivores and their predators, and critical ecological processes such as plant productivity and the cycling of minerals like carbon. Unfortunately, according to the authors of a report entitled "Human Alteration of the Global Nitrogen Cycle: Causes and Consequences," the addition of excessive nitrogen can pollute ecosystems and alter both their ecological functioning and the living communities that they support.

Additional nitrogen can enter the biological realm of the nitrogen cycle only after it is "fixed," that is, removed from the atmosphere and bonded to hydrogen or oxygen to form inorganic compounds, principally ammonium (NH_4) and nitrate (NO_3). The major natural sources of additional nitrogen are nitrogen-fixing terrestrial organisms (some bacteria and cyanobacteria, or blue-green algae) and lightning. Terrestrial organisms probably fix atmospheric nitrogen at an annual rate of between 90 and 140 teragrams or Tg (equal to a million metric tons of nitrogen). Lightning fixes a much smaller amount, perhaps less than 5 Tg per year. Thus, it is reasonable to assume that the upper bound of natural nitrogen fixation on land is on the order of 140 Tg annually.

An understanding of the natural rate of nitrogen fixation enables us to better appreciate the impact of human activities upon the global nitrogen cycle. Three such activities are particularly significant. The industrial fixation of nitrogen for use as fertilizer — far and away the largest cultural contributor of new nitrogen to the global cycle — currently amounts to about 80 Tg annually. The planting of nitrogen-fixing crops, such as soybeans, peas and alfalfa, adds an estimated 32 to 53 Tg each year; for the sake of simplicity, the panel uses an average of 40 Tg per year. (Remember, it's not the plants themselves that fix the nitrogen, but rather the symbiotic nitrogen-fixing bacteria that live in nodules on the roots of these plants.) The third major cultural source of new nitrogen is the combustion of fossil fuels. As coal, oil, and natural gas are burned, nitrogen that was previously fixed is liberated from long-term storage and released back to the atmosphere in the form of nitrogen-based trace gases, such as nitric oxide. Collectively, automobiles, factories, power plants, and other combustion processes add more than 20 Tg annually to the atmosphere. (Technically, fossil

these services, and their valuation. Their report, like "Human Alteration of the Global Nitrogen Cycle: Causes and Consequences" (see *Focus On: The Impact of Human Activities on the Nitrogen Cycle*), is one in a series of *Issues in Ecology*, a public-education effort supported by the Ecological Society of America.

The role and importance of some ecosystem services are obvious, such as the production of ecosystem goods (fuelwood, fiber, pharmaceuticals, fish, game, fruits, and vegetables), natural pest control, mitigation of floods and droughts, pollination of crops, and the provision of aesthetic beauty and intellectual and spiritual stimulation. The role and importance of others may be less obvious, but they are no less essential; these include generation and maintenance of biological diversity, seed dispersal, and the services supplied by soil, which moderates the water cycle, retains and delivers nutrients to plants, aids in the decomposition of dead organic mat-

fuel combustion does not add "new" fixed nitrogen to the global cycle, since the nitrogen was actually fixed millions of years ago. However, since human activities release this nitrogen from long-term storage, making it available once again to the biota, it must be taken into consideration when looking at human alteration of the global nitrogen cycle.) Thus, the panel concludes that human-driven nitrogen fixation has at least doubled the rate of nitrogen fixation on land. Moreover, they caution that human activities which mobilize nitrogen in long-term biological storage may also contribute significantly to changes in the global nitrogen cycle. Nitrogen bound up in tree trunks and soil organic matter, for example, is liberated when forests are burned, wetlands are drained, and lands are cleared for crops. Although there is considerable scientific uncertainty about the impact of these activities (in terms of both the quantity and fate of the nitrogen they liberate), it is believed that they may add as much as 70 Tg of nitrogen to the land-based biological nitrogen cycle — half again as much as the amount fixed by natural sources.

Scientists are certain that human alteration of the global nitrogen cycle has had four serious consequences. They are: (1) Increased global concentrations of atmospheric nitrous oxide, a greenhouse gas, as well as increased regional concentrations of other nitrogen oxides, such as nitric oxide (NO), which contribute to the formation of photochemical smog; (2) Depletion of soil nutrients, such as calcium and phosphorous, that are necessary to maintain the long-term fertility of soil; (3) Substantial acidification of soils, lakes, and streams in several regions; and (4) A sharp increase in the transport of nitrogen by rivers into estuaries and coastal waters, where it acts as a major pollutant. In addition to these four, scientists believe that two other problems have likely occurred as a result of human-driven change of the global nitrogen cycle. The first is an increased loss of biological diversity, particularly among plants that are adapted to nitrogen-poor soils and, consequently, the microbes and animals dependent upon such plants. The second is changes in the plant and animal life of estuarine and near-shore ecosystems, changes in the ecological processes of those systems, and subsequent long-term declines in all-important coastal marine fisheries.

To address the consequences of human alteration of the global nitrogen cycle, societies must slow the rate at which human activities accelerate nitrogen fixation on land. Of highest priority are efforts to curb and gradually decrease the production and use of nitrogen fertilizer. Increasing the efficiency of fertilizer can help to reduce the amount used. Management practices that enhance fertilizer efficiency include dissolving it in irrigation water, delivering it below the soil surface, and timing multiple applications to meet the needs of specific crops. These practices have the added benefit of lowering the cost of food production. Slowing nitrogen fixation that results from the combustion of fossil fuels requires an increase in the efficiency of fuel combustion and improved technologies to intercept the airborne by-products of combustion. It is essential that improvements in both fertilizer efficiency and fossil fuel combustion be transferred to developing countries as their economies and industries grow.

ter and wastes, recycles nutrients, and helps to regulate the Earth's major element cycles. The panel highlights the services provided by the interaction between climate and life: "The relative influence of life's stabilizing and destabilizing feedbacks remains uncertain; *what is clear is that climate and natural ecosystems are tightly coupled, and the stability of that coupled system is an important ecosystem service*" (our italics). This statement is particularly telling given the scientific consensus on the causes and possible impacts of human-induced climate change (see Chapter 12, Air Resources).

Ecosystem services are threatened by the ongoing destruction of natural habitats, invasion of nonnative species, loss of native biodiversity, overfishing, alteration of the Earth's biogeochemical cycles, degradation of farmland, squandering of freshwater resources, toxification of land and waterways, and unsustainable harvesting of renewable resources such as forests. All of these are driven by the rapid, unsustainable growth in the scale of the human enterprise — in population size, per capita consumption, and the environmental impacts of technologies and institutions — and the frequent mismatch between short-term individual economic incentives and long-term societal well-being.

In general, ecosystem services are greatly undervalued. There are a number of reasons for this. Many aren't traded or valued in the marketplace, and many serve the public good rather than individual landowners. Moreover, private property owners usually have no way to benefit financially from the ecosystem services that their land supplies to society. To the contrary, economic subsidies often encourage the conversion of such lands to other, market-valued activities. The panel concludes that "(t)he human economy depends upon the services provided 'for free' by ecosystems. The ecosystem services supplied annually are worth many trillions of dollars. Economic development that destroys habitat and impairs services can create costs to humanity over the long-term that may greatly exceed the short-term economics of the development. These costs are generally hidden from traditional economic accounting, but are nonetheless real and are usually borne by society at large. Tragically, a short-term focus in land-use decisions sets in motion potentially great costs to be borne by future generations. This suggests a need for policies that achieve a balance between sustaining ecosystem services and pursuing the worthy short-term goals of economic development."

The global community has begun to realize the importance of incorporating the value of ecosystem services into policy and decision-making. In 2001, the United Nations Environment Programme established the four-year Millennium Ecosystem Assessment (MEA) program. The MEA is an international collaborative effort to study how past, present, and future changes in ecosystem services affect human well-being. The ultimate goal of the MEA is to provide decision-makers with resources and tools that can be used to improve ecosystem management in a way that simultaneously alleviates poverty and ensures the conservation and sustainability of the world's ecosystems.

To assist policymakers in striking a balance between sustaining natural systems and promoting short-term economic development, some researchers have already attempted to estimate the economic value of ecosystem services. In "The Value of the World's Ecosystem Ser-

vices and Natural Capital," Robert Costanza, a zoologist and ecological economist, and twelve co-authors argue that ecosystem services and the natural capital stocks that produce them "are critical to the functioning of the Earth's life-support system. They contribute to human welfare, both directly and indirectly, and therefore represent part of the total economic value of the planet." Using published studies and some original calculations, the authors estimated the economic value of 17 ecosystem services for 16 biomes. They calculate that these services are worth at least $33 trillion dollars annually. Marine systems contribute nearly 63 percent of this total, with most of that originating in coastal areas. Terrestrial systems (chiefly forests and wetlands) contribute approximately 38 percent of the estimated value.

According to the authors of the report, one way to think about the value of ecosystem services "is to determine what it would cost to replicate them in a technologically produced, artificial biosphere. Experience with manned space stations...indicates that this is an exceedingly complex and expensive proposition." The City of New York appears to agree. In September 1996, the City announced that it had reached an agreement with the State of New York, the U.S. Environmental Protection Agency, and the upstate watershed communities from which the City draws its water. The agreement enables the City to preserve an ecosystem service — the maintenance of a clean water supply — at lower cost than is possible through technological means. From 1997-2006, the City is required to invest $260 million and solicit the owners of 355,000 acres of land to acquire additional watershed areas and conservation easements, which will provide a buffer around reservoirs, critical streams, and wetlands. By 2001, it had spent nearly $58 million to purchase an additional 17,800 acres of land. The City also is spending $82 million to upgrade the last three of the City's nine wastewater treatment plants in the watershed. Thanks to this agreement, the City can avoid building a new plant to filter water from two of its three upstate watershed systems. Construction costs and annual operating expenses for the plant — estimated at $4 billion to $6 billion and $300 million, respectively — would have doubled or tripled residents' water bills. New York discovered, as Costanza *et al.* point out, that the Earth "is a very efficient, least-cost provider of human life-support systems."

Summary

The biotic and abiotic components of the biosphere are inseparable, bound together by a complex and delicately balanced web of biological and physical processes that regulate the flow of energy and the cycling of materials. The gross primary productivity of an ecosystem — the total amount of energy fixed by autotrophs over a given period of time — represents energy that can be used as food for the producers themselves. After the autotroph's own energy needs are met, the amount of energy that is stored, and thus available to other organisms in the community, is the net primary productivity.

Food chains and food webs represent the feeding relationships and the movement of energy and materials among the organisms of the biotic community. Producers occupy the first trophic level in a food chain or web; primary consumers (herbivores), which feed directly on producers, occupy the second. Secondary consumers (carnivores) occupy the third trophic level, followed by tertiary consumers (top carnivores). An organism may operate at more than one level. Omnivores, for example, consume both plants and animals. At each trophic level, decomposers break down dead and decaying organic matter, thereby returning nutrients to the ecosystem to be used once again. Feeding relationships based on decomposing plant and animal material or animal wastes are known as detritus food webs.

Pyramidal relationships are used to depict the flow of energy and production of biomass in an ecosystem. The pyramid of energy represents the capture, use, and transfer of energy from one trophic level to another. Generally, only about 10 percent of the available energy is transferred to the next successive trophic level; the rest is lost to the environment as low-quality heat. The pyramid of biomass represents the total amount of living material at each trophic level, and the pyramid of numbers depicts the relative abundance of organisms at each trophic level. In general, lower levels contain a greater amount of biomass and a greater number of organisms than successive levels.

Unlike energy, materials cycle through ecosystems and are used over and over again by the biotic community. Because the processes by which materials cycle involve living organisms as well as geologic and chemical processes, they are known as biogeochemical cycles. The hydrologic cycle describes the movement of water between the air, seas, and land. Gaseous cycles include the carbon and nitrogen cycles. Major sedimentary cycles include phosphorus and sulfur.

The flow of energy and the cycling of materials are two important ecosystem services, functions or processes of a natural ecosystem that provide benefits to human societies. Others include the production of fuelwood, fiber, pharmaceuticals, fish and game, fruits and vegetables, and other goods; the control of pest populations; the mitigation of floods and droughts; and the generation and maintenance of biological diversity. These and other ecosystem services face a number of serious threats, among them the ongoing destruction of natural habitats, alteration of the Earth's biogeochemical cycles, degradation of farmland, and unsustainable harvesting of renewable resources such as forests. The driving forces behind these threats are unsustainable growth in the human population, high per capita consumption of resources in many countries, and increasing environmental impacts of technologies and institutions. Until ecosystem services are

Discussion Questions

1. Refer to Table 4-1. In the United States, coastal development and the destruction of wetlands are pressing environmental concerns. Explain, in terms of productivity, why that is so. What are some of the likely effects, on both humans and other living organisms, of the continued development of coastal regions and the continued conversion of swamps and marshes?

2. Construct a diagram to illustrate an ecosystem, then explain how it works. Be sure to include energy, biomass, and number relationships.

3. Author and ecologist Barry Commoner once wrote, "There is no such thing as a free lunch in nature." What does this mean in ecological terms?

4. What would be the environmental, social, and economic effects if people in the United States consumed most of their food at a lower trophic level?

5. Explain the role of decomposers in the cycling of materials through the ecosystem.

6. Explain why terrestrial food chains are seldom longer than three or four links.

7. In what way(s) have human activities interfered with biogeochemical cycles? Use specific examples, with respect to specific cycles, to support your answer.

(continued on next page)

(continued from previous page)

8. Describe at least three ways in which ecosystem services support human societies. For each one, discuss the measures and/or technologies that would need to be taken or developed to replace these services if they were no longer provided to us by natural systems. Are there some ecosystem services that cannot be replaced at any cost?

valued adequately by human societies, the threats to these processes will remain. Currently, efforts are being made to determine market valuations for many of the most critically important ecosystem services.

KEY TERMS

10 percent rule	estuary	nitrogen fixation
aerobic respiration	food chain	pyramid of biomass
anaerobic respiration	food web	pyramid of energy
biogeochemical cycles	gaseous cycle	pyramid of numbers
biomass	gross primary productivity	respiration
detritus food web	hydrologic cycle	sedimentary cycle
ecosystem services	net primary productivity	trophic level

CHAPTER 5

Ecosystem Development and Dynamic Equilibrium

It took hundreds of millions of years to produce the life that now inhabits the Earth – eons of time in which that developing and evolving and diversifying life reached a state of adjustment and balance with its surroundings.

Rachel Carson

So almost every corner of the planet, from the highest to the lowest, the warmest to the coldest, above water and below, has acquired its population of interdependent plants and animals. It is the nature of these adaptations that has enabled living organisms to spread so widely through our varied planet.

David Attenborough

Nature is not static; ecosystems change. In this chapter we will examine the development of ecosystems through a process called succession and also explore the concept of dynamic equilibrium — the idea that ecosystems maintain stability by constantly responding to internal and external change (Figure 5-1).

Learning Objectives

When you finish reading this chapter, you will be able to:

1. Explain how ecosystems develop over time through the process of succession.

2. Distinguish between primary and secondary succession.

3. Describe what ecologists mean by dynamic equilibrium.

4. Explain how feedbacks, species interactions, and population dynamics help to maintain dynamic equilibrium.

FIGURE 5-1: This wetland bog on the Huslis River in the Koyukuk National Wildlife Refuge, Alaska, maintains overall balance through dynamic equilibrium.

What Changes Ecosystems?

Individuals, species, populations, communities, ecosystems, and the biosphere itself are subject to continual environmental change: differences in amount of rainfall, daily and seasonal temperature and wind fluctuations, and differences in amount of sunlight. Natural systems can usually compensate for such variability without a drastic alteration in the biota. Some changes, however, can dramatically disturb ecosystems, either reducing the number or type of species present or making communities less productive. Such disturbances include fire, flood, drought, hurricane, volcanic eruption, and human activities like acid precipitation, deforestation, contamination by toxic wastes, and the draining and filling of wetlands.

It may be a long time after a major disturbance before the community of organisms present at the time of the disturbance again occupies that habitat. On Mount St. Helens, for example, a site of volcanic activity in 1980, the forest ecosystem was destroyed. Although both plant and animal life have returned, it will be hundreds of years before Mount St. Helens once again is blanketed by mature forests (Figure 5-2). Similarly, Hurricane Hugo's assault on South Carolina's coast in September 1989 had a devastating effect on the Francis Marion National Forest. Over 100,000 acres (40,469 hectares) of the forest were leveled. Living in the forest were approximately 500 breeding pairs of the endangered red-cockaded woodpecker. Since the woodpecker prefers to nest in cavities of live trees — trees that are no longer available — the species may not survive during the time that it takes the forest to recover.

If a disturbance is severe enough, an ecosystem may be altered beyond its ability to return to its former state. For example, what picture comes to mind when you

FIGURE 5-2: Mount St. Helens immediately after the eruption in 1980 and again in 1999. (a) The volcanic eruption drastically altered the site. (b) Nineteen years later, the forest ecosystem continues its recovery.

think of the Mediterranean country of Greece? Most likely, it's the image common to travel brochures: A sun-drenched land of whitewashed villages clinging to bare, rocky slopes that rise high above a glistening blue sea. But this image is a relatively recent one. In fact, much of the Mediterranean basin — which began to be inhabited by human ancestors some 500,000 years ago — was originally cloaked in forests of pine and oak. Then, at the end of the last ice age (about 10,000 years ago), the climate became increasingly arid. About the same time, human cultures in the region began the shift from nomadic, hunter-gatherer bands to agricultural communities. These agrarian communities laid the foundation for Ancient Greece, a sea-faring culture that came to dominate the Mediterranean region. The area's plentiful forests were cleared to accommodate cropland, to provide pasture for goats and sheep, and to obtain timber for ship building and mine smelters. As the human population expanded, people were forced to move out of the fertile valleys and plains into the uplands; logging of the steep hillsides caused severe erosion. Over time, human-induced changes in the landscape, coupled with the warmer climate, pushed the natural systems of the Mediterranean past a threshold where forests could not recover. Consequently, the forests that had characterized the region for millennia failed to regenerate and were replaced by the scrubby vegetation that we see today.

How Do Ecosystems Develop?

Ecological succession is the process by which an ecosystem matures; it is the gradual, sequential, and somewhat predictable change in the composition of the community. While the term succession is commonly accepted, it is more accurate to think of the maturation process as **ecosystem development**, which takes into account the accompanying modifications in the physical environment (such as microclimate and soil type) brought about by the actions of living organisms (Table 5-1).

In immature or developing ecosystems, when there is not much biomass to support, most of the energy captured through photosynthesis goes into growth or production of new biomass. In maturing ecosystems, less energy is available for production because increasing amounts of energy are required to maintain the biomass through respiration. As an ecosystem develops, the species composition of the community changes

TABLE 5–1: Comparison of Developmental Stages and Mature Stages in Ecosystems

Ecosystem Attribute	Developmental Stages	Mature Stages
Gross primary productivity	Increasing	Stabilized at moderate level
Biomass	Low	High
Production/ respiration	Unbalanced	Balanced
Growth	High	Low
Maintenance	Low	High
Use of primary production	Mostly via linear grazing food chains	Mostly via detritus food webs
Diversity	Low	High
Stability (resistance to stress)	Low	High

until the association of organisms best adapted to the physical conditions of the area is reached. In this community, production equals respiration; there is little net production and no further increase in biomass (Table 5-2).

Succession occurs in both terrestrial and aquatic habitats and has been observed in all habitats that support living organisms. There are two types of succession, primary and secondary. **Primary succession** is the development of a new ecosystem in an area previously devoid of organisms. The most common examples of primary succession are communities that develop on bare rock, after glaciers recede, after wind-blown sand stabilizes, and after volcanic islands form.

Secondary succession is the regrowth that occurs after an ecosystem has been disturbed, often by human activity. Although some organisms are still present, the ecosystem is set back to an earlier successional stage. Typically, secondary succession occurs more rapidly than primary succession, chiefly because the soil is usually already in place, eliminating the long process of soil building. Examples of secondary succession are the new growth in abandoned plowed fields or gardens, forests that have been burned or altered by storms, and lowlands that have been affected by floods or hurricanes.

TABLE 5–2: Production and Respiration as (kcal/m²/yr) in Growing and Climax Ecosystems

	Alfalfa Field (USA)	Young Pine Plantations (England)	Medium-Aged Oak-Pine Forest (NY)	Large Flowing Spring (Silver Springs, FL)	Mature Rain Forest (Puerto Rico)	Coastal Sound (Long Island, NY)
Gross primary production	24,400	12,200	11,500	20,800	45,000	5,700
Autotrophic respiration	9,200	4,700	6,400	12,000	32,000	3,200
Net primary production	15,200	7,500	5,100	8,800	13,000	2,500
Heterotrophic respiration	800	4,600	3,000	6,800	13,000	2,500
Net community production	14,400	2,900	2,000	2,000	Little or none	Little or none

Source: *Adapted from* Fundamentals of Ecology, *3rd Edition, by Eugene P. Odum. Copyright © 1971, by W.B. Saunders Company. Reprinted by permission of Holt, Rinehart and Winston, CBS College Publishing.*

Primary Succession from Glacial Till or Bare Rock to Climax Forest

An ecosystem develops as the physical environment is modified by each successive community. Figure 5-3 illustrates the primary succession that occurred on glacial till or bare rock in North America after the Wisconsin glacier receded northward some 12,000 years ago.

Stage I: Lichen Pioneer Community. The first community to occur in primary succession is composed of **pioneer species**, hardy organisms such as lichens and microbes that are capable of becoming established on bare rock and beginning the soil-building process.

Almost immediately after the glacier receded, development began when resistant spores of algae and fungi invaded the glacial till, the lifeless deposits of sand, clay,

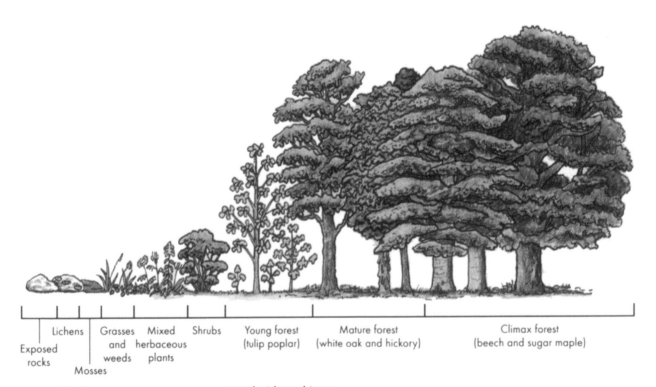

Exposed rocks | Lichens | Mosses | Grasses and weeds | Mixed herbaceous plants | Shrubs | Young forest (tulip poplar) | Mature forest (white oak and hickory) | Climax forest (beech and sugar maple)

Figure 5-3: Primary succession in a temperate deciduous biome.

gravel, and boulders left by the glacier. The invading organisms came from already established communities that had not been affected by glacial activity. Lichens, a resilient association of algae and fungi, began to colonize exposed rock surfaces. Neither the algae nor fungi would have done well alone in that inhospitable environment, where temperature fluctuation was great and water in short supply. But in association, these organisms, a producer and a decomposer, were capable of withstanding the periodic dry and harsh conditions that occurred for long periods after the glaciers receded. Over time, the fungus secreted acidic compounds that began to dissolve the rock. Slowly, the organic detritus of these pioneer organisms mixed with bits of weathered rock and minerals to form soil. Small arthropods and protozoans soon made their homes in the lichens, and their waste further added to the soil. This primitive soil held moisture and nutrients well enough to support the second stage, mosses. Thus, the lichen community modified the habitat, making it suitable for the moss community.

Stage II: Moss Community. As mosses became more successful, they crowded out the lichens. They, too, were autotrophs able to withstand harsh conditions. Acids formed by the mosses, weathering, and alternate freezing and thawing of water in small cracks in the rock combined to break down the rock into small particles, which slowly accumulated with the decaying organic matter to further build the soil.

The dense mats of moss attracted insects and other animals, which added more organic matter to the soil. Just as the lichens changed the environment and made it more suitable for the mosses that replaced them, mosses changed the environment and made it more suitable for other, larger plants. As the soil accumulated and matured, the communities that helped to form it were preparing it for the next community.

Stage III: Herbaceous Community. By stage III, the soil accumulated enough nutrients, organic matter, and moisture to support larger herbaceous plants such as goldenrods, asters, milkweed, primrose, ragweed, and wild carrot. These attracted more insects and burrowing and grazing animals such as snails, moles, voles, and mice. We are describing a location in which the temperature and moisture conditions were right for the eventual development of a temperate deciduous forest; in such a location the dominant organisms at this stage were dicot plants. In a different physical location, with drier climatic conditions, monocot prairie grasses might have dominated.

Stage IV: Shrub Community. Soil formation went on for several thousand years before the soil was able to support a shrub community. As the glacier receded, the climate was changing, becoming warmer and more moist. Because more soil, organic matter, and moisture were available, taller, woody shrubs became established and outcompeted the herbaceous plants of the previous stage. The community began to assume a pronounced vertical structure. Sumac, black locust, wild rose, blackberry, and mulberry proliferated and increasingly shaded the soil. The seeds of these plants were often first brought into the community in bird droppings. Deer, opossum, a variety of rodents, and red fox were successful in the shrub community.

Stage V: Tree Community. Over time, the soil became further enriched and was able to support sapling trees. After thousands of years, there was enough organic matter in the soil to support the first tree stages: red maple, willow, and cottonwood in wet areas; locust, tulip poplar, oak, and hickory in drier areas. Changes in the vegetation spurred changes in the physical factors of light, temperature, and moisture. Shade-tolerant varieties of plants developed on the forest floor. Early-blooming spring wildflowers were prevalent. Squirrels, raccoons, and chipmunks inhabited the early forest community.

Stage VI: Climax Forest or Equilibrium Community. The soil matured and, in the absence of climatic change, will remain essentially constant. Ecosystem development has resulted in the **climax community**, the association of organisms best adapted to the physical conditions of a defined geographic area. In general, conditions in the temperate location favored the seed germination, growth, and maintenance of large trees such as sugar maple and beech. However, microclimates within the climax forest community, characterized by different moisture, light and temperature conditions, favored different species, such as oak and hickory.

The climax community is usually dominated by a few abundant plant species, which give the community its name, such as the beech-maple forest or the tall-grass prairie. In a beech-maple forest, beech and maple trees might account for 80 percent of the number of trees. The other 20 percent might be composed of 10 to 12 other varieties, such as tulip poplar, black cherry, sweet gum, ash, oak, and hickory. Even though the climax community is named for its dominant plant species, it contains a unique assemblage of animals, microbes, and fungi.

Fields of Gold

Richard B. Fischer

It is late summer and the sun shines high above fields of gold — goldenrod fields (Figure 5-4). Although there are over a hundred species of goldenrod in the United States, they are most numerous in the Northeast. The apparent monotony of the goldenrod field is deceiving, for it teems with a diversity of life. If you are fortunate enough to live near a goldenrod field, you will find that it is a microcosm, a world within a world, which offers hours of interesting, and often surprising, observation.

A variety of insects feed on the goldenrod's foliage, pollen, nectar, flower, and sap. Look closely at the plant's leaves, and you will likely find evidence of *Trirhabda canadenis* and *Trirhabda virgata*, two species of beetle that feed primarily on goldenrod foliage. The adults of many species of beetles feed on the plant's pollen, while their young feed elsewhere. For example, the larvae of the locust borer (*Megacyllene robiniae*) tunnel in living black locust trees, but the adults are generally found on goldenrod blossoms. The eggs of short-horned grasshoppers nourish the larvae of the adult black blister beetle, which feeds on goldenrod pollen. The polistes wasp and the honeybee, two members of the order Hymenoptera, feed on the plant's pollen as well.

The pollen eaters illustrate an important point about goldenrod. This plant, with its heavy pollen, is insect-pollinated rather than wind-pollinated. Little of the heavy pollen gets into the air. But because goldenrod is in full bloom at the same time as the common ragweed, many people mistakenly assume that its pollen causes hayfever.

Goldenrod bloom in late summer, when most other wildflowers have gone to seed. Thus, for pollen-loving insects such as honeybees, goldenrod are the last big source of nectar before the first frosts of autumn. Many other species in the order Hymenoptera, including bumblebees, carpenter bees, solitary bees and assorted wasps, also sip goldenrod nectar, as do flower flies, or hover flies, which resemble small yellow jackets but are true flies belonging to the family Syrphidae. Goldenrod nectar is also an important food source for the migrating monarch butterfly, providing this beautiful insect with energy for its journey southward to the forests of Mexico and Latin America.

Several insects eat the entire goldenrod blossom, consuming both nectar and pollen. These are the adult Japanese beetle and the adult *Acullia angustipennis*, a moth whose colorful larva also feeds on goldenrod flowers. If you look closely at the stems of goldenrod, you might notice small, active creatures feeding on the plant's sap. Among the insect families that take advantage of this liquid food source are those belonging to the orders Hemiptera and Homoptera. Like other sap suckers, they have piercing and sucking mouthparts.

With so many herbivores feeding on various parts of this special plant, it is little wonder that the goldenrod also hosts many organisms at higher trophic levels which feed on the herbivores. Most of these predators are insects. The ambush bug, resembling a discolored bloom, hides among the flowers, lying in wait for pollen eaters or nectar sippers. It grasps its prey with its powerful forelegs and injects a proteolytic enzyme into the victim. Though the ambush bug is smaller than a Japanese beetle, it can capture insects as large as honeybees. The flower heads host other carnivores as well, including the assassin bug, damsel bug, and stink bug.

Many spiders feed on goldenrod herbivores. The banded argiope, the black and yellow argiope, and the harvestman ("daddy long-legs") prey on sap-feeding aphids. A very unusual arachnid is the crab spider. Hiding among the blossoms, it ambushes herbivorous insects ranging in size from small flies to large bumblebees.

While examining a goldenrod, you might notice strange swellings on the plant's roots, stems, leaves, or flowers. These are galls, the products of an odd assemblage of flies and moths. Although gall making involves many plant families, the goldenrod family bears more galls than any other. A female fly or moth of a particular species deposits an egg on the growing tissue of a particular species of goldenrod. The resulting larva secretes a minute amount of liquid which radically alters the tissue's genetic expression. A gall often results, sheltering the larva. Because different species of goldenrod produce different types of galls, if you can identify the gall, you can more easily identify the goldenrod itself. The most conspicuous gall belongs to the Canada goldenrod. It is a round swelling on the plant's stem made by the fly *Eurosta solidaginis*. By late summer, when the galls are fully developed, you might cut open a gall to discover the plump, cream-colored maggot. The maggot overwinters in its protective case, emerging as an adult in May. But not all gall-enclosed maggots survive. If, when you open the gall, you find a small, brown object resembling a seed, you will have found the pupa of a minute, highly specialized wasp that preys only on the maggot of the gall-making fly. The pupal case that ensconces the wasp is produced by the fly in response to the larval predator's attack! If you find a small hole drilled into the gall, you won't find anything after cutting it open. The hole is the mark of the downy woodpecker. During the winter months, maggots are a high-energy food source for these woodpeckers.

Not surprisingly, plant ecologists find goldenrod a rich and interesting object of study. They recognize one phase of old field or secondary succession as the goldenrod phase. Researchers have long been interested in the disproportionate length of time for which an area will remain in the goldenrod phase. Abandoned pastures may exist as goldenrod fields for 50 years or more! Research conducted by Jack McCormick and associates in Waterloo Mills, Pennsylvania, revealed the key to the mystery of the goldenrod field's longevity. Certain goldenrods and asters release chemicals that suppress the germination and seedling growth of other plants. Apparently, goldenrod inhibit the next stage of succession. In doing so, they help to maintain habitat diversity. The goldenrod field you study and enjoy today may continue to intrigue observers for many years.

Adapted from Richard B. Fischer, "Goldenrod: An Ecological Goldmine," The American Biology Teacher 47 (October 1985): 7.

Richard B. Fischer is a professor emeritus of Cornell University.

Figure 5-4: Goldenrod fields provide important nesting habitat for goldfinches; cover for quail, pheasants, sparrows, shrew, mice, rabbits, and deer; and food for a wide diversity of insects.

In temperate regions, rainfall of over 40 inches per year (100 centimeters) is needed to maintain a climax tree community, which in turn acts as a sponge to hold water. Where the average annual rainfall is just 10 to 20 inches (25 to 50 centimeters), the climax community might be a grassland prairie.

A climax community is dynamic. Some areas are set back by disease, fire, storms, or insect damage. These stressed areas eventually return to the climax stage. Human activities can also interrupt the process of succession. Today, most of the original forests that once blanketed the eastern United States are gone, having been replaced by towns, cities, roads, and millions of acres of farmland. Worldwide, natural communities that took thousands of years to mature are converted to less complex cultural systems, which are not as effective at maintaining biological diversity, retaining groundwater, and anchoring soil.

Secondary Succession from Old Field to Climax Forest

When an agricultural field or garden is abandoned and left to natural processes, it begins to change almost immediately. Figure 5-5 illustrates secondary succession from old field to climax forest in a temperate deciduous biome.

Stage I: Annual Weed Community. Invasion of an abandoned corn field usually begins before the last corn cob has been picked! In fact, the seeds of opportunistic plants, such as pigweed, ragweed and wild mustard, are present in the soil in what scientists call a weedy seed bank. Once those species are no longer kept under control by human efforts (weeding and the application of herbicides, for example), they begin to take over the field. These early plants are generally dominated in numbers and in species by annuals, which have a one-year life cycle and generally have small, windblown seeds.

Stage II: Perennial Weed Community. The annuals may dominate for a year or two. The next community, dominated by perennials like goldenrod, are better able to tolerate the amount of resources that are available in the soil after the annuals have taken what they need. Goldenrods inhibit other communities from getting started, perhaps by chemical suppression of other species, and may dominate the community for decades. (See *Focus On: Fields of Gold*, page 92.)

Stage III: Shrub or Young Tree Community. Eventually, shrubs and trees enter the old field with the perennials. Perhaps disease, predation, or shade on the field edges sufficiently changes the abiotic charac-

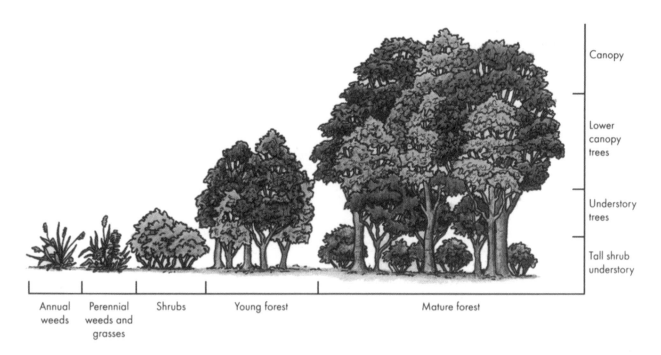

FIGURE 5-5: Secondary succession in a temperate deciduous biome.

teristics to allow the larger woody plants to get a start. Honeysuckle, multiflora rose, black raspberry, eastern red cedar, ash, and locust seem to be most successful at this stage.

Stage IV: Young Forest Community, Stage V: Climax Forest Community. For stages IV and V of secondary succession, the community tends to follow the changes described for stages V and VI of primary succession. It takes hundreds of years to establish a forest after succession begins in an old field.

Succession in Ponds, Lakes, and Wetlands

Succession also occurs in aquatic habitats. The maturation or natural aging of a lake or pond is called eutrophication. An aquatic ecosystem undergoes eutrophication as it becomes enriched with nutrients, which encourage the growth of aquatic plants. In deep lakes, eutrophication is usually slow. However, it is rapid in farm or beach ponds or shallow lakes that receive high amounts of nutrients and sediments from the surrounding land. In these cases, the aquatic system proceeds to dry land fairly quickly. It may pass through marsh, swamp, or boglike conditions before dry land succession commences. Presque Isle, a recurring sand spit that juts out into Lake Erie at Erie, Pennsylvania, affords an excellent opportunity to study pond succession (Figures 5-6 and 5-7).

Usually, an aquatic ecosystem undergoing succession proceeds to the climax terrestrial community of that particular area. But if a pond or lake is in a wetland habitat, the climax stage may be a marsh or swamp community — as long as water dynamics remain constant. In the Everglades ecosystem of South Florida, the type of climax community varies from site to site, depending mostly on the amount and seasonal availability of water. Marshes covered with saw-grass sedges

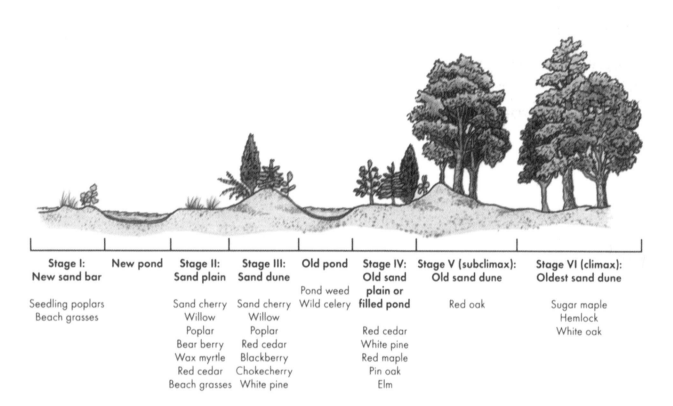

Stage I: New sand bar	New pond	Stage II: Sand plain	Stage III: Sand dune	Old pond	Stage IV: Old sand plain or filled pond	Stage V (subclimax): Old sand dune	Stage VI (climax): Oldest sand dune
Seedling poplars Beach grasses		Sand cherry Willow Poplar Bear berry Wax myrtle Red cedar Beach grasses	Sand cherry Willow Poplar Red cedar Blackberry Chokecherry White pine	Pond weed Wild celery	Red cedar White pine Red maple Pin oak Elm	Red oak	Sugar maple Hemlock White oak

Figure 5-6: Succession on sand at Presque Isle on Lake Erie at Erie, Pennsylvania. On the new sand bars that form near the water's edge, beach grasses and seedling poplars take root; they are the pioneer organisms that begin the process of succession. The sand plain represents the second stage of succession; there we might find sand cherry, willow, poplars, bear berry, wax myrtle, and red cedar. By the third stage, the sand dune, the trees have matured, joined by blackberry, chokecherry, and white pine. Older dunes may host other tree species as well, including red maple, pin oak, elm, and red oak. The climax community is represented by sugar maple, hemlock, and white oak. Note that as succession proceeds, vegetation gradually invades and fills in the oldest ponds — those farthest from the beach.

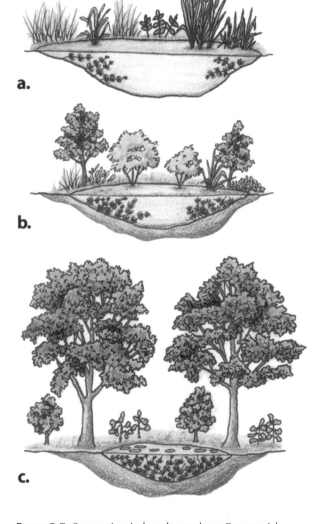

FIGURE 5-7: Succession in beach ponds on Presque Isle on Lake Erie at Erie, Pennsylvania. Careful observation reveals how the ponds have changed, proceeding toward dry land. (a) Young ponds are colonized by bulrush, chara, and cattail. (b) Older ponds, about 50 years of age, feature cottonwood, bayberry, and willow. (c) Ponds 100 years old and older are characterized by myriophyllum, yellow water lily, pondweed, and cottonwood.

comprise the glades for which the area was named. Within these broad "rivers of grass," as they have been described, are hammocks — drier "islands" that support hardwood trees. Shrub communities and cypress swamps are important too. To learn more about this unique ecosystem and the ongoing effort to preserve it, read *Environmental Science in Action: The Everglades* by going to www.EnvironmentalEducationOhio.org and clicking on "Biosphere Project."

Ecotone

The stages of succession are not discrete; rather, one blends into another. Another type of blending occurs where different communities, such as forest and grassland, meet and compete for resources and space. Each community has an area in which its species composition dominates. Where the different communities meet, a zone of transition and intense competition, called an **ecotone**, or edge community, occurs. Ecotones usually contain species from both communities as well as some transitional species that appear to be most successful in the ecotone. They are interesting areas of greater diversity than the adjacent habitats and are characterized by intense competition between community types undergoing succession or between climax communities. To encourage diversity, wildlife managers often create and maintain numerous ecotones in nature preserves.

What Is Dynamic Equilibrium?

Change and response to change are characteristic of all living systems, from the individual all the way up to the biosphere. Yet, changes in healthy living systems are constrained by opposing forces that interact in predictable patterns.

Ecosystems are dynamic, that is, they constantly go through changes of various kinds. Organisms die and new generations arise. Temperatures rise and fall with the season. One community is replaced by another during succession.

Ecosystems also possess equilibrium, a condition of balance or stability. A new generation of individuals within any species is not vastly different from the previous one. Maple trees lose their leaves for the cold months and put out new ones in the spring. While fire pushes back succession in one community, succession advances somewhere else, creating an overall balance of community types within the ecosystem. As these examples illustrate, ecosystems respond to changing conditions in ways that tend to preserve an overall pattern of plant, animal, and microbial life. This property of constant adjustment to change, maintaining an overall balance, is called **dynamic equilibrium**. The result of dynamic equilibrium is that the landscape is like a quilt composed of patches in different successional stages. A climax forest, then, may include not only stands of mature forest but also relatively bare patches where the trees have been recently destroyed by fire or insect damage, patches dominated by herbaceous plants or shrubs, and still others dominated by young trees.

What Factors Help to Maintain Dynamic Equilibrium?

In an ecosystem that is in equilibrium, organisms balance each other's inputs and outputs, the long-term average of the population of each species remains at a level that the environment can support, and nutrients are continually recycled by biogeochemical processes. As energy flows and materials cycle through the ecosystem in countless ways, equilibrium is maintained by many checks and balances. Several factors help maintain ecosystems in dynamic equilibrium; the most important of these are feedbacks, species interactions, and population dynamics. Diversity also may enable ecosystems to maintain equilibrium.

Feedbacks

Dynamic equilibrium is at least as complex in ecosystems as it is within individual organisms. Like the internal dynamic equilibrium of an individual, the dynamic equilibrium of an ecosystem depends on feedbacks. A **feedback** is any factor which influences the same trend that produced it.

To illustrate, let's look at the growth of bacterial populations after a fire. At first, there is little food for decomposers, which cannot utilize burned plant matter. Also, the loss of leafy plants on a burned site exposes the soil to sunlight and drying, unfavorable conditions for bacteria. Thus, bacterial populations are low. But the regrowth of plants provides some food, moisture, and shade for bacteria, so their populations increase. As bacteria decompose plant matter, enriching the soil, they make the site more favorable for plants. The growth of more plants makes the site even better for bacteria, and so on. In this system, enhanced plant growth is an "output" of the bacterial population growth trend. An output that promotes a trend is termed a **positive feedback**. The bacteria and the plants are related to each other in a positive feedback system.

As the preceding example illustrates, progressive ecological change promoted by positive feedback can contribute to the resiliency of ecosystems. **Resiliency** refers to the capacity of an ecosystem to undergo change and return to a similar, but not exact, system configuration. In their constant promotion of change, however, positive feedback systems alone are destabilizing. Recall our earlier discussion of the changes that took place, beginning about 10,000 years ago, in the Mediterranean region. Hundreds of years of farming, grazing, deforestation, and drought caused serious erosion that led to a decline in vegetation. With fewer trees to hold soil in place on the region's steep hillsides, erosion increased further, making it more difficult for trees to regenerate, and so on.

Negative feedbacks lend stability to ecosystems. A **negative feedback** is any output that interferes with the trend which produced it. Population trends of some owl species operate in negative feedback systems. When there are few owls in an area, rodent populations may increase rapidly. In response to the abundance of food, female owls lay more eggs. After a few generations, predation by the larger owl population (and other predators) reduces the rodent population. With less food available, female owls lay fewer eggs. With fewer owls, rodent populations expand once again, and female owls again lay more eggs. In this example, an output of the trend toward more owls is a lower food supply. The food supply acts as a negative feedback, stabilizing owl populations within a moderate range rather than promoting an ever-growing number of owls. This negative feedback system contributes to the inertia of the ecosystem to which the owls belong. **Inertia** refers to the ability of an ecosystem to resist change.

Species Interactions

Species interactions influence the composition and structure of communities. Community structure, in turn, influences ecosystem inertia and resiliency. That's why ecologists spend so much time trying to understand species interactions — besides, they're fun to learn about!

Species interact in numerous ways. Some of these are well known. The weasel stalks the rabbit, the butterfly pollinates the flower (Figure 5-8a), the flea sucks blood from the rat. Other species interactions are less familiar. The oxpecker, a bird native to East Africa, feeds on lice and other small insects found on rhinos. In coral reefs, the cleaner wrasse fish cleans the teeth or gills of other fishes, receiving a free meal for its trouble. Likewise, the clownfish finds safe haven amid the stinging tentacles of the sea anemone (Figure 5-8b). The familiar acorn is home to and food for an almost bewildering array of herbivores, predators, parasites, and decomposers. Weevils, ants, filbert worm moths, acorn moths, sap beetles, tachinid flies, braconid wasps, earthworms, termites, sow bugs, fungus gnat larvae, wire worms, slugs, millipedes, centipedes, and snails are just some of the animals that live in or use acorns for food. In addition, more than 80 species of North American birds

Figure 5-8: Species interactions. (a) An admiral butterfly pollinates a common daisy. (b) This two-banded clown fish finds protection by hiding among the tentacles of the sea anemone. The animal's sting is deadly to most fishes.

and mammals eat acorns. It's a wonder that enough acorns are left to become oak trees!

Four significant types of species interactions are cooperation, competition, symbiosis, and predation.

Cooperation. Intraspecific cooperation is best exemplified by the social structure of bees and termites and the behaviors of herds and packs — cooperation among members of the same species. Cooperative behaviors among members of the beehive, termite hill, bison herd, or wolf pack improve the chances for survival of the group as a whole.

Competition. Competition occurs when two or more individuals vie for resources (food and water), sunlight, space, or mates. Individuals that can outcompete others survive, reproduce, and pass on their attributes — a restatement of Darwin's theory of evolution by natural selection. (Evolution takes place at the population level, but the process goes on within the context of the ecosystem.) In this way, populations adapt over time to changes in the environment. Understanding competitive interactions helps us to understand how populations react to change, thus affecting the structure of communities.

Basically, two types of competitive interaction take place: intraspecific and interspecific. **Intraspecific**

competition is competition between members of the same species. When population densities (numbers per unit area) are low and there are adequate nutrients, sunlight, space and mates, little competition takes place. When population densities are high or resources are scarce, however, competition becomes keen.

Several examples illustrate the effects of intraspecific competition. Experiments have shown that frog tadpoles reared at high densities suffer higher mortality, experience slower growth, remain tadpoles longer, and metamorphose at smaller sizes than tadpoles reared at low densities. When plant densities are high and resources become limited, plants often experience a change in their reproductive strategies, producing fewer seeds and vegetative offspring. In animal populations, competition often leads to the establishment of social hierarchies and defended territories and the development of behaviors that affect the individual's success within the population (Figure 5-9).

Interspecific competition is competition between members of different species. The greater the similarity in their needs and the smaller the supply of resources, the keener the competition. Interspecific competition is most intense when different species occupy similar ecological niches. **Niche** is the complete ecological description of an individual species, including all the physical, chemical, and biological factors that the spe-

FIGURE 5-9: A wolf pack illustrates intraspecific cooperation and competition. These red wolf pups will learn social behaviors from older members of the pack. A high position in the social order gives the individual an advantage in the competition for food, mates, and territory.

FIGURE 5-10: Southern Grass-pink (*Calopogon tuberosus*). This delicate flower occupies a specialist niche throughout the southeastern United States and the Bahamas.

cies needs to survive and fulfill its role in the community. Niche includes the organism's interactions with other species; its trophic level position; its place, time, and method of reproduction; its physical requirements (temperature, moisture, salinity, light, and others); and its methods of protection and food gathering. A niche defines the functional role of an organism within its community; the habitat of a species (where it lives) is analogous to its address, while its niche is analogous to its occupation.

A biotic community can be thought of as a single unit composed of the niches of various organisms. Some organisms, like the Southern Grass-pink, occupy a specialist niche. This beautiful orchid, found throughout the southeastern United States and the Bahamas, favors wet habitats, including wet grasslands, pitted rock flats, marshes, and low flooded pineland (Figure 5-10). Other organisms, like the Canada thistle, occupy a generalist niche and can be found in many different habitats.

Rarely does interspecific competition lead to the elimination of a species. More commonly, the competing species eventually coexist at reduced densities by partitioning the available resources (Figure 5-11), or the less dominant species is forced to use other space or resources. In the tropical forests of New Guinea, as many as four species of pigeons, differing primarily in weight, rely on a particular fruit tree as their sole food source. The successively smaller and lighter birds have adapted to feed on the fruit that grows on successively smaller branches outward from the trunk.

Different species with similar requirements sometimes compete to the elimination of one of them, a phenomenon known as the **competitive exclusion principle**. This principle has been demonstrated in laboratory experiments with two closely related species of protozoans, fruit flies, mice, flour beetles, and annual plants (Figure 5-12). It is more difficult to observe in the field, since ecologists may not have all the information they need, particularly about species' life-history requirements, to prove conclusively that the exclusion of one species is due to competition with another.

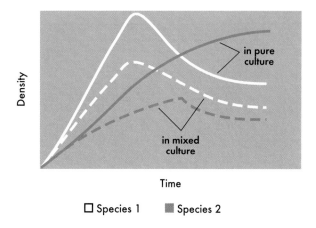

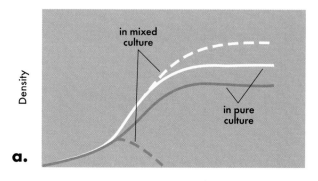

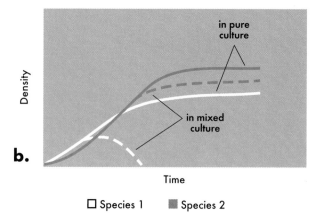

☐ Species 1   ■ Species 2

FIGURE 5-11: Competing species usually coexist at reduced densities. Researchers at the University College of North Wales conducted experiments with clover that showed the closely related species *Trifolium repens* (species 1) and *T. fragiferum* (species 2) are able to survive in mixed culture, although at lower leaf densities than when each exists in pure culture. Their results indicate that competing species can coexist despite crowding and competition for limited resources as long as there are differences in some significant factor, for example, the timing of growth; a requirement for nutrients, water, or light; or a sensitivity to grazing or toxic substances.

☐ Species 1   ■ Species 2

FIGURE 5-12: Other effects of interspecific competition. (a) In the 1940s and 1950s, investigators at the University of Chicago carried out a series of competition experiments with laboratory cultures of flour beetles. They found that *Tribolium castaneum* (species 1) excludes *T. confusium* (species 2) under hot and wet climatic conditions (93°F or 34°C, 70 percent relative humidity), even though both species can thrive under these conditions separately. (b) The opposite is true under cool and dry conditions (75°F or 24°C, 30 percent relative humidity). *T. confusium* (species 2) excludes *T. castaneum* (species 1), even though both do well at these conditions in pure culture.

Symbiosis. The intimate association of two dissimilar species, regardless of the benefits or lack of them to both species, is known as **symbiosis**. It can take one of several forms: mutualism, commensalism, or parasitism. Lichens, for example, are symbiotic associations of fungi and algae. Carbohydrates produced by the algae are used by both the algae and the fungi, and the fungi provide moisture and protection for the algal cells. Some biologists believe that the lichen association might be characterized as parasitism of the algae by the fungi, but others believe it is mutually beneficial.

Mutualism is an association of two species in which both benefit. Eighty percent of plants have mutualistic relationships with specific kinds of fungi known as mycorrhizae, from the Greek words for "fungus" and "roots." Mycorrhizae aid in the direct transfer of nutrients from the soil into the roots, while the plant provides organic carbon to the fungus. Mutualism is also illustrated by the association between termites and protozoa. While termites can chew through wood, they cannot digest the cellulose in it. Digestion of the cellulose is accomplished by protozoa that live in the guts of termites.

Commensalism is an association of two species in which one benefits and the other neither benefits nor is harmed. Remora fish attach themselves to sharks or rays with suckers and benefit from the sharks' or rays' feeding activities, but neither help nor harm the animals. Water that collects in bromeliads of the South American rain forests provides a place for tree frogs to breed, but the plants neither benefit nor are harmed by the frogs' activities.

Parasitism is an association of two species in which one benefits and the other is harmed. The organism that benefits from the activity is called the parasite; the other organism is the host. The host acts as the environment for the parasite, which derives food, shelter, and protection from the host. Parasites that live outside the host, such as the deer tick, are called ectoparasites; parasites that live inside the host's body cavity, organs or blood, such as tapeworms and flukes, are called endoparasites. Typically, the parasite does not kill its host, but the host may weaken and die from complications associated with the parasite's presence.

FIGURE 5-13: The predator-prey relationship. (a) A starfish pries apart a mussel's protective shell. (b) A cheetah, the world's fastest land animal, with its kill.

Predation. Perhaps the most widely recognized form of species interaction is **predation**, in which one species consumes another (Figure 5-13). Predators are generally larger than their prey, live apart from their prey, and consume all or part of their prey. Some, like the hyena, hunt in packs, where cunning and cooperation enable them to attack larger prey. Prey species have evolved behavioral, reproductive, physical, and chemical characteristics to help them avoid being eaten (Figure 5-14a). Predators, in turn, have evolved mechanisms to enhance capture of prey (Figure 5-14b). Predators may be herbivores, omnivores or carnivores, and prey may be producers, herbivores, omnivores, or carnivores. Predation (consumption) is the mechanism by which energy flows through all ecosystems.

Nonhuman Population Dynamics

Species interactions help ecologists to understand trophic level interactions, food webs, community structure, and the complex interactions between the biotic and abiotic components of ecosystems that help regulate nonhuman populations. In Chapter 3, we learned that a population is a group of organisms of the same species that occupies a particular habitat and has definable group characteristics. Seed dispersal, germination, and death rates are major characteristics of plant populations. Birth, migration, and death rates are major characteristics of animal populations. Once organisms obtain energy, they use much of it to reproduce. At the population level, this results in population growth.

If it were possible to construct a utopian environment, that is, an environment with no limitations, a population could experience unlimited growth. As long as the environment remained utopian, the population could continue to grow in an unrestricted manner. In just 24 hours, one yeast cell could give rise to several thousand cells, and one *Escherichia coli* bacterium could give rise to 40 septillion bacteria; in a single year, two houseflies could produce 6 trillion individuals! But environments are not utopian; all populations fluctuate over time and have an upper limit (set by the number of organisms)

FIGURE 5-14: The use of camouflage. Camouflage is an evolved mechanism that (a) helps hide the eggs of the least tern from predators and (b) enhances the ability of the green vine snake to capture prey.

their environment can support. In controlled laboratory experiments, populations exhibit similar growth patterns. From studies of bacteria, fruit flies, mice, and rats that were provided with unlimited food and water, scientists have identified five common phases of population growth: I, lag phase; II, slow growth phase; III, log, or exponential growth, phase; IV, dynamic equilibrium phase; and V, decline (Figure 5-15). (Caution must be exercised when making conclusions about population growth based on these experiments, because only single populations were studied and environmental conditions were carefully controlled.)

The experimental populations remained in phase I for some period of time. During this time, organisms were adjusting to their new environments and few were mature enough to reproduce, so the population did not grow. In phase II, the population grew slowly as more individuals began to reproduce. In phase III, many organisms were capable of reproducing, and the population grew rapidly until it reached phase IV, dynamic equilibrium. During phase IV, the population fluctuated around a particular level as reproduction basically equaled mortality. In all the systems studied, the population went through phase V, decline, eventually reaching zero. If the size of the experimental world was increased, that is, if the organisms were provided with additional space, maximum populations were larger, but once again all stages were observed.

The utopian growth represented in phase III cannot be maintained in a world of limited resources, as studies of naturally occurring populations have shown. These populations exhibit S-curve, or oscillating, growth patterns (Figure 5-16). It is as if phase IV of our experimental population pattern were magnified

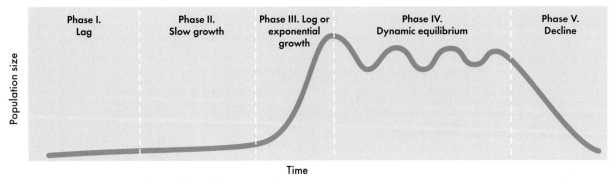

FIGURE 5-15: Five common phases of population growth.

for careful scrutiny. All natural populations fluctuate over time around the number of individuals the environment can best support. In other words, all known nonhuman populations thus far studied have an upper limit to growth. Many scientists believe that there is also an upper limit to human population growth, an idea we explore in Chapters 7 and 8.

The maximum growth rate that a population could achieve, given unlimited resources and ideal environmental conditions, is known as its **biotic potential**. The environment, through limiting factors known as **environmental resistance**, exerts a controlling influence on population size. Those organisms that are able to adapt to changing environmental conditions survive and reproduce. Over time, populations are regulated by the interaction between biotic potential and environmental resistance. This dynamic equilibrium, when birth and death rates are equal, is referred to as the **carrying capacity**, that is, the population size that can best be supported by the environment over time.

Environmental resistance can be separated into two categories: density-dependent factors and density-independent factors. **Density-dependent factors** are biotic; their effect is greater when the population density is high. They include a wide range of intra- and interspecies interactions. Some density-dependent factors are competition, predation, disease, spontaneous abortion, embryo absorption, infanticide, cannibalism, starvation, and stress. **Density-independent factors** tend to set upper limits on the population. They are abiotic and independent of population size. For example, whether there are 10 individuals in a population or 10 billion, they might all be killed in a volcanic eruption. Density-independent factors include a wide range of interactions between the physical environment and the population such as climate, weather, and availability of nutrients, water, space, den or nest sites, and seed germination sites.

For any given nonhuman species, it is likely that a combination of density-dependent and density-independent factors operate simultaneously to regulate populations. Certain rodent populations (mice and rats) seem to be regulated by available space (density-independent) and the stress caused by overcrowding (density-dependent). Possibly, stress-induced hormonal changes cause abnormal social behaviors and decrease the rodents' resistance to disease, thus lowering birth rates and increasing death rates. Because reproduction would then be less successful, population numbers would decline to a level where the stress caused by overcrowding would no longer be a major limiting

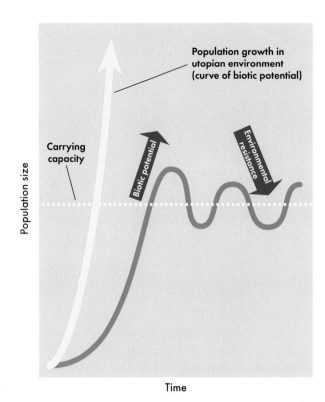

FIGURE 5-16: S-curve of population dynamics. The diagram illustrates the relationship of biotic potential, environmental resistance, and carrying capacity. The population in dynamic equilibrium oscillates around the level that can best be supported by the physical environment.

factor. However, for certain populations, some limiting factors may be more important than others. For lynx and hare populations, the prime regulator might be predation by the lynx, a density-dependent factor, or it might be the number of nest sites available for the hare, a density-independent factor (Figure 5-17). Rainfall and temperature (density-independent factors) are important regulators of insect populations.

A caution is in order. The results of population dynamics studies on nonhuman organisms should not be directly applied to humans. Conclusions cannot be made about the effect of environmental resistance on human populations in the absence of reproducible studies using human populations.

Species Diversity

For many years, it was generally accepted that the greater an ecosystem's diversity of species, the more easily it would recover from a disturbance. So long as the disturbance was not too severe, this assertion seemed reasonable. By the 1980s, however, scientists had begun to question the presumed link between diversity and

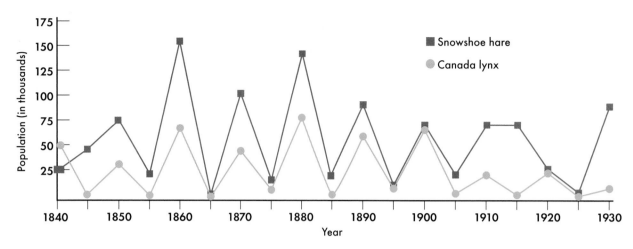

Figure 5-17: Snowshoe hare and lynx populations, northern North America. The graph shows the 9- to 10-year cyclic fluctuations that occur in the lynx and snowshoe hare populations.

stability. Discoveries about the inability of the most diverse of all ecosystems, tropical rain forests, to recover from deforestation, suggested that ecosystems vary in their response to stress and that factors other than diversity — factors not yet identified or understood — may contribute significantly to equilibrium. The high rate of tropical deforestation, and the subsequent loss of untold numbers of species, created a sense of urgency. As investigators studied the links between diversity, stability and ecosystem processes, several schools of thought emerged. One is centered on the **diversity-stability hypothesis**, which suggests that biodiversity promotes resistance to disturbance because species differ in their traits, that more diverse ecosystems are likely to contain some species which can thrive during a given environmental disturbance, and that those species should be able to compensate for species which are adversely affected by the disturbance. A related view, known as the **rivet hypothesis**, likens species in an ecosystem to the rivets that hold together an airplane: Some of the rivets (species) may be redundant, but others are not; therefore, removing rivets (species) beyond a threshold number may cause the airplane (ecosystem) to collapse. The **redundancy hypothesis**, in contrast, contends that there is considerable overlap in the roles that species play within an ecosystem; thus, some species can be lost with no adverse effect on the system as long as others which play the same role persist. Accord-

ing to this view, only minimum diversity is needed for an ecosystem to function properly. A fourth and final view promotes what is known as the **null hypothesis**: Ecosystem functions or processes are insensitive to the addition or deletion of species.

After decades of research and debate, most ecologists agree that there is no simple relationship between diversity and equilibrium. The scientific literature contains evidence supporting each of the various hypotheses about the diversity-stability link, and it may be that diversity enhances productivity and stability in some ecosystems but not in others. The bottom line is that there is still too much that we do not understand about how ecosystems function and change. However, it seems likely that an ecosystem's resilience — its ability to weather a disturbance and then return to a similar configuration — probably depends not just on the properties of individual species but also on the intricate ecological relationships between various species and between species and the physical environment. If human activities disrupt ecological relationships beyond a certain threshold, the system may become less resilient and return to a new and — from a human standpoint — less desirable community. Because we do not understand the complexities of ecosystem relationships well, the prudent course of action is to preserve the diversity in all ecosystems.

Summary

All ecosystems undergo succession, the gradual, sequential, and somewhat predictable changes in the composition of the biotic community. Succession, or ecosystem development, occurs in terrestrial and aquatic habitats. Primary succession is the development of a new community in an area previously devoid of organisms; secondary succession is the development that occurs after an ecosystem has been disturbed. Successional stages are not discrete; they tend to blend together. A similar blending occurs where different types of communities meet and compete for resources and space. The transitional zone between the communities is known as an ecotone, or edge community.

Ecosystems continually react to change and disturbance, thereby maintaining a dynamic equilibrium. Ecosystems maintain a dynamic equilibrium either by resisting change, known as inertia, or by restoring structure and function after a disturbance, known as resiliency. Numerous factors contribute to ecosystem stability. Of particular importance are feedbacks, species interactions, and population dynamics. Feedbacks can be positive or negative; positive feedback continues a particular trend, and negative feedback reverses it.

Four types of species interactions are cooperation, competition, symbiosis, and predation. Cooperation can take place between individuals of the same species or between those of different species. Intraspecific cooperation is exemplified by insect societies and animal herds and packs. Intraspecific competition takes place between individuals of the same species; interspecific competition occurs between individuals of different species. Different species with similar requirements sometimes compete to the exclusion of one of them, a phenomenon known as the competitive exclusion principle. Interspecific competition also may result in coexistence of two species at reduced densities. Organisms with a wide range of tolerances enabling them to compete successfully in different areas are said to occupy a generalist niche in the community. Organisms with a narrow range of tolerances allowing them to compete successfully in specific conditions occupy a specialist niche.

Symbiosis is the intimate association of two different species, regardless of the benefits or lack of them. Mutualism, commensalism, and parasitism are forms of symbiosis. Mutualism is an association of two species in which both benefit. Commensalism is an association of two species in which one benefits and the other neither benefits nor is harmed. Parasitism is an association of two species in which one benefits (the parasite) and the other is harmed (the host).

Predator-prey relationships are perhaps the most familiar type of species interaction. Prey species have evolved behavioral, reproductive, and physical and chemical characteristics to help them avoid being eaten; predators have evolved mechanisms to enhance capture of prey.

Population dynamics contribute to ecosystem stability. In laboratory studies, scientists have found that populations of organisms commonly exhibit five phases of growth: lag; slow growth; log, or exponential, growth; stability, or dynamic equilibrium; and decline. In contrast, natural populations fluctuate over time around the carrying capacity, the number of individuals the environment or habitat can best support. This number is lower than the population size that would result if the population achieved

Discussion Questions

1. Differentiate between primary and secondary succession. Give examples of successional stages. What maintains succession at its climax stage?

2. Explain the concept of dynamic equilibrium. What does it mean?

3. Discuss how feedbacks, species interactions, and population dynamics help to maintain an ecosystem in dynamic equilibrium.

4. Differentiate between commensalism, mutualism, and parasitism. Give an example of each.

5. Explain the concept of carrying capacity. If you were the manager of a nature park or preserve, how would an understanding of carrying capacity enable you to better manage the land under your control?

6. What might happen to population dynamics at each trophic level in an ecosystem if the top carnivores are eliminated?

7. Discuss an environmental problem in your locale or region that is caused or exacerbated by a positive feedback. What negative feedback is needed to slow or reverse this problem?

8. Of the four hypotheses concerning the link between diversity and ecosystem stability, which do you think seems most likely? Explain that hypothesis and your reasons for favoring it.

its biotic potential, that is, the maximum growth rate possible given unlimited resources and ideal environmental conditions.

Limiting factors, collectively known as environmental resistance, prevent a population from realizing its biotic potential. Environmental resistance factors may be density-dependent or density-independent. Density-dependent factors are biotic; their effect is greater when the population density is high. Density-independent factors are abiotic; generally, they set upper limits on the population.

Whether biological diversity contributes to ecosystem stability is a question studied by ecologists. The diversity-stability and rivet hypotheses suggest that biodiversity enhances an ecosystem's resistance to disturbance, while the redundancy hypothesis suggests that only minimum diversity is needed for ecosystem health. The null hypothesis maintains that there is no relationship between diversity and ecosystem stability. Despite these differing points of view, most ecologists agree that there is no simple relationship between diversity and equilibrium, and that, in fact, the relationship may vary from ecosystem to ecosystem.

KEY TERMS

biotic potential	ecosystem development	null hypothesis
carrying capacity	ecotone	parasitism
climax community	environmental resistance	pioneer species
commensalism	feedback	positive feedback
competitive exclusion principle	inertia	predation
density-dependent factor	interspecific competition	primary succession
density-independent factor	intraspecific competition	redundancy hypothesis
diversity-stability hypothesis	mutualism	resiliency
dynamic equilibrium	negative feedback	rivet hypothesis
ecological succession	niche	secondary succession
		symbiosis

CHAPTER 6

Ecosystem Degradation

We stand guard over works of art, but species representing the work of aeons are stolen from under our noses.

Aldo Leopold

The worst thing that can happen — will happen — is not energy depletion, economic collapse, limited nuclear war or conquest by a totalitarian government. As terrible as these catastrophes would be for us, they can be repaired within a few generations. The one process...that will take millions of years to correct is the loss of genetic and species diversity by the destruction of natural habitats. This is the folly our descendants are least likely to forgive us.

E. O. Wilson

Learning Objectives

When you finish reading this chapter, you will be able to:

1. List and define the five Ds of ecosystem degradation.

2. Describe what is meant by ecosystem damage and define pollutant.

3. Explain how human activities disrupt natural systems.

4. Identify the primary cause of ecosystem destruction.

5. Define desertification and identify three human activities that contribute to it.

6. Define deforestation and identify its primary causes.

When environmental conditions are altered in such a way that they exceed the range of tolerances for one or more organisms in the biotic community, the ecosystem becomes degraded. It loses some capacity to support the diversity of life forms that are best suited to its particular physical environment. Ecosystems can be degraded by natural catastrophes, like the eruption of a volcano, or by human activities, like the disposal of wastes into a river. In this chapter, however, we focus only on human-induced, or **anthropogenic**, ecosystem degradation. Our reason for doing so is simple: We cannot control nature (volcanic eruptions, earthquakes, hurricanes, and the like), but we can control our individual and societal activities (waste disposal, dam construction, the draining of wetlands, and so on).

Although ecosystem degradation is complex, it is helpful to classify it into categories we call the five Ds of natural system degradation: damage, disruption, destruction, desertification, and deforestation. These are very broad classifications; there is some overlap in these categories, and some forms of degradation could be placed in more than one category. Further, desertification and deforestation are really subsets of the first three (and deforestation can be a cause of desertification). However, we feel that these two forms of degradation are so serious that they warrant special attention and thus categories of their own.

What Is Ecosystem Damage?

Ecosystem damage is an adverse alteration of a natural system's integrity, diversity, or productivity. Pollution is the major cause of ecosystem damage. As we learned in Chapter 1, a **pollutant** is a substance or form of energy, such as heat, that adversely alters the physical, chemical, or biological quality of natural systems or that accumulates in the cells or tissues of living organisms in amounts that threaten their health or survival (Figure 6-1).

Pollution affects ecosystems virtually everywhere. You are probably very familiar with examples of pollution in your state or region. You may be less familiar with the increasing toll that pollution is taking on other ecosystems, such as the world's coral reefs, the marine equivalent of the tropical rain forests. Reefs are subject to various forms of pollution: oil spills, sewage, toxic pollutants, and warm water emitted from power plants. Those located near development projects may be damaged by sedimentation and runoff from construction sites. The collection of corals for sale to tourists and jewelers also damages the delicate reefs. Coral reefs are not isolated ecosystems; they are important to the overall stability and productivity of marine systems. They are connected to the grasslands and mangrove swamps along the shore, the rivers of the land, and the open ocean. For example, fish that spawn in the marshes and mangroves live as adults in the coral reefs.

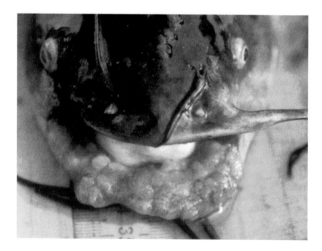

FIGURE 6-1: Catfish with tumor. Pollutants accumulate in the sediments of rivers and lakes, where they may adversely affect aquatic life. Toxic pollutants may cause tumors and malignancies in organisms that feed or live in the bottom sediments. This catfish, a bottom-feeder, was taken from Ohio's heavily contaminated Black River.

Worldwide, there are approximately 110,000 square miles (284,300 square kilometers) of coral reefs. In the *Status of Coral Reefs of the World: 2002* report, the Global Coral Reef Monitoring Network reported that 27 percent of the world's reefs have been effectively lost. The report finds that, while attempts to protect coral reefs still lag behind the rate at which they are being lost, some progress in reef conservation and awareness has been made. Thus, there is hope of preserving coral reefs, which are estimated to provide the world with $375 billion in goods and services and are, per unit area, amongst the world's most valuable ecosystems.

Four factors determine the damage done by pollutants: the effect of the pollutant, how it enters the environment, the quantity discharged to the environment, and its persistence.

Effect of the Pollutant

Acute pollution effects occur immediately upon or shortly after the introduction of a pollutant, and they are readily detected. For example, in October 2000, a coal waste pond near Inez, Kentucky collapsed into an abandoned mine, releasing 250 million gallons (945 million liters) of coal washing by-product, called coal slurry, into waterways used for drinking water and recreation. The slurry, with a volume twenty times greater than the amount of oil lost by the Exxon *Valdez*, was contaminated by arsenic, mercury, lead, copper and chromium, and killed all aquatic life in more than 70 miles (112 kilometers) of West Virginia and Kentucky streams, including approximately 385,000 fish. The spill polluted drinking water supplies, damaged water treatment systems, ruined septic systems, and covered yards with slurry, in some cases to a depth of eight feet. The U.S. Environmental Protection Agency (EPA) called the spill "one of the worst environmental disasters" in the history of the southeastern United States.

Chronic pollution effects act in the long term; they are not noticed until several years or decades after the introduction of the pollutant. This delay, known as lag time, makes it particularly difficult to predict what effect the pollutant may have at the ecosystem level.

At Lake Tahoe, the body of water that forms a partial border between Nevada and California, the popularity of motorized personal watercraft (PWC) is causing concern about chronic pollution. PWC, such as jet skis and small motorboats, are powered by the incomplete combustion of gas and oil, and their engines discharge as much as one-third of the mixture unused into the air and water. The discharged oil and gas, which can find their way into municipal drinking water reserves, contain more than 100 compounds, many of which, such as benzene and toluene, the EPA considers toxic to humans. Additionally, PWC emit hydrocarbons — organic compounds containing hydrogen and carbon — in their engine exhaust. Hydrocarbons are absorbed into the tissues of aquatic organisms over time, and their build-up can trigger harmful effects, including stunted growth and even death. In the presence of sunlight, the damage caused by such pollutants increases, a phenomenon called **phototoxicity**. Phototoxicity is of special concern at Lake Tahoe because the water's exceptional clarity and high altitude permit more intense levels of ultraviolet radiation below the surface. According to research begun in 1997 by ecotoxicologist Dr. James Oris, increased use of PWC at Lake Tahoe appears to contribute to greater phototoxic effects in fish and zooplankton, microscopic animals that are integral to many aquatic food chains. In response to this impact, the use of PWC on Lake Tahoe, as well as on waterways in the majority of U.S. national parks and in at least 25 states, has been restricted.

The effects of a pollutant are sometimes the result of interactions with other substances. As pollutants enter and move through ecosystems, they may interact and form new combinations of chemicals that are more harmful than the separate components. Such pollutants are said to have a **synergistic effect**, that is, their combined effect is greater or more harmful than the sum of their individual effects. Air contains a mix of thousands of chemicals. Some are believed to be safe; others are known to be toxic. For example, when hydrocarbons and nitrous oxides mix in the atmosphere and are exposed to sunlight, synergistic interactions can occur that produce photochemical smog containing new, more harmful compounds. Synergistic interactions may occur in aquatic or terrestrial habitats as well. Chlorine gas, which can be dangerous at certain levels, is used in small amounts to purify water. It may react with organic substances sometimes found in water to form chloramines, substances that may be toxic or carcinogenic. Synergistic interactions also can occur in landfills, resulting in new, potentially dangerous compounds that may then contaminate groundwater.

Bioaccumulation is the storage of chemicals in an organism in higher concentrations than are normally found in the environment. For example, nitrogen and phosphorus are usually found in lower concentrations in aquatic environments than are needed by phytoplankton. In order to obtain sufficient amounts for growth and reproduction, phytoplankton transport these nutrients through their cell membranes. Fat-soluble chemicals such as dichloro-diphenyl-trichloro-ethane (DDT) and polychlorinated biphenyls (PCBs) also pass through the cell membranes. Because such chemicals dissolve in fats (or lipids, as they are known), they tend to accumulate in lipid materials of the phytoplankton. If they were water soluble, the chemicals would be flushed from the cells and excreted. The fat-soluble chemicals may be present in the environment of the phytoplankton in small quantities, but biological activity concentrates them and they accumulate in the organisms.

Another example of bioaccumulation is the amount of tributyltin (TBT), a chemical in nautical paint, found in Pacific oysters along the coast of California in the late 1980s. During this time, Pacific oysters, which absorbed and stored the chemical in their tissues, exhibited shell thickening and chamber malformations that scientists attributed to TBT exposure. In some instances, oysters taken from central California's Moss Landing and Monterey Harbor had concentrations of TBT 30,000 times higher than concentrations found in the water.

Biomagnification is the accumulation of chemicals in organisms in increasingly higher concentrations at successive trophic levels. Consumers at higher trophic levels ingest a significant number of individuals at lower levels, along with the fat-soluble pollutants stored in their tissues. Biomagnification occurs in both aquatic and terrestrial habitats (Figure 6-2). Carnivores at the top of food chains, such as large fish and fish-eating birds (gulls, eagles, pelicans) in aquatic habitats or robins in terrestrial habitats, may accumulate poisons in concentrations high enough to prevent their eggs from hatching or to cause deformities or death. The concentrations of poisons in carnivores can be a million times higher than the concentration of the chemical in the water or soil.

Synergism, bioaccumulation, and biomagnification make it more difficult to estimate or assess the effect of a particular pollutant and its long-range cost — both the cost to the environment and the economic cost of cleanup or removal.

How the Pollutant Enters the Environment

Point-source pollution is emitted from an identifiable, specific source or point. Pipes that discharge sewage into streams and smokestacks that emit smoke and fumes into the air are point sources of pollution (Figure 6-3). Past regulatory efforts in the United States focused on

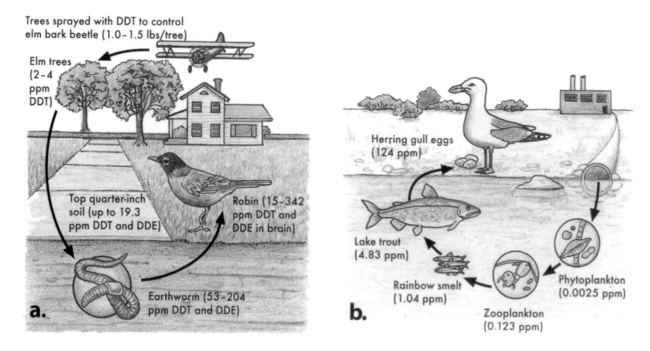

FIGURE 6-2: Biomagnification in terrestrial and aquatic habitats. Arrows show the movement of the substances through the food chain. (a) DDT was sprayed on trees to control the elm bark beetle, the insect responsible for the spread of Dutch elm disease. The insecticide entered the soil, where earthworms accumulated it in their tissues. The concentration of DDT was biomagnified in the tissues of earthworm-eating robins. (b) PCBs dumped into the Great Lakes move through the food chain. The concentration of PCBs increases with each step in the chain, and the highest levels are reached in the eggs of fish-eating birds such as herring gulls.

eliminating or reducing point sources because they are usually easy to identify.

Nonpoint-source pollution cannot be traced to a specific source but rather comes from multiple generalized sources, such as farm lands, parking lots and construction sites, in a wide area. It is much harder to identify than point-source pollution and, consequently, much harder to control. Puget Sound, framed by the Olympic and Cascade mountains in the State of Washington, is an ecosystem under siege by nonpoint-source pollution. Runoff from farm fields, pastures, parking lots, and streets contains large amounts of silts mixed with fertilizers, herbicides, insecticides, animal wastes, oils, fuel additives, solvents, and cleaners. The runoff enters streams and rivers that feed into the sound. In 2004, 40 percent of the sound's shellfish beds were closed due to pollution. Like Chesapeake Bay and San Francisco Bay, which are also greatly affected by nonpoint-source pollution, Puget Sound does not flush, or naturally cleanse itself, very rapidly.

Certain contaminants, called **cross-media pollutants**, affect an ecosystem by moving from one medium (air, water, or soil) to another. They can move from the air into soils or streams, seep from landfills into groundwater, or evaporate from surface water into the air. These pollutants can change their chemical character as they react with the chemicals present in the medium into which they move.

Acid precipitation is perhaps the best known cross-media pollutant. However, synthetic compounds and their by-products, such as solvents and cleaners, might eventually become our most troublesome cross-media pollutants because they are so widely used. Much of the pollution affecting New Bedford Harbor, near Cape Cod, Massachusetts, has entered the ecosystem as cross-media pollutants. Researchers at Woods Hole Oceanographic Institute report that, since 1940, two million pounds (900,000 kilograms) of PCBs have entered the harbor system from leaky landfills, illegal hazardous waste dumps, and unregulated industrial discharges to the air.

Quantity of the Pollutant

Some waste products, such as fertilizer runoff, human sewage, livestock waste, waste from food processing plants, carbon monoxide (from automobiles) and solid wastes, are produced in such quantities that natural

Figure 6-3: This aerial photograph shows pollution entering the Potomac River.

The bay became an "organic soup" laced with high concentrations of toxic metals, including silver, lead, zinc, and mercury. Bacteria, viruses, and trace metals accumulated in the tissues of clams, oysters, mussels, and fish such as striped bass, bluefish and white perch, presenting a danger to the organisms that fed on them. Before legislation banning all ocean dumping was passed in 1988, over 32 million tons (29 million metric tons) of sewage had been dumped at the Bight's deepwater municipal site. Although water quality in the Bight has improved, a 2002 report by the Natural Resources Defense Council states that in 1999 alone, more than 35 million pounds (16 million kilograms) of pollutants still were dumped into the Bight. Additionally, scientists report that the toxins dumped into the Bight have accumulated in sediments in the bay, causing lingering pollution problems.

In the late 1980s, Boston Harbor was known as the nation's dirtiest harbor. Wastewater from approximately 2.5 million people and over 5,000 businesses and industries was dumped into the harbor. Boston's sewage treatment facilities did not treat the toxic compounds in the industrial wastewater, meaning nearly one ton of heavy metals entered the harbor each day. However, a nearly $4 billion cleanup program by the Massachusetts Water Resources Authority (MWRA) has significantly improved the environmental quality of the harbor, in large part because of the MWRA's increased enforcement of industrial requirements to pre-treat wastewater and improved treatment of Boston sewage. Today, most of the toxic metals that enter the harbor are actually from household, rather than industrial, sources.

Persistence of the Pollutant

Materials that can be broken down and rendered harmless by living systems are called **biodegradable**. For example, if processing methods reduce sewage so that the concentration of materials in the water leaving the treatment facility does not alter the population of organisms present in the river or stream, then the system is biodegrading, or recycling, the sewage. A substance is **nonbiodegradable** when it enters a system in a form unusable by the organisms present in that system. As the cases of the New York Bight and Boston Harbor illustrate, even biodegradable substances, in large amounts, can adversely affect the productivity and species diversity of ecosystems.

A pollutant that accumulates in natural systems over time is called a **persistent pollutant**. Persistent pollutants include those that are nonbiodegradable and

systems cannot efficiently or readily process them. Americans generate, on average, about one ton of solid waste each year, from four to five pounds each day. A city the size of New York generates approximately 1.3 billion tons (1.18 billion metric tons) of solid commercial and residential waste and 1.35 billion tons (1.22 billion metric tons) of construction and demolition debris each year, as well as 1 to 1.2 billion gallons (3.8 to 4.6 billion liters) of raw sewage a day!

The disposal of massive quantities of wastes into a body of water eventually damages the ecosystem, with deleterious effects on the life it harbors. For example, from 1938 to 1987, the New York Bight, a bay between Long Island, New York, and Cape May, New Jersey, served as a sewage and industrial waste dumping site for New York City and neighboring communities.

those that are only slowly biodegradable. Although some insects and bacteria begin to break down the insecticide DDT almost as soon as it is released to the environment, it may persist in natural systems for 20 to 25 years. Moreover, its breakdown products (called metabolites), DDD and DDE, may persist in the environment for as many as 150 years. Plastics and other synthetic chemicals take hundreds, perhaps thousands, of years to degrade, and some radioactive substances persist in dangerous forms for hundreds of thousands of years. Future archaeologists may find artifacts of our culture that are too dangerous to handle. Some of our present-day pollutants will persist for longer periods than human cultures have been active on Earth!

What Is Ecosystem Disruption?

An **ecosystem disruption** is a rapid change in the species composition of a community that can be traced directly to a specific human activity. Changes in species composition have been linked to the use of persistent biocides (pesticides and herbicides), the introduction of new species (usually accidentally) into an area, the construction of dams or highways, and the overexploitation of resources. Often, the disruption caused by our activities results because we do not completely understand how ecosystems function or because we fail to adequately evaluate the long-term effects of our actions. To illustrate this point, consider an infamous incident that took place 50 years ago in the East Indies.

In 1955, the World Health Organization (WHO), in a massive campaign to reduce the incidence of malaria on North Borneo, sprayed dieldrin (similar to DDT) to kill disease-carrying mosquitoes. The campaign was a success; the dieldrin killed most of the mosquitoes, and malaria was nearly eliminated from the island. But the biocide also killed many other insects. Small lizards that fed on the dead insects died as well. Soon, cats that fed on the lizards began dying. Previously, cats had kept the rodent population in check, but as their numbers decreased, rats began to overrun the villages, and the people faced a new threat: sylvatic plague, carried by fleas on the rats. The situation was brought under control when the British air force, on instructions from the WHO, parachuted cats into Borneo.

The episode caused the villagers other troubles as well. The population of a particular type of caterpillar, which either avoided or was unaffected by the dieldrin, rose sharply when the populations of wasps and other insects, which were its natural predators, decreased.

Caterpillar larvae ate their way through one of their favorite foods — the leaves that formed the thatched roofs of the villagers' homes.

In many areas where dieldrin and DDT have been sprayed extensively to control mosquitoes, populations of dieldrin- or DDT-resistant mosquitoes have developed. How does this happen? Diversity or variation in genetic structure normally exists in living organisms. Some mosquitoes were naturally resistant to the levels of dieldrin or DDT sprayed to control their population numbers. Resistant mosquitoes were more likely to survive and breed with other resistant mosquitoes and produce entire populations of mosquitoes resistant to that level of the biocide. When levels of dieldrin or DDT were increased in the sprays, the same thing happened again (an example of a positive feedback), until it became too dangerous to increase the insecticide level any further. In some parts of the world, "super" mosquitoes now exist that are resistant to high levels of many types of insecticides. In fact, after years of using chemicals in an attempt to control pests, not one pest has been eradicated. Instead, our efforts have produced resistant populations of many human, animal, and plant pests and disease-causing microorganisms. These stronger pest strains and disease organisms cause tremendous human misery throughout the world. Rats, cockroaches, fire ants, fleas, staphylococcus bacteria, streptococcus bacteria, and the bacteria that cause syphilis and gonorrhea all show increasing resistance to our control efforts.

When a species is introduced to an area in which it is not native, it rarely has natural predators that can keep its population under control. Because the Japanese beetle, accidentally introduced to the United States in 1911, had no natural predators, its population grew wildly. It continues to cause extensive damage to vegetation in many regions of the country. But it is just one of a long list of nonnative species that have become, or may become, established in this country. Four excellent examples of the potential disruption to local ecosystems are the entry into Latin America and southern Texas of Africanized bees, popularly (but erroneously) called "killer" bees; the accidental establishment of an eastern blue crab colony in San Francisco Bay; the escape from gardens of the plant kudzu in the southeastern United States; and the spread of the plant purple loosestrife in the wetlands and marshes of the Great Lakes and midwestern states (Figure 6-4). (Ironically, purple loosestrife, a perennial, is sold in some greenhouses and nurseries in the Midwest for use in landscaping.) In each case, the introduced species competes more successfully

FIGURE 6-4: Purple loosestrife (at back), an invasive plant, encroaches on a field of native wildflowers at Montezuma National Wildlife Refuge, New York State.

than native species that occupy the same niche, resulting in the displacement of the native species.

Human constructions sometimes impede the migrations and movements of species and disrupt ecosystem processes. Dams block fish that are migrating to spawning grounds, oil and gas exploration interferes with the seasonal migrations of caribou and elk, cooling water intakes for electric-generating power plants suck in millions of planktonic larval fish and shrimp, and highway constructions interfere with animals' movements and complicate courtship or breeding behaviors. In *Song for the Blue Ocean*, conservation biologist Carl Safina asserts that the salmon populations of the U.S. Pacific Northwest are being decimated by the combined impact of such human interferences. Pacific salmon have disappeared from 40 percent of their breeding range in Washington, Oregon, Idaho and California, in large part due to dams on the Columbia River; intense irrigation of farm fields; and water pollution from pulp mills, pesticides, and livestock yard wastes. All major species of Pacific salmon — including chinook, coho, pink, chum, and sockeye — have been affected; Safina estimates that over 300 salmon populations are at moderate to high risk of extinction in the region.

Overexploiting resources alters the species composition of the community and can seriously hamper the functioning of ecosystems. Overgrazing, for example, promotes the spread of undesirable weed species over range grasses. Forests that are logged often are replaced with pine plantations or tree farms, stands consisting of one or two commercially valuable species. These stands, like croplands, require high inputs of energy and fertilizers. Overharvested areas no longer benefit from the basic functions that forests perform, including air and water cleansing, local climate moderation, and erosion control. This is especially true when riparian zones, vegetation zones near streams and rivers, are not maintained during timber harvests or when steep slopes are harvested, promoting rapid erosion.

Finally, driving a species to extinction disrupts natural systems. When species disappear or when drastic population changes occur within a community, other species — and the ecosystem in general — are likely to be affected as well. Hence, ecosystem disruption and decreasing diversity form a vicious cycle. To examine the impact of this cycle on whale populations, read *Environmental Science in Action: The Blue, Gray, and Humpback Whales* by going to www.EnvironmentalEducationOhio.org and clicking on "Biosphere Project."

What Is Ecosystem Destruction?

Human activities can and do disrupt the species composition of ecosystems, but more often human activities destroy ecosystems. **Ecosystem destruction** is the conversion of a natural system, such as a wetland, to a less complex human system, such as a farm. When a natural ecosystem is destroyed, we lose the benefit of the ecosystem services that it provided (such as protection of the watershed).

Obviously, we must convert some natural systems so that we have places to live, room to grow crops, and space for our schools, hospitals, businesses, roads, theaters, and ballparks. At issue is not whether we should convert any natural systems, but rather, how much of the biosphere we should convert and how much we should leave unaltered. As the human population increases, we take more and more habitat away from other species. Habitat loss is the primary cause of species extinction. In *The Biophilia Hypothesis*, E.O. Wilson, a highly-respected entomologist and naturalist, estimates that as many as 50,000 species may be going extinct annually. By the year 2025, the world could lose a quarter of its species. While extinction is a natural process, Wilson warns that the current extinction rate — driven largely by habitat loss — is 1,000 to 10,000 times higher than in prehuman times.

A second important issue is the uses to which we put particular lands. Natural systems are destroyed in order to accommodate five basic land uses: agriculture, urbanization, transportation (paving and the land required for freeways), industry, and recreation. Land suitable for agriculture also can be used for the other four purposes; consequently, there is tremendous pressure on good land. Each year, fertile farmland is lost to housing developments and other nonagricultural uses; some becomes less productive due to erosion. The Worldwatch Institute reports that, over the past half-century, land degradation has reduced food production by 13 percent on cropland and 4 percent on pasture. Also, 10 to 20 percent of the world's cropland suffers from some form of degradation, while over 70 percent of the world's rangelands are degraded.

As prime cropland is lost to urban and industrial uses, areas that are less suitable for agriculture are being lost to the plow. Over half of the original wetlands in the contiguous 48 states have been drained and developed since colonial times. The National Wildlife Federation (NWF) estimates that the United States loses wetlands at the rate of 12 football fields an hour, every day of the year, to croplands and urban projects.

Agricultural and urban conversion simplifies ecosystems and makes them highly susceptible to disturbances caused by insect pests, pollution, or soil erosion. The conversion of freshwater wetlands can cause floods to become more severe and eliminates important feeding habitat for waterfowl and birds. The draining of coastal wetlands (saltwater marshes, mangrove forests, and estuaries) seriously affects marine species that use these areas as nurseries, which in turn damages the fishing industry. When coastal wetlands are lost, the shoreline loses a natural protective barrier against storm waves.

The rate of wetland loss has slowed in recent years because of increased regulations at both the state and federal levels. In 2004, the National Resources Inventory reported that annual wetland loss to agriculture had dropped from 26,000 acres (10,500 hectares) annually between 1992 and 1997 to 10,000 acres (4,050 hectares) annually between 1997 and 2002. In April 2004, President George W. Bush announced a new national goal that moves beyond the previous goal of "no net loss" of wetlands to an overall increase of wetlands each year. However, a report by the NWF notes that, without full support from the U.S. Congress, the goal of net wetlands gain will not be realized.

What Is Desertification?

Desertification is land degradation in arid, semi-arid, and dry subhumid regions resulting mainly from adverse human impact (Figure 6-5). It poses the greatest threat to rangelands, where the native vegetation (mostly grasses; forbs, or wildflowers; and shrubs) is well-suited to forage by wildlife or livestock. However, desertification is also a problem on some rain-fed and irrigated croplands.

True deserts and grasslands are healthy ecosystems; the biota are adapted to the sequence of wet and dry periods (including droughts) common to these regions. When rainfall is plentiful, native plants flourish; when it is scarce, they rely on adaptive mechanisms to sustain them. But overgrazing, overcultivation, deforestation, and poor irrigation practices can tip the balance for native vegetation, such that it is no longer able to survive the stresses brought on by lack of moisture. Signs of desertification include degraded vegetation, wind erosion, water erosion and **salinization**, the build-up of

Hyperarid
Arid
Semiarid
Dry subhumid
Humid
Cold climates

Figure 6-5: Aridity zones.

salts in the soil. The latter is caused by poor irrigation practices; in the absence of adequate drainage, salts and minerals, which are naturally present in irrigation water, are left behind when the water evaporates. Over time, they may accumulate to levels that can impede or retard the growth of plants.

Desertification is a problem worldwide (Figure 6-6). According to the United Nations Convention to Combat Desertification (CCD), as of 2004, one-fifth of the world's population and one-third of the Earth's surface, over 15 million square miles (38 million square kilometers), are threatened by desertification. This includes over two-thirds of the African continent, over 30 percent of the land in the United States, 25 percent of the land in Latin America and the Caribbean, and 20 percent of the land in Spain. The CCD also reports that between 1990 and 2004, 14.8 million acres (6 million hectares) of productive land were lost every year due to land degradation, resulting in income losses worldwide of $42 billion per year.

In the western United States, desertification comes mainly from sustained, intense grazing by cattle and sheep, suppression of the natural fire cycle, and improper irrigation. Intense grazing is not necessarily harmful. Wandering herds of animals, such as the American bison before the European settlement of the American West, usually graze an area intensely and then move on, allowing the land to recover. Short, intense periods of grazing can be beneficial, but unrelenting pressure by cattle and sheep, as now occurs on drylands throughout much of the western United States, has contributed greatly to desertification.

In some nations, the landless poor are forced by political, economic, and ethnic factors to farm or graze herds on arid or semi-arid lands. During the past 50 years, 161 million acres (65 million hectares) in the southern fringes of the Sahara Desert alone have become nonproductive. For the desperate people trying to live in these marginal areas, famine and misery are the rule. Their plight was brought to the world's attention by the Sahel disaster in the 1970s and the Ethiopian disaster in the 1980s. In both instances, periods of better-than-average rainfall were followed by severe overcultivation and overgrazing on marginal lands. The overused lands were then subjected to normal, periodic but severe drought, and the rangelands became degraded.

Desertification can be halted through preventive measures: reducing livestock pressure on rangelands, planting hardy grasses and windbreak trees, encouraging the full use of good cropland, using agricultural techniques that minimize erosion, and providing adequate drainage for irrigation water. To be effective, preventive measures must be site-specific, matching local land, climate, economic, and cultural conditions. For example, bench terracing, an agricultural technique used on steep slopes to prevent water erosion, is very effective in areas where soils are moderately deep and

Figure 6-6: Cattle grazing in Kenya, 1996. Desertification threatens once productive semi-arid regions, and wildlife, livestock, and humans suffer.

labor is inexpensive. It has been used successfully for centuries in many parts of Asia.

What Is Deforestation?

Nearly one-half of the forests that once cloaked the Earth are gone, replaced by farms, pastures, and other human systems. **Deforestation** is the cutting down and clearing away of forests. According to the United Nations Food and Agricultural Organization, the world's total forest cover fell from 9.76 billion acres (3.95 billion hectares) in 1990 to 9.54 billion acres (3.86 billion hectares) in 2000.

Worldwide, the primary cause of deforestation is conversion to agricultural use. Other causes include conversion to pastureland; demand for fuel, timber, and paper products; and construction of roadways (Figure 6-7). In areas where natural gas, oil, and electricity are not available or are too expensive to use, such as in rural Haiti and much of Africa and Asia, women and children spend most of the day searching for fuel wood. Because wood and charcoal for cooking and heating are

becoming scarce and expensive, people cut or gather any available trees, branches, and fallen leaves.

All forests, from the tropical rain forest to the temperate woodland to the boreal coniferous forest, perform many beneficial functions. Healthy, intact forests ensure a steady supply of clean air and water. Often called the "lungs of the planet," forests help to regulate the carbon, nitrogen, and oxygen cycles. They help to regulate temperature and rainfall. The root systems of trees hold soil in place, thus preventing **siltation**, the process by which soil erodes from land surfaces and accumulates in streams and lakes. Siltation impedes water flow; moreover, the increased amount of sediments suspended in the water prevents light penetration to phototrophs, reducing primary productivity. The sediments also may settle on fish eggs, preventing them from obtaining sufficient oxygen. Forests also prevent flooding. Deforested hillsides are vulnerable to the erosive action of rains and spring thaws. Rain and melting snow rushing down hillsides can erode the soil and cause flashfloods in the valleys below. Moreover, forest habitats are vital to the continued existence of many species. Because they produce biomass quickly,

FIGURE 6-7: Surui Indian children watch a logging road being cut through their reservation, the Amazon.

forests are huge storehouses of food and shelter for a great diversity of organisms. In addition, forests produce humus, thus building up the soil and helping to keep the ecosystem productive.

Many forests are cathedral-like, serene, majestic, and beautiful; some are **frontier forests** — expansive tracts of contiguous forest largely untouched by human activities. Because of their vast size and undisturbed character, frontier forests function as intact ecosystems, providing several ecological services that make life on the planet possible. For example, they help regulate the Earth's climate by taking up and storing tremendous amounts of carbon dioxide. In addition, they provide a safe haven for species such as large predators that require a sizable range.

Like outright deforestation, fragmentation of frontier forests is a growing problem. As chain saws and bulldozers cut deep into frontier forests, the result is **fragmentation** — a patchwork of cropland, logging roads, and smaller, discrete forest areas. Fragmented forests do not appear to function as intact ecosystems do, and often,

they are not large enough to accommodate wide-ranging predators, such as big cats or bears. Other species also suffer, particularly those on the fringes of the forest patches. As heat rises from the adjacent cleared land, trees maladjusted to the temperature change may lose leaves and die prematurely. Moreover, fragmentation can hinder species interactions and therefore threaten the biodiversity that frontier forests protect. In the tropical rain forest, every species of fig is pollinated by a particular species of wasp; if the pollinating insect disappears from a fragment, the dependent fig can become locally extinct. Like tropical rain forests, temperate and boreal forests are being degraded by fragmentation. In 1998, the World Resources Institute reported that 76 countries have lost all of their frontier forest, and less than five percent remains in another 11 nations. Of the remaining frontier forests worldwide, 48 percent are located in boreal zones; 44 percent are in tropical zones; and only a tiny fraction is in the temperate zone. Nearly 40 percent of remaining frontier forests are currently threatened. In the last decade, fragmentation and deforestation have been particularly controversial issues in the rain forests of tropical and coastal temperate regions. Let's look at these areas more closely.

Tropical Rain Forests

On average, tropical forest areas decreased 33.4 million acres (13.5 million hectares) per year between 1990 and 2000 (Figure 6-8). As a result, during the 1990s, the world lost about 4 percent of its tropical forests. Deforestation on this scale could reduce tropical forests by half by 2050. Because tropical forests help to moderate the global climate, it is uncertain to what degree weather patterns and overall temperatures will be affected if we continue to destroy them.

By 2000, tropical forests covered an estimated 3.6 billion acres (1.47 billion hectares). Nearly half the remaining forests are in South America, 28 percent are in Africa, and 20 percent are in Asia. Brazil and Indonesia alone harbor 35 percent of the world's remaining tropical forests. The regions of the vast rain forests of the Amazon, Congo and Southeast Asia, which lie between the Tropic of Cancer and the Tropic of Capricorn, encompass some of the least industrialized countries with some of the highest birthrates. Over two billion people live in these regions, in countries with rapidly growing populations. The need to clothe, feed, house, and provide fuel for those populations contributes to massive deforestation, as does the economic need to export wood products and beef to developed countries.

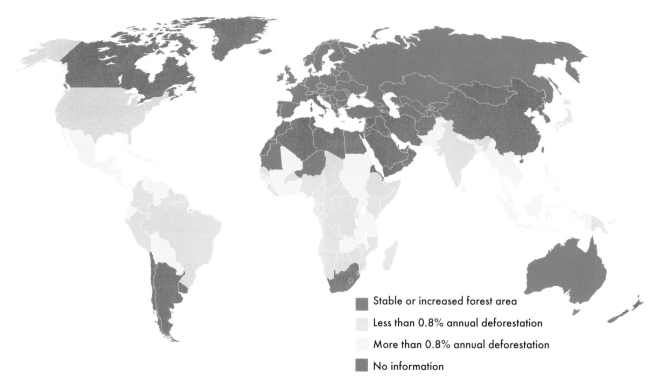

FIGURE 6-8: Estimated annual forest change rates, 1990-2000.

- Stable or increased forest area
- Less than 0.8% annual deforestation
- More than 0.8% annual deforestation
- No information

Tropical forests contain from 3 to 30 million species of plants, animals, and microorganisms — few of which have been discovered, named, or studied. Over 200 different tree species can be found in a single hectare of the Amazon rain forest. Compare this to the 10 or so usually found in a temperate forest. That same hectare of tropical forest is home to an immense array of insects, spiders, birds, and mammals. Half of all known species (and probably a much higher proportion of all species, known and unknown) live in tropical rain forests, yet these forests account for just seven percent of the Earth's land mass. Over half of the 650 bird species that breed in North America migrate to and from the tropical forests of Central and South America. For many of them the time spent in the north is minimal, yet we claim them as ours. What will be the result when migrating birds no longer find conditions suitable in the areas where they spend most of their adult lives? Declines in "our" bird populations have been linked to habitat destruction in tropical forests, but habitat loss in the United States and Canada likely plays a significant role in species decline as well.

Ironically, although conversion to cropland and pastureland is the major cause of tropical deforestation, many of the soils do not respond to traditional agriculture and cattle ranching. Most of the soils are shallow because 75 percent of the nutrients are tied up in the lush foliage, above ground. Leaves or trees that fall are quickly converted to nutrients, which are then recycled back to the foliage. Once an area is deforested, the exposed soil may be productive for a few years and then lose its fertility. Also, when the tree canopy is removed to create a field, the heavy tropical rains tend to wash away the soil, even in areas where it is relatively thick. Much of the rain that falls in tropical rain forests is generated by evaporation and by transpiration from the foliage itself; if large tracts of forest are removed, rainfall in the area may decrease markedly. In an intact forest ecosystem, about 50 percent of the precipitation that falls either evaporates from, or is taken up and then transpired by, the vegetation. Twenty-five percent of the precipitation is absorbed by the soil, and another 25 percent runs off the land into nearby surface waters. When the land is deforested, 75 percent of the precipitation ends up as runoff, carrying sediments from the denuded land into streams, rivers and oceans, while most of the remaining 25 percent evaporates from the soil. Because the deforested land returns less water vapor to the atmosphere, rainfall is generated less frequently and in smaller amounts. Often, the clearing of large tracts is preceded by the development of logging roads, which penetrate deep into the forest, allowing easier access to the lands. More trees are then cleared to accommodate farming and cattle ranching.

Fragmentation of tropical forests may be as problematic as deforestation; besides threatening species that live in the degraded forest edges, fragmentation disrupts the ecological services provided by intact forests. Based upon a 30-year study conducted by National Institute for Research in the Amazon, scientists concluded that frontier forests in tropical regions moderate levels of global atmospheric carbon dioxide through increased growth rates. The study's senior research scientist, Dr. William F. Laurance, and his associates monitored more than 100,000 trees in unfragmented forests and found that the total mass of living trees increased by an average of 37,000 pounds (17 metric tons) per acre. This striking increase in biomass may indicate a direct response to the larger quantities of carbon dioxide being released into the atmosphere from human activities, primarily fossil fuel combustion and deforestation. Because trees use carbon dioxide to synthesize their own food, an increase in carbon dioxide likely means more and faster tree growth *in unfragmented forests.* In a March 1999 *Natural History* article, Laurance points out that only large tracts of forests — from 2,000 to 15,000 acres (4,900 to 37,100 hectares) — can absorb the extra carbon dioxide. He estimates that frontier forests in the Amazon basin alone could be absorbing over one billion tons (910 million metric tons) of carbon dioxide each year — a strong indication that without the assistance of unfragmented forests, global climate change might be even more dramatic than it is now.

The causes of tropical deforestation and fragmentation — and the strategies necessary to eliminate those causes — are complex and could easily fill a book. Political, social, and economic factors all come into play. Many countries, for example, rely on their exports of beef and timber to raise cash to pay for their national debt. But debt burdens are often due largely to oil imports and military spending. Short-term economic gains, which benefit only a wealthy few, and disproportionate consumption of resources by developed countries have also been implicated.

Coastal Temperate Rain Forests

While tropical rain forests have garnered the lion's share of attention in recent years, another type of rain forest is also under increasing pressure. Coastal temperate rain forests share three common characteristics: proximity to the ocean, the presence of mountains, and high rainfall throughout the year (the latter a result of atmospheric interaction between the first two). In coastal temperate rain forests, terrestrial, freshwater, estuarine, and marine ecosystems interact in complex ways, particularly in the cycling of water. Abundant precipitation falls throughout the year, much of it in the form of fog, drizzle, or light rain. Coupled with little variation in temperature, the abundant precipitation results in fertile growing conditions. The combination of heavy rainfall and low temperatures produces high runoff rates, which lead to rapid rock wearing and soil formation, but which also can cause landslides and other erosion.

Coastal temperate rain forests produce more biomass per unit of area than any other terrestrial ecosystem. As you might expect, they harbor some of the largest and oldest trees in the world. One ancient Sitka spruce in the Carmanah Valley on Vancouver Island, British Columbia, is over 300 feet tall and 10 feet thick at its base! The diversity of lichens, mosses, and liverworts in coastal temperate rain forests may compare with that found in tropical rain forests. These majestic temperate forests are home to unique communities of plants and animals, and they sustain some of the world's most productive shellfish beds and spawning grounds for commercially valuable fish species.

Relative to other forests, coastal temperate rain forests have always been rare. While temperate deciduous forests currently encompass about 5 billion acres (2 billion hectares), coastal rain forests, even at their greatest extent, cloaked just 75 to 100 million acres (30-40 million hectares), an area equivalent to the size of Germany. They were found on the western margins of North America, New Zealand, Tasmania, Chile, Argentina, the Black Sea coast of Turkey and Georgia, Norway, Scotland, Ireland, and Iceland. The world's largest contiguous coastal temperate rain forest is in North America; it stretches over 1,800 miles (3,000 kilometers) from the Alaska Peninsula south through British Columbia and the State of Washington to Oregon's Siuslaw River. This frontier forest contains great old-growth trees — very old, very large trees, including western hemlock, Sitka spruce, Douglas fir, western red cedar, and coastal redwood. Old-growth trees are generally considered to be those over 200 years old; some may be as much as 500 to 1,000 years old!

Like tropical rain forests, coastal temperate rain forests are now fast disappearing; at least 55 percent have been logged or cleared for other uses. Currently, they comprise just 35 million acres (14 million hectares). The primary reason for the loss of these forests is global demand for wood products, especially the prized timber of the alerce and monkey puzzle (South America) and Sitka spruce and Douglas fir (North America). Since 1950, over half of the most productive rain forest in Alaska's Tongass National Forest has been clear-cut. Only about 10 percent of the original forest in Washington and Oregon remains (Figure 6-9). In Chile, logging by Japanese and other mulitnational companies threatens that country's remaining rain forests.

In the temperate zones, the fate of the coastal rain forests has become a matter of great concern. Nowhere is the debate more heated than in Canada, where, according to the World Resources Institute, nearly 500,000 acres (200,000 hectares) of forest are destroyed each year, about 2.5 acres (one hectare) every two minutes. On Vancouver Island, two-thirds of the original ancient groves have been logged; the remainder is fragmented or found on steep slopes, where road building and logging cause erosion and landslides and damage fish-bearing streams. Vancouver's Clayoquot Sound, a spectacularly diverse area that is home to abundant fish, seabirds, marine mammals, migratory waterfowl, shorebirds, bears, cougars and Orca whales, is also the site of the largest civil disobedience action in Canadian history. Since the early 1990s, hundreds of protesters have been arrested trying to halt logging in the area.

The fate of Vancouver's remaining old-growth, like that of ancient temperate rain forests everywhere, will

FIGURE 6-9: A clear-cut in a coastal temperate rain forest, Coos Bay, Oregon.

hinge on our ability to find ecologically sound solutions within the context of our political and economic systems. In other words, whether we preserve or log these forests, or find some acceptable combination of the two, is ultimately a question of values.

Summary

Ecosystems become degraded when human activities alter environmental conditions in such a way that they exceed the range of tolerances for one or more organisms in the biotic community. A degraded ecosystem loses some capacity to support the diversity of life forms that are best suited to its particular physical environment. The five Ds of ecosystem degradation are damage, disruption, destruction, desertification, and deforestation.

Damage occurs when the integrity of natural systems is altered. Pollution is the most common form of environmental damage. A pollutant is a substance or form of energy, such as heat, that adversely alters the physical, chemical, or biological quality of natural systems or that accumulates in living organisms in amounts that threaten their health or survival.

Four factors determine the damage done by pollutants: the effect of the pollutant, how it enters the environment, the quantity discharged to the environment, and its persistence. Acute effects occur immediately upon or shortly after the introduction of the pollutant to the system, and they are readily detected. Chronic, or long-term, effects may not be noticed for several years or decades after the introduction of the pollutant, rendering it particularly difficult to predict what effect the pollutant may have on the ecosystem. The interaction of two or more pollutants may have a synergistic effect, that is, their combined effect is greater than the sum of their individual effects. A pollutant that bioaccumulates becomes stored in an organism in concentrations higher than those normally found in the environment. A pollutant that biomagnifies appears in increasingly higher concentrations in organisms at successive trophic levels. Biomagnification occurs as consumers at higher trophic levels ingest a significant number of lower-level individuals that have accumulated persistent pollutants; the ingested pollutants are added to the persistent pollutants that have previously accumulated in the fatty tissues of the higher-level consumers.

Pollutants enter the environment through point or nonpoint sources. Cross-media pollutants, such as acid precipitation, move from one medium to another. Some otherwise harmless substances may be discharged to the environment in such large amounts that natural systems are unable to process, or biodegrade, them. Other pollutants may be nonbiodegradable, that is, they cannot be broken down by natural systems, no matter how small the amount. Persistent pollutants accumulate in the environment over time; they include nonbiodegradable pollutants as well as those that only degrade slowly.

A rapid change in the species composition of a community that can be traced directly to a specific human activity is known as a disruption. The use of persistent biocides, the introduction of new species into an area, construction, and overexploitation of resources have all been linked to changes in species composition.

Ecosystem destruction is the conversion of a natural system to a less complex human system. The major causes of destruction are agriculture, transportation, and urbanization. The most serious consequences of natural system destruction are habitat loss (the primary cause of species extinction) and the loss of the functions and services provided by natural systems, functions that human systems cannot duplicate.

Discussion Questions

1. Differentiate among the five Ds of environmental degradation and give an example of each.

2. What factors determine how a pollutant damages an ecosystem? How can soil erosion entering a stream, nutrient loading to a lake, and the discharge of a toxic synthetic chemical to the atmosphere all be considered pollutants?

3. Scan each issue of your local newspaper for one week. Identify several cases of pollution or potential pollution in your area. Relate each case to the factors that determine how pollutants affect ecosystems.

4. Differentiate between bioaccumulation and biomagnification. How is the concept of trophic levels related to biomagnification?

5. Identify three nonnative animal or plant species that are commonly found in your area. Research how their introduction altered the native community.

6. Suggest ways in which habitat loss might be reduced or minimized. Include suggestions that are appropriate for your own area and suggestions that are appropriate for such species-rich areas as the tropical rain forests and coral reefs.

(continued on following page)

(continued from previous page)

7. Some people argue that the decline or extinction of a single species does not degrade an ecosystem. What do you think? How might the biotic community of the Columbia River be altered if the chinook, coho, pink, chum, or sockeye salmon becomes extinct?

Desertification is the degradation of drylands (arid, semi-arid, and dry subhumid regions) resulting mainly from adverse human impact, especially overgrazing, overcultivation, deforestation, and poor irrigation practices. Rangelands and agricultural lands are most susceptible to desertification.

Deforestation is the cutting down and clearing away of forests; worldwide, it is caused primarily by conversion to agricultural land. Fragmentation results in further ecosystem degradation by segmenting an intact frontier forest into smaller, less functional patches. The demand for pastureland, fuel, timber and paper products, as well as the construction of roadways, also leads to deforestation and fragmentation.

KEY TERMS

acute pollution effects	desertification	persistent pollutant
anthropogenic	ecosystem damage	phototoxicity
bioaccumulation	ecosystem destruction	point-source pollution
biodegradable	ecosystem disruption	pollutant
biomagnification	fragmentation	salinization
chronic pollution effects	frontier forests	siltation
cross-media pollutant	nonbiodegradable	synergistic effect
deforestation	nonpoint-source pollution	

UNIT III

An Environmental Imperative
Balancing Population,
Food, and Energy

CHAPTER 7

Human Population Dynamics

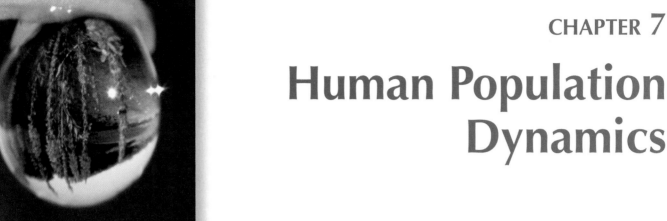

Learning Objectives

When you finish reading this chapter, you will be able to:

1. Explain the basic arguments of each side in the Great Population Debate.

2. Define demography and discuss why it is important.

3. Explain the three ways in which populations are measured.

4. Identify and discuss the factors that affect growth rates.

5. Discuss the relationship between population growth and the environment.

6. Discuss the relationship between population growth and quality of life.

7. Discuss the relationship between population growth and the status of women.

We are not faced with a single global population problem but, rather, with about 180 separate national population problems. All population controls must be applied locally; local governments are the agents best prepared to choose local means. Means must fit local traditions. For one nation to attempt to impose its ethical principles on another is to violate national sovereignty and endanger international peace. The only legitimate demand that nations can make on one another is this: "Don't try to solve your population problem by exporting your excess people to us."

Garrett Hardin

Why…have so few ecologists seriously applied their knowledge, skills, and talents to the question of the limits to human population?…The phenomena of human population growth and its impacts are all too apparent; is the ecological community willing to ignore the most pressing social and scientific issue of all time?

H. Ronald Pulliam and Nick M. Haddad

The topic of human population growth usually prompts a heated debate. Many people are alarmed about rapid population growth in critical regions of the world, particularly the tropics, and the sheer numbers in the world's most highly populated countries, China and India. Others point out that people in stable or slow-growing developed countries typically consume far more resources and generate far more pollution that those in less-developed countries; they maintain that many environmental problems would be nonexistent if resources were more equitably distributed (Figure 7-1). And the debate that always ensues in these discussions is the same debate that rages among ecologists, economists, philosophers, and politicians — a debate that has been raging for centuries. (See Box 7-1: *The Great Population Debate.*) In simple terms, the topic of the debate is, "Is human population growth a problem?" Chapters 7 and 8 are designed to help you decide where you stand in the Great Population Debate. In Chapter 7, we describe the human resource in detail; in Chapter 8, we look at how populations have been — and are being — managed.

FIGURE 7-1: Uneven distribution of space among the world's people. a) Traffic in Dhaka, Bangladesh. More than a million rural Bangladeshis have moved to the city in search of jobs. b) Children pause to enjoy an open vista in North Dakota.

BOX 7–1: **The Great Population Debate**

Whether in a classroom, a scholarly journal or the United Nations, passions run high when people ask the question, "Is human population growth a problem?" The answer depends largely on beliefs and values. The beliefs and values of the different groups depicted below dictate their positions in the Great Population Debate. Groups with different ideologies may take the same position on the population issue, but for different reasons.

Where do you stand in the Great Population Debate?

"There Is No Population Problem"

Cornucopians
Some people argue that people are the world's ultimate resource, and the more of them, the better. They adopt a cornucopian view of population which sees continued growth as a positive factor that can recreate the mythical "horn of plenty." Cornucopians contend that a growing population can actually help a country to become more prosperous, since additional people mean both more consumers and more producers. Moreover, they steadfastly maintain that there is no such thing as a "population problem," because human inventiveness will solve any problems — environmental, economic, social, or otherwise — that may beset continued population growth. Cornucopians believe that a free market society should be the driving force to accomplish population stability. In general, this is the

"unofficial" policy of the United States. Cornucopians emphasize that education is critical to understanding the relationship between free enterprise, development, and population growth.

The economists Julian Simon and the late Herman Kahn are probably the most well-known proponents of this view. In a 1981 interview in *Forbes* Magazine, Simon, author of *The Ultimate Resource,* asserted, "I have no reason to think that population growth is not a good thing. ... Population growth is a moral and material triumph. That's the flip side of the thing — what an enormous success for the human species. If the flies took over the world, the biologists would say, 'What a successful species!' But we get more and more people, and they say, 'What a failing species!' What a strange twist!"

Marxists
Marxists argue that poverty is the result of the unequal distribution of resources rather than unchecked population growth. They see the problem as one of power — the haves versus the have-nots. Consequently, Marxists do not acknowledge a population problem. Instead, they maintain that efforts should focus on distributing resources to all peoples more evenly. In China, for example, the stated goal of family planning efforts is to produce healthier, happier children; one way to do this is to limit the number of children born so that resources can be shared among a smaller population. Such a situation is good for the children, good for families, and good for China.

"There Is a Population Problem"

Malthusians
Malthusians are proponents of the beliefs credited to Thomas Malthus, an eighteenth-century parson, who believed that humans tend to overreproduce out of an innate drive to procreate. Because population increases geometrically, but food increases only arithmetically, he reasoned, population growth is greater than the Earth's power to sustain it. War, famine, and pestilence act as negative feedbacks that help limit population growth. According to Malthus, these factors affect the poor in greater proportion than the rich, since the wealthy can avoid service in armies and can afford medical attention and an adequate supply of food.

Neomalthusians
Malthus was convinced that food resources are the limiting factor on the size of the human population. Some contemporary thinkers argue that although food may not be as important as Malthus believed, his ideas about the link between poverty and family size are correct. They claim that the "Malthusian checks" (war, famine, pestilence) are more common in poorer nations. Those who espouse this adapted Malthusian view are known as neomalthusians.

Neomalthusians concede that cornucopians have fashioned a powerful image — that of each child as a potential gift to humanity. But, they counter, how can we expect to find answers to the many problems facing the human population among the starving masses in the developing world? Can those who have little food and water, no home, and a bleak future fulfill their human potential? Neomalthusians see continued population growth as a threat to environmental quality, political stability, international relations, and economic development. They contend that even if the rate of population growth could be decreased immediately, it would take many years to slow and halt the total increase in the absolute numbers of human beings on Earth. In the meantime, the human population would continue to stress and damage the Earth's life support systems, with dire consequences for the biosphere and its human and nonhuman inhabitants.

Advocates of Zero Population Growth

Some neomalthusians translate their beliefs about the consequences of population growth into recommendations for public policy. Because they believe that poverty is the result of high fertility, they recommend that birth control and mandatory family planning programs be introduced to achieve zero population growth (ZPG) as soon as possible. They predict that, unless zero population growth is reached, natural checks on population will cause widespread misery and death.

ZPG, Inc., is a group started by Drs. Paul and Anne Ehrlich, authors of *The Population Bomb* and *The Population Explosion*. The Ehrlichs' goal is to convince couples to have no more than two children through an extensive educational program to help people understand the relationships among population growth, poverty, environmental degradation, and standard of living. In an excerpt from *The Population Explosion*, the Ehrlichs contend that "Even some environmentalists are taken in by the frequent assertion that 'There is no population problem, only a problem of distribution.'...Unfortunately, an important truth, that maldistribution is a cause of hunger now, has been used as a way to avoid a more important truth — that overpopulation is critical today and may well make the distribution question moot tomorrow."

"The 'Population Problem' Is a Complex Issue"

Many people contend that most of the problems associated with human population growth (hunger, landlessness, poverty, pollution) involve distribution of resources and resource consumption patterns, not simply population numbers and growth rate. People in the United States, Canada, Japan, and Europe consume far more resources than do people in less-developed countries. In "The Grim Payback of Greed," Alan Durning, senior researcher at the Worldwatch Institute in Washington, D.C., writes "Overconsumption by the world's fortunate is an environmental problem unmatched in severity by anything but perhaps population growth."

Describing the Human Resource

In this chapter, we look closely at the human population. First, we examine the discipline of demography and why we study it. We discover the different ways in which populations are measured and the factors that affect one of those measures, the growth rate. We then look closely at the relationships between population growth and the environment, quality of life, and women's status.

What Is Demography?

All of us are born, age, and die. Most of us marry and have children and grandchildren. Many of us live in more than one location during our lifetimes; some of us migrate from one country to another. These are personal events; indeed, decisions about bearing children and raising a family are among the most intensely personal ones a human can make. But personal decisions are important on a societal level as well, for the sum of thousands and millions of individual acts has a tremendous impact on societies and the environment.

Demography is the scientific study of the sum of our individual acts as they affect measurements of the population. Demographers study population statistics, or **vital statistics**, such as births (natality), deaths (mortality), and migration. Population can be studied at many levels: all the individuals on Earth; the distinct populations occupying continents or regions of the world (Europe or North America, the Northern or Southern hemisphere); the populations of distinct political nations (Germany or Mexico); the populations of regions, states, or republics within countries or continents (Bavaria, Germany or Sonora, Mexico); the residents of cities (Munich or Mexico City); or the populations of smaller units, such as towns, counties, and townships.

The questions posed by demographers determine the level of population that they study. Those questions vary widely: How long do the people of a particular society live? How many children do parents want to have? How many people are younger than 15 and older than 65? How often do people move? When they move, where do they choose to go? What factors (economic, social, political) drive that choice? (One important question demographers do not ask is: How many people can a specific environment support?)

Demographers also try to understand the causes and consequences of changes in populations. Their research helps us to make social, economic, political, and resource use decisions for individuals, municipalities, states or provinces, and countries. For example, demographic information can help us to plan for health care facilities, schools, and transportation. It also can improve international relief efforts, such as extending aid to other countries. Finally, demographics enables us to understand the environmental and cultural consequences of our decisions and can help us implement the decisions necessary for the quality of life we want for our children and our children's children.

To understand the significance of demographic statistics, we must have some understanding of the culture they describe, especially its dominant beliefs and value systems. For example, in cultures where parents show a strong preference for boys over girls, parents may continue to have children in order to have the boy(s) they want, they may decide to abort a female fetus (when the sex of the child is known), or they may resort to infanticide after the birth of a baby girl.

Throughout this chapter, we must not lose sight of the fact that impersonal numbers, statistics, and percentages can obscure the lives and aspirations of the people they represent. Relating population statistics to our personal decisions about birth, death, marriage, and migration helps us realize that each statistic represents a person whose life, goals, and dreams are just as real as our own. It helps us to remember that the real population story is not numbers; it is people.

How Are Populations Measured?

A population can be measured in three ways: the number of people currently in the population, the rate at which the population is growing (or decreasing), and the time it will take for the population to double in size. Each of these measures provides useful information to demographers, government officials, and others interested in current and future demographic conditions in a country, a region, or the world.

Numbers of People

In 2004, the world population rose to nearly 6.4 billion. Over 5.1 billion people live in the developing world and over 1.2 billion in the developed world. The most populous countries are China with over 1.3 billion; India, 1.1 billion; and the United States, 293.6 million. Table 7-1 presents demographic data for selected nations.

Demographers calculate the change in a population over a period of time (usually one year, from mid-July to mid-July). World population growth, expressed as an absolute number, is determined by subtracting the total number of deaths worldwide from the total number of live births. In any given year, as long as live births outnumber deaths, world population will grow in absolute numbers.

For a country or region, population growth (or decline) is affected by **migration**, movement from one country or region to another for the purpose of establishing a new residence. Migration into a country or region is called **immigration**; migration out of a country or region is **emigration**. (In countries with substantial illegal immigration, accurate immigration data may not be available.) Assuming that the necessary data are available, the actual increase (or decrease) in absolute numbers for a specific country is determined by the formula: (births + immigration) – (deaths + emigration).

Growth Rates

Demographers also are interested in the rate at which a population is increasing or decreasing. This informa-

TABLE 7–1: Basic Demographic Data, 2004, Selected Countries

	Population Estimate, Mid-2004 (millions)	Crude Birth Rate	Crude Death Rate	Annual Natural Increase (%)	Population Projection, 2025 (millions)	Total Fertility Rate
World	6,926.0	21	9	1.3	7,934.0	2.8
More developed countries	1,206.0	11	10	0.1	1,257.0	1.6
Less developed countries	5,190.0	24	8	1.5	6,677.0	3.1
Less developed countries (excluding China)	3,890.0	27	9	1.8	5,201.0	3.5
Afghanistan	28.5	48	21	2.7	50.3	6.8
Australia	20.1	13	7	0.6	24.2	1.7
Bangladesh	141.3	30	9	2.1	204.5	3.3
Bolivia	8.8	28	9	1.9	12.2	3.8
Botswana	1.7	27	26	.1	1.1	3.5
Canada	31.9	11	7	0.3	36.0	1.5
China	1,300.1	12	6	.6	1,476.0	1.7
Congo, Dem. Rep. of	58.3	46	15	3.1	104.9	6.8
Estonia	1.3	10	13	-0.4	1.2	1.4
Germany	82.6	9	10	-0.2	82.0	1.3
Guatemala	12.7	34	7	2.8	19.8	4.4
Haiti	8.1	33	14	1.9	11.7	4.7
Hungary	10.1	9	13	–0.4	8.9	1.3
India	1,086.6	25	8	1.7	1,363.0	3.1
Iraq	25.9	36	9	2.7	41.7	5.0
Italy	57.8	10	10	-0.1	57.6	1.3
Japan	127.6	9	8	0.1	121.1	1.3
Kenya	32.4	38	15	2.3	39.9	5.0
Libya	5.6	28	4	2.4	8.3	3.6
Mexico	106.2	25	5	2.1	131.7	2.8
Niger	12.4	55	20	3.5	25.7	8.0
Russia	144.1	10	17	-0.6	136.9	1.4
Sweden	9.0	11	10	0.1	9.9	1.7
Thailand	63.8	14	7	0.8	70.2	1.7
United States	293.6	14	8	0.6	349.4	2.0

Note: *Figures for growth rate are sometimes rounded and thus do not exactly equal (birth rate – death rate).*
Source: *Population Reference Bureau, 2004 World Population Data Sheet.*

tion enables them to predict future population size, and that prediction allows others to assess the population's effect on the economic, political, and social structure of the society.

To determine how fast a population is growing, demographers must have information about birth and death rates. The **crude birth rate** is the number of live births per 1,000 people. It is calculated as follows: (number of live births) ÷ (total population at mid-year) x 1,000. The **crude death rate**, the number of deaths per 1,000 people, is: (number of deaths) ÷ (total population at mid-year) x 1,000. Thus, the **natural rate of increase** or **growth rate** is the surplus (or deficit) of births over deaths in a population in a given time period. It is determined as follows: (crude birth rate) – (crude death rate), with the difference expressed as a percent (that is, as a fraction of 100).

It's important to note that crude birth and death rates are rounded, so that the growth rate is an approximation. In 2004, the crude birth rate was 21/1,000 and the crude death rate was 9/1,000, but the growth rate was an estimated 13/1,000 (rather than 12/1,000) to account for rounding. When expressed in decimal form, 13/1,000 is 0.013. To express this figure as a percent, we multiply by 100: 0.013 x 100 = 1.3 percent. Thus, in 2004, world population grew at an approximate rate of 1.3 percent. Bear in mind that, for a country or region, the natural rate of increase does not account for the effects of migration, which may offset the natural increase (or decrease).

Growing at a rate of 1.3 percent, the global population adds nearly 80 million people annually, roughly equal to the population of Germany. Most demographers assume that the growth rate will continue to decline slowly, as it has since reaching a historical high of 2.2 percent in 1964.

Some people point to the decrease in the growth rate as evidence that there is no population problem, but that argument is misleading. Some regions and nations are experiencing rapid growth; the population of Africa — the world's fastest growing continent — is increasing at a rate of 2.4 percent. On a regional basis, growth rates are highest in middle and western Africa, both at 2.8 percent; eastern Africa, 2.3 percent; Central America, 2.1 percent; and northern Africa and western Asia, both at 2.0 percent (Figure 7-2). Moreover, while the decrease in the world growth rate is encouraging, we must not lose sight of the fact that a positive rate of any amount translates into an annual increase in absolute numbers. Thus, although the growth rate has slowed in the past several decades, it is still positive, and the population, in absolute numbers, continues to increase rapidly. While some less-developed countries (LDCs) have made considerable progress in lowering their growth rate — Kenya's growth rate, for instance, was almost 4.0 percent in the early 1980s but has since fallen to 2.3 percent — they are still growing rapidly.

The confusion over the meaning of falling growth rates versus increasing absolute numbers is complicated by the fact that demographers use rates per thousand for

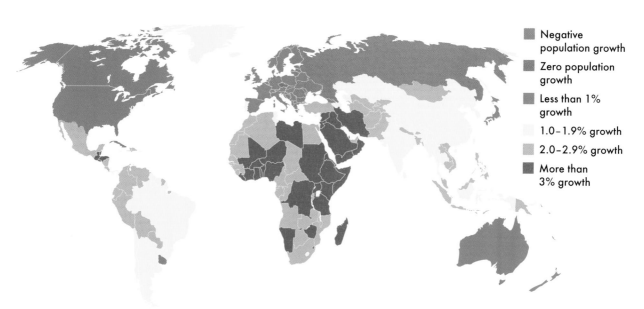

FIGURE 7-2: World population growth rates.

birth and death rates and rates per hundred (percent) for growth. The important thing to remember is that as long as births outnumber deaths, the growth rate is positive, and absolute numbers increase. When births are equal to deaths, the growth rate is zero, and absolute numbers remain the same. This is the condition known as **zero population growth**, or ZPG. (ZPG also assumes zero net migration, that is, immigrants equal emigrants.) In 2004, a number of European countries, including Greece and Poland, had growth rates of 0.0 percent and thus stable populations. In many other European countries, including every eastern European nation, deaths outnumbered births, a condition that causes **negative population growth**, or NPG, and thus a decline in absolute numbers. Europe as a whole had a negative growth rate, -0.2 percent, in 2004.

Doubling Times

A third measure of population is known as the **doubling time**, the number of years it will take a population to double, assuming that the current growth rate remains constant. Doubling times are calculated by the **rule of 70**: 70 divided by the current annual growth rate yields the time it will take to double the population, if the growth rate remains constant (Figure 7-3). The faster the growth rate, the more quickly a population will double in size. Most LDCs, where growth rates are high, have shorter doubling times than do more-developed countries (MDCs). At 1.3 percent growth

rate, the world population would double in 55 years (70/1.3), reaching about 12.8 billion in 2059.

There are two things to keep in mind concerning doubling times. First, the effect on resources and the environment depends on the absolute size of the initial population. Consider the United States and Australia. In 2004, both countries had a 0.6 percent rate of natural increase, but the U.S. population of about 294 million was nearly 15 times larger than that of Australia's. Under current conditions, if the U.S. population were to double, it would reach 588 million in 2120. Australia's population of about 20 million, meanwhile, would double to just 40 million. Second, the calculation of doubling times assumes that current growth rates will remain constant in the future. Usually they do not, and so doubling times may not accurately predict the real growth of populations. Nevertheless, the doubling time is a valuable measure because it allows us to visualize what future environmental and social conditions would be like if present growth rates were to continue. For example, with 3.5 percent annual growth (among the world's highest), the population of the West African nation of Niger, currently at 12.4 million, would double in 20 years and quadruple in 40 years. In 80 years, it would be 16 times its current size, reaching nearly 200 million! Demographers, economists, and ecologists agree that this rate could not be sustained. The stress exerted by such unchecked growth would overwhelm the natural resource base and further destabilize the region's political, social, and economic structures.

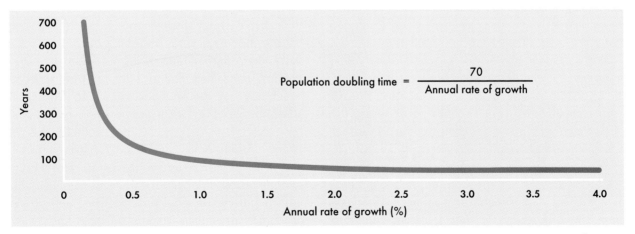

Figure 7-3: Doubling time and annual growth rate. A population's doubling time decreases dramatically as its growth rate rises. Even a small increase in growth rate results in a much shorter doubling time.

What Factors Affect Growth Rates?

Among the factors affecting growth rates are fertility, age distribution, and migration. The importance of any one factor in determining growth varies among countries. Fertility and age distribution are particularly important when considering future growth in the developing world, while migration is one reason why the United States is one of the fastest growing nations in the industrialized world.

Fertility

For many countries and regions of the world, crude birth and death rates are the only demographic data available. Crude rates are so called because they do not include information about age and sex. Consequently, they do not enable us to make accurate predictions about the future dynamics of the population. For example, men, children, and the aged are included in the calculation for crude birth rate. But only women bear babies, and women of certain ages are more likely to bear children than others. Thus, measures of **fertility**, the actual bearing of offspring, are more accurate indicators of the potential for future population growth. The most important of these are the general fertility rate, age-specific fertility rates, and the total fertility rate.

The **general fertility rate** is the number of live births per 1,000 women of childbearing age per year. Ages 15 to 49 are considered by many demographers (including the Population Reference Bureau or PRB, the nonprofit educational organization cited throughout this text) to be the childbearing years. A more helpful indicator of potential future growth is the **age-specific fertility rate**, the number of live births per 1,000 women of a specific age group per year, such as girls aged 15 to 19. The **total fertility rate** (TFR) is the average number of children a woman will bear during her life, based on the current age-specific fertility rate, assuming that current birth rates remain constant throughout the woman's lifetime. The TFR can be thought of as the average family size.

FIGURE 7-4: A young mother holds her child in Cochabama, Bolivia.

A TFR of 2.1 children per woman is generally considered to be the replacement level fertility for the more-developed countries (MDCs). The **replacement fertility** is the fertility rate needed to ensure that each set of parents is "replaced" by their offspring. A fertility rate of 2.1 births per woman enables each woman to replace herself and her mate and allows for some mortality, that is, children who fail to survive to adulthood and reproduce. In LDCs, the replacement fertility is 2.5, a reflection of the greater risk of mortality faced by children in those countries.

In 2004, the world TFR was 2.8, but the TFR for the developed world was just 1.6. As a whole, Europe had a TFR of only 1.4. Nearly every European nation had a TFR of 1.9 or lower; the only exceptions were Albania, 2.1; Iceland, 2.0; and Ireland, 2.0. It's little wonder, then, why Europe's population is declining! In the United States, the "baby boom" period (approximately 1946 to 1963) peaked in 1957 at a TFR of about 3.8 births per woman. By 1976, the TFR had dropped to 1.7. Since then, it has remained at or below the replacement level. In 2004, the TFR for the United States was 2.0.

The TFR for the developing world in 2004 was 3.1. Excluding China, which has pursued an aggressive population control program for a number of years (see Chapter 8), the TFR for the developing world was 3.5. On the continent of Africa, which had an overall TFR of 5.1, 17 nations had TFRs of 6.0 or higher, led by Niger, 8.0; Guinea-Bissau, 7.1; Somalia, 7.1; and Mali, 7.0.

Because statistics can paint a bleak picture for the future, it's important to realize that some regions and countries are making real progress in limiting growth. Throughout South America, birth rates have fallen over the past 30 years, a result of economic growth, urbanization, medical improvements, and expanding educational opportunities. In 1970, women in Bolivia had an average of 6.5 children; in 2004, that figure was 3.8 (Figure 7-4). Indonesia, the world's fourth most-populous nation with nearly 219 million people, has reduced its average family size to 2.6; in Thailand, it is 1.7; and in South Korea, 1.2. Even in Africa, there are hopeful signs. Through an aggressive program to control growth, Zimbabwe has lowered its TFR to 4.0, well below the average sub-Saharan rate of 5.6.

Seemingly small changes in fertility have a tremendous impact on growth rates and population sizes. The United Nations' projections of world population size in 2050 are based on three scenarios — low growth, medium growth, and high growth (Figure 7-5). These projections are 7.4, 8.9 and 10.6 billion, respectively, but the difference between the low and high projections is just one child per woman! If total fertility rates fall rapidly, until women average about 1.5 children each, the world population will rise to just 7.4 billion by 2050. On the other hand, if total fertility plateaus at 2.5, the population by mid-century will be 10.6 billion and growing. The UN's medium projection — a total world population of 8.9 billion people in 2050 — is based on the assumption that fertility will fall to about 2 children per woman.

Age Distribution

The **age distribution** of a population is the number of individuals of each sex and age (from birth through old age). It is represented by an age structure histogram, or **population profile**, which is constructed from census data. (A census is the periodic collection of demographic information by a government about its citizens.) Horizontal bars are used to depict the number of males and females of each age group. Typically, each age group is plotted at five-year intervals. Every five years, the bars are moved up one position. New bars are added to the bottom to represent new births and the uppermost bar is removed as the elderly die. Other upper bars are usually slightly reduced to reflect mortality related to aging.

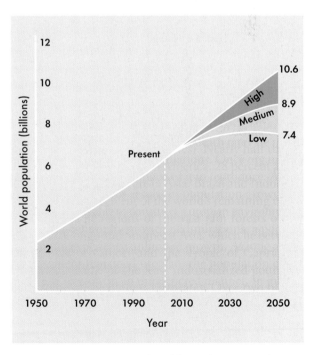

FIGURE 7-5: UN estimates of world population growth: Low, medium, and high projections.

In general, population profiles for LDCs have a pyramidal shape, a depiction of the rapid growth these nations are experiencing. In Nigeria and many other African nations, for example, about half of the population is under age 15; just two percent or so is age 65 or older (Figure 7-6). Throughout the developing world, high fertility, coupled with reduced mortality rates among infants and children, results in relatively high growth rates. As more people survive to their childbearing years, they produce an even larger next generation and a larger, younger population. In contrast, older age groups have far fewer members, a reflection of both higher mortality rates among infants and children in the past and the increased incidence of mortality that comes with aging.

The rapid growth of developing countries is largely a result of a decline in the death rate. Beginning after World War II (WWII), death rates fell throughout the developing world as a result of improved medical care and sanitation. Before long, birth rates far outpaced death rates in developing nations, a trend that continues in most LDCs today. Demographers generally attribute high fertility in the developing world to a complex mix of social factors, particularly a lack of access to sex education, family planning services, and safe and reliable means of birth control. Economic security — or rather, the lack thereof — also plays a

role. That's because, while death rates among infants and children in the developing world have fallen over the past 60 years, they remain high. Consequently, parents are compelled to have child after child because of the reality that many will die as infants or youngsters. In the absence of insurance, welfare programs and other measures to care for the aged, ill and injured, children become their parents' social security. In many parts of the world, sons are seen as particularly desirable because they are expected by custom to care for aged parents. In rural areas, children help their mothers with planting and harvest, childcare, and all household activities. In urban areas, they help support the family; children may work as peddlers (selling small trinkets and goods), scavengers (scouring dumps for cast-off items that can be used by the family or sold), or prostitutes. Bangkok, Thailand, for example, is notorious for its high number of child prostitutes.

The population profile for slow-growing MDCs like the United States and Canada is constrictive (Figure 7-6). A post-WWII baby boom means that the 43-to-60 age groups are larger than other age groups; the children of these "baby boomers" make for age cohorts that are large relative to the over-65 age groups, which comprise just 12 and 13 percent of the U.S. and Canadian populations, respectively. Population growth in these countries will stabilize and begin to decrease as

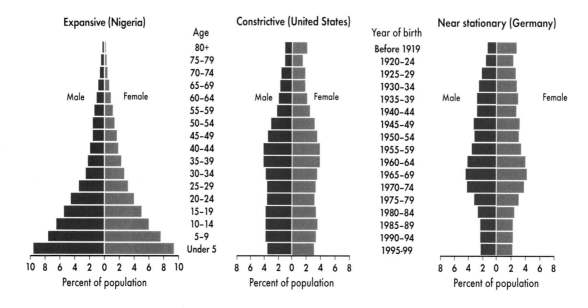

FIGURE 7-6: 2005 population profiles for Nigeria, the United States, and Germany. As the profiles show, the population of Nigeria (and LDCs in general) is expansive. Africa, with the highest growth rate of any continent or region, has the youngest age structure and the greatest potential for future growth. Growth in the United States is constrictive, or slow; in Germany and some other developed countries, it is stable or declining.

the baby boomers move into old age categories. This should happen in about 20 years as long as birth and death rates remain equal or current low fertility rates are maintained.

MDCs with stable or decreasing populations, such as Germany, have distinctive profiles (Figure 7-6). As a result of the nation's falling birth rate, the base of the population profile is narrower than the middle and upper portions. If the TFR rises and stabilizes at the replacement level TFR of 2.1, its age and sex structure gradually would assume a rectangular shape as the population becomes distributed evenly among the various age groups.

Age Distribution and Future Growth. Age distribution is an important indicator of future growth for individual nations, regions, and the world as a whole. For example, in 2004, 51 percent of Uganda's population was under 15 and just two percent was 65 or older. Uganda's young age distribution, coupled with a TFR of 6.9, means that there is a great potential for future growth because the majority of its population has not yet reached reproductive age. Even in LDCs with far lower fertility rates, a young age distribution ensures continued growth for many years. For example, China's population of 1.3 billion is expected to rise to nearly 1.5 billion in 2025, despite the fact that the country's TFR is 1.7, well below replacement level. Its large proportion of young people — 22 percent of Chinese are under age 15 — will fuel the country's continued growth.

Nowhere is the growth potential of a young age distribution more profound than in the developing world. Almost two billion people — *about one-third of the world's current total population* — are young people below the age of 15 who live in LDCs. They are currently, or soon will become, the world's newest group of parents. This "critical cohort," as it is called, ensures that there will be a long period of world population growth — even if birth and fertility rates fall dramatically — simply because more and more young people are reaching childbearing age each year. In fact, 98 percent of the people born in the coming decades will live in LDCs, where life is undoubtedly harshest. For a look at *Environmental Science in Action: Kenya and Zimbabwe*, which examines the causes and consequences of rapid population growth in developing countries, go to www.EnvironmentalEducationOhio. org and click on "Biosphere Project."

The developing world illustrates the phenomenon of **population momentum**, which occurs when there are large numbers of children living as fertility rates begin to drop; each successive generation that enters the childbearing years is larger than the preceding generation and bears more children overall, even if each set of new parents has fewer children than did parents of the previous generation. Because the number of children born is greater than the number of deaths among older generations, the population grows. Even when fertility drops below the replacement level, momentum can cause the population to grow for as long as a hundred years. During that time span, the population can undergo tremendous growth in absolute numbers.

Age Distribution and Dependency Load. Age distribution determines a nation's **dependency load**, the number of dependents — those under age 15 or over age 65 — in the population. People in these age groups are generally assumed to be too young or too old to be in the formal work force and to contribute to the country's gross national product. Worldwide, 30 percent of people are under 15 and seven percent are over 65. For MDCs, 17 percent are under 15 and 15 percent are over 65; not only are populations in MDCs growing slowly, they also are aging. In contrast, in LDCs, 33 percent of the population is under 15 and five percent is over 65. Populations in the developing world are growing quickly and, as a whole, becoming younger. Nowhere is this more obvious than in Africa, where 42 percent of the population is under 15 and just three percent is over 65.

Migration

The **actual rate of increase** for a country or region differs from the natural rate of increase because it takes into account migration. Immigration can have a significant effect on the rate of population growth. For example, immigration raises the United States's growth rate from 0.6 percent (the natural rate of increase) to about 1.0 percent. According to the Population Reference Bureau, during the 1980s and 1990s, Asians (a diverse group that includes Pacific Islanders) constituted the fastest growing minority group in the United States. This group more than doubled between 1980 and 1997, increasing from 3.8 million to 9.6 million; most of that increase was due to immigration. Asian-Americans now account for about 3.6 percent of the U.S. population. Assuming that they will exhibit medium levels of fertility, mortality and net immigration, Asian-Americans are expected to exceed 32 million by 2050, when they will account for eight percent of the country's population.

Latinos (individuals whose cultural heritage can be traced to a Spanish-speaking country in Latin America) are yet another fast-growing minority group. In 1900, there were approximately 500,000 Latinos in the United States; today, there are about 35 million. Most of that growth has occurred quite recently; between 1980 and 2000, the U.S. Latino population more than doubled, accounting for 40 percent of the growth in the total population. In 2003, the U.S. Census Bureau designated Latinos as the nation's largest minority group.

Migration also occurs within nations. In the United States, the South, West, and Southwest are experiencing significant growth, while the populations of some states in the Midwest and Northeast are declining. Cities such as Atlanta and Santa Fe are expanding as more and more people throng to the nation's "sun belt." Rising populations in these areas exert greater pressure on the environment and the local infrastructure (roads, water supply systems, sanitation systems, schools, hospitals, etc.).

How Are Population Growth and the Environment Related?

Before we answer this question, let's review a bit. In Chapter 1, *The State of the Biosphere,* we learned that all organisms live at the expense of their environment. In Chapter 5, *Ecosystem Development and Dynamic Equilibrium,* we learned that nonhuman populations are subject to numerous checks and balances. All natural populations have a maximum number that their environment can support. This upper limit, the maximum population size, is determined by the carrying capacity of the environment, that is, the number of a particular species that the environment can support year after year without degradation. Typically, populations do not exist for extended periods at the maximum population size. Rather, they fluctuate over time around the number of individuals the environment can best support, known as the optimum population size. Populations remain at or near the carrying capacity when a balance is struck between the biotic potential of the species and environmental resistance.

The concept of carrying capacity can be applied to humans *by analogy only* and with several considerations. First, as for nonhuman populations, there is no standard, universally agreed upon equation to calculate the carrying capacity for human populations.

Second, human populations are typically classified according to geopolitical units, and such cultural classifications are meaningless in ecological terms. Rivers, for example, often cross national borders. The amount of water available to people in one country (for drinking, sanitation, agriculture, and industrial use) depends on the actions (particularly diversions) of other nations upstream.

Third, unlike nonhuman populations, humans are able to raise the carrying capacity of an environment in several ways. Technologies exploit otherwise unavailable resources. Mining processes, for instance, have been so refined that ores containing smaller and smaller percentages of desired minerals can be used in production. Moreover, humans can bring in resources from outside their immediate environment to sustain themselves and their culture. The Middle East, for example, supplies about one-half of the oil needed to meet U.S. demand.

Fourth, it seems reasonable to measure the carrying capacity for nonhuman populations in terms of the resources needed for survival, but most people agree that this strategy is not acceptable for people. Consequently, the most important factor that distinguishes the calculation of carrying capacity for humans from that for nonhumans is **quality of life** or **standard of living**. Thus, the **cultural carrying capacity** is the optimal size that the environment can sustain in perpetuity, with a given level of technology, standard of living, and associated patterns of resource use.

Determining cultural carrying capacity for the biosphere or selected countries is a complex and inexact science, but some attempts have been made. *Limits to Growth* (1972) presented the results of a global modeling study that attempted to determine the Earth's carrying capacity. Using systems analysis, computer projections and a wealth of data, researchers determined that the planet could support a population of six billion at a European standard of living. In Chapter 8, we discuss how China determined its carrying capacity based on available fresh water and food.

Standards of living vary widely from nation to nation and from region to region (Figure 7-7). Less than 20 percent of the world's population (those living in MDCs) uses a disproportionately large share of the Earth's resources, including 60 percent of fossil fuels, 75 percent of paper, and 80 percent of metals. Because of their high resource use, each inhabitant of MDCs does approximately 7.5 times more damage to the biosphere than does each inhabitant of LDCs. If everyone used resources at the same rate as the inhabitants of MDCs, our fuels and raw materials would be depleted much more quickly, and far more pollution would be generated.

When a population exceeds the carrying capacity, the environment is eventually degraded, and environmental resistance increases. Certain factors increasingly come into play: hunger, starvation, disease, resource scarcity, and lack of space. There is simply less to go around — less drinkable water, less arable land, less food, less suitable living space. Consequently, the quality of life decreases. Unless resources are brought into the area from outside, the population eventually will be reduced to a size that can be supported by the environment.

In most nations where environmental degradation has reduced the carrying capacity, food and medical aid have been brought in to supplement (or replace) local resources. Some experts believe that this aid is misguided, for it falsely "props up" the carrying capacity of the land. The population continues to grow, and each year more and more people compete for fewer and fewer resources. Environmental stress increases, and the land's carrying capacity is further reduced, exacerbating and prolonging the misery of the entire population. When the situation in an area becomes intolerable, residents often flee (usually to cities or aid camps) in search of food. They become **environmental refugees**, people forced to abandon their homes because the land can no longer support them. In the absence of continued and accelerated relief aid, a rapid population decline is inevitable. During the 1980s, famine, drought, and an ongoing civil war combined to make environmental refugees of thousands of Ethiopians. Although a massive campaign of foreign aid was launched to help displaced Ethiopians, conflict between the country's warring factions prevented much of the aid from reaching those most in need. The result was prolonged and severe misery for many.

How Are Population Growth and Quality of Life Related?

Clearly, population growth affects the environment and ultimately a nation's or region's standard of living. To better understand this relationship, we again turn to demographic statistics, which reveal important

FIGURE 7-7: A woman washes her family's laundry in the Nyaderi River, Zimbabwe. This river inlet serves the needs of many families from several surrounding communities.

TABLE 7–2: Demographic Statistics Related to Quality of Life, 2004, Selected Countries

	Infant Mortality Rate (per 1,000)	Population Under Age 15/over 65 (%)	Life Expectancy at Birth (years)	Urban Population (%)	Married Women Using Contraception Total/Modern Methods (%)	Per capita GNI PPP, 2002 (U.S. $)
World	56.0	30/7	67	48	59/52	7,590
More-developed countries	7.0	17/15	76	76	69/58	23,690
Less-developed countries	62.0	33/5	65	41	57/51	3,850
Less-developed countries (excluding China)	66.0	36/4	63	41	48/40	3,630
Afghanistan	165.0	45/2	43	22	5/4	—
Australia	4.7	20/13	80	91	67/65	27,440
Bangladesh	66.0	37/3	60	23	54/43	1,177
Bolivia	54.0	39/5	63	63	58/53	2,390
Botswana	62.0	40/4	36	54	40/39	7,740
Canada	5.2	18/13	79	79	75/73	28,930
China	32.0	22/7	71	41	83/83	4,520
Congo, Dem. Rep. of	100.0	48/3	49	30	31/4	630
Estonia	6.0	17/16	71	69	70/56	11,630
Germany	4.1	15/17	78	88	75/72	26,980
Guatemala	39.0	44/4	66	39	43/44	4,030
Haiti	80.0	43/4	51	36	28/22	1,610
Hungary	7.3	16/15	73	65	77/68	13,070
India	64.0	36/4	62	28	48/43	2,650
Iraq	102.0	42/3	60	68	—/—	—
Italy	4.8	14/19	80	90	60/39	26,170
Japan	3.0	14/19	82	78	56/48	27,380
Kenya	78.0	44/4	51	36	39/32	1,010
Libya	28.0	35/4	76	86	49/26	—
Mexico	25.0	35/5	75	75	68/59	8,800
Niger	123.0	50/2	45	21	14/4	800
Russia	13.0	16/13	65	73	67/49	8,080
Sweden	2.8	17/18	80	84	—/—	25,820
Thailand	20.0	23/7	71	31	72/70	6,890
United States	6.7	21/12	77	79	76/72	36,110

Source: *Population Reference Bureau, 2004 World Population Data Sheet.*

information about the quality of life of the majority of people in a society. Five demographic factors are particularly telling: income levels, population density, urbanization, life expectancy, and infant and childhood mortality (Table 7-2).

Income Levels

Population growth affects economic conditions for a society at large and for the average individual within that society. To see how, let's take a closer look.

At the societal level, there is a direct relationship between population growth and economic development. For every one percent increase in population growth, the gross national product (GNP), an indicator of economic health, must increase three percent in order for the nation's economy to grow. This is a tall order for governments whose predominantly young populations are increasing at high rates. A growth rate of 2.4 percent — the average for Africa — means that GNP must increase 7.2 percent a year in order to provide basic services (schools, housing, hospitals or medical clinics, and roads) and employment opportunities to all citizens. It is virtually impossible for real, sustainable economic development to occur when the government is overwhelmed by the difficult task of providing health care, education services, and employment opportunities for a rapidly growing population. Yet in the absence of sustained (and sustainable) economic development, illiteracy and unemployment become more widespread, strengthening the cycle of poverty.

In general, per capita income in fast-growth countries is far lower than per capita income in slow- or no-growth countries. Demographers use a measure known as the gross national income in purchasing power parity (GNI PPP) to compare economic conditions in various nations. The GNI PPP refers to a nation's gross national income converted to "international" dollars using a conversion factor. International dollars represent the amount of goods and services an individual could buy in the United States with a given amount of money. In 2004, the annual per capita GNI PPP for Sierra Leone, a fast-growing country (2.1 percent) in West Africa, was $500, the world's lowest. In contrast, the slow-growing population of Luxembourg (0.3 percent) enjoyed the world's highest annual per capita GNI PPP, $53,290 — 105 times that of the average resident of Sierra Leone.

While LDCs are struggling to meet the needs of their young and growing populations, MDCs are faced with problems of their own. A large proportion of older citizens can strain social service networks, health care systems, nursing homes, and even public transportation systems. In the next several decades, the growing proportion of older citizens in the United States is expected to tax the country's already overburdened Social Security system. If fertility remains relatively low, the responsibility for the care of the aging baby boomers (as a sector of the population) may fall on an ever-shrinking proportion of younger laborers, unless changes are made in the way the country funds care for the aged.

Population Density

Humans are not uniformly distributed over the Earth. A nation's **population density**, how closely people are grouped, is an important demographic factor, especially with respect to the degree of resource use within defined geophysical or political boundaries. Consider, for example, that almost 141 million Bangladeshis crowd into an area about the size of Arkansas, which has a population of about 2.7 million (Figure 7-8). India has more than 3.5 times the population of the United States in about one-third of the space. Because both of these densely populated countries have a smaller resource base

FIGURE 7-8: Bangladeshi victims of the April 1991 cyclone await handouts of rice in the town of Pekua. The storm killed more than 125,000 people and left many others homeless. Its high population density renders Bangladesh susceptible to environmental problems and strains the nation's ability to respond to natural catastrophes.

than that of the United States, their people have a much smaller per capita resource base. In general, density in LDCs is more than double the density in MDCs.

Often, countries that are densely populated perceive the problem as one of an inadequate resource base rather than too large a population. When this perception is widely accepted, population pressures may be manifested in the desire to annex adjacent lands, colonize distant lands, acquire resources from outside the country, or encourage emigration. History is replete with examples of expansionist policies associated with population pressures: the Roman Empire, western European colonization of the Americas and Africa from the 1600s to the 1800s, Adolf Hitler's Germany, and the Japanese Empire of the 1930s and 1940s.

Urbanization

One indicator of increasing population density is **urbanization**, a rise in the number and size of cities. In 1970, about 37 percent of the world's population lived in cities. In 2004, nearly half (48 percent) inhabited the planet's sprawling cities, and by 2025, over 60 percent will be urban dwellers.

Currently, most people in MDCs (about 76 percent) live in urban areas, while the majority of people in most LDCs still live in rural areas. The developing world is rapidly changing, however. In 1975, only about one-fourth of the population of LDCs was urban; by 2004, that proportion had risen to 41 percent; and by 2015, well over one-half will live in cities.

Tokyo, at over 26 million people, heads the list of the 10 most populous megacities, followed by Sao Paulo, Brazil; Mexico City, Mexico; Mumbai (Bombay), India; New York, United States; Dhaka, Bangladesh; Delhi, India; Calcutta, India; Los Angeles, United States; and Jakarta, Indonesia. According to the United Nations, seven of the world's 10 most populous megacities are in LDCs. Of the world's 30 most populous cities, 23 are in LDCs.

Urban areas place a strain on the environment far greater than the absolute number of people. The concentration of people means that pollution is also concentrated, often to such a degree that natural processes are unable to cleanse the environment of contaminants. Rural dwellers are typically engaged in food production, while urbanites are strictly consumers. Finally, urban areas tend to isolate people from nature. Consequently, urban dwellers are at greater risk of forgetting that all life, including human life, is dependent upon the continued health of natural systems.

Several factors are responsible for the rapid growth of cities in poorer nations. Migrants from rural areas find more economic opportunities in urban areas; many are young people who are willing to work for low wages, which benefits employers and helps fuel the economy. For instance, through much of the 1990s, the manufacturing sector grew rapidly in the newly industrializing economies (NIEs) of Asia; at the same time, the agricultural sector was modernized. As a result, the demand for cheap labor rose in the manufacturing sector and dropped in the agricultural sector. Urban areas in the NIEs are expanding as migrants (many of them women) move to the region's cities to take advantage of the opportunities available.

In many cases, migration from rural to urban areas is caused by desperation. In Africa, environmental degradation and droughts have reduced soil fertility, making it difficult for farmers to earn a living from the land. In Latin America, a relatively small percentage of the population owns most of the land; wealthy landowners are able to buy out small landowners or charge such high land rent that peasant farmers are unable to make a living. They must move to the cities in search of work. Inequitable land ownership (and by extension the migration it fosters) is one reason that Latin America has the highest proportion of urban population (75 percent) in the developing world.

In general, the urban poor of LDCs are better off than the rural poor — at least they are able to find or scavenge food, make a living (however meager), and take advantage of services (such as maternal and child health programs and better immunization coverage) that may be nonexistent in rural areas. Even so, they face a multitude of hardships; some sleep under makeshift cardboard shacks in alleys or on sidewalks; others find shelter in one of the many squatter's areas or shantytowns that invariably ring cities in the developing world. Shantytowns are typically crowded and dirty, with nonexistent or inadequate sewage facilities and contaminated water supplies. Children of the poor may not attend school; often, they work to help contribute to the family's income.

Rapid urbanization in LDCs places a significant strain on governments' financial resources. Poor nations simply do not have the money or the infrastructure to provide adequate housing, education, transportation, and health and sanitary services to their expanding urban populations; they are overwhelmed by a ceaseless tide of humanity. Saddled with heavy debt burdens, they must allocate a large portion of their financial resources to pay the interest on overdue foreign loans.

To earn needed cash in the short run, they are often forced to take actions that deplete the countries' natural resource bases. The problems faced by governments of LDCs can be traced, in part, to the colonial period. Many LDCs were once colonies of industrialized nations; administrative activities were concentrated in a few urban areas, which also served as conduits for the movement of natural resources (raw materials) to the "mother country" and from thence to the rest of the industrialized world. While a small proportion of the population in LDCs prospered, the majority did not. They continued to work in the agricultural or informal sectors. After gaining independence, most LDCs have attempted to expand their industrial base and strengthen their urban infrastructure, but these attempts have been stymied by rapid population growth.

Governments in the developed world also are struggling to meet the needs of their urban poor. High rates of unemployment, increasing numbers of high school dropouts, and violent crime are all too common in the inner cities of the United States. In addition, the nation's urban areas appear to be experiencing a resurgence of racial unrest and ethnic violence as the economic gap widens between America's wealthy and her poor. Certainly, there are many differences between cities in LDCs and those in the United States, but sadly, the similarities are more important. Crowding, crime, pollution, sickness, economic deprivation, and desperation are the common experiences of the urban poor, whether those poor live in a shantytown in Mexico City, makeshift dwellings in Calcutta, or the slums of Los Angeles or Detroit (Figure 7-9).

In recent years, there has been a growing movement in the United States and other MDCs to revitalize cities, making them sustainable and ecologically healthy. By doing so, urban activists hope to also alleviate some of the social ills that plague cities. Many urban centers are revitalizing their waterfronts, reclaiming degraded riverbanks and lakefronts. In Toronto, Canada, an effort is underway to restore the Don River by reestablishing its large delta marsh, destroyed years ago by pollution and landfill. Other cities are reclaiming abandoned areas and converting them to greenbelts featuring diverse vegetation, bike paths, and park benches. In inner-city areas, urban activists and community groups are working together to transform vacant lots into parks, tree nurseries, and vegetable and flower gardens.

Making cities more livable may help to stem the flight of people to the suburbs, a trend that has been ongoing for the past several decades. Rapid suburbanization, common in the United States, is a low-density land use pattern that consumes prime farmland and wildlife

FIGURE 7-9: Homeless woman and child, Calcutta, India.

habitat such as forests and grassland. The suburb, with its maze of side streets and cul-de-sacs, is characterized by single-family homes; suburbanites depend on their cars for transportation to and from work (often in the city), schools, and shopping malls. Traffic congestion at rush hour in any big city is caused chiefly by commuters on their way to or from the suburbs. Compared to city dwellers, suburbanites use far more land, energy, and resources, and generate more pollution (principally because of auto exhaust). Concern over suburbanization has led some cities to take action to preserve the surrounding countryside.

Life Expectancy

Life expectancy at birth, the average number of years a newborn can be expected to live, is a good indicator of standard of living. Worldwide, life expectancy in 2004 was 67 years. For both men and women, life expectancy is higher in the developed world, 76 years, than in the developing world, about 65 years (Table 7-2). Among continents, North America has the world's highest life expectancy, an average of 78 years. Canadians enjoy a slight advantage over their southern neighbors — 79 years compared to 77 years for citizens of the United States. Life expectancy is also high, 75 and 74 years, in Oceania and Europe, respectively. The average life expectancy in Europe, however, masks some striking regional differences. In western, southern and northern Europe, life expectancy is 79, 78 and 78 years, respectively, but in eastern Europe — home to many of the former Soviet-bloc nations — life expectancy is 68 years (Figure 7-10).

Within the developing world, Africa has the lowest life expectancy, an average of just 52 years. As in Europe, significant regional differences exist on the continent. Contrast northern Africa, 67 years, with southern Africa, 52 years; western Africa, 51 years; middle Africa, 47 years; and eastern Africa, 46 years. National differences are even greater. In 2004, a newborn in the North African nation of Libya could be expected to live to age 76, while life expectancy for a newborn in Zambia was a mere 35 years.

Disparities are evident in Asia and Latin America as well. Life expectancy in East Asia is 72 years; the region includes Japan, which boasts the world's highest life expectancy, 82 years. In contrast, life expectancy in south central Asia is 62. South Central Asia includes the war-torn nation of Afghanistan, with a life expectancy of 43 years (the continent's lowest), but it also includes the Maldives, with a life expectancy of 73.

In general, life expectancies among LDCs are higher in the western hemisphere than in the eastern hemisphere. The average life expectancy for nations in Central America, the Caribbean, and South America is 74, 69 and 71 years, respectively. Haiti is the unfortunate exception. Plagued by political turmoil and corruption, violence, an epidemic of Acquired Immune Deficiency Syndrome (AIDS), poverty and unemployment, Haiti has a life expectancy of just 51 years.

What factors contribute to low life expectancy? Among the most visible are war and hunger. Ongoing strife is a factor in the lower life expectancy of citizens of Sierra Leone, 35 years; Angola, 40 years; Mozambique, 40 years; and Rwanda, 40 years. Disease also plays a role; in many African nations, the AIDS epidemic has reversed the decades-long trend of increasing life expectancies. By the late 1980s, life expectancies throughout sub-Saharan Africa had climbed to the low 60s; just 15 years later they had fallen by nearly half. A child born in 2004 in Botswana, for example, could expect to live to just 36 years. Sadly, life expectancy in many countries will continue to fall. According to the Population Reference Bureau, a Botswanan born in 2010 will have a life expectancy of 27 years.

According to estimates by the United Nations Programme on AIDS, sub-Saharan Africa is home to nearly two-thirds of the world's AIDS sufferers, some 25 million people. The epidemic is most virulent in southern and eastern Africa. In Swaziland, about 39 percent of the adult population (ages 15-49) is infected with the virus that causes AIDS (the Human Immunodeficiency Virus, or HIV). Other countries with high HIV infection rates include Botswana (37 percent), Lesotho (29 percent), Zimbabwe (25 percent), South Africa (22 percent), and Namibia (21 percent).

Despite the ravages of war, hunger and AIDS, the most significant factors that keep average life expectancy for the developing world so far below that of the developed world are infant and childhood mortality.

FIGURE 7-10: An elderly woman in Bulgaria relaxes in the shade of a late afternoon. Throughout eastern Europe, life expectancy is 10 years less than in other regions of the continent.

Kazi Ya Mwanamke

The Work of a Woman

Betsy A. Beymer

Kazi ya mwanamke is a kiswahili phrase meaning "the work of a woman." The following essay is a fictional, but nonetheless accurate, account of daily life for a woman in sub-Saharan Africa. It is a compilation of observations and notes made during my fieldwork in the West Usambara Mountains of Tanzania in the summer of 2003. The circumstances and challenges that it portrays mirror those faced by rural women throughout the developing world.

Imagine, for a moment, that you are a young woman in sub-Saharan Africa. Somewhere in your 20s (no one in your village keeps records on births and deaths), you are responsible for a household that includes five children, ages 11, 9, 8, 3, and 1. Your husband works in a mine a long way from your village, returning home only six or eight times a year. With little or no education and living in a remote rural area, there is no such thing as a "career choice." You and your children depend for your livelihood on what you can grow and gather. You do not have the money to hire anyone to help you; rather, you must rely on your children. Sound difficult? Well, that's just the beginning. Imagine that you have no choice over the type of crops you grow for cash income, if there is any cash income to be made; those decisions are made by someone else. Nor do you have any choice about the number of children you will bear or the timing of their births; you do not know about contraceptives, and even if you did, you have no access to them. Moreover, the land that you work everyday — a small plot on a rocky stretch some distance from your home — is held in your husband's name only. You cannot own property. If your husband dies, there is a very real chance that the land will be taken away from you and your children.

Given these circumstances, what is your life like? Imagine....

Although the sun has not yet risen, it's time for you to begin the day. So, at about 5 a.m., you rise quietly and leave your hut to collect water for the day's first meal and household chores. You bring along your youngest child, and he nurses as you walk. Your oldest child, an 11-year-old girl, is responsible for her siblings while you are gone. The water hole is more than a mile from your home, but you are so accustomed to the journey, you give the distance little thought. You fetch up to five gallons of water in an old plastic jug that you balance upon your head. Once home, you prepare tea for the rest of the household, then begin the daily ritual of washing the dirt floors with a wet rag in order to rid your house of insects. Soon after, you begin the day's laundry, which means another trip to the water hole. Without a washer and drier, washing clothes is a labor-intensive chore and takes you several hours. As you work, you talk with some of the women from your village who also have come to do their wash; their company makes the tedious task more enjoyable. After you finish, you return home to feed the livestock and tend to the crops. You perform all agricultural tasks — preparing the ground, plowing, planting the seeds, weeding, and harvesting — by hand. As you work, you again nurse your youngest son then listen to his cooing and babbling before he drops off to sleep.

While in the fields, you gather maize to prepare ughali, the staple of your diet. Returning home, you greet your four older children; your eldest son, nine, complains of hunger. The younger children play while their older sister helps you prepare the noon meal. Even food preparation is time-consuming and requires physical effort; you must shuck the maize, pick the kernels off the husks, and grind the kernels using a large, heavy pestle and mortar. You make enough ughali to last for several days; you eat meat only when you are able to catch a rat or one of the village men kills a bushpig. Once the maize is ground, you stoke the cook fire, transfer water from your container into a pot, and bring the water to a boil. You then begin the difficult task of stirring the ground maize into the water. As the ughali thickens, you feel the muscles in your back tense.

During the meal, you notice that your three-year-old isn't eating; her forehead feels warm and she is listless. There is no doctor in the village and the nearest clinic is more than 25 miles away. You tell your nine-year-old to go and find the village medicine man, a well-known healer. Perhaps he can advise you on what to do. Worried, you hold the three-year-old close and she rests, drowsing, against your side. With your other arm, you hold the baby, who again nurses greedily. Meanwhile, your older daughters begin to clear away the remains of the meal. Your son soon returns. The medicine man has gone to a nearby village; he will be gone for several hours. The three-year-old is sleeping again, though her face is flushed and hot. You worry, wondering if the fever that took her six-year-old brother's life several months ago will now take hers. And yet there is nothing more you can do for her right now, and so, leaving her in the care of her older siblings, you turn to go. There is work to be done.

And thus begins the most physically demanding part of your day: the search for firewood. Three or four times each week, you and your eldest daughter join a group of women and girls and set out for the nearest woodlot, a sparse stand of trees and brush more than two miles from the village. Usually, you look forward to the chance to visit with your friends, but today you are quiet, preoccupied with concern for your three-year-old. Your baby rides on your back; in one hand, you carry an axe dulled by years of use. Arriving at the woodlot, you begin the arduous task of cutting firewood. You collect as much wood and kindling as you can carry, then bundle it tightly and evenly with old twine. You sit for a few moments while the baby nurses yet again. After taking his fill, he fusses, wanting something more than milk in his belly. But you have nothing here for him, and so you pack him on your back once again and ready yourself for the return trip. You carry your bundle, which weighs about 80 pounds, on your head. Your daughter carefully balances a smaller bundle on her head. Anxious to be home, you set off as quickly as you can.

Like rural women throughout the developing world, you carry a heavy burden, literally and figuratively. Women produce 60 to 80 percent of the supply of basic foodstuffs in sub-Saharan Africa and the Caribbean. In India, they produce between 70 and 80 percent of food crops. Similarly, women perform over 50 percent of the labor involved in intensive rice cultivation in Asia, and in Indonesia and Central and South America, the household gardens tended by women are some of the regions' most complex agricultural systems. In Africa, as elsewhere, women and girls gather 60 to 80 percent of all domestic fuel supplies. Women in the developing world also are primarily responsible for collecting water

Infant and Childhood Mortality

The **infant mortality rate** (IMR) is the annual number of infants under age one who die per 1,000 live births. It is widely considered the single best indicator of a society's quality of life. The **childhood mortality rate** is the annual number of children between the ages of one and five who die per 1,000 live births. Both rates are sometimes expressed as percentages.

In 2004, the average infant mortality rate worldwide was 56/1,000, or 5.6 percent (Table 7-2). The rate for MDCs was 7/1,000 (0.7 percent), while the rate for LDCs was 62/1,000 (6.2 percent). The rate for LDCs excluding China was 66/1,000 (6.6 percent). Worldwide, infants in Singapore had the best chance of surviving their first year; 2.2 per 1,000 (less than 0.4 percent) died before their first birthday. Iceland, Japan, and Finland also boasted low IMRs — 2.4, 3.0, and 3.2 per 1,000, respectively. Prospects were grimmest for infants in Sierra Leone; 180 per 1,000 (18 percent) never saw their first birthday. Many African nations (especially sub-Saharan nations) had IMRs

needed for drinking, cooking, cleaning, and bathing. And they are almost solely responsible for childcare. They birth their children without trained medical assistance, sometimes to tragic results; childbirth is a leading cause of death for women in poorer countries. They also tend ill children without the help of medical professionals, and they bury those children they cannot save. Women do all this every day of the year. There are no holidays, no vacations: The children must eat and drink every day; the fire must be kindled; the fields must be tended. Even childbirth does not interrupt the daily pattern of life. In many cases, a woman resumes her normal activities within 24 hours after giving birth.

As you near the village, you hear your youngest daughter crying, and upon entering your hut, you find your neighbor spooning medicine into the child's mouth. She is treating the illness with a homemade remedy made from local plants that she gathered from the forest. As your three-year-old drifts back to a feverish sleep, you send your eldest daughter and son for more water. Before the evening is over, you stoke the fire, boil water, prepare the meal, feed the baby, clean up after dinner, and sponge-bathe the children and yourself. By the light of a kerosene lamp, you listen to your children as they tell you of their day's adventures, and you pray silently for their sister's recovery. Later, after the children are asleep, you lie near the little girl so that you can check on her through the night. You keep the baby close at hand as well, allowing him to

nurse himself to sleep. You touch your daughter's forehead, hoping for some sign that the fever has broken. Still warm. You lie awake, wondering. What will you do if the fever worsens? The minutes pass, and the night sounds fill the air outside your hut. You again touch your daughter's forehead. Perhaps, you think, she feels a little cooler. At the very least, the fever has not worsened. Finally, you close your eyes. It is after 10 p.m. You will yourself to rest, to sleep. You must rise tomorrow before the sun.

My time in Africa left me with an immense respect for the women with whom I worked. They possess an intimate knowledge of their environment — the soil, water, wild plants, crops, and firewood —for it is source and sustenance. And yet they have little control over basic decisions that affect the environment. I witnessed extreme hardship and poverty and a complex mix of social and environmental problems. There is no doubt in my mind that rich nations could — should — do more to help poor ones. Our collective security depends upon it. And yet when I recall my experiences, I think as often of what those women had as what they lacked. Strength. Resilience. Knowledge. Burdened with responsibilities and constricted by circumstances, they still found joy in living, and they clearly loved their children and their community. They even found the time and energy to be concerned for a young American far from home.

For that selfless concern, I am forever grateful.

well over 100/1,000, with the continent as a whole averaging 90/1,000. Other nations with IMRs exceeding 10 percent included Afghanistan (165/1,000), East Timor (129/1,000), Laos (104/1,000), and Iraq (102/1,000).

The West African nations of Sierra Leone and Togo illustrate how infant mortality rates affect a population's growth rate and life expectancy. In 2004, both had similarly sized populations (5.2 and 5.6 million, respectively) and high TFRs (6.5 and 5.5, respectively). But Togo grew at a rate of 2.7 percent, while Sierra

Leone — even with its higher TFR — grew at just 2.1 percent. What caused Sierra Leone's slower growth? An IMR of 180/1,000, more than double Togo's 72/1,000. As you would expect, its high IMR also affected Sierra Leone's average life expectancy, which, at 35 years, was far below Togo's (54 years).

Though infant mortality rates are uniformly higher in LDCs, they vary widely among countries in both the developed and developing world. The infant mortality rate in the United States, for example, is three times as high as that in Singapore. In fact, the United States,

Figure 7-11: A young boy drinks from a well that provides clean, safe water for the people of his village.

with an IMR of 6.7/1,000, ranks behind many other industrial nations, including Canada (5.2/1,000). While it's true that there have been improvements in recent years (the U.S. IMR was over 10/1,000 as recently as the 1980s), the nation's relatively high IMR remains a cause for concern. That's because the IMR for the poor and some minority groups is much higher than the national average. The IMR for African-American children, for example, is approximately 14/1,000, about 2.5 times higher than the rate for Caucasian infants. Many people argue that the average IMR would be markedly lower if the United States had a national health care system or better care (especially prenatal care) for its poor. The U.S. situation underscores the fact that, as Dr. Marsden Wagner of the World Health Organization has said, "Infant mortality is not a health problem; it is a social problem."

It seems almost bizarre that an IMR of one percent causes such controversy in the developed world while over six percent of infants routinely die in developing nations each year. Why do so many infants and children in the developing world die? Naturally, famine and malnutrition contribute to high infant and childhood mortality rates. In 1985, in some Ethiopian villages in the war-torn region of Eritrea, 60 percent of children died before their first birthday. But famine, while undeniably a relentless killer, is not the most common cause of infant and childhood mortality. More children die each year because they are afflicted with diarrhea (which can cause life-threatening dehydration when left untreated), are not immunized against common childhood diseases (particularly measles), or do not receive treatment for curable illnesses and infections

(such as malaria, whooping cough, pneumonia, and tetanus).

Often, the culprit in children's illnesses is disease-infested water. While a growing share of the global population enjoys clean drinking water and adequate sanitation, many people, particularly those in rural areas of LDCs, still lack access to these basic services. Sadly, parents generally have no choice but to continue to use contaminated water. Particularly in areas where wood is in short supply, parents feel they cannot "waste" fuel by boiling water that *looks* clean. Thus, even if children with water-borne diseases receive adequate medical attention and recover, they will repeatedly pick up parasites and bacteria. Providing safe, adequate sources of clean drinking water can help to reduce the incidence of disease and mortality, lengthening and improving people's lives (Figure 7-11).

A second cause of childhood mortality is inadequate nourishment. Children may be weaned, or withdrawn, from mother's milk too early if the mother does not sufficiently space the birth of her young, and may not be able to get sufficient nourishment from a meager adult diet. In many areas in Africa, however, children are weaned too late; their only source of nourishment may be breast milk until they are 12 to 18 months of age. When they are finally weaned, the children are given adult food, which may be difficult for their systems to digest.

Other factors that contribute to high infant mortality are poor maternal health and inadequate prenatal care. Low birth weight (like malnutrition) is one of the most highly reliable indicators of poor infant health. Low birth weight can be caused by a number of things, including the young age of the mother, poor maternal health and lack of prenatal care, all of which are linked to the low status of women.

The key to saving young lives is to educate women and provide them with the tools they need to better care for their children. Of the more than 10 million children who die each year from the combined effects of poor nutrition, diarrhea, and disease (especially measles, pneumonia, whooping cough, tetanus, and HIV/AIDS), most could be saved through simple measures. Oral rehydration therapy (ORT), for instance, can save the life of a child dangerously ill from diarrhea-induced dehydration; ORT salts cost just about ten cents a pack. But the mother who does not know why her child is dying, who does not have access to medical care, and who cannot afford to pay for care even if it is available, will not be able to prevent her baby's death.

How Are Population Growth and Women's Status Related?

In 1994, at the landmark International Conference on Population and Development (ICPD), representatives from 180 nations met in Cairo, Egypt and agreed to a new approach to population issues. They developed a 20-year program of action that called on governments to promote human development and stabilize population growth. The program emphasized investments to improve individuals' health, education, and rights — particularly women's rights — and for family planning services to be provided in the context of comprehensive reproductive health care. With the program of action developed by the ICPD, the global community acknowledged that improving women's status is key to creating greater economic prosperity, reducing poverty, improving overall health, and stabilizing population growth. Women's status is reflected in education, work, health care, and cultural traditions.

Education

According to the World Bank, the education of females is the "single most influential investment that can be made in the developing world." Unfortunately, despite some gains over the past 10 years, females still enjoy fewer educational opportunities relative to males (Figure 7-12). In the developing world, 76 percent of all boys attend primary school compared to 70 percent of all girls. By secondary school, the gap is greater; 57 percent of boys in LDCs are enrolled in secondary school compared to 48 percent of girls. (Latin America and the Caribbean are exceptions; in these regions, girls attend both primary and secondary school at rates equal to or greater than boys.) Girls discontinue their schooling for various reasons: household responsibilities, early marriage and childbearing, cultural perceptions that education is more beneficial for boys than for girls, parents' worries about their daughters' safety as they travel to schools away from their villages, and limited job opportunities for females in sectors that require higher education. Given the obstacles facing schoolgirls, it's little surprise that women account for two-thirds of the 860 million illiterate adults worldwide.

Work

Like education, formal employment is an important indicator of the status of women. Employment enables

FIGURE 7-12: A young schoolgirl, Eritrea. While the proportion of girls enrolled in school has increased significantly in the past several decades, it still lags behind the proportion of enrolled boys, particularly at the secondary level, and especially in parts of Africa and Asia.

females to earn income, assume greater control of resources, and improve decision-making skills; it also can increase their involvement in public life. About half of women worldwide participate in the formal paid workforce, compared to 82 percent of men. Women are most likely to be employed in the agricultural or service sectors, and — in every nation except Australia — they earn less than men for their labor. In South Korea, for example, women earn just 60 percent of men's wages; in the United States, they earn 75 percent of what men do. Bear in mind that while more and more women are working outside the home, they still shoulder the burden of housework (cooking, cleaning, and child care).

Health Care

Access to adequate health care is one of the chief determinants of women's status, and for women, quite often, health care means *reproductive* health care. According to the United Nations Population Fund, about 529,000 women die each year because of inadequate reproductive health care — more than one death every minute. Ninety-nine percent of these deaths occur in the developing world, with Africa, 251,000, and Asia, 253,000, accounting for the overwhelming majority. About four percent of maternal deaths, 22,000, occur in Latin America and the Caribbean and less than one percent, 2,500, occur in MDCs. For every woman who dies, an estimated 30 to 50 more suffer injuries, infections, or disabilities during pregnancy or childbirth — at least 15 million women annually. Serious complications to the birthing process include hemorrhage, obstructed labor, and infection. According to the United Nations, one of the most neglected issues in reproductive health care is obstetric fistula. A fistula is a hole that forms between a woman's vagina and bladder or rectum; it is caused by prolonged and obstructed labor. Usually, the baby dies. The underlying causes of obstetric fistula are early childbearing, malnutrition, and limited access to emergency obstetric care. A woman with obstetric fistula suffers chronic incontinence; the odor of urine or feces is constant and humiliating. Many sufferers become social outcasts — blamed for their condition and rejected by their husbands and communities. More than two million girls and women worldwide live with obstetric fistula; most are in sub-Saharan Africa, South Asia, and some Arab states.

In general, childbirth poses the greatest risk to the very young (those under 19), women over 35, and those in poor health. Early childbearing increases the risk of complications associated with pregnancy; adolescent mothers are twice as likely to die in childbirth as are women in their twenties. Early marriage is also linked to family size; in countries where most adolescent girls marry, women bear three times as many children as do women in countries where late marriage is more common.

The **maternal mortality ratio** is the number of women's deaths due to pregnancy and childbirth complications per 100,000 live births in any given year. Currently, the worldwide maternal mortality rate is over 400 deaths per 100,000 live births. While the rate is down slightly, the total number of women dying from complications of pregnancy has actually increased since the number of pregnancies has increased (with the continued rise of the global population). The maternal mortality ratio varies widely from country to country and from region to region. In the developed world, the maternal mortality ratio is 20 per 100,000; for the less-developed world, the figure is 440 deaths per 100,000. Maternal mortality is highest on the continent of Africa, where 830 women die for every 100,000 babies born alive. But as with other population statistics, ratios vary widely even across Africa, from 160 per 100,000 in Algeria to 2,000 per 100,000 in Sierra Leone (the latter being the world's highest maternal mortality ratio).

Access to family planning services, community-based maternity care, and essential obstetric care can help lower maternal mortality rates. Women (and couples) who have access to family planning services are able to prevent unintended pregnancies and unsafe abortions, protect themselves from sexually transmitted diseases, and avoid births that pose a threat to the health of the mother. Basic maternity care curbs maternal mortality by improving women's health, detecting and treating potential problems, providing trained assistance during delivery, and referring women to sources of additional help when complications arise.

Cultural Traditions

The backbone of demographic studies — the reliance on statistics — can be a curse; numbers can be numbing. To truly understand the social and environmental impact of the status of women, let's look at some of the practices faced by women in countries where the status of females is low. In many developing nations, women and girls eat only after men and boys have eaten (a practice that undermines the health of girls and their mothers when so little food is available to begin with), in spite of the fact that women often work longer hours and harder than men (Figure 7-13). *Focus On: Kazi Ya Mwanamke* details daily life for a woman in Tanzania.

In some Islamic nations, female heirs legally may inherit only half as much as male heirs. Women are not considered autonomous individuals in the most conservative Islamic countries; they must have a designated male "guardian," typically a father, brother, or husband. In many of these countries, very conservative interpretations of Islam forbid a woman to divorce her husband, but a man may easily divorce his wife (in Egypt, as well as some other nations, he need not even notify her), and men generally retain custody of the children. In the most conservative Islamic areas, women can be stoned to death for having sex outside marriage. In Nigeria in 2004, two women were sentenced to death by stoning; the sentences were to be carried out after the births of the babies. (No action was taken against the men involved.) In Brazil, battered women have little legal recourse, and husbands who murder their wives "to protect their honor" frequently do so with little fear of legal reprisal. And in Africa, genital mutilation (clitoridectomies) remains common for young girls. It is deemed necessary to keep women chaste and faithful, and in the most conservative regions a woman is not considered marriageable unless she has undergone such an "operation."

Perhaps the saddest indicator of the low status of women is the expanding sex trade, particularly in child prostitution. In parts of the developing world, especially Asia, children are bought, stolen, sold, or traded to provide for the sexual gratification of brothel patrons, many of them businessmen from developed countries such as Japan and Germany. After the devastating tsunami of December 2004, officials throughout Southeast Asia feared that young children orphaned by the disaster (or simply temporarily separated from their families) would be stolen and sold into the sex trade.

There are an estimated 800,000 child prostitutes in Thailand, 400,000 in India, and 60,000 in the Philippines. Brazil has 250,000 to 500,000 children trapped in the sex trade. Most child prostitutes are female, and most are younger than 16, although a few areas cater to pedophiles who prefer young boys. Sri Lanka, for example, has 20,000 to 30,000 child prostitutes, most of whom are male. The rapid growth in the child sex industry is partly due to increased concern over AIDS and other sexually transmitted diseases. Customers at brothels request young girls in the belief that they are less likely to be infected.

The majority of girls who end up in the sex industry are from remote villages. Some villages in northern Thailand, northeast Brazil, and other poor, rural areas have lost all of their teenagers. Some are turned out of their homes because their impoverished families cannot support them. They migrate to the city in search of economic opportunities. The United Nations reports that girls aged 10 to 14 have the developing world's highest rate of rural to urban migration. Arriving in the city, a young girl is easy prey for pimps and sex traffickers. Some girls are simply stolen from their homes. Others are sold by families who are tricked into believing that the child will become a maid or housekeeper for a rich family in the city. Tragically, even when parents know that their daughter will become a prostitute, they may be so desperately poor that it is impossible for them to refuse the cash payment offered by sex traffickers.

Parents who sell or trade a daughter may expect that she will return in a few years, prosperous and in good health. They probably do not realize the fate that awaits her. In Brazil, some young girls wind up in the hands of wealthy ranchers, and in what has become a Saturday night ritual, they are gang-raped to death. For those girls, the sex trade brings terror and agony, then a quick end. For most child prostitutes, the terror turns into despair then numbness as each day mirrors the day before. Locked in tiny cubicles, they service a dozen or more clients each day. They are told that they must keep "working" until they have earned enough to pay back the cash payment made to their parents. Eventually, many contract a sexually transmitted disease. According to the Children's Rights Protection Center in Thailand, the HIV infection rate among that nation's child prostitutes is about 50 percent. If a girl is found to have the AIDS virus, she is usually sent back to her home village, where she probably will not be able to access medical care. Because people in remote areas have had little experience with HIV, the virus may end up spreading further.

FIGURE 7-13: Women terracing in Burkina Faso. Using blocks of wood, these women beat the earth into ridges to control soil erosion. Such physical labor is only part of their daily routine, which also includes carrying water, collecting fuelwood, cooking, cleaning, and caring for children. (Note the child strapped to the back of several women in the foreground.)

Discussion Questions

1. Do you think that continued growth of human population is a problem? Why or why not?

2. Using Nigeria to illustrate your answer, explain how the biotic potential of the human species and environmental resistance factors might interact to raise or lower the carrying capacity of the land.

3. Suggest ways that the carrying capacity of the Earth might be increased for humans. Suggest ways that the quality of life (or standard of living) for all humans could be increased. Can both of these objectives be met simultaneously? Why or why not?

4. Referring to the demographic statistics in Tables 7-1 and 7-2, compare population dynamics and the quality of life in more-developed and less-developed countries.

5. Some people argue that the status of women is not relevant to population growth and is not an appropriate topic for environmental science. Do you agree or disagree and why?

(continued on following page)

Summary

The scientific study of the sum of individual population acts is called demography. Basic demographic data include the crude birth rate and crude death rate: the number of births or deaths, respectively, per 1,000 people. The crude birth/death rate is equal to the number of births/deaths divided by the total population at mid-year multiplied by 1,000. The growth rate is determined by subtracting the crude death rate from the crude birth rate, with the difference expressed as a percent; in 2004, the world growth rate was 1.3 percent. When births are equal to deaths, the growth rate is zero, a condition known as zero population growth, or ZPG. Negative population growth, or NPG, occurs when deaths outnumber births.

Doubling times are calculated by using the rule of 70: assuming that the current growth rate remains constant, 70 divided by a nation's current growth rate yields the time it will take for that nation to double its population, assuming that growth rates remain constant. Usually, they do not.

Growth rates are affected by fertility, the population's age distribution, and migration. The general fertility rate is the number of live births per 1,000 women of childbearing age for any one year. The age-specific fertility rate is the number of live births per 1,000 women of a specific age group for any one year. The total fertility rate is the average number of children a woman will bear during her life, based on the current age-specific fertility rate and assuming that the current birth rate remains constant throughout her lifetime. The replacement fertility — the fertility rate needed to ensure that each set of parents is "replaced" by their offspring — is 2.1 for more-developed countries (MDCs) and 2.5 for less-developed countries (LDCs).

The age distribution of a population can be represented by a population profile. In general, profiles for LDCs have a pyramidal shape, a result of the rapid growth rate these nations are experiencing. Slow-growing MDCs have constrictive profiles, and MDCs with decreasing populations have distinctive profiles with a narrowing base. Population momentum occurs when a population continues to grow, despite falling fertility rates, because it has a large number of children entering the childbearing years. Populations that have a large proportion of young people (under 15) or old people (over 65) are said to have a high dependency load, that is, a high number of people who are not in the formal work force and do not contribute to the country's gross national product. Lastly, migration — movement into (immigration) or out of (emigration) a country or region — can have a significant effect on a nation's growth rate.

Populations remain at or near the carrying capacity when a balance is struck between the biotic potential of the species and environmental resistance. Several considerations must be taken into account when applying the concept of carrying capacity to humans; one of the most important is quality of life. Demographics can tell us a great deal about the quality of life in a particular country; five indicators are particularly telling: income levels, population density, urbanization, life expectancy, and infant and childhood mortality.

As a rule, income levels are much higher in MDCs than in LDCs. Population growth affects economic conditions for a society at large and for the average individual within that society.

Population density, how closely people are grouped, gives us some idea of the pressure exerted on the natural resource base of a geographic or political area. Population density in LDCs is more than twice that in MDCs. Urbanization, a rise in the number and size of cities, indicates that population density is increasing. The greatest increases in urbanization will occur in LDCs, where over one-half of the population will live in cities by 2015.

Life expectancy at birth is the average number of years a newborn can be expected to live. Factors that contribute to low life expectancy include war, hunger, disease (especially AIDS), and infant and childhood mortality.

The infant mortality rate (IMR) is the annual number of children under age one who die per 1,000 live births; the childhood mortality rate is the annual number of children between the ages of one and five who die per 1,000 live births. The IMR is widely considered the single best indicator of a society's quality of life. In 2004, the average infant mortality rate worldwide was 56/1,000, or 5.6 percent; the rate for MDCs was 7/1,000 (0.7 percent), while the rate for LDCs was 62/1,000 (6.2 percent). Infant mortality rates, whether in LDCs or MDCs, could be lowered if all people had access to clean water, sanitation and adequate health care, and if mothers were given the education and health services needed to care for their children.

Improving women's status is key to reducing poverty, improving overall health, and stabilizing population growth. Women's status is reflected in education, work, health care, and cultural traditions. To truly understand the social and environmental impact of the status of women, the cultural traditions and practices of women in countries where the status of females is low needs to be addressed.

(continued from previous page)

6. A 21-year-old college student (a citizen of the United States) in 2006 will be 80 years old in 2065, the year when the human population could reach nearly nine billion. Speculate on what life will be like for that person. Consider housing, job opportunities, cost of living, availability of resources, and recreational activities such as wilderness camping and travel. Also speculate on what kind of world that person will inhabit. Are relations with other countries likely to be stronger and more secure or weaker and more precarious than they are now?

KEY TERMS

actual rate of increase	environmental refugees	population density
age distribution	fertility	population momentum
age-specific fertility rate	general fertility rate	population profile
childhood mortality rate	growth rate	quality of life
	immigration	replacement fertility
crude birth rate	infant mortality rate	rule of 70
crude death rate	life expectancy	standard of living
cultural carrying capacity	maternal mortality ratio	total fertility rate
demography	migration	urbanization
dependency load	natural rate of increase	vital statistics
doubling time	negative population growth	zero population growth
emigration		

CHAPTER 8

Managing Human Population Growth

Learning Objectives

When you finish reading this chapter, you will be able to:

1. Briefly recount how the human population has grown historically.

2. Describe the demographic transition and identify the factors needed for the transition to occur.

3. Explain what is meant by the demographic trap and demographic fatigue. Identify the factors, events, and policies — in both MDCs and in the international community — that contribute to demographic fatigue.

4. Explain what a population policy is and describe how policies can be used to encourage or discourage growth.

5. Summarize the arguments given in favor of controlling or limiting population growth.

6. Define family planning and clearly describe the goal of family planning programs.

7. Distinguish between preconception and postconception birth control methods and give at least two examples of each.

Stabilizing population is an absolute need. Let me make it really clear — that if we do not stabilize population with justice, with humanity and with mercy, that it will be done for us by nature, by natural processes, and it will be done brutally and without pity, and it will be done whether we wish it or not, and if that occurs we will leave a ravaged and mutilated world for subsequent generations. The future is to a considerable extent discretionary, and we have it in our power and in our hands, this generation, to alter that future.

Henry Kendall

Meeting the challenges of the future — providing the opportunity for a quality life for all people while protecting the living systems upon which we depend — requires a limit to population growth. In this chapter, we look at how human populations have grown historically, the demographic transition and the demographic trap, the arguments in favor of controlling population growth, the policies used to influence growth and how those policies are changing, and the technologies used to control births (Figure 8-1).

FIGURE 8-1: Two elderly Chinese women stroll down a busy street. As the world's most populous nation, China is a focal point in discussions of population growth and management.

FIGURE 8-2: Cave drawing in Lascaux, France. Beautiful paintings dating from 20,000 B.C. on the walls of this cave suggest that life was not as brutish for our ancestors as was once believed.

How Has the Human Population Grown Historically?

Population control — for some people the words bring to mind a futuristic, Orwellian society. But the idea of limiting population growth is not new. Anthropologists and historians seeking to unravel the tale of our species have shown that population control, in various forms, has taken place throughout human history.

The fossil record indicates that hominid (humanlike) species existed three million years ago, perhaps longer. Some 40,000 to 50,000 years ago, our own species, *Homo sapiens,* made its appearance. For most of its tenure on Earth, *Homo sapiens* has existed by gathering wild plants and hunting wild animals. The early hunter-gatherers were nomadic and probably lived in small bands of several dozen people. Dominated by the forces of nature, they possessed a strong sense of the Earth — the location and growth of plants, the habits of animals, and seasonal weather patterns.

It is popularly believed that life for these tribal peoples was harsh, a struggle against the elements, predation and disease, with high death rates keeping populations relatively low. However, anthropologists offer a different view of our early ancestors, maintaining that life for them was far less brutish than is widely believed (Figure 8-2). They contend that early hunter-gatherers were relatively healthy. The mainstay of their diet was meat — the large mammals, including reindeer, mammoths, horses, bison, and wild cattle that roamed lush, grassy plains. They practiced intentional population control through a variety of means, including abstinence from sexual intercourse, birth spacing, and infanticide. The global human population grew slowly. Up until about 10,000 years ago, when people first began to practice agriculture, the Earth probably supported about five million people, a size equivalent to the population of Finland.

Anthropologists believe that humans developed agriculture out of necessity rather than out of any innate desire to establish permanent dwellings. About 10,000 to 12,000 years ago, the last ice age came to an end. As the glaciers receded northward, the climate began to warm significantly and forests of evergreens and birches invaded the grassy plains. Loss of habitat, along with human predation, led to the extinction of numerous species of large animals, including the woolly mammoth, steppe bison, and giant elk. Accordingly, both the diet and lifestyle of prehistoric peoples underwent a significant change. For the first time, humans began to cultivate their own food, relying more on plants than they had previously. For many populations, the nomadic life style became a thing of the past, and as it disappeared, so did many of the reasons for limiting births. Having numerous young children did not present the same practical difficulties as it had for nomadic women. In fact, with the onset of agriculture, labor — to maintain homes and gardens, collect firewood and edible plants, and perform other necessary chores — became a strong impetus for having children. The rise of agriculture, then, led to increased human population pressures and impacts on the environment.

The rise of cities also dramatically increased the human impact upon the biosphere. Food produced in the countryside was consumed in the city. Food wastes were no longer returned to the soil, and the soil became less productive. Also, with urban populations, human wastes were concentrated, sometimes in amounts too great for rivers and streams to decompose effectively. The problems resulting from urbanization grew over the centuries. The urban poor often lived in horrendous circumstances, suffering from the ill effects of crowding, poverty, and hunger.

In thirteenth- and fourteenth-century England, as well as other medieval societies, infanticide was a means of managing population. "Overlaying" was common — the instance of a nursing mother who falls asleep and (presumably accidentally) rolls over onto her baby, suffocating it. Beginning in 1348, the Black Death, or bubonic plague, greatly reduced the populations of Europe and Asia, perhaps by as much as one-half. Crowding in cities contributed to the spread and severity of the plague. Within several hundred years, however, populations had rebounded. By the sixteenth century, infanticide was again common. Overlaying, drugging a baby with gin or opiates, and outright starvation were used to kill unwanted children. Others were left at "foundling" hospitals, which sprang up in the eighteenth century to handle the huge number of abandoned babies. Children left at such institutions rarely survived their first year of life.

During the early phases of the Industrial Revolution (around the late 1700s), children came to be viewed as valuable sources of cheap labor (for employers) and income (for parents). At the same time, advances in agriculture made it possible to produce more food with less labor; some people left rural areas to find work in cities. For the first time, the effects of exponential growth become apparent (Figure 8-3).

The population in those areas we now refer to as the developed world (Great Britain, Europe, the United States, and Canada) began to grow more rapidly. By the 1820s, the human population reached one billion for the first time. Medical advances and improvements in sanitation and hygiene helped to control diseases and

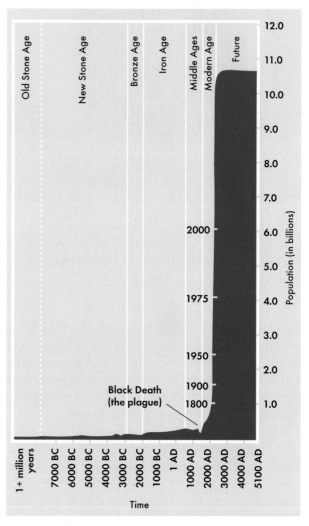

FIGURE 8-3: Growth in human population over time. The human population is currently increasing exponentially, with potentially catastrophic consequences for civilization and the biosphere.

epidemics. These advances in death control ushered in the current period of rapid growth in the global human population, as births began to outstrip deaths.

By 1900, another shift in human population growth occurred. The birth rate in the industrialized world dropped markedly, slowing the growth rate. Three factors contributed to this lowered growth. First, the Industrial Revolution led to a rise in living standards for many people, and lower growth rates typically accompany rises in the standard of living. Second, safe and inexpensive means of birth control (such as condoms made of sheep gut, vaginal douches, and vaginal plugs) were introduced. Third, an increase in the cost of child rearing (with the introduction of child labor laws and mandatory education statutes) meant that having more children yielded fewer material benefits. Thus, while the human population continued to grow, it grew far more slowly for most industrial nations.

What Is the Demographic Transition?

The **demographic transition** is the movement of a nation from high population growth to low population growth, as it develops economically. Historically, the economic development that spurred nations' demographic transition came about through industrialization.

Demographic transition consists of four stages (Figure 8-4). In Stage 1, birth and death rates are both high.

Birth rates are actually fairly constant, but death rates fluctuate as seasonal and cyclic factors temporarily raise or lower the carrying capacity of the environment. In Stage 2, death rates fall but birth rates remain high, and thus the population undergoes rapid growth. In Stage 3, as economic development improves the standard of living, birth rates fall, and the nation's growth rate declines. The population has made the transition from high growth to low growth. In Stage 4, the growth rate reaches zero, and the nation's population stabilizes. Some nations that reach Stage 4 of the demographic transition may even have negative growth rates and thus decreasing populations.

Industrialized nations are in Stage 3 or 4 of the demographic transition, as are a number of former developing nations. For example, through successful early efforts to lower fertility rates, South Korea and Taiwan created a self-reinforcing cycle in which the trend toward smaller families led to improvements in living standards, which further lowered fertility rates. In fact, both South Korea and Taiwan have a total fertility rate of 1.2, well below replacement level. Demographers expect their populations to decline, by eight and three percent, respectively, by the year 2050.

What Is the Demographic Trap?

Unfortunately, most developing countries are stalled in Stage 2 of the demographic transition. In effect, they are caught in a **demographic trap**, a situation

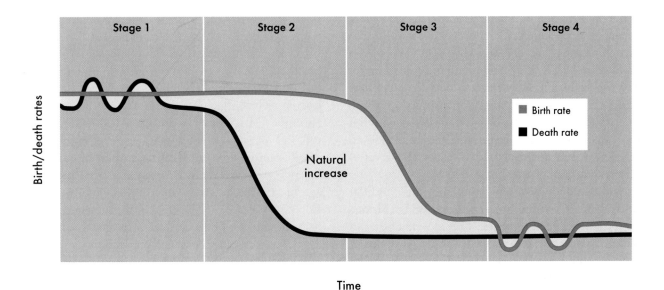

FIGURE 8-4: Demographic transition. ***Source:*** *Population Reference Bureau.*

that occurs when falling living standards reinforce the prevailing high fertility, which in turn reinforces the decline in living standards. As poverty deepens, people feel increasingly powerless to shape their own lives. In this situation, they tend to rely on more children to provide them with economic security.

For nations caught in the demographic trap, high birth rates and low death rates mean explosive population growth. In Africa, Asia and Latin America, death rates dropped dramatically throughout the 1950s and 1960s, thanks largely to advances in health care. At the time of the first World Population Conference, held in Bucharest, Romania in 1974, industrialized countries, including the United States (which had supported efforts to slow growth in developing nations since the Kennedy administration) argued that unchecked population growth would impede economic development and make it difficult to raise standards of living. Controlling population growth, they maintained, was the key to fostering economic growth. Developing nations, however, resisted calls to slow population growth, which some interpreted as racism on the part of the developed world. Countries throughout Latin America, francophone Africa, and parts of Asia insisted that the key to slower growth was economic development. They argued that once economic development took hold, population growth would slow and eventually stabilize.

For a time, conditions did improve in much of the developing world, but during the 1970s and 1980s, economic development did not keep pace with rapid growth. Governments throughout the developing world struggled to educate ever greater numbers of school-age children; create jobs for an expanding workforce; provide basic services (including sewage treatment facilities, roads and bridges, and healthcare facilities); secure adequate energy sources for industrial and business use; and combat environmental problems exacerbated by rapid growth (including deforestation, desertification, and the pollution and depletion of water supplies). Their decades-long struggle left many developing nations in a state of **demographic fatigue**, a condition characterized by a lack of financial resources and an inability to deal effectively with threats such as natural catastrophes and disease outbreaks.

At the second world population meeting, the 1984 International Conference on Population, held in Mexico City, less-developed countries (LDCs) were ready and willing to accept the developed world's assistance in order to slow population growth. But under the administration of President Ronald Reagan, the United States did an about-face, declaring that population

growth is not necessarily a problem, there is no negative relationship between population growth and economic development, and free market economics are the answer to development problems. In what became known as the Mexico City Policy, the Reagan administration eliminated funding for population activities in countries whose population programs were deemed coercive and for organizations involved in abortion-related activities. As a result of the Mexico City Policy, organizations such as the U.N. Population Fund, which had relied heavily on U.S. contributions for several decades, were forced to curtail or eliminate programs throughout the developing world. Unfortunately, many of the eliminated programs did not involve abortion or coercion; rather, they provided couples with education in human sexuality, information on planning and achieving a desired family size, and access to modern contraceptives. The United States did not restore funding for population activities until 1993, under President William Clinton. In 2002, President George W. Bush abruptly cut off funding once again, in response to allegations by a Virginia-based group that claimed U.S. funds were being used to support abortion activities in China. A committee investigating the allegations — established by President Bush — found no evidence to support the charges. Indeed, it verified that the U.N. Population Fund had set up a separate accounting system to track and control the use of U.S. funds in order to ensure that the agency complied with U.S. wishes. Even so, against the advice of the committee and then-U.S. Secretary of Defense Colin Powell, President Bush refused to restore funding for international population control efforts.

Sadly, some countries suffering demographic fatigue have slipped back into Stage 1 of the demographic transition, in which both birth and death rates are high. They soon may reach zero population growth, but they will do so at a terrible price. In much of sub-Saharan Africa, life expectancies are falling dramatically as civil conflict and disease, chiefly the Acquired Immune Deficiency Syndrome (AIDS), take their toll. In Botswana, where about 37 percent of the adult population is infected with the human immunodeficiency virus (HIV), which causes AIDS, the average life expectancy has fallen to 36 years. By 2004, Botswana's death rate (26/1,000) was nearly equal to its birth rate (27/1,000) and was expected to continue to rise. As a result, Botswana's population is expected to fall from 1.7 million in 2004 to 1.1 million in 2025. The populations of South Africa and Swaziland, with HIV infection rates of 22 and 39 percent, respectively, also are expected to decline in the next 20 years. The HIV/AIDS pandemic shows little sign of slowing, since most governments in sub-Saharan Africa — unlike those in the developed world — are unable to provide their citizens with the costly retroviral drugs needed to treat the disease.

In contrast to most potentially fatal diseases, HIV/AIDS affects primarily young adults, including professionals such as teachers, engineers, agronomists, and businesspersons — the very people so essential to sustained economic development! The epidemic is also creating a large contingent of orphans, a phenomenon that will further stress the political, social, and economic life of affected nations. In 2001, 34 million children in sub-Saharan Africa were orphans, one-third of them — 11 million — due to AIDS (Figure 8-5). By 2010, the number of AIDS orphans in the region will double to nearly 20 million.

Why Control Population Growth?

If you took a poll, and asked people, "Should we control human population growth?," you would likely find that those who answer "Yes" fall into one of two groups: those who oppose rapid population growth and those who oppose population growth of any kind. The first group sees rapid growth in LDCs as a major contributor to many social, economic, and environmental problems. They argue that, to solve those problems and free nations from the demographic trap, growth rates must be slowed and population sizes stabilized.

The second group, in contrast, opposes any population growth. They argue that slow-growth countries like the United States, Japan, and Canada exact a greater toll on the biosphere than do LDCs because of their high rate of per capita resource consumption. Opponents to growth maintain that a larger human population will ultimately lead to environmental degradation; degradation eventually will increase environmental resistance, resulting in worsened living conditions and a subsequent lowering of the carrying capacity. In addition, they point out that violent conflict and population growth are strongly interrelated. Often, resource scarcity fosters aggression; population pressures and conflicts over resources were at least partially responsible for World War I and World War II. Many links have been established between increasing population density in an area and antisocial behavior. Abundant evidence exists to link increased violence and mental disorders to the crowded conditions that occur in and around large cities.

FIGURE 8-5: A South African boy orphaned by AIDS holds a "memory box" filled with family photos, important documents, and positive memories. *Secure the Future*, an alliance between the Bristol-Myers Squibb Company and several African nations, creates memory boxes to help children safe-keep family history and learn how to grieve for parents lost to HIV/AIDS.

Whatever your position on human population growth, most people can agree that it is a worthwhile goal to provide all humans — regardless of the size of the global population — with the opportunity for a life of quality and dignity. That can appear to be an impossible task. Population growth may seem to be a problem beyond the reach of the individual, but there are things that each of us can do, from encouraging smaller families to lessening our own impact, and that of our families, on the environment (see *What You Can Do: Population Growth*, page 163).

What Is a Population Policy?

A **population policy** is any planned course of action taken by a government designed to influence its constituents' choices or decisions on fertility or migration. Approximately 90 countries (most of them in the developing world) have official policies on population growth or size.

Population policies are developed as a result of the way population changes are perceived: Is continued growth or lack of growth a problem? Are present fertility and growth rates too high or too low? Is the population size too large for the resource base or too small to defend itself from real (or perceived) aggression? For countries with rapidly growing populations, could the standard of living be improved if the population were smaller? For countries with markedly declining populations, what measures should be taken to cope with the consequences of an aging population and a smaller workforce? How we answer these questions and what we choose to do about them are very much a matter of values.

Pronatalist Population Policies

Throughout the developed world, where fertility rates are near or below replacement level and where populations are stable or declining, many governments are concerned about aging populations and the implications for national social security networks, healthcare systems, and labor pools. Other issues are also of concern: immigration, a loss of national identity, a loss of tradition, and (in some countries, such as Bulgaria) emigration and "brain drain," as young adults leave their native country to seek better prospects in other developed nations. These concerns are especially apparent in Europe, where fertility rates are very low, but even in the United States, slower population growth underlies many issues, explicitly or implicitly. For example, the country's declining fertility rate is at the heart of the ongoing debate over the future of Social Security: What, if anything, should be done to shore up Social Security, given that the retirement — and longer life expectancy — of the baby boom generation is certain to strain the national safety net?

Many nations with slowly growing or declining populations have a **pronatalist policy**, one that encourages increased fertility, a higher birth rate, and larger family size. Worldwide, 88 countries offer cash payments to families on the birth of a child to compensate them for a loss of income or an increase in expenses. In Italy, for example, mothers who bear a second child receive 1,000 euros. The United States and other nations provide tax credits or benefits to help defray the costs of children, thereby implementing an unofficial (or *de facto*) pronatalist population policy.

Many European nations have adopted pronatalist measures that promote not just childbearing but *child-rearing*. These measures include generous maternal and paternal leave on the birth of a child; parental leave during the early years of a child's life; childcare assistance, including free or subsidized daycare; and modified working arrangements, including flexible hours, part-time work, and family-related leave. Among developed nations, France has perhaps the strongest pronatalist policy: The government awards mothers of each new baby almost 800 euros, grants financial support to parents who leave their child (under the age of three) with a daycare mother, offers a child-raising allowance for parents who take a full or partial job break to stay home with children under three (if at least two children are cared for), and provides a subsidy to working parents with children under six who employ a daily helper.

Antinatalist Population Policies

Throughout the developing world, high growth rates, rather than low ones, are a cause for concern. As a result, many LDCs have developed an **antinatalist policy**, one that encourages lower fertility, a lower birth rate, and a smaller family size. Unfortunately, even when an official policy is in place, it may or may not be implemented due to a lack of funds.

Antinatalist population policies have proven successful in many countries and regions, including Cuba, the Indian state of Kerala, and Peru. In 1990, Peruvian President Alberto Fujimora instituted a National Population Program that includes free distribution of contraceptives. The best-known (and arguably the most controversial) antinatalist policy is that instituted by the world's most populous nation; see Box 8-1: *China's One-Child-Per-Family Policy*, page 164.

Although developing nations often take very different approaches to slowing growth, those that are successful share a commitment to family planning. Broadly defined, **family planning** encompasses a wide variety of measures that enable parents to control the number of children they have and the spacing of their children's births; it includes education in human sexuality, hygiene, prenatal and postnatal care, and preconception and postconception birth control measures. But while family planning is a key element of policies that effectively slow population growth, its goal is not necessarily to limit births; rather, it is to enable couples to have healthy children, to care for their children, and to have the number of children they want (Figure 8-6).

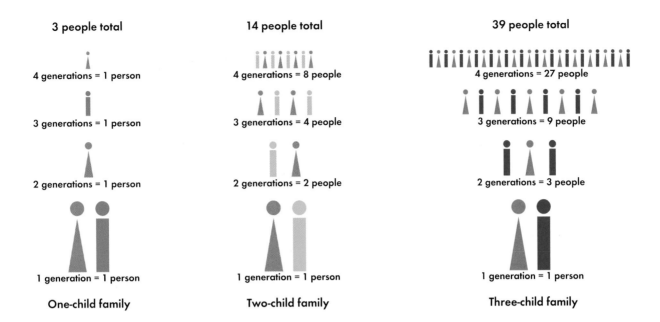

FIGURE 8-6: The consequences of family size across generations. Imagine that three couples decide to begin families. Couple A wants just one child. If their child and grandchild also have just one child, after three successive generations, a total of three people will have been born. Couple B wants two children, as do their children and grandchildren. After three successive generations, 14 people will have been born. Couple C decides to have three children, a decision their children and grandchildren share. After three successive generations, 39 people will have been born.

How Are Population Policies Changing?

During the twentieth century, population policies — both pronatalist and antinatalist — sometimes led to serious abuses. Perhaps the most infamous of these was the German population policy of the 1930s, designed to purify the Aryan "race" and eliminate "undesirable" groups. In addition to financial incentives (such as lower taxes, allowances, loans, and subsidies) for childbearing given to couples of German heritage, the Nazi government also awarded a "Mother's Cross" — in gold, silver, or bronze — to women who bore eight, six, or four children who were "of German blood and hereditarily healthy." A 1933 law authorized the compulsory sterilization of over 400,000 individuals labeled as having hereditary defects, ranging from mental illness to alcoholism. Abuses also occurred during the dictatorships of Spain's Francisco Franco (beginning in the 1930s) and Romania's Nicolae Ceausescu (beginning in the 1960s). In more recent decades, individuals, nongovernmental organizations, and other nations have criticized the antinatalist policies of China, and to a lesser degree, India (particularly under Indira Gandhi's first term), characterizing them as coercive and extreme.

By the late 1980s, women's health and rights advocates began to express grave concerns about policies that emphasize government-set targets for population size, contraceptive use, and family size. Such policies, they argued, focus too narrowly on delivering contraceptives and related information; they fail to meet the needs of women and their families, and in the most egregious cases, they violate a woman's reproductive freedom. Gradually, these arguments began to change the way demographers, nongovernmental organizations, and governments view population issues and develop population policies. Those changes were evident at the world's third international population conference — the largest gathering of its kind — the International Conference on Population and Development (ICPD).

The International Conference on Population and Development

Convened in Cairo, Egypt, in 1994, the ICPD, or Cairo Summit, is credited with fundamentally changing the way governments view population issues. Delegates from 179 nations and over 1,000 nongovernmental organizations agreed that without slower population growth, it will be virtually impossible to free poor countries from the demographic trap, with potentially dire results for the environment and humanity. They resolved the decades-old debate about which comes first, slower population growth or economic development. The answer: Both — population and development go hand-in-hand and cannot be separated. Moreover, in a reversal from past population conferences, the ICPD focused on individuals, human rights, and personal aspirations. The ICPD consensus: Responding to individual needs is a more humane and effective way to slow population growth than the old models that focused on population targets and narrow interpretations of family planning.

ICPD delegates crafted an ambitious Programme of Action that calls on governments to promote human development and stabilize world population by 2015. It includes over 200 recommendations in the areas of health, education, development, and social welfare. A key recommendation urges nations to provide comprehensive reproductive health care, including family planning, safe pregnancy and delivery services, abortion where legal, prevention and treatment of sexually transmitted diseases (including HIV/AIDS), information and counseling on sexuality, and the elimination of harmful practices against women (such as genital mutilation and forced marriage). Other recommendations concern empowering women and girls in the economic, political, and social arenas; removing gender disparities in education; reducing infant mortality rates; fostering economic and social development for the poor, disadvantaged, and elderly; increasing male participation and responsibility in family and community life; reducing consumption of goods and services, chiefly in the developed world; and removing or eliminating "target" family sizes from national population policies (Figure 8-7).

Ten years after the ICPD, progress has been made in numerous areas. Nearly three-fourths (131) of participating countries have changed national policies or laws, or made institutional changes, to recognize reproductive rights. The constitutions of South Africa and Venezuela, for instance, now include reproductive rights as fundamental human rights. Some developing nations have made basic philosophical changes to their national population policies; India, for example, eliminated mandated targets for contraceptive use and instead now emphasizes comprehensive reproductive and child health care services, with a focus on family health and welfare. In the many nations that now offer broader reproductive health and family planning

FIGURE 8-7: Schoolchildren in Kenya, 1996. The Programme of Action crafted by delegates to the 1994 ICPD includes a strong emphasis on increasing educational opportunities for children in LDCs.

services, the abortion rate has fallen. For example, in the early 1990s, abortion was the principal form of birth control in Kazakstan, Uzbekistan and the Krygyz Republic, but by 1999, abortion rates had dropped by as much as one-half while the use of modern contraceptives had increased 30 to 50 percent. Seventeen nations have outlawed female genital mutilation (a practice once deemed too sensitive to even discuss in an international setting), and many communities are taking action to eliminate it. And finally, there has been increased awareness and action regarding issues such as early marriage, obstetric fistula (a condition that occurs during obstructed labor, when a caesarean section is unavailable), unsafe abortion, gender-based violence, and trafficking in women and children.

Despite the progress, much work remains. More than 350 million couples do not have access to a full range of family planning services, and demand for such services will increase by 40 percent by 2025. In the developing world, one-third of women receive no health care during pregnancy, most deliveries (60 percent) take place outside of health care facilities, and trained personnel assist only half of all deliveries. (In many developing nations, particularly in sub-Saharan Africa, the proportion of women who receive no health care during pregnancy is much higher; and the proportion who deliver their child outside healthcare facilities, without the assistance of trained personnel, is much lower.) Rates of maternal and infant mortality are terribly high throughout the developing world; even in the United States, they remain unacceptably high among some demographic groups. Little progress has been made in slowing HIV/AIDS. In 2003, there were an estimated 5 million new HIV infections — an average of 14,000 per day; women accounted for 40 percent of new infections; children, 20 percent. Approximately 38 million people are presently living with the disease. Nations struggle to meet the social service needs of underserved rural communities while coping with a growing tide of urban immigrants. And everywhere, the stress on the global environment is inescapable: climate change, deforestation, water scarcity, degraded farmland, and diminishing biological diversity. These and other environmental problems make it more difficult to alleviate poverty, foster sustainable development, and address gender inequality.

The cost of implementing the "Programme of Action" was estimated at $17 billion by 2000. Two-thirds of the money was to come from LDCs; one-third was to be provided by more-developed countries (MDCs). Unfortunately, the MDCs, notably the United States,

failed to meet their financial commitment to the plan. For the Programme of Action to succeed, the international community must meet its obligations and provide promised resources to the developing world. If humanitarian reasons alone do not convince MDCs to assist LDCs, perhaps pragmatism will: 98 percent of population growth is now occurring in the developing world. Can we in the richer countries of the world afford not to invest in their well-being and security? Can we afford the political instability, social tensions, and environmental degradation that will almost certainly accompany continued rapid growth?

What Methods Are Used to Control Births?

Throughout most of human history, men and women had few options for controlling births. They relied on natural forms of contraception. Breast-feeding, for example, stimulates the production of hormones that may reduce the likelihood of ovulation; if a woman breast-feeds for two years or so, she may be able to delay pregnancy. The rhythm, or periodic abstinence, method is based on a woman's menstrual cycle. A couple abstains from intercourse during the time when the woman ovulates. Unfortunately, like breast-feeding, rhythm has a high failure rate; it is only about 76 percent effective at preventing pregnancy. Moreover, for it to be as effective as possible, a couple must have a fairly sophisticated knowledge of the body processes on which it is based, and it requires a degree of cooperation between husband and wife that is unusual in more traditional cultures. Although rhythm is not reliable, it is popular among certain groups (worldwide, 10 to 15 million people practice rhythm). Many Catholics, especially in the developing world, consider rhythm the only acceptable form of birth control, but religious beliefs do not always override the desire for control over family size. In Catholic Latin America, sterilization and the pill, though prohibited by the church, are the most popular methods of birth control. Many Catholics in the United States use the pill or other methods of contraception; Italy has one of the highest abortion rates in the world.

To meet the demand for reliable and safe birth control, researchers have developed a wide variety of modern contraceptive methods. These can be broadly categorized into two groups, those that prevent conception (preconception) and those that act after conception has occurred (postconception). The effectiveness of all methods depends heavily on factors beyond the control of science: the availability of medical services and reliable sources of birth control devices, adequate education in the use of the method, religious beliefs, social conventions (especially regarding the status of women), and the laws of each particular country.

Preconception Birth Control Methods

Preconception birth control methods include barrier methods, hormonal methods, and sterilization. Barrier methods include the condom, vaginal sponge, diaphragm, and spermicides.

Condoms, generally made of latex and thus commonly called rubbers, are from 70 to 90 percent effective depending on their quality and how they are used. Using a condom in combination with a spermicide, a foam or fluid that kills sperm, increases its effectiveness to about 95 percent. The condom has become increasingly popular in recent years, primarily because it is the only contraceptive that effectively prevents the spread of sexually transmitted diseases such as syphilis, gonorrhea, and especially AIDS (Figure 8-8).

Barrier methods for female use include spermicides, the vaginal sponge, and the diaphragm. These prevent passage of the sperm into the uterus and thus prevent fertilization. Spermicides are 75 to 82 percent effective, with foams more effective than creams, jellies, and suppositories. The sponge and diaphragm are typically used with spermicides, which increase their effectiveness to 83 percent (sponge) and 87 percent (diaphragm).

Barrier methods are generally not as effective as hormonal contraceptives, which interrupt the reproductive cycle, preventing the cyclic maturation and release of eggs. The pill, an oral contraceptive, is the third most popular birth control method in the world. This contraceptive can be 99 percent effective if taken consistently. Women with inconsistent access to the pill and little education as to the way it works often have significantly lower rates of success; 4 to 10 percent of users worldwide become pregnant because of improper use. For these reasons, the pill is generally not considered a good choice for women in LDCs, where supplies may be erratic. Moreover, many women in LDCs hesitate to take the pill because of its side effects (which pose a greater health threat in the absence of routine medical care).

Other hormonal methods, such as injections or implants, may be more efficient than the pill because they are less expensive and need to be administered

FIGURE 8-8: Poster encouraging condom use, Papua New Guinea. Translated, the text reads, "I am not afraid. I have protection."

much less frequently. Depo-Provera is an injectable hormonal contraceptive; it is effective for up to three months. Norplant is a hormonal contraceptive that is implanted under the skin of the upper arm. Effective for up to five years, Norplant must be removed by medical personnel after the duration of effectiveness expires. There are numerous side effects associated with these contraceptives: menstrual disorders, headaches, weight gain, depression, loss of libido, abdominal disorders, and delayed return in fertility after use of the method has been stopped. Some preliminary studies suggest that Depo-Provera also may be linked to an increased long-term risk of cervical cancer.

Sterilization, the most reliable method of birth control, is the most popular method throughout the world.

➤ Cherish fewer children. Support relatives and friends who decide to have just one or two children or none. Avoid pressuring your children to bear children. And don't believe the stereotype that says single children and single adults are unhappy — it isn't true!

➤ Spread the love around. If you've got a strong parental urge, consider adopting children rather than having your own. Make enriching the lives of other people's children a part of your life.

➤ "Onlies" are OK. If you decide to have children, consider having only one or, at most, two. Each child born in the United States has an enormous impact on the environment due to our heavy consumption of water, energy, and goods.

➤ Mandate equal opportunity for women. Where women have better educational and economic opportunities, the birth rate has declined.

➤ Make contraceptives available globally. During the next two decades, three billion young people will enter their reproductive years. Currently, only about half of fertile women have access to contraception. Encourage your congressional representatives to support expanded family planning programs in the United States and abroad.

➤ Work to eliminate the need for abortion. Outlawing abortion does not improve family planning or make for happy, wanted children; rather, it leads to dangerous illegal abortions, increased mortality rates for women, and unwanted children. Most people agree that eliminating the need for abortion is a worthier goal. Support family planning programs and efforts to make contraceptives widely available.

➤ Work to encourage an attitude of respect for human life — the unborn, the poor, the homeless, the parentless — so that everyone can enjoy a life of dignity.

➤ Limit development. Use your vote to promote land-use policies that preserve open space and farming, not only as a means of production, but as a way of life.

➤ Buy local produce or grow your own. When developing nations use scarce cropland to grow food for export, they deprive their populations of that land. Feed your family with foods produced from your area — even start your own garden.

➤ Sponsor a foster child in a developing country. For a small monthly fee, reputable organizations such as Plan USA link caring people in the United States with needy children and their families overseas. The programs help families and communities become self-sufficient. Write Plan USA, 155 Plan Way, Warwick, RI, 02886 or call 1-800-556-7918 toll-free (in RI, 401-738-5600).

Source: Adapted from a series on personal ecology by Monte Paulsen, editor and publisher of Casco Bay Weekly, *in Portland, Maine.*

Both men and women can be sterilized: Vasectomy is the cutting or tying of the tubes by which sperm leave the body, and tubal ligation is the tying of the oviducts by which eggs reach the uterus. Sterilization is becoming increasingly common in many LDCs. In China and India, for example, large numbers of women have undergone sterilization for contraceptive purposes.

Sterilization procedures, which are usually irreversible, are sometimes used in family planning programs that limit, rather than extend, the reproductive freedom of the individual. Among the countries and territories where abuses have been reported, Puerto Rico has been singled out by women's groups as having the most abusive sterilization program. The government began promoting sterilization in the mid-1940s, and by 1965, Puerto Rico had the highest sterilization rate in the world. Fully one-third of all women who had ever been married were sterilized, 40 percent of them before the age of 25. Many women were not told that the procedure was irreversible, and some women were sterilized unknowingly, while under anesthesia for other procedures.

BOX 8–1 China's Program: The One-Child-Per-Family Policy

China offers the best-known example of a population control program, but the Chinese wouldn't call it that. Because China follows a strong Marxist ideology, the Chinese do not officially acknowledge population growth as a problem. Instead, they maintain that the program was designed to "maximize the health and well-being of each child born." According to official government policy, births are to be controlled because healthy, happy children are good for the state, and what is good for the state is good for all of the people (Figure 8-9). Even though China's delegation to the 1974 Bucharest population conference formally came out against population limitations, their country had already instituted one of the strongest family planning efforts that had ever been attempted.

Why did the Chinese feel it was necessary to initiate this program?

The first census taken of the Chinese population revealed that in 1953, China had a population of over 580 million people. By the late 1970s, that figure had reached over one billion people. Fully 20 percent of the world's people were trying to live on seven percent of the world's arable land. After years of terrible famine (perhaps as many as 30 million people died), poverty and political turmoil, the government became convinced that China could adequately support no more than 650 to 700 million people. The availability of fresh water and food was especially critical in the determination

of carrying capacity. Because the population already exceeded the carrying capacity by more than 300 million people, a stringent family planning effort was needed to slow population growth and eventually reduce population size. The government estimated that, assuming growth rates continued to decline, it would take about a hundred years to reduce the population to 700 million. Because of population momentum, even if the growth rate fell to replacement level, China's population was predicted to peak between 1.2 and 1.5 billion people. China became the first nation to have as an official goal the end of population growth and the subsequent lowering of absolute numbers by a significant amount. Beginning in 1969, China achieved a transition from high to low birth rates. Let's examine how they did it.

In the late 1960s, the government instituted a family planning program that reached into every village across China. This program made birth control, including birth control pills, accessible to all. An extensive health care system was initiated that for the first time paid attention to China's rural poor. This important development would not have been possible without the major political change that took place in China. It was a change that stressed family planning as a way to increase the standard of living for all, gave women increased access to jobs and education, redistributed wealth to the masses, and provided for social security in the

form of old age pensions and food cooperatives.

The 1970s became known as the "later, longer, fewer" years. A mass educational effort, employing slogans, posters and radio and television programs, encouraged people to marry later, wait longer before starting a family, and have fewer children. The age for legal marriage for women was increased from as young as possible to the early twenties; contraceptives, abortions, and sterilizations were provided free of charge; and supplementary payments, longer maternity leaves, free education and health care for children, preference in housing, and retirement incomes were granted to couples who conformed to the goals of the program. Couples who did not conform lost benefits, paid heavy fines, and risked the loss of job promotions.

In 1979, China's leaders felt that population growth was not falling fast enough, and the emphasis switched to one child per family. Thus began the most restrictive family planning program ever attempted. The outside world deemed it extremely coercive. Intense pressure was brought to bear on women who became "unofficially" pregnant. Few women were able to resist this pressure; many had abortions or were sterilized. But even as the policy seemed to be having its intended effect — slower population growth — opposition to it increased, especially in rural areas, where children are needed to help with chores and food production. Critics claimed that the "one child per family" policy was unduly coercive. The government countered that the drop in population growth was the result of education, not coercion, and that in the long run, the policy would help to prevent widespread misery by offering a higher standard of living to a smaller population. The Chinese people, the government said, had agreed to put aside individual desires for the good of the society, at least temporarily.

Throughout the 1980s and 1990s, the Chinese economy grew at a healthy pace, spurred in part by economic reforms. The government began to encourage individual initiatives over collective agriculture and changed its guaranteed policies toward employment and old age. As personal incomes increased, families (particularly rural ones) began to have extra children, especially boys, in spite of the official policy. The government also relaxed its policy on marriage age. By 2004, China's population of 1.3 billion was growing at a much slower rate — just 0.6 percent — and was projected to reach nearly 1.5 billion by 2025.

Now, in the twenty-first century, China finds itself at a crossroads in its quest to gradually reduce its population size. One road is the controversial but proven path of population control that the country traveled during the strictest years of the "one child per family" policy. The other road is the path of the traditional demographic transition, where economic well-being leads to smaller families, slower growth, and eventual population stabilization. Whether or not this latter path can work quickly enough to slow the growth of such a large population is unknown. Economic prosperity appears to be slowing the growth rate. Moreover, the country is under mounting pressure from human rights activists — both internally and abroad — who maintain that the "one child per family" policy violates individual rights. That pressure helped bring about an agreement, in 1998, between the Chinese government and the United Nations Fund for Population Activities (UNFPA) that allowed the agency to set up a pilot program promoting voluntary family planning in 32 of China's 2,700 counties. In those counties, the "one child per family" policy was eliminated. In its place, the UNFPA established a comprehensive program that offers a full range of modern, safe and effective family planning methods adapted to the age, family, and personal circumstances of the individual concerned. Key elements of the program include maternal health services; assisted delivery; prenatal and postpartum care; counseling and treatment services for sexually transmitted diseases (with a focus on HIV/AIDS prevention); and numerous activities to raise the status of women, including literacy efforts, training in the management of small enterprises, loans for economic activities, and advocacy for gender equality. The success of this voluntary program may well determine which of the paths toward population stability China chooses to travel in the years to come.

Figure 8-9: The goal of Chinese family planning — a healthy, happy child. A baby enjoys his day sunning on a sidewalk in Beijing, China, in the summer of 1990.

Research on reversible sterilization is currently underway and appears promising. If reversible procedures are developed, they would be as effective as permanent sterilization, but more acceptable and less threatening.

Postconception Birth Control Methods

Postconception birth control methods include the intrauterine device and abortion. An intrauterine device (IUD) is a small plastic or metal object inserted into the uterus that prevents implantation of the fertilized egg. Next to sterilization, the IUD is the most popular form of birth control in use worldwide. Chinese women account for three-quarters of all IUD users. The IUD is 95 percent effective; when used with a spermicide, it can be 98 percent effective. Some people prefer the IUD to the pill because it does not have to be administered every day. However, there is a possibility of side effects, including pelvic inflammatory disease and infertility. Moreover, skilled medical care is needed to insert the IUD properly and to conduct periodic checkups. In LDCs where medical services are inadequate, such a method is not a popular or practical option.

Abortion is the termination of a pregnancy after the embryo has become implanted in the lining of the uterine wall. An abortion can be performed surgically or medically. In a surgical abortion, the lining of the uterus is removed along with the developing embryo. In a medical abortion, a woman takes a combination of drugs (typically, mifepristone and misoprostol) within the first seven to nine weeks of her pregnancy. Mifepristone, formerly known as the RU-486 pill, is taken first; the drug blocks progesterone, a hormone needed to maintain pregnancy. As a result, the uterine sheds its lining (as it does in menstruation) and thus the developing embryo. Approximately two days later, misoprostol is taken; this second drug stimulates uterine contractions, expelling the embryo. Mifepristone and misoprostol are used widely in numerous developed countries and in China. After a 10-year struggle to win approval by the Food and Drug Administration, mifepristone was approved for use in the United States in September 2000. (Misoprostol was already used in the United States as a treatment for ulcers.) By early 2004, more than 130,000 U.S. women had used mifepristone to terminate a pregnancy.

Worldwide, about 46 million pregnancies end in abortion each year. Nearly one-half (20 million) of these are illegal, carried out mainly in LDCs. Because many women prohibited from legal abortion attempt to induce abortion themselves (using knitting needles, wire coat hangers, and poisons) or go to unskilled practitioners, the risk of death from an illegal abortion is 30 times greater than from a legal abortion. The World Health Organization attributes roughly 200,000 women's deaths per year to illegal abortions.

In the United States, the right to a legal abortion is a controversial and extremely divisive issue. The controversy revolves around the conflict between religious and moral beliefs about the status of the fetus — Is it a living human being? At what point does it gain human status? — and the right of women to choose whether or not they will bear children.

Some people argue that the right to an abortion ought to be legally ensured; they maintain that the option to terminate a pregnancy is essential to a woman's reproductive freedom. In several European countries where abortions are legal and where contraception and family planning are widely accessible, the number of abortions has dropped. It has been shown that where abortions are illegal, the number of abortions remains about the same; what changes is the incidence of mortality for the mothers.

Many who are personally opposed to abortion are uneasy about intervening (or allowing the state to intervene) in a decision that is so personal and private. They feel that abortion is a matter between a woman and her conscience, and they are reluctant to impose their own values on anyone else. But for many anti-abortionists, the sanctity of human life supersedes a woman's right to determine whether or not she will bear her unborn child. For those who believe that life begins at conception, it is a moral imperative to try to prevent what they perceive as murder.

The abortion debate is complicated by the arguments and tactics of some people on both sides of the issue. The murder of personnel at abortion clinics, the bombing of such facilities, and the harassment of female patients make a mockery of the pro-life movement's stance of love and compassion for fellow human beings, and indeed, most people who oppose abortion condemn such activities. Similarly, certain pro-choice arguments — particularly, that women alone can and should make the decision regarding the fate of their unborn children — effectively eliminate male participation in what is clearly the concern of two people.

Sadly, using abortion as a means of birth control can lead to abuses. Some women in the former Soviet Union, for example, had an average of four or five abortions throughout their childbearing years, a reflection of the fact that contraceptives were difficult to obtain. In certain Asian cultures, "son preference" is so strong that abortion frequently is used to end a pregnancy if the fetus is discovered to be female. In Mumbai (Bombay), India, researchers found that of 8,000 fetuses aborted after amniocentesis (a test designed to discover genetic defects, but sometimes used to ascertain the sex of a fetus), all but one were female. There is a terrible irony in this situation: Abortion, hailed by many as a means of liberating women, is sometimes used to prevent the birth of females.

Given the controversial nature of abortion and the abuses it can invite, most people agree that preventing pregnancy is preferable to terminating a pregnancy through abortion.

Contraceptive Use Worldwide

Worldwide, contraceptive use varies widely; people in developed nations enjoy easy access to birth control, while those in developing nations, in general, do not. About 59 percent of married women worldwide used some form of contraception in 2004; 52 percent used modern contraceptives. In MDCs, 69 percent used some form, with 58 percent using modern methods. In LDCs (excluding China, where modern contraceptives are readily available), 48 percent used some form, with 40 percent using modern methods. The U.S. Agency for International Development (AID) estimates that, in order to stabilize populations in developing countries, 80 percent of women of childbearing age should use birth control.

Contraceptive use also varies widely within the United States. Teenagers (especially unwed teens) and poor women are the groups least likely to use birth control. Although the U.S. teenage pregnancy rate fell 28 percent between 1990 (when it reached its peak) and 2000, it remains a serious problem. The United States has the highest rate of teen pregnancy and births in the industrialized world: More than 820,000 young women under the age of 20 (about one-third of that age cohort) become pregnant each year; nearly 20,000 of them are girls under the age of 15. The majority (80 percent) of teen pregnancies are unintended; 57 percent result in birth and one-third (33 percent) end in abortion. While the teenage abortion rate is still very high, it has fallen considerably from its 1998 high of 46 percent.

Some researchers believe that teen pregnancy is a reasonable choice for poor women in the United States, given the fact that they are likely to have family support available and the probability that their health will deteriorate throughout their twenties. However, many serious problems are associated with teen pregnancies. Pregnant teenagers often do not have adequate diets and do not get the prenatal or postnatal care they need. The rate of low-birth-weight babies born to teen mothers is higher than that of babies born to older mothers, and the survival rate for babies born to teens is lower. Moreover, maternal age has some effect upon the emotional, social, and intellectual development of the child — the younger the mother, the more likely her child will not be as bright or adjust as well as other children while growing up. Teen parents (mothers and fathers) do not complete as many years of schooling, on average, as those who delay parenthood. This consequence of teen parenthood may haunt young parents throughout their lives, since lack of education decreases occupational options and earnings potential. Society loses, too, when teen pregnancy rates are high: Teen childbearing translates into greater demand for public health and welfare services and thus greater public expenditures. Citizens who are less educated and less well-adjusted cannot achieve their full human potential, and thus may not contribute as significantly to society.

Various factors contribute to high teen pregnancy rates. Premarital sex among teens has become common; according to a study by researchers at Johns Hopkins University, more than one-quarter of all 15-year-olds in the United States reported having intercourse, as did about one-third of all 16-year-olds and almost one-half of all 17-year-olds. In addition, contraceptive use among teens remains low. Only about one-third of all sexually active teenage girls (between the ages of 15 and 19) use birth control.

The low rate of contraceptive use by teenagers in the United States explains, in part, why the rates of teen pregnancy and abortion are much higher in this country than in all other developed nations — at least double rates in Canada, England, and France and seven times higher than rates in the Netherlands. Teens in other MDCs are as sexually active as teens in the United States, and socioeconomic conditions in these nations are similar to those in this country. However, in other MDCs, sex education and free or low-cost contraceptive services are universally available.

1. Describe the demographic transition. Which stage do you think the United States is in and why? Which stage do you think Nigeria is in and why? For developing nations that are attempting to free themselves from the demographic trap, what are the implications of the HIV/AIDS epidemic?

2. Explain the difference between pronatalist and antinatalist policies. What factors cause a country to adopt one or the other? Do you think the United States should adopt an official population control policy? Why or why not? If you answer yes, what policy should it adopt?

3. What are the goals of family planning? How is family planning different from policies and programs that aim for zero or negative population growth?

4. List some important features of China's "one child per family" policy. Do you think such a policy would work in India, the United States, Canada, and Mexico? Why or why not?

5. List at least five methods of birth control. Which ones are most effective in LDCs and in MDCs? Explain.

Summary

For most of its tenure on Earth, beginning some 40,000 or 50,000 years ago, *Homo sapiens* existed by gathering wild plants and hunting wild animals. Up until about 10,000 years ago, the Earth probably supported about five million people. With the rise of agriculture, populations began to grow more rapidly and the environmental impact of human activities increased. By the beginning of the nineteenth century, the human population had reached one billion. Advances in death control ushered in a period of rapid growth. Birth rates in the western world began to fall with the onset of the Industrial Revolution and subsequent rising standard of living, the introduction of safe and reliable means of birth control, and an increase in the cost of childrearing. However, the population continued to grow (though more slowly) because there were so many more people in total. Currently, growth rates for most industrial nations are either low, zero, or negative.

The population path followed by the industrial nations, the demographic transition, describes the movement of a nation from high growth to low growth. It consists of four stages. In Stage 1, birth and death rates are both high. In Stage 2, death rates fall, but birth rates remain high, and thus the population undergoes rapid growth. In Stage 3, birth rates begin to fall, and the growth rate declines until it eventually nears zero. In Stage 4, the growth rate is at or below zero.

Much of the growth in the human population in the past 50 years has occurred in less-developed countries (LDCs), as a result of falling death rates and constant or slightly rising birth and fertility rates. Rapid growth and the environmental deterioration it causes are fueling a downward spiral in the standard of living. Some developing nations are caught in a demographic trap, unable to break out of Stage 2 of the demographic transition. Their governments are in a state of demographic fatigue, worn down and financially strapped after decades of struggling to combat the consequences of rapid population growth. Some LDCs have even slipped back into Stage 1.

Any planned course of action taken by a government designed to influence its constituents' choices or decisions on fertility or migration can be considered a population policy. A pronatalist policy encourages fertility and a higher birth rate; an antinatalist policy discourages fertility and encourages a lower birth rate. Pronatalist or antinatalist policies are developed as a result of the way population changes are perceived. Family planning encompasses a wide variety of measures that enable parents to control the number of children they have and the spacing of their children's births. The goal of family planning is not to limit births, but to enable couples to have healthy children and to have the number of children they want.

China became the first nation to have as an official goal the end of population growth and the subsequent lowering of absolute numbers by a significant amount. Beginning in 1969, China achieved a transition from high to low birth rates by implementing the strongest measures ever attempted. These included free birth control, higher marriage ages for women, economic incentives to have fewer children, education, and media campaigns. The "one child per family" policy was effective but highly controversial. China began to relax its policies somewhat in the 1980s, due in part to an expanding economy and pressure from human rights activists.

Birth control can be achieved through various preconception and post-conception methods. Natural forms of contraception, such as rhythm, have a high failure rate. Other forms of birth control include barrier methods (the condom, spermicides, vaginal sponge, and diaphragm) and hormonal contraceptives. Sterilization, the most reliable method of birth control, is the most popular method of contraception in the world. It is usually irreversible. The IUD is the most popular form of reversible contraceptive in use worldwide. The pill, an oral hormonal contraceptive, is the third most popular birth control method globally. Other hormonal methods, such as injections or implants, may be more efficient than the pill because they are less expensive and need to be administered much less frequently. Abortion continues to be widely used worldwide as a form of birth control. Abortions can be performed surgically or medically. The abortion controversy revolves around the conflict between religious and moral beliefs about the status of the fetus.

Worldwide, contraceptive use varies widely. People in more-developed countries (MDCs) enjoy easy access to birth control methods, while those in LDCs, in general, do not. Contraceptive use also varies widely within the United States. Teenagers (especially unwed teens) and poor women are the groups least likely to use birth control. The rates of teen pregnancy and abortion are much higher in the United States than in all other developed nations.

KEY TERMS

abortion

antinatalist policy

demographic fatigue

demographic transition

demographic trap

family planning

population policy

pronatalist policy

CHAPTER 9

Food Resources, Hunger, and Poverty

Learning Objectives

When you finish reading this chapter, you will be able to:

1. Identify the critical components of a healthy diet.

2. Identify the major foods relied upon by the global human population and summarize the current status of food production.

3. Describe the various manifestations of hunger and explain how hunger affects human health.

4. Discuss how food consumption patterns vary worldwide.

5. Describe the relationship among hunger, poverty, and environmental degradation.

6. List various reasons why hunger continues to be a problem.

7. Briefly describe how agriculture has changed in the past fifty years in both the developed and the developing worlds.

8. Explain the strategies that could be used to increase the global food supply.

One evening a gentleman came to our house and told us there was a Hindu family with many children which had not eaten in several days. He asked us if we could do something. So I took some rice and went to them. When I got there I saw the hunger in the shallow eyes of the children, real hunger. The mother took the rice from my hand and divided it in two and left the room. She said simply, "Next door they are hungry, also."

Mother Teresa

We could save the lives of the world's poorest people at the cost of just 10 cents for every $100 of income. Since we can obviously afford it, how can we in good conscience refuse to accept this duty, especially given that the misery afflicting the poor is now washing up on our own shores in so many ways?

Jeffrey Sachs and Pedro Sanchez, Earth Institute, Columbia University

Like air and water, food is an absolutely essential resource; humans must consume enough food in order to obtain the energy and nutrients needed to grow and develop, to do work, and to regulate life processes. Whether we harvest from cultivated fields and pastures or wild forests, lakes, and oceans; whether we glean from bush, vine, and shrub or collect from fish and fowl, we depend upon the intricately interwoven fabric of the biosphere for sustenance and nourishment (Figure 9-1).

In this chapter, we look closely at food and the related issues of hunger and poverty, and we examine efforts to manage food resources.

FIGURE 9-1: Farmer working his fields, southern Colombia.

Describing Food Resources

We learn what constitutes a healthy diet, which foods humans rely on most, where our foods come from, and the current status of food production. We discover what real hunger is and why it continues unabated in our world, and we explore the relationship between hunger, poverty, and environmental degradation.

What Is a Healthy Diet?

A healthy diet is sufficient in both calories and nutrients. Daily caloric requirements vary with gender, age, body size, level of physical activity and, to some extent, climate. As a general guideline, the average person should consume about 2,400 calories daily. The range for men (ages 19 to 51+) is approximately 2,300 to 2,900 calories; for women (ages 19 to 51+), 1,900 to 2,200 calories. Pregnant or lactating women need extra calories. There is a normal variation in individual needs of plus or minus 20 percent.

Carbohydrates, fats, and proteins are known as macronutrients because they are needed by the body in large amounts. Carbohydrates and fats, which are composed primarily of carbon, hydrogen and oxygen, are the major sources of the energy required to maintain the body and to perform work. Fats also are needed to construct cell membranes, protect internal organs, and act as insulation. Proteins are needed for tissue building; they form the substance of muscles, organs, antibodies, and enzymes. Proteins are essential for growth and development, particularly fetal development. Small units called **amino acids** combine in various ways to form larger protein molecules. There are 20 naturally occurring amino acids. Eleven can be synthesized by the body; the others either cannot be synthesized or cannot be synthesized at a sufficient rate and must be obtained through diet. These nine are known as essential amino acids. A complete protein, such as mother's milk, contains all the essential amino acids in approximately the correct proportions to meet human needs.

A healthy diet is high in complex carbohydrates with an adequate amount of protein and fats. As a rule, an individual should obtain 60 percent of their daily caloric intake from complex carbohydrates, 25 percent from fats, and 15 percent from proteins.

Micronutrients are substances such as vitamins and minerals that the body needs in small, sometimes trace, amounts. Found in a wide variety of foods, micronutrients are necessary to regulate life processes such as transporting oxygen in the blood and maintaining the nervous and digestive systems.

What Foods Do We Rely On?

Although there are 80,000 potentially edible plants on Earth, global agriculture is dependent upon a few plant species. We derive 95 percent of our nutrition from just 30 plant species, and a mere eight crops supply 75 percent of the human diet. The top four are wheat, rice, maize (corn), and potatoes. These crops are known as **staples** because of their importance in many people's diets. In many less-developed countries (LDCs), people also rely on other food crops, including barley, sweet potato, sorghum, oats, rye, and soybeans (Figure 9-2).

Wheat, rice, corn, barley, oats, rye, sorghum, and millets are grasses. Their fruits, or grains, supply the human population with over half of its calories. Grains are a rich source of carbohydrates; they also contain some protein, oil, vitamins, and minerals. Because of their importance, the cereal grains (a name derived from Ceres, the Romans goddess of crops) have been called the staff of life.

Humans rely on just a small number of animals. Nine domesticated animals — cattle, pigs, sheep, horses, poultry, mules/asses, goats, camels, and buffalo — supply the bulk of the protein (meat, eggs, milk) from livestock. In order to produce foods and other products from animals, we use over seven billion acres (three billion hectares) of grazing area, about double the available cropland area. Although people in developing countries possess about half of the world's livestock, they consume only about 20 percent of the meat and milk they raise. The rest is exported to developed countries. People in more-developed countries (MDCs), in fact, consume three to four times as much meat as people in LDCs. Nearly 40 percent of the grain produced worldwide is used to feed livestock, but in MDCs that figure can reach 70 percent.

Wheat is the most important cereal worldwide, the staple for over one-third of the human population. Its protein content varies between 8 and 15 percent. Wheat is grown primarily in temperate climes and some subtropical zones.

Potatoes, grown best in cool, moist, temperate climes, are a staple carbohydrate in many developed nations.

Rice is the leading tropical crop in Asia. Because of wet-rice cultivation, which allows for continuous cropping, rice can support high densities of population. Its protein content ranges between 8 and 9 percent.

Maize, with a protein content around 10 percent, is a staple crop for the people in South America and Africa. The United States is the world's largest maize producer, but the bulk of the crop is fed to livestock.

Barley is the fourth most important cereal crop. It is used primarily for animal feed and malting for beer and whiskey. For people in parts of Asia and Ethiopia, it is an important food crop.

Cassava is an important food crop in Africa. It has low protein content, but is drought-resistant.

Sweet potatoes, grown in moist tropical regions, are generally used as a secondary food. Their chief value is as a source of starch.

Sorgum and millet are cereal grasses. They are staples in many drier parts of Africa and Asia. Because these grains lack gluten, they cannot be used to make bread.

Oats and rye are suited to cool, damp climes. Oats are grown chiefly for animal feed; rye is used principally for bread flour.

Pulses, especially **soybeans**, have high protein content, between 30 and 50 percent. In many poor regions, they may be the people's chief source of protein.

Figure 9-2: Important food crops worldwide.

In and of itself, eating meat or meat products in conservative amounts is neither nutritionally nor environmentally unsound. Meat is an important source of amino acids, vitamins, and minerals. The consumption of grazing animals can be sustainable as long as they feed primarily on grasses, crop residues, and vegetation and are not allowed to overgraze an area. The efficiency of grazing animals is increased when their wastes are used to fertilize the soil. In this situation, grazing animals in effect "harvest" their own food and help to replenish the soil, restoring its fertility. Unfortunately, meat consumption can generate problems for human health and the environment when it is practiced to the extent and in the manner common in the United States. When livestock are fed grains, the efficiency of food production is greatly reduced, because a significant amount of energy must be invested to plant and harvest crops. Synthetic fertilizers, used to replenish the soil and keep yields high, increase the "energy investment" of meat production. The overgrazing of range and pasture by livestock (especially cattle and sheep) also incurs substantial environmental costs, particularly soil erosion and the pollution of nearby waterways.

Where Does Food Come From?

We harvest our food from both land and sea. Of the Earth's total ice-free land surface, 3.7 billion acres (1.5 billion hectares), about 11 percent, can be used easily to produce food. The rest, about 32 billion acres (13 billion hectares), is too wet, dry, cold, hot, poor in nutrients, or shallow in soil to cultivate easily. With intense management and a significant financial investment, another 3.7 billion acres (1.5 billion hectares) of land could be brought into food production. Unfortunately, expansion of cropland would require undesirable losses in forests, wetlands, and grazing lands. Moreover, farming marginal land incurs greater environmental damage (such as soil erosion) and requires higher inputs of irrigation water, fertilizers, and so forth.

We harvest most of our aquatic foods from four biologically productive marine ecosystems: estuaries, offshore continental shelves, coastal wetlands, and coral reefs. Though they represent only a small part of the total water area of the Earth, these four ecosystems are 16 to 26 times more productive than the open oceans. (Because of the biological diversity they harbor, coral reefs and coastal wetlands are discussed in detail in Chapter 15, *Biological Resources.*)

An **estuary** is a shallow, nutrient-rich, semi-enclosed body of water; most estuaries have a direct connection to the ocean. Tides constantly mix salt water with fresh water from the surrounding land. Estuaries serve as habitats for larval stages of many ocean fishes and shellfishes. These larvae are washed out to sea with outgoing tides, where they feed and grow to adults. Estuaries produce over 80 million tons (73 million metric tons) of fish and shellfish per year, more than any other ocean ecosystem. Representing just two to three percent of total ocean areas, estuaries alone are responsible, either directly (through catch of adults) or indirectly (as nurseries for young), for 75 to 90 percent of all fish and shellfish caught.

Near-shore areas over continental shelves are also very productive. Where cold currents from the polar regions meet warm tropical currents, upwellings of turbulent, nutrient-rich waters nourish billions of phytoplankton. Phytoplankton are the base of a complex food web that includes zooplankton, fishes, birds, and marine mammals. Although upwelling systems comprise only 0.1 percent of the oceans, they contribute roughly half of the global marine fish catch. The major upwelling systems are found off the west coasts of North and South America and Africa and off the coasts of Arabia and Somalia in the Arabian Sea.

What Is the Current Status of Food Production?

For the last half of the twentieth century, global food production rose steadily — and in some cases, rapidly — either keeping pace with, or exceeding, human population growth. By the late 1990s, however, production in several sectors leveled off. A number of long-standing trends are contributing to the global loss of agricultural momentum: the cumulative effects of soil erosion (which reduces land productivity), the loss of cropland to desertification, and the conversion of cropland to nonfarm uses. Several more recent trends also are taking a toll on food production: falling water tables (which contributes to slower growth in the use of irrigation water), diminishing returns in the use of fertilizers, and rising global temperatures. Below, we look closely at production and consumption trends in grain, fish, and meat.

Grain

Food security is the ability of a nation to provide, on an ongoing basis, enough food to keep its people alive and healthy. Grain is used as a measure of food security for two reasons. Consumed directly, it provides half of human food energy intake; consumed indirectly (through eggs, meat, and dairy products), it supplies part of the remainder. Also, because grain is less perishable than fruit and vegetables, it can be stored for use during the cold season (in the higher latitudes) or dry season (in areas with pronounced wet/dry periods).

Wheat, rice, and corn are the three major cereal crops, accounting for 85 percent of the world's grain harvest. Global grain production is concentrated in China, India, North America, Europe (including the former Soviet states), Australia, and Argentina. China and India, respectively, are the two largest producers of wheat and rice. The United States is the leading producer of corn; the U.S. crop makes up at least one-third of the global corn harvest.

Between 1950 and 1996, the global grain harvest nearly tripled, but from 1997 through 2003, it leveled off, showing little or no increase. In 2004, stronger grain prices at planting time and the best weather in a decade produced a significantly larger yield, approximately 2 billion tons (2.04 billion metric tons). The 2004 harvest was about 8.5 percent higher than that in 2003.

Compared to the grain harvest, per capita grain production is a more revealing — and more disturbing — indicator of food security because it takes into account population growth. From 1950 to 1984, per capita grain production rose 38 percent, from 543 pounds (247 kilograms) to an all-time high of 752 pounds (342 kilograms) per person. Since that time, growth in the grain harvest has fallen behind population growth. Even with a bumper crop, in 2004 the amount of grain produced per person was just 679 pounds (308 kilograms).

There are important regional differences in per capita grain production. In recent years, Western Europe's output per person has risen, the result of slow population growth (0.1 percent) and high support prices that compel farmers to adopt advanced agricultural technologies. African farmers face a different situation: Population growth on the continent is the highest worldwide (2.4 percent), and the technologies that can be used in semiarid regions are limited. In many parts of sub-Saharan Africa, unrelenting drought reduced crop yields in the early 2000s. Throughout the continent, poverty has slowed production further, since poor farmers are unable to afford fertilizers. The Acquired Immune Deficiency Syndrome (AIDS) pandemic is also taking a terrible toll: Already, Africa has lost seven million farmers to the disease. By 2020, in southern Africa alone, it will claim the lives of 20 percent of farm workers. Not surprisingly, per person grain production in Africa continues to decline.

Another important indicator of food security is world **carryover stocks** of grain, the total amount in storage when the new harvest begins. Stocks are reported in either millions of metric tons per year or days of consumption, a measure that accounts for population size. Many countries rely on grain from grain-exporting nations like Canada and the United States to supplement their own harvests. A certain amount of time is needed to transport grain from farms in North America to flour mills and bakeries in grain-importing countries. Large quantities of grain are thus needed to keep the supply lines between producers and consumers operating. Carryover stocks must be at a minimum of 70 days in order to ensure food security. When carryover stocks fall below 60 days of consumption, there is barely enough to keep the pipeline filled. With the flat harvests of 1997-2003, production fell below consumption, drawing down stocks. By 2003, carryover stocks had fallen to their lowest level in 30 years, just 59 days of consumption. Even the bumper crop of 2004 was not large enough to begin rebuilding the world's carryover stocks (Figure 9-3). However, grain production in 2005 is expected to exceed consumption, leading to the first increase in stocks in at least five years.

Whenever carryover stocks fall, approaching the critical 60-day level, grain markets become highly volatile and prices rise. In early 2004, wheat and corn

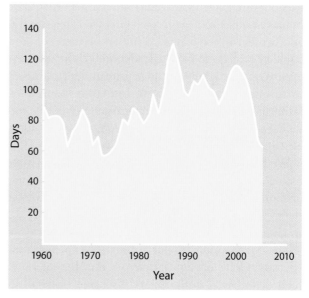

Figure 9-3: World grain stocks as days of consumption, 1960-2004. **Source:** *U.S. Department of Agriculture.*

prices reached seven-year highs, while rice prices climbed to five-year highs. High grain prices translate into higher prices for flour, bread, and grain-based livestock products like meat, milk, and eggs. Volatility in the marketplace means more than just higher food prices, however; it can spell tragedy. From 1972 through 1974, as carryover stocks reached an all-time low, wheat and rice prices more than doubled. In an effort to keep domestic food prices under control, exporting countries like the United States restricted exports and denied requests for food aid. Hundreds of thousands of the world's poorest people, in countries like Ethiopia and Bangladesh, starved to death. The problem was not a scarcity of food; the problem was that they were simply too poor to buy the available food.

Grain markets are especially anxious about climate patterns in North America. In 1988, record heat and drought reduced the U.S. grain harvest by 27 percent and the Canadian harvest by 31 percent. A major crisis was averted because carryover stocks were near the record high of 1987. From an agricultural perspective, global climate change is particularly worrisome; as a rule of thumb, with each one degree Celsius rise in temperature above the optimum during the growing season, the yields of wheat, rice, and corn fall by 10 percent. Warmer temperatures and less precipitation could dramatically decrease production in the North American breadbasket, the region that encompasses the Great Plains of the United States and Canada and the U.S. cornbelt. While it's true that agriculture likely would expand northward in both Canada and Russia, the trade would not be an even one. Those regions do not enjoy the same fertile soils of the North American breadbasket, nor do they have the existing infrastructure to support large-scale agriculture.

In general, grain prices fell throughout the twentieth century due to increased production. That trend could be reversed, however, if demand begins to outstrip supply, a scenario suggested by trends in the 1990s. While world grain production slowed throughout that decade, demand continued to climb, due partly to annual population growth and partly to rising wealth in Asia, especially China. As the nation becomes increasingly affluent, the diets of many Chinese are changing. People are eating higher up the food chain, demanding such grain-intensive products as pork, poultry, beef, eggs, and beer. Thus, where once the Chinese relied on grain itself (rice or wheat) as the staple of their diet, they now consume a larger share of grain in the form of meat or drink.

In the 1980s and 1990s, China's grain production rose steadily, peaking at 432 million tons (392 million metric tons) in 1998. In recent years, however, production has dropped, falling to 366 million tons (322 million metric tons) in 2003. That decline — 77 million tons (70 million metric tons) — exceeds the entire grain harvest of Canada. It is directly linked to a decline in the amount of land devoted to cereal production. In 1998, grains were cultivated on 222 million acres (90 million hectares); just five years later, that figure was 188 million acres (76 million hectares), a drop of more than 15 percent.

Five problematic trends underlie the reduction in land devoted to grain production: the loss of irrigation water; the expansion of deserts (primarily due to over plowing and over grazing); the conversion of arable land to nonfarm uses; the shift to higher value fruits and vegetables; and (in more prosperous areas) the loss of rural labor needed for double cropping (a labor-intensive practice that consists of quickly harvesting one grain crop and immediately preparing the ground for the next crop). None of these trends can be reversed easily. For example, water tables are dropping and lakes, rivers, and wells are running dry under the North China Plain, which produces one-half of the nation's wheat and one-third of its corn.

Thus far, China has covered the shortfall in its grain harvests by drawing down its once impressive stocks, but by 2004, those stocks were nearly depleted. Moreover, domestic wheat and rice prices were up by 20 percent or more, and China began to turn abroad for grain. If, as expected, China's grain production falls an estimated one-fifth by 2030, the Chinese will face a huge grain deficit. A conservative estimate of the shortfall is some 238 million tons (216 million metric tons); other estimates put the figure at closer to 440 million tons (400 million metric tons).

Even if China has enough foreign exchange to import grain, it's unlikely that any nation could meet its demand. Since 1980, average annual world grain exports have totaled about 220 million tons (200 million metric tons). Moreover, demand is also expected to increase in other regions. By 2030, Africa is expected to require grain imports of 275 million tons (250 million metric tons). Imports also are expected to rise on the Indian subcontinent and in fast-growing countries like Iran and Mexico. As demand rises around the world, grain prices could soar, further widening the nutrition and health gulf between the wealthy and the poor.

Fish

Fish is the leading source of animal protein worldwide; fish consumption exceeds that of beef and chicken combined. For over one billion people in Asia, the coastal nations of Africa and island countries, fish is the primary source of animal protein. About 25 percent of the annual marine catch is not used to feed humans directly; instead, it is ground for meal to supplement the diets of livestock, pets and zoo animals, or it is processed to yield fish oil products.

Total world fish production has risen steadily since World War II and in 2003 stood at 145.7 million tons (132.2 million metric tons). The marine wild catch, or capture, was 89.6 million tons (81.3 million metric tons); the inland (freshwater) capture was 10 million tons (9 million metric tons). **Aquaculture**, the controlled production of fish, shrimp, shellfish, and seaweeds in ocean pens and inland ponds or tanks, contributed an additional 46.1 million tons (41.9 million metric tons). Marine aquaculture yielded 18.4 million tons (16.7 million metric tons); inland aquaculture yielded 27.7 million tons (25.2 million metric tons). Currently, aquaculture accounts for about one-third of total world fish production, a figure expected to rise to one-half by 2030.

As with grain, the per capita marine capture is a more revealing indicator of food security; like per capita grain production, it signals trouble ahead for the world's hungry. Between 1950 and 1988, the marine capture increased by four and half times, so that, despite rapid world population growth, the per capita marine capture still doubled, increasing from less than 18 pounds (eight kilograms) to 37 pounds (17 kilograms). But since 1988, the marine capture has leveled off, fluctuating between 94 and 105 million tons (85 and 95 million metric tons). As a result, the per capita marine capture has declined about nine percent since 1988. Moreover, less valuable species now account for a growing share of the marine capture, offsetting declines in the stocks of commercially profitable species like cod and haddock.

Clearly, the oceans account for the vast majority of total world fish production. According to the U.N. Food and Agriculture Organization (FAO), however, we have reached the global potential for marine capture. In 2003, just over 50 percent of the world's main fish stocks were exploited fully, producing catches close to or at sustainable levels; another 25 percent were overexploited, depleted, or recovering from depletion and in need of rebuilding. Only about 24 percent of fish stocks were underexploited or only moderately exploited, so that they could accommodate increased harvesting. Fisheries in serious decline include the northwest Atlantic, the southeast Atlantic, and the Mediterranean and Black seas. Most fisheries experts believe that more efforts are needed to rebuild depleted stocks and prevent the decline of those that currently are being harvested at or near maximum levels.

Pollution (particularly near estuaries and other coastal areas) and habitat destruction (especially of coastal wetlands) have taken a toll on ocean fisheries, but the chief culprit in declining stocks is overharvesting. Bigger and faster boats and more advanced hunting technologies (such as sonar to locate schools) have made fishers more efficient — and, in some cases, more wasteful (Figure 9-4). "Dragging," for example, is a technique in which a powerful trawler drags a net and metal door across the ocean floor; groundfish (such as cod and haddock), starfish, sharks, and rocks all are swept into the net. Undesirable or undersized fish are thrown back into the water, but many of these die because they are out of the water too long.

As fishers become more efficient, fewer are needed to bring in the catch. But according to the FAO, the world fishing fleet doubled between 1970 and 1990,

FIGURE 9-4: Tuna boats and nets. The large nets used to capture tuna can also ensnare non-target species such as dolphins and sea turtles.

and the capacity of the world's fishing fleet is now two and a half times the sustainable yield of fisheries. In part, government subsidies — for fuel, vessel construction, insurance, and other expenses — encourage fleet overcapacity. When fish populations are healthy, fishers invest in new boats and equipment, often with government support. When those populations fall, however, either because of overfishing or as part of a natural cycle, the fishing industry seeks government help to protect jobs and investment. Worldwide, fishing subsidies total over $15 billion annually. Unfortunately, subsidies can enable overextended fishers to stay in business and encourage new fishers to enter the market, thus maintaining overcapacity.

Fleet overcapacity also occurs because some fisheries allow open access. Regulators who want to limit the total catch must estimate the potential take of all boats and adjust the length of the season accordingly. The season then becomes a race for fishers, who compete against one another to try and get the biggest catch in the allowable time. The season becomes shorter as the number of fishers or their capacity increases. The "season" for Alaska's halibut industry is limited to just two or three 24-hour periods per year.

Because much of the global fish catch comes from fully exploited stocks, continued harvesting will cause them to decline. In the past, we have witnessed the collapse or near collapse of numerous fish or shellfish species due to a combination of overexploitation and climatic or ecological changes. Examples include the collapse of the Peruvian anchovy fishery, the demise of California sardines, the depletion of the cod and haddock fisheries off the coast of eastern Canada, and the severe reductions of Alaskan king crab, Atlantic striped bass, and Atlantic herring.

The decline of marine fisheries has several ramifications. Because so many people in the developing world depend on fish as their sole source of animal protein, fewer fish may translate into reduced protein consumption for those who can ill afford it. As the fish supply tightens, seafood prices are likely to rise; rising prices give developing nations a strong incentive to increase exports to industrialized countries, thereby gaining needed foreign exchange. If this happens, people in LDCs will face a reduced domestic supply and rising prices. Essentially, they will have to compete on the international market for a commodity they rely on as a dietary staple. Finally, the decline in marine fisheries will continue to adversely affect the world's 15 to 21 million fishers. Already, over 100,000 fishers worldwide have lost their source of income in just the past few years.

Although the income from marine fisheries accounts for only about one percent of the global economy, it is essential to many coastal and island regions. In the early 1990s, the collapse of eastern Canada's cod and haddock fisheries led to massive layoffs in the fishing and fish processing industries. Nearly 50,000 Canadian fishers lost their jobs in 1992 and 1993. In the United States, fishing communities in the Northeast and the Pacific Northwest also have suffered due to declines in stocks of groundfish and salmon.

Meat

As with grain and fish, world meat production increased substantially in the latter half of the twentieth century. Between 1950 and 2002, meat production soared five and a half times, increasing from 48.4 million tons (44 million metric tons) to 267 million tons (242 million metric tons). On a per capita basis, it more than doubled, rising from 37.4 pounds (17 kilograms) to 86 pounds (39 kilograms). Much of the increase occurred in the western industrialized nations and Japan, but since about 1980, meat production has increased rapidly in East Asia (particularly China), the Middle East, and Latin America.

Three types of meat — pork, poultry, and beef — are particularly important to the world food supply. Mutton is a distant fourth, although it is an important protein source in New Zealand, Australia, Kyrgyzstan, and Kazakhstan.

Pork accounts for about 40 percent of world meat production — 102 million tons (93 million metric tons) in 2002. From 1950 until 1980, beef and pork production rose in concert, but in the wake of the economic reforms in China, where pork is the dominant meat, pork production soared ahead of beef production, nearly doubling in less than two decades. Not surprisingly, China is the world's largest pork producer, accounting for about half of the global total. It is now also the largest consumer of red meat.

Similarly, world poultry production has increased rapidly in recent years, more than doubling between 1987 and 2002, when it reached 79 million tons (72 million metric tons). In the late 1990s, poultry production overtook beef production. Worldwide, the United States produces and consumes the most poultry; other major producers are China, Brazil, and France.

Worldwide, beef production reached 66 million tons (60 million metric tons) in 2002. Not surprisingly, nations with plentiful grazing land produce and consume the most beef. Brazil is the world's largest producer of

Figure 9-5: A hog factory farm, Missouri. The barns in the foreground house almost 9,000 individuals; there are eight other such sites on this particular factory farm. The manure lagoon to the left of the hog barns holds 20 million gallons of wastes. Each of the nine sites has its own manure lagoon.

beef and second largest consumer, after the United States. Other nations that play an important role in beef production include Argentina, Uruguay, Australia, Canada, and New Zealand.

In and of itself, eating meat or meat products *in conservative amounts* is neither nutritionally nor environmentally unsound. Meat is an important source of amino acids and minerals. The consumption of grazing animals can be sustainable as long as they feed primarily on grasses, crop residues, and vegetation and are not allowed to overgraze an area. The efficiency is increased when the animals' wastes are used to fertilize the soil. In this situation, grazing animals in effect "harvest" their own food and help to replenish the soil, restoring its fertility. Unfortunately, if too many animals are allowed to graze an area, they may degrade the pasture or range. Degraded range is prone to desertification and soil erosion, which can pollute nearby waterways.

In recent decades, with rangelands worldwide at or near their carrying capacity, many producers have turned to industrial farming to raise cattle, pigs, and fowl. In 2002, 43 percent of the world's beef and half of its pork and poultry were produced on so-called "factory farms." As the name implies, factory farms are huge operations, and they raise serious environmental and health concerns (Figure 9-5). Feedlots (for beef

cattle) and industrial hog and chicken farms generate tremendous amounts of manure, which can contaminate surface- and groundwater, cause air pollution, and give off an ever-present stench. Crowded conditions facilitate the rapid spread of disease, contributing to the overuse of antibiotics. In 2002, Russia temporarily closed markets to U.S.-produced chicken and turkey due to concerns over drug residues in poultry. Factory farms also depend on the use of grains (especially corn), supplemented with high-protein soy meal, to fatten livestock. Nearly 40 percent of the grain produced worldwide is used to feed livestock, but in MDCs that figure can reach 70 percent.

When livestock are fed corn and soy meal, a significant amount of energy must be invested to plant and harvest the grain and soybeans. Synthetic fertilizers, used to replenish the soil and keep yields high, increase the "energy investment" of meat production. The relative conversion efficiencies of various animals also come into play. It takes much less grain to produce a pound of weight gain per chicken or turkey than it does to produce a pound of weight gain per hog or steer. A tightening grain supply, then, favors the production of chicken (and, to a lesser extent, pork) over beef. The surge in poultry production since 1995 can be attributed in part to its relatively favorable conversion efficiency.

What Is Hunger?

At first glance, hunger seems easy to define. Even a young child knows that being hungry means wanting something to eat. For most of us, hunger is a sensation, our body's way of signaling that it is running short of food. For the chronically hungry, however, **hunger** is the inability to acquire or consume an adequate quality or sufficient quantity of food through non-emergency food channels, or the uncertainty of being able to do so. According to the U.N. World Food Programme (WFP), approximately 852 million were chronically hungry in 2002, an increase of seven percent over the levels of the late 1990s. That's more than the combined populations of the United States, Canada, Japan, and Europe. Given that nearly one in seven people worldwide do not have enough food to be healthy and to lead an active, productive life, it's no surprise that hunger and malnutrition together are the number one risk to health worldwide — greater than AIDS, malaria, and tuberculosis combined.

Few of us have known real hunger — an aching, gnawing desire for food that remains unabated day after day, week after week. Few of us have felt the effects of hunger — the sharp pain in the belly that gradually dulls but never leaves, the growing weakness, the increased susceptibility to disease and illness, the slow deterioration of mind and body. Indeed, few of us can even imagine chronic and persistent hunger. Perhaps that is part of the reason that hunger exists in our world, for how can we fight an enemy that we do not know? Before we can eradicate hunger, then, we must learn to recognize it, and hunger is an enemy with many faces.

Starvation and Famine

Starvation, the most easily recognized manifestation of hunger, is suffering or death from the deprivation of nourishment; a person does not consume enough calories to sustain life. **Famine**, or widespread starvation, is typically the result of many interrelated factors, including natural disasters such as floods, droughts, and earthquakes and human activities such as war and the inequitable distribution of land. For example, the Ethiopian famine of the mid-1980s was a product of prolonged drought, a civil war that ravaged the country, and the production of cash crops. Although famines are widely reported by the international community and media, they are only one aspect of hunger: The starving account for just eight percent of hunger's victims.

Undernutrition

Hunger is primarily a quiet killer. Of the more than nine million people who will die this year because of hunger, most will die from diseases that afflict them because they are undernourished or malnourished (Figure 9-6). **Undernutrition** is the consumption of too few calories and protein over a prolonged period; it occurs when an individual consumes fewer than 2,000 calories daily for weeks, months, or years. From 2,000 to 2,400 calories, undernourishment may also occur, depending on a person's activity level. To compensate for the lack of energy, the body slows down both physical and mental activities. Consequently, those who are chronically undernourished find it difficult to concentrate, study, work, or play. They become weaker and less productive; they may appear listless, and their weakened bodies are less able to ward off (or recover from) illness and disease. Hungry children are especially vulnerable; common infections like measles

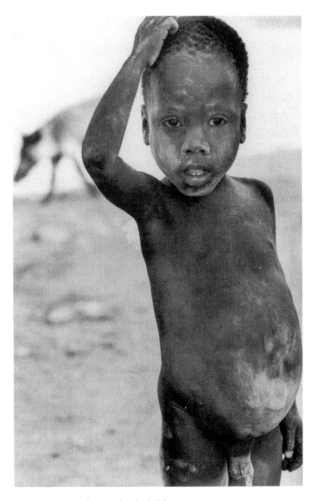

Figure 9-6: Malnourished child, Haiti.

and diarrhea kill millions of children annually. Chronic undernutrition is a slow but steady — and therefore often unnoticed — decline toward death.

Malnutrition

Malnutrition is the consumption of too little (or, more infrequently, too much) of specific nutrients essential for good health; it is more about the *quality* than the *quantity* of food. Even if people consume a sufficient number of calories, they will become malnourished if their food does not provide the proper amount of protein or micronutrients to meet daily nutritional requirements. Malnourished mothers give birth to malnourished children. Each year, approximately 17 million children are born underweight, a consequence of inadequate nutrition both before and during the mother's pregnancy.

Malnutrition can result from inadequate dietary intake, infection, or both. It may not kill, but malnutrition can maim and deform. **Malabsorptive hunger** often accompanies malnourishment and undernourishment. The body loses the ability to absorb nutrients from the food consumed. Malabsorptive hunger can be caused by parasites in the intestinal tract or by a severe protein deficiency.

To identify a malnourished child, health workers compare physical measurements of the body (weight or height) to age. Three determinants are particularly important: stunting, wasting, and underweight.

Stunting reflects shortness-for-age; an indicator of chronic malnutrition, it is calculated by comparing the height-for-age of a child with a reference population of well-nourished and healthy children. According to the FAO, almost one-third of all children are stunted.

Acute malnutrition that causes a recent and substantial weight loss results in **wasting**. It is calculated by comparing weight-for-height of a child with a reference population of well-nourished and healthy children. Wasting is usually associated with starvation and/or disease. Because it is strongly related to mortality, wasting is often used to assess the severity of food emergencies.

Underweight is a measure of a child's weight-for-age compared to a reference population of well-nourished and healthy children. According to the WFP, underweight contributes to 60 percent of child deaths each year.

There are various forms of malnourishment, depending on what nutrients are deficient in the diet, for how long, and at what age. Some result from a deficiency of macronutrients (fats, carbohydrates, and proteins);

others from a deficiency of micronutrients (vitamins and minerals).

Macronutrient Deficiencies. For children, insufficient protein is a particularly serious and common health problem. A deficiency of protein inhibits mental and physical development. Severe protein insufficiency may cause mental retardation, wasting, or death. Protein insufficiency contributes to kwashiorkor and marasmus, two of the most common childhood diseases related to a nutritionally deficient diet.

Kwashiorkor is associated with a low protein diet, a diet composed primarily of starches, often maize or sweet potato. Kwashiorkor, which means "displaced child," occurs when a child, typically from one to three years old, is weaned from mother's milk so that a new infant can be nursed. The child is deprived of a source of complete protein, and even though the child may consume a sufficient number of calories, protein insufficiency brings on the disease. Symptoms include skin rashes or discolored patches on the knees and elbows; swelling of the hands, feet, and face; liver degeneration; possible permanent stunting; hair loss; diarrhea; mental apathy; and irritability.

Marasmus is caused by a diet deficient in both protein and overall number of calories. It is most common among infants or very young children who are not breast-fed or who are fed an inadequate, high-starch diet after they are weaned. The child with marasmus typically has emaciated arms and legs, shriveled skin, wide eyes, and a gaunt, aged-looking face and suffers from diarrhea, muscle deterioration, and anemia. Kwashiorkor and marasmus can cause irreversible brain damage and mental retardation, especially in the very young, and both can be fatal. Fortunately, both can be reversed if the deficiency is not prolonged.

Annually, malnutrition contributes to the deaths of approximately 10 million children under the age of five. According to estimates by the U.N. Children's Fund (UNICEF), many of these deaths could be prevented through programs that promote breast feeding, oral rehydration therapy (ORT) to counteract diarrhea, and overall improved child care. The cost: $5 per child, per year. Combating diarrhea, a leading cause of death in children, is particularly important; untreated episodes can lead to malnutrition and retarded growth and development.

Micronutrient Deficiencies. According to the WFP, combating dietary deficiencies in micronutrients is one of the most difficult nutrition challenges for the twenty-first century. Over one-third of the world's populations do not take in sufficient quantities of iron, iodine, vitamin A and zinc, with serious consequences, particularly for women and children.

Worldwide, iron deficiency is the most widespread form of malnutrition, affecting an estimated 1.7 billion people. Iron forms the molecules that carry oxygen in the blood; not surprisingly, those with an iron deficient diet are often tired and lethargic. In children, iron deficiency can also impede cognitive development; an estimated 40 to 60 percent of children in LDCs are likely affected. A severe iron deficiency results in anemia, which affects 45 percent of women in the developing world. Every day, anemia contributes to the deaths of 300 women during childbirth. In India alone, 22,000 mothers die annually as a result of anemia.

Worldwide, 780 million people suffer iodine deficiency. The most obvious symptom is a swelling of the thyroid gland, called a goiter. The most serious effect, however, is on fetal development. According to U.N. researchers, approximately 20 million children each year are born mentally impaired because their mothers' diets were deficient in iodine. In severe cases, children may suffer cretinism, a condition characterized by mental retardation and physical stunting.

Vitamin A deficiency is a leading cause of child blindness in LCDs; it also weakens the immune systems of young children, increasing their vulnerability to disease. A diet deficient in Vitamin A elevates a child's risk of dying from diarrhea, measles, and malaria by 20 to 24 percent. According to the WFP, Vitamin A deficiency affects more than 140 million children in 118 countries and more than seven million pregnant mothers.

A weakened immune system and a failure to grow are the effects of a zinc-deficient diet in young children. A zinc deficiency is linked to a higher risk of diarrhea and pneumonia, resulting in an estimated 800,000 deaths each year.

Seasonal Hunger

For many people, **seasonal hunger** is a way of life, a part of the annual cycle. Each year there is a time before the new harvest when the reserves from the previous year's harvest run out. The people go hungry, perhaps for days or weeks, but sometimes, if the previous harvest was small or if the new crop is late in maturing, for months. Seasonal hunger is so common in parts of Africa that the names of the months before harvest time reflect the expected lack of food. The Iteso people of eastern

Uganda, for instance, refer to May as "the month when the children wait for food."

The United States has its own version of seasonal hunger. It is the period, from a few days to a few weeks, when the money runs out at the end of the month or when the food stamps are used up and there is little or no food. A particular kind of seasonal hunger can occur with a new political administration, which may eliminate or severely reduce assistance programs.

Who Are the Hungry and Where Do They Live?

Children account for about 40 percent of the world's hungry; most of the rest are women. In fact, 70 percent of those who suffer chronic hunger are women and girls. Most of the hungry live in the developing world. Of the world's 852 million chronically hungry people, the vast majority — 815 million — live in LDCs. Another 28 million live in countries making the transition to an industrialized society, and 9 million live in MDCs. Most citizens of LDCs consume about 10 percent less than the minimum calories required for good health. In contrast, many individuals in MDCs consume about 30 to 40 percent more than necessary. On average, a person in the developing world consumes only two-thirds of the calories and one-fifth of the animal protein consumed by a person in the developed world. If we include all the grain fed to livestock, people in the developed world consume three times as much food, on average, as their LDC counterparts. Pets in Europe and North America receive more food (meat and calories) each day than many people in LDCs.

Most of the world's hungry live within the "great hunger belt," a region encompassing various nations in Southeast Asia, the Indian subcontinent, the Middle East, Africa, and the equatorial region of Latin America. Just five nations — India, Bangladesh, Nigeria, Pakistan, and Indonesia — are home to at least half of the world's hungry. Within the great hunger belt, hunger is a daily reality. The FAO and World Bank estimate that 500 million people consume less than the minimum critical diet needed to remain healthy and maintain body weight, even with little physical activity. Of these, more than two-thirds live in Asia; most of the rest live in Africa and Latin America. In other parts of the world, hunger is less visible and less widely recognized, but it is nonetheless real. In the United States, nearly 36 million Americans — including 12.9 million children — fall below the poverty line and are at risk from hunger and associated diseases. For a look at *Environmental Science in Action: Hunger in the United States*, go to www.EnvironmentalEducationOhio.org and click on "Biosphere Project."

Statistics can tell us how widespread hunger is and where it occurs, but the hungry are not just numbers. They are people who love and feel pain, people with diverse personalities and talents, people who laugh too seldom and weep too often, people who suffer and hope. Most of all, they are people who want not just food, but the means to make a living; people who want not just bread, but the land to grow wheat; people who want not just to survive the present, but to realize a better future for themselves and for their children.

Why Hunger in a World of Plenty?

Food production, especially of grains, is sufficient to feed the world's people. Moreover, if grain and root crops presently used to feed animals were instead used for human consumption, perhaps as many as twice the present global population could be adequately fed. Yet hunger continues to be a deadly and intractable problem. In part, natural environmental conditions account for inequities in food production. Some areas are more fertile and have better growing conditions, while other regions are less productive and more vulnerable to degradation. But the chief cause of widespread, intractable hunger is poverty. Other contributing factors include environmental degradation, war and civil unrest, natural disasters, and faulty economic practices. These factors are interrelated and often act synergistically.

Poverty

On December 26, 2004, an earthquake under the Indian Ocean near Indonesia spawned tsunamis that swept throughout the region and as far away as the eastern coast of Africa. The devastation was incomprehensible. Within hours, at least 175,000 people perished; three months later, another 143,000 people were still missing and presumed dead. The killing waters orphaned tens of thousands of children, leaving them alone and vulnerable. Homes, schools, hospitals, roads, bridges, and wells were destroyed. More than 1.2 million people were left homeless, having lost nearly everything to the deadly waves. In the following days, relief agencies warned that, without prompt and generous aid, many more people might die due to dehydration,

disease (especially cholera), illness (such as diarrhea), and hunger. Stunned by the disaster, people around the world opened their hearts and their wallets. The unprecedented response from the global community enabled relief workers to provide quick and effective humanitarian aid, thus averting secondary crises and beginning the long, difficult process of rebuilding.

As we write this chapter in March 2005, we are heartened by the global response to the terrible tragedy in Southeast Asia. At the same time, we are saddened by humanity's collective failure to respond to a crisis of even greater proportions: global hunger.

Imagine that the week after the December tsunamis, a second earthquake of a similar magnitude again struck, with the same awful results. Then imagine that another struck the following week and the week after and the week after that, for an entire year: 52 disasters, each one killing at least 175,000 people, so that the total loss of life was more than nine million. The scenario is so horrible that the mind rebels; it is impossible to fathom suffering and loss on such a scale. And yet, *each year*, an equal number of people — over nine million — die of hunger or malnutrition. *Every day*, hunger and malnutrition claim 25,000 lives. *Every five seconds*, somewhere in the world, another child dies for lack of food. Yet too few voices are raised in disbelief, in protest, in outrage. In fact, hunger rarely makes the evening news; rather, it is a silent tsunami, a relentless force that does not ebb. Periodically, famine precipitates a humanitarian crisis — Ethiopia in the mid-1980s, the Sudan in the early 1990s, and more recently, the Democratic Republic of the Congo, North Korea, and the Darfur region of the Sudan — and the world is roused from its apathy. Briefly, we see them: the emaciated infants lying in their mothers' arms; the listless young children too weak to play or smile; the men and women without hope, resigned to a fate they cannot escape. We are moved to sympathy perhaps, or to pity. But as a global community, we are not moved to sustained action. The sympathy, the pity, wanes; other concerns distract us. The moment passes, the global media focus on more recent events, and the hungry are forgotten again.

There are striking parallels between the December 2004 tsunamis and global hunger — particularly in terms of the depth of suffering and the effect on vulnerable populations such as children. But there is one critical difference: The tsunamis were spawned by an earthquake, and we cannot prevent earthquakes. Hunger, in contrast, is caused primarily by poverty, and we can prevent poverty.

Of the world's population of 6.3 billion, nearly half are poor. Poverty comes in degrees: extreme or absolute poverty, moderate poverty, and relative poverty. According to the World Bank, **extreme poverty** is defined as subsisting on less than $1 a day; it is known as "the poverty that kills." Those in extreme poverty cannot meet their basic needs for survival. Some lack even the most rudimentary shelter. Without enough income to buy sufficient food, they are chronically hungry; they lack access to safe drinking water and sanitation; and they are unable to purchase basic items like clothing. Many (perhaps most) are ill, but they cannot afford medical care; their poor health further hampers their ability to work and earn income. They cannot send their children to school, and without an education or job training, these children face a similarly bleak future (Figure 9-7). In a world where more than 150 billionaires and 2 million millionaires enjoy lives of incredible wealth, an estimated 1.2 billion people live in extreme poverty, all of them in the developing world. Three-quarters live in rural areas, but those who inhabit urban slums are the fastest-growing segment of the absolute poor.

Moderate poverty is defined as living on $1 to $2 a day; one's basic needs are met, but just barely. There is no "margin of error" for those in moderate poverty; one poor harvest, for example, can spell ruin. Those in **relative poverty**, defined as a household income level below a given proportion for the national average, lack material items and comforts that the middle class takes for granted.

Those who suffer the most from poverty are children. Most of the world's poor are under the age of 15; one-third of the world's children who live in poverty die before the age of five; many who survive are physically or mentally impaired by the age of six months. Poverty creates an insidious cycle. It leads to undernutrition and malnutrition, which in turn lead to decreased resistance to disease and decreased energy. Illness and lack of energy lead to decreased ability to learn and work, which perpetuates poverty. Poverty and hunger undermine national productivity, lowering the gross national product of LDCs by as much as two to four percent. Poverty also results in shortened life expectancy. Unless something is done to break the cycle of poverty, poor children are doomed to a life of hardship and want simply by virtue of having been born (Figure 9-8).

Relief funds and health measures can provide immediate help to alleviate some symptoms of poverty, but such measures alone are not enough. Unless poor,

FIGURE 9-7: A child of a homeless family holds her doll on the streets of Sao Paulo, Brazil.

hungry, and sick people eventually are able to feed and care for themselves, unless they are given the tools to become self-sufficient, the poverty and hunger trap will never be eliminated. Economic development is imperative. In those parts of the developing world where economic development has taken hold, particularly eastern and southern Asia, the number of people living in poverty is expected to drop significantly. Other areas are expected to show only modest decreases in the number of poor, and in sub-Saharan Africa, it is estimated that the number of poor will increase markedly. In these regions, emergency relief efforts will again be urgently needed.

The plight of the world's poor and hungry stands in direct contrast to the "scandal of the waistline," the situation in which the wealthy are able to obtain so much excess food or so many rich foods with little nutrient value that they become seriously overweight. Illness caused by overeating or poor food choices is common in the United States and some other developed countries. A high-fat diet low in nutrient value is

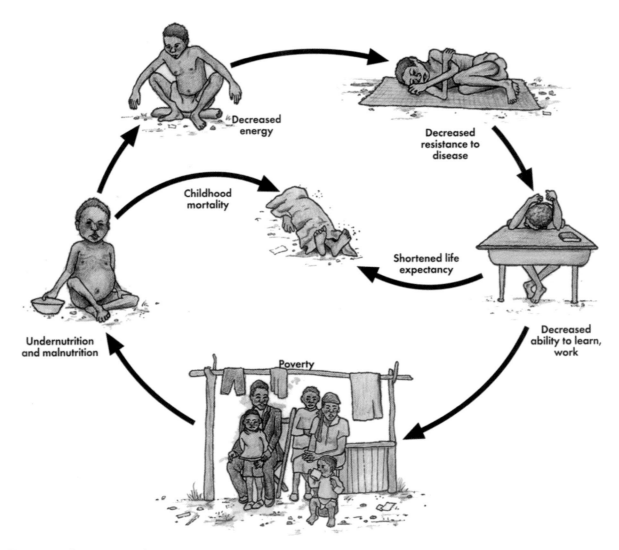

Figure 9-8: The poverty cycle.

linked to heart disease, colon cancer, high blood pressure, and other illnesses that shorten lives. About 20 percent of all adults in the United States, some 33 million Americans, are significantly overweight. But the scandal of the waistline also occurs within other nations. In LDCs, the wealthy class or large landowners often consume an excessive amount and variety of foods while their fellow citizens starve or get by on meager diets. This widening gap often skews the statistics for average dietary consumption. In many LDCs, the wealthiest 20 percent of the population are often 10 to 20 times richer than the poorest 20 percent. Consequently, there is an unequal distribution of food, with the wealthy receiving an inordinate share.

In our world of stark and sobering contrasts, ironies abound. India exports wheat, despite the fact that 221 million of its people are chronically hungry. Annually, Americans spend billions of dollars on special diets that promise to help them "control" their appetite and "shed unwanted pounds," while 500 million people worldwide are so undernourished that their bodies and minds are wasting away. And North Americans spend far more on pet food each year ($25 billion) than they spend on foreign aid ($18 billion).

Environmental Degradation

Lacking arable lands or decent paying jobs, the poor are forced to try to squeeze out a living on marginal lands where overworked soils cannot sustain cultivation; to overgraze grasslands and pastures; and to clear forests for fuel, food, or marketable resources. The result is erosion, desertification, increased salinity in the soils, and the contamination of local water supplies. The degraded environment offers less and less to support its human inhabitants, resulting in still more hunger and poverty. According to the WFP, overfarming has destroyed more than 15 percent of the Earth's fertile farmland.

In LDCs, poverty and environmental degradation are inseparable. Hungry people's concern for protecting the environment is outweighed by their immediate need to survive. Consequently, they will search for food and fuel wherever they can find it. When the food or fuel is gone or the soil is depleted, they will move on to a new area if they have the strength or die in their homelands if they do not. Forty percent of Guatemala's arable land has been lost to erosion. In Haiti, little topsoil remains for cultivation. In Bangladesh, Nepal and Java, severe pressure on hillsides for fuel and growing crops has caused much erosion and flooding in lowland areas (Figure 9-9). And throughout sub-Saharan Africa, in the period from 1980 to 2001, environmental degradation contributed to a five percent decline in agricultural production; simultaneously, the number of hungry increased by 50 percent.

FIGURE 9-9: Severe soil erosion in the Genangan area of Indonesia is a consequence of planting cassava every year without rotating crops.

Civil War and Unrest

Civil war and unrest cause some of the world's worst hunger crises. Particularly in Asia, Africa and Latin America, conflict forces millions to flee their homes, precipitating a food emergency. Often, warring sides will use food as a weapon, seizing and destroying crops and livestock and disrupting or wrecking local markets. They may mine fields and contaminate wells, so that residents cannot return to the land even after the fighting stops or moves elsewhere. In early 2002, conflict had displaced nearly 20 million people worldwide, a number greater than the population of Australia.

Natural Disasters

Prolonged drought, floods, tropical storms, and insect pest outbreaks are among the most widely recognized causes of hunger. In fact, drought is the single most common cause of food shortages worldwide. In the early 2000s, drought reduced crop yields worldwide, including in MDCs like the United States, Canada, and Australia. It was most problematic in LDCs, however, where a poor harvest can force those living on the edge into chronic undernutrition or malnutrition. In recent years, prolonged or recurrent drought precipitated crop failures and heavy livestock losses in LDCs like Guatemala, India, Ethiopia, Kenya and Uganda, among others. Unfortunately, scientists warn that global warming is altering the climate, causing an increase in the number and severity of weather-related natural disasters. (These ideas are explored in detail in Chapter 12, *Air Resources.*) Such "unnatural disasters" exact an increasingly heavy toll on the world's most vulnerable populations.

Economic Forces

A number of economic factors exacerbate poverty and hunger; among the most significant are cash crops, inequitable land distribution, and faulty government policies.

Cash crops are those grown for export rather than domestic consumption; typically, they are nonessential items such as coffee, tobacco, or cotton. The high prices these commodities often command compel farmers to abandon food crops in favor of cash crops. As long as the market for cash crops remains strong, the farmers earn more money by growing crops for export than by growing food crops. With their earnings, they can purchase the foods they need as well as tools and technologies to improve production. The country benefits, too, since the exports bring in critically important foreign exchange earnings, which bolster the economy and aid development. But if the market for cash crops weakens, exports drop, and foreign exchange earnings plummet. Farmers are left with no cash and little or no food crops with which to feed their families.

The inequitable distribution of land also contributes to chronic hunger. In many LDCs, the best lands are controlled by a small number of wealthy landowners. Poor peasants must work for the wealthy, rent good land at high prices, or try to make a living from marginal croplands. Cash crops tend to favor or support the establishment of large plantations. The lure of large profits encourages wealthy landowners to buy up as much acreage as possible and convert it to cash crop production. Under these circumstances, small, independent farmers find it difficult to hold onto their land. In the Philippines, for example, sugarcane plantations are quite profitable, but the poor who work them are paid little and are often underfed and malnourished. Those who harvest the sugarcane do not reap the benefits of the sugarcane industry. In the Sahel region of northern Africa, high yields of cash crops, especially coffee, were recorded even as famine killed millions in the 1980s.

Faulty economic policies also perpetuate hunger and poverty. On average, LDC governments earmark just 5 to 10 percent of their budgets for agricultural development in an effort to boost local food supplies, even though most people in the developing world live in rural areas. They fail to develop and support the infrastructure — roads, warehouses, and irrigation systems — needed to support and promote agricultural production. As a result, high transport costs, lack of storage facilities, and unreliable water supplies can reduce agricultural yields and access to food. In addition, government agricultural policies too often focus on export markets at the expense of local needs. According to the WFP, 54 LDCs do not produce enough food to sustain their populations.

To learn more about the role of economics and politics in perpetuating hunger, read *Focus On: Beyond Hunger – Extending Democracy*, by Frances Moore Lappe, author or co-author of 14 books, including the 1971 best-seller, *Diet for a Small Planet*. To find the essay, go to www.EnvironmentalEducationOhio.org and click on "Biosphere Project."

Managing Food Resources

As we've seen, food resources, hunger, and poverty are inextricably linked. The actions we take, or fail to take, to increase food supplies can alleviate — or exacerbate — hunger. Similarly, addressing the root causes of hunger — particularly poverty and environmental degradation — can enhance food security. Failing to address those causes, however, will likely increase food insecurity worldwide and precipitate an ever-greater number of food emergencies. In this section, we look closely at how we can promote food security and eliminate hunger. To do so, we first must understand how food resources have been managed in the past, and so we begin this section with a brief summary of the history of agriculture.

What Is the History of Agriculture?

Domestication is the result of the long-term selection of traits useful for survival in captivity over those traits necessary for survival in the wild. Early farmers domesticated certain species by selecting and planting the seeds or tubers of those plants that had desirable characteristics, such as a particular taste or high yield. The earliest domestications of plants occurred in **centers of diversity** (or Vavilov centers, after the Russian botanist Nikolai I. Vavilov, who first recognized their existence), where the genetic diversity for a particular species was the greatest.

Domestication gave way to **agriculture**, the purposeful tending of particular plant and animal species for human use. This system of deliberate food production began over 10,000 years ago, in a series of river basins around the globe: the Nile in Africa, the Euphrates-Tigris in the Middle East, the Ganges-Brahmaputra in India, the Indus in Pakistan, and the Yangtze in China. These river basins enjoyed year-round warmth and a constant and ample supply of water. Farmers' deliberate selections of certain strains resulted in **land races**, plant varieties adapted to local conditions such as climate and soil type. Similarly, their careful interbreeding of the best animals yielded unique strains of livestock that thrived under local conditions. In some parts of the world, a small number of traditional farmers continue to grow land races and raise traditional livestock breeds, much as their ancestors did (Figure 9-10).

Figure 9-10: Beans and corn for sale in a marketplace, Tananarive, Madagascar. In some areas, people still cultivate the land races that they have tended for generations.

Historically, farmers increased yields simply by cultivating more land. Agriculture was labor intensive; the power of humans and livestock was used to till and prepare the land, plant the seed, care for the growing plants, and harvest the crop. Machines were used, but they were relatively simple. Farmers grew a variety of crops, a strategy known as **polyculture**, and they allowed fields to lie fallow periodically in order to restore soil fertility.

Around 1950, agriculture began to undergo rapid changes. In the United States (and much of the developed world), agriculture became energy intensive, or more specifically, fossil-fuel intensive. In 1950, an amount of energy equivalent to less than half a barrel of oil was used to produce a ton of grain. By 1985, the amount of energy needed to produce a ton of grain had more than doubled. To increase production, farmers began to rely heavily on high-yielding hybrid crop strains and inputs of water, synthetic fertilizers, and pesticides (many of which are petroleum-derived

products). In some areas, especially the drier regions of the Southwest, irrigation projects allowed marginal lands to be cultivated. Farmers began to concentrate on producing only one or two profitable crops, a strategy known as **monoculture** farming.

At about the same time that food production began to change in the developed world, significant changes took place in the developing world as well. The **Green Revolution**, which began in Mexico about 1943, was based on the process of hybridization, or selective breeding — the crossing of one or more varieties of a particular plant or animal species to produce a hybrid with qualities considered desirable. The hybridization process can lead to a loss of reproductive capabilities; parents must be continually crossed to get the desired hybrids. (Plants grown from the seeds of the hybrid are more likely to resemble one or the other parental type than the desired hybrid, or in some cases the hybrids will be sterile.)

Special genetic strains of hybrid wheat were developed that enabled Mexico to double production by 1958 and then double it again during the 1960s. By 1970, 90 percent of Mexico's wheat crop was planted in the new high-yield varieties (HYVs). HYVs typically have short stalks and an expanded planting season, and mature earlier than traditional varieties. Many countries were able to double food production in 30 to 40 years using Green Revolution varieties.

Despite its successes, the Green Revolution did not end hunger in LDCs. For one thing, the Green Revolution virtually bypassed Africa altogether; its greatest impact was in Latin America and Asia (Figure 9-11). Moreover, productivity has biological limits, and recent world harvests seem to indicate that present HYVs have reached their peak of production. In addition, the use of HYVs poses significant problems for the farmers who rely on them. To achieve their full potential, HYVs require large amounts of fertilizer, carefully controlled irrigation, and pesticides. Thus, HYV farming is high input and expensive. The seeds, too, are expensive, and they typically require an extensive knowledge of appropriate farming techniques. Consequently, these varieties are out of reach of the poorest farmers. In many countries, the use of HYVs has meant that the rich have become richer, and the poor have become poorer and hungrier. And with the onset of the Green Revolution, many farmers stopped planting different types of crops (such as legumes) and land races in favor of the high-yield strains of cereals. Thus, the rapid spread of HYVs has diminished the diversity of crop species. Finally, because HYVs are genetically more

Figure 9-11: Rice paddies in the Philippines.

similar than the native varieties they replace, they are vulnerable to disease caused by pests and insects.

How Can We Promote Food Security?

Future improvements in the global food supply will result from changes in consumption patterns and the way we manage food resources, or through the increased use of underutilized resources. Of these, the most promising strategies include modifying the diet of people in the developed world, improving management of ocean fisheries, adding new plants to the human diet, and preserving the genetic diversity of food crops and livestock. Potentially, aquaculture and biotechnology also may improve the global food supply, but both of these strategies carry significant environmental and social risks and therefore must be pursued carefully.

Modify the Diet of People in MDCs

As we've seen, most people in MDCs eat very differently than most people in LDCs. The typical developed-world diet is meat-rich, and most of our foods are grown or raised in places far away — often with methods that are environmentally damaging. To put it another way, our diets are energy-intensive — in production, packaging, and distribution. But they needn't be; we can choose to minimize the negative environmental and social impacts of our diets. *What You Can Do: Managing Food Resources*, page 190, offers simple, concrete suggestions for adopting a diet that is healthy, environmentally sound, and socially just.

Improve Management of Ocean Fisheries

Managing ocean fisheries in an environmentally sound manner is essential if we are to get the most from today's catch while protecting stocks into the future. Addressing the various causes that give rise to overfishing, particularly fleet overcapacity, can help protect healthy fisheries and may help rebuild stocks that are overexploited. By increasing our understanding of different species, we may be able to selectively harvest only those schools that are large enough to sustain it. Improved management also means eliminating the waste associated with current harvesting techniques. For example, tremendous waste also occurs when non-target fish are tossed overboard from factory trawlers or are thrown away at processing plants. According to the FAO, between 17 million and 39 million tons of fish are thrown overboard each year because they are too small or are not in demand. Strict adherence to mesh size would reduce the catch of small fish. Additionally, creative thinking could enable us to utilize less desirable species, but before we do so, it's important that we first understand the consequences for ocean ecosystems. A critical question is: What will be the effect on marine food webs and species interactions if we significantly increase our harvest of certain species?

Add New Plants to the Human Diet

With research, we will surely discover many plants to add to the human diet. For example, researchers are investigating the potential of plants in the genus amaranth, members of the cockscomb family. These wild plants vary in taste, appearance and other traits, but all provide a protein-rich grain. Quinoa (pronounced "keen-wha"), a member of a related family, is a major source of protein for millions of people living in the Andean highlands. The protein in the quinoa grain contains a better amino acid balance than the protein in most true cereals. In fact, its protein is of such high quality that, in nutritional terms, it can take the place of meat in the diet. Quinoa is made into flour for baked goods, breakfast cereals, beer, desserts, and livestock feed. It also can be prepared like rice or used to thicken soups. The malted grains and flour can be used as a weaning food for infants. The grain holds particular promise for improving life and health for the poor in marginal upland areas, including highland regions in Ethiopia, Southeast Asia, and the Himalayas.

Preserve the Genetic Diversity of Food Crops and Livestock

The discovery of "new" plants to supplement the human diet presupposes that wild lands and their complement of species will remain intact. Unfortunately, environmental degradation, habitat fragmentation, and the widespread use of HYVs are contributing factors to the loss of species and the reduction of genetic diversity. The demand for increased productivity to feed ever-growing populations makes HYVs necessary, but at the same time, we cannot afford to lose the characteristics possessed by native plant varieties. Among these are adaptations to local climate and soil conditions, resistance to drought, resistance to diseases and pests, and nutritional qualities such as high protein, mineral, and vitamin content. The value of traditional plant and animal varieties lies primarily in their genetic code. Unfortunately, since 1900, about 75 percent of the diversity of agricultural crops has been lost. Similarly, about 1,000 domestic animal breeds were lost in the last century, and another 30 percent of existing breeds are close to extinction.

In recent years, the international community has recognized the urgent need to protect the remaining genetic diversity of crop plants and livestock (Figure 9-12). An important component of that effort is the protection of wild lands within the centers of diversity. Preserving these lands can help to protect the wild relatives of crop plants as well as other potentially important, but currently uncultivated, crops. Because wild lands are disappearing so rapidly, however, protecting remaining undisturbed lands cannot adequately ensure the preservation of genetic diversity. For that reason, researchers and farmers alike are adapting an age-old tactic practiced by many traditional societies, including the Kayapo Indians of the Brazilian Amazon. For

Figure 9-12: The traditional Red Spotted cattle breed recently faced extinction in eastern Europe. Heifer International organized a multinational program to support the breed and improve its genetics.

centuries, the Kayapo have tended hillside gardens that contain representative samples of various food crops, particularly tuberous plants. The gardens' location protects the collections from flooding, thus preserving this important and valued resource. Nations that establish gene banks are doing much the same thing that the Kayapo do: preserving a commodity upon which society depends.

A **gene bank** is a place in which the germplasm (genetic material) of plant or animal species is preserved for future use. Samples of useful or economically valuable species are gathered and then maintained under controlled conditions. Plant genetic material is usually preserved outside of its natural environment in seed banks, although field gene banks, especially in the tropical regions, are becoming more common.

To maintain a seed sample in long-term storage, the seeds are dried to reduce their moisture content, preventing damage to the tissue. The sample is then stored at -4°F (-20°C). Under these conditions, samples can remain viable for up to 100 years. To test for viability, a subsample is periodically taken and germinated. If less than 85 percent of the seeds germinate, the entire sample is regerminated in order to maintain and protect

WHAT YOU CAN DO ▷ Managing Food Resources

There are many things you can do to help foster food security and end hunger. Here are some suggestions:

➤ Keep a "food diary" of everything you eat for three days. Try to identify when, where, and how each item was produced. Then, modify your diet in order to eat more foods grown or produced locally.

➤ A first step: Shop at your local farmers market for produce, eggs, meats, and other foods. The food is fresher, you avoid excessive packaging, and your money stays in your community.

➤ Take it a step further: Identify and patronize local *organic* farmers who grow fruits and vegetables without the use of chemical fertilizers or pesticides and who raise livestock without the use of growth hormones. Ask local supermarkets to carry organic produce and meat.

➤ For foods that aren't produced locally (for many people, coffee, tea, and chocolate are three good examples), support "fair trade" programs, which ensure that farmers in LDCs are paid a price that exceeds their production costs. Some programs allocate a portion of profits to support community improvement projects, such as the construction of wells, sanitation systems, and schools. To learn more, visit www.fairtradefederation.org or www.transfairusa.org

➤ Cut down on the amount of animal products in your diet. Learn how to combine grains, legumes, and other foods to assure adequate consumption of high quality proteins. Frances Moore Lappe's *Diet for a Small Planet* is an excellent reference.

➤ If you love seafood, limit your menu choices to fish and shellfish that are relatively abundant and that are caught or raised using methods that cause little or no damage to habitat and other wildlife. To obtain a *Seafood Miniguide* for use at the market and in restaurants, visit the Blue Ocean Institute at www.blueoceaninstitute.org

➤ Find out more about hunger in your area, your country, and around the world. Then translate your knowledge into action: Volunteer at a local food bank, soup kitchen or homeless shelter, or support an organization dedicated to eliminating hunger.

➤ At the next gift-giving holiday, consider charitable, alternative gifts that directly support anti-hunger efforts. One excellent source for socially conscious gifts is Heifer International; for more information, visit www.heifer.org (Figure 9-13).

➤ Learn as much as you can about issues related to food production and hunger so that you can make informed decisions at the grocery store, in your own garden, and in the voting booth. Contact organizations concerned with improving agriculture, conserving natural resources, and eliminating hunger. Ask them for publication listings; subscribe to newsletters so that you can keep up-to-date on developments such as proposed legislation or promising research.

➤ Write your state and national lawmakers to express your concern about hunger, food security, and agriculture. Share your views; encourage them to support legislation that promotes equitable trade and meaningful foreign assistance and that discourages environmentally harmful farming and fishing practices.

Figure 9-13: A youngster in Kenya with a kid — a baby goat — provided by Heifer International (HI). For over 60 years, HI has worked with communities to end hunger and poverty and to care for the Earth.

its genetic variability. Because some genes are extremely rare in a population, even a minimal deterioration in a sample's viability could mean the loss of potentially valuable genes.

Not all plants can be preserved as seed in a gene bank. Root crops and plants that have recalcitrant seeds, which must remain moist and thus do not survive the drying process, can be preserved in one of three ways: field gene banks, in vitro preservation, and cryopreservation.

Field gene banks are plots in which plants to be preserved are grown. In a field gene bank, the plants are germinated annually. There are two major disadvantages in preserving plants in a field gene bank. First, a large space is required to grow the plants and, second, the labor costs necessary to maintain the field can be high.

In vitro preservation, in which plant tissue is stored in a test tube, allows storage of many species for up to two years before the culture must be renewed. Samples saved in this manner are kept in windowless, air-conditioned rooms. Cool temperatures retard growth, and artificial lighting allows the operator to control day length. The major advantage of *in vitro* preservation is that many more samples can be stored in far less space than that required by field banks. For example, a 10-acre field is required for 6,000 potato plants at the Huancayo substation of the International Potato Center (CIP, Centro Internacional de la Papa) in Peru. But at CIP headquarters in Lima, a duplicate collection is housed in facilities that require only 0.1 percent as much space.

In vitro preservation does have two major drawbacks. Certain materials appear to be genetically unstable in tissue culture and so cannot be stored reliably in this manner, and it is only a temporary means of storage. **Cryopreservation** — storing plant and animal materials in liquid nitrogen — may prove to be a longer term and safer method of preservation for certain species. The extremely low temperature of liquid nitrogen, -321°F (-196°C), should suspend genetic instability and change.

Increase Yields through Aquaculture

Nearly all aquaculture products are used to directly feed humans (compared to the one-quarter of the marine catch that is used to make fishmeal and fish oil); consequently, 25 percent of all the fish and shellfish now consumed by humans is the result of aquaculture, making it an increasingly important element of the global food supply. Aquaculture allows for optimal use of feedgrains because the conversion efficiency of fish is high; it takes slightly less grain to produce a pound of weight gain per fish than it does for chicken or turkey. China, India, and Japan are the largest producers of farm-raised fish, with China producing almost half of the world total. In the United States, operations included cage-raised catfish in the South (especially Mississippi and Arkansas), trout and salmon in Washington, Oregon and Idaho, and crayfish in Louisiana.

Despite its potential as a source of much-needed protein, aquaculture has some negative impacts. These include the easy spread of disease among fish raised in close quarters, the spread of disease to wild fish by farmed fish that escape from pens, inbreeding and genetic weakening of wild stocks, and pollution caused by fish wastes. Perhaps the most troublesome impact of marine aquaculture is its effect on coastal habitats. In Bangladesh, Honduras, Thailand and elsewhere, a significant portion of mangrove forests were destroyed to make artificial shrimp ponds. Mangroves and other coastal wetlands are important nurseries for wild fish and shellfish stocks, so their destruction adversely affects marine ecosystems — and ultimately the oceanic catch so vital to humankind. Marine aquaculture has adverse social effects too. In developing countries, coastal residents do not benefit from aquacultural activities, since the farmed seafood is destined for industrialized countries. In fact, they may be hurt by such activities if the wild stocks that rely on coastal nurseries decline.

For the poor in developing countries, the most promising fish farming is freshwater aquaculture. Freshwater farms produce less expensive species, such as carp and tilapia (an African river and lake fish), which are within the economic reach of more people. And when fish ponds are integrated with crop production, as in China, waste from the ponds can be used to fertilize crops, thereby reducing pollution.

Increase Yields through Biotechnology

According to the 1992 U.N. Convention on Biological Diversity, **biotechnology** is any technological application that uses biological systems, living organisms or derivatives thereof, to make or modify products or processes for a specific use. While this broad definition encompasses even traditional techniques used to make wine and cheese, modern biotechnology typically refers to the modification of living organisms (plants, livestock animals, and fish) through the manipulation of genes.

Biotechnological processes consist of two main types. **Marker-assisted selection** uses genetic information to speed up and improve conventional plant and animal breeding. Essentially, researchers analyze crop varieties to identify genetic lines that possess a desired characteristic, such as cold-tolerance, drought-resistance, or high protein content. Those lines are then used in conventional breeding programs.

Genetic modification (GM) is the direct transfer of genes from one organism to another in order to convey useful characteristics on the recipient organism. For example, maize and cotton have been genetically modified with a bacterial gene to yield varieties that produce an insect-killing toxin, thus reducing the need for pesticides. The most common GM crops are soybeans (a variety was produced that is resistant to herbicides), maize, cotton, and canola.

Genetic modification has numerous potential benefits — and an equal number of serious risks. On the positive side, GM could allow us to increase the nutritional value of food staples, raise fish yields, and produce crops that tolerate poor environmental conditions (such as cold, drought, or salinized soils). For example, researchers have developed a variety of rice containing genes that make the plant produce beta carotene, which the body converts into Vitamin A. This "golden rice" has the potential to dramatically reduce the incidence of vitamin A deficiency. On the negative side, with GM, allergens can be transferred inadvertently from an existing organism to a target organism, and genes inserted into a species artificially may cross accidentally to an unintended species. For instance, herbicide resistance might spread from a GM crop into weeds, rendering them herbicide-resistant also. In addition, GM crops might have unforeseen consequences for farming systems — by depleting the soil of nutrients, for example, or by requiring a greater amount of water than unmodified varieties. GM fish that escape into the wild could displace native fish populations. And finally, a growing reliance on GM crops could mean the continued disappearance of land races and traditional livestock breeds, further eroding genetic diversity.

For biotechnology to enhance food security worldwide, a number of economic and social issues must be addressed. First, most of the money to finance costly biotechnological research comes from multinational corporations that anticipate making huge profits by providing the fruits of their research to the world's farmers. But impoverished farmers cannot afford costly plant varieties — certainly not every year. Currently, patent law prohibits farmers from "saving seed" from one year's harvest to plant the next year, a practice farmers the

world over have used for centuries. If biotechnology is to play a significant part in alleviating widespread hunger, the financial interests of large corporations must be balanced by the desperate need of impoverished farmers. To ensure fair access by the poor and hungry, the public and non-profit sectors will need to play a greater role in research and development.

A related issue concerns the goals of biotechnological research. At present, more money is being spent on developing technologies to improve animals than on plants. Plant biotechnologists seem to be concentrating on improving flavors and sweetness in fruits and vegetables. To promote food security in those areas that need it most, more money must be committed to developing crops that are resistant to pests and diseases and that are adapted to less-than-ideal growing conditions.

How Can We End Hunger?

Ending hunger means addressing its root causes, chiefly poverty and environmental degradation. At the turn of this century, the global community committed to doing just that. Under the auspices of the United Nations, the countries of the world established the Millennium Development Goals (MDGs), a set of internationally-agreed upon targets for reducing poverty, hunger, disease, illiteracy, environmental degradation, and discrimination against women by 2015. The Millennium Project, an independent advisory group, was commissioned by the U.N. Secretary-General to advise the international body on strategies for achieving the MDGs.

Jeffrey Sachs, a Columbia University economist who directs the school's Earth Institute, heads the U.N. Millennium Project. According to Dr. Sachs and his Millennium Project colleagues, the world is wracked by instability arising from "failed states," nations plagued by hunger, disease and death, where young people face poverty, mass unemployment, lack of education, and hopelessness. The norm, in these societies, is a lack of health care clinics, a shortage of schools and teachers, a lack of rural roads and other basics. These nations desperately need major investments in social services and infrastructure, but they do not possess the resources to make those investments alone. It is the cycle depicted in Figure 9-8 on a national scale: Poverty begets poverty, because the basic investments needed to overcome it are beyond the means of LDCs, while the degree of financial assistance from MDCs is too limited to enable poor countries to break free. Take the United States, for instance; the nation spends about $450 billion each year for the military to defend it against global threats but only about $13 billion to fight the underlying causes — poverty, disease, and despair — that give rise to those threats.

In 2005, the Millennium Project released its report, *Investing in Development*, which contains specific cost-effective measures that together could cut extreme poverty in half and radically improve the lives of at least one billion people in poor countries (Figure 9-14). Highlights include measures designed to address the health crisis in poor nations, launch a Green Revolution in Africa, and invest in basic rural infrastructure. (For more information and to read the complete Millennium Project report, please go to www.unmillenniumproject.org)

FIGURE 9-14: School feeding program, Malawi. Such programs have proven highly effective in improving child health and well-being.

Address the Health Crisis in Poor Nations

The health crisis is a major reason why the poverty trap continues to plague sub-Saharan Africa and other impoverished regions. Large numbers of the poor are sick and dying, and sick people are unable to earn income or pay taxes. In many LDCs, bankrupt governments and impoverished households are the rule; under these conditions, health systems have collapsed and epidemics are running unchecked. It will be impossible to break this vicious cycle without increased assistance from the world's richest countries. According to the Millennium Project, for an annual investment of about $25 billion, the industrial world could finance a serious, concerted attack on AIDS, tuberculosis, malaria, vaccine-preventable diseases, unsafe childbirth, and other killer conditions. While $25 billion is a considerable sum, it is just one one-thousandth of the combined annual income — about $25 *trillion* — of the United States, Canada, Western Europe, Japan, Australia, and New Zealand. The monies would be used for specific, proven measures, including anti-malarial drugs and bed nets and antiretroviral medicines for AIDS sufferers.

Launch a Green Revolution in Africa

The Green Revolution in Africa should be tailored to the unique conditions and constraints of that impoverished continent, particularly in sub-Sahara. Five science-based components are key. First, African farmers need soil nutrients. Inexpensive "green fertilizers" — in the form of manure, nitrogen-fixing trees, and cover crops — are especially desirable. Second, they need reliable water sources, particularly in regions where rains often fail. The answer: water-harvesting technologies and small-scale irrigation (which will avoid disrupting ecosystems and displacing people, as happens with giant hydroelectric projects). Third, African farmers need effective agricultural extension services so that they can adopt small-scale irrigation and other appropriate, state-of-the-art technologies, including low-till or no-till agriculture (to reduce soil erosion) and integrated pest management (to limit the use of herbicides and pesticides). Fourth, they need improved germplasm — crop seeds, tree varieties, and livestock breeds — that are highly productive, adapted to local climates, and resistant to pests and disease. And fifth, supplementary feeding should be provided for the groups most vulnerable to malnutrition — pregnant and nursing mothers, children younger than age two, and those in school. The best option is to use locally purchased foods, thereby increasing market demand while providing villagers with a balanced diet comprised of the foods they prefer. Food aid shipments should be limited to emergency starvation situations where foods cannot be procured locally.

Invest in Basic Rural Infrastructure

To eliminate extreme poverty in the world's most troubled regions, wealthy nations must help poor countries make major investments in schools, roads, power, water, sanitation, and more. Again, the Millennium Project recommends specific, proven measures: new wells, pit latrines, grain storage facilities, feeder roads to carry farm goods to the cities, market information systems and value-added processing facilities for high-value products, including fruits and medicines from newly domesticated plants.

According to the Millennium Project, approximately $50 billion is needed to address the related challenges of education, social services and infrastructure, for a total of some $75 billion annually. About half of that total, $35 billion, should come from the United States. That translates into 35 cents for every $100 of U.S. Gross National Product (GNP). To put that figure into context, $35 billion per year is about one-seventh of President Bush's tax cuts — one-half of annual U.S. military spending in Iraq — and one-twelfth of the total yearly U.S. military budget.

It may surprise you to learn that, when measured as a percentage of GNP, U.S. contribution to development aid is currently the lowest of the world's 22 donor countries. Sweden, the most generous nation, gives 0.87 percent of its GNP, while the United States gives about 0.13 percent. To put it another way, the United States donates about 13 cents per every $100 of income. By increasing our share another 35 to 50 cents per every $100 of income, we could help end extreme poverty and chronic hunger and save tens of thousands of lives each year.

For a moment, let's set aside humanitarian reasons for increasing assistance to LDCs. What benefits will the rich world realize from helping poor nations meet their health, development, and social needs? Without a doubt, the most significant benefit is a more peaceful world. By ending extreme poverty and hunger, we eliminate breeding grounds for violence and terror, disease and misery, unwanted mass migration, and drug trafficking. In their place, we foster healthier, sustainable communities — societies that become good neighbors and trading partners with the more developed world. As Sachs and Pedro Sanchez (a Millennium Project colleague) have written, "We are surely the first generation in human history that could actually bring about an end to extreme poverty on the planet. Whether or not we will seize this glorious opportunity, this historic chance to fundamentally transform the human condition, depends on the strength of our human will, and on our willingness to do the right things."

Summary

Carbohydrates and fats are the major sources of the energy required to maintain the body and perform work. Proteins are the "building blocks" of the body, forming the substance of muscles, organs, antibodies, and enzymes. Small units called amino acids combine in various ways to form larger protein molecules. Of the 20 naturally occurring amino acids, 11 can be synthesized by the body at a sufficient rate; the remaining nine, the essential amino acids, must be obtained through diet.

Of the 80,000 potentially edible crops on earth, global agriculture is dependent upon just a few plant species. The top four, called staples, are wheat, rice, maize, and potatoes. Our agricultural system has also relied on just a small number of animals. Nine domesticated animals supply the bulk of the protein (meat, eggs, milk) from livestock.

Food security is the ability of a nation to provide, on an ongoing basis, enough food to keep its people alive and healthy. Grain is used as a measure of food security because it provides (both directly and indirectly) well over half of human food energy intake and because it can be stored for use during cold or dry seasons. Per capita production and carryover stocks are the most important measures of the status of the grains that are our "staff of life." In both cases, trends in recent years are worrisome: per capita production is declining (although total production continues to grow slowly) and carryover stocks are low.

Along with grain, fish and meat are the most important categories of human foods. Ocean fisheries account for the largest share of total fish production, but aquaculture, the controlled production of fish, shrimp, shellfish, and seaweeds in ocean pens and inland ponds or tanks, accounts for the fastest growing share. As with grain, the growth in total fish production has slowed in recent years, falling behind population growth; the factor most responsible for this slowed growth is overharvesting of marine fisheries. Not surprisingly, the per capita marine catch declined throughout the 1990s. Only per capita production of meat has increased, in part due to increased demand for meat in the developing world, especially China. Meat production can cause serious environmental problems when it is practiced to the extent and in the manner common in the developed world. When livestock are fed grains, the efficiency of meat production is greatly reduced. Also, the overgrazing of range and pasture by livestock (especially cattle and sheep in the American West) causes soil erosion and water pollution.

Hunger is the inability to acquire or consume an adequate quality or sufficient quantity of food through non-emergency food channels, or the uncertainty of being able to do so. Worldwide, about 852 million people are chronically hungry — a number greater than the combined populations of the United States, Canada, Japan, and Europe. With nearly one in seven people unable to consume enough food to be healthy and to lead an active, productive life, it's no surprise that hunger and malnutrition together are the number one risk to health — greater than AIDS, malaria, and tuberculosis combined.

The most extreme and dramatic effect of hunger is starvation, which occurs when a person does not consume enough calories to sustain life. Famine is widespread starvation. Each year, of the deaths attributed to

Discussion Questions

1. Describe the current status of food production worldwide, emphasizing grain, fish, and meat production. Next, describe current patterns of consumption worldwide, again emphasizing grain, fish, and meat. What conclusions can you draw about the sources of production and consumption?

2. What measures do you think are appropriate to help restore and protect marine fisheries? Now suppose that you live in a small fishing community in the northeastern United States or the maritime provinces of Canada — do those measures still seem appropriate to you? If not, what do you think should be done to help the fisheries?

3. Define hunger. Name at least three factors that cause hunger and explain how they are related. Which do you think is most important and why?

4. Imagine that you are one of the world's hungry people. What might your life be like? What factors might have contributed to your situation?

5. What are some benefits of growing high-yield crops? What are some disadvantages? Should the world's poorest farmers be allowed to save the seeds of high-yield crops from one year to another? Why or why not?

(continued on following page)

(continued from previous page)

6. Of aquaculture and bio-technology, which do you feel has more potential to increase food production? What are the drawbacks of each? Would one be more effective than the other under different circumstances? Explain.

7. Explain why it is essential to preserve the genetic diversity of food crops and livestock.

8. Is there hunger in your community or city? What resources are available for the hungry in your area? What measures do you think should be taken to end hunger in the United States?

9. According to the Millennium Development Project, an annual investment of $70 billion is needed from the world's richest nations in order to eradicate extreme poverty, end chronic hunger, improve healthcare, and foster sustained development in the world's most impoverished regions. Do you think it is a wise investment? Use specific arguments to defend your answer.

10. Many people argue that it is in the best interest of rich nations to alleviate poverty and suffering in less developed countries. Do you agree? Why or why not?

hunger, most come not from starvation but from undernutrition, malnutrition, and hunger-related diseases. Chronic undernutrition results when an individual consumes too few calories and protein over an extended period of time. Malnutrition occurs when an individual consumes too little (or infrequently, too much) of specific nutrients essential for good health. In children, signs of malnutrition include stunting, wasting, and underweight.

There are various forms of malnutrition; some result from a deficiency of macronutrients (fats, carbohydrates, and proteins); others from a deficiency of micronutrients (vitamins and minerals). Many children in LDCs suffer from kwashiorkor, caused by a diet deficient in protein, and marasmus, caused by a diet deficient in both protein and overall number of calories. One of the most pressing nutritional problems worldwide is dietary deficiencies of micronutrients, particularly of iron, iodine, and vitamin A. Micronutrient deficiencies have serious health consequences, particularly for women and children. Malabsorptive hunger — which occurs when the body is no longer able to absorb nutrients from the food consumed — often accompanies undernutrition and malnutrition. Seasonal hunger is the lack of adequate food for a period of time (days, weeks, or even months) after the reserves from the previous year's harvest have run out but before the new crops have been harvested.

Children account for about 40 percent of the world's hungry; most of the rest are women. Seventy percent of those who suffer chronic hunger are women and girls. Most of the hungry live within the "great hunger belt" — a region encompassing various nations in Southeast Asia, the Indian subcontinent, the Middle East, Africa, and the equatorial region of Latin America. In other parts of the world, hunger is less visible but nonetheless real. In the United States, nearly 36 million Americans — including 12.9 million children — fall below the poverty line and are at risk from hunger and associated diseases.

The chief cause of widespread, intractable hunger is poverty. Other causes include environmental degradation, war and civil unrest, natural disasters, and faulty economic practices. These causes are interrelated and often act synergistically. Extreme poverty is defined as subsisting on less than $1 a day; those in extreme poverty cannot meet their basic needs for survival. An estimated 1.2 billion people (all of them in the developing world) live in extreme poverty. Moderate poverty is defined as living on $1 to $2 a day; one's basic needs are met, but just barely. Those in relative poverty, defined as a household income level below a given proportion for the national average, lack material items and comforts that the middle class takes for granted. As with hunger, those who suffer the most from poverty are children; most of the world's poor are under the age of 15. Poverty leads to undernutrition and malnutrition, which in turn lead to decreased resistance to disease and decreased energy. Illness and lack of energy lead to decreased ability to learn and to work, which perpetuates poverty. Unless something is done to break the poverty cycle, poor children are doomed to a life of hardship and want simply by virtue of having been born.

For most of our history, humans were hunters and gatherers. Eventually, however, humans began to domesticate plants. Domestication is the result of the long-term selection of traits useful for survival in captivity over those traits necessary for survival in the wild. Domestication gave way to agriculture, the purposeful tending of particular plant species for human use. Farmers' deliberate selections of certain strains resulted in land races, varieties of plants adapted to local conditions such as climate and soil type.

In the past 50 years, food production in the United States has come to rely heavily on inputs of energy, water, synthetic fertilizers, herbicides, and pesticides. At about the same time that food production began to change dramatically in the developed world, the Green Revolution began in Mexico. Special high-yield strains of hybrid wheat were developed that enabled Mexico to more than double wheat production. However, despite its successes worldwide, the Green Revolution has not ended hunger in the developing nations.

To promote food security worldwide, four strategies are particularly important: modifying the diet of people in the developed world, improving management of ocean fisheries, adding new plants to the human diet, and preserving the genetic diversity of food crops and livestock. Aquaculture and biotechnology also may improve the global food supply, but both of these strategies carry significant environmental and social risks.

To end hunger, we must address its root causes, chiefly poverty and environmental degradation. In 2005, the Millennium Project, an independent advisory body to the U.N. Secretary-General, released a report, *Investing in Development,* which contains specific, proven, and cost-effective measures that together could cut extreme poverty in half and radically improve the lives of at least one billion people in poor countries by 2015. Highlights include measures designed to address the health crisis in poor nations, launch a Green Revolution in Africa, and invest in basic rural infrastructure.

KEY TERMS

agriculture	food security	moderate poverty
amino acids	gene bank	monoculture
aquaculture	genetic modification	polyculture
biotechnology	Green Revolution	relative poverty
carryover stocks	hunger	seasonal hunger
cash crops	kwashiorkor	staples
centers of diversity	land races	starvation
cryopreservation	malabsorptive hunger	stunting
domestication	malnutrition	undernutrition
estuary	marasmus	underweight
extreme poverty	marker-assisted selection	wasting
famine		

CHAPTER 10

Energy: Fossil Fuels

*If one has cut, split, hauled, and piled his own good oak,
he will remember much about where the heat comes from,
and with a wealth of detail denied to those who spend the
weekend in town astride a radiator.*

Aldo Leopold

*Since the Industrial Age the world has increasingly
depended on fossil fuels. Modern civilization is actually
based on non-renewable resources. This puts a finite limit
on the length of time our civilization can exist.*

The Gaia Peace Atlas

Learning Objectives

When you finish reading this chapter, you will be able to:

1. Define energy and explain how it is measured.

2. Explain how energy resources are classified.

3. Explain briefly how fossil fuels were formed and differentiate among proven, subeconomic, and indicated resources.

4. Describe the current status (uses, availability, and environmental effects) of coal, petroleum, natural gas, oil shales, and tar sands.

5. Describe how energy consumption varies worldwide.

6. Summarize how energy consumption has changed in the United States in the past 200 years.

7. Give examples to illustrate how industries, communities, and individuals can increase energy efficiency and conserve fossil fuels.

The cover of the June 2004 edition of *National Geographic* is one of foreboding. Superimposed over a photograph of evening rush hour traffic on the congested Washington, D.C. Beltway, in bold yellow letters, reads the headline "The End of Cheap Oil" (Figure 10-1). Historically, fossil fuels have been an important energy source, particularly in the industrialized nations, enabling us to enjoy a high standard of living and everyday conveniences uncommon in many countries. But there is a dark side to the lifestyle that runs on fossil fuels: Escalating environmental impacts; the growing scarcity of reserves; volatility in world politics and economics; and intensified fears of compromised national security.

Take a moment to consider the following two statements: The amount of fossil fuels that humans use in just one year took about a *million years* to form. And in the United States, fossil fuels have supplied over 90 percent of the energy consumed. Although there is debate on when, exactly, the reign of cheap oil will end, researchers agree that supplies are rapidly dwindling. Perhaps, then, the most pressing question isn't *when* will the peak and decline occur but rather *will* we be prepared for it?

We begin this chapter by defining energy, energy resources, and energy efficiency. We then turn specifically to fossil fuels, describing each of the major fuels — coal, petroleum (also called crude oil), natural gas, oil shales, and tar sands — and examining efforts to manage them.

Defining Energy

Silent and unseen, energy fuels our bodies, our machines, our societies, and our planet. It is the common denominator among widely diverse activities. The beat of a hummingbird's wing, for example, like the fantastic journey of the Voyager-2 spaceship, is possible only through the conversion of energy from potential to useful forms. Similarly, a wildflower and a large urban shopping mall are both the product of harnessed energy. As we learned in Chapter 3, *Ecosystem Structure*, energy is part of the abiota, the nonliving or physical component of the biosphere. In this section, we further define what energy is and examine how it is measured and classified. We also discuss the calculation and implications of energy efficiency.

Figure 10-1: Evening rush hour moves at a crawl on the Washington, D.C. Beltway.

What Is Energy and How Is It Measured?

Energy is the ability to do work. All energy is either **kinetic** — energy due to motion or movement, such as falling water or an electrical current — or **potential** — energy in storage, that is, energy which can be converted to another form. For example, an automobile engine converts the potential energy stored in the chemical bonds of gasoline into kinetic energy in the form of heat, eventually causing the car to move.

Energy can take different forms, such as heat, light, chemical, or electrical energy. Various units are used to measure the quantity of different forms of energy. For example, heat energy can be measured in calories, British thermal units (btu), or therms; electrical energy is measured in kilowatts. Table 10-1 explains some important units of energy.

In addition to measuring the quantity of a specific form of energy, such as heat, we also measure the amount of a specific type of energy resource. Petroleum, for example, is measured in barrels, while natural gas is measured in cubic feet and coal is measured in short tons. Table 10-2 presents some common measures of energy resources.

TABLE 10–1: Selected Units of Energy

Btu (British thermal unit) is the standardized measure of quantity of heat produced by various sources of energy. It is the quantity of heat required to raise the temperature of 1 pound of water 1° F. One Btu is the approximate energy equivalent of one burning wooden match tip.

Therm is a measure of quantity of heat derived from natural gas; 1 therm is equal to 100,000 Btus. It is roughly equivalent to the energy in 87 cubic feet of natural gas.

Kilowatt (kw) is a measure of electrical power; 1 kilowatt is equal to 1,000 watts; 1 megawatt (Mw) is equal to 1,000,000 (1 million) watts; 1 gigawatt (Gw) is equal to 1,000,000,000 (1 billion) watts. Watt is an electrical unit of power or rate of doing work or the rate of energy transfer equivalent to 1 ampere flowing under the pressure of 1 volt.

Kilowatt-hour (kwh) is a measure of electrical use, the energy equal to that expended by 1 kilowatt in 1 hour. For example, 1 kwh is the amount of electricity than an operating 100-watt light bulb would consume in 10 hours. It is equivalent to 3,411 Btus.

Quad is a measure of quantity of heat; 1 quad is equal to 1,000,000,000,000,000 (1 quadrillion) Btus. It is equivalent to the heat produced by 171.5 million barrels of oil.

TABLE 10–2: Selected Measures of Energy Resources

Barrel is a measure of amount of crude oil or other liquid petroleum products; 1 barrel is equal to 42 gallons. Each barrel of petroleum motor gasoline produces 5.25 million Btus; each gallon produces 12,500 Btus.

Cubic foot (cu ft) is a measure of amount of dry natural gas. One cubic foot of dry natural gas produces approximately 1,031 Btus of energy.

Short ton is a measure of amount of coal; 1 short ton is equal to 2,000 pounds. (For comparison, 1 metric ton is equal to 2,204 pounds.) One short ton of coal equals approximately 3.8 barrels of crude oil and 21,000 cubic feet of dry natural gas. There are 1.102 short tons in 1 metric ton.

How Are Energy Resources Classified?

Energy resources are generally classified as nonrenewable, renewable, or perpetual. **Nonrenewable resources** exist in finite supply or are renewed at a rate slower than the rate of consumption. Fossil fuels — coal, oil, and natural gas — best exemplify nonrenewable energy resources. Nuclear energy, as it is widely used, is also nonrenewable.

In 2005, 94 percent of the energy consumed in the United States was supplied by nonrenewable resources (Figure 10-2); worldwide, the figure was 92 percent.

How long the world's supply of fossil fuels will last depends on many factors: market price, energy demand, development of improved extraction techniques, discovery of new deposits, use of more energy-efficient technologies (especially automobiles), conservation practices, and development of other energy resources.

As the term implies, **renewable resources** are replenished at rates faster than or consistent with use; consequently, supplies are not depleted. Renewable energy resources, such as biomass (wood, plant residues, and animal wastes), can become exhausted — and thus nonrenewable — when they are used faster than they can be regenerated. This is the case in some less developed countries, where wood has become so scarce that people are forced to use crop residues or animal dung for fuel. See Chapter 11, *Alternative Energy Sources,* for more information on perpetual and renewable resources and nuclear energy.

Perpetual resources originate from a source that is virtually inexhaustible, at least in time as measured by humans. Examples include the sun, tides, falling water, and winds. Perpetual and renewable resources often are called **alternative resources** because they offer us a choice other than fossil fuels. They supply about 10 percent of the world's energy, but the proportion varies from two to five percent in many industrialized nations to 50 percent in some nonindustrialized countries. While most alternative resources are renewable or perpetual, our nation's reliance on fossil fuels is so high that we believe almost any other energy resource, including nuclear energy, can be considered an alternative.

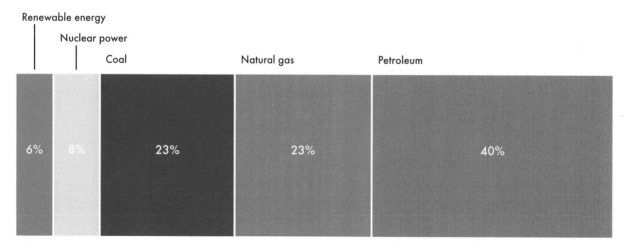

Renewable energy	Nuclear power	Coal	Natural gas	Petroleum
6%	8%	23%	23%	40%

FIGURE 10-2: U.S. energy consumption, 2005. Nonrenewable resources meet almost all of the nation's energy demand. Petroleum is used largely in the transportation sector; natural gas in the industrial and residential sectors; and coal and nuclear power in the generation of electricity. **Source:** *U.S. Energy Information Administration (2005).*

What Is Energy Efficiency?

According to the first law of thermodynamics, also known as the law of the conservation of energy, energy can be neither created nor destroyed, but it can be changed or converted in form. Although the same amount of energy exists before and after the conversion, not all of it is in a useful form. The conversion of energy from one form to another always involves a change or degradation from a higher quality form to a lower quality form, a principle contained in the second law of thermodynamics. Ultimately, all forms of energy are degraded to low-quality, diffuse heat.

For any system, device, or process that converts energy, the energy efficiency of the conversion can be determined. **Energy efficiency** is a measure of the percentage of the total energy input that does useful work and that is not converted into low-quality, diffuse heat. Because heat is given off every time energy is converted from one form to another, the energy efficiency of any device or process is always less than 100 percent.

The electricity produced by a coal-fired power plant illustrates a chain of three energy conversions and their associated efficiencies. When coal is burned, about 60 percent of its potential chemical energy is converted to intense heat or thermal energy used to produce steam; the rest is converted to diffuse heat and given off to the environment. As the steam is used to drive a turbine, its thermal energy is converted to mechanical energy. The efficiency of this conversion is 47 percent. The turbine, in turn, drives an electric generator; the conversion efficiency of mechanical energy to electrical energy is very high, 99 percent.

The **net efficiency** of a process or system that includes two or more energy conversions is found by determining the efficiency of each conversion. For every 100 units of potential energy in a specific amount of coal, only 60 units of thermal energy are produced. Only about half of the thermal energy in the steam actually drives the turbine, so a little less than 30 units of mechanical energy are produced. Nearly all of that, however, is converted to electrical energy by the generator. Thus, the 100 units of potential energy in the coal results in about 30 units of electrical energy. The net efficiency of the process is slightly less than 30 percent.

Our calculation of net efficiency does not include all the energy that is required to extract, transport, and pulverize the coal, nor does it include pollution control devices and other environmental costs. You can see that getting energy in a form that is convenient and useful — available at the flip of a light switch — is an inefficient undertaking! Moreover, when the electricity is used to provide light in a home, the energy efficiency of the entire process is much less. In conventional incandescent light bulbs, for example, up to 95 percent of the electrical energy is converted to heat and only five percent to light. That's why the bulbs become so hot. "Cool lights," compact fluorescent bulbs, are much more efficient; using 13 to 15 watts of electricity, they deliver the equivalent light of a 60- or 75-watt incandescent bulb, an energy savings of about 80 percent.

Stop and think about all the devices found in the typical home that run on electricity (such as kitchen appliances, power tools, gardening equipment, grooming tools, toys, computers, and entertainment equipment). A list of such devices might total a hundred or more. Using the most energy-efficient devices available can produce a significant energy savings and conserve nonrenewable fuels. The purchase price of energy-efficient devices is usually higher than that of conventional models, but they are economical in the long run because they have a lower **life-cycle cost**: (initial cost + lifetime operating costs).

Describing Fossil Fuels

In this section, we discuss how fossil fuels were formed and where they are located. We also examine each of the major fuels, particularly their use, current and projected availability, and environmental effects.

What Are Fossil Fuels?

Fossil fuels include coal, petroleum, natural gas, oil shales, and tar sands. The term **fossil fuels** suggests that these resources are ancient in origin, and indeed they are; they are the fossilized remains of organic matter. Fossil fuels are composed chiefly of carbon and hydrogen. When a fossil fuel is burned, the chemical bonds that bind the carbon and hydrogen molecules are broken, and energy is released. This energy, whether it is contained in a tankful of gas or a lump of coal, was once sunlight captured by phototrophs. Each of the fossil fuels possesses unique characteristics and plays a different role in industrialized societies, but all originated from living plants or animals millions of years ago.

How Were Fossil Fuels Formed?

The greatest period of coal formation occurred some 300 million years ago during the Carboniferous era. Giant ferns and other plants died and settled on the bottom of swamps and wetlands, forming thick layers of organic matter rich in carbon. Anaerobic decomposition, coupled with warm temperatures and the significant pressure exerted by the water, transformed the organic matter into a low-quality fuel called peat. The peat then was buried and compressed beneath layers of sediments. Additional pressure from settling sediments compacted the peat further, squeezing more water from the buried matter. Over millions of years, the heat, along with the weight and pressure of the overlying sediments, turned the peat into coal. Similar processes were responsible for the production of petroleum and natural gas. But while the organic source of coal was largely plant material, petroleum originated from microscopic aquatic organisms (animals and plants) that died and settled to the bottom of shallow, nutrient-rich seas. Over time, sediments accumulated over these deposits. Heat from the Earth's interior and the pressure of the sediments enhanced the oil's formation. Increasing amounts of heat, pressure, and anaerobic decomposition were required to form natural gas, which is derived from both plant and animal organic matter.

The processes that formed fossil fuels continue today. (For example, decaying plant materials currently are accumulating in coal-forming environments like Florida's Everglades.) But because those processes occur on a geologic time scale, the world's present deposits of coal, petroleum, and natural gas are all we can ever expect to use. Fossil fuels thus are classified as nonrenewable and finite in supply.

Where Are Fossil Fuels Located?

Fossil fuels are not distributed evenly beneath the Earth's surface. Some areas are rich in deposits of coal; others are rich in deposits of petroleum or natural gas. This uneven distribution of fossil fuels exists because the conditions that gave rise to the preservation and fossilization of organic matter did not occur everywhere.

In the search for rich deposits, geologists use their knowledge of the origin and formation of fossil fuels. They explore areas that were exposed to the conditions favorable to the formation of coal or petroleum. Many of the areas most likely to yield deposits of fossil fuels already have been explored. Arid deserts, particularly those located on the edges of drying seas, often have been the site of major oil reserves. In geologic history, such areas likely were covered by shallow, nutrient-rich seas, which fostered an abundance of aquatic organisms. In recent years, oil exploration has focused on the continental shelves, especially those located along partially landlocked waters such as the Baltic Sea, the Gulf of Mexico, the Arctic Ocean, and the Bering Sea. Geologists believe that upwelling currents may have been present at these sites in the past 200 million years. Upwelling currents enrich seawater with nutrients, providing an ideal habitat for plankton and other microorganisms.

The location of fossil fuels has very important implications because whether or not a deposit will be extracted depends on where it is found and if the technology exists to extract it at a profit. The exploration and measurement of fossil fuels are ongoing processes. Estimates of the amount of each fuel are revised as new information about various deposits becomes known. **Proven** or **economic reserves** are deposits that have been located, measured, and inventoried. The fuel located in a proven reserve can be or currently is being extracted at a profit. In contrast, **subeconomic reserves** are deposits that have been discovered but cannot be extracted at a profit at current prices or with current technologies. **Indicated** or **inferred reserves** are deposits that are thought to exist and are likely to be discovered and available for use in the future.

What Is the Current Status of Fossil Fuels?

Coal, petroleum, and natural gas are the world's major fossil fuels; oil shales and tar sands are comparatively minor energy sources. Although they share a similar biological origin, fossil fuels vary widely in their status (the ways in which they are used and their availability) and in the extent to which they affect the environment.

Coal

Coal is a solid composed primarily of carbon (55 to 90 percent by weight) with small amounts of hydrogen, nitrogen, and sulfur compounds. Other elements or compounds may be present in trace amounts.

Coal is derived from deposits of plant matter. Depending on the amount of time that deposits were exposed to conditions of high pressure and temperature, different types of coal were formed. **Anthracite**, or hard coal, has the highest carbon content and lowest moisture content of all types of coal and consequently is the most efficient, releasing the largest quantity of heat per unit weight when burned. Anthracite's efficiency, coupled with the fact that it is the cleanest burning coal, makes it the preferred coal for heating homes and commercial buildings, but it comprises the smallest portion of U.S. reserves, less than one percent, making it very expensive.

Bituminous, or soft coal, is the most common coal, accounting for nearly half of U.S. reserves. It has a heating value slightly lower than that of anthracite. Bituminous coal has long been preferred for electric power generation and the production of coke, a hard mass of almost pure carbon used to make steel. When heated in air-tight ovens, coal is transformed into coke. Coke then is burned with iron ore and limestone to produce the pure iron required to make steel. Over 50 percent of bituminous coal reserves have a medium to high level of sulfur content, resulting in more air pollution.

Together, **subbituminous** and **lignite** coals account for approximately 45 percent of our nation's coal reserves. Because both have low heating values, they must be burned in large amounts in order to heat effectively. One advantage of subbituminous and lignite coals is that they contain very little sulfur; all of the subbituminous and 90 percent of the lignite coals have a sulfur content of less than one percent. To meet state and federal pollution standards, many utility companies have switched to subbituminous or lignite coal.

Use. In general, coal consumption has declined over the last century, largely because it is a dirty fuel with serious environmental impacts. Recently, however, coal use has increased in the face of global uncertainty in the petroleum market and mounting concerns over the safety of nuclear power. Currently, coal accounts for 23 percent of the United States' and 24 percent of the world's energy budget. At current rates, world consumption of coal — 5.3 billion short tons (4.8 billion metric tons) in 2001 — will increase to 7.6 billion short tons (6.9 billion metric tons) by 2025. In order, China, the United States, and India will drive much of the increase (Table 10-3).

Worldwide, coal is used primarily to produce electric power (Figure 10-3). In 2004, approximately 92 percent of the coal used in the United States went to electric utilities. Coal also is used heavily in industry,

TABLE 10–3: Coal Consumption, 2001-2025 (in Millions of Tons)

Region	2001	2025 Projected	Percent Change
China	1,383	2,757	99.3%
United States	1,060	1,567	47.8%
India	360	611	69.7
Japan	166	202	21.7
Former Soviet Union	446	436	-2.2
Western Europe	574	463	-19.3
Rest of the world	1,274	1,538	20.7
Total World	**5,263**	**7,574**	**43.9**

Source: U.S. Energy Information Administration, 2004.

principally to purify iron and produce steel. While fuel oil and natural gas have been replacing coal as a heating source for residential and commercial structures throughout western Europe, the rising cost of these fuels is causing similar buildings in the United States to convert back to coal.

Availability. Coal is the most abundant fossil fuel on Earth. Current world reserves are estimated to be approximately 1,083 billion short tons (983 billion metric tons), representing about 92 percent of the world's fossil fuel resources. The United States, the former Soviet Union, and China have the majority of the world's coal reserves. U.S. coal reserves represent about 90 percent of the nation's energy resources. In terms of potential energy, they are comparable to Saudi Arabia's known oil reserves. The majority of these reserves lie in three broad regions: western or Rocky Mountain, Appalachian, and the interior plains of the United States. Wyoming leads all states in coal production, followed by West Virginia and Kentucky.

Estimates developed by the coal industry predict that, at present consumption rates, world reserves should last about 210 years. However, coal use is likely to increase as reserves of other fossil fuels, especially oil, diminish. Naturally, increasing consumption would shorten the life expectancy of coal reserves. Some experts believe that if we were to become completely dependent on coal, the United States' proven reserves would last only 47 years and total estimated reserves only 75 years.

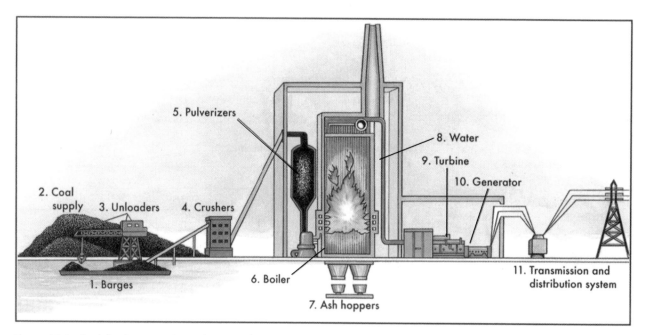

FIGURE 10-3: Coal-fired power plant. Electric utilities consume approximately 90 percent of the coal used in the United States. Coal supplies are brought in by barge or rail (1,2), unloaded (3), then crushed and pulverized to form a fine dust (4,5). The dust is blown into a furnace area inside a boiler (6), where it burns at very high temperatures. Waste ash collects in hoppers (7). The boiler is a huge structure made up of 30 miles of steel tubing. Purified water constantly circulates through the tubing (8) and is converted to steam by the intense heat of the furnace. The high-pressure steam turns the fanlike wheels of a turbine, spinning its shaft (9). The shaft turns the generator rotor (10), a large electromagnet, to produce electricity in coils of wire in the enclosure around the rotor. The electricity is then sent through a distribution system (11) to consumers.

Petroleum

Petroleum, also called crude oil, is a dark, greenish-brown or yellowish-brown liquid composed largely of hydrocarbon compounds, which account for 90 to 95 percent of its weight; the remaining 5 to 10 percent is a mixture of oxygen, sulfur, and nitrogen compounds. Typically, crude oil is found mixed with salt water (brine) and gas in the pores and cracks of sedimentary rock. When a well is drilled into oil-bearing rock, the pressure of the gas trapped in the oil forces part of the oil to the surface. "Gushers" produced by the pressure were spectacular proof of the great oil fields of the United States (Figure 10-4).

Use. Petroleum is perhaps the most versatile fossil fuel. A gooey, dark liquid when pumped from the ground, it is refined to yield many different materials such as propane, gasoline, jet fuel, heating oil, motor oil, kerosene, and road tar. One-quarter of petroleum resources are used in the production of nonfuel substances and chemicals, which are used in the manufacture of plastics and medicines. Each day, over $3 billion is spent worldwide on approximately 3,000 petroleum-derived products such as ink, mascara, insect repellent, heart valves, and dishwashing liquid.

Given petroleum's versatility, it is not surprising that global consumption is approximately 80 million barrels per day. The United States consumes a full quarter of this amount — 20 million barrels per day — even though it accounts for just five percent of global population. Gasoline is by far the most consumed petroleum product in the United States. Annually, over 60 percent of the nation's oil is consumed by the transportation sector in the form of gasoline for automobiles, diesel fuel, and jet fuel.

Availability. In 1950, proven reserves of petroleum amounted to 76 billion barrels. This figure quickly grew to 664 billion barrels by 1973, an 874 percent increase in just 23 years. Since 1973, however, global proven reserves have increased by a paltry five percent. This drastic slowdown in proven reserves has generated concern about future supplies. Over 80 percent of all the oil discovered to date on the North American continent has already been burned. At current consumption rates, proven U.S. reserves — estimated at 25 billion barrels — would meet domestic demand for a mere 4.5 years (to learn more about U.S. petroleum reserves, read *Environmental Science in Action: The Alaska Pipeline* by going to www.EnvironmentalEducationOhio.org and clicking on "Biosphere Project").

FIGURE 10-4: First oil well at Titusville, Pennsylvania, in 1859. Edwin L. Drake (at right, in top hat) was responsible for the drilling of the world's first commercial oil well. The age of petroleum had begun.

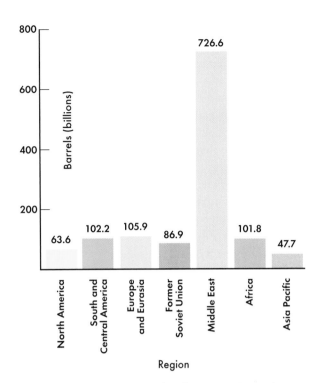

FIGURE 10-5: Total proven crude oil reserves, December 31, 2003. The Middle East is the region with the largest oil reserves, most of which are controlled by members of the Organization of Petroleum Exporting Countries (OPEC). *Source: BP Statistical Review of World Energy 2003, Oil Proved Reserves.*

Estimates of proven global reserves are slightly more than one trillion barrels. Two-thirds of these reserves are located in the Middle East (Figure 10-5). Other regions that have considerable oil deposits are Europe and Eurasia (about 10 percent of global reserves), Latin America (about 9 percent of global reserves), and Africa (also about 9 percent of global reserves).

Even though global reserves seem substantial, they cannot last long given the world's appetite for oil. If the present annual global consumption rate remains constant, proven reserves would last approximately 39 years. Estimates that include subeconomic and indicated reserves are more optimistic. No one can predict precisely when our oil supplies will become exhausted. What is certain, however, is that the world's oil supply is being drained from a pool that will eventually be emptied.

Natural Gas

Natural gas is composed of several different types of gases. Methane is by far the most abundant, approximately 85 percent of the typical deposit; other gases include ethane (10 percent), propane (3 percent), and small amounts of butane, pentane, hexane, heptane, and octane.

In most parts of the world, natural gas occurs in deposits along with petroleum; in this situation, it is known as **associated gas**. In the United States, however, 70 to 80 percent of natural gas comes from deposits not associated with petroleum. Such **nonassociated gas** seeps through sedimentary rock until it becomes trapped against impervious rock layers.

Use. In the early days of petroleum exploration and exploitation, natural gas that was found in the same deposits as crude oil was burned off as a waste product at the wellhead. Later, technological advances in the handling of gas — welded pipelines to move gas effectively, improved storage systems, and methods for liquefying and shipping gas in tankers — made it an important resource on its own. The main reasons for the recent emphasis on natural gas are a recognition of its relative cleanness as a fuel and a growing awareness of the limits of our oil supplies.

Since 1970, the most common application of natural gas in the United States has been in the industrial sector. Five key industries — chemicals, refineries, steel, paper, and cement — account for 32 percent of total gas use. The residential sector accounts for 23 percent, with gas being used for heat and to power sundry home appliances such as hot water heaters, stoves, and dryers.

Worldwide, electricity markets are increasingly dependent on natural gas. Global use of natural gas for electric generation is projected to double by 2025, as technologies for gas-fired generation continue to improve and ample gas reserves are exploited.

Natural gas systems are desirable in part because of their high conversion efficiency. A unit of natural gas can be produced and delivered at an efficiency of nearly 89 percent. In contrast, the efficiency of the production and delivery of an equivalent unit of coal is only about 30 percent. Increasingly, utilities also must consider the environmental aspects of power generation. The combustion of natural gas produces small amounts of ash or sludge by-products. In contrast, a 1,000-megawatt power plant operating on high-sulfur coal produces 700,000 tons of sludge and 250,000 tons of ash per year.

Natural gas holds significant potential in the transportation sector. According to the Natural Gas Vehicle Coalition, over two million natural gas-powered vehicles (NGVs) are in use worldwide. Natural gas offers fuel cost savings, greater efficiency, and lower air emissions than conventional gasoline or other alternatives such as propane and methanol. An estimated 130,000 NGVs are cruising the roads and highways of the United States, mostly in utility service fleets and urban transit.

Availability. Many countries are investing large amounts of capital and resources to determine the location and extent of their natural gas reserves. Much of this gas exists in unconventional sources such as sands, shales, coal seams and geopressurized zones, and we currently lack the technology to extract it economically. Instead, we must rely on deposits of conventional associated or nonassociated gas.

Global reserves of conventional natural gas are relatively abundant, with the largest deposits in the Middle East, Eastern Europe, and the former Soviet Union. Together, these regions account for over 70 percent of known reserves, whereas the largest gas consumers — North America and Western Europe — have only seven percent (Figure 10-6).

Unlike petroleum, most of the gas consumed in the United States is produced in North America. Substantial deposits of natural gas are known to exist in about half of the 50 states. However, just five states (Texas, Louisiana, Alaska, New Mexico, and Oklahoma) hold more than half of the country's reserves.

At current consumption rates, the world's proven natural gas reserves are estimated to last between 60 and 70 years. However, advancements in exploration and extraction technologies, coupled with growing global demand, make it difficult to accurately estimate the life expectancy of proven reserves.

Oil Shales and Tar Sands

Oil shales and tar sands are deposits of fossil fuels that have played a minor role as energy sources. Despite their tremendous energy potential, commercial production of these resources generally has been considered uneconomical.

Oil shales are the result of an interruption in the process that formed oil. In fact, the term is a misnomer; the shales do not contain oil. Rather, **oil shales** are fine-grained, compacted sedimentary rocks that contain varying amounts of a waxy, combustible organic matter called **kerogen**. Apparently, oil shales were not subjected to enough heat to complete the conversion process that produced petroleum.

The mined shale is crushed and then heated to about 900°F (482°C). At these temperatures, kerogen is vaporized, and the vapor then is condensed to yield oil. A heavy, slow-flowing liquid, shale oil must be refined

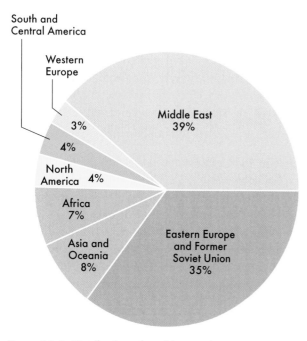

Figure 10-6: Distribution of world natural gas reserves, 2005. The recent rise in the consumer price of natural gas in the United States is attributed, in part, to challenges in transporting the resource from gas-rich nations. The United States has only four ports capable of importing natural gas from the countries with the greatest reserves. ***Source:*** *U.S. Energy Information Administration (2005).*

to increase its flow rate and hydrogen content and to remove impurities. The resulting synthetic crude oil then is further refined to produce gasoline, heating oil, and other products.

Oil shale reserves in the United States are vast — an estimated two trillion barrels of kerogen. Most of these reserves were identified over a century ago, and in 1912, President William Howard Taft established the Naval Petroleum and Oil Shale Reserves to oversee extraction, production, and use of these resources. Since that time, however, most commercial attempts to exploit oil shales in the United States have failed for an array of economic reasons, including high costs associated with insufficient technology and competition with relatively low petroleum prices. Yet, in light of national security concerns in the twenty-first century, oil shales likely will be re-examined as a potential supplement to domestic supplies of other fossil fuels.

Tar sands are sandstones that contain **bitumen**, a thick, high-sulfur, tar-like liquid, within their porous structure. High-pressure steam is used to force the bitumen from the sandstone. The bitumen then is purified and upgraded to synthetic crude oil. Tar sands generally are extracted using surface mining techniques; at present, underground mining is neither technically nor economically feasible.

The world's largest deposits of tar sands are found in Venezuela and Alberta, Canada. The Venezuelan Orinoco deposit alone has reserves estimated at 1.8 trillion barrels of bitumen. Deposits throughout Alberta — most notably the Athabasca tar sands — total approximately 1.7 trillion barrels. Of this amount, the government of Alberta has determined that 174 billion barrels can be tapped economically. When added to Canada's proven reserves, these accessible tar sands catapult the nation to second place in the ranking of oil-rich countries, behind Saudi Arabia and ahead of Iraq, Iran, and Kuwait. Smaller tar sand deposits also have been located in Trinidad, Russia, and western Kentucky.

What Are the Environmental Effects of Fossil Fuels?

All energy use affects the environment to some degree, but the extraction, production, and consumption of fossil fuels have negative impacts greater than most other energy sources. In the following paragraphs, we examine the environmental effects of each of the fossil fuels more closely.

Coal

Of all energy sources in use today, coal has the most serious environmental impacts. Coal is extracted from the Earth through various mining techniques. (The effects associated with mining are discussed more thoroughly in Chapter 18, *Mineral Resources*, but a brief synopsis is included here.) Underground or subsurface mining produces **acid drainage**: When air and water come into contact with sulfur-bearing rock and coal, the sulfur is oxidized to form sulfuric acid. The acid can drain into groundwater and eventually make its way into streams and rivers. Additionally, as underground voids created by mining collapse, the land may subside, or sink, making buildings situated in mining areas prone to structural damage.

The environmental damage associated with surface mining is even more severe. To strip mine an area, the overlying vegetation, soil and rock layers, collectively known as the **overburden**, are removed. Wind and rain acting on the exposed overburden can cause erosion and sediment runoff, which may find its way into nearby streams and rivers, increasing their sediment loads. As with underground mining, sulfur-rich rocks exposed to the air undergo oxidation, forming sulfuric acid and acidic runoff. The steeper the slope, the greater the danger of severe erosion and runoff. Environmental laws require that strip-mined areas be reclaimed or restored after mining has ceased. Reclamation is difficult in some areas, particularly mountainous terrain, but is possible if undertaken seriously.

Coal also contains trace elements of heavy metals and other substances that pose serious environmental threats and health hazards (Figure 10-7). Arsenic, lead, mercury, cadmium, selenium and uranium, for instance, can be toxic to plants and animals. These substances can contaminate both air and water near coal mining sites.

The combustion of coal has serious environmental impacts. Coal contains varying amounts of sulfur, which has been implicated in acid precipitation. Coal from the eastern United States, in particular, has a high sulfur content. Technologies have been developed to remove sulfur from coal before or during combustion or from the flue gases that form after combustion. These technologies are expensive, but they are being used by coal-fired power plants, especially those that rely on high-sulfur coal.

Carbon dioxide — an end product of combustion and the primary greenhouse gas responsible for climate change — may create a more serious problem than sulfur because carbon dioxide emissions can not be

FIGURE 10-7: Vietnamese women at work at a coal mine. Workers who lack protective gear are at increased risk from exposure to toxic substances present in coal dust.

controlled with current technologies. The combustion of coal produces proportionately more carbon dioxide than the combustion of either oil or natural gas. If coal consumption rises as predicted, especially in China, the United States and India, carbon dioxide emissions also will increase.

Petroleum

Oil extraction, production, and use adversely affect the environment in myriad ways. Like other fossil fuels, petroleum releases carbon dioxide when burned. Unless pollution control devices are in place, the combustion of crude oil products also releases carbon monoxide, sulfur oxides, nitrogen oxides and hydrocarbons, all of which are primary air pollutants. (Chapter 12, *Air Resources*, discusses these pollutants in greater detail.)

Perhaps the most disturbing aspect of oil development is the risk of oil spills, such as the 1989 *Exxon Valdez* disaster. In the largest oil spill in North American history, the Exxon oil tanker rammed into the well-marked Bligh Reef in Alaska's Prince William Sound, spewing 11 million gallons of crude oil into the cold Arctic waters. The spill affected over 1,300 square miles of beaches, marshes, wetlands, estuaries, salmon streams, islands, bays, and fjords. Tens of thousands of animals died in the days and weeks following the accident. Human inhabitants suffered as well, as the Alaskan government closed the sound's $12 million herring fishery. Just three months after the *Exxon Valdez* disaster, smaller but still substantial spills occurred in Texas, Delaware and Rhode Island, all in the space of one weekend. As destructive as these spills were, they were dwarfed by the spill of crude oil into the Persian Gulf, an environmental casualty of the Persian Gulf War (1991). However, the oil well fires during the Persian Gulf War and, more recently, the War in Iraq (2003) may have an overall greater environmental effect because of the addition of large amounts of carbon dioxide and sulfur to the atmosphere.

Natural Gas

Environmental concerns provide a compelling reason for shifting from coal and oil to natural gas. When burned, coal and oil release 40 percent and 30 percent more carbon dioxide, respectively, than does gas. In fact, gas is the cleanest of all fossil fuels. Utilities using natural gas produce far less nitrogen oxides and sulfur oxides than do utilities burning coal and oil. Even so, natural gas is not an entirely clean fuel. Its combustion does release some carbon monoxide, nitrogen oxides, and carbon dioxide.

Oil Shales and Tar Sands

As with other fossil fuels, several problems are associated with the development of oil shales and tar sands. Extracting and refining these resources are energy-intensive processes. The energy equivalent of about one barrel of conventional crude oil is needed to produce one barrel of shale oil and three barrels of heavy oil from tar sands. Consequently, the net useful energy yield of these resources is negative, or, at best, low.

Large amounts of water also are needed to extract the heavy oils and to process and refine them. Producing one barrel of shale oil, for example, requires two to six barrels of water, which is then contaminated. The world's richest deposit of oil-bearing shale is found in the Green River formation in Colorado, Utah, and Wyoming — a region with scarce water supplies.

The extraction of both oil shales and tar sands also disturbs large land areas and produces enormous amounts of waste for each unit of oil produced. For example, many potential sites for oil shale extraction in the United States are located on significant tracts of federal public lands or in wilderness areas. Development likely would destroy the wild quality of these areas. Likewise, the development of tar sand deposits leave vast landscapes marred with mines and tailing ponds, which hold wet sand after the bitumen is extracted. Given that over two tons of sand must be strip-mined for each barrel of bitumen produced, these tailing ponds are a considerable environmental concern.

Managing Fossil Fuels

In this section, we examine how energy consumption varies worldwide and how fossil fuels have been used historically. We also discuss how modern events — including the 1973 oil embargo, conflicts in the Middle East, and debate over the Arctic National Wildlife Refuge — continue to shape energy consumption in the United States. We close the section by describing two strategies for reducing the nation's dependence on imported oil: enacting a long-term energy policy and increasing energy efficiency.

How Does Energy Consumption Vary Worldwide?

Commercial energy — to power industry and provide services such as heating homes, fueling automobiles, and generating the electricity that drives personal computers, televisions, and refrigerators — is supplied primarily by fossil fuels. Commercial energy consumption varies widely with economic activity and living standards.

Consider the following comparison: With four times the population of the United States, China has eight personal motor vehicles (such as cars and trucks) per thousand people; the United States, 780 per thousand. This comparison is particularly striking because it not only emphasizes Americans' relentless consumption of commercial energy but also indicates a troubling trend: As developing nations continue to industrialize, fossil fuel resources will become increasingly strained. According to the Worldwatch Institute, global consumption of fossil fuels rose by 1.3 percent in 2002. Petroleum use in the developing world alone has quadrupled since 1970, and China now ranks third for global oil consumption, while Brazil is sixth. Such trends in energy consumption simply cannot be sustained. If the average Chinese consumer used as much oil as the average American, China would require 90 million barrels per day — 11 million more than daily global production in 2001!

Taking into account both commercial energy and traditional fuels (such as wood, plant residues, and animal dung), less-developed countries (LDCs) — with over 80 percent of the global population — consume just 30 percent of the global energy budget. For most people in LDCs, energy is a daily concern, similar to food and shelter, and they are quite aware of the origins of their fuels (Figure 10-8). They may spend a major portion of their day collecting traditional fuels or buying and bartering for the fuel needed to power stoves and lanterns. Because they are so directly involved in the search for fuel, these people have an intimate knowledge of the sources of energy upon which they depend.

For the most part, we in more-developed countries (MDCs) are indifferent to the origins of the energy we use. Except for a brief period in the 1970s, when an embargo by the Organization of Petroleum Exporting Countries (OPEC) disrupted oil supplies, an

FIGURE 10-8: A woman collects fuel wood, Burkina Faso. Two or three times a week, she walks for an hour to an area where she can find wood. The World Bank estimates that fuel wood consumption has reached four and a half times the level of sustainable production.

inexpensive and plentiful supply of energy always has been available. Think about your own energy supplies. From where does the electricity that illuminates your home, classroom, or office originate? What fuel heats your home? Where does it come from? Recall the last time you filled your car's gas tank. From what part of the world was that petroleum extracted — the Gulf of Mexico's continental shelf, Saudi Arabia, Alaska?

These questions may seem irrelevant, but they are not. Ignorance of our energy sources has caused people in MDCs to make several dangerous assumptions. Chief among these assumptions is that energy will always be available at prices we can readily afford. Also, we generally are unaware of the origins of our energy supplies and of the processes used to locate, extract, and produce them. Thus, we are equally unaware of the environmental effects of energy exploration and development. Moreover, as debate over energy policy and its relationship to national security intensifies in the United States, it is imperative that voters equip themselves with knowledge rather than relying on easy assumptions.

How Have Fossil Fuels Been Used Historically?

The discovery that coal can be burned to produce heat likely was made independently by humans in various parts of the world during prehistoric times. By the A.D. 300s, the Chinese had developed a coal industry. They mined coal from surface deposits and used it to heat buildings and smelt metals. Commercial coal mining in Europe lagged behind development in China. Commercial mines were started in England and in the area now known as Belgium in the 1200s; the coal was used chiefly for smelting and forging metals. Europeans were slow to exploit coal because they considered it a dirty fuel; they preferred to use wood and wood-based charcoal. By the start of the seventeenth century, however, severe wood shortages in western Europe forced communities and factories to increase their coal use. By the late 1600s, England accounted for 80 percent of total world coal production.

In North America, the Pueblo Indians were the first to use coal, digging the combustible material from hillsides to bake pottery. In the early eighteenth century, European colonists began using coal; the first commercial operation was established in Virginia in 1730. The first recorded commercial shipment of American coal, 32 short tons (29 metric tons), was transported from Virginia to New York in 1759. All large-scale early mining was done underground. The coal was extracted with picks and bars from solid beds and transported up from the mines in wheelbarrows, baskets, and buckets.

Like coal, petroleum has a long history. About 3000 B.C., the Sumerians of Mesopotamia used it to waterproof ships and wicker baskets and as a bedding compound for mosaic and inlay work. They also mixed it with sand for use in architecture and road construction. But perhaps their most interesting use of petroleum was its medicinal applications. The Sumerians softened bitumen with olive oil and rubbed the mixture on sores, open wounds, and rheumatic joints. They drank bitumen mixed with beer as a cure for numerous ailments and burned the pitch-like substance to ward off evil spirits. Belief in the medicinal qualities of petroleum also was widespread in the days of the ancient Roman Empire, and it endured over the intervening centuries. In the early part of the nineteenth century, enterprising persons collected and bottled oil that flowed from springs in western Pennsylvania into a tributary of the Allegheny River. They marketed the substance as "Seneca Oil," referring, most likely, to its medicinal use by the Seneca Indians. Despite petroleum's long history of use, little more than a century has passed since humans began to exploit its energy potential.

How Have Fossil Fuels Been Used in the United States?

For more than 250 years after the first English settlement was established in North America, wood was the nation's primary energy source. By 1885, however, wood gave way to a more powerful fuel — coal. The Industrial Revolution, which dawned in the United States in the early 1800s, increased coal's importance in both manufacturing and transportation. The steam engines that powered factories, ships, and railroads required large amounts of coal to fire their boilers. As industry and transportation grew, so did the production of coal, and by the late 1800s, the United States had replaced England as the world's leading coal producer. In the early 1900s, coal met 90 percent of U.S. energy needs.

The twentieth century brought rapid change to U.S. society, increasing the country's demand for more energy. Coal's years of dominance were numbered as the popularity of oil began to rise, in large measure because of the invention of the "horseless carriage" in 1892. Previously, the primary derivative of petroleum was kerosene, used for heating; indeed, much of the oil typically was discarded as useless. But with the development of the first American gasoline-powered engine, many found that this "new" fuel was the most practical for automotive transportation. Manufacturers, quick to appreciate the fuel's advantages, made gasoline-driven cars at a rapid rate. The increase in demand for gasoline caused the first severe petroleum shortage in the United States, from 1903 to 1911. As techniques for locating, recovering, and refining petroleum improved between the two World Wars, supplies rose. The resulting glut of oil was so great that at times the selling price was driven to absurdly low levels, and many people became convinced that the supply of petroleum was limitless.

In addition to spurring the development of the nation's modern transportation system, petroleum revolutionized agriculture. Farmers quickly found that horses and mules could not compete with the higher productivity of tractors and mechanical harvesters. Because of their convenience, oil and gas also began to displace coal in the small industry and home heating markets.

In the decade from 1937 to 1947, more petroleum was taken from the Earth than had been extracted in all previous history. With the end of gas rationing after World War II, Americans took to the roads with vigor, and motor vehicle registrations climbed by 50 percent. The use of farm tractors also increased by 50 percent, and the number of domestic oil burners doubled.

As the consumption of all petroleum products rose to record levels, the rate of production began to outpace the rate of new oil discoveries. In 1947, for the first time ever, the United States became a net petroleum importer. President Dwight D. Eisenhower realized the importance of this transition and pledged in his 1952 inaugural address to make the nation energy self-sufficient, a pledge that Eisenhower — as well as all of his successors — failed to meet.

With increasing oil consumption, the use of natural gas also increased. It became a major fuel for home heating and for industry in the production of glass and other commodities. By 1946, oil and natural gas together displaced coal as the chief source of the nation's energy supplies. By 1950, oil alone outdistanced coal to become the primary energy source.

In the years since 1960, oil, coal, and natural gas collectively have played the dominant role in U.S. energy consumption. Along with nuclear power, they account for 94 percent of all energy consumed in the United States.

What Was the OPEC Oil Embargo?

Into the mid-twentieth century, the United States' escalating use of oil as its major energy source made the nation increasingly reliant on imports, yet few people realized the implications of dependence on foreign suppliers. The illusion of an abundant supply of cheap energy vanished abruptly in 1973, when the Organization of Petroleum Exporting Countries (OPEC) placed an embargo on the United States, several western-European nations, and Japan. The embargo was the result of a complex play of international politics and economics.

In March 1973, President Richard Nixon attempted to boost a sluggish economy and curb rising inflation by devaluing the U.S. dollar in world markets. Because the global oil market was priced in terms of U.S. dollars, the fall in the value of the dollar brought about a concurrent fall in the price of oil. In October of the same year, conflict erupted when Egypt and Syria attempted to overtake Israeli fortifications in the Golan Heights, around the Suez Canal, and on the Sinai Peninsula. The "October War" lasted approximately

WHAT YOU CAN DO ▷ To Conserve Energy

To Save Energy at Home

➤ The chief use for energy in the home is for space heating and cooling. Prevent or minimize heat loss by adding insulation, caulking, installing storm windows or covering windows with plastic. These simple measures can reduce annual energy consumption by 25 percent or more.

➤ Use energy-efficient compact fluorescent light bulbs in place of incandescent bulbs. A 15- to 18-watt compact fluorescent bulb (costing $11 to $20) replaces a 75-watt incandescent bulb. The compact fluorescent will save about $40 in electrical costs and, because it lasts longer, additional savings are realized in replacement costs for incandescents. Because it requires less energy, over the course of its lifetime the compact fluorescent will prevent emissions of up to 2,000 pounds of carbon dioxide and 20 pounds of sulfur dioxide (from a coal-fired plant) or 25 milligrams of plutonium waste (from a nuclear-powered plant). It will also save the energy equivalent of about 500 pounds of coal or 1.3 barrels of oil.

➤ If building a new home, superinsulate by doubling the amount of insulation and using airtight liner in walls. After superinsulating, some new homes in Minnesota required 68 percent less energy for heat.

➤ Use energy-efficient appliances rather than conventional models for tremendous energy savings. The chart below compares average conventional models, new energy-efficient models, and models that use the best available technology for the greatest energy efficiency and the most significant energy savings.

To Save Energy in Transportation

➤ Walk or bicycle when traveling short distances. (Not only is it more energy-efficient to do so, it's also healthier!) For longer trips, especially those you make routinely (like trips to school or work), use public transit or organize car pools. Public transit and car pools also curb pollution and reduce congestion in crowded metropolitan areas. If those effects seem inconsequential, consider that Los Angeles freeways are so crowded that

the average speed in 1988 was just 35 miles per hour. That figure is expected to drop to 19 miles per hour by 2010.

➤ Consider driving a hybrid automobile. Powered partly by a traditional combustion engine and partly by a battery-charged electric motor, some hybrid cars average 55 miles per gallon for combined city/highway driving. Although hybrids comprised less than one percent of the 17 million new vehicles sold in the United States in 2004, some analysts predict that they could comprise 30 to 35 percent of the market by 2015.

To Reduce Indirect Energy Consumption

➤ Reuse and recycle! Enormous amounts of energy are used to produce consumer goods. You can minimize energy waste by reusing or recycling goods such as paper products, aluminum and bimetal cans, glassware, and plastics.

Appliance	Conventional	New Energy-Efficient	Best Available Technology	Potential Energy Savings [a]
Refrigerator[b]	1,500	1,100	750	87%
Central air conditioner[b]	3,600	2,900	1,800	75%
Electric water heater[b]	4,000	2,900	1,800	75%
Electric range	800	750	700	50%
Gas furnace[c]	730	620	480	59%
Gas water heater[c]	270	250	200	63%
Gas range[c]	70	50	40	64%

[a] The potential savings listed is the percent reduction in energy consumption realized when the best available technology is used rather than today's conventional model.

[b] Kilowatt-hours per year.

[c] Therms per year.

three weeks, during which time the United States and several allies issued diplomatic warnings against Egypt and Syria.

Displeased with the United States' attempt to control oil prices and with intervention in the October War, OPEC responded by withholding much of its oil exports to western nations. U.S. imports of oil from OPEC countries dropped from 1.2 million barrels a day to 19,000. The retail price of a gallon of gasoline rose from a national average of 39 cents in May 1973 to 55 cents in June 1974. At the embargo's onset, oil and gas accounted for three-quarters of the United States' energy supply. Although domestic production of the two fuels peaked during the embargo, it did not prevent severe fuel shortages, mile-long gas lines, and large price increases in consumer products. This situation was repeated during the Iranian revolution in 1978-1979.

The shortages of the 1970s forced the United States to think more realistically about its energy future. At the federal level, the "energy crisis" raised concerns over national security, trade deficits, and environmental degradation. Closer to home, rising energy costs resulted in worries about the family budget, transportation to and from work and school, and the heating and cooling of homes and offices. For the first time in peacetime, Americans were forced to drastically limit their energy consumption.

Events in the 1970s created, in effect, a **conservation revolution** in the United States. Conservation means reducing or eliminating wasteful use of a resource — in this case, energy — in order to protect or extend supplies. Concern over the energy crisis brought policy revisions, community action and individual efforts, all designed to conserve energy and discover workable ways to mitigate the threat to the American lifestyle. The success of the conservation revolution gave the nation time to adjust to reduced energy consumption. Between 1972 and 1981, the industrial sector reduced consumption by six percent. The 1975 Energy Policy and Conservation Act established corporate average fuel efficiency (CAFE) standards which mandated fuel efficiency levels for all new automobiles. These standards saved billions of gallons of gasoline. Energy-efficient technologies were developed for heating and lighting, and people began to form car pools, turn out lights not in use, and lower thermostats. Conservation efforts saved about six million barrels of oil daily.

What Are the Implications of U.S. Reliance on Imported Oil?

Since the last oil shortage in 1978-1979, supplies have been uninterrupted, despite significant price fluctuations. While a steady supply of oil allows us to maintain the American lifestyle, it also impedes conservation. Since 1982, oil consumption has risen steadily. In 2005, the United States consumed 20 million barrels of oil per day, 54 percent of which was supplied by foreign sources (Figure 10-9). By 2025, U.S. reliance on imported petroleum is expected to climb to 68 percent, leaving the nation and its economy vulnerable to disruptions in supply.

The United States' growing dependence on imports is a matter of economics. A barrel of oil, which sold for just over $2 in 1973, cost $34 by 1982, providing consumers with a powerful incentive to conserve. By the late 1980s, however, the price had declined to about $10 per barrel. An oil glut driven by OPEC was responsible for the falling prices, which slowed the conservation effort and lulled consumers into a false sense of security — and complacency — once again. Many people abandoned energy conservation measures, and less fuel-efficient vehicles again became popular as energy consumption slowly crept back to the levels of the early 1970s.

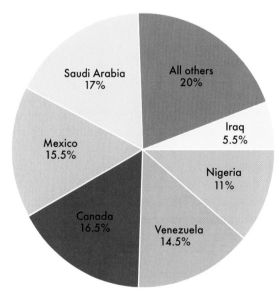

FIGURE 10-9: Crude oil imports to the United States, 2005. To quench its thirst for oil, the nation relies on imports from dozens of countries. ***Source:*** *U.S. Energy Information Administration (2005).*

Even the Persian Gulf War did not reverse the trend of falling prices and increasing imports. Although the price of petroleum rose sharply after Iraq's invasion of Kuwait in August 1990 (from $16 in July to $55 in November), it soon dropped as Saudi Arabia increased production to compensate for the disruption in supplies. Throughout most of the 1990s, the price of a barrel of oil fluctuated between $16 and $24, but by early 1999, it had plummeted to just less than $10, its lowest price since the early 1970s. U.S. oil producers found it nearly impossible to compete with the flood of inexpensive foreign oil; as a result, many were forced to curtail their operations, and domestic production fell to about six million barrels daily, the lowest level since the Eisenhower presidency. Some U.S. producers charged that the cheap oil was merely a tactic designed to dismantle the national oil industry and increase reliance on foreign suppliers. Such increased reliance, they warned, would lead to greater control over the U.S. economy by foreign sources and, eventually, to rising oil prices.

Less than a decade later, some would argue that these predictions have held true, as oil prices rose to over $52 a barrel by May 2005, leaving consumers faced with higher prices for petroleum-based goods (Figure 10-10). The United States, again at war in the Middle East, is engaged in an increasingly heated discussion over why oil prices are rising and how the situation should be addressed. After the events of September 11, 2001, a wellspring of public opinion emerged about the connection between national security, oil resources, and the conflicts in Afghanistan and Iraq. Some believe that escalating oil prices represent an attempt by OPEC to influence U.S. foreign policy. Others say the prices were timed to promote proposed legislation that would open restricted federal lands, such as areas of the Arctic National Wildlife Refuge (ANWR), to exploration and extraction. Market analysts, however, cite strong demand in Asia and bottlenecks in refining capacity as primary reasons for the increase in oil prices.

Regardless of why oil prices are rising, most Americans concur that continued reliance on foreign energy sources leaves the United States in a vulnerable position. After this point of agreement, however, opinions once again diverge about how best to reduce this reliance. To illustrate differing viewpoints, let's examine the controversy over granting land leases that would allow fuel companies to drill for oil and natural gas in ANWR. ANWR encompasses 19.6 million acres on the magnificent north coast of Alaska. In 2003, Congressional debate began over whether a 1.5-million-acre

FIGURE 10-10: a) In 1999, gas prices across the nation reflected the abundant supply of inexpensive imported oil. b) By 2005, however, prices had increased by 250-300 percent, leaving the nation to question its reliance on foreign energy sources.

tract of ANWR's coastal plain — known as the "1002 Area" — should be opened to oil and gas exploration. Proponents of allowing access to the 1002 Area argue that increasing domestic oil production is the fastest, most efficient way to reduce the nation's dependence on imports. Opponents counter that drilling is a short-term solution at best, given the relatively small amount of extractable oil estimated to be in Area 1002. In a 2004 analysis of ANWR, the Energy Information Administration (EIA) concluded that 10.4 billion barrels of technically recoverable oil are in the coastal plain. At the current consumption rates, this amount would fuel the nation's oil habit for just under one-and-a-half years. Further, some opponents argue that no amount of oil is worth the degradation of this pristine habitat, home to 45 species of land and marine mammals (including the polar bear, black bear, wolf, moose, musk ox, caribou, and bowhead whale), 36 species of fish, and 180 species of birds.

In April 2005, Congress passed a $2.6 trillion concurrent budget resolution by a narrow margin in both houses. (The House of Representatives passed the measure by a vote of 214-211; the Senate, 52-47). While the resolution does not mention ANWR by name, it does require the joint Congressional Budget Conference Committee to revise the bill by increasing revenue projections by $2.4 billion — approximately the same amount as the estimated federal share of revenue from ANWR land leases. Many believe that the resolution will be revised to include drilling in ANWR, and if passed in a final Congressional vote, it then will go before President George W. Bush for signature or veto.

As the debate over the fate of ANWR demonstrates, there is no easy solution to reducing U.S. reliance on foreign energy sources. The nation must broadly assess its energy outlook, examining not only short-term "quick fixes" but also strategies that address future challenges. Two such strategies are to enact a comprehensive, farsighted energy policy and to use energy more efficiently.

What Is the Status of U.S. Energy Policy?

For people born after 1970, the energy crises that characterized the decade — and the inflation that those crises helped spur — are no more than historical footnotes. In the ensuing decades, U.S. consumers have enjoyed a steady and ample supply of inexpensive energy anchored by fossil fuels. The nation's energy policy reflects a short-sighted optimism that the supply of fossil fuels will remain indefinitely plentiful, even though many scientists predict that world oil production will peak by the middle of the twenty-first century. Every presidential administration from Ronald Reagan to George W. Bush shares the dubious distinction of putting forth a national energy policy built around continued reliance on fossil fuels and increased production of domestic oil reserves. Instead of using these times of plenty to develop strategies and technologies that would secure long-term, cleaner energy alternatives, the United States maintains a "business as usual" stance.

Indeed, the federal government's unofficial strategy concerning energy reserves has been to "drain America first." For example, the Clinton Administration's 1998 national energy plan advocated increasing domestic gas production and diversifying — not reducing — import sources. Similarly, the 2003 plan put forward by the Bush Administration encourages diversification, as well as significant investment ($2 billion over 10 years) in "clean coal" technologies. If successful, these technologies will reduce air emissions from coal combustion, but they will do nothing to address coal's other serious environmental impacts or move the nation away from its continued use.

As the United States progresses through the twenty-first century, it can continue along the path of careless optimism, or — heeding the adage, "Those who do not remember the past are doomed to repeat it" — the nation can forge a new path. Early planning is critical to developing and implementing a national energy policy that is both effective and farsighted. It takes many years to phase in a new energy source. If the nation is to move smoothly from a dependence on oil and gas to a reliance on alternative fuels, the transition must begin now — before oil production peaks. On the other hand, if the nation chooses to rely on fossil fuels and nuclear power, it must be willing to risk environmental damage, devise means to mitigate that damage, and develop methods to enhance the safety of nuclear energy.

How Can Energy Efficiency Be Increased?

Researchers at the Rocky Mountain Institute, an energy think tank, estimate that the United States wastes about $300 billion a year because of energy inefficiency. Drafty doors and windows, gas-guzzling cars and appliances, and poorly insulated buildings are three major culprits. Energy efficiency — obtaining greater productivity while using less energy — is our greatest untapped "source" of domestic oil. The energy efficiency campaigns of the future will include reduced energy requirements of electric appliances, automobiles, and other technologies and more efficient and environmentally sound industrial processes. A simple axiom can characterize the twenty-first century: *Conservation and efficiency mean building better things and doing things better, not doing without.* The following sections illustrate some of the creative, environmentally sound ways industries, communities, and individuals are making processes and products more efficient.

Industries

The industrial sector offers myriad opportunities to conserve and extend energy supplies by increasing efficiency. Energy-efficient transformers use 70 to 90 percent less electricity than conventional transformers. Since most electric power passes through at least two transformers between the power plant and the consumer, efficient transformers can yield significant savings. Savings also can be realized through the use of improved combustion technologies. For instance, through fluidized bed combustion, huge fans keep powdered coal suspended in mid-air so that it burns cleaner and loses less energy to the environment. A variation developed in Sweden captures waste heat to achieve an energy efficiency of 85 percent, with sulfur oxide and nitrogen oxide emissions that are just one-tenth and one-sixth, respectively, of those allowed in U.S. plants.

Despite such progress, numerous opportunities for increasing efficiency in the industrial sector remain untapped. For example, using a variety of tax incentives and disincentives, the government could encourage industries and utilities to increase the efficiency of their processes and seek alternatives to coal and oil. Improving the efficiency of electric motors is another important option available in many industries, including paper, chemical, and cement-making. In the United States, electric motors consume 80 percent of all electricity used for industrial purposes. By decreasing energy consumption, U.S. industries could reduce production costs and make their goods more competitive at home and abroad.

Because the transportation sector accounts for over 60 percent of all oil consumed in the United States, conservation and efficiency measures in this area can result in substantial savings. Raising CAFE standards is the easiest and most significant measure the nation could take to conserve fuel, decrease dependence on foreign oil, and improve air quality. Increasing fuel efficiency by just one mile per gallon would cut carbon dioxide emissions by approximately 40 billion pounds, an amount roughly equivalent to closing six coal-fired power plants.

Further measures can be taken in the transportation sector; one of the most significant is to raise the gasoline tax to reflect the product's real worth. In many countries, including several European nations, Japan and Argentina, the tax exceeds the price of the fuel itself. In the United States, however, tax rates register about $0.40 per gallon, much less than the price of regular gas. Finally, a long-term way to promote efficiency is to provide incentives that encourage mass transit in urban areas — perhaps even a national rail system similar to those used across Europe.

Communities

Communities can increase energy efficiency and enhance quality of life in diverse ways. Consider Osage, Iowa, a farm-based community of 3,800 people. Like many other agricultural communities, it has struggled to survive in recent years, hit hard by the recession of the 1970s, the farming crisis of the early 1980s, and the drought late in that same decade. But Osage is on sound economic footing, primarily because of a unique economic development program based on energy efficiency. Since 1974, publicly owned Osage Municipal Utilities has led the state in establishing programs that help its customers save energy, thereby reducing the amount of electricity and natural gas the town must buy from out-of-state suppliers.

One major program offers free infrared scans of citizens' houses. The scans spot energy leaks, allowing residents to insulate their homes more effectively. The improved insulation has resulted in savings of up to 50 percent on many heating bills. A second program requires that new homes meet minimum efficiency standards before they are hooked up to natural gas lines. These requirements spurred construction companies to improve insulation in attics and side walls and to install energy-efficient windows. Yet another program distributes free water-heater jackets and compact fluorescent light bulbs. Over their lifetime, these bulbs save the equivalent of 120 short tons (109 metric tons) of coal. Trees and a tree spade are donated to anyone who wishes to plant shade trees around homes. The trees are expected to reduce the need for air conditioning, especially in peak months.

Local businesses soon joined the utility company's approach to Osage economic development. Steele's Super Value, the local grocery store, put all its refrigerator compressors together in an insulated compartment and installed two fans and an insulated duct to carry the waste heat from the compressors into the main part of the store. The store saves $600 each month on heating, enough to keep prices low enough that Osage residents will not be tempted to take their business to a larger supermarket 15 miles away. The owners also are experimenting with new energy-efficient lighting that appears to use 22 percent less energy than the older system.

These programs have yielded significant benefits for the community, increased knowledge about energy conservation, and enhanced community spirit.

The local Jaycees, for instance, donated time to help lower-income residents install weather stripping and water-heater jackets. Financially, efficiency measures have saved the community an estimated $1.2 million each year. Osage has not had to issue any additional bonds, and the community has been able to retire all its old debts. The utility company itself has passed the savings on to consumers, cuttings its rates five times over the last 10 years. Moreover, it has not had to add to existing generation capacity in 15 years.

Communities everywhere could take a lead from Osage and institute a number of conservation and efficiency measures for the commercial and residential sectors. Chief among these measures is the revision of building codes to mandate efficiency standards for heating and lighting in new buildings. Although super-insulated buildings cost 5 to 10 percent more to build than conventional buildings, they are less expensive in the long run due to lower utility bills. Similarly, public service campaigns could alert consumers to the economic savings of home insulation, compact fluorescent bulbs, and energy-efficient appliances.

Individuals

Can individual efforts really make a difference in the drive to conserve energy? Consider the following sobering facts: The amount of heat that leaks out of American windows and doors each year is equivalent to the amount of energy that flows through the Alaskan pipeline annually. Throwing away an aluminum beverage can instead of recycling it wastes as much energy as half-filling the same can with gasoline and dumping the fuel on the ground. The same amount of energy is wasted each time a daily edition of the *New York Times* or *Washington Post* is tossed into a garbage can rather than a recycling bin. On an individual level, great energy savings could be realized by making relatively simple changes in home use, transportation, and indirect energy consumption (see *What You Can Do: To Conserve Energy*, page 212).

Summary

Energy, the ability to do work, is part of the abiota, the nonliving or physical component of the biosphere. All energy is either kinetic — energy due to motion or movement — or potential — energy in storage, that is, energy which can be converted to another form. Energy can take different forms, such as heat, light, chemical, or electrical energy. Various units are used to measure the quantity of different forms of energy. For example, heat energy can be measured in calories, British thermal units (btu), or therms; electrical energy is measured in kilowatts. In addition to measuring the quantity of a specific form of energy, such as heat, we also measure the amount of a specific type of energy resource. Petroleum, for example, is measured in barrels, while natural gas is measured in cubic feet and coal is measured in short tons.

Energy resources generally are classified as nonrenewable, renewable, or perpetual. Nonrenewable energy resources, such as fossil fuels, exist in finite supply or are renewed at a rate slower than the rate of consumption. Renewable resources are resupplied at rates greater than or consistent with use, so that supplies are not depleted. Perpetual resources originate from a source that is virtually inexhaustible, at least in time as measured by humans. Nuclear, renewable, and perpetual resources are called alternative resources because they offer us an alternative to fossil fuels.

In any energy conversion, although the same amount of energy exists before and after the conversion, not all of the energy remains useful. The conversion of energy from one form to another always involves a change or degradation from a higher quality form to a lower quality form. The energy

Discussion Questions

1. What is the difference between energy efficiency and net efficiency? How is net efficiency determined?

2. List the pros and cons of (a) increasing the nation's reliance on coal; (b) developing domestic oil supplies; and (c) switching to natural gas for as many applications as possible. In your opinion, should the United States expand its use of coal, oil, or natural gas (or any combination of the three)? Why or why not?

(continued on following page)

(continued from previous page)

3. Imagine that a mining company has asked for permission to strip mine coal near your community. Would you allow it? If so, what conditions, if any, would you place on the company to minimize environmental damage?

4. Why should less-developed countries worry about saving energy when their most important concern is generating economic growth.

5. Imagine that you have been asked to help develop a national energy policy for the United States. What factors do you need to consider? What will you recommend?

6. Develop an energy efficiency program for your community. What major components would you include? What incen-

efficiency of a system, process, or device is a measure of the percentage of the total energy input that does useful work and is not converted into low-temperature heat. Heat is given off with every energy conversion; as a result, the efficiency of any device or process is always less than 100 percent. The net efficiency of a process or system that includes two or more energy conversions is found by determining the efficiency of each conversion. Because they use less energy, efficient devices have a lower life-cycle cost (initial cost + lifetime operating costs) than inefficient ones.

Coal, petroleum, and natural gas are the fossilized remains of organic matter, as are oil shales and tar sands, two comparatively minor energy sources. All fossil fuels are composed chiefly of carbon and hydrogen. The processes that formed them continue today, but because those processes are extremely slow, fossil fuels are classified as nonrenewable and finite.

Fossil fuels are not distributed evenly beneath the Earth's surface. Proven or economic reserves are deposits that have been located, measured, and inventoried. Fuel in proven reserves can be or is currently being extracted. Subeconomic reserves have been discovered but cannot yet be extracted at a profit. Indicated or inferred reserves are thought to exist and are likely to be discovered and available for use in the future.

Coal is a solid composed primarily of carbon, with small amounts of hydrogen, nitrogen, sulfur, and other trace elements. The four major types of coal are anthracite, bituminous, subbituminous, and lignite. Coal is the most abundant fuel on Earth. Coal consumption declined over most of the last century, but it has increased in recent years as a result of the relative scarcity of oil and natural gas and growing concern over the safety risks associated with nuclear power.

Of all energy resources in use today, coal has the most serious environmental impacts. Underground or subsurface mining produces acid drainage, which can contaminate groundwater, streams, and rivers. Land subsidence makes buildings situated in mining areas prone to structural damage. Strip mining can cause erosion and sediment runoff. Coal combustion releases sulfur — which has been implicated in acid precipitation — and carbon dioxide — the primary greenhouse gas responsible for climate change.

Petroleum, also called crude oil, is a liquid composed largely of hydrocarbon compounds. When refined, petroleum yields propane, gasoline, jet fuel, heating oil, motor oil, kerosene, and road tar. Even though global resources seem substantial, they may not last long, given the world's appetite for oil.

Oil production and use adversely impact the environment in numerous ways. Perhaps the most disturbing risk is the potential for oil spills. In addition, petroleum, like other fossil fuels, releases pollutants such as carbon dioxide, sulfur oxides, nitrogen oxides, and hydrocarbons.

Natural gas is composed mostly of methane. Worldwide, natural gas usually occurs with petroleum and is known as associated gas. The United States, however, taps most of this resource from deposits of nonassociated gas. Natural gas has taken on increasing importance because of the growing awareness of finite oil supplies. Natural gas systems also have a high conversion efficiency and produce minimal by-products. Deposits are found on every continent, but many of them occur in unconventional resources such as sands, shales, coal seams, and geopressurized zones.

Oil shales are fine-grained, compacted sedimentary rocks that contain varying amounts of kerogen. Tar sands are sandstones that contain bitumen — a thick, high-sulfur, tar-like liquid — within their porous structures. Significant inputs of energy are necessary to extract and refine the oil and tar from shales and sandstones. Extraction and processing also disturb large land areas, produce enormous amounts of waste, and require large amounts of water.

Humans have used fossil fuels for thousands of years. In the United States, wood was the nation's primary energy source until about 1885, when it gave way to coal. In the twentieth century, the popularity of oil began to rise. With increasing oil consumption, the use of natural gas also increased.

In the twenty-first century, continuing dependence on oil has made the nation increasingly vulnerable to foreign suppliers. A 1973 embargo by OPEC caused global concern about energy sources. This embargo, along with other events in the 1970s, created a conservation revolution, leading to national policy revisions, local community action, and individual efforts to conserve energy. However, since 1982, oil consumption has inched upward steadily, and prices rose to over $52 per barrel in May 2005. Increased prices have spurred much debate over the connection between national security, oil resources, and continuing conflicts in the Middle East.

Debate also ensues over opening the Arctic National Wildlife Refuge for oil exploration and extraction. Proponents argue that opening a 1.5-million-acre tract of the Alaskan coastal plain is the most efficient way to reduce U.S. dependence on foreign oil. Opponents counter that drilling is a short-term solution that could lead to serious environmental degradation.

The U.S. government historically has put forth national energy policies built around continued reliance on fossil fuels and increased production of domestic reserves. This attitude must change now — before oil production peaks — if the nation is to move smoothly into relying on alternative and sustainable fuels.

KEY TERMS

acid drainage

alternative resources

anthracite

associated gas

bitumen

bituminous

coal

conservation revolution

economic reserves

energy

energy efficiency

fossil fuels

indicated reserves

inferred reserves

kerogen

kinetic energy

life-cycle cost

lignite

natural gas

net efficiency

nonassociated gas

nonrenewable resources

oil shales

overburden

perpetual resources

petroleum

potential energy

proven reserves

renewable resources

subbituminous

subeconomic reserves

tar sands

CHAPTER 11

Energy: Alternative Sources

Learning Objectives

When you finish reading this chapter, you will be able to:

1. Identify eight alternative energy resources, summarize their present and future use, and explain the advantages and disadvantages of each.

2. Briefly summarize the role that alternative energy sources have played historically.

3. Describe how alternative energy resources might be developed to complement and eventually replace fossil fuels.

The expanded use of renewables and a greater commitment to energy efficiency are the most environmentally sound and cost-effective approaches to mitigating many seemingly intractable problems.

Worldwatch Institute

In the ongoing effort to create a sustainable society, many renewable and perpetual energy resources and several nonrenewable resources represent alternatives to fossil fuels. They include nuclear energy, solar energy, wind power, hydropower, geothermal energy, ocean power, biomass energy, and solid waste.

Alternative resources are a diverse lot: Geothermal energy is nonrenewable, while wind power is perpetual. The energy of the sun has been used by humans for millennia; the energy of the atom has only recently been harnessed. Humans have understood and exploited hydropower extensively; researchers have only begun to explore the applications of energy generated by the world's oceans. Nuclear, solar, wind, hydro, and geothermal power are physical forces; biomass energy is derived from once-living things; and solid waste is a by-product of human activity.

In all their diversity, alternative fuels offer many advantages. They can supplement and eventually replace fuels that are dwindling or that will be too costly in the future. They can broaden our fuel base, allowing us to avoid dependence on any one fuel or on foreign suppliers. And, while alternatives do affect the environment to some degree, their impact is usually much less detrimental than that of fossil fuels.

Yet, there are limitations associated with alternative energy resources. Most are site-specific; that is, unlike coal and oil, they cannot be transported from one area or nation to another. Further, solar and wind resources are interrupted regularly, and hydropower and geothermal energy are not evenly distributed. Therefore, we must be able to adapt or compensate for their variability if we are to make good use of them. Although renewable and perpetual resources are essentially free, the technologies needed to harness them are not. The costs associated with alternatives should drop, however, as technologies are improved and as fossil fuels become increasingly scarce and expensive. In this chapter, we describe alternative energy sources and examine efforts to manage them.

Describing Alternative Energy Sources

In this section, we look closely at eight alternative energy sources. We discuss the present and future use and advantages and disadvantages of each source.

What Is Nuclear Energy?

Nuclear energy is the power contained within the nucleus of the atom. The constituents of the nucleus — protons and neutrons — are held together by an attractive force called the **binding force**. The strength of the binding force varies from atom to atom. If the force is weak enough, the atom will split, or undergo **fission**, when bombarded by a free neutron. As the atom splits and bonds are broken, it produces smaller atoms, more free neutrons, and heat. The release of particles initiates a chain reaction as the free neutrons strike other nearby atoms, which also split (Figure 11-1).

Nuclear reactors are designed to sustain the fissioning process. The fuel that powers a nuclear reactor is uranium 235 (U-235), a readily fissionable isotope of the element uranium. An **isotope** is one of several forms of an element that have the same number of protons but different numbers of neutrons. U-235 is a rare isotope of uranium, making up less than one percent of natural uranium. U-238, an isotope that does not fission easily, is the common form and accounts for over 99 percent of natural abundance.

Before uranium can be used in a nuclear reactor, the impurities must be removed and the concentration of U-235 raised to increase the incidence of fissioning. The ore is milled to a granular form, and an acid is added to dissolve the uranium out of the granules. The uranium is converted to a chemical form that exists as a gas slightly above room temperature. The gas is purified, and the concentration of U-235 is enriched to approximately three percent. The enriched uranium then is converted into powdered uranium dioxide. A highly stable ceramic material, uranium dioxide powder is subjected to high pressures to form fuel pellets, each about five-eighths-inch long and three-eighths-inch in diameter. About five pounds of pellets are placed into a fuel rod, a 12-foot-long metal tube made of zirconium alloy that is highly resistant to heat, radiation, and corrosion.

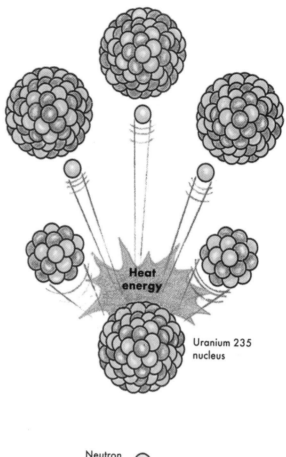

FIGURE 11-1: Fission reaction.

Fuel rods are grouped together to form a fuel assembly measuring about 14 feet high and weighing about 2,000 pounds. Graphite or water around the fuel rods slows the speed of the neutrons and increases the likelihood of fission. The fuel assembly also contains 16 vacant holes for the insertion of metal rods that absorb neutrons. By raising or lowering these control rods, operators control the rate of fission inside the reactor core. One hole within the assembly is left vacant for a probe that monitors temperature and neutron levels. Hundreds of fuel assemblies comprise the **reactor core** of a nuclear power plant, which is housed within a thick-walled **containment vessel**.

Uranium atoms contain tremendous potential energy and can drive the production of significant amounts of electricity. The fissioning of one uranium atom releases about 50 million times more energy than the combustion of a single carbon atom. A fuel pellet composed of trillions of atoms can release an enormous amount

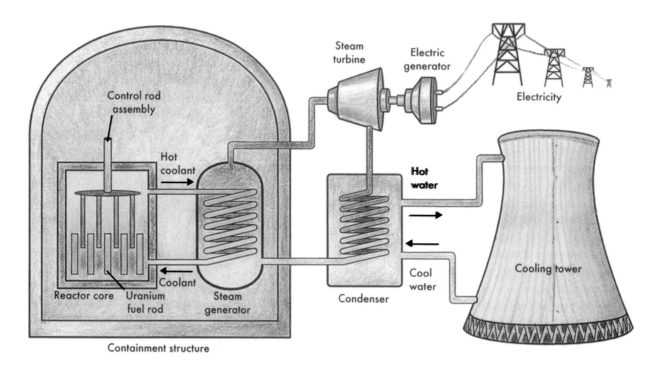

FIGURE 11-2: Essential components of a pressurized water reactor. Energy produced in the fuel rods within the reactor core heats water to steam, which then drives a turbine that powers an electric generator.

of heat energy. The heat energy is used to boil water and create steam, which drives turbines that produce electricity (Figure 11-2). Three pellets generate approximately the same amount of electricity as 3.5 tons of coal or 12 barrels of oil.

Several different types of nuclear reactors are in use today. In a light-water reactor (LWR), the reactor core is immersed in water, which acts both as a moderator to slow the rate of fission and as a coolant to transfer the heat that drives the turbines. Light-water reactors are the most common nuclear reactors in the United States and worldwide. Reactors with graphite moderators are used mainly in Russia. The United Kingdom and France primarily have gas-cooled reactors, which use a gas such as helium to transfer heat from the core to the turbine. Canada uses heavy-water reactors, which operate on nonenriched uranium. Using nonenriched uranium is much less expensive, but the incidence of fissioning is decreased. To compensate, deuterium oxide, or heavy water, which does not readily absorb neutrons, is used as a coolant.

Present and Future Use

Worldwide, the nuclear energy industry has grown slightly, increasing seven percent between 1992 and 2004. Nuclear power now accounts for approximately 23 percent of the world's electricity and six percent of total energy output. While the United States has more nuclear power plants online than any other nation (104 as of 2004), over a dozen countries rely more heavily on nuclear energy for electric generation (Figure 11-3).

In the early 1970s, before the oil embargo of the Organization of Petroleum Exporting Countries (OPEC), nuclear energy accounted for about five percent of electric production in the United States. In 2004, it accounted for about 20 percent, making nuclear power the second largest source of the nation's electricity, after coal. Because of the nation's interconnected power system, virtually every American gets some electricity from nuclear energy. Most nuclear plants are in the East, with Vermont, Connecticut, and South Carolina relying most heavily on nuclear energy (Figure 11-4).

Even so, growth in the U.S. nuclear industry has slowed and even declined since peaking in the 1980s. A number of factors are responsible for this decline: lower-than-predicted demand for electricity; increased costs due to longer lead times for licensing and construction of plants; higher financing expenses due, in part, to uncertain economic conditions; rising interest rates; inability to secure insurance against accidents; and increased concerns about safety and disposal of nuclear wastes. Indeed, widely publicized accidents at Pennsylvania's Three Mile Island facility (1979) and

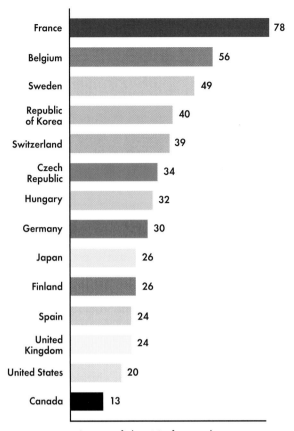

Figure 11-3: Countries with the highest reliance on nuclear energy for electric production.
Source: *Nuclear Energy Agency (2004).*

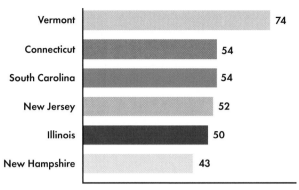

Percent of electricity from nuclear energy

Figure 11-4: States with the highest reliance on nuclear energy for electric production.
Source: *Nuclear Energy Institute (2003).*

the former Soviet Union's Chernobyl plant (1986) have spurred continuing debate over the use of nuclear power.

Advantages and Disadvantages

A primary advantage of nuclear power plants is that — unlike plants that use fossil fuels — they do not contribute to air pollution through emissions of carbon dioxide, sulfur and nitrogen oxides, and particulates. Each year, nuclear plants in the United States reduce emissions of greenhouse gases by 128 million tons, including a five-million-ton reduction in sulfur dioxide and a two-million-ton reduction in nitrogen oxides. During the 1980s, as France tripled its nuclear energy production in response to concerns over dependence on foreign oil, total pollution from the nation's electric power system fell by 80 to 90 percent. Proponents of nuclear energy maintain that the United States could realize the same benefits if it increased its reliance on this energy source. In answer to concerns over safety,

supporters argue that modern nuclear power plants are inherently safe. The concentration of fissionable fuel (three percent) is far less than that needed for a nuclear explosion. (Nuclear weapons have a 97 percent concentration of U-235.) Further, improved reactor design and overlapping safety systems decrease the risk of accidents. Proponents also insist that increased reliance on nuclear energy means decreased dependence on foreign supplies of oil and gas.

Those who oppose further development of nuclear energy argue that it is expensive, dangerous, and environmentally unsound. In the United States, nuclear plants generate electricity at a total cost of more than 13 cents per kilowatt-hour, twice the prevailing rate. Much of this higher cost is due to stricter environmental regulations. Largely because of economics, the United States has not approved the construction of any new plants since 1978.

Concern over safety is also a major issue. Opponents argue that no matter how safely a reactor is designed, human error cannot be eliminated, and because the consequences of a serious accident can be so grave, nuclear energy is a risk that we should not take.

A serious environmental concern is the large amounts of cooling water nuclear plants use. Most plants use ponds, cooling towers, or spray systems to discharge waste heat. However, when warmed water is released back to its source, it can adversely affect the ecosystem. Likewise, when cooling water is pumped from lakes, rivers and oceans, aquatic organisms may be trapped and killed.

Perhaps the most serious environmental concern is the disposal of nuclear wastes — spent fuel from both functioning and decommissioned plants (nuclear power plants have a maximum life of 30 to 40 years).

Permanent storage sites for nuclear wastes still are not available, and temporary on-site storage areas are approaching capacity. High radiation levels in reactors make the retirement of plants a complex and expensive process. According to the Worldwatch Institute, retiring a plant costs an estimated $50 million to $3 billion. (These and other important issues surrounding nuclear resources are discussed in detail in Chapter 19.)

What Is Solar Energy?

Solar energy, radiated from the sun to the Earth, provides the energy that all life needs to survive; the energy base to transform organic material to fossil fuels and renewable biomass; and the energy that powers abiotic cycles in the atmosphere, hydrosphere, and lithosphere. Solar-generated heat creates winds which power windmills. Solar energy also moves the hydrologic cycle, feeding the streams and rivers that can be harnessed to drive water mills and the turbines of hydroelectric plants.

Present and Future Use

Because much of the energy consumed in the United States is used to heat space and water and to dry materials, solar power holds enormous promise. The heating of residential and commercial buildings alone accounts for one-third of all energy consumption; according to studies by the U.S. Department of Energy (DOE), advanced solar technologies have the potential to meet up to 80 percent of a building's energy requirements. Similarly, industry consumes 40 percent of the nation's energy to power manufacturing processes; the DOE estimates that by 2025, 20 percent of industrial energy needs could be met by solar devices that achieve and maintain higher temperatures than conventional means.

Solar energy can be used to heat space and water through passive or active systems. A **passive solar system** relies only on the natural forces of conduction, convection, and radiation to distribute heat. Passive systems incorporate specific design features into buildings — such as strategically placed windows, overhangs, and insulation — to capture maximum solar radiation in the winter and minimum in the summer. The features absorb radiation during the day and distribute it throughout the building at night. According to the National Information and Resource Service, under optimal conditions, "a well-designed, well-insulated

home can obtain as much as 90 percent of its heating and cooling needs through the use of passive solar design." Passive solar buildings save as much as 50 percent on heating bills for only one percent more in construction costs.

An **active solar system** uses fans or pumps driven by electricity to enhance the collection and distribution of the sun's heat (Figure 11-5). Air, water, or other fluids are pumped through solar collectors, where they absorb energy from the sun. The heated substance can be distributed throughout the building immediately or transferred to storage units for later use. For example, hot water produced in rooftop collectors can be pumped through radiators and used for heat, or it can be stored in an insulated tank and used for washing and bathing.

Solar energy also holds great potential for the generation of electricity. If only one percent of the Earth's deserts was used to harness solar energy, more electricity could be generated than currently is produced worldwide by fossil fuels. Ninety percent of the world's installed solar capacity is represented by solar thermal electric systems equipped with "power tower" technology. Power tower technology relies on a central receiving system, usually a tall tower. Reflector fields located around the tower focus the sun's rays on a gas- or fluid-filled receiver set atop the tower. The high concentration of sunlight produces temperatures reaching 2,000°F (1,093°C). The heated substance then is pumped through insulated pipes to a central power plant, where it drives conventional turbines. In the United States, solar thermal electric systems currently generate more than 400 megawatts of power, enough to meet the residential needs of Seattle, Washington. There are over 700 megawatts of solar thermal electric systems currently deployed worldwide. By 2010, the market for these systems is estimated to exceed 5,000 megawatts, enough to serve the residential needs of seven million people and to save the energy equivalent of 46 million barrels of oil per year.

Another method used to produce electricity is the **solar pond**, a lined cavity filled with water and salt. Because salt water is denser than freshwater, it sinks to the bottom of the pond, where it remains trapped by the overlying layer of freshwater. The salt water absorbs and stores heat; because it cannot rise and evaporate, it can reach its boiling point even in the coldest winters. Heated water drawn from the bottom of the pond is sent through insulated pipes and used to generate electricity. For example, a solar pond located near the Dead Sea in Israel covers nearly 62 acres and produces

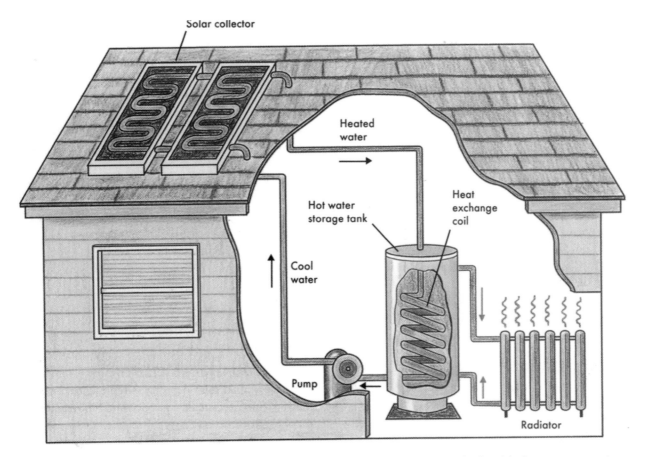

Solar collector

Heated water

Hot water storage tank

Heat exchange coil

Cool water

Pump

Radiator

FIGURE 11-5: This active solar system employs two separate loops of circulating water. In the first (black arrows), water is moved by means of an electric pump through solar collectors, where it absorbs energy from the sun. As this water enters the heat exchange coil in the hot water storage tank, heat energy is exchanged between it and the water circulating through the radiator-storage tank loop (green arrows). Ultimately, the heat dissipates across the radiator surface to warm the air in the house.

up to five megawatts of electricity during peak demand periods. Currently, the United States' largest solar pond is near Chattanooga, Tennessee; it covers one acre, is 10 feet deep, and contains 2,000 tons of salt. In southern California, plans are being developed to build the world's largest solar pond. With a capacity of 48 megawatts, it would supply some 40,000 households with electricity.

Perhaps the most exciting and promising development in solar technology is the **photovoltaic (PV) cell**. This technology generates clean, affordable electricity directly from sunlight, with no boilers, turbines, generators, pipes, or cooling towers. The cell relies on a process that begins when light hits its semiconductors, light-sensitive materials usually made of water-thin slices of crystalline silicon. Atoms within the semiconductors absorb sunlight energy (photons) and liberate electrons. The movement of electrons produces a direct electrical current that is funneled into a wire leading from the cell. Originally, PV cells were expensive, av-

eraging $79 per watt in 1975, but cost has decreased significantly, down to $3.65 per watt in 2000.

Photovoltaic systems are based on technology discovered as long ago as 1839, but it was not until the mid-1950s that the first practical solar cells were manufactured. They were used to supply small amounts of power in such devices as remote weather equipment. As the technology advanced, its applications increased, and by the mid-1960s, PV cells were used to power communication satellites. In 1973, they constituted the major power source for the orbiting Skylab. The technology now is used widely for other applications, such as hand-held calculators, watches, and portable lights and battery chargers.

One especially intriguing development is the possibility of using PV cells for the large-scale generation of electricity. In 1994, the DOE's National Renewable Energy Laboratory collaborated with the Centro de Pesquisas de Energia Electrica to implement a PV technology project in Brazil. The project uses demonstrations of PV power systems in a large array of applications, including rural home lighting, public lighting systems (such as street lights), and a water pumping system. Researchers also installed a meteorological station that allows them to collect solar radiation data.

In addition to the Brazilian project, the DOE is incorporating PV technology in several other experiments, including a 25-kilowatt irrigation system and a 60-kilowatt plant for an Air Force radar station. Further, to encourage increased PV research, the United States has enacted the Photovoltaic Demonstration Program of 2005, which allocates $80 million to states that demonstrate and encourage the use of advanced PV technologies.

Worldwide, PV cells increasingly are used to electrify homes. In remote regions of Mexico, Indonesia, South Africa, Sri Lanka and other developing nations, rooftop PV systems have been added to homes; such systems power lights, water pumps, and appliances. Rooftop PV systems also are being used in more developed nations. In Norway, PV cells provide electricity for 50,000 country homes, and Finland has equipped 20,000 summer homes similarly.

Advantages and Disadvantages

Solar energy offers many important advantages. It exists in unlimited supply, is available everywhere (albeit intermittently in some areas), is non-polluting, conserves natural resources, and is technologically available for widespread use.

Yet, solar power is not without disadvantages. Because solar energy is intermittent, it requires some means of storage. It is costly and difficult to build a solar energy system large enough to store adequate heat to last through several days of cloudy weather. Therefore, most systems are installed together with on-demand sources of heat, such as woodburning stoves or conventional furnaces. Other economic drawbacks include the initial costs of building solar facilities, which can be high, although the life-cycle cost of a solar home, for instance, may be lower than that of a conventional, fossil-fuel-dependent home.

Additionally, using solar power to generate significant amounts of electricity requires a large land area. This requirement might be considered solar power's most serious adverse effect on the environment. Generating 50 megawatts of electricity requires one square mile (2.6 square kilometers) of collectors. To supply electricity to the city of Pittsburgh, Pennsylvania, a 30-square-mile (78-square-kilometer) collector area would be needed. Considerable space also is needed for solar

fields of PV cells, although the requisite area will drop as the efficiency of solar cells increases.

Finally, while solar technologies emit virtually no pollutants during operation, some toxic chemicals are used in the manufacture of certain types of PV cells. Some newer technologies use gallium arsenide and cadmium cells, which contain minute quantities of toxic gases.

What Is Wind Power?

The sun's unequal warming of the Earth's surface and atmosphere causes the regional differences in pressure that are responsible for initiating winds. These flowing rivers of air can be harnessed to provide **wind power** — a safe, clean source of perpetual energy.

Present and Future Use

Like other renewable and perpetual energy resources, wind power has enjoyed a revival in popularity during the past several decades. During the 1970s, dozens of small manufacturers sprang up, producing over 10,000 wind machines, or **wind turbines,** in 95 countries. These units were generally small and were used either to charge batteries or to produce minuscule amounts of electricity, usually less than 100 watts. In the 1980s, the market expanded into larger designs capable of producing significant amounts of electricity.

Spurred by falling costs, wind energy generating capacity reached nearly 32,000 megawatts by the end of 2002, an increase of 27 percent over 2001. At this time, wind power provided enough electricity to meet the residential needs of 35 million people worldwide. Moreover, the DOE has set a goal of generating five percent of U.S. electricity from wind by 2020.

Wind turbines typically require an average minimum wind speed of 13 miles per hour to produce electricity efficiently. Much research has focused on overcoming the limitations of wind speed through design modifications. Advances in aerodynamics and airfoil design have yielded a variety of experimental windmills of nontraditional design. Hundreds of research projects are also concerned with reducing costs and developing economical ways to store energy produced by wind turbines. A **wind farm** is a cluster of wind turbines in a favorable geographic location, that is, where wind speeds average 14 to 20 miles per hour (Figure 11-6). The world's largest wind farm is in California's Altamont Pass, where more than 7,000 turbines have the capacity to provide one billion kilowatt-hours of electricity.

Figure 11-6: Wind turbines generating electricity at a wind farm in the Netherlands.

The largest markets for wind power are Germany, where 38 percent of the world's wind capacity is found (approximately 12,000 megawatts), Spain (4,830 megawatts), the United States (4,685 megawatts), and Denmark (2,880 megawatts). In recent years, the most significant growth has occurred in Asia. India added 250 megawatts to total 1,702 megawatts, and Japan's capacity rose 36 percent to about 470 megawatts with 1,800 megawatts of wind capacity in the development pipeline.

Advantages and Disadvantages

Wind power is a clean and perpetual energy source, but it does have some disadvantages. Unreliable and intermittent, it varies with climate, season, daily weather condition, geography, and topography. As in any region, average wind strength varies in different parts of the United States, but the national average is just 10 miles per hour. The equipment used to harness wind power is expensive, though its cost should decline with further development. Another much-publicized drawback is the problem of bird kills. Traits that characterize a good wind farm location often are also attractive to birds. For example, mountain passes are frequently windy and provide excellent potential for wind farms; however, these mountain passes are also the preferred migratory routes for birds. At the Altamont Pass wind farm, 165 raptors, including 27 golden eagles, were killed in 1993. While this amount is small compared

to bird mortality from glass buildings, pollution, power lines and pet cats, many opponents of wind power claim that any bird deaths from wind turbines are unacceptable. Finally, wind turbines produce some noise and are very visible (covering entire hillsides in some parts of California, for example), so there is an aesthetic cost to their construction.

What Is Hydropower?

Hydropower is the energy of falling water. Like wind power, it is an indirect form of solar energy. For centuries, people have used the power of water cascading downstream and over natural falls or dams to turn paddles or turbines. Water at the top of the falls or dam is in a higher state of gravitational potential energy than water at the bottom. As the water drops from top to bottom, its potential energy is converted to kinetic energy; when the water strikes the blades of a turbine and spins the turbine shaft, the kinetic energy is converted to mechanical energy. The mechanical energy can be harnessed directly to run machinery or used to drive a generator to produce an electrical current.

Present and Future Use

First constructed in more-developed countries (MDCs) in North America and Europe, hydroelectric plants traditionally are large in scale and output and are located at ideal sites — areas with a steep, narrow gorge through which water falls. By 1980, 59 percent of potential hydropower sites in North America and 36 percent in Europe already had been developed. Most of the remaining sites were excluded from development as parkland or because of their natural beauty.

Some less-developed countries (LDCs) now appear to be following the pattern set by MDCs, relying on large dams to supply the energy needed to fuel economic development. China and Brazil have the most ambitious hydroelectric programs underway. Brazil's Itaipu Dam, completed in 1991, has the highest generating capacity of any hydropower installation in the world. Five miles long and half as high as the Empire State Building, the dam generates 12,600 megawatts — a capacity equal to 12 large coal or nuclear power plants. However, the Itaipu Dam's generating capacity could be surpassed by China's Three Gorges Dam, which is scheduled for completion by 2013. Once completed, the dam will produce 18,200 megawatts of energy by harnessing the power of the Yangtze River.

Other developing nations are taking a different route. Many are installing generators thousands of times smaller on remote rivers and streams. Such plants usually have a capacity of 15 megawatts or less and provide power to sparsely populated communities and agricultural processing plants that are remote from electric utility power lines. In Burma, Costa Rica, Guatemala, Guinea, Madagascar, Nepal, Papua New Guinea and Peru, small hydropower potential exceeds the installed generating capacity of all other energy sources. China also uses smaller generators; with some 90,000 turbines supplying electricity to rural areas, China leads all nations in this regard.

Hydropower generates approximately 24 percent of electricity worldwide and 10 percent in the United States. It is responsible for the greatest proportion of electric-generating capacity of all renewable energy sources. However, in the United States, the proportion of hydroelectric production to total electric production has declined in the past several decades. The proportional decline is a consequence of several factors, notably the lack of new sites to develop and growing resistance to large dam projects from environmental groups and the general public.

Advantages and Disadvantages

Proponents of large-scale projects argue that dams offer many advantages. Hydropower provides a cleaner way to produce electricity, as the process emits no pollutants to the atmosphere. Dams are also multipurpose. In addition to generating electricity, they provide impounded water for recreational facilities, municipal water supplies, irrigation, and flood control.

But opponents favor free-flowing rivers and decry the flooding of lands to create reservoirs. Many communities have banded together to fight proposed dams, which they contend will ruin local streams and waterfalls and destroy fishing in the area.

Damming a river does radically alter surrounding ecosystems. For example, the reservoir of Egypt's Aswan High Dam on the Nile River provided a habitat for river snails that allowed their number to increase markedly compared to the population present before the dam was built. Snails are an intermediate host for the parasitic worm *Schistosoma*, or liver fluke, which causes schistosomiosis, a serious disease in humans. The flukes move from the snails to people walking, swimming, or bathing in the impounded water. Mosquitoes, carriers of malaria and yellow fever, also breed in the still water.

Dams also collect silt. Sediments that normally would provide nutrients for downstream organisms or settle on agricultural floodplains instead accumulate behind turbines. Large hydropower projects affect the temperature and oxygen content of downstream waters, altering the mix of aquatic and riparian (or streamside) species. Moreover, dams displace people and wildlife. Globally, an estimated 30 to 60 million people, the majority in China and India, have lost their homes and agricultural fields — their way of life — to these projects.

Because of such economic, environmental and social concerns, the United States and many other MDCs are contending with legal disputes over whether dams should be maintained or decommissioned. In 2001, the National Park Service filed suit in the Colorado courts on behalf of the Black Canyon of the Gunnison National Park to restore its historic water flows, which have been altered by a series of upstream dams. One of the nation's most spectacular gorges, the Black Canyon is known for its natural beauty and unparalleled vistas. The National Park Service claims that, as a designated national park, the Black Canyon of the Gunnison should be preserved in its natural condition as much as possible. Many residents in the nearby Uncompahgre Valley and town of Montrose, however, fear that removing the dams will strain water supplies used for electricity and irrigation, discourage new development, and compromise river recreation. Both sides predict that, as with nearly all legal disputes over water rights, the suit will take years to resolve.

Hydropower offers many advantages, but large dam projects also require sound management that considers the entire watershed. Typically, various components of watershed management are parceled off to different agencies, each more concerned with fulfilling its own departmental responsibilities than with the complexity of the entire watershed. According to Professor Donald Worster of Brandeis University, "Everyone wants a piece of rivers, wants to siphon them off, dump waste into them, drink from them, or move barges along them, but no one has ever been given overall charge of protecting them and their renewability." Environmentally sound hydropower cannot be realized until myriad concerns — including flood control, irrigation, transportation, power production, forestries management, land use, fisheries management, and sanitation — are coordinated within the primary goal of maintaining healthy and productive rivers.

What Is Geothermal Energy?

Geothermal energy is heat generated by natural processes occurring beneath the Earth's surface. Fifteen to 30 miles below the Earth's crust lies the mantle, a semimolten rock layer. Beneath the mantle, decaying radioactive elements and molten rock of nickel and iron cause intense pressure that helps warm the planet's surface. Generally, the heat source lies too deep to be harnessed, but in certain areas, where the molten rock has risen closer to the Earth's surface through massive fractures in the crust, underground deposits or reservoirs of dry steam, wet steam, and hot water form. These deposits can be drilled, like oil deposits, and their energy used to heat space and water, drive industrial processes, and generate electricity.

Dry steam is the rarest and most preferred geothermal resource. It is also the simplest and cheapest form for generating electricity. Steam released from a hole drilled into a reservoir is filtered to eliminate solid materials. The filtered steam is piped directly to a turbine, where it is used to produce electricity.

Wet steam, which consists of a mixture of steam, water droplets and impurities such as salt, is more common than dry steam, but it is also more difficult to use. The water in a wet steam deposit is under such high pressure that it is superheated, that is, its temperature is far above the boiling point of water at normal atmospheric pressure. When wet steam is brought to the surface, a fraction of the water vaporizes instantly because of the dramatic decrease in pressure. The steam and water mixture then is spun at high speed in a separator to remove impurities, and the steam is used to drive a turbine to generate electricity (Figure 11-7).

Hot water deposits are the most common source of geothermal energy. Iceland's tremendous geothermal resources provide 85 percent of the residential heat used by its population. Nearly all the homes and commercial buildings in its capital, Reykjavik (with a population of over 112,000), are heated by hot water geothermal deposits that lie deep beneath the city. Boise, Idaho, has used geothermal energy for space heating for over 80 years. The city's geothermal district heating system

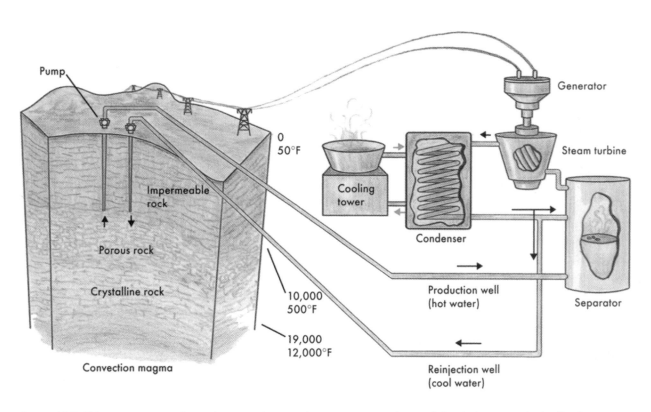

FIGURE 11-7: This wet steam geothermal power plant employs two separate loops of circulating water. In the first loop (black arrows), superheated water is extracted from the Earth and piped to a separator, where impurities are removed and steam is used to turn a turbine and generate electricity. In the second loop (green arrows), the steam reverts to water as is passes through the condenser, where residual heat moves across the coil to water circulating through the cooling tower. The cooled water is then reinjected into the Earth.

consists of over 400 wells which tap the 104° to 230°F (40° to 110°C) water. The system heats city and state buildings as well as nearby residences.

Present and Future Use

People have always used hot springs for purposes such as heating, bathing, cooking, and therapeutic treatment. Today, 20 countries produce 8,000 megawatts of electricity from geothermal resources at a cost competitive with conventional power plants (five to eight cents per kilowatt-hour). In the United States, geothermal plants produce approximately 2,800 megawatts of electricity — enough to supply about 2.8 million average U.S. homes and save four million barrels of oil per year that otherwise would be used for electric production. According to the Renewable Energy Policy Project, a public information and analysis group, global reliance on geothermal energy increased by over 40 percent in the 1990s. Use is expected to rise well into the twenty-first century, primarily for direct heating in MDCs and for electric generation in LDCs.

Advantages and Disadvantages

Geothermal resources are generally more environmentally benign than fossil fuels or nuclear energy. Costs are moderate since most of the requisite technology already has been perfected by the oil industry. Advances in the design of drill bits should produce equipment better able to tolerate high temperatures and pressures. Research currently is underway to reduce the effects of corrosion on equipment caused by salts and silica picked up from subterranean rocks.

There are several major drawbacks to geothermal energy. Like fossil fuels, some geothermal deposits are nonrenewable on a human scale, although reinjecting the fluids extracted from a well can help conserve the resource's energy and extend the life of the site. Moreover, there are relatively few easily accessible deposits, and emptying underground reservoirs may affect land stability. Finally, substances dissolved in the steam and water may affect air and water quality. A variety of gases may be released into the atmosphere when deposits are tapped. Hydrogen sulfides, which smell like rotten eggs, cause the major complaints from people living near geothermal plants. However, scrubbers, similar to those used on coal stacks, are rated as 90 percent efficient.

What Is Ocean Power?

The oceans are a relatively untapped energy resource. Continued research may prove that many forms of **ocean power** — energy derived from the seas through means such as harnessing tides and thermal currents — can be significant, perpetual sources of clean energy. For instance, the flow and ebb of tides in and out of restricted areas can be harnessed to turn turbines for electrical energy and to drive water wheels for mechanical energy. Turbines also can be powered by the movement of waves, by the temperature differential between the sun-heated surface waters and the cold waters below, by floating and stationary "windmills" that use the natural airflow of onshore and offshore breezes, and by ocean currents, such as the Gulf Stream and Japanese Current. Researchers already have developed the means to use the near-surface heat of the Earth's core to spin land-based electric-generating turbines; therefore, the near-surface heat under the oceans also might be used to generate electricity.

Present and Future Use

Of the many forms of ocean power, only two have been used commercially: tidal energy and wave energy. Tidal electric power plants consist of barriers, or dams, built across inlets or estuaries in which a series of turbines is housed. As the tide rises and the water moves inland, it flows through gates in the barrier and is directed into the blades that turn the turbines. At the peak of the tide, the gates are closed and the turbine blades are reversed. As the tide recedes, the gates are reopened and the water flows back over the turbine blades.

Other designs are based on capturing the water of the rising tide. The rising tide is allowed to fill a basin behind the dam. Once the basin is filled and the tide has peaked, the receding water is channeled through turbines to generate electricity. In this type of design, electricity is generated only during the ebb tide.

Wave energy plants, which came into commercial use in the 1980s, use a variety of methods for converting wave energy to electricity. Among the most widespread of these technologies is the oscillating water column (OWC). Devices based on OWCs use the rise and fall of waves to compress air in a vertical pipe. This air is forced through a turbine to generate electricity. Japan was one of the first countries to test wave-power generators based on OWCs. The first two wave-power stations in the world were built in Norway and came online in 1986. One was based on OWC technology, while the other used a tapered-channel system called the Tapchan, in which water funnels through a tapering channel into an elevated reservoir and then flows back to the sea through a turbine.

Another form of ocean power, ocean thermal energy conversion (OTEC), shows promise for not only power generation, but also desalination and aquaculture. OTEC involves harnessing the temperature difference between warmer surface waters and colder, deeper layers by alternately vaporizing and condensing a working fluid (such as warm seawater or ammonia). Once OTEC is more fully developed, numerous regions could benefit, including the Gulf Coast of the United States, the Caribbean, and coastal areas of Africa, Asia, and the Pacific Islands.

Advantages and Disadvantages

Ocean power is non-polluting, renewable, and, like other renewables, essentially "free." However, large facilities both inside and outside of tidal reservoirs have several potentially adverse environmental effects, including reduced tidal range (which would dry out the perimeter of the basin), reduced tidal current flow, altered sea levels, and the death of migratory fish and other aquatic species.

What Is Biomass Energy?

Biomass energy, harnessed from organic matter used as fuel, is one of the oldest and most versatile sources of power. Derived directly or indirectly from photosynthesis, biomass is capable of providing high-quality gaseous, liquid, and solid fuels. Forms of biomass that may serve as fuel are wood and wood processing residues; crop residues; animal waste products; and crops such as switchgrass, seaweed, and kelp that are grown specifically to provide energy.

Present and Future Use

Worldwide, biomass provides over 14 percent of all energy consumed, with LDCs obtaining about 35 percent of their energy from biomass and MDCs, just three percent. In some LDCs, biomass provides up to 90 percent of all energy consumed, with wood generally the principal fuel. For example, Brazil produces charcoal from eucalyptus trees for industrial purposes.

Crop residues are an increasingly important source of biomass. Examples include sugarcane and cotton stalks, rice husks, coconut shells, peanut and other nut hulls, fruit pits, and coffee and other seed hulls. Rice-husk steam power plants currently are operating in India, Malaysia, the Philippines, Suriname, Thailand, and the United States. In Punjab, India, 20 tons of husks are burned each hour to fuel a 10.5-megawatt power plant.

Portions of the sugarcane industry currently are being subsidized by the sale of electricity generated from crop residues. Consider, for example, the sugarcane industry in Hawaii. In the wake of falling sugar prices in the late 1970s, sugar companies began installing 150-megawatt-capacity plants which burn bagasse, the residue after the juice is extracted from the cane. Approximately 50 percent of the electricity produced is used to operate sugar-processing plants, and the remainder is sold to electric companies. The Hawaiian sugar industry supplies 10 percent of the state's electricity, saving about 2.7 million barrels of oil a year.

Using bagasse to generate electricity is fairly inefficient. For instance, a moderately efficient facility can produce only about 20 kilowatt-hours for every ton of cane. But with newly developed technology — a combined bagasse gasifier and steam-injected gas turbine — to cogenerate heat and electricity, electricity production could soar to 460 kilowatt-hours per ton of cane, 23 times greater than before.

The production of clean-burning alcohol fuels is another application of biomass energy technology. One of the most important biomass fuels is ethyl alcohol, or ethanol, which also can be used to make chemicals, solvents, detergents, and cosmetics. The use of alcohol as a fuel source dates back to the late nineteenth century. Ethanol fueled the engines of the first automobiles in the 1880s. To power tractors and other machinery, farmers in the 1920s mixed gasoline with ethanol, a mixture which became known as agrifuel. With the oil shortages of the 1970s, agrifuel resurfaced under the name of gasohol, a mixture of 90 percent gasoline and 10 percent ethanol. Brazil leads the world in ethanol production. More than four million cars in Brazil run on pure ethanol or ethanol mixtures. Most of this ethanol is made from domestic sugarcane, which can be grown almost year-round and takes up only one percent of Brazil's cropland. However, the tremendous success of the ethanol program has resulted in a shortage, forcing Brazil to import ethanol to meet the demand.

Alcohol fuels are considered a potentially significant supplement to the United States' gasoline supplies. According to some estimates, ethanol and other alcohol-based fuels could replace as much as seven percent of the nation's gasoline consumption. These fuels also are gaining support because of concerns about worsening air pollution in urban areas. Ethanol generates 90 percent less carbon dioxide (the leading greenhouse gas) and 70 percent less sulfur dioxide (the leading precursor of acid rain) than does reformulated gasoline. Colorado now requires motorists in its major cities to use gasohol during the winter, when air pollution is worst.

The gases released by decaying plant matter and animal waste, collectively called **biogas**, also can be captured and used as a boiler fuel. Biogas is actually a mixture of gases produced by the anaerobic microbial decomposition of organic materials. Methane is usually the chief component (typically 50 to 70 percent). Other gases include carbon dioxide (30 to 48 percent) and hydrogen sulfide. Several countries are successfully using biogas for diverse applications (Figure 11-8). In China, for example, more than eight million biogas digesters convert manure and other organic waste into methane. These are especially useful in southern China, where higher temperatures stimulate decomposition.

Advantages and Disadvantages

Biomass is a readily accessible and fairly inexpensive resource. It is also a relatively clean fuel; the carbon dioxide released by burning plant matter can be offset by replanting. In some areas, however, biomass fuels

FIGURE 11-8: Household biogas digester in Nepal. Biogas systems enable families to turn biomass into fuel. The manure of two to four pigs can supply a family of five with enough energy for lighting and cooking needs.

have been exploited to such an extent that the supply of the resource and the land's fertility are threatened. The threat is particularly apparent in some LDCs, where wood is such a highly prized resource that deforestation has become a serious problem. Tree-planting ventures by local communities have become more common as people have become aware of the need to reforest and protect their lands. But deforestation is not the only problem. Removing residues from agricultural lands deprives the soil of the nutrients needed to maintain fertility, thereby reducing yields.

What Is Solid Waste?

Solid waste, or refuse, is material that is rejected or discarded as being spent, useless, worthless, or in excess. On average, every U.S. resident produces between four and five pounds of solid waste each day. Much of this refuse is derived from biomass, such as paper and food waste, and like biomass, it can be tapped as an energy source. Using solid waste to produce energy is called **trash conversion**.

Present and Future Use

One way to use solid waste as a source of energy is to sort out noncombustible materials, such as glass and metal, and then shred, screen, and pelletize the combustible refuse into a usable fuel. The pellets can be mixed with an equal amount of a fossil fuel and burned to supply heat and electricity to industries and surrounding communities. The Netherlands and Denmark are leaders in the use of municipal solid waste as fuel; in contrast, the United States exploits this resource very little. Trash conversion has the potential to meet a much larger proportion of the nation's electricity demand. A $165-million, 47-megawatt plant on the Hudson River in Peekskill, New York, converts about half a million tons of refuse into electricity each year. Because it is able to generate twice the electricity the community needs, the plant sells the rest to Consolidated Edison Company, which channels the electricity into its power grid.

Burning refuse is not the only source of energy. Biogas digesters now are being installed in some landfills. These mountains of solid waste gradually decompose anaerobically, forming pockets and veins of methane gas. Pipes drilled into the landfills siphon the gas into storage tanks. As of 2004, approximately 300 methane-recovery facilities were in operation at landfills across the United States.

Advantages and Disadvantages

Using solid waste as an energy source reduces the amount of material that must go into landfills. Further, tapping the methane produced in landfills yields an energy supplement that could save the equivalent of 20,000 to 85,000 barrels of oil per day. Another advantage of recovering methane is not so obvious. As the gas is siphoned off, the landfill settles and becomes smaller, allowing room for additional refuse and extending the useful life of the fill.

There are economic and environmental drawbacks to using refuse as a fuel. To be economically feasible, a plant is limited in location to an area where large amounts of refuse are produced. Further, if a municipality relies on trash conversion to generate electricity, local recycling and resource recovery efforts will be hampered, even though more energy can be saved through resource recovery than can be generated by burning recyclable materials. Also, emissions of mercury, lead, and other harmful substances sometimes result from trash incineration.

Managing Alternative Energy Sources

In this section, we offer a brief synopsis of the use of alternative energy sources throughout the centuries. While not intended to be a comprehensive overview, the following paragraphs do provide a basic understanding of the political, economic, and social factors that drive the development — or lack thereof — of sources of alternative energy.

For centuries, people have used radiation from the sun; have tapped heat from the Earth; and have harnessed the power of winds, water, and ocean tides. The Egyptians took advantage of wind power when they sailed their crafts down the Nile River some 5,000 years ago. The ancient Persians used water wheels for turning and grinding. In 85 B.C., the Greek poet Antipater celebrated the development of a water-powered gristmill, noting that it had liberated Greek maidens from the arduous task of grinding grain. Hot water from thermal springs heated pools used by the Romans when they ruled England. The Japanese applied the basic mechanics of harnessing water in the seventh century. The ancient cliff dwellings of ancestral Pueblo Indians, in Mesa Verde, Colorado, are excellent examples of the simplicity and effectiveness of passive solar use (Figure 11-9). In Hamburg, Germany, a tidal-powered sewage pump functioned until 1880. In the 1700s, New Englanders built mills powered by tidal motion. Windmills, for centuries a source of mechanical power in parts of Europe, also were used by farmers and ranchers in the United States, particularly on the plains of the Midwest, to operate water pumps.

In the twentieth century, people began to experiment with alternative energy sources to generate electricity. The world's first commercial hydroelectric plant started producing electricity at Niagara Falls on the border between the United States and Canada in the early 1900s. In the 1930s, Europeans began to investigate the use of wind power for large-scale electric generation. Soon after, the world's largest windmill was constructed near the town of Rutland, Vermont. A giant propeller-driven model, it had a 175-foot sweep and was capable of generating 1,250 kilowatts. Although it fed electricity into the local commercial power system for three years, Rutland's windmill was doomed by the same phenomenon that cut short interest in other perpetual and renewable energy sources — the age of fossil fuels.

The widespread availability of inexpensive supplies of oil, natural gas, and coal spurred utilities, industries, and home owners to rely increasingly on fossil fuels to meet almost all energy demands. Only hydroelectric power, probably the least environmentally benign of all renewable sources, managed to grow during the age of fossil fuels. It was joined, in the latter half of the twentieth century, by a promising new resource: nuclear energy.

The world first witnessed the awesome power of the atom at the close of World War II, with the attacks on Hiroshima and Nagasaki. But with the 1954 Atomic Energy Act, the United States government stated its commitment to developing civilian nuclear capabilities. This act essentially opened the nuclear industry to the private sector, although government research and activities still remained off-limits. The first major developments within the civilian sector were in the areas of medicine, biological research, and agriculture. Then, as the 1950s came to a close, interest began to turn toward converting nuclear energy into electric power. This interest was spurred by lowered estimates of finite fossil fuel reserves, a growing demand for electrical power, and the 1956 war over the Suez Canal, a major route for transporting oil. By 1957, the world's first commercial nuclear reactor, the Shippingport nuclear plant, came online at Pittsburgh, Pennsylvania.

Figure 11-9: Ancestral Pueblo Indians used the principles of passive solar heating to warm and cool their cliff dwellings. Massive rock overhangs provide shade from the high, hot summer sun, but permit the rays of the lower winter sun to penetrate the dwellings and bathe them in warm sunlight.

The OPEC oil embargo of 1973 and the resulting oil shortages renewed the United States' interest in perpetual and renewable resources, especially for home heating and similar uses. Government funding for development of these alternative fuels rose (though not to the levels enjoyed by the nuclear industry) and peaked by 1980. That year saw the election of an administration which favored increased development of domestic oil reserves and nuclear energy over continued support for renewables. Funding dropped precipitously and has remained well below funding levels of the late 1970s.

As with other alternative fuels, interest in nuclear energy increased worldwide during the 1970s and early 1980s, as governments sought to reduce their dependence on foreign oil. Nuclear energy's share of electric production grew significantly during the period. Despite support for nuclear energy among governments worldwide, a small but growing number of people continued to voice concerns about the safety and wisdom of increasing reliance on nuclear energy. By the 1990s, the issue of what to do with the by-products of nuclear power became a matter of great debate in the United States as the government searched for a geologically stable — and politically acceptable — storage site for the nation's mounting nuclear wastes. With the accidents at Three Mile Island and Chernobyl, many people became convinced that nuclear energy is not the solution to the world's energy problem, and public support for nuclear energy fell throughout North America and Western Europe.

Summary

Numerous energy sources may prove to be important alternatives to fossil fuels. These include nuclear energy, solar energy, wind power, hydropower, geothermal energy, ocean power, biomass energy, and solid waste. Nuclear energy is the energy contained within the nucleus of the atom. When an atom splits into smaller particles, it releases neutrons and emits heat. The released neutrons may bombard other fissionable atoms, causing a chain reaction. To harness the energy of the atom, nuclear reactors are designed to sustain the fissioning process. Uranium 235 (U-235), a relatively rare uranium isotope, is the fuel that powers a nuclear reactor. The energy from the nuclear reactor heats water to produce steam; the steam, in turn, runs a turbine which generates electricity. Nuclear energy accounts for a modest proportion of the United States' and the world's total energy supply. In terms of electric generation, however, nuclear power's share is about 20 percent in the United States and 23 percent worldwide. The significant growth of the nuclear industry in the United States during the 1970s and early 1980s has tapered off in recent years, primarily because of safety concerns and less-than-anticipated demand for electricity.

The sun provides the energy that all life forms need to survive. Currently, solar energy is used primarily for space and water heating. Passive solar systems rely on natural forces to distribute the heat; active systems use fans or pumps driven by electricity to enhance the collection and distribution of the sun's heat. All solar systems incorporate design features of buildings to capture the maximum amount of radiation from the sun in winter months and a minimum amount of radiation in summer months. In addition to space heating, solar energy also holds potential for electric generation. Photovoltaic cells generate electricity directly from sunlight. Atoms within semiconductors absorb sunlight energy and liberate electrons, which produce a direct electrical current.

The sun's unequal warming of the Earth's surface and atmosphere causes regional pressure differences that produce winds. Winds can be harnessed to provide a safe, clean, perpetual source of energy. Wind turbines clustered in favorable geographic locations make up wind farms.

Hydropower is the energy produced as flowing water strikes the blades of a paddle or turbine. The energy can drive an electric generator or run machinery. Currently, hydropower generates almost one-fourth of the world's electricity, a contribution slightly greater than that of nuclear power. Worldwide, it is responsible for a greater proportion of electric-generating capacity than all other renewable sources of energy.

Geothermal energy is heat generated by natural processes beneath the surface of the Earth. Dry steam deposits, the rarest geothermal resource, are also the simplest and cheapest for generating electricity. Wet steam deposits are more common than dry steam deposits, but they are also more difficult to harness. Hot water deposits are the most common type of geothermal energy. For all three forms, a hole is drilled into a reservoir to release deposits, which are filtered as necessary and then piped to a turbine.

Ocean power is a relatively untapped energy resource. Researchers believe that the oceans contain numerous different energy sources, including the tides, waves, deep ocean currents, and on- and off-shore winds. Of these, only tidal and wave power have been used commercially.

Discussion Questions

1. Explain, in general terms, how a nuclear power plant generates electricity.

2. What are some major advantages and disadvantages of each of the perpetual or renewable energy resources described in this chapter? Which seem to be most desirable from an environmental standpoint? Why?

3. Imagine that a new power plant is to be built to serve your area. What type of energy would be best suited to your area's resources and needs? What kind of plant would you be most willing to live near? Why? Consider both economic and environmental factors in your response.

4. Imagine you are planning a new home for yourself. What kind of energy sources would you want to use? What energy-saving features would you incorporate into your design? Explain how these features function.

5. If you were in charge of crafting a U.S. energy policy, which alternative sources — if any — would you promote? Why?

Biomass energy, harnessed from organic matter used as fuel, is one of the oldest and most versatile sources of power. Derived directly or indirectly from photosynthesis, biomass is capable of providing high-quality gaseous, liquid, and solid fuels. Among its primary sources are wood and wood processing residues; crop residues; animal waste products; and crops such as switchgrass, seaweed, and kelp. Another application of biomass energy technology is the production of clean-burning alcohol fuels, which are considered a potentially significant supplement to gasoline supplies. The gases released by decaying plant matter and animal waste, called biogas, also can be captured and used as a boiler fuel. Biomass holds significant potential as a renewable energy resource, especially in LDCs. It is a clean, readily accessible, and fairly inexpensive resource. In some areas, however, biomass fuels have been exploited to such an extent that the land's fertility is threatened.

One unique alternative fuel, solid waste, is a direct result of human activity. Some communities have experimented with burning solid waste to produce energy, commonly called trash conversion. Biogas digesters, used to collect the methane gas that results from the decomposition of solid waste, also are being installed in urban landfills.

For centuries, people have harnessed energy through radiation from the sun; heat from the Earth; and the power of winds, water, and ocean tides. In the past several centuries, however, the widespread availability of inexpensive supplies of coal, oil, and natural gas spurred utility companies, industry, and residential users to rely on fossil fuels to meet almost all energy demands. In the latter half of the twentieth century, the emergence of nuclear power, the OPEC oil embargo of 1973, and an awareness of the dwindling supplies of fossil fuels stimulated renewed interest in some perpetual and renewable sources of alternative energy.

KEY TERMS

active solar system	isotope	solar energy
binding force	nuclear energy	solar pond
biogas	ocean power	solid waste
biomass	ocean thermal energy conversion	trash conversion
containment vessel		wind farm
fission	passive solar system	wind power
geothermal energy	photovoltaic cell	wind turbine
hydropower	reactor core	

UNIT **IV**

An Environmental Necessity

Protecting Biospheric Components

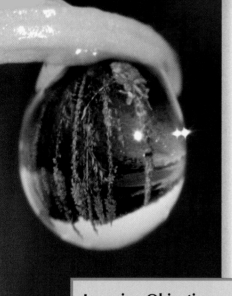

CHAPTER 12

Air Resources

Now I see the secret of the making of the best persons.
It is to grow in the open air, and to eat and
sleep with the earth.

Walt Whitman

The global climate is an essential foundation of
natural ecosystems and the entire human economy. If
we are entering a new period of climate instability, the
consequences could be serious indeed, affecting virtu-
ally all of Earth's ecosystems, accelerating the
pace of extinction, and leaving few areas
of economic life untouched.

The Worldwatch Institute

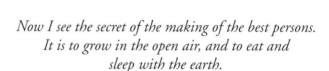ir is a vital resource, nourishing both body and spirit. With the exception of a small number of anaerobic species, all living things require oxygen. Humans can survive for weeks without food and for days without water, but if we are deprived of oxygen, we will die within minutes. Air also feeds our spirits; although invisible to the human eye, it helps create the sights that stop us in our tracks: sublime sunsets, dazzling rainbows, and spectacular meteor showers. Air carries thousands of scents, both subtle and pungent: the fresh smell of approaching rain, the salty tang of an ocean breeze, the fragrant perfume of blooming flowers, and the rotten-egg odor of sulfurous gases. Air also bears the sounds of reflection and comfort: the rustle of fall leaves, the laughter of children, and the tea kettle's whistle on a lazy weekend morning. Even with our relatively poor sense of smell, humans can detect dust or strong chemicals carried along by wind currents. And finally, air brings tactile pleasure, from a cooling breeze on bare skin to a gusty wind laden with rain or snow. Many other animals, from snakes to deer, also rely on the sensory cues they pick up from the air to hunt prey — or to avoid being hunted. But while air surrounds us, we seldom are aware of it unless it is degraded by pollution, natural or human-induced. In this chapter, we describe air resources and examine efforts to manage them.

Describing Air Resources

In this section, we look closely at the atmosphere and how it helps to sustain life on Earth. We examine the major types of air pollution, their effects on environmental and human health, and factors such as weather and topography that affect pollution levels.

What Is the Atmosphere?

The layers of the atmosphere and the chemical composition of air are the chief physical characteristics of air resources. The **atmosphere** consists of four layers of gases that surround the Earth (Figure 12-1). The bottom layer, called the **troposphere**, extends to about seven miles (11 kilometers) above the Earth's surface and contains the gases that support life. The **stratosphere** extends from roughly seven to 31 miles (50 kilometers) above the Earth's surface. **Ozone** gas (O_3) present in the stratosphere acts as a shield, preventing harmful amounts of ultraviolet radiation from penetrating the troposphere and reaching the Earth's surface. The **mesosphere** extends from the top of the stratosphere to about 56 miles (90 kilometers) above the Earth; the **thermosphere** extends from 56 miles to outer space.

Even though wind patterns in the upper atmosphere influence weather conditions, weather occurs in the troposphere. **Weather** is the day-to-day pattern of precipitation, temperature, wind (direction and speed), barometric pressure, and humidity. In contrast, **climate** is the long-term weather pattern of a particular region. Climatic factors such as temperature and precipitation are expressed as averages observed over time, typically 30 years. Although short-term predictions of weather trends (up to five-day forecasts) have become increasingly accurate, long-term patterns and subtle shifts in climate remain difficult to predict.

Clean, dry air is a mixture of gases containing 78 percent nitrogen and 21 percent oxygen. The remaining one percent is composed of 0.03 percent carbon dioxide (CO_2) and rare gases such as helium, argon, and krypton. But air is rarely clean or dry. Air over the tropics may contain as much as five percent moisture, and air over the Earth's poles may contain less than one percent. Typically, air contains varying amounts of water (in the form of vapor, droplets, hail, or snow-flakes), tiny particles of solids such as dust and soot,

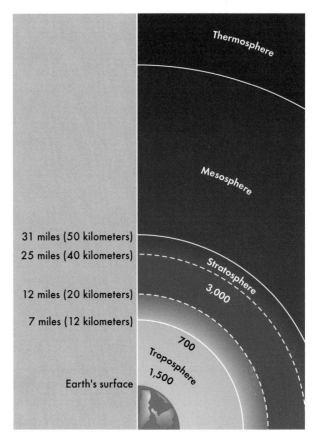

FIGURE 12-1: The atmosphere. The Earth's atmosphere consists of the troposphere, stratosphere, mesosphere, and thermosphere. The numbers given indicate the concentration of ozone (in billion molecules per cubic centimeter of air). Atmospheric ozone is found at its densest concentration, about 3,000 billion molecules per cubic centimeter, in the stratosphere at 12 to 25 miles above the Earth.

other gases and elements such as methane and sulfur, living organisms such as bacteria and molds, and reproductive cells such as pollen and spores.

How Does the Atmosphere Help to Sustain Life on Earth?

A vital — and as far as we know, a *unique* — relationship exists between the Earth's atmosphere and the life that the planet harbors. The atmosphere helps to sustain life by regulating the Earth's temperature, thereby maintaining a global climate hospitable to organisms. In turn, living organisms, particularly green plants and algae, affect the composition of the atmosphere in ways that benefit life. Let's examine this relationship more closely.

Much incoming solar radiation actually is reflected away from Earth by particles, water vapor, and ozone in the atmosphere and by the reflectivity, or **albedo**, of the Earth's surface. Lighter-colored areas of the Earth — the polar ice caps, deserts and rangelands, and areas with heavy cloud cover — have a high albedo. A low albedo is associated with dark surfaces, such as forests and ocean areas rich in phytoplankton and other plant life. Because areas with a high albedo reflect more solar energy into space than do areas with a low albedo, they retain less heat. Dark areas absorb more solar radiation than they reflect; the absorbed radiation is degraded to long-wave infrared radiation, or heat, thus raising the temperature of the areas. Eventually, this long-wave heat is radiated back into space. The capacity of a surface to radiate heat is its **emissivity**. It is equal to the amount of solar radiation that is absorbed, degraded to heat energy, and radiated back into space. The combined effects of the Earth's albedo and emissivity help to maintain the average global temperature at about 59°F (15°C).

Temperature also is determined by the atmospheric levels of water vapor, CO_2, and other greenhouse gases. As we learned in Chapter 1, these gases trap reradiated heat energy and warm the Earth, a phenomenon known as the **greenhouse effect** (Figure 12-2). To understand the effect of greenhouse gases, we need only look at our planetary neighbors. Venus, nearer the sun and with a much higher proportion of greenhouse gases, is far too hot to sustain life, with an average temperature of 837°F (447°C). The thin atmosphere of Mars, although predominantly composed of CO_2, does not retain enough of the sun's heat, and Mars is far too cold to sustain life; its average temperature is -63°F (-53°C). Earth benefits from the Goldilocks effect: Its concentration of greenhouse gases is "just right" to enable life to flourish.

So then, how do living organisms affect the composition of the atmosphere? During the process of photosynthesis, green plants, blue-green algae, and phytoplankton remove CO_2 from the atmosphere and give off oxygen. In this way, they help to maintain the proper atmospheric concentration of greenhouse gases. Undisturbed forests play a particularly important role because they act as "carbon sinks," removing and storing large amounts of atmospheric CO_2 for long periods of time. According to the Gaia hypothesis, developed by British atmospheric scientist James Lovelock, the atmosphere — an abiotic component of Earth — and its living organisms — the biota — work together to create and maintain the conditions (including climate) necessary for life.

What Is Air Pollution?

An **air pollutant** is any substance present in or released to the atmosphere that adversely affects environmental or human health. Pollutants emitted by natural sources, such as volcanoes, forest fires and living organisms, are part of the complex biogeochemical cycles that regulate the cycling of materials among land, sea, and air. A significant and growing share of pollutants is emitted by cultural, or anthropogenic, sources. In the United States, transportation — which relies chiefly on the combustion of petroleum — accounts for over half of all

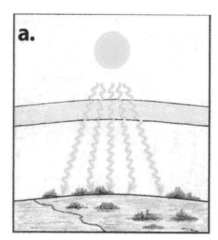
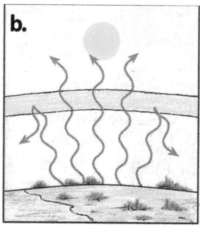
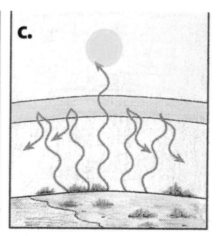

FIGURE 12-2: The greenhouse effect. (a) Most of the solar radiation that penetrates the Earth's atmosphere consists of visible light energy. (b) Objects on Earth heated by the sun's rays eventually reradiate this heat towards space in the form of infrared rays. Some outgoing heat is trapped by greenhouse gases in the atmosphere, keeping the Earth's surface temperature moderate. (c) When greenhouse gases build up in the atmosphere, more heat is trapped near the Earth's surface.

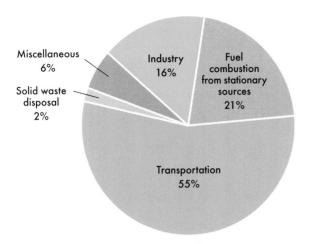

Figure 12-3: Cultural sources of air pollutants in the United States. Transportation is the largest single source of cultural pollutants, with fuel combustion from stationary sources, such as coal-fired electric generating plants, the second-largest source. Clearly, the combustion of fossil fuels is by far the most significant source of anthropogenic, or human-caused, air pollution.

anthropogenic air pollutants. Other significant polluters include stationary sources such as coal-fired electric power plants and industrial sites (Figure 12-3).

There are two categories of air pollutants, primary and secondary. **Primary pollutants** are emitted directly into the atmosphere, where they have an adverse impact on human health or the environment. In the 1970s, in the face of overwhelming evidence linking certain primary pollutants to grave health and environmental impacts, the U.S. Environmental Protection Agency (EPA) began to regulate emissions of carbon monoxide, sulfur dioxide, nitrogen dioxide, hydrocarbons, particulates, and lead. (Importantly, the EPA did not — and still does not — regulate emissions of CO_2, a primary pollutant emitted in large quantities.) Those regulations have improved the nation's air quality, but as we'll see below, they have not eliminated the health and environmental threats posed by primary pollutants.

Once in the atmosphere, primary pollutants may react with other primary pollutants or atmospheric compounds such as water vapor to form **secondary pollutants**. When sulfur or nitrogen oxides react with water vapor in the atmosphere, for example, they give rise to acid precipitation, a secondary pollutant; nitrogen oxides are also a precursor of photochemical smog, commonly known as ground-level ozone.

Effects of Primary Pollutants

Table 12-1 summarizes the sources, health effects, and environmental impacts of the EPA-regulated primary pollutants. In general, the groups most at risk from exposure to air pollution are children, the elderly, and those who suffer from respiratory illness or cardiovascular disease.

The dangers posed by acute exposure to very high levels of primary pollutants have been recognized widely for many decades. However, until recently, little was known about the long-term effects of exposure to pollutant levels routinely found in many urban areas. Increasingly, research indicates that chronic exposure to air pollution has serious health effects, and not just for those belonging to at-risk groups. For example, in a long-term study conducted between 1993 and 2001, scientists tracked the levels of major air pollutants in 12 southern California communities while following the pulmonary health of nearly 1,900 children as they progressed from grades four through 12 (ages 10 to 18). The communities included some heavily polluted sites in greater Los Angeles and several low-pollution sites outside the area. The researchers found that the lungs of many children routinely exposed to polluted air are five times more likely to have clinically low lung function — less than 80 percent of the lung function expected for a person of his or her age. (Lung function increases steadily throughout one's childhood, peaking at about age 18 in women and in the early 20s in men.) Low lung function has both short- and long-term implications. For instance, a person with low lung function may exhibit more severe cold symptoms and may take longer to recover from an illness. In the long term, low lung function increases the risk of respiratory disease and heart attack and is second only to smoking as a risk factor for mortality.

Of the pollutants examined in the California study, fine particulates showed the strongest correlation between air levels and deficits in lung development, although correlations also were shown for nitrogen dioxide and acid vapor. Some fine particulates are emitted directly from combustion sources, but most are created by chemical processes in the atmosphere. Fine particulates penetrate into and are retained in the walls of small airways; the finest particles can pass through the lungs to the blood. The resulting damage is similar to that found in the lungs of cigarette smokers — markedly higher levels of fibrous tissue and microscopic evidence of particle accumulation in the respiratory bronchioles. Indeed, mounting evidence shows a strong relationship between particulate pollution, illness, hospitalization,

Table 12–1: Sources, Health Effects, and Environmental Impacts of Selected Primary Pollutants

Pollutant	Description
Carbon monoxide (CO)	Major source is internal combustion engines used in transportation (77 percent); minor sources include charcoal grills, improperly adjusted oil and gas heaters, and cigarette smoke. Interferes with the oxygen-carrying capacity of the blood because it has a greater affinity than oxygen for the binding sites on the hemoglobin molecule. At greatest risk are people who suffer from cardiovascular disease, especially those with angina or peripheral vascular disease. Healthy individuals may be affected at higher levels. Exposure to elevated CO levels slows reflexes; causes headaches and unconsciousness; impairs visual perception and manual dexterity; may affect fetal development and mental development. Precursor of photochemical smog.
Sulfur dioxide (SO_2)	Major source is the combustion of fossil fuels, especially high-sulfur coal, to generate electricity. Exposure to high concentrations of SO_2 may aggravate existing respiratory and cardiovascular disease; lead to respiratory illness; cause difficulties with breathing; and alter the lungs' defenses. Irritates nose, nasopharynx, and bronchi; increases likelihood of cancer. At greatest risk are children, elderly, asthmatics, and allergy sufferers. SO_2 also can damage tree foliage and crops; makes plants more susceptible to disease. Precursor of acid precipitation.
Nitrogen dioxide (NO_2)	Major sources are fuel combustion and transportation. Can irritate the lungs and lower resistance to respiratory infections such as influenza. Continued or frequent exposure to elevated NO_2 levels may cause increased incidence of acute respiratory disease in children. At greatest risk are children and those who suffer from asthma, emphysema, and chronic bronchitis. Precursor of acid precipitation and photochemical smog.
Hydrocarbons (HC)	Major sources are transportation, fuel combustion, petroleum refineries, petrochemical plants, paint and printing solvents, and cigarettes (benzene). Benzene is a suspected carcinogen; HCs adhere to particulates and enter lungs. Precursor of photochemical smog.
Particulates	Major sources are diesel vehicles, residential wood combustion, coal-fired power plants, agricultural tilling, construction, and unpaved roads. Particulates, especially fine particles that are able to reach the lower regions of the respiratory tract, affect breathing; aggravate existing respiratory and cardiovascular diseases; alter the body's defense systems against foreign materials; and damage lung tissue. Particulates also may be carcinogenic and can absorb and deliver gaseous pollutants (such as SO_2) directly to the lungs. At greatest risk are children, the elderly, and persons with chronic obstructive pulmonary or cardiovascular disease, influenza, and asthma. Particulate matter is responsible for atmospheric haze and soils and damages materials.
Lead (Pb)	Major sources are transportation (vehicles that use leaded fuels) and stationary sites such as smelters, battery plants, and solid waste disposal facilities. Residential sources include paint in old homes and lead solder in old pipes. In the body, lead accumulates in blood, bone, and soft tissue. It is not readily excreted, and can affect the kidneys, liver, nervous system, and blood-forming organs. Continued exposure may cause neurological impairments (seizures, mental retardation, and/or behavioral disorders). Even at low doses, exposure to lead is linked with changes in fundamental enzymatic, energy transfer, and homeostatic mechanisms in the body. At greatest risk are fetuses, infants and children, all of whom may suffer central nervous system damage after exposure to low lead levels. Lead may also be a factor in high blood pressure and subsequent heart disease in middle-aged white males. At the environmental level, lead bioaccumulates in food chains.

and premature death. Each year, fine particle exposure is responsible for an estimated 40,000 premature deaths from respiratory illnesses (such as emphysema and chronic bronchitis), heart attacks, strokes, and lung cancer; the elderly are particularly at risk. Exposure to fine particulates causes children to become ill more frequently and experience greater respiratory problems, such as difficult and painful breathing. For both children and adults, exposure to fine particulates also aggravates asthma and increases hospital admissions and emergency room visits. In addition to their human health effects, fine particulates are the major cause of reduced visibility in many areas of the United States, including numerous national parks.

Unfortunately, unhealthy air is a problem worldwide. Over 1.4 billion people, most of them in the developing world, breathe dirty air every day. In some of the most polluted cities, such as Mexico City, Mexico; Beijing and Shanghai, China; Tehran, Iran; and Calcutta and Rajkot, India, chronic respiratory disease is a leading cause of death. Sadly, few less-developed countries (LDCs) regulate air pollution, which is seen as a necessary by-product of economic activity. In addition, a lack of zoning regulations means that in many LDCs, industrial and residential areas are in close proximity (Figure 12-4). In general, the air quality problems that plague developing nations are the same ones that developed nations began tackling decades ago: lead, particulates, industrial smog, and ground-level ozone. In fact, as many as 80 percent of children between the ages of three and five living in the developing world have blood lead levels that exceed the standard set by the World Health Organization (WHO).

FIGURE 12-4: Children at their school in Lanzhou, China. Some of the children wear masks to protect themselves from pollution emitted by nearby factories. Residential and industrial areas are in close proximity in Lanzhou, as they are in many cities in the developing world.

On an encouraging note, there are signs that developing nations are beginning to address issues of air quality. Between 1998 and 2000, for example, China and other Asian nations phased out the use of leaded gasoline in order to combat childhood lead poisoning. The nations of central and eastern Europe must do the same in order to become members of the European Union, a status seen as vital to the countries' economic development. Even in Africa, where leaded gasoline is ubiquitous, a phase-out appears imminent. Many African nations have acknowledged the need to end the use of leaded fuels by 2006.

Yet lead is just one air pollutant, and in the long run, it may be the easiest to eliminate. Developing nations must begin to tackle more difficult air quality issues, notably the pollution caused by coal combustion. In this respect, they find themselves in a position similar to that of developed nations, which have yet to solve the problems caused by a heavy reliance on fossil fuels. Moreover, the nations of the world must work together to address regional and global air quality issues. In the following pages, we examine six issues that warrant special attention: climate change, stratospheric ozone depletion, acid precipitation, smog, air toxics, and indoor air pollution.

What Is Climate Change?

Greenhouse gases play a vital role in maintaining the conditions necessary for life on Earth, but when human activities cause the concentrations of these gases to rise — as has been happening for approximately the past 140 years — the greenhouse effect is enhanced, increasing the likelihood that the average global temperature will rise several degrees or more. This phenomenon is popularly known as global warming, but it is more accurately described as **climate change**, a shift in the Earth's long-term weather patterns caused by an anthropogenically enhanced greenhouse effect. Scientists use the term climate change to emphasize the fact that a higher average global temperature is likely to affect *many* aspects of the global climate, including precipitation patterns and the occurrence and severity of storms. Before we discuss the effects of climate change, however, let's look at the trends in greenhouse gases and the average global temperature and at forecasts of expected climate change. We conclude this section with a discussion of some of the scientific studies that document the wide-ranging effects of climate change on natural systems.

Trends in Greenhouse Gases and Average Global Temperature

Carbon dioxide (CO_2) is responsible for about 50 percent of climate change. The United States alone accounts for nearly one-quarter of global CO_2 emissions; other countries with high emissions include China, Japan, Russia, and most European nations. Additional greenhouse gases include chlorofluorocarbons, chemicals found in aerosol propellants, coolants and solvents, which account for about 20 percent of climate change; methane, about 16 percent; ozone, about eight percent; and nitrous oxide, about six percent.

Since the dawn of the Industrial Revolution (circa 1750), human activities have slowly but steadily increased the atmospheric concentrations of greenhouse gases. The most problematic of these activities is the combustion of fossil fuels, which releases CO_2 and nitrogen oxides (NO_x) and produces ozone. Carbon dioxide also is released when forests are burned or when land is cleared and the vegetation left to decompose. Deforestation has a double impact: Once cleared, the area no longer removes and stores additional amounts of atmospheric CO_2. Finally, decomposition also releases methane, as does the digestive activity of ruminants (like cows and sheep) and termites.

The atmospheric concentration of CO_2 has increased an estimated 31 percent since 1750, rising about 18 percent since 1960 alone (Figure 12-5). In addition, since pre-industrial times, the atmospheric concentrations of methane and NO_x have risen by 151 percent and 17 percent, respectively. Because many greenhouse gases remain in the atmosphere for a long time (for CO_2, about 100 years; for NO_x, about 120 years), their warming effect is long-term.

The steady build-up of greenhouse gases appears to be correlated with an observed rise in the Earth's average surface temperature over the last century. The average global temperature has increased by about 1°F (0.56°C) since the late nineteenth century and 0.27°F (0.15°C) since the 1960s. In general, nighttime temperatures over land have increased more than daytime temperatures. Recent years have been among the warmest on record, despite the cooling effect caused by the eruption of Mt. Pinatubo in 1991. In fact, since record-keeping began in 1866, all of the 10 warmest years on record have occurred in the 1990s and 2000s, and the 1990s are the warmest decade on record. The year showing the highest temperature increase was 1998, which brought a range of extreme weather events attributable only in part to an unusually strong El Niño phenomenon. Those events included droughts and rare fires in tropical and subtropical forests in Indonesia, Florida, and Mexico; historic floods in China and Bangladesh; severe storms in Africa and the Americas; a devastating ice storm in the northeastern United States and Canada; and withering heat waves in the United States, India, and southern Europe. According to the Worldwatch Institute, economic losses associated with the year's weather-related disasters are estimated at $89 billion.

After 1998, the next three warmest years on record were 2002, 2003, and 2004. According to Dr. James

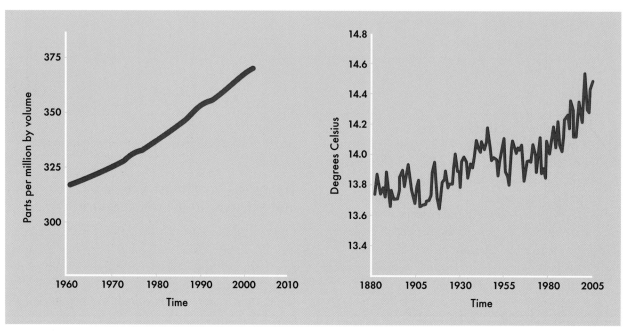

Figure 12-5: Atmospheric concentration of carbon dioxide, 1000 to 1997, and average temperature at the Earth's surface, 1866-1998. The concentration of carbon dioxide in the atmosphere (left) has risen sharply since the late 1800s. Records from the past 100 years also show an increase in the average global temperature at the Earth's surface (right) of about 1°F (0.56°C), which may indicate that the increasing concentration of carbon dioxide and other greenhouse gases is causing a warming of the planet's climate.

E. Hansen of the Goddard Institute for Space Studies, National Aeronautics and Space Administration (NASA), the global temperature in 2005 is likely to exceed those of 2003 and 2004.

Forecasts of Expected Climate Change

Given current knowledge, the global climate is expected to continue to warm in the future. Computer models differ in the projected degree of change, depending on the assumptions used. These models project an increase in mean surface temperature of 1.8 to 6.3° F (1 to 3.5° C) by 2100. While the actual changes that occur over specific years or decades will show considerable natural variability, models indicate that, over the next century, the average rate of warming will be greater than any seen in the last 10,000 years.

Some people dispute the concept of human-induced climate change. They argue that because we have records of atmospheric CO_2 and temperature levels for only the past 100 years, the planet's rising temperatures could be the result of natural cycles rather than human actions.

Most scientists, however, believe that human-induced climate change is now or soon will be occurring. In 2001, two major reports were released that confirmed this strong consensus within the scientific community. A distinguished panel of scientists appointed by the National Research Council (NRC) issued the first report, which focuses on the science of climate change and its impacts on the United States. The second report, issued by the United Nations Intergovernmental Panel on Climate Change (IPCC), summarizes the international scientific community's position. The 2001 IPCC report is the panel's third assessment and builds upon the previous two while incorporating new climate research. Over 2,000 experts participated in drafting and reviewing the IPCC report. Both the NRC and IPCC reports come to the same conclusions, which including the following: 1) Global temperatures have warmed over the past century, resulting in changes in the Earth's climate system, including changes in atmospheric and oceanic circulation; 2) emissions of greenhouse gases due to human activities continue to alter the atmosphere in ways that are expected to affect climate; 3) there is new and stronger evidence that most of the warming observed over the last 50 years is attributable to human activities; 4) human influences will continue to change atmospheric composition throughout the twenty-first century; 5) global average temperature and sea level are projected to rise, further altering the climate system and resulting in changes such as increased precipitation, exacerbated drought, and more frequent flooding; and 6) anthropogenic climate change will persist for many centuries.

Another important conclusion of both the NRC and IPCC reports is that scientists are becoming increasingly proficient in using computers to model climate change. As a result, they have greater confidence in the ability of models to understand and predict the factors driving this change. Computer models allow scientists to more fully understand the roles that ocean currents, cloud cover, and vegetation play in modifying the warming effect. Even so, scientists acknowledge that additional work is needed to accurately detect, attribute and understand climate change; reduce uncertainties; and develop projections of future warming.

It is important to realize that scientific uncertainties affect only the *degree* to which the climate will warm; they do not alter the fact that most scientists believe that the climate is changing as a result of human activities. In fact, in October 2003, over 1,000 scientists in the United States signed and submitted a letter to the U.S. Congress expressing the clear consensus among top experts in the field that human activities are altering the global climate.

Effects of Climate Change

It does not take a dramatic shift in average temperature to have a substantial effect on the Earth's climate. The average global temperature of the last ice age was only about 16°F (9°C) lower than the average global temperature today. Among the most likely and worrisome effects of a warming climate are changes in rainfall patterns, growing seasons, and arable land. Southeast Asia, for instance, is projected to be wetter, while parts of North America and Asia could be much drier. If rainfall patterns do change, parts of North America's grain-producing areas might suffer extended drought. The United States might not be able to produce as much grain as it currently does, while Canada, the former Soviet states, and northern Europe might produce more. In general, then, changing rainfall patterns might shift agricultural regions northward. Plants and crops in currently temperate regions would need to be heat resistant, even if enough irrigation water could be supplied.

Warmer temperatures also should produce a more vigorous hydrologic cycle, which could mean more severe droughts or floods in some places and less severe droughts or floods in others. Some computer models also indicate an increase in the intensity of precipitation, which might translate into more extreme rainfall events. Sustained rapid climate change has the potential to shift the competitive balance among species. It could even lead to forest die-back, altering the terrestrial uptake and release of carbon, with further effects on the global climate.

The impact of climate change is not restricted to terrestrial systems. Over the last 100 years, ocean levels worldwide have risen by approximately four to 10 inches (10 to 25 centimeters), which is attributable chiefly to two factors. The first is thermal expansion of the upper layers of the seas, since water expands as it warms. The second is the melting of the polar ice caps, which adds more water to the oceans. Because the oceans are so vast, sea levels will rise gradually. The oceans presently are rising at 0.4 inch (about one centimeter) per decade, lagging behind the CO_2 increase by about 20 to 25 years. As with average global temperature, projections for the rise in average sea level vary according to the assumptions used in different computer models. They range from a low of 5.85 inches (15 centimeters) to a high of 37 inches (95 centimeters) from the present until 2100. Beyond 2100, sea levels would continue to rise at a similar rate even if greenhouse gases and average global temperature were stabilized.

A sea-level rise of 37 inches would cause widespread coastal erosion and flooding of low-lying areas, greatly affecting cities such as New Orleans, states such as Florida, and countries such as the Netherlands. Vast stretches of coastal wetlands would be inundated and groundwater reserves would become contaminated by salt water encroachment. According to the Organization for Economic Cooperation and Development, the economic damages associated with rising sea levels could reach as high as $970 billion by 2100. As extraordinary as this amount is, it would be dwarfed by the human toll. As many as 200 million people could be made homeless; those at greatest risk are people living in the low-lying and densely populated river deltas of Bangladesh, China, Egypt, and Nigeria.

A warmer climate also affects mountain glaciers; biological diversity; the composition and geographic distribution of ecosystems; the energy, industry, and transportation infrastructure; and human health. Some of the effects on human health are directly related to climate change: higher cardiorespiratory mortality due to an anticipated increase in the intensity and duration of heat waves; higher incidence of death, injury, psychological disorders, and exposure to contaminated water supplies following an increase in extreme weather; and fewer cold-related deaths as temperatures warm in colder regions. Other human health effects are indirect. As the geographic range and season expand for disease-carrying organisms, the potential increases for transmission of infectious diseases such as malaria, yellow fever, and viral encephalitis. For example, one computer model shows that, in response to global temperature increases, the geographic zone of potential malaria transmission will rise from approximately 45 percent of the world population to 60 percent by 2100.

Scientific Evidence Supporting Climate Change

Increasingly, scientific studies are providing evidence of a warming climate. For example, since 1984, Dr. Mauri Pelto and colleagues at Nichols College in Dudley, Massachusetts, have monitored 117 glaciers in the North Cascades of Washington State. Their research shows that the glaciers are in increasingly poor health, becoming both thinner and smaller in area. In fact, seven of the glaciers have disappeared completely and the rest are receding. The retreat of the North Cascades' glaciers is mirrored in mountain ranges from the Arctic to the equator to the Antarctic: in the Pyrenees and Alps of Europe, the Northern Caucasus of Asia, and the Andes of South America (Figure 12-6). Glaciers provide essential ecosystem services. In the spring, they store snowmelt, reducing runoff and helping to prevent or moderate flooding. During summer, when runoff from rainfall and alpine snowmelt is low and evaporation is high, glacial runoff peaks, providing a significant portion — from 20 to 50 percent in many cases — of total streamflow. Worldwide, glacial runoff accounts for 10 percent of surface water, and many municipalities and agricultural areas rely on it. In the North Cascades, glaciers contribute 20 to 25 percent of the region's total summer water supply, nearly all of it used for irrigation, fisheries, and power generation. In the past, the region enjoyed adequate water supplies regardless of climate, but continued rapid development, especially around Puget Sound, has increased demand. At the same time, the summer water supply has declined, resulting in shortages and forcing water resource managers to consider the changing contribution of glaciers.

Dr. Jonathan Patz, of the Johns Hopkins School of Hygiene and Public Health in Baltimore, and his colleagues found that a local outbreak of malaria in Rwanda in 1987 was precipitated by climate change. Reported cases of the disease rose by 501 percent in high altitude areas where malaria previously had been

FIGURE 12-6: Pasterze Glacier, (a) 1875 and (b) 2003. Located in Carten, Austria, the glacier still extended into the valley as recently as the 1920s.

rare and was best explained by the rise in minimum (nighttime) temperature that has occurred over the past three decades. This explanation suggests that climate change can impact the occurrence and distribution of malaria by shifting the geographic range of the mosquito that transmits the disease.

In 1996, Dr. Camille Parmesan, of the University of California at Santa Barbara, reported that the range of Edith's checkerspot butterfly (*Euphydryas editha*) has shifted northwards and upwards altitudinally. Over a four-year period, Parmesan censused 151 populations of Edith's checkerspot throughout its range, from northwestern Mexico to southwestern Canada. She found that the butterfly is declining at the southern portion of its range, with populations in Mexico four times as likely to be extinct as those in Canada. Additionally, Parmesan found that populations above 7,872 feet (2,400 meters) were significantly more persistent than those at all lower elevations. She contends that other factors — such as differences in initial population isolation or subsequent land-use changes — are not likely to be responsible for the range shift of the Edith's checkerspot. Moreover, earlier detailed studies found climate-caused extinctions in this butterfly. These findings provide the clearest evidence to date that global climate change already is influencing species' distributions.

What Is Stratospheric Ozone Depletion?

Ozone has a split personality as far as air quality is concerned. It is a serious threat to human health when it occurs near ground level, but in the upper atmosphere, it acts as a shield, protecting the Earth from the sun's lethal ultraviolet (UV) radiation. When chemicals capable of destroying ozone accumulate in the upper atmosphere, they can reduce its concentration, resulting in the phenomenon known as **stratospheric ozone depletion**.

Ozone occurs naturally in the upper atmosphere up to 37 miles (60 kilometers) above the surface of the Earth, but it is most dense in the stratosphere, between 12 and 25 miles (20 and 40 kilometers). Even at its highest concentrations, only one molecule in 100,000 is ozone. If we could collect all of the stratospheric ozone and spread it evenly over the Earth's surface, it would form a layer just 0.12 inch (three millimeters) thick. Thus, even small changes in the concentration of ozone in the stratosphere can profoundly alter its ability to screen UV radiation.

Although stratospheric ozone is continually produced and destroyed by natural processes, anthropogenic pollutants increase the rate of destruction. The most important of these pollutants is chlorine; others include nitrogen oxides, fluorine, and bromine. Atmospheric chlorine and fluorine are the breakdown products of chlorofluorocarbons (CFCs), and bromine is a breakdown product of halons, chemicals most frequently used in fire extinguishers.

CFCs are versatile chemicals. First developed as coolants, they were used as aerosol propellants during World War II to spray pesticides in the fight against malaria. After the war, CFCs were used in many products: as propellants in aerosol sprays, as coolants in home and automobile air conditioners, as solvents in the electronics industry, and as agents in molding polystyrene plastic foams such as Styrofoam.

The chemical stability that makes CFCs so useful in industry also enables them to destroy stratospheric ozone. It takes about eight years before CFCs reach the stratosphere; once there, they may persist for decades before they break down. CFC-12, for example, which accounts for 45 percent of ozone depletion, has an atmospheric lifetime of 111 years. Eventually, however, the sun's UV energy destroys the CFC molecule, releasing chlorine and fluorine atoms. Chlorine reacts with ozone, converting it to ordinary oxygen. Because chlorine acts only as a catalyst and does not undergo permanent change in the reaction with ozone, a single chlorine atom can catalyze the destruction of as many as 100,000 ozone molecules.

F. Sherwood Rowland and Mario Molina, of the University of California at Irvine, first discovered the destructive potential of CFCs in 1974. Other researchers and the CFC industry initially resisted their findings, but Rowland and Molina publicized their results, and by 1978, CFCs had been banned from use as aerosol propellants in the United States, Canada, Norway, and Sweden. Despite this measure, the use of CFCs actually increased because of their versatility in other products. International regulations have since resulted in the phasing out of most CFC use; however, an active black market in CFCs still exists. Further, one of the classes of chemicals legally being used in place of CFCs is hydrochlorofluorocarbons, or HCFCs, which contain chlorine and thus deplete stratospheric ozone, though to a lesser extent than do CFCs.

Effects of Stratospheric Ozone Depletion

The depletion of stratospheric ozone has serious consequences. This blue-tinted gas is the only atmospheric gas capable of screening out UV rays. The UV radiation that does reach the Earth's surface is responsible for cataracts, sunburn, snow blindness, and aging and wrinkling of the skin. It is the primary cause of skin cancer, which claims some 12,000 lives each year in the United States alone. Exposure to UV radiation also suppresses the immune system, enabling cancers to become established and increasing susceptibility to diseases such as herpes. In addition, UV radiation slows plant growth, delays seed germination, and interferes with photosynthesis. Phytoplankton show a particular sensitivity to UV radiation, which also damages the larval development of crabs, shrimps, and some fish. Because increased UV radiation affects aquatic organisms at their most vulnerable developmental stages *and* reduces their food supply, it could disrupt marine eco-

systems, with devastating effects on world fisheries.

The destructive impact of CFCs and halons is greatest over Antarctica, where researchers first noted a significant thinning of stratospheric ozone in 1985, a depletion that became popularly known as the "ozone hole." The Antarctic ozone hole recurs each year during spring in the Southern Hemisphere, when temperatures and winds maximize ozone depletion. Typically, about 40 to 50 percent of the ozone layer over the Antarctic is depleted. In the upper stratosphere, the decreases are small, but in the lower stratosphere, about 95 percent of the ozone is destroyed. Thus, while spring levels of ozone over the Antarctic should normally be about 300 Dobson units (DU), levels in recent decades average about 150 DU or lower. (A Dobson unit is a measurement related to the physical thickness of the ozone layer if it could be compressed and brought to the Earth's surface. A layer that is 300 DU thick is equivalent to two pennies stacked atop each other; a layer 150 DU thick is equivalent to one dime.) The Antarctic ozone hole reached record size in 2000, extending over approximately 11.5 million square miles (29.8 million square kilometers), an area larger than the entire North American continent (Figure 12-7).

In contrast to Antarctica, the depletion of Earth's protective ozone shield is much smaller over the mid-latitudes. However, it is particularly troublesome because these regions are both heavily populated and agriculturally important. Since the mid-1960s, globally averaged ozone losses have totaled about five percent, with cumulative losses of 10 percent in the winter and spring and five percent in the summer and autumn over Europe, North America, and Australia. As a result, these areas have seen increases in UV radiation of 6 to 14 percent. In another sobering sign, a springtime ozone hole over the Arctic that first appeared in 1995 has recurred in six of the following nine years, with ozone losses of up to 30 percent.

What Is Acid Precipitation?

In the 1960s, researchers began to link an alarming array of environmental impacts: the die-off of large tracts in the Black Forest of Germany and the Adirondack and Appalachian mountain ranges of the United States; the decline or disappearance of fish populations in streams and lakes in Scandinavia, the northeastern United States, and Canada; and the rapid deterioration of centuries-old monuments and buildings throughout Europe. The common link, researchers discovered,

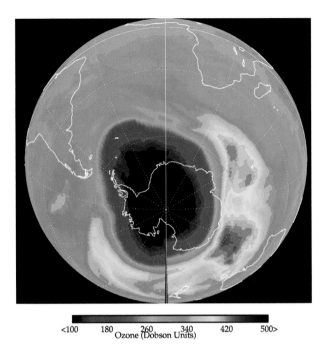

<100   180   260   340   420   500>
Ozone (Dobson Units)

Figure 12-7: The Antarctic ozone hole, 2000. At its peak, the hole was larger than the continent of North America.

was **acid precipitation** — rain, snow, fog, or mist that contains enough sulfuric acid, nitric acid, or their precursors to raise the acidity of the precipitation above normal. Although normal precipitation is slightly acidic, it is not harmful to plant or animal life. In contrast, acid precipitation can damage natural systems and adversely affect human health. It also can corrode marble, limestone, sandstone and bronze, destroying statues, monuments, gravestones, and buildings.

Acidity is measured on the pH scale, which spans from 1 to 14. Substances with a pH from 1 to 6 are called acidic; from 8 to 14, alkaline; and 7, neutral. Sulfuric acid has a pH of 1 (a strong acid); tomato juice, 4 (a moderate acid); cow's milk, 6 (a weak acid); baking soda, 8 (a weak alkali); and lye, 14 (a strong alkali). Because the pH scale is logarithmic, each change in whole numbers represents a tenfold increase or decrease in acidity. For example, a solution with a pH of 4 contains 10 times more acid than one with a pH of 5, and 100 times more acid than a solution with a pH of 6.

Precipitation with a pH as low as 5.6 to 5.9 is considered normal, but precipitation with a pH of 5.5 or lower is labeled acidic and is cause for concern. Readings taken in pristine, remote areas of the world and in core samples of ice from glaciers in Greenland and Antarctica suggest that prior to the Industrial Revolution, the pH of precipitation was rarely lower than 5.0. In the United States in the early 2000s, the most acidic rain had a pH of about 4.3, with much lower values recorded at certain sites. For example, acid fogs of pH 3.0 have been recorded in Maine's Acadia National Park, and in the highest elevations of the Great Smoky Mountains National Park, the pH of clouds that blanket the region's sensitive spruce-fir forest can be as low as 2.0 — as acidic as vinegar.

Formation and Transport of Acid Precipitation

Industrial processes are believed to be responsible for the formation of highly acidic precipitation. Sulfuric and nitric acids are formed in the atmosphere when sulfates (derived from sulfur dioxide) and nitrates (derived from nitrogen oxides) react with water vapor and fall to the Earth as wet deposition (Figure 12-8). Sometimes the sulfates or nitrates fall to the Earth as dry deposition and then combine with water in a stream, lake, or pond to form acids. The major sources of sulfur dioxide (SO_2) and nitrogen oxides (NO_x) are electric utilities (chiefly those that burn coal), motor vehicles, and industrial and manufacturing processes.

Acid precipitation was one of the first environmental problems shown to have a large-scale regional impact; pollutants generated in one area can be transported hundreds of miles by winds before they are flushed out of the atmosphere in precipitation or fall to the ground as dry deposition. If sulfur and nitrogen emissions are deposited far from their source, a much larger amount of acid has a chance to form. For example, after three days, as much as 50 percent of atmospheric SO_2 is converted to acids. The long-distance transport of pollutants is made easier by the construction of tall smokestacks because pollutants that enter the atmosphere at greater heights are likely to travel farther before they are deposited. Beginning in the 1960s, many utilities and industries attempted to alleviate local pollution problems by constructing extremely tall smokestacks, which emit gaseous pollutants 500 feet (150 meters) or more into the atmosphere.

Over 80 percent of the SO_2 and NO_x released over North America each year originates in the United States. Emissions from Ohio, Pennsylvania, Indiana, West Virginia, and Kentucky account for nearly 40 percent of all the SO_2 produced in the United States each year, the bulk of which comes from coal-fired power plants. Ohio emits 10 percent of all SO_2, more

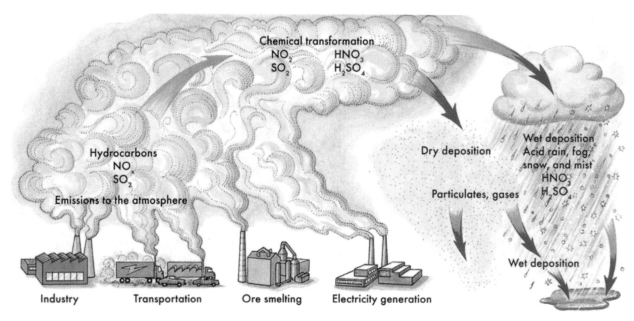

FIGURE 12-8: Formation of acid precipitation.

than any other single state, and more than all of the New England states combined. Prevailing winds out of the Mississippi and Ohio valleys pick up SO_2 and NO_x in Missouri, Ohio, Kentucky, Indiana, Illinois, West Virginia, and Pennsylvania and carry them north and northeast, where they are deposited over New York, New Jersey, New England, and Canada. Tracer experiments have shown that just half of the SO_2 in the Northeast is produced locally, with the rest produced in the Midwest. The same is true of SO_2 in Canada; about half originates in Canada and half in the United States. Because the SO_2 not produced locally has been aloft for several days, it tends to produce about 75 percent of the actual acid precipitation that falls in the Northeast and Canada.

Controls imposed since 1990 have greatly reduced emissions of SO_2 in many industrial areas; in Europe, for example, emissions fell approximately 45 percent by 2000. Total U.S. SO_2 emissions dropped 39 percent between 1980 and 2002, from 17.3 million tons (15.6 million metric tons) to 10.2 million tons (9.2 million metric tons). In 2003, they rose slightly, to 10.6 million tons (9.6 million metric tons), partly due to the increased production of electricity by coal-fired plants.

Over half (56 percent) of the NO_x released into the atmosphere each year comes from transportation. Industrial procedures that involve high-temperature combustion and electric utilities account for most of the remaining NO_x emissions. In the western states, NO_x emitted in automobile exhaust and metal-smelting operations are the major precursor of acid precipitation. The EPA has identified acid-sensitive areas in every western state but Arizona; affected areas include 11 national parks and millions of acres of wilderness and forests. Acid fogs of pH 1.7, 10,000 times more acidic than normal precipitation, have been measured over Los Angeles. The pH of this fog is comparable to the pH of commercial toilet bowl cleaner.

Despite the progress to date, acid precipitation remains a serious problem worldwide. Annual rainfall pH values in polluted areas of Scandinavia, central Europe, and Japan can be as low as 3.5. Acid precipitation affects much of European Russia and has taken a significant toll on historic buildings, forests, lakes, and streams in the former German Democratic Republic (East Germany), Poland, the Czech Republic, Slovakia, and Hungary. In Africa and South America, the burning of grasslands and forests to prepare the land for agriculture and the lack of controls on industrial emissions contribute to acid precipitation. The problem is particularly acute in China, which relies heavily on high-sulfur coal. After the Chinese government instituted market reforms in 1978, coal use skyrocketed as the nation sought to fuel economic growth as cheaply as possible. About 40 percent of China's agricultural land is affected by acid precipitation, which has been linked to large-scale die-offs in the nation's southwestern forests. But the problems are not China's alone. About one-third of Japan's sulfur deposition originates in China and is carried by winds across the Sea of Japan.

Effects of Acid Precipitation

Emissions of SO_2 have fallen in many industrial countries in recent years, but forests, lakes, and other natural areas have not rebounded as scientists expected, suggesting that the damage caused by acid precipitation may be more extensive and long-lasting than once believed. In fact, scientists still do not completely understand how acid precipitation and other air pollutants act and interact to damage natural systems and living organisms. Nevertheless, their understanding is growing, and in the paragraphs that follow, we highlight some of what is known about the biological effects of acid precipitation.

To simplify this complex subject, we'll work from the ground up — literally. Soil minerals are a major factor in determining how severely acid precipitation affects a region. Most of the midwestern United States and much of southeastern England and northeastern China, for example, are able to resist acidic precipitation because they have alkaline soils; the calcium in the soil buffers, or neutralizes, the acids, rendering them harmless. Lakes with limestone or sandstone beds have a similar buffering capacity. In contrast, thin glacial soil and thick slabs of granite are unable to buffer acidic fallout. These sensitive areas are the most severely affected. For example, southern Norway, where the government estimates a loss of more than half the fish population in 30 years, and Sweden, where 18,000 lakes are acidified, are downwind from the major industrial and urban centers of Britain and Germany. The granite bedrock, soil, and water of these affected areas cannot neutralize acid precipitation (Figure 12-9). Similarly, a 2002 report by the U.S. National Park Service found that high elevation ecosystems in the Rocky Mountains, Cascades, Sierra Nevadas, southern California, and upland eastern states are generally the most sensitive to atmospheric deposition due to their poor ability to neutralize acid. Thousands of lakes in the Northeast are too acidic to support life, and acid precipitation has caused extensive damage in the forests of the Adirondack Mountains, which lie atop thin granite soils. (To read *Environmental Science in Action: Acid Precipitation in the Adirondack Mountains*, go to www.EnvironmentalEducationOhio. org and click on "Biosphere Project.")

While some soils and waters have a natural buffering capacity, they may become degraded by a prolonged assault by acid deposition. Acid precipitation leaches as much as 50 percent of available calcium and magnesium from forest soils, altering their chemistry. As soil pH falls, there are lasting effects on the biotic community. The microbial community, for example, shifts from bacteria to fungi, some of which are more tolerant of acid conditions. One exception is the mycorrhizae, a group of fungi whose numbers decline as the soil becomes more acidic. Potentially, this decline is very serious because mycorrhizae help transfer soil nutrients into plant roots. Studies also show greatly reduced numbers of some soil organisms (especially short-lived species, such as earthworms, which are closely linked to the surrounding soil chemistry) in areas receiving high doses of acid precipitation. Researchers who compared total biomass and respiration of soil organisms along an east-west gradient from Ohio to Illinois found that the highest-dose site (in Ohio) had only 30 percent of the animal life found in a reference site (in Illinois). Half of the animal species at the highest-dose site had become locally extinct. Although some studies show that as acidity increases the total number of *organisms* may remain essentially the same (because a few acid-tolerant species become more abundant), every study done to date shows that the total number of *species* declines.

Like soils, forests around the world are showing the effects of decades of exposure to acid deposition. Conifers, which tend to grow at higher altitudes than hardwoods, are particularly affected because precipitation is most likely to form when air is lifted and cooled as it travels over mountain ranges (Figure 12-10). Many mountain peaks are shrouded in acid mists, covered by acid snows, and pelted by acid rains throughout the year. Some scientists believe that this phenomenon

FIGURE 12-9: Liming an acidified lake, Sweden.

FIGURE 12-10: Forest damaged by acid precipitation, Poland.

contributes to mountain forest decline and fish mortality in mountain lakes.

Like forests, aquatic ecosystems can be damaged or destroyed by acid precipitation. An acidified lake or stream undergoes changes in vegetation, food chains, and fish populations. Small invertebrates are often the first affected; when they begin to disappear, the organisms that feed upon them — including salamanders, fish, and frogs — suffer. Populations dwindle as the food supply decreases. Moreover, various species of fish stop breeding at certain levels of acidity. Sensitive fish species such as salmon, trout, minnows, and arctic char are directly affected by heavy metals leached out of soils and washed into streams and lakes. Amphibians, which spend part of their time in water and part on land, may be especially sensitive to acidic waters and heavy metals because they respire through their skins. In addition, amphibians lay their eggs directly in streams, ponds and lakes, where they are exposed to influxes of acid deposition.

Acid surges are periods of short, intense acid deposition in lakes and streams. A spring snowmelt or a rainfall after a prolonged drought can transfer acids previously locked in snow or soil into nearby waters, resulting in a sudden surge of acidity. This temporary rise can have a devastating effect on animal species that reproduce in the spring. Often, the acid surge is enough to deform or kill young fish and newly hatched salamanders and frogs. Acid surges also can kill adult fish and decimate populations of aquatic insects, crayfish, and snails that serve as food for fish.

What Is Smog?

Smog is a generic term that literally means "smoky fog." It was coined to describe the sooty emissions from nineteenth-century factories. Today, we recognize two main types of smog: industrial and photochemical.

Industrial smog, or smoke pollution, consists chiefly of sulfur oxides and particulates (primarily smoke, soot, and ashes) and is emitted to the air from industrial and manufacturing facilities. Typically grayish in color, industrial smog is a problem in urban areas, where it reduces visibility and contributes to unhealthy air. In the United States, it is particularly troublesome in the Northeast and around the Great Lakes, where industries are heavily concentrated.

Photochemical smog is produced when volatile organic compounds (VOCs) react with nitrogen oxides and oxygen in the presence of sunlight to form chemicals such as ozone and peroxyacetyl nitrate. Photochemical smog is sometimes called "brown smog" because of its characteristic color, given to it by nitrogen dioxide, a brown gas. The incomplete combustion of fossil fuels, particularly from automobiles, is the primary cultural source of VOCs, or hydrocarbons, so it is not surprising that photochemical smog cloaks many urban centers, including Los Angeles and Houston; Ankara, Turkey; New Delhi, India; Melbourne, Australia; Mexico City, Mexico; and Sao Paulo, Brazil.

The primary component of photochemical smog is ground-level ozone, or ozone that occurs anywhere from a few feet to a few miles above the Earth's surface. Ozone levels are generally higher in the afternoon, after radiation from the sun begins to take effect. Smog build-up is also greater when air is trapped close to the ground for long periods of time and in valleys or basins, areas where large numbers of people tend to live and work. The 12 million inhabitants of greater Los Angeles, for example, live in a valley nearly surrounded by mountain ranges. The city has over eight million cars; 10 million buses, trucks, and trains; and thousands of stationary sources of pollution, including numerous oil refineries. Its climate is hot, with little air-cleansing rainfall each year. Consequently, Los Angeles has the highest ozone levels of any city in the United States, and in 2004 it was the only region of the country to be listed in the EPA's "severe" pollution category (Figure 12-11). While ground-level ozone is a year-round problem in warmer climates, it also occurs seasonally in temperate zones, reaching its highest levels in the Northern Hemisphere between April and September.

Effects of Smog

Industrial smog is responsible for some of history's most deadly air pollution episodes. In 1911, 1,150 people died when coal smoke blanketed the city of London, England; four decades later, in 1952, 4,000 Londoners died when unusual weather conditions again held industrial smog in place over the city. In 1930, pollutants from coke ovens, steel mills, sulfuric acid plants, blast furnaces, zinc smelters, and glass factories accumulated in Belgium's industrial Meuse Valley, killing 63 and sickening 600. In 1948, in Donora, Pennsylvania, smog from a local steel mill, zinc smelter, and sulfuric acid

plant killed 20 people; another 6,000 residents — 43 percent of the town's population — fell ill. And in December 1962, adverse weather conditions in normally well-ventilated New York City gave rise to a lingering smog that was blamed for nearly 300 deaths.

Given the principle components of industrial smog, it is not surprising that its health effects and environmental impacts are similar to those of SO_2 and particulates. Smog aggravates existing conditions, especially respiratory and cardiovascular diseases; worsens respiratory illnesses like pneumonia and bronchitis; causes difficulty breathing; and irritates the nose, nasopharynx, and bronchi. Thus, most deaths attributed to acute, episodic industrial smog are the result of increased susceptibility to disease or illness rather than to air pollutant poisoning. Groups at greatest risk include asthmatics, people with chronic respiratory or pulmonary and heart disease, and the elderly. Children are another high-risk group; Box 12-1: *Air Pollution and Human Health* describes the factors that make children particularly susceptible to air pollution.

Like industrial smog, photochemical smog adversely affects both human health and the environment. A principal component, ozone, is toxic; less than one part per million in air is poisonous to humans. It irritates eyes and skin, dries out the mucous membranes of the nose and throat, and interferes with the body's ability to fight infections. Even when inhaled at very low levels, ozone can inflame lung tissues, leading to acute respiratory problems and increased hospital admissions

FIGURE 12-11: The photochemical cloak in Los Angeles. (a) Los Angeles has the worst air quality in the United States. (b) On a clear day, however, it is easy to imagine that the area was once beautiful.

and emergency room visits. Ten to 20 percent of all summertime hospital visits for respiratory problems in the northeastern United States are linked to ground-level ozone pollution.

Ground-level ozone has significant effects on vegetation and ecosystems, as well. It reduces photosynthesis, plant growth, and the survival of tree seedlings; lowers crop yields and the productivity of commercial forests; and increases the susceptibility of plants to environmental stresses, among them diseases, pests, and harsh weather. According to the EPA, crop yield losses in the United States attributed to ozone damage amount to several billion dollars annually. Moreover, research suggests that ground-level ozone has contributed to forest decline in the eastern half of the United States, especially in the Appalachian Mountains. In the Great Smoky Mountains and Shenandoah national parks, ozone has affected 95 plant species, including sassafras, yellow poplar, sugar maple, milkweed, and black cherry. Species that rely on these plants, such as the monarch butterfly, whose larvae feed exclusively on milkweed leaves, are likely to be affected also.

What Are Air Toxics?

Hazardous air pollutants, commonly called air toxics, are a varied group of chemical substances known or suspected of causing cancer, other serious human health effects, or ecosystem damage. Examples include heavy metals like mercury and lead and organic chemicals like benzene perchloroethylene. By volume, hazardous air pollutants are the largest source of human exposure to toxic substances.

The range of sources that emit hazardous air pollutants is mind-boggling; it includes large stationary industrial facilities (such as mining operations), electric power plants, smaller area sources (such as neighborhood dry cleaners and metal-plating operations), and mobile sources (such as automobiles). Obviously then, air toxics are a problem nationwide. Atmospheric deposition may account for 20 percent of toxic water pollution in the United States. Almost one-third of the heavy metals entering the Chesapeake Bay come from the air. Air toxics also are one of the largest sources of contaminants for all of the Great Lakes. For example, toxaphene is an insecticide used primarily on cotton and secondarily on soybeans, peanuts, and coffee. Although 86 percent of the croplands treated with toxaphene are in the South, the chemical has been detected in Great Lakes fish since the 1970s. Coal-fired power plants are the single largest source of atmospheric mercury, which contaminates inland lakes and rivers. Once present in waterways, mercury is converted to methylmercury by bacteria in soils and sediments. Tiny aquatic plants and animals take up the methylmercury; these organisms then are eaten by small fish; larger fish consume the small fish, and so on. Through bioaccumulation and biomagnification, methylmercury can reach levels harmful to humans, especially young children. Because of widespread contamination, 45 states have issued fish advisories for mercury for children and women who are pregnant (or may become so) or nursing.

Just as air toxics can pollute water systems, contaminated water supplies can give rise to toxic air pollution. Each day, sewer systems receive thousands of gallons of industrial and household cleaners. These and other toxins can vaporize at sewage treatment plants and enter the air.

Effects of Air Toxics

About 80,000 chemicals are produced in the United States, most of which have not been tested adequately for their effects on human health; even less is known about their environmental impact. Nonetheless, evidence links many of these chemicals to serious health effects, including disruption of the endocrine and reproductive systems, cancer, and neurodevelopmental disorders. For example, mercury and other heavy metals may cause autoimmunity, in which a person's immune system attacks its own cells. Autoimmunity can lead to joint diseases, such as rheumatoid arthritis, and diseases of the kidneys and circulatory and nervous systems. Fetuses, infants, and young children are particularly vulnerable to the toxic effects of metals since their bodies are developing so rapidly. Mercury and lead easily cross the placenta, damaging the fetal brain. The U.S. government estimates that as many as 300,000 newborns each year may be at risk of learning disabilities associated with in utero exposure to methylmercury. And the risk does not end at birth; an infant or child exposed to mercury or lead may suffer learning disabilities, impaired memory, nervous system damage, and behavioral problems. At high doses, mercury and other heavy metals can cause irreversible brain damage or death.

Chapter 20, *Unrealized Resources: Solid and Hazardous Wastes,* explores in greater detail the health effects and environmental impacts of toxic substances.

BOX 12–1 **Air Pollution and Human Health**

Each time we breathe, we take about a quarter of a gallon (one liter) of air into our lungs. That translates into some 7,800 gallons (30,000 liters) of air that pass through the lungs daily. When the lungs are functioning properly, they take in oxygen and expel carbon dioxide. Most of the time, the body's defense system can cope with pollutants that enter the lungs. Mucus traps foreign particles, protecting delicate lung tissue. Cilia, minute, hairlike structures that line air passageways, help to move the mucus away from lung tissue. Coughing expels the mucus and impurities. Microorganisms and chemicals that escape or overwhelm initial defense mechanisms may cause infections or cell damage. In these cases, the body's immune system helps to fight the invaders and restore health. Pollutants too powerful for the body's defenses can affect health in a number of ways, including choking, labored breathing, burning and watery eyes, and, in the most severe cases, death.

Although the link between air pollution and mortality is difficult to establish definitively, many experts believe that air pollution is responsible for as many as 50,000 premature deaths in the United States each year. Most of these deaths are believed to be caused by pollutants released when fossil fuels are burned. The effects on human health depend on the type of pollutant, its concentration, and length of exposure. An EPA study suggests that up to 1,700 cancer deaths each year are attributable to toxic air pollution. This study covered only a small fraction of the airborne toxins released to the air; it did not consider the synergistic effects of combined pollutants, accidental releases of chemicals, or secondary pollutants. Over long periods of time, even mildly polluted air can have damaging effects: lung cancer, chronic bronchitis, emphysema, asthma, and lead or other heavy metal poisonings of the blood, nervous system, and kidneys (Figure 12-12).

Air pollution takes the greatest toll on the very old, very young, and persons already suffering from respiratory or circulatory disease. Researchers at the University of California at Irvine have shown that the risk of health problems due to air pollution is six times greater for children than for adults. Because they have smaller air passages and must breathe more air per unit of body weight to maintain their metabolism, children tend to collect more pollutants than do adults. Further, children often breathe through their mouths and so do not benefit from the filtering effect of the nasal passages.

Air pollution also has a greater effect in urban and suburban areas than in rural areas and on cigarette smokers than on nonsmokers. A smoker living in a large city is at greatest risk, especially if that person works in a dusty environment or has impaired respiratory functions. Indoor air in improperly vented buildings, as well as air polluted by improperly operated furnances or wood-burning stoves, also can adversely affect health. Good indications of dangerous pollutant levels are often breathing difficulties; increased susceptibility to respiratory infection; headaches; irritation of the nose, throat, eyes and sinuses; nausea; drowsiness; colds; coughs; and impaired judgment.

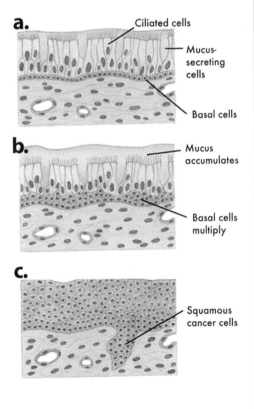

FIGURE 12-12: A brief history of lung cancer. (a) In a healthy lung, the lining of the respiratory passages includes both ciliated and mucus-secreting cells. In a smoker's lung, the cilia become partially paralyzed and mucus accumulates on the irritated lining. (b) In an early cancerous state, the underlying basal cells divide rapidly and displace the columnar ciliated and mucus-secreting cells. (c) In later stages, most of the normal, columnar cells are replaced by the cancerous cells. If the disease reaches an advanced stage, clusters of cancerous cells may be carried away by the lymphatic system and spread to other parts of the body.

a.
Ciliated cells
Mucus-secreting cells
Basal cells

b.
Mucus accumulates
Basal cells multiply

c.
Squamous cancer cells

What Is Indoor Air Pollution?

Indoor air pollution, airborne contaminants present inside homes, offices and other buildings, became a topic of serious concern in the 1980s. Like outdoor air pollution, indoor air pollution poses a health threat. In fact, the EPA considers indoor air pollution to be in the top five public health threats in the United States, and a chief factor in the rising number of asthma sufferers. According to the Asthma and Allergy Foundation of America, both the prevalence of asthma and asthma death rates have increased since 1980 across all age, gender, and ethnic groups. For children under the age of 19, the asthma death rate has risen nearly 80 percent.

The chief indoor contaminants are formaldehyde, radon 222, tobacco smoke, asbestos, combustion products from gas stoves and poorly vented furnaces (including carbon monoxide, nitrogen oxides, sulfur dioxide, and particulates), chemicals used in building and consumer products (such as pesticides and household chemicals), and disease-causing organisms (such as bacteria and viruses) or spores (Figure 12-13). Inadequate ventilation can lead to carbon monoxide and nitrogen dioxide levels that exceed the standards set for outdoor levels of these gases. In the homes of cigarette smokers, levels of carbon monoxide, particulates, and benzene may be 50 to 100 times higher than the standards set for outdoor air. Mobile homes are particularly susceptible to indoor air pollution because they often have lower air exchange rates than conventional homes, are more likely to use propane for heating and cooking, and feature more plywood and other synthetic construction materials, many of which are made with formaldehyde.

In 1988, the EPA published the findings of its first in-depth study of indoor air pollution in the workplace and public buildings. Among the chemicals found were benzene, a known carcinogenic substance emitted by synthetic fibers, plastics, and some cleaning products; trichlorethylene, a carcinogenic organic solvent used

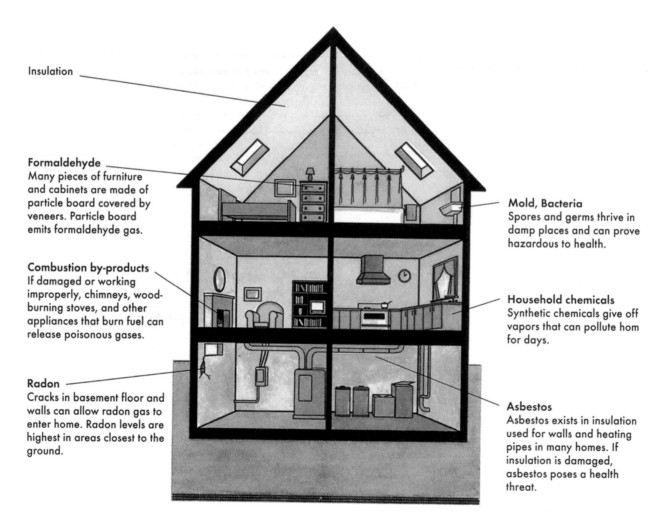

Insulation

Formaldehyde
Many pieces of furniture and cabinets are made of particle board covered by veneers. Particle board emits formaldehyde gas.

Combustion by-products
If damaged or working improperly, chimneys, wood-burning stoves, and other appliances that burn fuel can release poisonous gases.

Radon
Cracks in basement floor and walls can allow radon gas to enter home. Radon levels are highest in areas closest to the ground.

Mold, Bacteria
Spores and germs thrive in damp places and can prove hazardous to health.

Household chemicals
Synthetic chemicals give off vapors that can pollute hom for days.

Asbestos
Asbestos exists in insulation used for walls and heating pipes in many homes. If insulation is damaged, asbestos poses a health threat.

Figure 12-13: Common sources of indoor air pollution.

for cleaning and degreasing; carbon tetrachloride, a carcinogenic substance found in cleaning solutions; and paradichlorobenzene, an animal carcinogen found in mothballs and air fresheners. Formaldehyde and chloroform also were found. Chloroform, known to cause cancer in laboratory animals, is vaporized in hot water sources such as showers. It is formed when chlorine, which is used to kill bacteria in water, interacts chemically with organic substances in the water.

Generally, federal and state health and safety standards regulate pollutants in workplaces, schools and public buildings, but there are no laws that regulate pollutants in homes. Ironically, buildings constructed to be energy-efficient often restrict air flow or continually recirculate the same air, conditions that allow pollutants to build up to dangerous levels. Most U.S. citizens spend 90 percent of their time indoors, thus dramatically increasing their exposure to indoor pollutants. *What You Can Do: To Protect Yourself Against Air Pollution* recommends ways that individuals can reduce their exposure to air pollutants and help safeguard air resources.

Effects of Indoor Air Pollution

Depending on the substance, indoor air pollutants affect human health in myriad ways. In this section, we focus on the health effects of four of the most common pollutants: formaldehyde, radon 222, tobacco smoke, and asbestos.

Formaldehyde. Commonly recognized as a preservative for biological specimens, **formaldehyde** also is used in numerous other products, largely because it prevents bacteria and mold growth: foam insulation, permanent-press clothes, carpets, toothpaste, shampoos, some paper products, some medicines, and pressed wood products. Formaldehyde emits fumes that can irritate the eyes, nose, and throat and cause shortness of breath, nausea, and headaches. This steady release of noxious fumes, called outgassing, is greatest when products are new, but materials can continue to release fumes for years. Because formaldehyde is so widespread, we may be exposed to it more frequently than to any other potentially dangerous chemical. A known carcinogen in rats, it also causes mutations in bacteria and cell changes in primates, and some evidence links formaldehyde to skin cancer in humans. The EPA has not regulated the use of formaldehyde in consumer products, despite recommendations to do so from health scientists.

Radon 222. A colorless, odorless, and inert radioactive gas, **radon 222** is formed by the disintegration of radium, itself a by-product of the decay of uranium 238. Uranium 238 is an element found naturally in the Earth's crust. Radon also occurs naturally and can be found in high concentrations in soils and rock containing uranium, granite, slate, phosphate, and pitchblende. It also can be found in soils contaminated from wastes of uranium or phosphate mining. Some of it filters to the surface through cracks in rocks and pores in the soil and passes into the atmosphere. Radon 222 also can seep through cracks in basement walls or around improperly sealed pipes.

Radon 222 is stable only for a few days, after which it undergoes a series of four changes to form chemically active solids. These solids can attach to dust particles or clothing. When inhaled, they can penetrate lung tissues, emitting dangerous radiation that may cause lung cancer. Next to smoking, radon exposure is the most common cause of lung cancer. EPA studies indicate that exposure to radon may contribute to or cause as many as 10,000 to 30,000 cases of lung cancer each year and that radon might be a threat in one of every 15 homes. In homes with increased radon levels, those at greatest risk are smokers, children, and people who spend a lot of time indoors.

Geographic locations having greater amounts of uranium typically emit increased amounts of radon 222. In Colorado, radon exposure has been traced to working in uranium or other hard rock mines and to proximity to uranium mill tailings, the waste rock left after the ore is processed. Uranium tailings have been used as fill under the concrete slabs on which homes are built, as backfill around foundations, as heat sinks in fireplaces, and as an accidental component of concrete or mortar. In east-central Pennsylvania, abnormally high levels of radon gas were first noted in 1984, when the radiation detection badge of an atomic energy employee registered high radiation levels from his home and not from his workplace. This finding led to the realization that radon was a threat in homes and not just near uranium mines.

In 1988, the EPA and the Surgeon General's Office issued a statement calling for radon testing in virtually all homes in the United States. Radon test kits are commercially available and easy to use. Corrective measures, such as cementing cracks and repairing seals around pipes and ventilators, are usually inexpensive to moderately priced.

Tobacco Smoke. Smoking poses many health risks. Cigarette smoke contains such toxins as carbon monoxide, nitrogen oxides, hydrogen cyanide and cadmium, as well as carcinogenic substances such as formaldehyde, vinyl chloride, nitrosamines, nickel carbonyl, and benzopyrene. Compared to nonsmokers, smokers suffer more respiratory illnesses and have been shown to have more complications, such as pneumonia, after surgery. They also suffer death rates that are twice as high as those of nonsmokers at any age.

A plethora of health effects are associated with smoking, among them cough, loss of appetite, shortness of breath, abdominal pains, headaches, weight loss, bronchitis, and insomnia. The two greatest dangers, however, are lung disease and heart disease. An estimated 85 percent of all lung cancer cases are attributable to smoking. Smokers are about twice as likely as nonsmokers to suffer an immediately fatal heart attack. Researchers speculate that carbon monoxide, which binds to hemoglobin in the blood, reduces the amount of oxygen to the heart muscle, forcing it to pump harder. Smoking is a major cause of emphysema, a disease in which the alveoli, tiny sacs found at the end of each bronchiole in the lungs, do not expel air as they should. Emphysema sufferers have difficulty expelling used air from their lungs in order to take a fresh breath.

Passive, or second-hand, smoke is also dangerous. Nonsmokers, including children, who are exposed to cigarette smoke are at higher risk for developing respiratory illnesses and lung cancer. A 1990 University of California at San Francisco report that summarized research on second-hand smoke for the EPA found that the health effects on nonsmokers include as many as 32,000 heart-disease deaths annually.

In recognition of the health risks of smoking, many governments have taken the initiative and put smoking bans into place. For example, in 2004, Ireland, Holland, and Norway banned smoking in all indoor workplaces — including bars and restaurants. The European Union is considering a similar, continent-wide smoking ban. In the United States, a handful of states, including Massachusetts, New York, Connecticut, Maine, Delaware and California, have enacted indoor smoking bans.

Asbestos. Found in insulation, vinyl floor tiles, cement and other building products, **asbestos** is a substance that causes lung cancer and asbestosis, an ailment in which breathing is made difficult because of the presence of asbestos fibers in the lungs. In general, asbestos poses a problem only in older buildings; newer building materials are typically made without asbestos. Because asbestos becomes a health threat only if the fibers are released to the air and inhaled, it is thought to be relatively safe as long as the materials containing asbestos are not disturbed.

What Factors Affect Air Pollution Levels?

Air pollutants are dissipated, concentrated, and transported as a result of the interactions of weather and topography. These factors may act singly or in combination to affect the air quality in a particular locale.

Weather

With respect to air pollution, bad weather is good and good weather is bad. Cool, windy, stormy weather with turbulent low-pressure cells and cold fronts favors the upward mixing and dissipation of pollution. In contrast, hot, sunny, calm weather with stagnating high-pressure cells and warm fronts usually favors the build-up of pollutants close to the ground, contributing to worsening air quality. The unusually warm summer of 1998 meant high levels of photochemical smog and generally unhealthy air for many U.S. cities. The prolonged heat wave and resulting poor air quality was caused by stagnating high-pressure systems over the Pacific, the Gulf of Mexico, and the Rocky Mountains.

Weather affects pollution levels in several ways. Precipitation helps cleanse the air of pollutants, although those pollutants then are transferred to the soil, lakes, or streams. Winds transport pollutants from one place to another, usually moving west to east, and sometimes traveling as far as 500 miles in a 24-hour period. Winds and storms may dilute pollutants with cleaner air, making pollution levels less troublesome in the area of their release. Air heated by the sun or bound in a low-pressure cell rises, carrying pollution with it. When wind accompanies the rising air mass, the pollution is carried aloft and diluted with clean, fresh air. Sometimes air sinks toward the ground, as in a high-pressure cell. In the absence of winds, pollutants are trapped and concentrated near the ground, often leading to serious air pollution episodes.

Weather also can affect pollution levels through chemical reactions. Winds and turbulence mix pollutants together in a sort of a giant chemical soup in the sky. Energy from the sun, moisture in clouds, and the proximity of highly reactive chemicals may cause the formation of secondary pollutants, many of which may be more dangerous than the original substances.

Topography

Topography, the shape of land formations, helps to direct pollution movements. For example, valleys can act as sinks for pollutants, and mountains can act as barriers to air flow. Cities like Pittsburgh, Pennsylvania, and Cincinnati, Ohio, suffer poor air quality because air masses often stagnate in the deep river valleys that these cities occupy. Large urban areas like Los Angeles and Mexico City, which are located in valleys ringed by mountains, experience even more serious pollution episodes. On average, Los Angeles has the worst air pollution in the United States, and Mexico City has the worst in the world — with high levels of sulfur dioxide, particulates, carbon monoxide, ground-level ozone, lead, and nitrogen oxide. Why? During the night and early morning, downslope winds push cooler, denser air from the mountains into the valley. Throughout the day, the sun warms the air in the valley, and the lighter, heated air rises, accompanied by upslope winds. But if the upslope winds are not strong enough to carry the air mass over the mountains during the day, the pollution that the air mass holds does not leave the valley. Instead, it keeps moving up and down with the rising and falling air masses, and each day the total pollutant load increases.

Mountains affect the movement of air masses and the deposition of pollution in other ways as well. When an air mass encounters a mountain, it begins to rise. As it rises, the air cools and precipitation forms. Often, the air mass will drop its moisture — and whatever pollutant load it carries — on the windward side of the mountain. Once it clears the mountain, the air mass has lost its moisture, and the lee side of the mountain receives little or no precipitation, a phenomenon known as the **rain shadow effect**. The rain shadow effect explains why one side of a mountain may experience severe damage from air pollution while the other side of the same mountain remains essentially unharmed.

Temperature Inversions

Usually, air temperature decreases with distance from the ground. Heated gases emitted from the surface and air heated by normal radiation rise until their temperature cools to the same temperature as the surrounding air. Occasionally, however, because of specific weather or topographic features, cooler air becomes trapped temporarily beneath a layer of warmer air (Figure 12-14). Any hot gases emitted at ground level pass through the cold layer next to the ground and into the warmer

Figure 12-14: Temperature inversion. Normally, hot gases released from stationary and mobile sources rise and disperse. Unusual weather patterns, however, can create a temperature inversion, in which a layer of lighter, warm air acts as a cap, holding cooler, denser air closer to the ground. Hot gases begin to rise through the cooler air but as they do, they cool and do not rise further. Consequently, the total load of pollutants near ground level increases.

inversion layer above. When their temperature equals the temperature of the air around them, the gases stop rising. In this way, the inversion layer acts as a cap to prevent the further upward dispersal of the pollutants. Typically, inversions last less than a day, and the pollution quickly disperses. Under special conditions, however, pollution can remain trapped near the ground for up to several weeks. A temperature inversion was responsible for 4,000 deaths attributed to smog in London in 1952.

Temperature inversions may be caused by radiational cooling, downslope movements in valleys, and high-pressure cells. Radiational cooling occurs after the sun sets. The surface of the ground reradiates heat faster than the air above it. Therefore, the ground tends to cool faster than overlying air, and lower air masses tend to cool faster than the air above them. This denser, colder air usually remains close to ground level. In the morning, before the sun has had a chance to warm the Earth's surface, the cool air mass remains under the warmer air mass above it. This type of radiational cooling is common in temperate zones during the cool, clear nights of early spring and fall.

Cold air descends from higher valley walls or mountain slopes during the night and tends to rise again during the day. Inversion layers caused by downslope air movements tend to be deeper and last longer than radiational layers. Consequently, larger amounts of pollution can accumulate in them and be trapped for longer periods of time. In the fall and early spring, a combination of radiational cooling and downslope air movements can intensify air pollution.

High-pressure cells are generally characterized by clear, calm weather. Air masses in high-pressure cells sink toward the ground, and air pressure at ground level increases. As air becomes compressed, its temperature rises, often trapping a layer of cold air between the compressing air and the ground surface. This gives rise to an inversion layer that is generally higher in the air column than other inversion layers. When high-pressure cells stagnate over an area for several days to a week, pollution can build up to dangerous levels (Figure 12-15). This type of weather pattern can occur in the fall, compounding the effects of radiational cooling and downslope air movements. More often, it occurs during summer when stationary weather systems trap pollutants close to the ground for extended periods of time. Because the effects of sunlight are greater during summer and weeks may pass without cleansing rain or winds, photochemical smogs can become even more dangerous.

Managing Air Resources

The Clean Air Act guides U.S. air pollution control efforts. In this section, we examine the Act's provisions and its impact on the nation's air quality. We also briefly summarize the Clear Skies Initiative, a controversial proposal by the Bush administration to amend the Clean Air Act. We close the chapter with a review of international agreements to address climate change and stratospheric ozone depletion.

What Is the Clean Air Act?

First enacted in 1963, the Clean Air Act called on states to develop air quality standards and to implement plans to curb pollutants. States and municipalities were slow to respond, however, and in 1970, 1977, and 1990, growing public concern led Congress to pass tough amendments to the Act. Today, the Clean Air Act protects air quality through measures and programs that address six criteria pollutants, air toxics, and acid precipitation.

Criteria Pollutants

Just three decades ago, America's skies were visibly troubled. Six primary pollutants, identified as criteria pollutants by the EPA, were particularly pervasive: lead, particulates, sulfur dioxide (SO_2), carbon monoxide (CO), nitrogen dioxide (NO_2), and volatile organic compounds (VOCs). These pollutants either directly affect human health, as in the case of lead and particulates, or give rise to secondary pollutants that do so. Nitrogen dioxide and VOCs, for example, are precursors of ground-level ozone. To regulate criteria pollutants, the Clean Air Act directed the EPA to establish standards for both emissions and air quality. Legally, the EPA is required to review these standards every five years and revise them if necessary to protect human health and welfare.

Emissions Standards. In 1970, the EPA set **emissions standards** for the six criteria pollutants — limits on the amount of each pollutant that could be emitted by specific point sources. New plants or factories could not be approved for use unless they were shown to

Figure 12-15: Pittsburgh, Pennsylvania, fall 1945. Before the introduction of air pollution controls, stagnating air masses translated into poor air quality for the residents of industrial towns such as Pittsburgh, home to many steel manufacturers. This photo was taken at 9 a.m., looking east on Liberty Avenue.

remove 55 percent of SO_2 emissions, 20 percent of NO_x emissions, and 70 percent of particulates. The idea was that as new plants replaced older plants, air quality would improve. (Unfortunately, because many older plants — which were exempt from emissions standards — remain in operation, they continue to pollute the atmosphere.) In addition, the EPA began a phase-out of leaded gasoline. This was a significant step because in 1970, vehicle exhaust was by far the major contributor of lead emissions to the atmosphere.

To meet the requisite standards, power plants, manufacturing operations, and the automotive industry developed technologies to remove pollutants from emissions or to convert pollutants in emissions to harmless substances. For stationary sources, two of the most significant developments were electrostatic precipitators and scrubbers, both of which now are used commonly in coal-fired power plants. Electrostatic precipitators remove particulates; scrubbers remove SO_2. Both of these technologies collect toxic liquids or solids that then must be disposed of properly.

Another way to reduce SO_2 pollution is to use "clean" coal technologies. Sulfur can be "washed" or removed from coal before it is used. New techniques for burning coal can make the combustion more efficient so that less coal is used, thus reducing the amount of sulfur produced. Clean coal technologies also allow older electric utility plants to retrofit their facilities, resulting in as much as twice the output of electricity with lower emissions. Power plants become less efficient after 30 years and must be replaced at great cost or retrofitted to extend their productive years.

Technological advances such as the catalytic converter help to reduce pollution from mobile sources.

Catalytic converters are attached to the exhaust systems of cars; as exhaust gases pass through the converter, CO and hydrocarbons are converted to water and CO_2. Catalytic converters made possible the significant cuts in automobile emissions required by the Clean Air Act. Future gains in emission reductions probably will be made through the use of refined converters, technologies that allow motor vehicles to burn cleaner fuels such as methanol, and particulate traps installed on new diesel-powered vehicles and retrofitted on older ones. Eventually, new technologies might lead to replacing diesel-powered buses in urban areas.

Air Quality Standards. To monitor the effectiveness of emission standards, the EPA established maximum allowable levels of lead, particulates, SO_2, CO, NO_2 and ground-level ozone, averaged over a specific time period in ambient air. Ambient concentrations are those in the outside air that people breathe. The recommended limits are called **national ambient air quality standards** (NAAQS, pronounced "nacks"). Primary air quality standards are designed to protect human health and welfare; secondary standards protect the environment. Primary standards have been set for all six of the criteria pollutants; secondary standards have been set for all criteria pollutants except CO. With the exception of SO_2, secondary standards are identical to the primary standards for all pollutants.

In 1997, under pressure from the public, healthcare professionals, and non-profit groups such as the American Lung Association (ALA), the EPA tightened the NAAQS for ground-level ozone and particulate matter (PM). The agency's action was based on mounting scientific evidence linking health problems with exposure to these pollutants at levels below the pre-existing standards.

With respect to exposure to ground-level ozone, the EPA eliminated the one-hour standard of 0.12 parts per million (ppm) and replaced it with an eight-hour standard of 0.08 ppm, in order to reflect the health consequences of more prolonged contact (particularly for children playing outdoors all day). However, the one-hour standard remains in effect for counties that failed to meet it.

With respect to particulate matter, the EPA established standards for fine particulates, those less than 2.5 microns in diameter ($PM_{2.5}$). (The agency maintained the pre-existing standards for coarse particulates, those between 2.5 and 10 microns in diameter.)

Continuous sampling enables the EPA to determine how often primary air quality standards are violated. Generally, the pollutant with the highest measured level is used to determine the **air quality index**, a measure of the frequency and degree of air pollution in a particular city or region. According to the ambient air quality standards, an index of 100 should not be exceeded more than once a year. An air pollution alert can be called when the index reaches 200 or higher.

During an alert, people with existing heart or respiratory illness are advised to reduce physical exercise and outdoor activities. As air quality worsens, people with health problems, the elderly, and the very young also are warned to curtail physical activity and stay indoors. At the emergency level, even healthy people may be included in the warning. If emergency conditions persist, stationary sources such as factories and foundries can be ordered by the EPA to curtail any or all activities that would add pollutants to the atmosphere.

Improvements in Air Quality. Inarguably, the Clean Air Act has improved the quality of the nation's air. By 2002, aggregate emissions of the six criteria pollutants had decreased by 48 percent. All six showed significant declines, with the largest reductions in emissions of lead, which fell 98 percent (Figure 12-16). Like emissions, ambient air concentrations of the criteria pollutants have decreased. Between 1983 and 2002, ambient concentrations of CO, lead, and SO_2 fell by more than half. These improvements occurred during a period of growth in both the nation's population and economy. From 1970 to 2002, the U.S. population grew by 38 percent, the total number of vehicle miles traveled rose by 155 percent, energy consumption increased by 45 percent, and the gross domestic product jumped by 164 percent.

Remaining Challenges. Despite the progress made in addressing the effects of criteria pollutants, serious problems remain. Of particular concern are ground-level ozone and particulate matter. The danger is widespread; according to the *American Lung Association State of the Air: 2004* report, 159 million Americans — over 55 percent of the population — live in counties with unhealthy levels of either ground-level ozone or particulate pollution.

In 2004, nearly half of the U.S. population (about 136 million people) lived in areas with elevated ozone levels. Only 19 states are in compliance with the new ground-level ozone standard; 31 states (encompassing 474 counties) are not. The EPA has taken various measures to combat the problem. Smaller industrial polluters, such as dry cleaners, printers and automobile repair shops, must limit emissions of smog-forming chemicals. Moderately or severely polluted urban areas

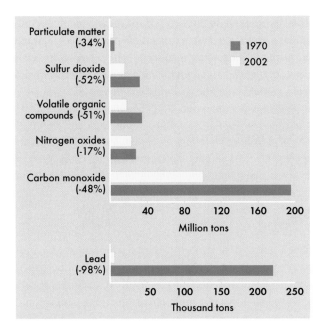

Figure 12-16: Air quality trends, 1970–2002. The Clean Air Act amendments of 1970 significantly reduced emissions of five primary pollutants.

must achieve annual reductions in ground-level ozone — 15 percent within six years and three percent each year after that — until the standards are met.

In 2004, to address the role of automobile exhaust in producing ground-level ozone, the EPA implemented the Tier 2 Vehicle and Gasoline Sulfur Program, modeled after legislation enacted in California in 1998. The program requires that all new passenger vehicles — including previously exempt models such as sport utility vehicles, pick-ups and light trucks, and vans — be held to the same emissions standards. Those standards apply whether the vehicle runs on gasoline, diesel fuel, or alternative fuel. Under the Tier 2 standards, the EPA estimates that new vehicles will be 77 to 95 percent cleaner than 2003 or older models. The program also requires that the sulfur content of gasoline be reduced by up to 90 percent. As a result of the Tier 2 Vehicle and Gasoline Sulfur Program, the EPA estimates that NO_x emissions will fall by 74 percent by 2030; emissions of VOCs and particulates also will fall. Consequently, progress should be made in addressing other air issues, including air toxics, acid precipitation, and regional visibility problems.

Like ground-level ozone, PM remains a persistent public health threat. The 2004 report from the American Lung Association found that about one-quarter of Americans (66 million people) live in areas with unhealthy year-round levels of particulate pollution; 28 percent live in areas plagued by unhealthy short-term

levels (those that last from a few hours to a few days). In the United States, the largest unregulated source of PM is wood-burning stoves. Because most stoves are airtight, they can burn wood slowly with minimum airflow. While burning this way tends to conserve wood, it produces more PM, CO, carcinogens, and creosote. On windless winter days or days when a temperature inversion occurs, wood smoke envelops many communities in a mass of particulates and gas.

As the EPA struggles to enforce compliance with existing NAAQS, it is in the process of reviewing those standards, as required by the Clean Air Act. A growing body of scientific evidence finds that ground-level ozone and PM pose significant health risks — including premature death — *even at levels below current air quality standards*. In 2004, the *Journal of the American Medical Association* published a landmark study that examined respiratory and cardiovascular deaths in 95 U.S. cities between 1987 and 2000; the researchers linked increasing levels of ozone air pollution to thousands of premature deaths annually. Their results mirror those of a large European study (published in the *American Journal of Respiratory and Critical Care Medicine*) that found equivalent health effects in 19 of 23 European cities. Similarly, a 2005 EPA "Staff Paper" concludes that the current NAAQS for PM do not protect public health. According to the document, "new studies support findings…on associations between PM and cardiorespiratory mortality, hospitalization and emergency department visits for respiratory disease, respiratory symptoms and decreased lung function." The EPA's own scientists have stated that "the available information clearly calls into question the adequacy of the current suite of $PM_{2.5}$ standards, and provides strong support for giving consideration to revising the current $PM_{2.5}$ standards to provide increased public health protection."

Air Toxics

The EPA's Toxic Releases Inventory (TRI) is the chief source of information on emissions of air toxics. Established by the Emergency Planning and Community Right to Know Act of 1986, the TRI requires major industrial producers to report how much of each toxic chemical they release to air, water, land, public sewers, and injection wells. Currently, the TRI covers 188 chemicals or chemical groups. Industries reporting the highest total releases are chemical products, primary metals, paper products, transportation equipment, and rubber and plastic.

Air emissions are the single largest source of toxic substances to the environment. In 2002, total emissions of hazardous air pollutants were 1.6 billion pounds (0.7 million kilograms). Top states for total releases include Texas, Louisiana, Ohio, Pennsylvania, Indiana, Tennessee, and Alabama (Figure 12-17). Increasingly, the Sunbelt states of the South and Southwest are replacing the Rustbelt states of the Northeast and Midwest as the largest emitters of toxic substances to the environment.

Many of the toxins emitted to the nation's air are done so legally. Although the Clean Air Act requires the EPA to regulate emissions of toxic substances, the agency has set emission standards for only a handful of chemicals, including asbestos, arsenic, mercury, vinyl chloride, and radioactive isotopes. Hundreds of dangerous chemicals remain unregulated. Moreover, the total amount of U.S. toxic emissions is probably much higher than indicated by the TRI. There are several reasons for this: First, the TRI requires emissions reports on less than one percent of the 80,000 chemicals produced and used in the United States. Emissions of the remaining substances go unreported. Second, only the largest manufacturing and processing facilities in a limited number of industries must report their releases; the TRI excludes government-owned plants, waste management firms, and manufacturing facilities with fewer than 10 employees. Third, mining operations, a major source of toxic substances, are not required to include waste rock, which is typically highly contaminated, in

their reports. And finally, TRI data are self-reported, and facilities are not required to perform any actual monitoring or testing to develop their TRI estimates. As a result, the accuracy of reported data likely varies on a yearly basis and among facilities.

Acid Precipitation

The 1990 amendments to the Clean Air Act established the Acid Rain Program, a major effort to reduce SO_2 and NO_x emissions from electric power plants that use fossil fuels. The hallmark of the Acid Rain Program is a market-based "cap and trade" approach. The program sets a permanent cap on the total amount of SO_2 that may be emitted by electric power plants. The cap is set at about one-half of the total SO_2 these sources emitted in 1980. To meet emissions limits, individual utilities can install scrubbers, switch to low-sulfur coal, or participate in the allowance trading program. This last option enables plants that reduce their emissions below required levels to sell their extra emissions permits to other plants. Alternatively, they can "bank" their unused allotments from one year and use them in another year when their emissions rise. Proponents argue that allowance trading minimizes compliance costs, maximizes economic efficiency, promotes strong economic growth and ensures self-enforcement, thereby reducing the need for government intervention.

In 2002, the nation's electric power plants had reduced SO_2 emissions by 41 percent compared to 1980 levels. They did so at about half the expected cost. Total SO_2 emissions from all sources (including smelters and sulfuric acid manufacturing plants) fell 39 percent between 1980 and 2002.

While the Acid Rain Program established tough restrictions on SO_2 emissions, it failed to set a cap on total NO_x emissions and required only modest emissions reductions from the largest coal-fired power plants. However, controls imposed by some states, primarily in the Northeast, contributed to an overall 33 percent drop in NO_x emissions between 1990 and 2002.

What Is the Clear Skies Initiative?

In 2002, President Bush announced the Clear Skies Initiative, a proposal to amend the Clean Air Act. According to the administration, the Clear Skies Initiative will rely on a "cap and trade" approach to reduce emissions of SO_2, NO_x, and mercury by approximately 70 percent each by 2018. Affected industries, particularly

FIGURE 12-17: Exxon refinery near Baton Rouge, Louisiana, one of the world's largest. It is located in what is often referred to as Cancer Alley, a stretch of land along the Mississippi River from Baton Rouge to New Orleans that is home to many petrochemical plants and industrial sites. Residents of Geismer, St. Gabriel and Carville, towns along Cancer Alley, suffer an unusually high cancer rate.

coal-fired power plants, support the Initiative and its market-based approach to pollution control. It is opposed by a broad coalition of public health organizations — including the American Lung Association, the American Public Health Association, Physicians for Social Responsibility, the American Nurses Association, and the Children's Environmental Health Network — and environmental groups.

Critics claim that "Clear Skies" is a misnomer; they argue that the proposal rolls back current standards (by delaying and reducing required cuts in pollutant emissions), thereby allowing industry to pollute at rates higher than those established by the Clean Air Act. They also maintain that the "cap and trade" approach favored by the administration will enable some power plants and industrial sites to pollute at much higher levels. The result: hot spots of pollution that undermine public health and the environment in affected communities.

Using the EPA's own internal assessments, opponents point out that, relative to the Clean Air Act, the administration's proposal allows for more than one-and-a-half times as much NO_x for nearly a decade longer (2010-2018), and one-third more NO_x even after 2018, twice as much SO_2 for nearly a decade longer (2010-2018), and one-and-a-half times as much after 2018, more than five times as much mercury for a decade longer (2010-2018), and three times as much after 2018. Moreover, because of the emissions "banking" provisions, the proposed reductions are likely to be delayed until as late as 2025.

In addition to the rollbacks in current standards, the Clear Skies Initiative will dismantle key provisions of the Clean Air Act. Local governments could no longer require state-of-the-art pollution controls in new plants of any type or in older plants slated to increase their pollution when they rebuild or expand their facilities. Also, states located downwind from other states with polluting utilities could no longer take legal action to protect their citizens. (Under the Clean Air Act, Northeast states took legal action to effectively fight pollution originating from power plants in the Midwest and the South.)

What Is the Kyoto Protocol?

The Kyoto Protocol is an amendment to the United Nations Framework Convention on Climate Change (1992), an international treaty concerning global warming. Nations that ratify the Protocol agree to reduce their collective emissions of greenhouse gases by 5.2 percent from 1990 levels. (This target represents a 29 percent cut from the emissions levels that would be expected without the Protocol.) In addition to CO_2, the Protocol focuses on five other greenhouse gases: methane, nitrous oxide, hydrofluorocarbons, perfluorocarbons, and sulfur hexafluoride. It specifies how much (if any) each developed nation must contribute toward meeting the overall 5.2 percent reduction. The European Union, for example, must cut emissions by eight percent; Japan, six percent; and Russia, zero percent. At present, developing nations that ratify the treaty are not required to reduce their greenhouse gas emissions.

Negotiated in Kyoto, Japan in December 1997, the Protocol took effect in February 2005. At that time, it had been ratified by 127 countries (including Canada and Mexico), representing 61 percent of 1990 global CO_2 emissions. The United States, under the Clinton administration, signed the U.N. Framework Convention on Climate Change, but the U.S. Senate has not ratified it. President George W. Bush does not support the Kyoto Protocol, and U.S. ratification of the treaty is unlikely.

The Kyoto Protocol is significant because it is the first international attempt to set binding limits on greenhouse gas emissions. Supporters argue that it is an important first step in the process of weaning the world away from its reliance on fossil fuels.

Although the Kyoto Protocol represents real progress in the effort to combat climate change, complex issues remain. One is the concept of "emissions trading" between developed nations. A nation whose emissions fall below its treaty limit can sell credit for its remaining emissions allotment to another nation, which can then use the credit to meet its own obligations under the Protocol. Proponents argue that it is a cost-effective way to curb greenhouse gases. However, developing a workable trading plan — one that involves six different gases moving across international borders — is a difficult task.

The most complex issue concerns the role of developing nations. The Kyoto Protocol did not set binding limits on emissions from developing nations, nor did it establish a timetable for them to assume such limits voluntarily. Carbon emissions are rising rapidly in many LDCs in concert with energy consumption, which is soaring in China, India, Indonesia, and Brazil as these nations attempt to develop economically. Critics of the Protocol, including the Bush administration and some U.S. senators, argue that the United States should not ratify the treaty until LDCs agree to binding limits on their emissions. They maintain that reductions in carbon emissions from the industrialized world will be for naught if emissions increase in the developing world. For example, with its heavy reliance on coal, China is expected to surpass the United States as the leading emitter of greenhouse gases by 2015. Further, they argue that the United States will be at a disadvantage in the marketplace if its goods must compete with goods produced in nations that do not have to install costly emissions-control technologies.

What Is the Montreal Protocol?

The 1987 Montreal Protocol on Substances that Deplete the Ozone Layer called on countries to reduce the production of ozone-destroying chemicals. The Protocol has been amended four times since its conception and, as of 2003, controls 96 different chemicals. Under the Protocol, developed countries are required to have phased out CFCs, carbon tetrachloride, methyl chloroform, and halons by 1996; methyl bromide by 2005; and 65 percent of hydrochlorofluorocarbons by 2010. LDCs have a ten-year grace period before they must start their phase-out schedules, in recognition that developed countries are responsible for the bulk of total emissions into the atmosphere and that they have more financial and technological resources for adopting replacements. The Protocol also established a special fund to enable LDCs to afford less polluting, but more expensive, technologies. By 2003, approximately $1.7 billion had been transferred to LDCs seeking to make the transition from ozone-destroying chemicals.

The Montreal Protocol is expected to yield significant benefits. According to Environment Canada, a government agency, over 19 million cases of nonmelanoma skin cancer will be avoided through 2060, as will at least $459 billion in damages to fisheries, agriculture, and plastic building materials. Even so, the number of skin cancer cases attributable to ozone loss is projected to rise. Skin cancer cases in the United States are expected to peak in the year 2060, at 10 percent above 1996 levels. This projection accounts for the delay between exposure to UV radiation and the occurrence of cancer.

There is a time lag between the production of ozone-destroying chemicals and damage to the ozone layer. According to a report by the World Meteorological Organization (WMO) and United Nations Environment Programme (UNEP), *Scientific Assessment of Ozone Depletion: 2002*, stratospheric abundances of chlorine were at or near their peak in 2001, while bromine abundances continue to increase. Thus, the ozone layer will remain particularly vulnerable in the next decade, even given full compliance with the Montreal Protocol. Moreover, there will be a delay between reductions in emissions of these chemicals and a restoration of the ozone layer. According to the WMO/UNEP report, the thinning of the stratospheric ozone layer will not show signs of improvement until 2010. Assuming the nations of the world continue to make rapid progress in phasing out ozone-destroying chemicals, the ozone layer should reach pre-thinning levels in 2050. However, several factors could hamper efforts to restore the ozone layer. The first is the illegal trade in contraband CFCs, estimated at $300 million annually; China and Russia are leading suppliers in the black-market trade in CFCs. The second factor is the pace at which Russia is meeting its obligations under the Copenhagen Amendments, enacted in 1994. Mired in an economic slump as it attempts to switch to a market economy, Russia has been slow to reduce its emissions of CFCs and continues to produce a large share of the global output. In fact, Russia failed to meet the 1996 deadline to halt production of CFCs. Finally, because the Copenhagen Amendments allow LDCs to produce CFCs for another decade, the demand for CFC-reliant products in newly affluent nations like China could slow progress in reducing global emissions of these chemicals.

WHAT YOU CAN DO ▷ **To Safeguard Air Resources and Protect Yourself against Air Pollution**

➤ Install air exchange systems to reduce indoor air pollution buildup in tight, energy-efficient homes.

➤ Put catalytic converters on wood-burning stoves.

➤ Avoid the use of toxic chemicals at home, in school, and in the office.

➤ Become aware of the danger of skin cancer from ultraviolet rays, a danger that increases as stratospheric ozone thins.

➤ Test your home for radon.

➤ Help improve the air quality in your home, school, or office: keep house plants! Some researchers believe that house plants — or more specifically, microbes specific to the plants' roots — can help filter air, lowering the risk of asthma, allergies, and "sick building syndrome." For more information, see *How to Grow Fresh Air: Fifty Houseplants that Purify Your Home or Office* by Bill Wolverton.

➤ Help improve global air quality by reforesting the Earth: plant trees!

➤ Let your legislative representatives know you think it is important to pass stricter ground-level ozone laws and laws to reduce reliance on fossil fuels.

➤ Find out about organizations that are active in attaining and maintaining air quality. Write to:
National Clean Air Coalition
530 Seventh Street, SE
Washington DC 20003
(202) 523-8200

Summary

The atmosphere is all the gaseous matter that surrounds the Earth. The innermost layer, called the troposphere, contains the gases that support life. The stratosphere contains a protective layer of ozone. The mesosphere and thermosphere are the two outer layers of the atmosphere. Weather, the day-to-day patterns of precipitation, temperature, wind (direction and speed), barometric pressure and humidity, occurs in the troposphere. The atmosphere plays an important role in influencing climate, the long-term weather pattern of a particular region.

Atmospheric gases allow some of the incoming solar radiation to pass through the atmosphere, but trap heat that would otherwise be reradiated back toward space. This phenomenon is known as the greenhouse effect. The greenhouse gases are essential to maintaining the conditions necessary for life on Earth, but increased levels of carbon dioxide (CO_2), methane, chlorofluorocarbons, and nitrous oxide can lead to global climate change.

Any substance present in or released to the atmosphere that adversely affects human health or the environment is considered an air pollutant. Primary pollutants are emitted directly into the atmosphere. They may react with other primary pollutants or atmospheric compounds, such as water vapor, to form secondary pollutants. Of the thousands of primary pollutants, six are of special concern because they are emitted in particularly

Discussion Questions

1. Explain the greenhouse effect. How is it related to global climate change? What measures, if any, do you think should be taken to prevent climate change? What could developed countries do to help developing countries reduce emissions of carbon while still raising living standards?

2. Distinguish between primary and secondary pollutants, and give three examples of each. What are the major sources of air pollution?

(continued on following page)

(continued from previous page)

3. What factors are contributing to the destruction of the stratospheric ozone layer? What effects could this destruction have?

4. Why is acid precipitation such a widespread problem? What political problems are involved with regulating it?

5. Imagine that your community faces an air pollution problem. What steps would you recommend to reduce the pollution? Explain your choices in terms of the size, location, and demographic composition of your community.

6. Summarize the air pollution problems caused by the burning of fossil fuels. What measures could we take to alleviate the unwanted consequences of fossil fuel combustion?

7. What do you think is the most pressing concern with respect to air resources? Explain.

large quantities: carbon dioxide, carbon monoxide, sulfur dioxide, nitrogen oxides, hydrocarbons, and particulates. In the 1970s, efforts at cleaning up air pollution in the United States focused on primary pollutants. Currently, increased attention is being given to six air pollution issues that pose a particular threat: global climate change, stratospheric ozone depletion, acid precipitation, smog, airborne toxins, and indoor air pollution.

Global climate change is expected to result from increased atmospheric concentrations of greenhouse gases, especially CO_2. Since 1750, atmospheric CO_2 has increased 31 percent. Since the late nineteenth century, the average global temperature has increased by about 1°F. Computer models show that the average global temperature will rise 1.8 to 6.3°F (1 to 3.5°C) by 2100. The most likely effects of global warming are rising sea levels, changes in rainfall patterns and growing seasons, and increased incidence of storms and weather-related disasters.

Ozone in the stratosphere plays a critical role in protecting the Earth from harmful ultraviolet radiation. The stratospheric ozone layer is extremely thin, so even small decreases are expected to cause significant increases in negative health effects among humans and decreases in ecosystem productivity. Chlorine, NO_x, fluorine, and bromine are responsible for ozone destruction. Atmospheric chlorine and fluorine have been traced to the chemical breakdown of chlorofluorocarbons (CFCs). Under the terms of the Montreal Protocol, the nations of the world have begun to phase out the use of CFCs and other ozone-destroying chemicals.

Acid precipitation is rain, snow, fog, or mist that contains enough sulfuric acid, nitric acid, or their precursors to raise the acidity of the precipitation above normal. Acid precipitation can damage lakes, forests, and crops; adversely affect human health; and corrode marble, limestone, sandstone and bronze, destroying statues, monuments, gravestones, and buildings. Acid surges are periods of short, intense acid deposition in lakes and streams. Damage caused by acid precipitation is widespread, both in the United States and globally.

Industrial smog is essentially smoke pollution; it consists chiefly of sulfur oxides and particulates and is emitted to the air from industrial and manufacturing facilities. Photochemical smog, or brown smog, is formed when hydrocarbons react with nitrogen oxides (NO_x) and oxygen to form chemicals such as ozone and peroxyacetyl nitrate. Photochemical smog is a serious problem in urban areas worldwide. Its primary component, ground-level ozone, is toxic; it can adversely affect human health in many ways and damage crops and forests.

Hazardous air pollutants, commonly called air toxics, are a varied group of chemical substances known to cause (or suspected of causing) cancer or other serious human health effects.

Major indoor air pollutants are formaldehyde, radon 222, tobacco smoke, asbestos, combustion products from gas stoves and poorly vented furnaces, chemicals used in building and consumer products, and disease-causing organisms and spores.

Air pollution takes the greatest toll on the very old, the very young, and persons already suffering from respiratory or circulatory disease. In general, it is worse in urban and suburban areas and can be concentrated in improperly ventilated buildings.

Weather and topography affect pollution levels. Energy from the sun, moisture in clouds, and the presence of highly reactive chemicals in the atmosphere may cause the formation of new pollutants. Valleys act as sinks for pollutants, and mountains act as barriers to air flow that could carry pollutants away from certain locations.

In the United States, air pollution is regulated chiefly by the Clean Air Act, first passed in 1963 and amended in 1970, 1977, and 1990. The Clean Air Act directed the EPA to develop national ambient air quality standards, emission standards that specify the quantities of air pollutants that can be emitted by specific sources. Other provisions concern smog, airborne toxics, and acid precipitation.

The Clear Skies Initiative is a proposal to amend the Clean Air Act. Supported by the Bush administration, the Initiative will rely on the "cap and trade" approach to reduce emissions of several major air pollutants.

The Kyoto Protocol is an amendment to the United Nations Framework Convention on Climate Change, an international treaty concerning global warming. Nations that ratify the Protocol agree to reduce their collective emissions of greenhouse gases.

The Montreal Protocol on Substances that Deplete the Ozone Layer calls on countries to reduce their production of ozone-destroying chemicals. The Protocol has been amended four times and currently controls 96 different chemicals.

KEY TERMS

acid precipitation	formaldehyde	primary pollutant
acid surge	greenhouse effect	radon 222
air pollutant	hazardous air pollutants	rain shadow effect
air quality index	indoor air pollution	secondary pollutant
albedo	industrial smog	stratosphere
asbestos	mesosphere	stratospheric ozone depletion
atmosphere	national ambient air quality standards	thermosphere
climate	ozone	troposphere
climate change	photochemical smog	weather
emission standards		
emissivity		

CHAPTER 13

Water Resources

Learning Objectives

When you finish reading this chapter, you will be able to:

1. List the properties of water that enable it to support life.

2. Identify how water resources are classified and briefly describe each classification.

3. Describe worldwide water consumption patterns.

4. Identify eight broad categories of water pollution.

5. Describe how both drinking water supplies and wastewater are treated.

6. List some of the current threats to groundwater, surface waters, and marine waters.

The sober citizen who would never submit his watch or his motor to amateur tamperings freely submits his lakes to drainings, fillings, dredgings, pollutions, stabilizations, mosquito control, algae control, swimmer's itch control, and the planting of any fish able to swim. So also with rivers. We constrict them with levees and dams, and then flush them with dredgings, channelizations, and the floods and silt of bad farming.

Aldo Leopold

Earth, the water planet. How incongruous it seems that "Earth" is both a word for soil and the name of our planet when oceans cover 71 percent of its surface! In addition to the water in the seas is that held in lakes, ponds, rivers, streams, marshes, bays and estuaries, as well as the water locked in polar ice caps and glaciers (Figure 13-1). Water so dominates the Earth's surface that it is responsible for the familiar image seen in satellite photos: a beautiful blue ball existing in stark contrast to the blackness of space. Vast amounts of water also exist in liquid form underground and as water vapor in the atmosphere. All of this water is linked in a continuous cycle of evaporation (from the surface to the atmosphere), precipitation (from the atmosphere to the surface), and runoff (from the land or beneath the land's surface). The hydrologic cycle (see Chapter 4, *Ecosystem Function*) is driven by the sun's energy and has been recycling the same water through the millennia of the Earth's existence. Indeed, the water we bathe in today might have quenched the thirst of dinosaurs millions of years ago. In this chapter, we describe water resources and examine efforts to manage them.

FIGURE 13-1: Diversity of water habitats. As a life-giving environment, water shows remarkable diversity: (a) the Little Colorado River; (b) a freshwater lake in Mancelona, Michigan; (c) Forked Lake in McCurtain County, Oklahoma; (d) the shoreline of Lake Erie's western basin; (e) the Red Sea's Gulf of Aqaba, Jordan; and (f) a winter scene in Clifton Gorge State Park, Yellow Springs, Ohio.

Describing Water Resources

In this section, we look at how water supports life, how much freshwater is on Earth, and how water is classified. We also examine how humans use water and how those uses contribute to water pollution.

How Does Water Support Life?

Water is the largest constituent of living organisms, and it is also a habitat for a great diversity of life on Earth. Water is able to accomplish these dual roles because of unique physical and chemical characteristics (Table 13-1). For most of the Earth's existence, living organisms have been able to capture some of the water in the hydrologic cycle to maintain their internal environments. The human body is composed primarily of water (about 65 percent); there are about 43 quarts of water in a 150-pound person. Water makes up 83 percent of blood, helps digest food, regulates salt and mineral balances, transports body wastes, and lubricates joints. The body's internal water supply must remain constant and free from impurities to maintain health. Like all living organisms, we are intricately bound to the hydrologic cycle.

How Much Freshwater Is On Earth?

Most of the Earth's water (97 percent) is salty; only a small portion (less than 3 percent) is fresh. Three-quarters of all freshwater is found in polar ice caps and glaciers, and nearly one-quarter, known as **groundwater**, is found underground in water-bearing porous rock or sand or gravel formations. Only a small proportion (0.5 percent) of all water in the world is found in lakes, rivers, streams, and the atmosphere.

How Is Water Classified?

Water is classified as either fresh or salt (marine), depending upon its salt, or saline, concentration. The saline concentration of marine waters is generally fairly consistent, about 35 parts per thousand (ppt).

TABLE 13–1: Physical and Chemical Properties of Water

Superior solvent. More materials can be dissolved in water than in any other solvent. Water can permeate living cell membranes; dissolved materials also diffuse through the membranes or are "pumped" through using respiration energy.

Strong attractive force. The strong attractive force between water molecules allows them to be transported through the spaces in soil, into roots, and through the conducting tissue in plants and then to be transpired through leaves and needles back into the atmosphere.

State changes. In the hydrologic cycle, the sun supplies the energy that vaporizes water to a gaseous state. When water vapor condenses to a liquid, energy is given off, helping to distribute the sun's energy across the Earth.

High specific heat. It takes substantial amounts of energy to raise the temperature of a body of water, and, conversely, water is slow to cool. This property protects aquatic environments against rapid temperature changes. The different heating and cooling rates of bodies of water and adjacent land masses help to circulate air, create winds, and establish weather patterns.

Expansion at freezing. Water is the only common substance that expands when it freezes. Other substances contract when they freeze.

Density at freezing. Water reaches its maximum density at 46° F (4° C), 14° fahrenheit above the freezing point. As water approaches freezing, it becomes lighter and moves toward the surface. When frozen, it floats on the surface as ice. Thus, streams, lakes, and ponds freeze from the top down, protecting aquatic environments from freezing solid in the winter.

Oxygen- and nutrient-holding capacities. More oxygen dissolves in cold water than in warm water, but warm water generally holds more nutrients. The largest populations of aquatic organisms are found where nutrient-rich warm waters and oxygen-rich cold waters mix.

Flow. Water tends to move easily downhill over land toward streams, rivers, lakes, ponds, and oceans. Bodies of water tend to flow downhill and toward larger bodies of water. Water also percolates through the ground, sometimes collecting in huge, slowly moving underground aquifers.

Aesthetic quality. Flowing water, shimmering water, cascading water, waves, small drops of dew collected on a spider's web, rainwater pattering on a roof or running down a windowpane, puddles, a single drop of water from a pond teeming with life, a tear: each affects the human psyche in ways science is unable to explain.

On average, the saline concentration of freshwater is 0.5 ppt. The saline concentration of freshwater tends to vary more than that of marine waters because lakes, rivers, and streams are much more dominated by local environmental conditions, such as the lands they drain and the rate of evaporation.

Freshwater

Freshwater is found on land in two basic forms: surface waters and groundwater. The two are not entirely distinct. Some water from lakes, streams, and rivers may percolate downward to groundwater supplies. Similarly, during dry spells when surface runoff may be unavailable, groundwater may help to maintain the flow and/or level of rivers, streams, or lakes.

Surface waters. Surface waters are bodies of water (such as lakes and rivers) recharged by precipitation that flows along land contours from high to low elevations as **runoff**. Runoff makes its way into streams, rivers, ponds and lakes, eventually reaching the oceans. The entire runoff area of a particular body of water is known as its **watershed**. A watershed contributes water and dissolved nutrients and sediments to the body of surface water toward which it drains. The fertility of surface waters, then, depends on the fertility of the land in the watershed.

Surface water ecosystems include both standing water habitats (including ponds, lakes, reservoirs, and in some cases, wetlands) and running water habitats (including springs, streams, and rivers). **Standing water habitats** are relatively closed ecosystems with well-defined boundaries; generally, they contain both inlet and outlet streams. Standing water generally has a lower oxygen level than running water because there is less mixing of water. Water does move, or flush, through ponds and lakes, but at a slow rate — from one year or less in small lakes to hundreds of years in large lakes such as Lake Superior. With slow movement of water, pollutants have more time to build up to dangerous levels and to settle in sediments.

Lakes may be categorized according to the amount of dissolved nutrients they contain (Figure 13-2). **Oligotrophic lakes** contain a relatively low amount of dissolved solids, nutrients, and phytoplankton and are thus clear and deep blue in color. They lie in infertile watersheds, are cold and deep, have a high oxygen content, and usually have rocky bottoms. The low nutrient level of oligotrophic lakes results in a low production

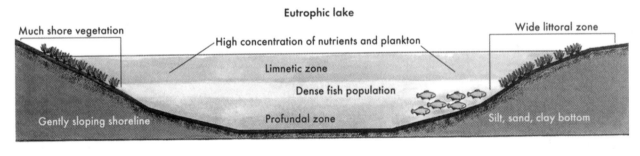

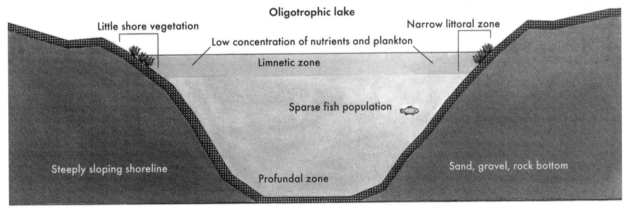

FIGURE 13-2: Eutrophic and oligotrophic lakes.

of organic matter, particularly phytoplankton, relative to total volume.

Compared to oligotrophic lakes, **eutrophic lakes** are warmer, more turbid, have a lower oxygen content, and often have muddy or sandy bottoms. They also are far more productive: Fertile watersheds supply eutrophic lakes with abundant nutrients, enabling them to support a large phytoplankton population and a rich diversity of organisms. As we learned in Chapter 5, *Ecosystem Development and Dynamic Equilibrium*, lakes undergo succession, becoming more eutrophic over time. Eventually, as eutrophic lakes and ponds grow warmer and shallower, they become weed-choked and more like marshes and dry land.

Lakes are composed of internal zones based on depth and light penetration. The **littoral zone** is a shallow, near-shore area where rooted plants grow because light can penetrate to the bottom. The **limnetic zone** is deeper water where light can penetrate and support populations of plankton. Below the limnetic zone is the **profundal zone**, into which light does not penetrate enough to allow photosynthesis to occur. Organisms of the profundal zone depend on food and nutrients filtering down from above.

Large lakes in temperate zones undergo **thermal stratification** (Figure 13-3). In the summer, surface waters heated by the sun become lighter and less dense and form a top layer, or **epilimnion**. Colder, denser water sinks to the bottom to form the **hypolimnion**. A sharp temperature gradient, or **thermocline**, exists between the upper and lower layers. Sometimes, especially in eutrophic lakes, all photosynthesis and oxygen production occur in the warm upper waters. The thermocline prevents the two layers from mixing, and the hypolimnion can become depleted of oxygen. Occasionally, late in the summer, the shallower areas of eutrophic lakes and ponds also can become completely devoid of oxygen. The lack of oxygen can stress fish populations and result in high fish mortality. Generally, this phenomenon can be traced to very warm weather, little wind to mix the surface water, and a high decomposition rate of algae or other organic matter. In the autumn, when seasonal changes help to cool the surface water, the thermocline disappears and the top and bottom layers mix, returning oxygen to the bottom layers. This mixing is known as fall turnover. If the winter is cold and snow covers the ice for long periods, oxygen also can be depleted in the deeper

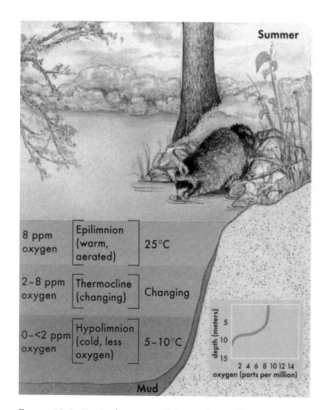
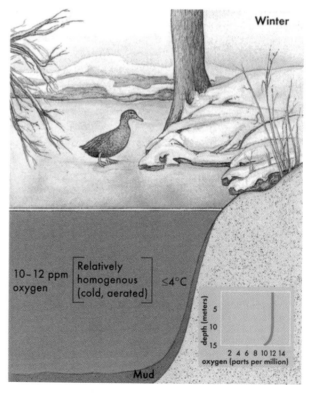

FIGURE 13-3: Typical pattern of thermal stratification in lakes. In summer, the thermocline separates the colder, oxygen-poor water in the hypolimnion from the warmer, oxygen-rich water in the epilimnion.

areas. Spring warming causes the water to turn over once again, returning oxygen to the deepest levels and circulating both nutrients and the overwintering stages of organisms to the warmer, light-filled waters.

Running water habitats include streams and rivers, continuously moving currents of water that cut channels or beds through the land's surface. The speed of the current influences both the composition of the channel (such as rock, sand, gravel, or mud) and the oxygen content — and, consequently, the composition of the organisms found there. Fine-grained particles tend to collect in pools or shallows, while large rocks and boulders are found on the streambeds of fast-flowing waters. The oxygen content of fast-flowing streams is higher than that of slow-flowing ones.

Temperature also affects the composition of organisms in streams and rivers. In general, the temperature of small streams tends to rise and fall with the air temperature. Rivers with large surface areas exposed to the sun generally are warmer than those with trees and shrubs on their banks. Often, when streams are channelized or when surrounding land is cleared for farming, the water goes through large changes in temperature with subsequent drastic changes in species composition.

A look below the surface of a river or stream reveals many interesting habitats and associations of living organisms. Fast-moving rivers and streams usually contain two kinds of habitats, riffles and pools. Riffles, with a high oxygen content, tend to house the producers of biomass, while pools tend to contain consumers and decomposers.

Groundwater. As its name suggests, groundwater is freshwater found underground in porous rock strata. Imagine that you could contain all of the fresh surface water (from lakes, rivers, ponds, and streams) in the United States in one lake. Our imaginary lake would contain only a fraction of the water that exists underground. It would have to be 20 times larger to hold all of the groundwater! This huge amount, some 30 to 60 quadrillion gallons (114 to 227 quadrillion liters), represents 96 percent of the United States' freshwater.

Groundwater percolates downward through the soil from rain, snow, or surface water and is stored in an **aquifer**, a water-bearing geologic formation composed of layers of sedimentary material such as sand, gravel, or porous rock. Water fills the crevices of the rock and the pores between the particles of sand and gravel. The depth at which the aquifer begins is known as the **water table**. Before water reaches the aquifer, it passes through an unsaturated zone, where pores contain both water and air. Plants remove some of this water; the rest continues to move downward to the saturated zone.

Aquifers vary widely in terms of length and thickness. They can cover a few miles or thousands of square miles. For example, the High Plains aquifer underlies about 174,000 square miles (450,660 square kilometers) in eight states (South Dakota, Wyoming, Nebraska, Colorado, Kansas, New Mexico, Texas, and Oklahoma) and encompasses the 134,000-square-mile (347,060-square-kilometer) Ogallala aquifer. Because the Ogallala was formed long ago and is not significantly recharged by present precipitation patterns, the water in the aquifer is known as fossil groundwater. Aquifers can be several feet to several hundred feet thick; they may occur several hundred feet underground or quite close to the surface, perhaps emerging as a free-flowing spring or contributing water to a stream, river, lake, or wetland. If the material above the aquifer is permeable, allowing water to move freely downward to the water table, the aquifer is said to be **unconfined**. If an impermeable layer exists above the water table, thereby restricting the downward flow of water, the aquifer is said to be **confined**. Lacking a protective, impermeable rock or clay layer, unconfined aquifers are susceptible to contamination.

Groundwater flow also is highly variable, percolating from higher to lower elevations sometimes very slowly and sometimes surprisingly fast — anywhere from a fraction of an inch to a few feet a day (Figure 13-4). Groundwater flow is determined by the slope of the water table as well as by the permeability of the material through which the water is moving. The steeper the slope and the more permeable the substrate, the more rapidly the water flows. Usually, but not always, groundwater flows parallel to the aboveground flow. This may not be true, however, in the case of a confined aquifer, where an impermeable layer separates the surface water from the groundwater.

Aquifers supply drinking water for half of the U.S. population, including almost all rural residents. While groundwater is used to some extent in every state, it is the major source of drinking water in about two-thirds of them. In addition to its use as a source of drinking water, groundwater supplies over 95 percent of total rural household needs, 40 percent of agricultural demand, and 26 percent of industrial demand. It also helps to maintain water levels and the productivity of streams, lakes, rivers, wetlands, bays, and estuaries. Furthermore, mineral-rich groundwater helps to supply nutrients that nourish such aquatic ecosystems as

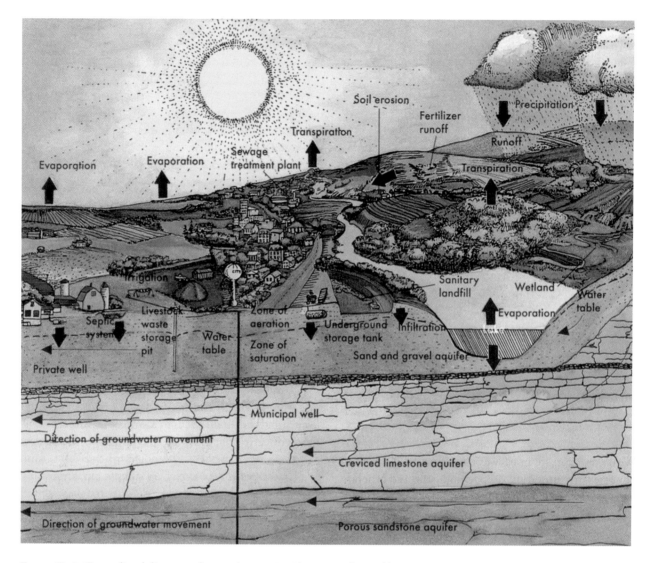

Figure 13-4: Generalized diagram of groundwater. Aquifers are recharged by precipitation that percolates through the ground or water that seeps down from lakes, rivers, streams, and wetlands. They are subject to many diverse sources of pollution.

Tarpon Springs and the Apalachicola River in Florida. These systems often support unique plants and animals and provide excellent fishing and recreation opportunities.

However, aquifers are slow to recharge after water is removed and slow to cleanse after they are contaminated. Contamination is becoming more significant as aquifers are increasingly threatened by pollution from sources as varied as municipal and industrial landfills, oil drilling, mining, urban and agricultural runoff, and leaking underground storage tanks for toxic substances.

Marine Waters

The oceans are one huge living system. Cradles of life, crucibles of diversity and nurturers of civilizations, they influence nearly everything we do. Constantly moving water currents operate under the influence of the sun's energy, the Earth's rotation, and the moon's gravitational pull. Currents move huge masses of water around the continents and across the vast open reaches, circulating nutrients that have washed in from the land. Further, the interaction of the oceans and the atmosphere affects heat distribution, weather patterns, and concentrations of atmospheric gases throughout the world. Because they are so expansive and deep, the oceans moderate climates and provide a sink for dissolved solids and gases. Beneath the ocean floor lies a rich storehouse of minerals, petroleum, and natural gas. Finally, hundreds of millions of people live near the ocean shores and on coastal plains. Even greater numbers depend upon ocean fisheries, energy, and minerals for their sustenance.

The ocean is divided into zones. As in a lake, the littoral zone is the area of shallow waters near the shore. The **euphotic zone** is the area of light penetration; here, phytoplankton exist in great abundance. Where light can penetrate to the ocean floor, attached plants like the giant kelps are found. The **neritic zone**, comprising seven to eight percent of the oceans, is the part of the euphotic zone over the continental shelf and near-shore islands. The greatest variety of life is found in the neritic zone, where ocean waters mix with waters from the land, and this zone supports most of the ocean fisheries. The **pelagic zone** is the deep water of the open oceans. Below it lies the **abyssal zone**, the deepest part of the ocean.

How Do We Use Water?

We use water in countless ways: to drink, bathe, dispose of wastes, irrigate crops, support industry, and generate power. Some of the world's most important transportation corridors are waterways; many play a vital role in global economics and politics. Further, water provides many of our favorite recreational opportunities. Finally, water is a source of inspiration for poets, writers, artists, and others whose souls are stirred by the sight and sound of this most basic of all compounds.

Globally, agriculture is the single largest drain on water supplies; approximately 70 percent of fresh water withdrawn from lakes, rivers, and aquifers is used to produce food. By 1995, the total amount of land under irrigation worldwide was 625 acres (253 million hectares), approximately equal to an area the size of India. Industry accounts for about 22 percent of water use worldwide, and municipalities and households account for 8 percent.

Water uses may be nonconsumptive or consumptive. Nonconsumptive uses remove water from a river or lake, use it, and return it — usually altered in some way — to its original source. Municipal uses (such as drinking and bathing) and industrial uses (such as cooling and power generation) are included in this category. About 90 percent of the water used by homes and industry is available for reuse, although not all of

this water will be suitable for use without extensive treatment. In contrast, consumptive uses remove water from one place in the hydrologic cycle and return it to another. For example, water withdrawn from a stream, lake, or aquifer to irrigate crops might not return to its original source; it might be "lost" through evaporation, or it might run off the land into a different stream or lake.

How Do Water Use Patterns Vary Worldwide?

Water resources are not evenly distributed; some areas have abundant water and others have little. In freshwater-poor areas, such as the Middle East and northern and eastern Africa, **desalination** — the process of removing dissolved salts from marine water or brackish groundwater — is used to provide citizens with water for drinking, cooking, and other needs. Even highly developed countries with seemingly adequate water resources can suffer from lack of rainfall and groundwater resources or from pollution that renders water supplies unfit for use.

The United States' current water consumption pattern is a mixed bag. In 2000, the nation as a whole withdrew 408 billion gallons (1.5 trillion liters) per day of freshwater; despite an increase in population, this amount is approximately three percent less than that withdrawn in 1990 and 14 percent less than in 1985. The U.S. Geological Survey attributes this decrease to better water efficiency in industry and agriculture. Yet, irrigation remains the largest use of freshwater in the United States; with 137 billion gallons (519 billion liters) per day withdrawn in 2000, it accounted for just over one-third of the year's total usage.

On an individual basis, water is becoming a resource Americans no longer can take for granted. Today, each citizen served by public supplies uses an average of 100 gallons of water (379 liters) per day, 70 percent of which is used in the home — mostly for flushing toilets. Figure 13-5 shows the daily water use of a typical U.S. family of four. If we calculate industrial and home use combined, each American consumes nearly 400 gallons (1,516 liters) of water per day.

The Pacific Institute for Studies in Development, Environment, and Security considers 15 gallons (57 liters) of water per day as the minimum standard for meeting four basic needs — drinking, sanitation, bathing, and cooking. Approximately 55 countries with a combined population of over one billion people cannot meet this

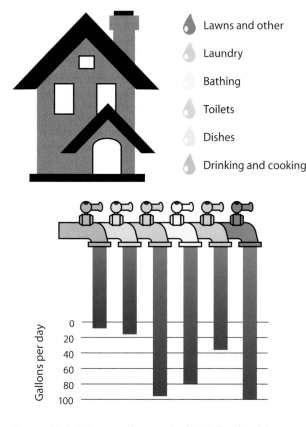

Lawns and other

Laundry

Bathing

Toilets

Dishes

Drinking and cooking

Gallons per day

0
20
40
60
80
100

FIGURE 13-5: Water use by a typical U.S. family of four.

FIGURE 13-6: A girl collects water from the Bishnumati River in Kathmandu, Nepal.

standard; in fact, many of the world's people do not have access to even five gallons. This statistic means that one flush of a conventional toilet in the United States uses more water than many people in the developing world use in an entire day. In many less-developed countries (LDCs), clean drinking water and adequate sanitation facilities are unknown luxuries. Many areas do not have toilets, latrines, or proper drains; wastes are disposed of near or in the same rivers, lakes, or wells used for drinking and food preparation (Figure 13-6). Because developing countries dedicate most of their water resources to agricultural expansion, little remains for domestic use. India, for example, devotes 90 percent of all water to agricultural purposes, seven percent to industry, and only three percent to household use. Out of 3,119 towns in India, only 217 treat their sewage — 8 fully, 209 partially. The untreated wastes threaten to contaminate surface waters with several waterborne diseases, such as cholera, typhoid, shigella, meningitis, and diarrheal disease. Worldwide, the lack of sanitary waste disposal and clean drinking water results in over 12 million deaths annually, most of which are in LDCs. Of the estimated four billion people who contract diarrheal disease (the major wa-

terborne disease) each year, three to four million die, most of whom are children.

To increase water availability and provide basic waste treatment requires large sums of money and cooperation among nations. But the results — better health, reduced infant mortality, and population stability — yield a much better return than dollars spent on medical treatment. Disease prevention through an adequate supply of clean freshwater is the key to improving health standards. The greatest successes have come in countries where grassroots movements have encouraged better education about sanitation, the development of community-based water programs, and the use of low-energy treatment technology.

What Kinds of Water Pollution Are There?

Water pollutants can be divided into eight general categories: organic wastes, disease-causing wastes, plant nutrients, toxic substances, persistent substances, sediments, radioactive substances, and heat. These broad categories — and the interactions among them — indicate the complex nature of water pollution. Consider that one source may be responsible for more than one type of pollutant. For example, improperly treated sewage may contribute organic wastes, disease-causing wastes, plant nutrients, toxic substances, and persistent substances to a waterway. Likewise, a pollutant may fit in more than one category. For example, mercury and polychlorinated biphenyls (PCBs) are both toxic and persistent. Further, one type of pollutant may enter water attached to another type. For example, organic chemical pollutants often adhere to sediments. Many water systems are assaulted by pollutants from all eight categories, compounding the difficulty of clean-up not only by the magnitude of total pollutants but also by the need to use several different techniques. Effective clean-up is essential because a single pollutant may not be found in amounts necessary to kill organisms, but many pollutants acting together, or synergistically, may be deadly.

Organic Wastes

Organic wastes are small pieces of once-living plant or animal matter. They usually are suspended in the water column but sometimes can accumulate in the sediments of lakes or streams. Most suspended organic matter comes from human and other animal wastes or plant residues. Aerobic bacteria pull dissolved oxygen (DO) from the water to break down organic matter, a process called decomposition. Anaerobic bacteria, which often operate in the sediments, do not require oxygen to break down organics; they may emit noxious gases, such as hydrogen sulfide or methane, as a by-product of decomposition.

Decomposition by aerobic bacteria removes DO from the water to the detriment of other aquatic organisms, such as fish and shellfish. As the oxygen is depleted, the species composition of the area may change dramatically, as higher aquatic organisms (such as fish, oysters, and clams) die or leave the area, and organisms that can tolerate low oxygen levels (such as sludge worms and rattailed maggots) proliferate. If the oxygen depletion is drastic enough, it may cause many fish to die, an event known as a fish kill.

Biological oxygen demand (BOD) is a measure of the amount of oxygen needed to decompose organic matter in water. A stream or lake with a high BOD will have a low concentration of DO because oxygen is being used by bacteria to decompose organic matter. Accordingly, a body of water with a high BOD is also high in organic matter. Because DO is a major limiting factor in aquatic habitats, oxygen depletion contributes to the degradation of a stream or lake through the loss of species diversity.

About half of oxygen-demanding pollutants come from nonpoint sources, the most conspicuous of which is animal waste used as fertilizer. Sewage treatment plants and the food-processing and paper industries are notable point sources.

Disease-Causing Wastes

Untreated human or other animal wastes that enter surface waters increase the chance that infectious organisms (such as bacteria, viruses, and protists) will spread disease to humans (Figure 13-7). Disease outbreaks occur commonly in LDCs, where human sewage is treated improperly, if at all, before it enters an aquatic environment. Since millions of people in LDCs have no basic sanitation, providing adequate supplies of disease-free water presents a major challenge. Cholera, typhoid fever, hepatitis, and dysentery are some major diseases transmitted through drinking water supplies. Even in the United States, where the chlorination of drinking water helps to prevent major outbreaks, contaminated water causes some cases of infectious disease each year. Bacteria from human and nonhuman wastes is the number one contaminant in rural water supplies.

Water quality in the United States is also an issue outside of residential areas. Giardiasis, a disease caused by the protist *Giardia lamblia*, is the country's most common waterborne disease. It is known as camper's-, hiker's-, or backpacker's disease because of its high incidence in those groups. Even the most pristine lakes

FIGURE 13-7: Aerial view of the destruction caused by the December 2004 offshore earthquake and tsunami that leveled parts of Indonesia, most notably the Aceh Province in Sumatra, and caused serious damage in other countries, including Sri Lanka, India, and Thailand. In this photograph, salinated plots of land replace once-thriving nurseries, paddy fields, and aquaculture operations. Following the immediate devastation, survivors and relief workers focused on preventing water-borne disease outbreaks that could result from the lack of clean drinking water and sanitation infrastructure.

and streams can harbor these infectious protistans, which enter the water via animal wastes.

Many other diseases also are transmitted by organisms in water. For example, mosquitoes (whose larvae live in water) transmit the protist that causes malaria, and snails transmit the fluke that causes schistosomiasis. Both of these diseases are serious threats to human health in LDCs.

The presence of fecal coliform, bacteria normally found in the large intestines of humans and other animals, indicates the likelihood of contamination by waterborne diseases. Coliform standards for drinking water are stricter than those for water used for recreation. The International Food Policy Research Institute recommends that drinking water have a coliform count of no more than 100 organisms per 100 milliliters, but many waterways do not meet this recommendation. For example, several rivers in Latin America have coliform counts of more than 100,000 organisms per 100 milliliters. In the United States, despite attention to proper sewage treatment, many beaches are closed each summer because of high coliform counts.

Plant Nutrients

Algal and aquatic plant growth is normally limited by the amount of nitrogen and phosphorus in the water. The large amount of these nutrients contained in sewage — including the phosphates in some detergents and agricultural and urban runoff — stimulates massive reproduction and rapid growth in algae, known as algal blooms. Algae impart a green color to water and form a "scum" on the surface and on rocks near shore. When algae die and decompose, more organic matter is added to the water, increasing the BOD. Fast-moving water generally is not as affected by algal growth as slow-moving water. Ponds, lakes, and bays are most affected through the process of eutrophication (to learn more about this process in the Great Lakes system, read *Environmental Science in Action: Lake Erie* by going to www.EnvironmentalEducationOhio.org and clicking "Biosphere Project").

Excess phosphorus generally is not a human health hazard, but excess nitrogen — in the form of nitrates — is. Found largely in fertilizers and organic waste from livestock feedlots, nitrates are soluble in water and do not bind to soil particles, making them highly mobile. As such, they wash into surface water supplies and percolate into groundwater. Nitrates in drinking water pose a significant health threat. For example,

in the intestinal tract of infants, they are reduced to nitrites, which oxidize the hemoglobin in blood, rendering it unable to carry oxygen. This condition, called methemoglobinemia, can result in brain damage or death. Nitrates and nitrites also can form toxic substances called nitrosamines, which have been found to cause birth defects and cancer in animals. Still other studies have established a link between high nitrate levels and stomach cancer in humans.

Sediments

By weight, sediments are the most abundant water pollutant. With every hard rain, muddied streams and rivers are a visible reminder of the millions of tons of sediments that wash from the land into aquatic systems (Figure 13-8). Particles of soil, rock, sand, and minerals run off the land, enter waterways, and fill in lake bottoms and river channels.

Sediments contain nutrients, and natural erosion thus helps to maintain the fertility of aquatic ecosystems. However, a dramatic increase in the sediment load — from poorly managed agricultural lands and urban construction sites, for example — can cause problems. While the threat to human health is minimal because filtration easily removes sediments from drinking water, aquatic life often is adversely affected. Excess sediments smother fish eggs and prevent light from penetrating to rooted aquatic plants. Further, chemicals may adhere to sediments and accumulate on the bottom of bodies of water, where they can contaminate aquatic habitats. Some dredged sediments are so laced with dangerous chemicals that they must be treated like hazardous wastes. For example, polyaromatic hydrocarbons (PAHs), first identified in cigarette smoke, are carcinogens that bind to sediment particles. Major sources of PAHs include emissions from coke ovens, creosote plants, and coal gasification plants; sewage containing petroleum products; automobile exhaust; and oil leaked from boats.

Toxic Substances

Toxic substances have the potential to cause injury to living organisms and their cells. Toxic organic substances include oils, gasoline, greases, solvents, cleaning agents, biocides, and synthetics. Thousands of organic chemicals enter aquatic ecosystems every day. Most are by-products of industrial processes or are present in countless commonly used products. Some organic

Figure 13-8: In this 1988 photo, a clear tributary flows into Peru's sediment-laden Palcazo River.

chemicals, such as dioxin, PCBs and trichloroethylene, are carcinogenic, meaning they cause (or are believed to cause) cancer.

Inorganic toxic substances include acids from mine drainage and manufacturing processes; salt from roads and irrigation ditches; brine from offshore oil and natural gas wells; and metals such as chromium, copper, zinc, lead, and mercury from industrial processes. Lead and mercury, in particular, have received a great deal of attention. In the United States, lead contamination from old pipes and solder joints caused the Environmental Protection Agency (EPA) to release an advisory explaining how consumers can reduce the lead content in their drinking water until new pipes and joints are installed. Also highly dangerous, mercury in water becomes methylated by bacteria; methylmercury accumulates in the fatty tissues of organisms such as fish and magnifies through food chains. In Japan, this process has led to a condition known as Minimata disease, which stunts the development of the brains and nervous systems of children who eat fish or shellfish contaminated with mercury.

Many rivers, lakes, and bays have thousands of toxic substances in their sediments. The increased incidence of cancers discovered in bottom-dwelling fish has been traced to contaminated sediments. Particularly affected in the United States are New York Harbor, Boston Harbor, Puget Sound, Ohio's Black River, and Michigan's Torch Lake. Further, the leaching of toxic chemicals from dumps and landfills into surface waters and groundwater supplies is a serious environmental problem in many industrialized countries.

Persistent Substances

As their name implies, persistent substances are pollutants that normally are not changed or degraded to harmless substances, instead persisting in their original form in the environment. Pesticides, such as DDT and chlordane, and organic contaminants, such as PAHs and PCBs, do not break down easily and tend to magnify through food chains. Consequently, organisms at higher trophic levels, such as the bald eagle and the polar bear, suffer the most serious effects of these persistent substances.

Many plastic products, such as bags, monofilament line and beverage six-pack rings, have a life expectancy of hundreds of years. Plastic pollution is particularly troublesome in marine systems (Figure 13-9). Plastic bags and balloons can become entwined in the stomachs and intestines of sea turtles and mammals, which mistake them for food such as jellyfish. Polystyrene plastic foam breaks up into pellets, which also resemble food. When consumed by sea turtles, the buoyant pellets can keep them from diving and clog their systems, causing them to starve to death. Parent seabirds also can feed these pellets to their young, inadvertently starving them.

FIGURE 13-9: Monofilament gill net around the neck of a California sea lion, *Zalophus californianus*. Because they do not biodegrade, plastics pose a significant threat to wildlife. For example, an estimated six million seabirds and 100,000 marine mammals die each year after becoming entangled in beverage six-pack holders.

Radioactive Substances

Radioactive isotopes, such as strontium 90, cesium 137 and iodine 131, enter water from several sources: the mining and processing of radioactive ores; the use of refined radioactive materials for industrial, scientific, and medical purposes; nuclear accidents; the production and testing of nuclear weapons; and the use of cooling water in nuclear power plants. Many of these isotopes magnify in food chains and can cause cell mutations and cancers. Water pollution from radioactive substances is so serious that when Soviet authorities found extensive groundwater contamination in the area around the 1986 Chernobyl nuclear accident, they sealed off over 7,000 wells within a 50-mile radius of the plant; drilled emergency wells into bedrock to provide water to bakeries and milk-bottling plants; and constructed an emergency pipeline from the uncontaminated Desno River.

Heat

Changes in water temperature can cause major shifts in the structure of biotic communities. Water is used for cooling purposes in manufacturing, industry, and electric power plants. The heated water then is returned to its source. Hot water holds less oxygen, speeds up respiration in aquatic organisms, and tends to accelerate ecosystem degradation. The relatively still waters of lakes and bays are particularly vulnerable to thermal pollution.

Managing Water Resources

In this section, we discuss problems surrounding the use of water worldwide, focusing on four "hotspots" around the globe. We also examine the legislation that guides the management of water resources in the United States. Finally, we look at how this country treats drinking water and wastewater, as well as how it manages and protects groundwater, surface waters, and marine waters.

What Problems Surround the Use of Water Worldwide?

In 1998, Population Action International reported that eight percent of the world's population (nearly half a billion people) living in 29 countries are affected by **water stress** — the episodic lack of renewable freshwater — or the more serious condition of **water scarcity** — the chronic lack of renewable freshwater. Just five years later, in 2003, the United Nations' World Water Assessment Programme (WWAP) published the *World Water Development Report,* which found that a full two billion people living in 40 countries are suffering water shortages. Depending on future rates of population growth, between 2.7 and 3.2 billion people may be living under water-stressed or water-scarce conditions by 2025 (Figure 13-10). By 2050, one in four people will live in countries affected by episodic or chronic shortages of freshwater.

Globally, water resources are the source of hundreds of political tensions, economic concerns, and environmental problems. For example, some countries must share access to rivers, which generates political unrest. Further, many areas have been subjected to prolonged droughts; as the water table falls, streams and waterholes dry up, as do sources of groundwater. Yet, while drought poses a serious problem, many difficulties associated with too little water are of our own making. The combination of intensive agriculture and population growth places a great strain on water availability in arid climates such as the African Sahel, the Middle East, vast areas of the Asian continent, and the American Southwest. We examine each of these regions more closely in the paragraphs below.

African Sahel

Perhaps nowhere is water scarcity more severe than in the three principle countries of the Nile River Valley: Egypt, Sudan, and Ethiopia. Locked in the vise of water scarcity, these nations have growing populations and rely heavily upon the Nile River for irrigation. In 2004, Egypt supported a population of 73 million people and garnered much of the Nile's water. With its population projected to reach 127 million by 2050, Egypt's demand for water will rise dramatically. Likewise, Sudan — whose 2004 population of 39 million is

expected to more than double by 2050 — is becoming increasingly dependent on the Nile River. Even more startling, however, is the situation in Ethiopia, which controls 85 percent of the Nile's headwaters. A country whose current population of 72 million is expected to explode to 173 million by 2050, Ethiopia plans to recover from years of civil unrest by relying on the Nile River to promote economic development. In fact, by the end of 1998, Ethiopia already had constructed 200 small dams and designed extensive irrigation and power generation projects. As Ethiopia continues to increase its reliance on the Nile, less water will be available to Egypt and Sudan — just as their populations demand more. Yet, the difference in per capita gross national product between Egypt ($1,080) and Ethiopia ($100) emphasizes the very real need of Ethiopia to increase its use of the Nile River.

Middle East

Like the African Sahel, the Middle East faces the prospect of severe water shortages within the next several decades. In this region of political turmoil, water may be the resource that forges closer ties — or it may be

Water-stressed and water-scarce countries in 2000

Additional water-stressed and water-scarce countries by 2025

No data available

FIGURE 13-10: Water-short countries in 2000 and 2025. Areas most severely affected by water stress and water scarcity include the African Sahel, the Middle East, India, Haiti, and Peru. ***Source:*** *Population Action International,* People in the Balance: Population and Natural Resources at the Turn of the Millennium *(2002).*

the match that sets off the powderkeg. Syria's plight, especially, illustrates the difficulties of securing clean freshwater in a dry and fast-growing area. To examine Syria's situation, we must look to Turkey, its neighbor to the north. Since the 1930s, Turkey has been planning the Southeastern Anatolia Project, an ambitious effort to develop the nation's eastern region through a series of dams and hydroelectric power plants. The project's cornerstone is the gigantic Ataturk Dam on the Euphrates River, and in 1990, Turkey began to fill the Ataturk Reservoir to reach a volume equal to the Sea of Galilee. To carry this water from the Ataturk Reservoir to eastern Turkey, a series of tunnels measuring over 16 miles in length and 25 feet in diameter was built. When the project is completed in 2010, 22 dams will hold back the waters of the Euphrates and Tigris rivers, both of which rise in eastern Turkey. The flow of the Euphrates could be reduced by as much as 80 percent, creating a near-crisis situation for the downstream nations of Syria and Iraq, both of which are heavily dependent on the river. Because upstream nations are under no legally binding obligation to provide water to downstream nations, the only recourse open to Syria and Iraq (short of war) is to claim that they have historical rights of use and urge Turkey to honor those rights. Syria and Iraq both claim historical use of the Euphrates; Iraq also claims use of the Tigris. In advance of the completion of the Southeastern Anatolia Project, the project's Council of Ministers has ordered the development of a guiding document drafted with input from other governments. However, a technical committee comprising representatives from Turkey, Syria, and Iraq has been unable to agree on how best to meet the water needs of all three countries.

Even now, Syria is beset by serious water-related problems. Its population growth rate of 2.4 percent strains the nation's ability to provide water, food, and power for its people. Syria already has exploited its western farmland, which is fed by rain, and consequently finds that it must farm the arid eastern steppes in order to grow enough food to ensure self-sufficiency. Water withdrawn from the Euphrates River at the Euphrates Dam, located in Tabaq, Syria, irrigates some 500,000 acres (202,347 hectares). But feeding the country's population in 2050 — an estimated 35 million people — will require cultivating millions of additional acres. Currently, there is only enough water in the Euphrates River to operate two of the eight turbines at the Euphrates Dam. One immediate result has been a reduction in power generation; the electricity in Syrian cities routinely is cut off for several hours each day. Although there was never enough water to power all the turbines, Syria maintains that it has even less water now, a consequence of the construction of the Ataturk Dam.

Asian Continent

Conflicts surrounding the distribution and use of water erupt not only between nations, but also within them. For example, the construction of the Three Gorges Dam across the mighty Yangtze River has caused internal discord among China's people. Though the dam is expected to supply one-tenth of China's energy output upon its completion in 2009, it also will flood prime farmland and thousands of villages, displacing more than 1.2 million people and threatening many other species, including the white-fin dolphin, the Chinese alligator, the finless porpoise, the white crane, and several species of monkey.

A major contributing factor to water stress and water scarcity, groundwater depletion is a problem in many regions of Asia. Parts of India and China, for example, are experiencing severe shortages as demand exceeds supply. Competition for the use of surface waters also has become increasingly intense. In 1960, the Aral Sea, located in Central Asia, was the fourth largest inland body of water (in surface area) in the world, behind the Caspian Sea, Lake Superior, and Lake Victoria. Now, it is about one-third its former size (in area) and one-eighth in volume, and its salinity has tripled. Wind blowing across the exposed sea bottom deposits salty grit throughout the region. The principal cause of the Aral Sea's decline is large-scale irrigation, which withdraws water from the sea's two major freshwater tributaries, the Amu Dar'ya and the Syr Dar'ya.

American Southwest

The water problems of the African Sahel, the Middle East, and areas of Asia are half a world away, but a growing population poses similar concerns in the American Southwest. Throughout history, cities in the United States have tried to balance population growth and water use by finding ways to increase quantity while still protecting quality. But today, the water wars in the American Southwest clearly indicate that water is a limited resource. Communities here always have had too little water, and continued growth only exacerbates the problem. The struggle over water often pits community against community, farmers against urban dwellers, and ranchers against wilderness preservationists. Water is

Figure 13-11: (a) Trickle drip irrigation system and (b) conventional irrigation system.

power, particularly in this area where, as Mark Twain once noted, "whiskey's for drinken [sic], water's for fightin." Today's water wars are fought with legal suits, research papers, and blueprints rather than with six shooters and dynamite. For example, El Paso, Texas, sued for the right to sink wells into aquifers that traditionally have served other interests in New Mexico. The Central Arizona Pipeline, completed in the early 1990s, was intended to relieve the state's serious groundwater shortage by diverting billions of gallons of Colorado River water to Phoenix and Tucson. However, the water delivered to many Tucson residents was plagued by poor color, taste, and smell due to a combination of factors, including new treatment methods and the corrosion of old plumbing systems. As a result, the city continues to deplete its groundwater resources while attempting to devise a solution to its water dilemma. The Colorado, a once-mighty river, now trickles to the sea, laden with silt, pollution, and salts. From Utah to California, cities and irrigation projects drink the river dry.

Consider that from the one-hundredth meridian to the Pacific Ocean, rainfall averages less than 20 inches per year (with the exception of the Pacific Northwest and high mountains). Westerners live in a near perpetual drought. While Chicago receives about 40 inches of rain per year, Los Angeles averages just 9, and Phoenix averages 8. Many western areas do not receive enough rain to sustain agribusiness or rapid increases in urban growth. Yet, cities like Palm Springs continue watering dozens of golf courses, Arizona farms grow alfalfa in the desert, and dry towns like Phoenix experience phenomenal growth because federal support for large development projects makes water accessible at a low consumer cost.

As long as the price remains low, consumers have little incentive to evaluate future projects in terms of the actual availability of water or to conserve current supplies. For example, drip irrigation systems are more efficient than conventional systems, but because they are costly to install, they are used less frequently (Figure 13-11). But as demand continues to escalate, so, too, will the cost of pumping and transporting water. Moreover, future generations will not have the easy options that have been made available to us: aquifers to tap, rivers to dam, and lakes to drain.

What Is the Safe Drinking Water Act?

Underlying all water management legislation in the United States is the attempt to address one perennial question: How do we increase the quantity and protect the quality of water supplies to keep pace with increasing population? This question took on an added dimension in 1974, when the EPA announced that high concentrations of 66 human-made chemicals, including six suspected carcinogens, were discovered in the drinking water of New Orleans, Louisiana, situated on the banks of the Mississippi River. Many of the contaminants were traced to petrochemical businesses upstream of the city, agricultural operations, and numerous other industries based along the river and its tributaries. These discoveries focused attention on an entirely new water purification problem: removing substances toxic to human health.

In part, the condition of New Orleans' drinking water helped spur the passage of the Safe Drinking

Water Act of 1974. The Act set national drinking water standards, called maximum contaminant levels, for pollutants that might adversely affect human health, and three years later, the first standards went into effect. It also established standards to protect groundwater from toxic wastes injected into the soil, once a common waste disposal practice.

In 1982 and 1983, the EPA established a priority list to set regulations for over 70 toxic substances that are likely to be found in drinking water. In 1986, when Congress reauthorized the Safe Drinking Water Act, it directed the EPA to monitor drinking water for unregulated contaminants and to inform public water suppliers of targeted substances. The 1986 reauthorization also instructed the EPA to set standards within 30 years for all 70 substances on its priority list. By 1995, the priority list had been expanded, and the EPA had set standards for 83 substances.

When the Safe Drinking Water Act was reauthorized again in 1996, the EPA was directed to regulate additional substances, including radon, sulfates, and arsenic. The amendments also instructed the EPA to identify and monitor at least five new contaminants every five years. To better familiarize residents with the water they drink, water suppliers must now provide customers with annual Consumer Confidence Reports that include information about several factors, including the sources of water, levels of regulated contaminants, and health concerns for any contaminants that are in violation.

What Is the Clean Water Act?

In 1972, President Richard Nixon signed the Federal Water Pollution Control Act, commonly known as the Clean Water Act. Amended in 1977, the Act divided pollutants into three classes: toxic, conventional, and unconventional. It stipulated that industries must use the best available technology (BAT) to treat toxic wastes before releasing them into waters. Conventional pollutants, such as municipal wastes, must be treated using the best conventional technology (BCT). All other pollutants, classified as unconventional, must meet BAT standards, though waivers can be granted for pollutants in this class. If the BAT will not protect certain waters, stricter standards (including "no discharge") must be enforced.

The Clean Water Act contained several stipulations pertaining to industry. It established pretreatment standards for industrial wastes that pass through sewage treatment plants. These standards require industries to handle toxic wastes rather than simply dump them into municipal sewers. In the past, such dumping resulted in the production of toxic sewage sludge. The Act also mandated that the EPA or state must grant a permit to allow the release of any pollutant into navigable waters. These permits define and limit the amount of pollutants that can be included in the wastewater and form the primary means for implementing the Clean Water Act. Discharges must meet the appropriate toxic, conventional, or unconventional standards. Unfortunately, the goal — to phase out toxic discharges by 1985 — was not met, largely because the EPA allows local municipalities to enforce compliance with discharge standards.

The Water Quality Act of 1987, a reauthorization of the original Clean Water Act, provided $20 billion to curb water pollution, primarily through the construction of wastewater treatment plants. The federal government provides 20 percent of the money needed; local sources provide 80 percent. (Even so, the cost of building or remodeling existing sewage treatment plants remains prohibitive for many small communities.) The 1987 Act also provided money for nonpoint-source pollution programs, which afforded appropriations to help state and local governments improve the water quality of bays and estuaries. Through a National Estuary Program, money was earmarked to help clean up and protect San Francisco Bay, Puget Sound, and Boston Harbor. In addition, special appropriations have been made to monitor and control pollutants entering the Great Lakes and Chesapeake Bay.

Another provision of the reauthorization granted the U.S. Army Corps of Engineers the authority to regulate the dredging and filling of wetlands. That provision has proven controversial, with some landowners maintaining that limits on the use of private land amount to a "taking" by the federal government. They contend that landowners should be compensated financially for a portion of the development value of the land. Those who support the wetlands provision point out that draining wetlands adversely affects the entire community because it interferes with the invaluable ecosystem services they provide, including water purification and flood control (Figure 13-12). Therefore, when a private landowner develops a wetland, he or she alone realizes the profit, but the entire community pays an ecological and economic cost. Wetlands and wetland regulation are discussed more thoroughly in

Chapter 15, *Biological Resources;* see especially *Focus On: Where Diversity Reigns: Tropical Rain Forests, Coral Reefs, and Wetlands.*

An extension of the Clean Water Act, the Great Lakes Legacy Act of 2002 was passed to address the growing problem of sediment contamination in the rivers, harbors, and other waterways that comprise the Great Lakes system. Persistently high concentrations of contaminants in sediments can make bottom-dwelling fish unsafe to eat and cause disease in aquatic and shore wildlife. The Act authorized $270 million for five years, beginning in 2004, to remediate contaminated sediments in the Great Lakes; this remediation also includes funds designated for research and public outreach.

FIGURE 13-12: Wetlands provide numerous ecosystem services for surrounding communities.

How Are Drinking Water Supplies Treated?

After water is withdrawn from a lake, river or aquifer, it is treated before being distributed to local destinations such as homes, businesses, schools, and hospitals. Most water treatment systems, large or small, include certain basic steps: A chemical such as aluminum sulfate (alum) is added to water supplies to create small gelatinous particles, called floc, which gather dirt and other solids. Gentle mixing of the water causes floc particles to join and form larger particles; floc and sediment fall to the bottom and eventually are removed as sludge. The water then is filtered through a granular material such as sand or crushed anthracite coal (carbon). Chlorine is added to kill bacteria and other microbes, and a chemical such as quicklime is added to reduce acidity and prevent corrosion in city and household pipes. Many municipal treatment plants also add fluoride to water supplies to prevent tooth decay.

Treated water is sent through a network of pipes to consumers. To ensure its quality, water must be monitored and tested throughout the treatment and delivery process. Generally, surface water is more complicated to treat than groundwater because it is more likely to be contaminated.

How Is Wastewater Treated?

In rural (and some suburban) areas that have suitable soils, sewage and wastewater from each home usually are discharged into a **septic tank**, an underground tank made of concrete or fiberglass. Solids settle to the bottom of the tank. Grease and oil rise to the top where they are trapped and periodically removed to prevent them from clogging the tank. Through anaerobic respiration, bacteria in the wastewater feed on the sludge and scum and liquefy the waste products. The treated wastewater then filters out into a drain field. This process requires time, and septic systems must be large enough to accommodate the expected flow from each home (Figure 13-13).

Wastewater in urban areas must be managed much differently. Underground sewers collect wastewater from residences, businesses, schools, hospitals, industrial sites, and other buildings and transport it to sewage treatment plants. These plants are designed to make wastewater safe for discharge into streams or rivers or for reuse.

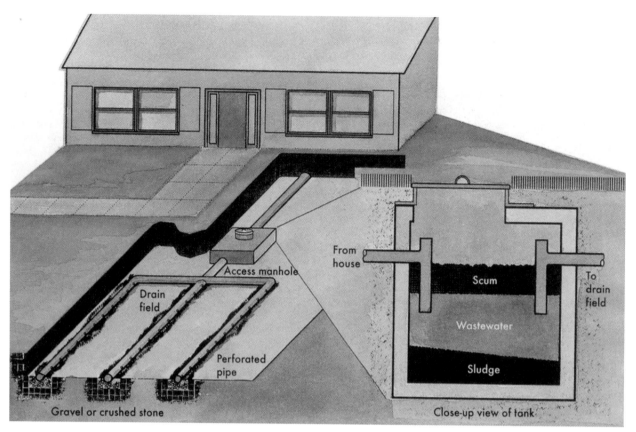

Figure 13-13: Septic tank system. In a properly functioning septic tank, only treated wastewater filters into the drain field. The drain field should be constructed so as to allow the water to drain properly and the soil to absorb the water and nutrients. For this reason, drain fields should not be placed in clay or sandy soils. When septic systems are not functioning properly, effluent (the water that leaves the tank) can be seen rising to ground level and drains or toilets operate slowly or not at all. Odors can often be detected near or below drain fields.

There are two kinds of sewer systems: combined and separate. A **combined sewer system** carries both wastewater and rainwater. Pipes that collect each type of water connect with an interceptor pipe that leads to a treatment plant. The interceptor pipe is large enough to hold several times the normal combined flow. But during storms, the rainwater flow might increase by a factor of 100. Since the combined sewer system is designed to protect the treatment plant, which can handle only a certain amount of flow, and to prevent flow from backing up into buildings, some of the combined flow (raw sewage and rainwater) may bypass the treatment plant and go directly into a receiving stream.

As the name implies, a **separate sewer system** consists of sanitary sewers, which carry only wastewater, and storm sewers, which carry only rainwater and melting snow. Rainwater follows a different route from that of wastewater, bypassing the treatment plant and going directly into a receiving stream. Separating sanitary and rainwater flow is costly, but it does significantly lower the volume of pollutants that enter a receiving stream after a storm. Unfortunately, separate systems have one significant flaw. Rainwater often picks up and transports sediments, oils, greases and other non-point-source pollutants, and because it is not treated, these pollutants are carried directly into the aquatic environment.

Once the flow enters a sewage treatment plant, it usually goes through a multistage process to reduce it to an acceptable effluent, the water that leaves the plant (Figure 13-14). Properly treating sewage maintains the integrity of the stream or river receiving the effluent.

Primary treatment is a physical process to remove undissolved solids. Screens remove sticks, rags, and other large objects. In some plants, the remaining sewage is chopped or ground into smaller particles, then passed to a grit chamber where dense material such as cinders, sand, and small stones settle out and are removed. Undissolved suspended materials, including greases and oils, are removed in a settling tank or primary clarifier. Greases and oils are skimmed off

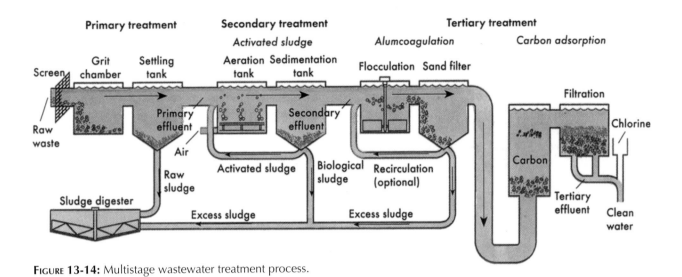

Figure 13-14: Multistage wastewater treatment process.

of the top; the undissolved organic material sinks to the bottom, where it forms a mass called raw sludge. This sludge is drawn off to a sludge digester where it is further reduced by anaerobic bacteria.

In plants that provide only primary treatment, the remaining liquid, called primary effluent, may be chlorinated and released to the receiver stream. Primary treatment is only about 50 percent effective at removing solids. It cannot remove excess nutrients, dissolved organic material, or bacteria.

Secondary treatment is a biological process. At sewage treatment plants, primary effluent is treated in one of two ways: trickling filters or activated sludge. Plants with trickling filters have tanks with beds of stones several feet thick. As wastewater passes, or trickles, through the beds, a diverse population of organisms living on the stones, including aerobic bacteria, protists, rotifers, algae, fungi and insect larvae, consume most of the dissolved organic matter. After leaving the filter, wastewater is allowed to settle in a secondary clarifier and then may be disinfected with chlorine and released to the environment.

In plants using activated sludge, primary effluent is pumped into an aeration tank (in which oxygen is added) and combined with biological sludge, creating a suitable environment for aerobic bacteria to digest the dissolved organic material still present in the wastewater. After aeration, the secondary effluent passes to a sedimentation tank or secondary clarifier where biological sludge is removed. Part of the sludge is recycled back to the aeration tanks, and the remainder is removed to the digester. The secondary effluent can then be chlorinated and released. Secondary effluent is usually considered to be 85 to 90 percent treated.

Tertiary treatment may be a physical or chemical process. It is designed to remove ammonia, nutrients, and organic compounds still remaining in the wastewater. Nearly all of the remaining solids or chemicals can be removed by sand filters, alum coagulation, or activated carbon **adsorption** — whereby organic compounds bind, or stick, to carbon particles but do not chemically combine with them. Tertiary effluent then can be disinfected and released to the environment. Tertiary effluent is considered to be 90 to 95 percent treated. In many instances, it is of a higher quality than the water in the receiver stream and is near drinking water quality. If phosphates are removed during this stage, such advanced treatment can render the water as much as 97 to 99 percent treated. However, because the cost of tertiary treatment is often prohibitive, most sewage treatment plants in the U.S. end the process with secondary treatment.

Proper sewage plant construction and operation are expensive, and for many towns and municipalities, the costs of upgrading are prohibitive. Large metropolitan areas also are struggling to pay for new plant construction to accommodate increased populations. Though the Clean Water Act aims to provide funds for maintaining treatment plants, more could be done on the national, state, and local levels to foster better wastewater management. Perhaps funding priorities should turn toward research and development: Effluent and properly treated sewage sludge can be resources used for constructive purposes. (Chapter 20, *Unrealized Resources: Solid and Hazardous Wastes,* explores ways in which effluent and sewage sludge can be used.) In addition, individuals should begin to realize that, once flushed, sewage is not merely "out of sight, out of mind." See *What You Can Do: To Protect Water Resources* (page 293) for ways to reduce the amount of wastewater leaving the home.

How Is Groundwater Managed?

Groundwater withdrawal in the United States has increased 300 percent since the 1950s. As demand increases, both the quantity and quality of this essential resource are threatened. Our past use of groundwater supplies was based on two popular misconceptions: (1) there is an inexhaustible supply of groundwater, and (2) groundwater is purified as it percolates through the ground. Let's examine both misconceptions more closely.

It is difficult to imagine that the total groundwater supply in the United States, some 30 to 60 quintillion gallons, could be exhausted, but because groundwater is not evenly distributed and because all aquifers do not recharge in the same way, many aquifers experience overdraft. An **overdraft** is the withdrawal of water from a source faster than it can recharge; it is akin to writing a check for more money than you have deposited in your bank account. Recharge also is adversely affected by draining wetlands, clearing forests and diverting streams, all of which can reduce the amount of water absorbed by the soil. Paving for buildings, parking lots, and roads prevents water from entering the soil, as do prolonged drought and flooding.

In 2000, the USGS found that 68 percent of all groundwater extracted in the United States is used for agricultural purposes, particularly irrigation. Irrigation uses seven times more groundwater than all of our city water systems combined. Especially in the western and southern states, irrigation strains groundwater resources. Severe overdrafts, called **water mining**, can deplete aquifers, lowering the water table so drastically that further extraction is no longer economically feasible. The High Plains aquifer of the U.S. West has suffered water mining largely as a result of accelerated pumping for irrigation. In the early 1980s, about 16,000 square miles (41,440 square kilometers) of the aquifer experienced water table declines of more than 50 feet (15 meters), and an additional 50,000 square miles (129,500 square kilometers) fell more than 10 feet (three meters). Floyd County, Texas, endured the most severe decline — 200 feet (60 meters). By 1994, water tables in parts of Texas, the Oklahoma Panhandle, and southwestern Kansas had declined by 140 feet (42 meters). Projecting continued population increases in the South and Southwest, the USGS forecasts widespread, severe depletions throughout the High Plains region by 2020.

Overdrafts can lead to land subsidence. Water pressure in the pores of aquifers helps to support the overlying material. When large volumes of water are removed without adequate recharge, the pores collapse, allowing the rock, sand, or clay particles to settle. If the affected area is large, the ground can collapse, or subside, and form sink holes, cracks, and fissures of varying sizes. Once the ground subsides, there is little chance for it to hold water again, as it generally becomes too tightly compressed to expand with new water. Dramatic effects of subsidence have occurred nationwide. For example, in parts of Florida, sink holes have swallowed trees, cars, and buildings (Figure 13-15).

In addition to problems of quantity, problems of groundwater quality also are becoming critical. For many years, we erroneously believed that groundwater was safe from surface pollutants. We thought that aquifers were deep enough so that contaminants would not reach them. Landfills were routinely located in abandoned gravel pits along rivers, enabling chemicals to leach into both surface and groundwater. Moreover, we thought that pollutants dumped on or near the ground's surface would be filtered out of the water as it percolated through soil, sand and rock layers, so that the water would be purified by the time it reached the water table.

FIGURE 13-15: Land subsidence in Florida. Pauline Bennett looks at what is left of her home in Frostproof, Florida, after a sinkhole opened up under the house.

Because groundwater usually moves slowly, the resulting pollution took a long time to show up. But now, groundwater contamination has reached such magnitude that surveys indicate it is a concern to people in all 50 states. Improperly treated sewage is a problem in rural areas nationwide, where septic tank systems are the norm. Thirty-four states have reported serious groundwater contamination from toxic chemicals. Pesticides, which have been found in groundwater in 25 states, are the major contaminants in agricultural states such as Iowa, Florida, Nebraska, Mississippi, Minnesota, Wisconsin and Idaho, all of which also are among the 10 states most dependent on groundwater for drinking water.

How Are Surface Waters Managed?

Pollution and development pose serious challenges to those responsible for managing the United States' network of lakes, rivers, and coastal areas. The Clean Water Act helped to protect the nation's tens of thousands of lakes by dramatically improving the treatment of wastewater (principally organic wastes), but toxic pollution remains a serious threat. Airborne toxins are thought to contribute significantly to the total contamination in the Great Lakes and are responsible for the acidification of lakes and streams in the Adirondack Mountains and parts of eastern Canada. Likewise, accelerated eutrophication is a problem in many lakes and reservoirs, which receive high inputs of nutrients from wastewater and agricultural runoff. In some lakes, such as Lake Erie, eutrophication has been slowed by banning phosphates in detergents, encouraging farming practices that reduce runoff, and improving wastewater treatment. Finally, development of coastal areas, especially along the Great Lakes and the Gulf of Mexico, can cause soil erosion, thereby contributing to water quality problems.

An extensive system of rivers crosses the United States. Thousands of rivers and streams flow year-round in the East. The Midwest also has many rivers, though some streams are dry for part of the year. In the West and Southwest, many waterways are dry beds for significant periods of time each year. Yet regardless of flow pattern, rivers throughout the country are besieged by numerous threats. In the West and Southwest, the Bureau of Reclamation began a desert reclamation project in 1902 to provide water storage and conveyance for irrigation projects in 17 western states. Proponents of the desert reclamation project maintained that it would pay for itself via increased use of public lands and a growing population. Unfortunately, low interest rates, water subsidies, and procedural loopholes delayed the payback, and the onset of cheap water discouraged conservation. In response, dams and water diversion became the order of the day across the country. On the positive side, dams control flooding, generate hydropower, and store water for a variety of uses. On the negative side, when unusually heavy rainfall causes a dammed river to rise, the build-up of energy can become so great that the river breaches levees and releases all of the energy — and water — in one devastating burst.

Dams, diversions, and channelizations have left few free-flowing rivers in the United States, largely the result of federal policy that puts power generation before environmental concerns. For example, the Federal Energy Regulatory Commission (FERC) grants licenses for hydroelectric plants on free-flowing rivers. FERC also can override states' designations of scenic rivers and permit the construction of dams to generate power. Further, in western states, water rights are guaranteed by law for hydroelectric, agricultural, municipal, and industrial use. Streams can be diverted to the extent that no water remains, leaving the streambed dry. In contrast, few water rights are granted for environmental purposes, including uses for wildlife, wilderness and human recreation, as well as the replenishing of groundwater.

Pollution from a variety of human activities has significantly altered most rivers and streams in the United States. Much of this pollution stems from non-point sources, including acid drainage from mining operations and agricultural and urban runoff. However, grassroots activists and responsible corporate citizens are rising to the defense of the nation's surface waters. For example, the two sewage treatment plants in Milwaukee, Wisconsin, flush approximately 200 million gallons (758 million liters) of wastewater into Lake Michigan each day. Included in that wastewater are hundreds of pounds of toxic chemicals generated by the more than 500 industries that use the Milwaukee District Sewage System. To address this issue, Milwaukee has developed a voluntary program that calls for companies to halt or minimize their toxic waste discharges. Participating companies realize economic savings while lessening their environmental impact. Likewise, in a first-of-its-kind action, the Pillsbury and Midbury Coal Mining Company donated its water rights (worth $7.2 million) for conservation purposes to help maintain flow through Black Canyon on the Gunnison River in Colorado. Black Canyon is considered one of the best natural trout streams in the West and is noted for its scenic beauty and populations of river otters and eagles. In addition, residents in many cities and towns are voting to enact river corridor management programs that create greenways and parkways, regulate flood plain development, and rejuvenate unused manufacturing areas. Similarly, communities in Oregon's Willamette River Basin are helping to protect the river and its watershed; to learn more, read *Environmental Science in Action: The Willamette River Basin* by going to www.EnvironmentalEducationOhio.org and clicking on "Biosphere Project."

How Are Marine Waters Managed?

Both pollution and overuse can result in abuse of the ocean ecosystem. Just as oceans are a vast sink for minerals, nutrients and gases, they also are a sink for pollutants. Eighty-five percent of ocean pollutants come from the land; 90 percent directly affect estuaries, coastal wetlands, coral reefs, and continental shelves. Indeed, humans dump thousands of chemicals into the most biologically productive ocean zones, where they slowly accumulate in aquatic food chains.

Our careless attitude toward the oceans — evident in our indiscriminate disposal of wastes, liberal use and release of toxic substances, and unwillingness to prevent erosion — has begun to exact a tremendous ecological toll on their vital life processes. In the United States, more than one-half of the coastal wetlands that existed at the beginning of European colonization have been dredged and filled, and entire communities have been eliminated from productive estuaries. We also have deposited billions of tons of sewage sludge, garbage, and dredge spoils into coastal and shelf waters. The magnitude of this pollution is so great that it often returns as "gifts from the sea," closing beaches and destroying shellfish beds all along the coasts.

Until recently, little thought was given to managing and preserving marine environments. However, the most promising trend in water resources management is the recognition that coastal zones and oceans are living ecosystems. Realizing the important role that each plays in maintaining the integrity of the biosphere, we are beginning to propose viable management strategies. For example, ongoing research that uses living organisms to biomonitor ocean and estuarine ecosystems appears promising. One such program uses mussels and oysters to record levels of pollution from heavy metals and petroleum hydrocarbons. Because these organisms concentrate and retain pollutants in their tissues, they can make detection easier and help researchers determine the effects of such substances on other constituents of the ecosystem.

Most scientists agree that we cannot afford to degrade our remaining coastal wetlands, estuaries, bays, and coastal zones (Figure 13-16); rather, we need to seek new ways to demonstrate their importance. In Florida, where the population increases by some 6,200 residents each day — most of whom want to live near the coast — almost unrestricted development has destroyed a third of the state's sea grass beds and more than half of its mangrove swamps. In response, Florida has taken several long-term actions to reverse these trends, including purchasing 70,000 acres (28,328 hectares) of shoreline for preservation and replanting mangroves in northern Biscayne Bay. In 1984, the Iron Bridge Sewage Treatment Plant, which serves parts of Orlando and neighboring communities, almost reached its capacity of 24 million gallons (91 million liters) a day, and the city faced building moratoriums unless capacity could be increased and effluent properly disposed. In response, the city established the Orlando Easterly Wetlands Reclamation Project, a 2.6-square-mile (6.6-square- kilometer) wilderness park that accepts treated effluent from the Iron Bridge plant. Already cleaned to secondary standards, the effluent filters through the reclaimed wetlands' three plant

FIGURE 13-16: A vital part of bays and estuaries, sea grass beds provide habitat for many fishes, crustaceans, and shellfish.

➡️ Locate and correct leaks around your home. An estimated 50 percent of all households have some kind of plumbing leak. Worn-out washers and faulty tank valves are the prime culprits.

➡️ To test for leaks in your toilet, add food coloring to the toilet tank. If there is a leak, color will appear in the bowl within 30 minutes.

➡️ Look for dripping faucets. Consider that it takes about 11,600 average-size droplets to fill a one-gallon container. If your faucet is leaking at the rate of one droplet per second, and those droplets are of average size, about 7.5 gallons (28 liters) of water will be wasted each day. Over a year, the leak will send 2,700 gallons (10,200 liters) of unused water down the drain. Leaks are costly: A hot water leak increases your utility bill, and since sewer charges are based on water consumption, leaks also increase your sewer charge.

➡️ Do not use detergents containing phosphates.

➡️ When landscaping, use indigenous plants. For example, if you live in a dry climate, forego green lawns in favor of plants that do not require constant watering.

➡️ Never throw garbage into a body of water.

➡️ Read your water meter before and after a period of hours when no water is used in the house. If the meter shows a change, there is a leak in the house. Reading meters is easy: Most record gallons much as a car's odometer records mileage. For meters that show cubic feet of water used, you can convert it into gallons by multiplying the figure shown on your meter by 7.5, the approximate number of gallons in one cubic foot.

➡️ Get to know a stream, river, pond, lake, estuary, or bay through observation and study; help to protect watershed or drainage basins.

➡️ Become involved with a group or organization involved in restoring a body of water near your home.

➡️ Make local and national politicians aware of your views.

➡️ Install water-saving devices in your home or business:

• Toilet dams block off a portion of the toilet tank, preventing water behind the dam from leaving the water closet. Properly installed, they reduce water use by about two gallons (7.6 liters) per flush.

• Water-saving showerheads have a flow capacity of about 3 gallons per minute, compared to 6 to 9 gallons (23 to 34 liters) for conventional showerheads. Water-saving showerheads can be installed easily with a pair of pliers.

• Low volume toilets use about 1.5 to 3 gallons (6 to 11 liters) of water per flush, compared to 5 or 6 gallons (19 to 23 liters) for conventional toilets.

• Flow restrictors reduce the size of faucet openings, conserving water while maintaining the same pressure. They are inexpensive and easy to install on most faucets and showerheads.

---

communities (deep marsh, mixed marsh, and hardwood swamp) and becomes "polished" before being discharged into the St. Johns River.

In recent years, the United States, as well as nations around the globe, witnessed a series of horrors on the oceans: hypodermic needles, catheter bags, and blood vials — some of which tested positive for infectious hepatitis and the Human Immunodeficiency Virus — washed up on beaches; fish with tumors and rotted fins; and marine mammals and sea birds maimed or killed by plastic debris. Perhaps these and other horrors best illustrate the immediate and ongoing need for individual, national, and international action to protect the Earth's life-giving waters.

Discussion Questions

1. Why is water so important?

2. More than 71 percent of the Earth's surface is covered with water, yet many ecosystems and people suffer from lack of water. Explain how this is possible.

3. Differentiate between surface waters and groundwater and explain how the two are related.

4. What are the major uses of freshwater? How do those uses affect water supplies?

5. List the eight categories of water pollution and briefly describe each.

6. Distinguish between water scarcity and water stress. Name three regions worldwide where water shortages are common.

7. Briefly describe the stages of wastewater treatment. Be sure to distinguish during which stages physical, biological, and chemical treatments take place.

8. Briefly describe the basic stages of municipal drinking water treatment, telling what substances each stage removes and how.

9. Name five things individuals can do to help protect water resources.

10. What is the primary legislation governing water resources in the United States?

Summary

Most of the Earth's water is salty. Only 0.5 percent of all water in the world is found in lakes, rivers, streams, and the atmosphere. Water is classified as either fresh or salt (marine), depending upon its salt content. Freshwater is found on land in two basic forms: surface water and groundwater.

Lakes are categorized according to the amount of dissolved nutrients they contain. Oligotrophic lakes are cold, blue, and deep; often have rocky bottoms; and have a high oxygen content. They contain low amounts of dissolved solids, nutrients, and phytoplankton. Eutrophic lakes are warmer and more turbid, often have muddy or sandy bottoms, and have a lower oxygen content. They are far more productive.

Lakes are composed of internal zones based on depth and light penetration. These include the littoral zone, the area where rooted plants grow; the limnetic zone, deeper water where light can penetrate and support populations of plankton; and the profundal zone, where light does not penetrate enough to allow photosynthesis to occur. Large lakes in temperate climates undergo thermal stratification. In the summer, surface waters heated by the sun become lighter and less dense, rise, and form the epilimnion. Colder, denser water sinks to the bottom to form the hypolimnion. A sharp temperature gradient, or thermocline, exists between the upper and lower layers. In the autumn, the thermocline disappears and the epilimnion and hypolimnion mix, a process known as fall turnover.

Running water habitats include streams and rivers. The speed and temperature of the current affects the kinds of organisms that live in a habitat. Fast-moving rivers and streams usually contain two kinds of habitats: riffles, with a high oxygen content, tend to house the producers of biomass, and pools tend to contain consumers and decomposers.

The ocean is divided into zones. In the euphotic zone, light penetrates easily and phytoplankton are abundant. The neritic zone is that part of the euphotic zone over the continental shelves and near-shore islands. The neritic zone hosts the greatest diversity of life and it supports most ocean fisheries. The deep waters of the open oceans comprise the pelagic zone; below them lies the abyssal zone, the deepest part of the oceans.

The United States has about 20 times more groundwater than surface water. Groundwater percolates downward through the soil from precipitation or surface water and is stored in an aquifer, a water-bearing geologic formation composed of layers of sedimentary material. The depth at which the aquifer begins is known as the water table. Groundwater is the major source of drinking water in about two-thirds of the United States. In addition, it helps to maintain water levels and the productivity of streams, lakes, rivers, wetlands, bays, and estuaries.

Water uses may be nonconsumptive or consumptive. Nonconsumptive uses remove water, use it, and return it to its original source. Consumptive uses remove water from one place in the hydrologic cycle and return it to another. Worldwide, agriculture (a consumptive use) is the chief use of water.

Water pollution can be divided into eight general categories: organic wastes, disease-causing wastes, plant nutrients, toxic substances, persistent substances, sediments, radioactive substances, and heat. Biological oxygen demand (BOD) is a measure of the amount of oxygen needed to decompose the organic matter in water.

Globally, water resources are the source of hundreds of political tensions, economic concerns, and environmental problems. Water stress (the episodic lack of renewable freshwater) and water scarcity (the chronic lack of renewable freshwater) are at the heart of these problems. The African Sahel and the Middle East face an especially difficult situation, as rapid population growth further strains water resources already severely limited by arid climate and shared access to rivers.

Legislation, or the lack thereof, determines how water is managed in the United States. The Safe Drinking Water Act of 1974 set national drinking water standards, called maximum contaminant levels, for pollutants that might adversely affect human health. It also established standards to protect groundwater from toxic wastes injected into the soil. The 1986 reauthorization of the Act instructed the EPA to monitor drinking water for unregulated contaminants and to inform public water suppliers of targeted substances. The 1996 reauthorization built upon the 1986 regulations, requiring the EPA to identify and monitor five new contaminants every five years and to provide residents with annual Consumer Confidence Reports that detail the status of the water they drink.

The 1972 Federal Water Pollution Control Act, commonly known as the Clean Water Act, divided pollutants into three classes: toxic, conventional, and unconventional. The act stipulated that industries must use the best available technology (BAT) to treat toxic wastes before releasing them and the best conventional technology to treat conventional pollutants, such as municipal wastes. Unconventional pollutants must meet BAT standards. The 1987 reauthorization of the Clean Water Act granted the Army Corps of Engineers the authority to regulate the draining and filling of wetlands, though that authority remains contested by many landowners. An extension of the Clean Water Act, the Great Lakes Legacy Act of 2002 was passed to address the growing problem of sediment contamination in the rivers, harbors, and other waterways that comprise the Great Lakes system.

After water is withdrawn from a lake, river or aquifer, it is treated before being distributed. In areas that have suitable soils, sewage and wastewater from each home usually is discharged into a septic system consisting of an underground tank and a drain field. Wastewater in largely urban areas must be collected in underground sewers and treated in sewage treatment plants. These plants are designed to make wastewater safe for discharge into streams or rivers or to make it acceptable for reuse. Most sewage plants employ a multistage process to reduce wastewater to an acceptable effluent.

Fresh and marine waters worldwide are besieged by numerous threats. Many aquifers experience overdraft, the withdrawal of water faster than the aquifer can be recharged. Overdrafts can lead to land subsidence. When groundwater becomes contaminated, it is a costly, slow, and sometimes impossible task to remove pollutants. Toxic pollution and accelerated eutrophication are serious threats to the United States' lakes. Threats to rivers include dams, diversions, channelization, and pollution. Abuse of ocean and coastal ecosystems can result from pollution and overuse. Until recently, little thought was given to effectively managing and preserving marine environments.

KEY TERMS

abyssal zone

adsorption

aquifer

biological oxygen demand (BOD)

combined sewer system

confined aquifer

desalination

epilimnion

euphotic zone

eutrophic lakes

groundwater

hypolimnion

limnetic zone

littoral zone

neritic zone

oligotrophic lakes

overdraft

pelagic zone

profundal zone

running water habitats

runoff

separate sewer system

septic tank

standing water habitats

surface waters

thermal stratification

thermocline

unconfined aquifer

water mining

water scarcity

water stress

water table

watershed

CHAPTER 14

Soil Resources

Soil is as lively as an army of migrating wildebeest and as fascinating, even beautiful, as a flock of flamingoes. Teeming with life of myriad forms, soil deserves to be classified as an ecosystem in itself — or rather, as many ecosystems.

Gaia, An Atlas of Planet Management

Moving together in a perfect rhythm, without a word, hour after hour, he fell into a union with her which took the pain from his labor. He had no articulate thought of anything; there was only this perfect sympathy of movement, of turning this earth of theirs over and over to the sun, this earth which formed their home and fed their bodies and made their gods....They worked on, moving together — together — producing the fruit of this earth — speechless in their movement together.

Pearl S. Buck

Learning Objectives

When you finish reading this chapter, you will be able to:

1. Describe the major components of soil and explain how soil is formed.

2. Describe a typical soil profile.

3. List the major uses for land and describe how they affect the soil and the environment.

4. Compare conventional and sustainable agricultural practices.

5. Discuss methods to prevent soil erosion and maintain soil fertility.

Like air and water, soil is essential for life, and when used wisely, it is a renewable resource. Unfortunately, most of us take this precious resource for granted. We are aware of it only when we track it into our homes, wash it from our clothes, or clean it off our automobiles. And yet most of the food we eat, the clothes we wear, the medicines we use, and the materials in our homes and vehicles originate, directly or indirectly, in the soil.

If there is one concept we hope to leave you with by the end of this chapter, it is this: First and foremost, healthy soil is a living entity. Billions of organisms live in soil and help to generate and maintain its fertility, enabling it to nourish the plants that, in turn, sustain all animals, including humans. The terrestrial web of life, then, is firmly rooted in the living soil beneath our feet. In this chapter, we describe soil resources and examine efforts to manage them.

Describing Soil Resources

In this section, we look closely at what soil is, how it is formed, and how living organisms help to maintain soil fertility. We also discuss how land is used and how those uses affect soil.

What Is Soil?

Soil is the topmost layer of the Earth's surface. It is an ecosystem composed of both abiotic and biotic components — inorganic chemicals, air, water, decaying organic material, and living organisms (Figure 14-1). As such, it is subject to the dynamics that operate in all ecosystems.

The abiotic composition of soil varies from one location to another, but on average it is 45 percent minerals (particles of stone, gravel, sand, silt, and clay), 25 percent water (the amount varying with rainfall and the soil's capacity for holding water), 25 percent air, and five percent **humus**, partially decomposed organic matter. Humus is an essential component of fertile soil. The humus content of soil accumulates over many hundreds, or even thousands, of years and remains at a fairly constant level unless it is carried away by erosion. Humus helps to retain water and maintain a high nutrient content, thus enabling soil to remain fertile.

Soil fertility refers to its mineral and organic content, while **soil productivity** refers to its ability to sustain life, especially vegetation. A primary factor in productivity is the availability of water. Desert soils, for example, have high fertility but low productivity largely because of a lack of water. Rain forest soils have low fertility, since the nutrients are bound up in the vegetation, but high productivity in part because of ample rainfall.

Maintaining soil productivity depends on maintaining soil as a living, renewable resource. Like air

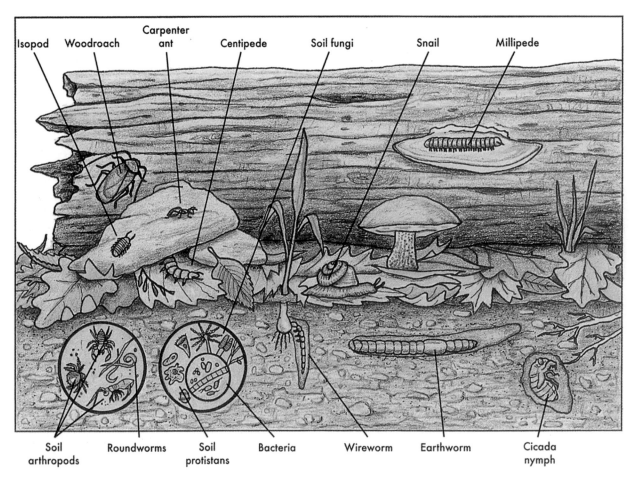

FIGURE 14-1: A soil ecosystem.

and water, soil can be degraded and depleted to such an extent that it becomes, for all practical purposes, nonrenewable. We degrade soils when we cover them with concrete and asphalt; poison them with pesticides, herbicides, and toxic wastes; or allow them to wash or blow away.

What Is Soil Texture?

Since soil is composed mostly of minerals, it is the mineral content that determines the texture, or feel, of the soil. The mineral content consists largely of sand, silt, or clay particles. Sand particles range from 0.002 inches to 0.08 inches, about the thickness of a paper clip, and feel gritty to the touch. Silt particles range from 0.00008 to 0.002 inches; rubbed between the fingers, silt particles feel like flour. Clay particles are even smaller in diameter; they feel like corn starch when dry, but are sticky when wet. Soil texture is determined by the proportions of sand, silt, and clay present (Figure 14-2).

Texture determines the oxygen-holding and water-holding capacity of a soil. There are three broad categories of soil based on texture: loams, clays, and sands.

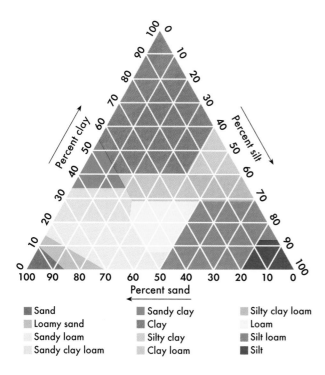

FIGURE 14-2: Soil texture classes. The U.S. Department of Agriculture classifies soils according to the percentages of sand, silt, and clay they contain. Loams are the best soils for cultivation.

Loam soils, comprised of about 40 percent silt, 40 percent sand and 20 percent clay, have the best texture for growing most crops. These soils provide adequate air spaces, and they allow good drainage while retaining enough moisture for plant growth. Clay soils can retain too much water; they also can become deficient in oxygen, either because the spaces between particles become filled with water or because the soil becomes compacted. Sandy soils are typically too porous to retain sufficient moisture for plant growth.

What Is Soil Structure?

The texture of a soil helps determine the soil's structure. Soil structure, or **tilth**, is the arrangement of soil particles, or how they cling together to form larger aggregates such as crumbs, chunks, and lumps. The structure of a particular soil is strongly influenced by the amount of clay and organic material it contains. Because of their physical and chemical properties, clay particles and organic matter are able to form links with other particles, thereby forming larger soil aggregates to which water molecules and nutrients adhere. Soils that lack clay or organic matter have an unstable structure and are likely to form dust or loose sand, which can easily blow or drift away.

How Is Soil Formed?

All soil types result from physical, chemical, and biological interactions in specific locations. Just as vegetation varies among biomes, so do the soil types that support them. Soils of the tundra and rain forest differ vastly from each other and from soils of the prairie and deciduous forest. Scientists typically recognize 10 major soil types, or orders, but there are many variations of these. An estimated 100,000 different soil types have been formed through a combination of five interacting factors: parent material, climate, topography, living organisms, and time.

Parent material, the raw minerals from which soil eventually is formed, has the greatest effect on the texture and structure of soil. Soil begins to form when rocks or other materials are broken down into smaller particles by weathering, abrasion, and dissolution. Weathering occurs when water from rain or snow seeps into the cracks of rocks, freezes, expands, and breaks the rocks into smaller pieces. The abrasive action of moving glaciers, ocean waves, rivers and streams, and wind also breaks down parent material. The minerals in parent material can be dissolved by the action of acids secreted by organisms or deposited on land from the atmosphere. For example, water and carbon dioxide form carbonic acid, which reacts with calcium to form carbonates, which are dissolved and carried away by water passing over the rocks.

Climate is an important factor in the formation of soil. The **macroclimate** of a region, or the average weather pattern including precipitation and temperature, affects the soil because it regulates both the amount of water entering the soil and the amount that can be evaporated from its surface. Within a region, soils may vary because of variations in microclimate. The **microclimate**, or the weather conditions just above the surface of the ground, influences soil temperature, rates of chemical and microbial activity, and plant growth.

The topography of the land influences the natural drainage and amount of erosion that occur in a particular region. Steep slopes, which are usually well drained, often have little topsoil because much is carried away by wind and water. Valleys, usually poorly drained, often contain thick layers of soil that have been washed or blown there from other areas.

Living organisms participate in soil formation in several ways. Pioneer lichens and mosses secrete mild acidic solutions that dissolve rock. The roots of herbs and shrubs penetrate cracks in rock, eventually expanding the cracks and breaking the rock into smaller particles. As organisms die, they add organic matter to parent material, improving moisture retention and nutrient availability. Eventually, complex food webs form that promote soil formation and fertility.

The final factor affecting soil formation is time. Because parent material changes into soil only very slowly, it may take hundreds or even thousands of years for mature soil to develop. Thus, there are places where soil has accumulated to a thickness of several feet or more and others where it is only a thin band less than an inch thick.

What Is a Soil Profile?

As soils develop, they form distinct horizontal layers called **soil horizons**, each of which has a characteristic color, texture, structure, acidity, and composition. A vertical series of soil horizons makes up a **soil profile**. Although soil profiles differ greatly from region to region, soils generally have three to five major hori-

zons. Their thickness depends on the interaction of the physical, chemical, and biological components of the soil as well as the climate, major vegetation type, and length of time the profiles have had to develop (Figure 14-3).

The top layer, or O horizon, consists mostly of surface litter and decaying organic matter. The O horizon protects the lower horizons from the compacting effect of rain and the drying effect of wind. When there is no O horizon, the exposed soil is extremely susceptible to erosion.

Directly below the surface litter is the A horizon. Commonly called the **topsoil**, the A horizon contains humus, living organisms, and some minerals. It is the most fertile part of the soil. Plants with shallow root systems obtain the nutrients they need directly from the topsoil, which usually also retains adequate water and oxygen for their needs.

The E horizon, or **zone of leaching**, lies between the fertile topsoil and the less fertile layers below. Water and soluble minerals pass through this layer. Not all soil types possess an E horizon.

The layer that lies below either the A or E horizon is the B horizon, commonly called the **subsoil**. Soluble minerals, such as aluminum and iron, may accumulate here. The B horizon is not as fertile as the topsoil, but deep-rooted plants can withdraw water, minerals, and oxygen that have leached into it. By transporting nutrients out of the B horizon, these plants keep the minerals in circulation in the ecosystem, since the minerals are returned to the topsoil once again after the plant has died and decomposed (assuming that the plant with its constituent nutrients is not harvested and removed from the ecosystem altogether).

The lowest soil layer is the C horizon; it consists mostly of parent material. Too far underground to contain any organic material, the C horizon lies above the impenetrable layer of bedrock, sometimes called the R horizon.

How Do Organisms Maintain Soil Fertility?

Soil teems with life. Every teaspoonful of topsoil contains billions of beneficial organisms. Most of these are bacteria, fungi, nematodes, and viruses. Nitrogen-fixing bacteria are microorganisms that convert the nitrogen gas in the air spaces of the soil into nitrates, a form that can be used by plants. Soil microorganisms maintain the fertility and structure of healthy soil as

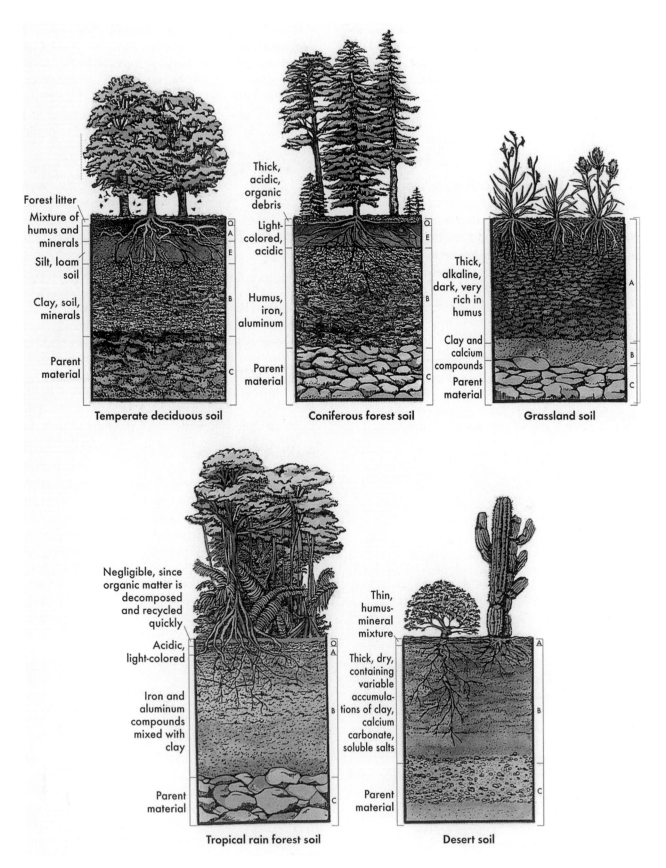

FIGURE 14-3: Generalized soil profiles found in five major ecosystems. Depending on soil type, the number, composition, and thickness of the soil horizons vary.

they decompose organic material, recycle the constituent nutrients, and produce humus. Some microbes, such as penicillin and streptomycin, even provide us with antibiotics.

Larger soil organisms also play an important role in soil fertility. Earthworms turn over an estimated 11 tons of soil per acre per year, mixing fertile topsoil with deeper, less fertile soils and helping with aeration. The digging and burrowing of small mammals, insects and their larvae, and other arthropods also aerate and drain the soil. Larger animals play a role in the nitrogen cycle, since their feces contain many forms of nitrogen that are readily converted into compounds plants can use. Additionally, the matter bound in animal tissues is a reservoir of nutrients that are released gradually as the organisms decompose.

How Is Land Used?

People use land for many different purposes — to grow plants and livestock for food and fiber; to build houses and communities; to pursue recreational activities; to dispose of wastes; and to build roads, highways, and airports. Very simply, the way in which a particular piece of land is used is known as its **land use**. Cities are a high-density land use; they accommodate many people per acre. Low-density land uses include suburbs and farmlands; they accommodate far fewer people per acre.

Depending on soil type and topography, there may be many potential uses for a specific parcel of land. The Soil Conservation Service has developed a system to classify land for different uses based on the limitations of the soil, the risk of damage, and the soil's predicted response to management techniques. Classes I to IV are suitable for cultivation or pasture. Class I land has no limitations to cultivation, while limitations on Class II, III, and IV lands are moderate, severe and very severe, respectively. Classes V to VIII are better suited to forestry, conservation, and rangeland. Rangeland differs from pasture in that its vegetation is sparser and the climate (and land) is drier. In the United States, approximately 50 percent of land is used for crops and livestock, 45 percent for forests and natural areas, and 5 percent for urban centers and transportation corridors.

How Do Land Uses Affect the Soil?

The purposes for which we use land have an immediate and often long-lasting effect on soil resources. Sometimes the effect is obvious: land paved for a parking lot, for instance, cannot be farmed. The spread of low-density residential areas, known as suburban sprawl or suburbanization, consumes two million acres (809,000 hectares) a year in the United States, much of it prime farmland (Figure 14-4). According to the Worldwatch Institute, only 18 percent of all rural land is classified as prime farmland, but a full 27 percent of land within 50 miles of the largest urban areas is prime. In 1997, the American Farmland Trust (AFT), the nation's largest farmland conservation group, published *Farming on the Edge*, the culminating report of a five-year study on farmland threatened by sprawling development. The report named the 20 most threatened agricultural

FIGURE 14-4: Effects of land use. (a) Conversion of farmland to nonfarm uses, such as subdivisions, poses a serious threat to agricultural production. (b) Suburban sprawl places increasing pressure on natural habitats, meaning less space for wildlife.

regions in the United States, all of which are in or next to counties with high population growth rates (Table 14-1). Though these 20 areas comprise just seven percent of all U.S. farmland, they produce a substantial portion of gross national agricultural sales, including 51 percent of fruit crops, 39 percent of vegetables, and 28 percent of dairy products. Working at the state and local levels, the AFT continues to spearhead conservation efforts in these regions by focusing on collaboration among governmental, agricultural, and environmental groups.

Urban growth and suburban sprawl also pose a long-term threat to soil resources worldwide, as the majority of future global population growth is expected to occur in and around cities. According to the Food and Agriculture Organization of the United Nations, fully 83 percent of the population in Latin American and the Caribbean will be urban dwellers by 2030. Similar growth is expected in Asia and the Pacific (53 percent in 2030) and in Africa (55 percent in 2030). Consequently, more and more land will be needed to accommodate housing, schools, hospitals, roads, and commercial development. Without careful management, it is likely that rapid urban and suburban growth will continue to result in the loss of productive soils — soils that increasingly will be needed to keep pace with the world's rapidly growing population and the ever-rising demand for food.

Perhaps the greatest threat to soil fertility and conservation is erosion. Erosion by wind and water is a natural process, and when unexposed, soil is usually replaced faster than it erodes. The amount of soil that can be lost through erosion without a subsequent decline in fertility is known as the **soil loss tolerance level** (T-value), or replacement level. Depending on the type of soil, land may have a T-value between two and five tons per acre per year, with natural processes compensating for that loss with the production of new topsoil. But natural processes cannot offset soil loss as it accelerates above the T-value, nor can they make up for the declining productivity that results.

Runoff from construction sites and urban areas, as well as deforestation, can substantially erode an area's soil resources. However, as much as 85 percent of eroded soil originates from croplands. Agricultural soil erosion is a serious problem worldwide, and scientists believe that as many as one-third of all croplands are eroding faster than natural processes can replace them. The two major causes of agricultural soil erosion are the cultivation of marginal cropland and the use of poor farming techniques on good cropland.

TABLE 14–1: The Top 20 "On the Edge" Agricultural Areas in the United States

1. Sacramento and San Joaquin Valley
2. Northern Piedmont
3. Southern Wisconsin and Northern Illinois Drift Plain
4. Texas Blackland Prairie
5. Willamette and Puget Sound Valleys
6. Florida Everglades and Associated Areas
7. Eastern Ohio Till Plain
8. Lower Rio Grande Plain
9. Mid-Atlantic Coastal Plain
10. New England and Eastern New York Upland, Southern Part
11. Ontario Plain and Finger Lakes
12. Nashville Basin
13. Central Snake River Plains
14. Southwestern Michigan Fruit and Truck Belt
15. Central California Coastal Valleys
16. Columbia Basin
17. Imperial Valley
18. Long Island-Cape Cod Coastal Lowland
19. Connecticut Valley
20. Western Michigan Fruit and Truck Belt

Managing Soil Resources

In this section, we examine the history of agricultural land management in the United States and legislation governing the use of soil resources. We also take a closer look at the differences between conventional and sustainable methods of agriculture.

What Is the History of U.S. Agricultural Land Management?

The first European settlers to North America used standard European farming practices such as rotating crops, fertilizing fields with livestock manure, and liming to neutralize acidic soils. For a time the soil remained fertile despite cultivation. But as land became increasingly scarce and the soils in some areas became less productive from year to year, settlers began to move westward in search of more land. By the late 1700s, agriculture moved through New York to Ohio, and by the mid-1800s, the wave of settlers had reached the Midwest.

To those settlers, the vast, fertile North American continent must have seemed an endless resource. Most farmers had access to plentiful land, but little manpower or horsepower. It was the nation's small population, in fact, that helped spur the development of labor-saving devices. The first horse-drawn reaper came into use in the 1830s; horses gave way to tractors between the two World Wars. The 1950s saw the rise of the combine, which could harvest and thresh crops at once. The average farmer, who could produce food for five people in 1870, could now feed 40 people.

When settlers reached the Great Plains, they encountered land far different from that to the east. The rich grassland soils were resistant to soil erosion and slow to lose their fertility. They continued to produce large yields year after year. But even these soils were not inexhaustible. As nutrients absorbed by the crops were not replenished, soil fertility gradually declined. A cycle of dry spells in 1890 and 1910 further decreased crop productivity and exposed more bare soil.

The Dust Bowl

To capitalize on high grain prices during and after World War I, farmers planted millions of acres in wheat in the short-grass environment of a 150,000-square-mile area in Kansas, Colorado, New Mexico, Oklahoma, and Texas. Further, to avoid losing out on a season's profits, farmers did not periodically allow the land to lie fallow. This practice, coupled with the fact that wheat is a medium grass not suited to the short-grass prairie, began to take its toll. Then, in 1931, an unprecedented drought struck the Great Plains and turned that 150,000 square miles into a "Dust Bowl." The bare, dry topsoil was easily swept up by the wind

and carried thousands of miles, darkening the skies of the eastern United States and rendering life impossible for livestock and people in many parts of the Great Plains (Figure 14-5). The drought severely damaged over 10 million acres (four million hectares) of farmland. In some places, as much as 12 inches of topsoil blew away, exposing the infertile subsoil. Thousands were forced to leave their homes in Kansas, Oklahoma, and Texas. In a land suffering from an extensive economic depression, this exodus of homeless farmers increased the hardships of all.

The devastation of the Dust Bowl brought the issue of soil conservation to the forefront and forced the government to take action. On August 25, 1933, Congress formed the Soil Erosion Service, later renamed the Soil Conservation Service (SCS). One of its first projects was to conduct a nationwide soil survey in order to produce maps and data for erosion control. Armed with this information, the SCS recommended the best conservation practices for particular areas and provided technical assistance when needed. In the following decades, the government implemented programs of economic incentives to keep erodible land out of production and to protect land under cultivation.

Increased Production

In the 1950s and 1960s, research efforts focused on raising food production, largely in response to rapid population growth worldwide. Agriculture became energy-intensive rather than labor-intensive, requiring high inputs of synthetic fertilizers and pesticides (which require the use of fossil fuels in their manufacture); large machinery (which also requires substantial amounts of fossil fuels); and hybrid strains of crops, especially grains like wheat and corn. These changes raised production dramatically. From 1790 to 1930, for instance, American farmers produced an average 22 to 26 bushels of corn per acre per year; that figure jumped to 80 bushels per acre by 1968. In just 20 years (between 1949 and 1969), American agriculture increased output by 50 percent.

But the technological advances that made this "agricultural miracle" possible rendered croplands less and less like natural ecosystems. Unlike natural systems, which are diverse and self-supporting (given energy from the sun), modern farms are often monocultures dominated by one or perhaps several crops that can be maintained only through external chemical inputs.

FIGURE **14-5:** A dust storm approaches Hooker, Oklahoma, on June 4, 1937.

The Farm Crisis of the 1980s

By 1980, a recession, growing inflation, high interest rates, and the soaring costs of energy-intensive agriculture made farming a precarious business. Many of the nation's farmers could not afford to continue without borrowing heavily; those who were unable to repay loans were forced to sell their farms or watch as the bank foreclosed on their property. Hard times had again visited the nation's farming communities.

Since 1960, the number of family farms in the United States has fallen by about half. Many have been purchased and consolidated by large corporations that cultivate just a few profitable crops. In 1985, 30 percent of farmland was controlled by a mere one percent of farm owners. Corporate or industrial agriculture has the capital to take advantage of economies of scale: The expensive, specialized equipment characteristic of conventional farming is most profitable when used on vast acreage.

Corporations involved in agriculture virtually have assumed control of the U.S. food production system — from growing the food to processing, distributing, and selling it. Fewer than 50 corporations account for more than two-thirds of all food processing in the United States. A few dozen firms dominate the food processing, manufacturing, and marketing industries. Since 1970, a handful of petroleum companies have taken over more than 400 small seed businesses, and only a few firms supply half of all hybrid seeds. In the past, small businesses offered a variety of seeds with widely differing characteristics suited to diverse environmental conditions, tastes, and prices. The petroleum

companies, however, concentrated their efforts on the production of crop seeds that require the intensive use of petroleum-based additives. Not coincidentally, these companies also control 75 percent of the chemicals and 65 percent of the petroleum-based products used in agriculture.

Several government programs have promoted the trend toward highly specialized, energy-intensive, and environmentally questionable farming practices. Perhaps the best examples are the federal commodity programs. Although some of these programs were being restructured by the late 1990s, many had been in place for over 50 years, and their impacts — both on farmers' livelihoods and on the land — likely will continue to be felt well into the twenty-first century. In previous decades, farmers enrolled in one such program received subsidies, called price supports, that rose when commodity prices dropped. In exchange for this source of consistent financial support, farmers agreed to allow the program to dictate which crops they could plant and in what amount. Planting acreage in a different crop disqualified a farmer from receiving government payments for that acreage. For example, because soybeans were not subsidized, farmers had to forego price supports on their acreage planted in soybeans. Additionally, to be eligible for subsidies, acreage had to be planted in the same crop for five years in a row, a stipulation that discouraged crop rotation, which bolsters soil fertility naturally. Other commodity programs encouraged farmers to reap the highest per acre yield each year, a short-term goal that necessitates the heavy use of fertilizers and pesticides. Finally, although price supports and quotas helped to ensure a steady food supply, they also may have encouraged farmers to produce more than they could sell in a true free market and to cultivate marginal lands.

In the 1980s, farmers received an average of 38 percent of their income from subsidies; today, farm subsidies account for approximately 20 percent of farm income. According to the Worldwatch Institute, in 2001, countries belonging to the Organization for Economic Cooperation and Development gave $311 billion in agricultural subsidies. Because subsidies are tied to production, payments are weighted toward the largest and wealthiest farmers. In the United States, for example, the largest 10 percent of farms are due to receive approximately two-thirds of an estimated $125 billion in farm subsidies over the next decade. Thus, smaller family farms are often at a disadvantage, unable to compete with large corporate farms.

The image of the family farm and the way of life it represents is a cherished part of the American culture. The qualities we associate with that way of life — hard work, independence, and resilience — are qualities we like to believe define our national character. The farm crisis in America, which had been building for years but came to its apex in the 1980s, saddens and frustrates people precisely because it shakes their beliefs in such a national identity. If the family farm and the small independent farmer are doomed to failure, then perhaps everything they represent may be lost as well.

What Is the 2002 Farm Bill?

Although federal commodity programs can adversely affect soil quality, the U.S. government also has taken steps to protect soil resources. Since the late 1970s, legislators have passed several "farm bills" that include provisions to protect farmland and conserve soil resources. The 2002 Farm Bill, formally called the Farm Security and Rural Investment Act of 2002, is a six-year endeavor that includes a $17-billion conservation title. The bill also reauthorized several existing programs through 2012; these programs include the Conservation Reserve Program (CRP), the Wetlands Restoration Program (WRP), and the Farmland Protection Program (FPP).

Originally part of the 1985 Food Security Act, the CRP was established as a voluntary program in which farmers receive annual "rent" payments of approximately $50 per acre to take highly erodible land out of production. Farmers agree to plant the land in trees, grasses, or legumes for 10 years, with the costs of planting to be shared by the federal government. Further, the Agricultural Stabilization and Conservation Service (ASCS) and the SCS, in conjunction with county agencies, help farmers develop and secure funds for their participation in the program.

The Food Security Act also established the WRP, popularly called "swampbuster," in recognition of the need to protect and restore wetlands that had been converted to farmland. Landowners participating in the WRP may sell or lease land to the U.S. Department of Agriculture (USDA); the USDA then pays all or part of the costs of restoring the farmland to wetlands.

Started under the 1996 Federal Agriculture Improvement and Reform Act (FAIR), the voluntary FPP provides matching grants to help environmental agencies purchase conservation easements from landowners who want to keep their acreage as farmland in perpetuity. A conservation easement is a legal agreement between an agency and a landowner; the easement stipulates that the owner retains property rights to the land but limits the ways in which it can be used. For example, a conservation easement might allow for a parcel of land to continue being farmed but restrict the construction of additional barns, grain silos, or animal lots.

The 2002 Farm Bill also created the Conservation Security Program, which rewards farmers for applying conservation practices to working lands, and the Grasslands Reserve Program, which offers ranchers rental payments in return for a commitment to not develop their land for a minimum of 10 years. Programs such as these play an important role in protecting soil resources and preserving the agricultural and environmental quality of farmland.

How Does Conventional Agriculture Affect the Environment?

Conventional agriculture adversely affects the environment in numerous ways, most notably through soil degradation, soil erosion, reliance on synthetic chemicals, groundwater depletion, poor irrigation practices, and loss of genetic diversity.

Soil Degradation

Soil degradation is one of the most serious problems associated with high-input farming. Degradation, a deterioration of the quality and capacity of the soil's life-supporting processes, usually stems from the hallmarks of conventional intensive farming: erosion, soil compaction, loss of nutrients and biotic activity, and a build-up of salts, acids, or bases in the soil.

All plant communities withdraw nutrients from the soil, but natural ecosystems also replenish nutrients. The organic matter from dead plants and animals stays near the soil surface, where microorganisms can break it down, releasing nutrients for growing plants to absorb. Modern agriculture generally removes most of the plant material from soil; because the nutrients cannot return to the topsoil, the farmer compensates for the loss with fertilizers. Natural systems also include nitrogen-fixing bacteria associated with the roots of certain plants; domestic varieties of such plants include the legumes — alfalfa, soybeans, and peas. As modern agriculture has become more specialized, fewer farmers rotate nitrogen-depleting crops with nitrogen-fixing ones. Rather, they make up the deficit by applying more fertilizer. Further, some farms have no livestock, so manure is not readily available, and synthetic fertilizers are the rule. The heavy machinery used to plow and harvest also compacts soil, lessening its ability to retain water, which then simply runs off the hard-packed surface.

Soil degradation is a key part of a difficult cycle: As land loses its fertility, farmers must find new land to cultivate. However, much of this land is not optimal for farming, so cultivation leads to rapid degradation. In many dry regions, marginal lands have been cultivated by practicing extensive irrigation. Over time, irrigated land can suffer from **salinization**, the buildup of mineral salts, which eventually ruins the soil's ability to produce.

Soil Erosion

Many of the practices that degrade soil also lead to erosion, which removes the fertile topsoil from the land (Figure 14-6). Soil experts estimate that arable land in the United States lost one-third of its topsoil between 1776 and 1976. The USDA reported in 2001 that 27 percent of cropland was considered "highly erodible," as characterized by steeper slopes and lower fertility. But even higher rates of soil erosion occurred in the 1970s, when the global economic climate favored heavy grain production. For example, in 1974, the SCS found that in Iowa, the heart of the Midwest cornbelt, some areas were losing 40 to 50 tons (36 to 46 metric tons) of soil per acre, with erosion reaching 100 to 200 tons (91 to

182 metric tons) per acre on unprotected slopes. Some topsoils once 12 to 16 inches thick are now just six to eight inches thick. A six-inch loss of soil can reduce crop productivity by as much as 40 percent. According to the Worldwatch Institute, for every inch of topsoil lost, yields of wheat and corn drop by about six percent. Think about it: If farms lose even one inch of soil per decade — an amount scarcely detectable on a daily basis — those with six inches left today will deplete their productive soil within one human lifetime.

Erosion due to wind and water continues to be a nationwide problem. Of the approximately 400 million acres (162 million hectares) of productive farmland in the United States, about 90 million acres (36 million hectares) are eroding at one to two times the replacement level, while an additional 100 million acres (41 million hectares) are eroding at two times the replacement level. Some croplands on highly erodible land are even being lost at four times the replacement level.

The Wind Erosion Research Unit, an interdisciplinary team supported by the USDA and Kansas State University, estimates that soil is lost to wind erosion at an annual rate of 2.5 tons (2.3 metric tons) per acre. Such erosion is costly. The loss of topsoil from a 1.2-million-acre (500,000-hectare) plot of land in southwestern Kansas can result in annual yield reductions of 339,000 bushels of wheat and 543,000 bushels of grain sorghum.

FIGURE 14-6: Water erosion on a farm field in northern Ohio. When farmers plow their fields in the fall and leave the fields bare, the soil is unprotected from erosion by wind, rain, and melting snow throughout the winter and early spring.

While wind erosion is most prevalent in the Great Plains states, it is also a serious problem on sandy coastal areas and alluvial soils along river bottoms. In arid areas like the western and central plains, severe wind erosion can lead to desertification. Throughout the United States, about five billion acres (two billion hectares) have been identified as being at risk of desertification. Not only does desertification ruin land, but the resulting dust damages buildings and chokes water sources.

Although experts say that soil is now disappearing 25 percent faster than in the days of the Dust Bowl, the loss is not attracting as much attention because more soil is being eroded by water than by wind. Water causes two-thirds of the erosion on U.S. farmland. Water erosion occurs on cropland chiefly because of a lack of plant cover and root systems; after a crop is harvested, many farmers plow the fields, leaving bare soil exposed. A heavy rain can wash away precious topsoil and even newly planted seeds or seedlings which have not put down firm roots. Sheet erosion occurs when rain falls faster than soil can absorb it. Heavy sheets of water, or runoff, can cut grooves, or rills, into the soil. As the runoff carries away more soil, the rills may form large gullies. Water erosion is also a serious problem in areas that have been deforested, particularly on steep slopes.

The practices that lead to soil erosion also can adversely affect wildlife. When soil fertility declines, more acreage often is brought under the plow to maintain output, resulting in diminished habitat for wildlife. Streams filled with silt from water erosion can reduce fish populations, and fall plowing eliminates ground cover and possible grazing material for land mammals. Research in Manitoba's pothole country, a region dotted with hundreds of small ponds and water holes produced by receding glaciers, showed that duck production was four times greater on fields that had not been tilled than on fields that had.

Reliance on Synthetic Chemicals

The most common synthetic chemicals used in conventional agriculture are pesticides and fertilizers. Agrichemicals that kill crop "pests" are broadly categorized as pesticides. However, many pesticides have specific purposes and can be classified more narrowly. For example, insecticides kill unwanted insects; herbicides, unwanted vegetation; and fungicides, unwanted fungi. Fertilizers, on the other hand, promote the rapid growth of crops.

Pesticide resistance is a serious problem associated with the intensive use of agrichemicals. In the 1950s,

as the use of pesticides became widespread, organisms began to show resistance to the chemicals. When exposed to a pesticide, most individuals in a pest population are killed; a few, possessing a genetic make-up that enables them to tolerate or detoxify the poison, are left to breed. They pass on their resistance to their offspring, and over time, applications of the original pesticide in equal amounts are increasingly ineffective. Pesticides also can kill off beneficial predator organisms, increasing populations of existing pests.

Agrichemicals also pose a threat to animal populations. Consider that each year, approximately 200 to 300 million pounds (90 to 135 million kilograms) of herbicides are applied to cropland in the midwestern United States alone to control weeds. When applied inefficiently, a large amount of these chemicals can enter the environment through agricultural runoff and erosion, posing a threat to nontarget species.

Most agricultural human health problems also occur as a result of direct contact with pesticides, such as when the chemicals are being applied to fields. Worldwide, approximately three million people are poisoned by pesticides annually, and 200,000 die. People exposed to pesticides, especially farmers and migrant workers, often complain of nausea, vomiting, and headaches. Certain pesticides are suspected of causing birth defects, damage to the nervous system, and cancer.

While less hazardous than pesticides, fertilizers can encourage the growth of algae and aquatic plants, hastening the eutrophication of lakes and ponds. Ultimately, the washing of fertilizers into waterways results in both economic and environmental costs by polluting water resources and destroying wildlife habitat.

Rice and cotton account for the largest share of the world agrichemical market; other crops that receive a significant amount of pesticides are corn, soybeans, and wheat. Industrialized countries consume most of the world's agrichemicals, although their use has leveled off in the United States. Health and environmental concerns have led many U.S. farmers to use pesticides more sparingly; some monitor climate, insect presence, and soil conditions year round, spraying only as a last resort. In contrast, agrichemical use is rising in the developing world, which presently purchases approximately 31 percent of global exports (Figure 14-7). Many of the substances exported to developing nations are banned or restricted in the countries in which they are manufactured. DDT, chlordane and heptachlor, all of which are banned for farm use in most industrialized countries, are still commonly used in the developing world.

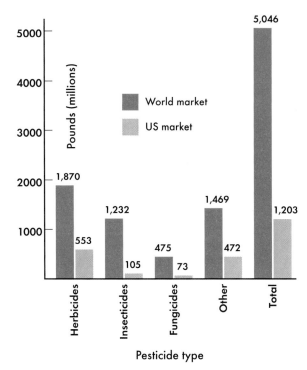

Figure 14-7: World Pesticide Market, 2001. The U.S. Environmental Protection Agency classifies pesticide types as herbicides, insecticides, and fungicides. Included in the "other" category are a variety of chemicals commonly used as pesticides, such as sulfur and petroleum oil.

Groundwater Depletion

Agriculture consumes the lion's share of the world's supply of freshwater. The Worldwatch Institute estimates that farming accounts for roughly 70 percent of global water use. In the United States, the use of groundwater for irrigation increased 300 percent in 30 years, between 1955 and 1985. Today, irrigation remains the major use of the nation's groundwater. Most irrigation systems are grossly inefficient; in many cases, 75 percent of the water is not taken up by the crop.

One-fifth of the nation's total irrigated area — about 10 million acres (four million hectares) — is watered by overpumping aquifers. The depletion of aquifers is particularly severe in eight food-producing states: Texas, California, Kansas, Nebraska, Idaho, Arizona, Arkansas, and Florida. Collectively, these states account for two-thirds of all groundwater pumping in the nation. See Chapter 13, *Water Resources,* for more information on groundwater supplies.

Poor Irrigation Practices

In some ways, modern irrigation has been a boon to agriculture, causing the desert to bloom and raising yields on farmlands from the southwestern United States to the Middle East to China. For example, a farmer in Nebraska can produce about 40 bushels per acre without irrigation water; with it, the yield rises to 120 bushels per acre. But poor management of irrigation water can be environmentally destructive. Irrigating farmland year after year without allowing it to lie idle can cause the soil to become waterlogged as air spaces in the soil fill with water. The excess water leaches away nutrients needed by crops, prevents roots from getting oxygen, kills soil organisms, and promotes mold growth.

Increased salinity is also a problem on poorly drained farmlands that are irrigated perennially. In dry climates, water on or near the soil surface evaporates, leaving mineral salts behind. Over time, the salts accumulate and the soil suffers from salinization. Salinization gradually leads to impoverished farmlands, since most plants cannot tolerate high levels of mineral salts. According to the International Irrigation Management Institute, of the world's 667 million acres (270 million hectares) of irrigated cropland, 49 to 74 million acres (20 to 30 million hectares) are severely affected by salinity, while another 148 to 198 million acres (60 to 80 million hectares) are damaged to a lesser extent. In the United States, about one-quarter to one-third of irrigated land in the United States suffers from salinity. Salinization also affects other important food producing regions, including 23 percent of China's irrigated lands and 11 percent of India's. In Egypt, where traditional irrigation systems were in place for centuries, the widespread use of perennial irrigation, introduced in the 1960s, has salinized one-third of the nation's farmland and waterlogged 90 percent of it.

Adequate drainage systems can minimize salinization, but such systems can be very expensive to install on the huge fields served by modern irrigation systems. And even effective drainage systems cannot prevent salinization indefinitely. The only way to avoid increased salinity is to allow the land to lie fallow periodically so that accumulated salts can be washed from the soil.

Loss of Genetic Diversity

As farmers rely on fewer and fewer hybrid strains of crop species, the risk that crops will be destroyed by pests or disease increases (Figure 14-8). In natural systems, a wide variety of plant strains helps to ensure that some will be resistant to specific pests or diseases. As natural habitats are disrupted, wild relatives of food crops often are destroyed, thus depleting the genetic diversity that could be used to strengthen agricultural strains. The value of wild strains became clear in 1970,

FIGURE 14-8: Monoculture: Corn crop.

when severe southern corn leaf blight struck the U.S. corn crop. At that time, 70 percent of the corn seed came from only six types. Because there was so little genetic diversity, much of the corn crop was vulnerable to the blight. One-seventh of the entire crop was lost, raising prices by 20 percent and resulting in a loss of approximately $2 billion. The damage was finally controlled by crossbreeding with blight-resistant germ plasm that originated in Mexico. As awareness of the importance of genetic diversity has grown, researchers have become more interested in preserving rare and endangered plant species in gene banks (see Chapter 9, *Food Resources, Hunger, and Poverty*). In addition, increased effort is being made to preserve older strains of crop species, known as heirloom varieties.

What Is Sustainable Agriculture?

The problems associated with conventional agriculture have spurred the development of, or return to, methods of **sustainable agriculture**, farming that protects and maintains the fertility and productivity of soil while ensuring a stable and healthy food supply. Sustainable agriculture safeguards soil resources; healthy soil, in turn, supports continued cultivation.

Sustainable agriculture encompasses a broad spectrum of techniques and practices, including growing a variety of crops rather than one or two, rotating crops, using organic fertilizers such as animal manure and crop residues, allowing croplands to lie fallow periodically, planting cover crops to protect the soil between crops or during the winter, and encouraging the natural enemies of crop pests. Farmers may embrace a few or all of these techniques. At one end of the spectrum are practitioners of low-input sustainable agriculture, who use synthetic inputs in reduced amounts. An estimated 30 to 40 percent of the nation's farmers have taken simple steps to decrease the use of chemicals and other inputs on their farms. For most, economics — the need to reduce costs — is the driving force behind the switch to low-input farming. At the other end of the spectrum are organic growers, who disavow the use of any synthetic fertilizers or pesticides. All forms of sustainable agriculture, however, share a common dual goal: to prevent erosion and to maintain soil fertility and structure.

Most of the practices that define sustainable agriculture are not new. Until the middle of the twentieth century, for instance, crop rotation and the use of manure and cover crops were common. However, with the development of pesticides, hybrid strains and other agricultural advances, many farmers abandoned these practices. Yet some groups, such as the Amish, retained sustainable methods, and their farms prospered throughout the farm crisis of the 1980s. The prosperity of Amish farms is due in part to their farming methods and in part to their accounting system. The Amish believe that land ownership is a privilege, unlike industrial agriculture, which counts the "cost of ownership" of the land as a fixed cost. The Amish hire no additional labor and consider their own labor as part of their profit, not their cost.

Measures to Prevent Excessive Soil Erosion

The most effective way to prevent excessive soil erosion is simply to not farm marginal croplands, especially slopes and soils with structures not suited to cultivation. Another effective measure is to plant a cover crop, which is grown when the land is not planted with a main crop. The roots of the cover crop anchor the soil, protecting it from the effects of rain, snow, and wind.

When it's time to plant the main crop, the cover crop litter and residues act as a soil amendment, or **green manure**, vegetation left on the ground or tilled under to fertilize the soil. Likewise, a **windbreak,** or shelterbelt, is a row or group of trees and shrubs planted along the windward side of fields in order to reduce wind erosion. The most effective windbreaks are wide and tall enough to deflect the wind upward and over the cropland while slowing its velocity (Figure 14-9).

Several alternative methods of tilling can help prevent soil erosion. **Ridge tilling** is planting a crop on top of raised ridges. Conservation tillage includes both low-till and no-till methods of planting. **Low-till planting** consists of tilling the soil just once in the fall or spring, leaving 50 percent or more of previous crop residue on the ground's surface. With **no-till planting**, crops are planted amid the stubble of the previous year's crop, which acts as a mulch to fertilize the soil and prevent it from drying out. A no-till planter or drill is used to sow the seeds without turning over the soil (Figure 14-10). One disadvantage of conservation tillage is that the previous year's mulch provides food and cover for insect pests and allows weeds to become established. Consequently, conservation tillage may require the use of pesticides and herbicides.

The risk of erosion is greatest on sloped land, but there are a number of remediation techniques. **Strip cropping** is the alternation of rows of grain with low-growing leaf crops or sod. It offers greater protection to the soil than conventional planting, and the strips of legumes or grasses also help to enhance soil structure and organic content. Strip cropping can be used in conjunction with contour plowing or terracing. **Contour plowing** is tilling the soil parallel to the natural contours of the land rather than in the straight rows and square fields characteristic of conventional fields. Contour plowing helps to keep the soil from washing down the hillside. **Contour terracing**, which is used on steep slopes, consists of broad "steps," or level plateaus, built into the hillside. Swales, or trenches, located at the edges of the terraces act as catch basins for rain, channeling it along the hillside. Swales planted with a grass, called grassed waterways, help to both slow the course of water and absorb excess water and eroding soil.

FIGURE 14-9: Windbreak. Note how the staggered height of the plantings deflects the wind up and away from the house (hidden behind the trees).

Figure 14-10: No-till technique, corn field.

Whether the land is flat or sloped, improving soil structure is imperative to preventing the long-term loss of soil. Soil with a healthy and stable structure is less prone to washing or blowing away because the soil particles tend to stick together. Mulching and fertilizing with crop residues and animal manure improve the structure of the soil and enhance its fertility. To learn more about how one region combated soil loss, read *Environmental Science in Action: The Minnesota River Basin* by going to www.EnvironmentalEducationOhio. org and clicking on "Biosphere Project."

Measures to Enhance Soil Fertility

Four of the most effective measures to enhance soil fertility are allowing fields to lie fallow periodically, rotating crops, using organic fertilizers, and practicing wise irrigation techniques.

A field that is allowed to lie fallow is planted in a cover crop such as hay or clover. **Crop rotation** is the practice of changing the type of crop grown on a field from year to year. Rotating a grain crop, which uses nitrogen, with a legume crop, which fixes nitrogen, is especially effective.

Organic fertilizers are an essential part of alternative agriculture, and they offer some important benefits over synthetic ones. The most obvious advantage is that many can be obtained free or at low cost. Organic fertilizers tend to become an integral component of the topsoil, while synthetics may wash away easily in rains. And, unlike synthetic fertilizers, organics stimulate the growth and proliferation of soil microorganisms, which are essential to maintaining the fertility of the soil over the long term. Although gathering and spreading manure requires more labor than synthetics, organic fertilizers can make up for this drawback with lower cost, better soil fertility, and less risk of environmental damage. However, because organic amendments release nutrients more slowly than synthetic fertilizers, they may not provide the "burst" of nutrients young plants need early in the season to become established and produce sufficient leaf surface to provide the energy needed for growth.

Trickle drip irrigation can protect soil fertility from the salinization that accompanies long-term traditional irrigation. **Trickle drip irrigation** is the delivery of water through permeable or perforated pipes directly onto the soil surrounding the base of the plants. It uses far less water than traditional irrigation methods, resulting in less evaporation and thus leaving fewer salts behind to salinize the soil. However, it is not appropriate for all croplands, such as the dry-land wheat fields of the Great Plains states.

Organic Farming

Although techniques for preventing erosion and enhancing soil fertility can be used on all types of farms, they work best when integrated into a program designed for overall sustainability, such as organic farming. Of all forms of agriculture, organic farms most closely resemble natural systems. They typically promote species diversity through polyculture, the planting of a variety of crops suited to the particular climate and soil of the area. Rather than synthetic fertilizers, organic farmers use animal manure, green manure, and fish emulsions to bolster soil fertility. Likewise, organic farmers control weeds through natural means, such as mulching between plant rows and maintaining biological controls. For example, some organic farmers keep geese and ducks, which eat weeds. Carefully timed planting and cultivating regimens also minimize problems caused by weeds. Killing weed seedlings in the spring with a rotary hoe, for instance, before undesirable plants germinate and go to seed, is far more effective than trying to weed a field after the plants have taken hold.

Insect pests are controlled through a variety of methods collectively known as **integrated pest management**. If the plot is small enough, pests can be picked off by hand. Traps (typically baited with pheromones or other substances) attract and capture pests. Biological control can be achieved in several ways. Microbes, insects, birds, and animals that feed on pests can be introduced to a field or encouraged to take up residence nearby. Some species of predatory insects can reduce pest populations without harming crops. The ladybug, for instance, preys upon crop-damaging aphids. Another simple preventive measure is to provide bird houses and water (a small pond, for example) to encourage the presence of martins and other insect-eating birds.

Organic farming offers several important environmental advantages over conventional farming. It improves the structure of the soil, making it easy to work and enhancing plant growth. Maintaining good soil structure also minimizes runoff, reducing the loss of top soil after heavy rains and the leaching of nutrients. The use of natural, instead of synthetic, fertilizers further enhances soil fertility. Finally, because pests and weeds are biologically controlled, organic farming diminishes the health and environmental risks associated with pesticides.

In addition to its environmental advantages, organic agriculture appears able to compete economically with conventional agriculture, especially in the long-term performance of the farm. The yields of crops with high nitrogen requirements, such as corn, wheat and potatoes, are usually somewhat lower when organic techniques are used. But for some crops, notably alfalfa, soybeans and oats, yields may be higher. In addition, crop rotation can result in higher yields by replenishing specific nutrients for a variety of crops. For example, potatoes yield best when planted on a plot in which corn was grown the year before. Figure 14-11 illustrates how crops might be rotated on a plot in succeeding years in order to maximize benefits to the soil and enhance the yields of various crops. Perhaps most importantly, because organic farmers have lower capital costs for equipment and machinery and do not have to pay for synthetic chemicals, their profits may exceed those of conventional farmers.

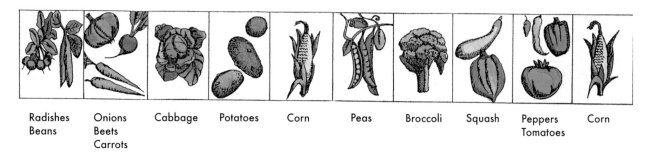

| Radishes Beans | Onions Beets Carrots | Cabbage | Potatoes | Corn | Peas | Broccoli | Squash | Peppers Tomatoes | Corn |

FIGURE 14-11: Ten-year crop rotation plan devised by organic farming expert Eliot Coleman. The diagram represents the first year of a 10-year rotation in 10 plots. Each succeeding year, the crops are moved one plot to the right, with the crop from the tenth plot moving to the first. Rotation is based mainly on the nutrient requirements of the plants. For example, corn, a heavy feeder with high nitrogen requirements, follows legumes (peas and beans), lighter feeders that fix nitrogen.

Despite its many advantages, organic agriculture still has far fewer practitioners than conventional, high-input farming. In 2001, the USDA reported that only 0.3 percent of the nation's croplands and pasture were certified as organic. One reason is economic: limited consumer demand. Many consumers have grown accustomed to picture-perfect produce and have a prejudice against buying produce with any blemishes, though slight imperfections are not harmful and do not signal inferior quality.

This attitude seems to be changing, however, as increasing concern over pesticides is helping companies and consumers to put the cosmetic appearance of produce in proper perspective. For example, by 1990, several supermarket chains announced that they would begin to test certain fruits and vegetables for pesticide residues. In a move of particular import to parents and childcare advocates, the H.J. Heinz Company announced that it would no longer make baby food from produce that contained traces of specific pesticides. A 1999 *Consumer Reports* article validated such action when it revealed the startling results of a study concerning the toxicity of pesticides found on conventionally grown fruits and vegetables. Conducted by the Consumers Union, the study analyzed pesticide-use data collected by the USDA between 1994 and 1997. Of the 27 foods tested by the USDA in those four years, seven — apples, grapes, green beans, peaches, pears, spinach, and winter squash — registered exceptionally high toxicity scores. This news is especially distressing for children, who often eat far more produce per pound of body weight than adults and whose developing nervous and endocrine systems are much more sensitive to the effects of pesticides. One primary way to minimize exposure to pesticides, the article advised, is to buy organic produce.

Organic farming also has fewer practitioners because it requires training, a willingness to deal with a new market, and time — an average of three to five years — to convert a farm from conventional techniques to organic practices. Legal definitions of "organic" farming vary from state to state, and it is not always clear just what the "organically grown" label means. To avoid the confusion, some states where organic farming is popular, such as California and Oregon, where organic farming is more popular, have established agencies to define organic standards. These agencies certify growers who meet their requirements and offer assistance to farmers attempting to make the transition from conventional farming. That assistance is important, for organic farming requires an intimate knowledge of the soil and a sophisticated understanding of ecological processes.

Polyculture

An innovative approach to making agriculture less environmentally harmful is polyculture. While most conventional farms grow just a few crops, farms practicing polyculture grow numerous crops. The diversity of a polyculture helps provide stability for a farm; if one crop fails in a year, the other crops may produce

---

WHAT YOU CAN DO ▷ **To Protect Soil Resources**

➡ Learn about the type of soil in your area and to what uses it is best suited.

➡ Support farmers who use organic and other alternative methods of agriculture by buying their fruits and vegetables or by encouraging your supermarket to buy from them.

➡ If you have a vegetable garden, try biological pest controls, crop rotation, and other alternatives to conventional pesticides and fertilizers. In addition, try planting heirloom crops rather than hybrid species and begin to save your seeds from year to year.

➡ Get involved in your area's land use planning meetings.

➡ Write to your legislative representatives to express your support for measures which will help protect soil resources as well as human health and the environment.

well enough to compensate for it. A polyculture is the agricultural equivalent of the saying, "Don't put all your eggs in one basket."

Perennial polyculture — growing a mixture of self-sustaining, or perennial, crops — is an extended area of polyculture research. Fields have a greater number of species than a conventional or even organic farm. Because perennial plantings do not require that the soil be turned over every year and instead maintain their root systems in the ground, they reduce soil erosion and allow soil to accumulate — much like natural prairies! Research on perennial polycultures is being conducted by Wes Jackson and his colleagues at the Land Insti-

tute in Salina, Kansas. Land Institute researchers are working to develop new strains of wild grain-producing perennials in order to create "domestic" prairies that closely mimic natural ones (Figure 14-12).

As Wes Jackson's research illustrates, managing soil resources is a challenge that concerns all citizens since agricultural health issues affect everyone — urbanites and rural dwellers, consumers and producers alike. (See *What You Can Do: To Protect Soil Resources.*) The goal of environmentally sound management — to sustain the soil ecosystem as a living resource — requires a comprehensive management approach that guides land use and safeguards invaluable arable land.

FIGURE 14-12: Interns at the Land Institute in Salina, Kansas, sample prairie vegetation for research on perennial polyculture.

Discussion Questions

1. Describe the texture, structure, and profile of a typical soil.

2. What factors affect the formation of soil? Briefly describe how these factors interact to form soil.

3. What effects can land use have on soil resources? On water resources? On air resources? What conclusions can you draw about these three resources?

4. Briefly describe some of the problems associated with high-input agriculture. How can sustainable farming techniques reduce or solve these problems?

5. Discuss the environmental implications of soil erosion and salinization.

6. What are some of the advantages and disadvantages of organic farming?

7. Explain the concept of perennial polyculture. What are the benefits of this form of agriculture?

Summary

Soil, the topmost layer of the Earth's surface, is an ecosystem composed of abiotic and biotic components. Humus, which consists of partially decomposed organic matter, helps to retain water and to maintain a high nutrient content, thus enabling soil to remain fertile. The mineral content of soil determines its texture; texture determines the soil's oxygen-holding and water-holding capacity. Soil structure, or tilth, is how soil particles are arranged, that is, how they cling together.

Scientists typically recognize 10 major soil orders and an estimated 100,000 soil types. The soil types have been formed through the interaction of five factors: parent material, climate, topography, living organisms, and time. As soils develop, they form distinct horizontal layers called soil horizons. These layers include the O horizon, or litter layer; the A horizon, or topsoil; the E horizon, or zone of leaching; the B horizon, or subsoil; and the C horizon, or parent material. The C horizon lies above the impenetrable layer of bedrock, sometimes called the R horizon. A vertical series of soil horizons is a soil profile.

Soil contains billions of organisms that help to maintain soil fertility by aerating it and adding nutrients.

The way in which a particular piece of land is used is known as its land use. Worldwide, urban growth and suburban sprawl pose a long-term threat to soil resources. An even greater threat to soil fertility and conservation is erosion. Erosion by wind and water is a natural process, and when unexposed, soil is usually replaced faster than it erodes. The amount of soil that can be lost through erosion without a subsequent decline in fertility is known as the soil loss tolerance level (T-value), or replacement level.

In the United States, no land use has been more important historically than agriculture. When Europeans first arrived in North America, they cleared seemingly inexhaustible forests for farmlands. As land became increasingly scarce and the soils in some areas became less productive, settlers began to move westward. The rich grassland soils of the Great Plains produced large yields for years, but soil fertility gradually declined. High grain prices during and after World War I and years of sufficient rainfall encouraged farmers to plant millions of acres with wheat and to practice continuous cropping. In 1931, drought returned to the Great Plains, severely damaging over 10 million acres of farmland and creating the Dust Bowl. In response, Congress formed the Soil Erosion Service, later renamed the Soil Conservation Service (SCS). In the following decades, the federal government implemented programs that paid farmers to keep erodible land out of production.

In the 1950s and 1960s, agricultural production was increased by the use of high inputs of chemicals, large machinery, and hybrid strains of crops. But high-input or conventional agriculture can cause serious problems, including soil degradation, soil erosion, reliance on fossil fuels and agrichemicals, groundwater depletion, overirrigation, and loss of genetic diversity.

In part, the Farm Crisis of the 1980s was a result of high-input practices. Coupled with an economic recession and high interest rates, the soaring costs of energy-intensive agriculture forced farmers to borrow heavily. Those who could not repay their loans faced the difficult decision of selling their farms or watching as the bank foreclosed on their land.

The Farm Security and Rural Investment Act of 2002 focuses on protecting farmland and conserving soil resources. The 2002 farm bill reauthorized existing programs and created several additional ones that encourage farmers to practice more conservation-friendly farming.

Various systems of alternative or sustainable agriculture can protect the soil and restore its fertility. Sustainable agriculture involves techniques such as growing several crops, rotating crops, using organic fertilizers, allowing croplands to lie fallow periodically, planting cover crops, and encouraging the natural enemies of crop pests. Alternative methods of tilling can help prevent soil erosion. Ridge tilling, or planting crops on top of raised ridges, minimizes the use of pesticides and fertilizers. Conservation tillage includes both low-till and no-till methods of planting. Low-till planting is tilling the soil just once in the fall or spring, leaving 50 percent or more of previous crop residue on the ground. No-till planting consists of sowing seeds without turning over the soil.

A number of techniques for sloped land can be used to reduce erosion, slow the flow of water, and add organic content. Strip cropping is planting alternating rows of grain and low-growing leaf crops or sod. Contour plowing is tilling the soil parallel to the natural contours of the land. Contour terracing is building broad steps or terraces into a hillside. Swales are trenches along the edges of terraces; when planted with grass, they slow water movement and prevent soil loss.

Of all forms of agriculture, organic farms most closely resemble natural systems. They are characterized by diversity. Soil texture and fertility are enhanced by organic soil amendments. Weeds are controlled by mulching and cultivating techniques. Insect pests are controlled through a variety of methods collectively known as integrated pest management. Biological pest control involves the introduction of insects, birds, or microbes that attack pests. Crop rotation prevents the depletion of soil nutrients and promotes higher yields. Perennial polyculture, a new agricultural approach in which a number of crops are established in an area, also renders agriculture less environmentally harmful.

KEY TERMS

contour plowing	no-till planting	strip cropping
contour terracing	perennial polyculture	subsoil
crop rotation		sustainable agriculture
green manure	ridge tilling	
humus	salinization	tilth
integrated pest management (IPM)	soil fertility	topsoil
	soil horizon	trickle drip irrigation
land use	soil loss tolerance level (T-value)	
low-till planting		windbreak
macroclimate	soil productivity	zone of leaching
microclimate	soil profile	

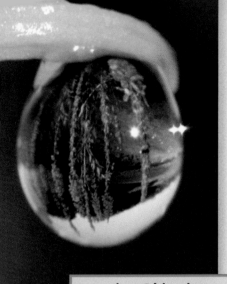

CHAPTER 15

Biological Resources

Learning Objectives

When you finish reading this chapter, you will be able to:

1. Define biological diversity. Differentiate between genetic diversity, species diversity, and ecosystem diversity.

2. Define extinction and explain the relationship between extinction and biological diversity.

3. Identify the major threats to biological diversity.

4. Explain why it is important to preserve biological diversity.

5. Briefly relate how plant and animal species have been managed historically.

6. Describe how species are currently managed both within and outside of their natural habitats, and explain the advantages and disadvantages of each.

7. Discuss the major legislation designed to preserve biodiversity.

For one species to mourn the death of another is a new thing under the sun. The Cro-Magnon who slew the last mammoth thought only of steaks .... The sailor who clubbed the last auk thought of nothing at all. But we, who have lost our last pigeons, mourn the loss.

Aldo Leopold

Each species, to put the matter succinctly, is a masterpiece. It deserves that rank in the fullest sense; a creation assembled with extreme care by genius.

E.O. Wilson

Earth is a planet defined by life, life that inspires awe and reverence, wonder and delight (Figure 15-1). Unfortunately, many of the life forms and living systems that grace this lovely planet are in danger of disappearing forever. Because it's difficult to care about something that we don't know — and may never have even heard of — we want to showcase just a small measure of the Earth's biological wealth. For that reason, we have lined the bottom of each page in this chapter with photos of various species and ecosystems — about 100 photos in all. If we were to include every species known to science, the chapter would have to be 425,000 pages long! It would probably be at least several million pages long if every species that exists on Earth were included! In this chapter, we learn about biological resources and efforts to manage these irreplaceable treasures.

(left–right) Wild Potato, Guadelupe Mountains National Park (Texas), Jellyfish, Mountain Gorilla

Figure 15-1: Two earthly treasures: a) Dewy dragonfly and b) Freshwater diatoms.

Biological diversity, or **biodiversity**, refers to the variety of life forms that inhabit the Earth. We often think of biodiversity at the level of **species diversity**, the millions of distinct species that share this planet with us. Scientists have discovered and described approximately 1.7 million species. (Many other species may be known and named by local peoples, but as indigenous cultures are assimilated into the dominant culture, the knowledge they possess can be lost.) Even so, we do not know how many species exist, not even to the nearest order of magnitude. Conservative estimates of the total number of species range from three to 50 million, most of them microbes, insects and tiny sea organisms, but some researchers believe the figure may be much higher. Terry Erwin, a biologist with the Smithsonian Institute, conducted a now-famous study in the Panamanian tropical rain forest in the early 1980s. He discovered a tremendous number of previously unknown beetle species, so many that he estimates the total number of insect species alone may be 30 million! Edward O. Wilson and Paul Ehrlich, both highly respected biologists and outspoken conservationists, estimate that there may be as many as 100 million species. Their estimate is based, in part, on Erwin's and others' work that indicates that the diversity of insects and other arthropods is probably much higher than had previously been estimated for the *entire* world flora and fauna. Moreover, little study has been devoted to nematodes, fungi, mites and bacteria, each of which is highly diverse, containing undescribed species that may total in the hundreds of thousands.

In addition to species diversity, we can measure biodiversity in terms of **ecosystem diversity**, the variety of habitats, communities, and ecological processes in the biosphere. Some entire ecosystems, such as rain forests

Describing Biological Resources

In the following pages, we carefully examine biological resources — what they are and where they are found. We discover how beliefs and attitudes affect biological resources, and we learn about the greatest threats to biological diversity. This section ends with a look at extinction, its relationship to biological diversity, and reasons for preserving biological resources.

What Are Biological Resources?

All species, from the smallest microorganisms to the great blue whales, are **biological resources**. Many scientists classify organisms into five kingdoms: Monera, Protista, Fungi, Plantae, and Animalia (Figure 15-2). Some biological resources, including crop plants, livestock and pets, are domesticated. They are discussed in Chapter 9, *Food Resources, Hunger, and Poverty,* and we mention them again here, but this chapter focuses primarily on wild biological resources, the Earth's most important and perhaps most threatened resource.

(left–right) Cardinal, Cicada, Greenland, Irises

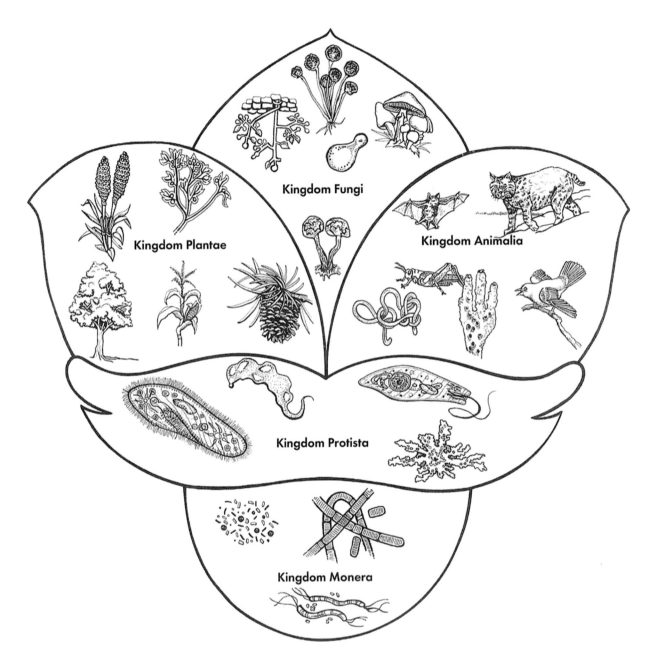

FIGURE 15-2: Five kingdoms of living organisms.

(left–right) Snapping Turtle, Black-eyed Susans, Wavy-browed Albatrosses, Millipede

and coral reefs, are just as endangered as the individual species that inhabit them.

On the other end of the scale from ecosystem diversity is **genetic diversity**, the variation among the members of a single population of a species. Each member has a unique **genotype**, the individual's complement of genes. A **gene** is the fundamental physical unit of heredity that transmits information from one cell to another and thus from one generation to another. The hereditary material of an organism is known as its **germ plasm**. The sum of all of the genes present in a population of organisms is known as a **gene pool**.

Which Habitats Have the Greatest Number of Species?

The planet harbors life everywhere, from the ocean floors to the rocky mountain summits, from high deserts to old-growth forests, from tropical pools to the icy waters of the Antarctic. Yet some areas are especially biologically rich, and some have an inordinate number of **endemic species** — species that are unique to the area and occur nowhere else. About 80 percent of Madagascar's plants, 50 percent of Papua New Guinea's birds, and 50 percent of the Philippines' mammals are endemic to those countries.

Rain forests are the most biologically rich of all ecosystems. Many of the megadiversity countries, so called because of the large numbers of species they contain, owe their biological wealth to rain forests. These countries include Brazil, which probably has more species than any other nation, Colombia, Indonesia, Peru, Malaysia, Ecuador, Zaire, and Madagascar. Second in terms of diversity are the coral reefs of the oceans, which have been called the rain forests of the sea. After the tropical rain forests, the land areas with the greatest diversity are those with a mild climate. The Mediterranean basin, coastal California, and the southern part of western Australia all have a large number of endemic species. Yet another important habitat, and one under increasing threat worldwide, is wetlands. Several of the planet's biologically rich areas are the subject of *Focus On: Where Diversity Reigns: Tropical Rain Forests, Coral Reefs, and Wetlands* (see page 324).

How Do Our Beliefs and Attitudes Affect Biological Resources?

Cultural attitudes and beliefs affect how people value, use, and manage biological resources. For example, **charismatic megafauna** are relatively large animal species, typically mammals or birds, that have symbolic value. The California condor, bald eagle, giant panda, harp seal, and humpback whale are charismatic megafauna. Because many of these species are endangered, their plight receives great media attention, and their management is well-funded, both by private groups and governmental agencies. Although the focus of preservation has shifted from species to habitats and ecosystems, the appeal of these "flagship" species remains high. They serve a useful function by alerting people to the need for preservation of biodiversity.

There are some species that some people have historically learned to hate, from the wolf to the dandelion. We label undesirable animals "**vermin**" and undesirable plants "**weeds**." Although we have spent a large amount of time and money trying to eliminate these misunderstood species, public attitudes toward some of them are beginning to change. The wolf, for example, is enjoying a surge in popularity in some areas of the United States, spurred in part by research and understanding of the animal and by books and films, such as *Of Wolves and Men* and *Never Cry Wolf,* that show the wolf in a sympathetic light. For many people, the wolf, once considered vermin, has come to symbolize the wilderness. This shift in attitude has led to positive action; where efforts were once directed toward exterminating the species, reintroduction programs are now underway in North Carolina, Idaho, Minnesota, and Wyoming's Yellowstone National Park. Although these programs

(left–right) Great Egret, Sea Urchin, Young Mexican Wolves, Pacific coastline

FIGURE 15-3: For most species, human indifference is the rule. Perhaps if we knew more about them, we would realize how interesting these charismatic minifauna are. (a) The brightly colored banana slug, shown here in Washington's Olympic National Forest, has both male and female reproductive organs. Banana slugs can grow as large as 10 inches and weigh as much as a quarter of a pound! (b) A gravid (pregnant) female Plain pocketbook, or *Lampsilis cardium*, a freshwater mussel native to the southeastern United States. This photo shows the mussel displaying its lure, a modified portion of the mantle that resembles a small fish (note the eyespot). Larvae of freshwater mussels must spend a period of time as parasites on the gills of fishes; the lure elicits an attack from a potential host fish, facilitating the transmittal of the larvae to the fish.

are controversial, the wolf is likely to become more highly valued as people begin to appreciate more fully the importance of our shrinking wilderness areas.

Species hunted for sport, such as Canada geese, are called **game animals**. They are highly valued by many different groups: people who hunt for recreation, manufacturers of hunting equipment, and people who regard the animals' presence and migration as a sign of continuity and stability. The management of game animals, like that of the charismatic megafauna, is well-funded, with much of the financial support coming from hunting and fishing license fees, taxes on hunting and fishing equipment, permits, and the sale of duck stamps and conservation stamps (many states require the latter to be purchased when you obtain a hunting or fishing license).

Nongame animals are species not hunted for sport. They include songbirds, butterflies, meadow voles, and turtles. Victims of indifference, they may not suffer the same problems that face misunderstood or disliked

species, but they do not arouse the strong sentiment for preservation that charismatic megafauna and game animals do. Some states have check-off programs that allow residents to donate a portion of their income tax refund to management programs for nongame species, but unfortunately, most such programs are not well funded. The value of these species cannot be neatly determined; they are not worth X amount of dollars annually to anyone and they have little emotional appeal. Even when a species' numbers are greatly reduced and its existence is threatened, as in the case of desert tortoises and certain warblers, there may be no group particularly interested in saving it, so little money is allocated toward effective management.

Like nongame species, most plants, even those listed as rare, threatened or endangered, do not generate much public support. In fact, most living organisms — plants, animals, and microorganisms — fall into this category. If they succumb to extinction, it will not necessarily be because of human greed, but because of our failure to

(left–right) Bullsnake, Aravaipa Canyon (Arizona), King Penguins, We really like this one.

recognize their true worth. We look forward to the day when these interesting but often overlooked creatures are cherished just as much as the giant panda or bald eagle. Perhaps then they will be known as **charismatic minifauna**, small and inconspicuous species valued for their intrinsic worth (Figure 15-3)!

What Are the Greatest Threats to Biological Resources?

Many human activities threaten biological resources and biodiversity. Among the most significant are habitat loss, overharvesting and illegal trade, selective breeding, introduced species, and pollution.

Habitat Loss

Currently, the single greatest threat to wildlife is loss of habitat, either through outright loss of areas used by wild species (such as when a grassland is converted to a housing development), degradation (such as when a plot of rain forest is logged, depriving species of food, shelter, and breeding areas), or fragmentation (when species are forced onto small patches of undisturbed land because surrounding areas are cleared to accommodate roads, agriculture, and other uses). Fragmentation has numerous adverse effects: It can interfere with ecosystem functions, such as the hydrological cycle. Native species that cannot persist in small, isolated pockets of habitat may be crowded out. And fragment edges may prove unsuitable for native plants and animals adapted to conditions in the forest interior; exposure to wind, sunlight, new predators, and other factors may stress species.

Habitat loss is difficult to measure, particularly on a global scale. But, assuming that human activity is associated with conversion of land to other uses, degradation and fragmentation, we can use human disturbance as a surrogate measure of habitat loss. A 1993 study

by Conservation International (CI) mapped areas over 98,000 acres (40,000 hectares) in size that have undergone low, moderate, and high levels of human disturbance. Using deforestation, forest degradation and human population density data, CI found that less than half of the world's vegetated land remains in relatively undisturbed areas (Figure 15-4). South America had the highest percentage of vegetated land in minimally disturbed areas (63 percent), while Europe had the lowest (12 percent). Nearly all of Europe's relatively undisturbed areas are found in Norway and Iceland. Bear in mind that the effect of habitat loss on global biodiversity is greater in the tropical regions because both microenvironments and entire species are more narrowly distributed in the tropics than they are in the mid- and increasingly higher latitudes.

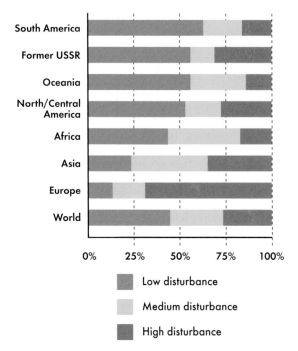

FIGURE 15-4: Degree of human disturbance, vegetated land area.

(left–right) Leopard, Maple leaf, Burrowing Owl, Honeybees

Where Diversity Reigns: Tropical Rain Forests, Coral Reefs, and Wetlands

Tropical Rain Forests

In the tropical rain forests, diversity reigns supreme. These forests are unparalleled in terms of biological richness. Covering just seven percent of the Earth's surface, they are home to nearly half of all known species. They harbor an estimated 45 percent of all plant and animal species, including 30 percent of all bird species and 96 percent of all arthropod species.

How species-rich are the tropical rain forests? According to the U.S. National Academy of Sciences, a typical patch of rain forest about four miles square contains as many as 1,500 species of flowering plant, 750 tree species, 400 bird species, 150 butterfly species, 100 reptile species, 60 amphibian species, and an unknown number of insect species (Figure 15-5). A single Brazilian river harbors more species of fish than all the rivers in the United States, and a single Peruvian wildlife preserve contains more bird species than does the United States. In one study, about 300 tree species were found in single-hectare plots in Peru; all of North America has just 700 native tree species. In that same study, just one Peruvian tree was found to host 43 species of ants, roughly equal to the total ant fauna of the entire British Isles.

Tropical rain forests vary according to altitude, precipitation and other factors, but all help to moderate both global and regional climates (Figure 15-6). Because carbon is stored in woody vegetation, tropical forests (and all forests) are an important global sink for carbon dioxide; intact and undisturbed, large stretches of forest may help to offset global warming. At the regional level, tropical forests play an important role in the cycle of precipitation, evaporation, and transpiration. Most rainfall is absorbed by plants and transpired, or it is evaporated from the surface of vegetation and soil. Consequently, most of the precipitation that falls in the

FIGURE 15-5: Tropical beetle, *Euchroma gigantia*. Insects comprise the majority of species found in the tropics (as they do worldwide). The golden color on the beetle's back is pollen.

tropical rain forests is returned to the atmosphere, where it condenses and again falls to the ground, beginning the cycle anew. Forests also hold the soil in place, minimizing erosion. Rain forests yield numerous products useful to humans: pharmaceuticals, woods, fibers, fruits, nuts, vegetables and other foods, spices, gums, resins, and oils. Unfortunately, these resources are typically ignored in economic assessments of forest use despite the fact that their value often outweighs that of timber.

The tropical rain forests have been reduced by almost one-half their original area. During the 1990s, the world lost about 43.6 million acres (14.6 million hectares) of forest. Deforestation was slightly offset by new forest growth and the establishment of plantations, so that the net annual rate of forest loss for the decade amounted to 28.1 million acres (9.4 million hectares). Some areas have faced especially intense pressure.

FIGURE 15-6: Tropical forests can vary widely. (a) Tropical rain forest, Costa Rica. (b) Cloud forests, such as this one on Negros Island, the Philippines, are found at high altitudes. Temperatures are cooler and much of the precipitation appears as mist or fog.

Ninety-eight percent of the tropical dry forest along the Pacific Coast of Central America is gone. Of all tropical regions, the Asian and Pacific regions have the least remaining forest. Malaysia, the Philippines, and other Asian countries have already lost much of their forests; recently, deforestation rates have risen sharply in Cambodia, Myanmar, Thailand, and Vietnam. Many other nations have lost essentially all their rain forests, including Benin, Côte d'Ivoire, El Salvador, Ghana, Haiti, Nigeria, and Togo. Where forests remain, they often occur in small fragments isolated by developed or degraded land. Although serious deforestation continues in Brazil, it has slowed somewhat since its

peak in 1987. By national and international standards, the Brazilian Amazon is relatively untouched; nearly 94 percent of its forests are still standing. However, deforestation in the area has increased in recent years, and government estimates predict that 25 percent of the forest will be lost by the year 2020. Other rain forests in Brazil have suffered even more; over 95 percent of Brazil's once extensive Atlantic coastal rain forests and the coniferous Araucaria forests in the south have been destroyed. The Atlantic Forest once covered more than 386,000 square miles along Brazil's coast with extensions into eastern Paraguay and northeastern Argentina. Today, less than seven percent of the original forest cover remains, and that has been fragmented into forest islands by centuries of clearing for agriculture and urban development.

Tropical deforestation is linked to several causes. The first is economic pressure. Exploiting the forest's timber resources or transforming it into pastureland for cattle ranching yields short-term benefits for a wealthy minority, but they are not sustainable and do not contribute to real, lasting economic development. Much of the timber is used to make cheap, disposable goods. Eighty percent of logs imported by Japan are used to make cheap plywood, much of which is used for concrete molding frames and scaffolding and is discarded after being used once or twice; tropical wood is also used to make disposable chopsticks and paper. In the United States and Europe, tropical wood is used mostly to make door and window frames, furniture, plywood, blockboard and veneer sheets, which could be manufactured from native hardwoods and softwoods harvested on a sustainable basis. The damage from logging is extensive; many non-target trees are damaged in order to remove target species. Logging also opens up previously inaccessible areas to poor farmers. The extensive network of roads and skid trails needed to remove logged timber destroys even more forest. Erosion from roads and unprotected land chokes rivers with silt.

Tropical forests tend to contain large quantities of minerals and oil, and the exploitation of these resources is contributing to deforestation in countries such as Ecuador and Brazil. The Grande Carajas program in Brazil will open up one-sixth of eastern Amazonia to industry—an area the size of Great Britain and France. Some industries will mine for coal and bauxite, while

FIGURE 15-7: A tract of cleared rain forest, Amazonia, Brazil. The forest was destroyed to make room for a tin mining operation.

others will be devoted to producing cash crops for export and sugar cane for the production of synthetic fuels. In the northwest Brazilian Amazon, thousands of "wildcat" miners have invaded, seeking gold, diamonds, uranium, titanium, and tin (Figure 15-7). These activities threaten the remaining homeland of the Yanomami Indians, who already have lost more than one-third of their territory to mining.

Forests are also being cleared for farming and ranching, typically agribusiness ventures conducted by the wealthy. The island of Negros in the Philippines, which once was almost completely forested, is now little more than a large sugar plantation. More than 50,000 acres (20,000 hectares) of the forested uplands of the island are being cleared every year as those who previously farmed the lower lands are driven onto these marginal lands to grow their crops. In Thailand, much forest land in the east and northeast of the country has been cleared in order to increase the production of cassava, which is mainly exported to feed livestock in Europe. The poor who are displaced by these cassava plantations are forced to clear more forest to grow the crops they need to survive. Since 1950, two-thirds of the lowland tropical forests in Central America have been cleared, mostly for pastureland, and cattle now outnumber people in several countries. Most of this beef is exported, with 80 to 90 percent going to North America.

Large amounts of land in Brazil also are cleared for cattle ranches, but for a different reason: land speculation. Foot-and-mouth disease, which is endemic in Amazonia, prevents much beef from being exported, and even with government subsidies, only three out of 100 large ranches make a profit from livestock. However, under Brazilian law, anyone who clears an acre of forest may claim the land — and the mineral rights below it — and cattle ranching enables large amounts of land to be claimed with minimal labor. Not surprisingly, the areas most frequently cleared are close to gold strikes. Gold, not beef, makes the owners rich; at least half of the large ranches in Amazonia have never even sent a cow to market.

FIGURE 15-8: Beetle with fungi. In tropical forests, decomposition begins almost as soon as an organism dies. Organic matter is quickly broken down by decomposers, such as these fungi, and the nutrients made readily available to plants. As a result, tropical soils tend to be nutrient-poor.

FIGURE 15-9: Cuvier's Toucan. Loss of habitat is the single greatest threat to biological diversity in the tropics and worldwide.

Another cause of deforestation in the tropics is slash-and-burn clearing for subsistence agriculture, which is directly related to population pressures and the inequitable distribution of agricultural land. In Central and South America, land ownership is concentrated in the hands of the wealthy elite. Less than 10 percent of the population typically owns nearly half the land; in non-Amazonian Brazil, just 4.5 percent of landowners control over 80 percent of the land. Their large holdings are often underused; land lies idle that could be more intensively cultivated to provide food crops and employment opportunities for the poor. The growing ranks of the poor, many of whom have been forcibly dispossessed of their own land to make way for development projects, must search elsewhere for the land and resources they need to survive; for many, that search has led to the rain forests. These settlers use slash-and-burn techniques to clear small plots of land. Unfortunately, the soils are not very fertile, since the nutrients are locked up in the vegetation (Figure 15-8). Thus, after only a few years, the soil in cleared areas is exhausted and poor farmers must move on to clear a new area, where the cycle is repeated.

Deforestation has many victims (Figure 15-9). The land itself is seriously and perhaps permanently degraded. The vast amounts of nutrients once bound up in the natural vegetation disappear, and the thin topsoil is susceptible to erosion. Nearby streams and rivers become choked with silt, killing many aquatic organisms. On the Philippine island of Palawan, erosion due to deforestation has almost destroyed coastal fisheries. Two-thirds of the rivers in Sarawak, Malaysia, are officially classified as polluted by soil erosion, and fish catches, which provide a major part of the native peoples' diet, have been drastically reduced. Flooding also becomes a serious problem. In 1998, heavy rains brought record-setting floods to many deforested regions, including areas in Bangladesh, India, and Mexico. Flooding along China's Yangtze River watershed, which has lost 85 percent of its forests to logging and agriculture, killed 3,700 people, displaced 223 million people, inundated 61 million acres (25 million hectares) of cropland, and cost economic losses of $30 billion. The poor who try unsuccessfully to eke out a living from crops planted in a cleared area must move every few years to a new, more productive area. Many Javanese farmers who were moved by the Indonesian government to the outer islands of the Indonesian archipelago in the mid-1970s and 1980s have been unable to survive in their new environment. Colonists were forced to clear more and more forests as previously cleared fields were degraded. In some areas, the soil is so poor that some colonists have starved, while others have returned to Java. The entire population of the region suffers if precipitation patterns are altered markedly. And finally, when wealthy landowners "develop" a tract of forest for timber or pastureland or when settlers clear an area to establish small farms, they displace the native peoples who originally lived there. These native peoples depend on the forest for food, shelter, medicines and clothing, and they have deep cultural and spiritual ties with the forest that gives meaning to their lives (Figure 15-10).

Slowing and preventing deforestation will require efforts in a number of areas. The first is to reform natural resource accounting methods so that nontimber goods and ecosystem services are included in economic assessments of forest use (Figure 15-11). For example, in order to calculate the value of a forest, one should consider not just the economic value of the timber, but also the value of the forest as a carbon sink, habitat for wildlife and indigenous peoples, and moderator of the climate. The second is to change tenure laws in order to grant title or legal use rights to forest inhabitants. Agrarian reform also can ease pressure on the rain forests; more equitable distribution of land would enable poor farmers to make a living on land that is suited for agriculture. Given an opportunity to raise their standard of living, they would have a real stake in the future. International actions that can be taken to slow deforestation include reducing the demand for tropical timber, using debt reduction to finance conservation, and expanding and reforming development assistance.

FIGURE 15-10: Basketweaver from an indigenous tribe, Borneo. Among the endangered resources of the world's tropical rain forests are the cultures of the indigenous peoples who dwell in these dwindling habitats.

FIGURE 15-11: In the El Cielo Biosphere Reserve, Tamaulipas, Mexico, a local resident holds the harvested leaves of palmilla, a nontimber forest product used in the cut foliage industry. Harvesting palmilla is a sustainable extractive industry; it provides residents with a reliable source of income but does not damage the forest.

Coral Reefs

Coral reefs have been called the rain forests of the oceans because of the abundance and diversity of life found there. Scientists estimate that one million species are probably associated with reefs, but little is known about most of them. Australia's Great Barrier Reef, which at 1,250 miles (2,000 kilometers) is the world's largest coral reef ecosystem, contains more than 3,000 animal species. The richest, most diverse coral reefs in the world are found in the Indo-West Pacific. They harbor over 2,000 species of fish, 5,000 species of molluscs, 700 species of corals, and countless species of crabs, sea urchins, brittle stars, sea cucumbers, and worms (Figure 15-12). The diversity of coral reefs is largely due to great numbers of fish species, many as yet unidentified. Yet it is the hermatypic, or reef-building, coral animals themselves that make this spectacular ecosystem possible.

Reefs consist of the limestone skeletons of dead corals cemented together by the action of single-celled algae, called dinoflagellates, that live in association with the

FIGURE 15-12: Blue tang and red-fringed sponge.

coral animals or polyps. The reef surface, which lies just below water level at low tide, is covered by a thin layer of live polyps. These small cup-shaped creatures, related to the sea anemones of temperate waters, have a ring of tentacles around a central mouth. Each individual secretes its own coral skeleton; these skeletons can be cemented together to form massive coral blocks. Most corals are small, and most live in colonies, but there are exceptions to both of these generalities.

Coral reefs are an aggregate of many coral colonies (Figure 15-13). The reefs grow rapidly; individual colonies grow at an estimated annual rate of 0.3 to one inch (8 to 25 millimeters). The symbiotic dinoflagellates that live inside the polyps accelerate the rate at which the coral skeletons are formed, and during photosynthesis they use carbon dioxide produced in the polyps' tissues. In turn, the algae excrete nitrogenous materials that the polyps require for growth. The shape of any one particular coral colony is a product of the water

FIGURE 15-13: Coral reef.

temperature, amount of wave action, and nutrient and light levels. These factors vary in different parts of a reef system.

Coral reefs develop in areas where the water is clear and the temperature does not fall below 68°F (20°C). These conditions typically occur in the coastal waters in the tropics. Hermatypic corals thrive in water between 72° and 82°F (22° to 28°C); they grow best in warm, shallow waters where light and oxygen are plentiful. They are rarely found at depths greater than 230 feet (70 meters). Hermatypic corals are not found in cool waters, such as those far north or south of the equator, and also do not occur on western continental coastlines, where zones of upwelling bring colder water to the surface. Coral reefs are also absent from areas where rivers transport fresh water and sediments to the ocean. Sediments smother the delicate corals and prevent light from reaching their surface, thus interfering with photosynthesis by the dinoflagellates.

There are three major classes of coral reef structures: fringing reefs, barrier reefs, and atolls. Fringing reefs grow as platforms from the shores of continents or islands. Barrier reefs are found offshore; deep lagoons separate the barrier reef from the coast. Atolls are found in the open ocean; they are circular reef structures that were once fringing reef structures around volcanic islands (Figure 15-14). Rising sea levels or the sinking of the land surface caused the island to disappear below the water's surface. The reefs continued to grow upward, however, until only the coral remained as a circular or oval structure. Sometimes, the reef is capped

FIGURE 15-15: Pale anemone.

by a sandy isle known as a cay in the Caribbean and a motu in the Pacific.

The waters in which coral reefs grow are nutrient poor, a fact that long puzzled scientists. We are beginning to understand, however, that the water is deficient in nutrients because they are locked in the biota. Nutrients are quickly removed from the water and cycled and recycled through the food webs of the reef, much as the nutrients in the tropical rain forests are quickly removed from the organic litter and returned to the vegetation.

Coral reefs provide habitats for many organisms that cannot live elsewhere (Figure 15-15). Much of the primary production on reefs is carried out by the dinoflagellates living on the coral itself, forming the basis for a very complex food web capable of supporting many different animals with many different lifestyles. Many reefs support commercially valuable fisheries; according to one estimate, the potential yield of fish and shellfish from coral reefs is nine million tons per year. Coral reefs protect coastal areas from being battered by waves and thus prevent coastal soil from eroding. The quiet waters between the reef and the coast, called a lagoon, provide a home for organisms too delicate to withstand direct battering by waves. Reefs also may help regulate the amount of mineral salts in the world's oceans. Every year, runoff deposits massive amounts of mineral salts in the

FIGURE 15-14: Talatea, French Polynesia.

sea. The lagoons and closed seas formed by coral reefs serve as natural evaporation basins that eventually dry out, leaving behind huge salt deposits. Thus, large quantities of salt are removed from the sea. Finally, to build their limestone skeletons, coral polyps remove carbon dioxide, an important greenhouse gas, from the atmosphere.

These unique ecosystems are threatened by a variety of human activities. The World Resources Institute reports that 60 percent of the world's reefs are at risk from human activities, with about 27 percent of reefs at high or very high risk. In the last few decades, over 35 million acres (14 million hectares) of coral reef in 93 countries have been destroyed by human activity. Coral reefs are especially vulnerable to any disturbance that stirs up sediment, which smothers coral polyps and interferes with photosynthesis by the dinoflagellates. Deforestation, especially of coastal areas and mangroves, contributes heavily to the sediment load of coastal waters. Logging in the watershed of Bascuit Bay in the Philippines has destroyed five percent of the coral reefs in the bay. Sewage and the runoff of fertilizers and pesticides have caused destructive algal blooms or poisoning of some reefs. The crown-of-thorns starfish, *Acanthaster planci*, has ravaged many reefs in the Red Sea and the Indian and Pacific oceans, and it is feared that it may enter the Caribbean through the Panama Canal. It is not known if the infestations are a result of anthropogenic change, but overfished reefs seem particularly vulnerable to invasion.

Many reefs have been mined for their limestone, which is used for construction. Coral miners dislodge coral using crowbars or explosive charges. Dynamite fishing is also extremely destructive, not only killing large numbers of fish and other animals, but also blowing apart the coral infrastructure. Dynamite fishing has destroyed some of the finest reefs in Kenya, Tanzania, and Mauritius.

Reefs also fall victim to the tourist industry. Coral and shells are collected in large quantities for sale, while some tourists also hack out pieces straight from the reef to take home as souvenirs. Boat anchors and chains also damage coral.

Nuclear testing is perhaps the most violent of human activities that affect coral. France has detonated about 100 nuclear devices on the Polynesian atoll of Mururoa, which is gradually sinking into the sea. The French are considering moving their testing site to another Pacific island, Fangataufa.

Coral reefs also are threatened by massive bleaching. When corals are subjected to high water temperatures (even two degrees above normal), cold water temperatures, chemical pollution or dilution by fresh water, the colorful algae that they harbor die. The disappearance of the algae exposes the white skeletons of the corals. If the temperature stress persists, the corals eventually die. Bleaching, considered by some to be the greatest threat to reefs worldwide, occurred without warning over the last several decades, affecting the Caribbean reefs most seriously. The worst year on record for bleaching was 1998, in the wake of El Nino, the weather phenomenon that brings warmer-than-usual water currents to the Pacific. Although the causes of bleaching are unknown, some scientists believe it is a harbinger of global climate change.

Wetlands

Wetlands, transitional zones between land and water, tend to be relatively high in species diversity. They are among the most biologically interesting and productive but least understood habitats, ranging in diversity from prairie potholes in Canada and the United States to papyrus swamps in Uganda. Many wetlands have formed along riverbanks or near river deltas, including the Mississippi in the United States, the Nile in Egypt, the Okavanga in Botswana, the Rhône in France, and the Brahmaputra in India.

The presence of water is the distinguishing characteristic of wetlands. It may vary from standing water several feet deep to waterlogged soil without standing water. The soil in wetlands differs from that of adjacent uplands; it is often saturated long enough to become anaerobic. Wetland vegetation must be able to tolerate both flooding and the lack of oxygen in the soil.

There are two major categories of wetlands: inland and coastal. Inland wetlands are freshwater ecosystems and include marshes, swamps, riverine wetlands, and bogs. Coastal wetlands may be either fresh or salt water and are affected by tides. Examples include tidal salt marshes, tidal freshwater marshes, and mangroves.

Freshwater marshes are areas of incredible diversity (Figure 15-16). Dominated by emergent grasses and sedges, including cattail, wild rice and bulrush, they

FIGURE 15-16: Leopard frog, a resident of freshwater marshes.

have a high pH, high levels of nutrients, high productivity, and high rates of decomposition. Marshes usually are recharged by both precipitation and lake or river flooding, and it is the flooding that serves as a source of nutrients. Some marshes remain wet year-round while others are wet only seasonally. Marshes are very important habitats for wildlife, particularly migratory birds and waterfowl. The marshes at Point Pelee, a provincial park in Ontario, Canada, offer refuge to a wide spectrum of waterfowl and other wildlife. Over half of the waterfowl harvested in North America are produced in the marshes of Canada.

"Swamp" — for many people, the name evokes feelings of repulsion or dread. Popular culture is at least partially responsible for the poor image these areas have. B-grade movies portray swamps as places of evil and terror. In reality, swamps are hauntingly beautiful. With standing water up to five or six feet deep throughout most or all of the year, there is little or no understory or emergent vegetation. Instead, swamps are dominated by woody vegetation, especially cypress, water tupelo and black gum, trees well-adapted to their watery environment. To obtain oxygen from the air (since it is typically in short supply in the often stagnant water), cypress and other trees send up pneumatophores, knobby aerial roots, or "knees," through which air is funneled to the submerged roots (Figure 15-17). Reptiles, amphibians, and fish are common denizens of swamps. Among the best-known swamps in the United States are the bayous of Louisiana and the Okefenokee Swamp in Georgia. Cienaga Grande,

the "Great Swamp" region of Colombia, is one of the richest fishing grounds in that country.

Riverine or riparian wetlands, commonly called floodplains, are lands adjacent to rivers that undergo periodic flooding. During flooding, they receive a high input of nutrients from the surrounding area. Since they are found alongside rivers, floodplains have a linear form. A floodplain system includes the river channel, its floodplain, and all oxbows. An oxbow is a river channel formed by the natural meandering pattern of a river that becomes cut off from the main channel (Figure 15-18). Severe flooding, for example, may carve a new and straighter channel through an area, isolating a bend or meander of a river. Although the oxbow becomes separated from the main river, it is recharged through precipitation and groundwater. Weston Lake in South Carolina is an oxbow lake formed from the Congaree

FIGURE 15-17: Cypress tree swamp, Mississippi.

FIGURE 15-18: Oxbows, Manu National Park, Peru.

River. The Congaree Swamp, a national monument, includes the river, the floodplain, and the oxbow lake. Designated a Biosphere Reserve, it contains 50 tree species, including several 300-year-old Loblolly pines, the largest of which measures 27 feet in circumference.

Riverine wetlands vary from areas that are almost always flooded to areas that are rarely flooded. The frequency and duration of flooding are a product of climate (precipitation, spring thaw), floodplain levels (higher plains are flooded less frequently than low-lying areas), drainage area (the larger the area drained, the longer the duration of the flooding), and the soil's ability to hold water. Hardwoods are the dominant vegetation, with the species composition depending on the frequency of flooding. Cypress, tupelo, and buttonbush are found in wetlands that are almost always flooded, while areas subject to progressively less frequent flooding host water locust, persimmon, sweetgum, box elder, sycamore, hickory, beech, and white oak. The soils of riverine wetlands are anaerobic part of the year, again depending on the frequency of flooding. One of the best-known riverine wetlands in the United States is

the wetland formed by the Mississippi River in Louisiana. The Pantanal, a seasonal wetland in Brazil, is an important food source for birds. Every year, rains cause the rivers flowing through the Pantanal to flood the surrounding flat ground. As the Pantanal dries out, many fish are trapped and forced into smaller pools, where they become easy prey for birds.

Historically, floodplains were very important to agriculture. For thousands of years, seasonal flooding of the Nile River in Egypt was used to enhance the productivity of agricultural fields. In late summer, as the river began to rise, water was channeled into basins and released onto fields, where it was allowed to stand for 40 days or more. Today, however, seasonal flooding is prevented by a series of dams, and productivity in agricultural fields in the Nile Delta must be boosted by expensive artificial fertilizers. Tribal pastoralists in southern Sudan also have developed a migratory way of life that revolves around vast swamps known as the Sudd. At the end of the rainy season, the people burn the grasslands surrounding the floodplains, encouraging the growth of new, rich grass. As the floodwaters recede, the people bring their herds of cattle in to graze on the rich toich grass. When the rains return, the people move to higher ground, where they practice agriculture.

Bogs are generally found in cooler regions such as the northeastern United States, Canada and the British Isles, but some tropical countries, such as Indonesia, possess rather extensive bogs. Although the water table in a bog is typically high, precipitation, not underground aquifers, is responsible for most water inflow. There is no outflow; the water is static, and the land is permanently waterlogged. The oxygen supply in standing water soon becomes depleted; coupled with the cool climate, the static water contributes to slow rates of decomposition. Deposits of peat, a brown, acidic material made up of the compressed remains of plants that have not decomposed, gradually accumulate. Because nutrients remain locked up in plant and animal matter and are not quickly recycled, the productivity of a bog is low compared to that of the surrounding terrestrial system. Not surprisingly, bogs are colonized by plants that are acid tolerant, such as

sphagnum moss, cranberry, pine, spruce, and tamarack. Eventually, enough peat may accumulate that it raises the level of the land making it less wet and suitable for other vegetation. Dried peat burns well and for centuries has been used for fuel; until recently, almost every household in western Ireland relied on peat for heating and cooking. Countries in which fuelwood is scarce, such as China and Indonesia, are being encouraged to exploit their peat resources. Peat is also harvested for use as humus to improve the soil or for use in potting compost. However, the destruction of peat bogs may have implications for global climate change. Active peat bogs "fix" carbon in the soil, acting as sinks; when peat is dug up, carbon is released into the atmosphere, where it contributes to global warming.

Like inland wetlands, coastal wetlands exhibit great variety. Tidal salt marshes occur along coastlines throughout the world in the mid-to-high latitudes, being replaced in the tropics and subtropics by mangroves. Salt marshes form wherever the accumulation of sediments is equal to or greater than the rate of

FIGURE 15-19: Flamingos, *Phoenicopterus ruber,* feeding at dawn in the Camargue, a tidal freshwater marsh in France.

land subsidence, as long as the area is protected from the destructive power of waves and storms. They are found near river mouths, in bays, along protected coastal plains, and around protected lagoons. The lower salt marsh is that area found in close proximity to the ocean; its water has the salinity of the adjacent seawater. The upper marsh is that area which is further removed from the ocean; its salinity is variable due to exposure and dilution by fresh water. Salt-tolerant grasses are the dominant vegetation in salt marshes. These wetlands are very productive ecosystems, but much of the primary productivity is washed out with the tides and is utilized by aquatic communities. In the United States, extensive salt marshes are located behind barrier beaches in Long Island, New Jersey, and North Carolina. Salt marshes along the Waddenzee of Germany and the Netherlands are extremely important habitats for waterfowl and many other animals and plants. Also, almost the entire population of North Sea herring is dependent on these marshes at some point in their life cycle. Unfortunately, the Waddenzee is now seriously threatened by pollution.

Tidal freshwater marshes combine features of tidal saltwater marshes and inland freshwater marshes. They are found where a river enters the ocean or where precipitation is high. The tide influences water level, as in a tidal salt marsh, but species diversity is very high, as in an inland marsh, because of the absence of the salt stress. Plant diversity is high; vegetation varies with elevation, with submergent aquatic vegetation dominating in low-lying areas and emergent vegetation dominating at higher elevations. Tidal freshwater marshes serve as vital nurseries for many species of fish and are used by more birds than any other type of marsh. For example, the Camargue, in France, is a large wetland formed where the River Rhône flows into the Mediterranean (Figure 15-19). Much of the Camargue, which contains very fertile soil, has been drained for agriculture, but the remaining area, which is the breeding ground for 15,000 to 20,000 greater flamingos and is visited by more than 300 species of migratory birds each year, has been designated a Biosphere Reserve. Approximately 280 species of birds are found in U.S. tidal freshwater marshes. Unfortunately, tidal freshwater marshes are often found near urban areas, since people tend to settle along river mouths, and are therefore subjected to significant stresses, especially pollution and development.

The mangrove swamp is the dominant type of coastal wetland in tropical and subtropical regions. It is a forested wetland, an association of salt-tolerant trees, shrubs, and other plants growing in brackish to saline tidal waters. Mangroves cover approximately 39 million acres (16 million hectares) worldwide, with the largest concentrations in tropical Asia. The largest stretch of mangroves in the world is the Sundarbans, along the Lower Ganges Delta in Bangladesh.

Mangrove is a general term referring to any tree that can survive partly submerged in the relatively salty environment of coastal swamps. These trees do not require a saline environment and may in fact grow better when the salt content is low, but, under such conditions, they tend to be outcompeted by other tree species. They have survived because they are able to adapt to a saline environment. Some mangrove species excrete excess salt through the action of special glands. Others isolate the salt in inactive tissues within the plant. Pneumatophores enable the tree to transport oxygen to its roots. Because they colonize shifting silt beds, mangroves must have some means of maintaining their stability. They do this through stilt roots, which act as buttresses (Figure 15-20). The mangrove roots also trap silt, and over time, mud banks accumulate.

FIGURE 15-20: Mangrove thicket, Puerto Rico.

As the mud banks rise above the water level, other plants are able to invade the area. Mangroves may be totally inundated or they may have no standing water. They are important exporters of detritus for nearby estuaries. Half of a mangrove's primary productivity may be exported.

Traditionally, wetlands have been misunderstood and underappreciated, but we are learning that these varied ecosystems provide essential services. Wetlands intercept and store runoff, helping to prevent flooding. The stored water can help to recharge shallow underground aquifers. As runoff flows through a wetland, suspended solids are precipitated out, reducing sediment loads to nearby waterways. Anaerobic and aerobic processes, such as denitrification, remove excess nutrients and other chemicals from the water, thus purifying it before it reaches the lake or ocean. For this reason, wetlands have been called the "kidneys" of the Earth. In one study of a Florida swamp, water hyacinth were found to remove 49 percent of the nitrogen and 11 percent of the phosphorus present in the water.

Coastal wetlands, particularly mangrove swamps, protect shorelines from the erosive force of ocean tides and minimize storm damage. In addition, mangroves protect coral reefs and other offshore areas from land-based pollution.

Both coastal and inland wetlands are valuable reservoirs of biological diversity. They are important nesting and feeding habitats for birds and waterfowl and nurseries for fish and shellfish. Draining wetlands for questionable short-term economic gain might prove to be one of the biggest ecological disasters humans have foisted on the biosphere. Further, it is a questionable practice on economic grounds. A study which originated at the National Center for Ecological Analysis and Synthesis at the University of California, Santa Barbara, estimates the value of one acre of wetland to be $36,533 per year. This estimate includes water and

FIGURE 15-21: Endangered prairie pothole, Alberta, Canada. Formed over time by the action of receding glaciers, water-fowl-breeding potholes in agricultural land are being rapidly lost to the plow.

soil, as well as processes such as climate regulation, crop pollination, and biological pest control.

In the past, wetlands have been drained and destroyed, without a thought to their importance, and converted to agricultural land or commercial development. Channelization of rivers in order to achieve better navigable depths or to prevent flooding has also had disastrous effects on riparian wetlands. Without protective wetlands, the straightened rivers are more prone to flooding, a fact dramatically demonstrated by the Mississippi River flood of 1993.

Wetlands throughout the world are threatened (Figure 15-21). Australia and New Zealand have lost over 90 percent of their wetlands. In the United States, over half of the wetlands that existed before the advent of European settlement in the seventeenth century have been lost. In some states and regions, the loss has been exceedingly high. California has the dubious distinction of leading the nation in this category, with 91 percent of its wetlands destroyed. Many of the marshes and swamps that once ringed the Great Lakes, especially lakes Erie and Michigan, were drained for conversion to farmland. Mangrove swamps have been significantly reduced in size in Africa, Latin America, and western Asia. Logging to produce pulpwood and charcoal and

conversion to aquaculture ponds are the chief threats. For example, the Indonesian state of Kalimantan plans to clear 95 percent of all mangroves in order to accommodate pulpwood production, despite the fact that the fisheries nursed by those mangroves earn about seven times as much in export revenue as all wood and charcoal production combined.

A more immediate threat to wetlands is their definition under national laws. While new advances in ecology have shown that wetlands are critical ecosystems, changes in the legal definition of "wetlands" can seriously hamper efforts to protect them. Currently, the Environmental Protection Agency, Army Corps of Engineers, U.S. Fish and Wildlife Service, and USDA Soil Conservation Service define "wetlands" similarly. Each organization requires positive indicators of hydrology, hydrophytic vegetation, and hydric, or saturated, soils. The Fish and Wildlife Service also includes nonvegetated areas such as mudflats, gravel beaches, rocky shores, and sandbars. However, this definition does not always cover drier or seasonally saturated wetlands, which are just as valuable as wetter wetlands. The belief that wetter wetlands are more valuable has resulted in pressure to exempt wetlands — at least drier ones — from protection under the Clean Water Act.

Tropical deforestation affects not only exotic species but also those more familiar to us. Many songbirds that summer in the United States and Canada are threatened by the deforestation that is reducing or eliminating their tropical wintering grounds. At the same time, development and urban encroachment are reducing their summer breeding grounds in the north (Figure 15-22). Other threatened habitats in North America include wetlands and grasslands, both of which are transformed for agriculture and development. Without immediate action, these habitats and their native inhabitants may go the way of the tallgrass prairie. Prior to the 1850s, tallgrass prairies extended over a quarter of a billion acres in central North America. Today, less than one percent of those prairies remain; the rest have been replaced by acre upon acre of corn and wheat.

FIGURE 15-22: Adult rose-breasted grosbeak. The grosbeak is one of many migrant songbirds whose populations are declining. Others include the olive-sided flycatcher, white-eyed vireo, blue-winged warbler, northern parula, chestnut-sided warbler, cerulean warbler, Canada warbler, and scarlet tanager.

Overharvesting and Illegal Trade

Butterflies, parrots, lizard skins, tortoise shells, orchids, corals, and cacti: all are part of the huge global trade in wildlife and wildlife products worth an estimated $10 billion annually. Much of the wildlife trade is legal, governed by national laws and international treaties, the most notable of which is the Convention on International Trade in Endangered Species. However, a significant portion of the trade is illegal. In fact, international law enforcement agencies report that the illegal wildlife trade is second in volume and profit only to the illegal drug trade. In the United States, the legal trade in wildlife, approximately $1 billion, is dwarfed by the illegal trade, estimated at $5 billion.

The wildlife trade is a serious drain on populations and a significant factor in driving species toward extinction. For instance, 40 of the world's 330 species of parrots face extinction because of trade and loss of habitat. In all, over 600 species of animals and plants worldwide face extinction as a result of the wildlife trade, while another 2,300 animals and 24,000 plants are endangered.

The largest suppliers of the wildlife trade are South America (especially Bolivia, Argentina, Brazil, Peru, and Guyana); Africa (Senegal, Tanzania, the Congo Republic, the Sudan, and South Africa); East Asia (Philippines, Taiwan, Thailand, Indonesia, Singapore, and Japan); and the United States, which exports cat skins, reptile skins, bear products, cacti, and ginseng. The major importers and consumers of wildlife and wildlife products are the United States, Europe, and Japan. Reptiles and reptile-leather products, primates, and tropical fish account for most U.S. imports. The leading European imports are exotic birds, reptile skins, monkeys, and small cats. Worldwide, the chief market for illegal wildlife products is Japan, where a clouded leopard coat may sell for over $100,000.

Social and cultural factors, especially fashion, custom and tradition, are responsible for the enormous wildlife trade. For example, in 1970, the estimated population

FIGURE 15-23: Poachers slaughtered this black rhino for its horn, Swaziland, Africa.

of black rhino throughout its range south of the Sahara was 60,000, making it the most numerous of the five species of rhino. During the 1970s, the value of rhino horn, which had been constant for decades, suddenly and dramatically escalated because of increased market demands. Rhino horn was valued for medicinal uses throughout the Orient and as an aphrodisiac in parts of Asia. In Yemen, it was used to produce high-quality handles for jambias, the traditional dagger. From 1972 to 1978, the North Yemenis imported 40 percent of all rhino horn which entered international trade, much of it from the black rhino. The increased demand for rhino horn created an upsurge in poaching all over Africa, and consequently, the number of black rhinos in the wild has plummeted to fewer than 2,500 (Figure 15-23). By 1993, international pressure prompted Yemen to ban internal trade in rhino horns and to encourage the use of water buffalo horn as a substitute in jambias. Taiwan and China also have banned rhino horn imports and internal trade, and China has banned the manufacture of all medicines containing rhino powder. The recovery of rhino populations may depend upon how well these bans are enforced.

(left–right) White Sands National Monument (New Mexico), Great Spangled Fritillary, Starfish, Chestnut-bellied Heron

Selective Breeding

In many areas of the world, modern crop hybrids have replaced naturally occurring strains and semidomesticated or traditional strains, those bred by farmers over centuries. Inarguably, hybrid strains have yielded many benefits, especially in terms of productivity. But the reduction in the number and frequency of wild and traditional strains has come at a high price — the loss of rare and potentially valuable genes that carry traits such as resistance to pests or disease.

The genetic erosion of crop plants and the widespread use of genetically similar hybrid strains are two of the most serious threats facing modern agriculture. The genes of wild and semidomesticated plants are our only defense against this threat. The productivity of cultivated crops cannot be maintained, much less improved, without constant infusions of germ plasm from wild relatives and primitive cultivars. Crop geneticists have observed that between five and 15 years after new crop strains resistant to certain diseases are introduced, the strains succumb to newly evolved diseases. The corn blight of 1974 in the United States dramatically illustrated both the danger of homogeneous strains and the value of traditional and wild strains. That summer, a blight swept through the country, threatening 70 percent of the corn crop. It eventually destroyed 15 percent of the crop; in some areas, the loss was as high as 50 percent. Weather changes finally reduced the threat that season, and the blight was eventually controlled by introducing a resistant strain from Mexico, the ancestral home of maize.

Wild biological resources constitute a genetic library of invaluable and irreplaceable characteristics. Genetic engineers are continually developing their ability to draw upon these characteristics to improve the quality of human life in myriad ways, including increasing crop yields and strengthening livestock strains. Unfortunately, the safety of this genetic library is in question. With every extinction — of both known and unknown species — we lose yet another valuable book (Figure 15-24). Most seed companies contribute to the problem by offering only a limited selection of modern hybrids, but some innovative entrepreneurs offer environmentally friendly alternatives.

Loss of genetic variability is a problem in domesticated animals as well as plants; in the United States, nearly half of the existing breeds of livestock are in danger of extinction. Breeds with economically desirable characteristics, such as high productivity (the most milk, the most meat per animal), are becoming more popular at the expense of other breeds. However, each of those other breeds also has valuable characteristics, such as disease resistance, adaptability to regional environments, low maintenance, or special qualities as food (taste, fat content). We never know when these characteristics may become vitally important. For example, commercial cattle had all but replaced the native breeds in Africa's Chad Valley in the early 1970s. However, the advent of the area's periodic seven-year drought wiped out the imported stock, leaving farmers to rebuild herds with the few remaining local cattle varieties, which had adapted to the drought cycles over the centuries. In the United States, the American Minor Breeds Conservancy works to preserve genetic diversity in livestock.

Introduced Species

Because they rarely have predators that can control their populations, introduced species may compete more successfully than native species that occupy the same niche and may eventually displace the native species. The introduction of exotic species was a contributing factor in 68 percent of the extinctions of North American freshwater fishes recorded from 1900 to 1984. In East Africa's Lake Victoria, the Nile perch — introduced in 1960 to "improve" fishing for food and sport — threatens many of the 200 species of freshwater fish of the family Cichlidae, which are endemic to the lake. The economic costs associated with introduced species, like the environmental costs, are often high. The zebra mussel, a native of the Caspian Sea, was

(left–right) Tiger Swallowtail and Daylilies, Glacier National Park (Montana), Giant Clam, Shoebill Stork

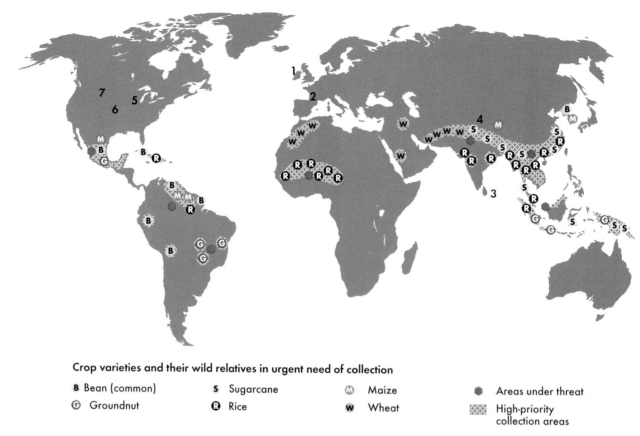

Crop varieties and their wild relatives in urgent need of collection

B Bean (common)   **S** Sugarcane   Ⓜ Maize   ⬡ Areas under threat

Ⓖ Groundnut   Ⓡ Rice   Ⓦ Wheat   ▨ High-priority collection areas

FIGURE 15-24: Ancestral areas of wild relatives of major crop species. Numbers indicate areas where disasters have occurred due to lack of genetic diversity: (1) 1840s, blight devastated Irish potato crops; two million people died; (2) 1860s, vine disease crippled Europe's wine industry; (3) 1870-1890, coffee rust destroyed Ceylon's coffee harvest; (4) 1942, rice crop in Bangladesh destroyed; millions died; (5) 1946, a fungus epidemic devastated U.S. oat crop; (6) 1950s, stem rust devastated U.S. wheat harvest; (7) 1970, maize fungus threatened 80 percent of U.S. corn acreage.

brought to the Great Lakes in the bilge water of ocean-going ships sometime in the early 1980s. Its population soared throughout the decade, and by 2000, it had spread down to the Arkansas River, where it is expected to wreak havoc on the Southeast's already declining native freshwater mussel fauna. In the Illinois River, zebra mussels have been blamed for wiping out almost a quarter of the native species. Millions of dollars have been spent in mitigating the damage to boats, docks, and pipes caused by the mussel and in attempting, thus far unsuccessfully, to control its population.

Pollution

Habitat can be "lost" to native species when it is rendered uninhabitable as a result of chemical contamination from hazardous waste dumps, industrial and municipal wastewater discharges, agricultural and urban runoff, and airborne toxins. Pollution may be especially detrimental to aquatic biodiversity; since human settlements tend to form along rivers, lakes and coastal areas, industrial activity and wastes are often concentrated in these areas. Pollution and the

  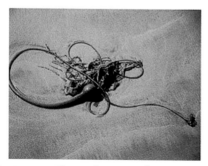 

(left–right) Common Horsetail, Grey Parrot, Kelp, Montane coniferous forest (Rocky Mountains)

FIGURE 15-25: Streamside salamander.

What Is Extinction?

Extinction is the elimination from the biosphere of all individuals of a group of organisms of any taxonomic rank (family, genus, or species). It is a natural process that has occurred throughout Earth's history — the fossil record shows that, on average, a species persists from one to 10 million years before one or more factors — climate change, persistent drought, natural catastrophe, the emergence of a new predator, or genetic mutation — cause it to become extinct. Over geologic time, extinction is estimated to occur at the rate of one mammal species every 400 years and one bird species every 200 years. However, since the start of the Industrial Revolution, the earth has lost 83 species of mammals, 113 species of birds, 23 species of amphibians and reptiles, 23 species of fish, about 100 species of invertebrates, and over 350 species of plants. Such high extinction rates appear directly related to human activity. In analyzing animal extinctions since 1600, the World Conservation Union, or IUCN (an independent global body that oversees efforts to conserve biodiversity), found that when the cause was known, 39 percent had resulted from the introduction of an exotic species, 36 percent from habitat destruction, 23 percent from hunting and deliberate extermination, and two percent from other factors.

Recorded extinctions over the past four centuries are surely far lower than actual extinctions, since they do not include plants, microbes, and animals other than mammals or birds. Further, recorded extinctions pertain only to known species. Considering how little we know about the number and variety of living organisms, it seems likely that many species have disappeared without ever being described. Moreover, a species is not officially classified as extinct until fifty years after its last sighting; for that reason, species that have disappeared in the past few decades are not accounted for in recorded extinctions. Finally, widespread habitat loss in the tropics has occurred only recently. Tropical spe-

deliberate poisoning of waterways have contributed to over one-third of North American freshwater fish extinctions. The recent worldwide decline in many amphibian species also may be linked, at least in part, to acid precipitation. Frogs, toads, and salamanders are particularly susceptible to pollutants for several reasons (Figure 15-25). First, they respire through their skin; its permeability also may render them vulnerable to pollutants and increased ultraviolet radiation (UV), a consequence of stratospheric ozone depletion. Second, they lay their eggs, unprotected, in ponds, lakes and streams, where they are exposed to pollutants, including acidic meltwater, and increased UV radiation. One study found that when the eggs of some frog and toad species are bathed in slightly acidic water, they either die or produce deformed tadpoles. Other studies have shown that increased acidity can slow growth rates in frogs; species at greatest risk are those that breed in temporary pools and must mature before the water evaporates.

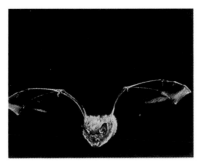

(left–right) White-lined Sphinx Moth and Verbena, Big Brown Bat, Heirloom sweet pepper harvest, Massasauga Rattlesnake

cies that have been adversely affected (by, for example, deforestation or habitat fragmentation) may persist for several generations, but if their numbers drop too low, they will become vulnerable to disease, predation and other factors, and may eventually succumb to extinction. The loss of tropical forest habitat is a key reason that Wilson estimates the current extinction rate to be 1,000 to 10,000 times higher than in prehuman times.

A 1997 analysis of about 240,000 plant species worldwide found that nearly 34,000 were threatened; 380 have gone extinct. These figures are probably conservative; relatively little work has been done in some species-rich regions of the world. The United States, Australia, and South Africa harbor the greatest number of at-risk plant species; for example, 4,669

plant species in the United States — 29 percent of the nation's flora — are threatened; 200 are already extinct (Figure 15-26). However, these countries may head the list simply because their flora is much better-studied compared to other species-rich countries, especially those in the tropics, such as Malaysia and Indonesia. U.S. wildlife is the subject of Box 15-1: *The State of U.S. Plants and Animals.*

Worldwide, about 23 percent of mammal species, 30 percent of amphibians, and 12 percent of birds are classified as threatened. For fish, the proportion ranges from 3-49 percent; for invertebrates it is 0.2-58 percent. This wide range for fish and invertebrates reflects the fact that they have been studied far less than mammals, birds, and amphibians. Animal species at greater risk include those that have a low reproduc-

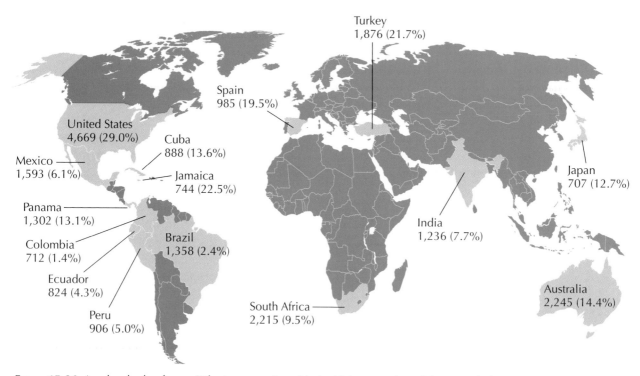

Turkey
1,876 (21.7%)

Spain
985 (19.5%)

United States
4,669 (29.0%)

Cuba
888 (13.6%)

Mexico
1,593 (6.1%)

Jamaica
744 (22.5%)

Panama
1,302 (13.1%)

Colombia
712 (1.4%)

Brazil
1,358 (2.4%)

Ecuador
824 (4.3%)

Peru
906 (5.0%)

South Africa
2,215 (9.5%)

India
1,236 (7.7%)

Japan
707 (12.7%)

Australia
2,245 (14.4%)

FIGURE 15-26: A paler shade of green? The ten countries with the highest number of threatened plant species are the United States, Australia, South Africa, Turkey, Mexico, Brazil, Panama, India, Spain, and Peru. The figures indicate the number of threatened plant species and the percentage of the country's total flora that they represent.

(left–right) Largemouth Bass, Silverweed, Basidiomycete fungus, Glacier Bay (Alaska)

BOX 15–1 **The State of U.S. Plants and Animals**

In its *1997 Species Report Card: The State of U.S. Plants and Animals*, The Nature Conservancy (TNC) finds that approximately one-third of U.S. plants and animal species are at risk of extinction. The organisms in greatest peril are those that rely on freshwater habitats, such as mussels, crayfish, fishes, and amphibians. The group with the largest proportion of at-risk species is freshwater mussels; over two-thirds (68 percent) of these species are of conservation concern. With more species of freshwater mussels than any other country in the world, the United States holds a biological treasure in its rivers and streams. That treasure is imperiled, however, by the construction of dams and reservoirs (especially in the Southeast), water pollution, and streambed alterations.

Also in trouble are the flowering plants, with over one-third, 5,144 species, at risk of extinction. Many wild plants are adapted to very specific soil types or microclimates and grow only in very limited areas. For this reason, they are especially vulnerable to direct human disturbances. For example, the construction of an interstate highway is believed responsible for the extinction of the Sexton Mountain mariposa lily, which was known to exist in only a single Oregon locality.

While every state in the union has suffered losses, Hawaii, Alabama, and California have recorded the greatest number of extinctions. Hawaii is the "extinction capital" of the United States, with 269 species known or presumed extinct. Alabama has lost 98 species, many of them species of freshwater fauna, especially mussels and snails, that succumbed to dams and other riverine alterations. In California, rapid human population growth and subsequent development have radically altered many of the Golden State's myriad unique habitats. The result has been 46 species extinctions, about half of which are flowering plants.

tion rate, specialized feeding habits, limited nesting or breeding habitats, a valuable fur or hide, fixed migratory patterns, or behavioral patterns that make them susceptible to exploitation. Species that are large, narrowly distributed, move across international borders, or require a large range are also more likely to become rare. The giant panda, for example, reproduces slowly, feeds only on bamboo, and prefers undisturbed terrain away from humans.

If habitat destruction in the tropical regions and coral reefs continues along present trends, from one to 11 percent of the world's species will be committed to extinction by 2015. Other estimates are even bleaker. Wilson and Ehrlich argue that as many as 100,000 species may become extinct each year. They point out that by 2025 we may lose one-quarter of all species now present on Earth. If present trends continue, the 150 years from the beginning of the twentieth century to the middle of the twenty-first century will witness the extinction of more species than any other similar time span since life began. And while it is true that there have been mass extinctions in the Earth's geologic past (such as occurred at the end of the Cretaceous period, when dinosaurs disappeared), even these took place more slowly. For example, at the end of the Permian period (about 250 million years ago), two-thirds of all marine species and almost half of all plant and animal families went extinct over a period of 10,000 years. The slower rate of past mass extinctions allowed speciation and evolution to replenish the Earth's biota.

(left–right) Claret Cup Cactus, Pronghorn Antelope, Great Smoky Mountains, Superb Starling

How Are Extinction and Biological Diversity Related?

The disappearance of an individual species often has far-reaching effects. Each species plays a distinct role in its ecosystem, and when it is gone, the organisms that depended on it may be threatened as well. For example, in the Amazonian rain forest, female euglossine bees pollinate the Brazil nut tree, and the males pollinate a particular species of orchid during their courtship displays. If the orchid becomes extinct, the bees may also die out. The Brazil nut trees might go unpollinated, rendering them unable to set seed and produce nuts. Extinction also disrupts food chains. As the links between species are ruptured, the entire ecosystem becomes more vulnerable to stresses. Thus, diminishing species diversity threatens ecosystem diversity, just as the degradation of an ecosystem threatens the chances of survival of its inhabitants and thus threatens both species and genetic diversity.

Decreasing genetic diversity in turn threatens species diversity by making species more vulnerable to extinction. If the size of a population of a species decreases, as individuals die or are removed, the gene pool decreases. If no new individuals from other populations join the group, the population will be less genetically diverse than the original population, even if it regains its former size through new births. Further, the genetic variability among individuals will decrease with each successive generation because of inbreeding, unless individuals from other populations are added to the group. The resultant loss of genetic variability is known as **genetic erosion**, a phenomenon that lessens a population's ability to adapt to changes in its environment.

Genetic erosion is of great concern to wildlife biologists, zoo curators, managers of botanical gardens, and plant and animal breeders. The **minimum viable population size** is the smallest population unit of a species that has a specified percent chance of surviving for a specified period of time, given foreseeable demo-graphic, genetic and environmental occurrences, as well as natural catastrophes. This figure has not been determined for most species. Extensive, time-consuming, and expensive research is needed to determine it for a single species. It is unlikely that we will be able to conduct the necessary research for some species before some of their populations fall below the minimum viable population size.

The current rate of extinction and the subsequent loss of biological diversity are sobering trends. And yet, there are reasons to hope, for all around the world, people are attempting to protect species, ecosystems, and wild places. In *Focus On: Four Rays of Hope*, world-renowned wildlife researcher and conservationist Jane Goodall explains why she is hopeful that humanity will change course and blaze a path toward a sustainable future for all life on Earth.

Why Should We Preserve Biological Diversity?

The arguments in support of preserving biological resources and maintaining maximum biological diversity fall into five categories: ecosystem services; aesthetics; ethical considerations; evolutionary potential; and benefits to agriculture, medicine, and industry.

Ecosystem Services

Ecosystems perform many beneficial services, such as chemical and nutrient cycling and soil generation. For example, each year, earthworms carry in their casings from 2.2 to 69 tons (2 to 63 metric tons) of soil to the surface of each hectare of land, effectively aerating the soil. Eliminating living resources may hinder the ability of ecosystems to perform vital functions. American oysters were once so plentiful in the Chesapeake Bay that they filtered all the water in the bay every three days. Since 1870, populations of the oyster have dropped by

(left–right) Roadrunner, Stick Insect, Colorful potato harvest, Wetland in Grand Tetons National Park (Wyoming)

Four Rays of Hope

This article by Jane Goodall originally appeared in the Winter 1999 edition of Orion Magazine. 195 Main Street, Great Barrington, MA 01230.

As we move toward the millennium, it is easy to be overwhelmed by feelings of hopelessness. We humans have destroyed the balance of nature: forests are being destroyed, deserts are spreading, there is terrible pollution and poisoning of air, earth, water. Climate is changing, people are starving. There are too many humans in some parts of the world, overconsumption in others. There is human cruelty to "man" and "beast" alike; there is violence and war. Yet I do have hope. Let me share my four reasons.

Firstly, we have at last begun to admit to the problems that threaten the survival of life on Earth. And we are problem-solving creatures. Our amazing brains have created modern technology, much of which has greatly benefited millions of people around the globe. Sadly, along with our tendency to overreproduce, it has also resulted in massive destruction and pollution of the natural world. But can we not use our awesome problem-solving ability to now find more environmentally friendly ways to conduct our business? Good news — it's already happening as hundreds of industries and businesses adopt new "green" ethics. And we must play our part — in our billions we must adopt less-harmful lifestyles. Refuse to buy products from companies, corporations, that do not conform to new environmental standards. We *can* change the world.

Secondly, nature is amazingly resilient. Given the chance, poisoned rivers can live again. Deforested land can be coaxed — or left — to blossom again. Animal species, on the verge of extinction, can sometimes be bred and saved from a few individuals.

My third reason for hope lies in the tremendous energy, enthusiasm, and commitment of young people around the world. Young people want to fight to right the wrongs, for it will be their world tomorrow — they will be the ones in leadership positions, and they

FIGURE 15-27: Jane Goodall and friend.

themselves will be parents. This is why the Jane Goodall Institute started Roots & Shoots, an environmental education and humanitarian program for youth. Roots creep under the ground to make a firm foundation. Shoots seem small but to reach light they can break brick walls. Hope — millions of roots and millions of shoots can break through, break all the problems humans have created, make change. Roots & Shoots groups, from kindergarten to college, work to make the world a better place for animals, the environment, and the human community. The central message of Roots & Shoots is that every individual matters, every individual has a role to play, every individual makes a difference.

My fourth reason for hope lies in the indomitable nature of the human spirit. There are so many people who have dreamed seemingly unattainable dreams and, because they never gave up, achieved their goals against all odds, or blazed a path along which others could follow.

So let us move into the next millennium with hope — with faith in ourselves, in our intellect, in our indomitable spirit. Let us develop respect for all living things (Figure 15-27). Let us try to replace violence and intolerance with understanding and compassion. And love.

99 percent; it now takes a year for the oysters to filter the bay's water. The role that the oyster played has been severely limited, one reason that the bay is muddy and oxygen-poor relative to its historical condition. (Ecosystems and ecosystem services are discussed throughout this book, especially in Chapter 4, *Ecosystem Function,* and other chapters in Unit II, *Ecological Principles and Applications*.)

Aesthetics

Many people feel that wild species and wild places imbue the Earth with much of its beauty, grandeur, and mystery. Most of us do not revel in the physical, chemical, and biological conditions that enable Earth to support life, although we recognize and appreciate those conditions. Instead, we revel in life itself, clover and condor, forest and field mouse. And perhaps it is this fundamental thing — the connection we feel with other living beings — that will compel us to take the actions necessary to protect biological diversity. *What You Can Do: To Preserve Species* offers some suggestions for personal action.

Ethical Considerations

Many people argue that biological resources have inherent worth and so have a right to exist for their own sake, independent of their value to humans. Proponents argue that moral and ethical grounds alone are reason enough for humans to preserve biological resources and protect diversity.

In *A Sand County Almanac*, Aldo Leopold asks, "If the biota, in the course of aeons, had built something we like but do not understand, then who but a fool would discard seemingly useless parts? To keep every cog and wheel is the first precaution of intelligent tinkering." Eliminate a species, and the health of the ecosystem may suffer. Scientists have no way of knowing how the loss of a single species will affect a given ecosystem. Many species play vital ecosystem roles

that are not proportional to their size or abundance. The loss of one or two of these keystone species may result in major changes in the functions of the ecosystem, just as the loss of the keystone in an arch would result in the toppling of the entire structure. Leopold maintains that a land ethic is necessary to "change the role of *Homo sapiens* from conqueror of the land community to plain member and citizen of it. It implies respect for his fellow members and also respect for the community as such."

Homo sapiens has always been a tinkerer. Long ago, our ancestors domesticated certain animal species and practiced selective breeding for particular plant strains. We, too, are tinkerers, but our greater numbers and powerful technologies render our activities far more dangerous and potentially destructive. We must become intelligent tinkerers, protecting what we do not understand rather than eliminating, in our ignorance, that which may be invaluable to us. We must remember that what we lose — or, more to the point, allow or cause to disappear — is irreplaceable.

Evolutionary Potential

In *Conservation Biology: An Evolutionary-Ecological Perspective*, biologists Michael E. Soulé and Bruce A. Wilcox write, "Perhaps even more shocking than the unprecedented wave of extinction is the cessation of significant evolution of new species of large plants and animals. Death is one thing — an end to birth is something else." All other arguments — ecosystem services; benefits to agriculture, medicine and industry; aesthetics; and ethical considerations — are strong and convincing reasons to preserve diversity now and in the foreseeable future. But what if we were to take on a more universal, timeless perspective, more like the perspective the Earth itself might have?

The time scale of the Earth is far different from our own. None of us, for instance, will ever see a new species arise from an ancestor. Seeing things from the planet's perspective means thinking of time in terms

(left–right) Water lilies (Lotus), American Bison, Basidiomycete fungus, Black-capped Chickadee

WHAT YOU CAN DO ▷ To Preserve Species

At Your Home, School, or Office

➡ Write your Congressional representatives and Senators to express your views on reauthorization of the Endangered Species Act.

➡ Take advantage of the non-game wildlife check-off on your state tax form.

➡ Vote for candidates who share your opinions on conservation and the preservation of biological diversity. If you do not know a candidate's opinion on a conservation issue, ask.

➡ Learn about gardening organically and try your hand at growing heirloom or traditional varieties suited to your region.

➡ Pull weeds instead of using herbicides.

➡ Let bird and insect predators take care of native garden pests.

FIGURE 15-28: In the United States, the butterfly known as *F. Lycaenidae Eumaeus atala,* or Florida atala, is restricted to the southern tip of Florida. Planting or preserving vegetation that butterflies require (either as adults or in their larval, or caterpillar, stage) helps protect biological diversity.

➡ Use beer traps for slugs instead of baiting with poisons.

➡ Don't use a lawn care service. If you must use one, find out what chemicals it uses and make sure it applies them correctly.

➡ Plant shrubs or trees around your home, school, or office to provide food and shelter for bats, birds, butterflies, and other creatures (Figure 15-28).

➡ Put up bird houses and bird baths.

➡ If you live in an arid region, landscape with plants that do not require a great deal of water.

➡ Landscape with plants that aren't prone to insect and fungus problems.

➡ Watch out for wildlife; give consideration to all living things. Slowing down just a few miles per hour while driving could help you to avoid hitting animals crossing the road. Be especially alert at night and during the first warm spell in late winter or early spring, when many frogs, toads, and salamanders migrate to and from their breeding grounds.

➡ Don't buy products that come from endangered animals or plants.

➡ Don't keep exotic pets or rare plants.

➡ Read books and articles on wildlife and environmental issues and watch nature programs on television.

➡ Teach children to respect nature and the environment; take them on a hike, help them plant a tree or build a bird house, buy them a nature book or a subscription to a wildlife magazine.

➡ Volunteer your time to conservation projects designed to increase or protect wildlife habitat.

➡ Check your lifestyle — think about the effects of your daily actions on the environment.

➡ Join an organization that works exclusively for the protection of biological resources and habitat. Some suggestions: The Nature Conservancy, National Wildlife Federation, the Wilderness Society, Conservation International, World Wildlife Fund, Defenders of Wildlife, and Cousteau Society.

Traveling

➡ Don't pick flowers or collect wild creatures for pets; leave them where you find them. If you want a remembrance, take pictures!

➡ Don't buy souvenirs or products made from wild animals or plants.

➡ When hiking, stay on the trail.

➡ Consider a vacation centered around wildlife observation and study. Choose an area or country with native wildlife in which you are particularly interested.

➡ Carefully research tourist businesses, such as safari tours, to make certain that you deal with a reputable firm.

Benefits to Agriculture, Medicine, and Industry

Biological resources benefit agriculture in many ways. The economic value associated with pollination, an ecosystem service provided for free by intact and healthy natural systems, is estimated at $400 billion annually (Figure 15-29)! The genes of wild crop relatives may hold valuable characteristics, which, when introduced into crop species, can dramatically increase yields. Wild species are also the source of "new" food crops. Of the 80,000 edible plant species on Earth, only a small percentage have been investigated for use as a food crop, and just eight supply 75 percent of the human diet. Similarly, most of the protein from livestock is derived from just nine domesticated animals. Thus, the global human diet is precariously dependent on a handful of species. Around the world, thousands of new beneficial food crops await discovery. Some of these may be useful only locally, while others may have the potential for wider cultivation. For example, the winged bean *Psophocarpus tetragonolobus* of New Guinea has many potential uses. Its seeds, pods, shoots, and tubers are all edible and nutritious; the plant is fast-growing (reaching 12 feet in just a few weeks); and its roots fix nitrogen, improving the quality of soil for other crops. Other plants, currently known only to indigenous peoples, may hold the key to bolstering the world's food supply in the twenty-first century.

Many important and familiar pharmaceuticals originated from biological sources. One-quarter of all pharmaceuticals dispensed in the United States contain active ingredients derived from plants, including digitalis, a heart stimulant derived from purple foxglove, and quinine, a malaria treatment derived from the bark of the cinchona, a tree native to the tropics. Many of these medications literally save lives: An extract of the rosy periwinkle, for example, has proven very effective in the treatment of leukemia. In 1960, a child suffering from leukemia had only a one-in-five chance of long-term survival; today, the odds are four-in-five. Animals

FIGURE 15-29: A honeybee on a sunflower. Pollination, an ecosystem service, is valued at an estimated $400 billion annually.

of millions, rather than tens or hundreds, of years. According to that time scale, the preservation of diversity is paramount because it alone can produce further diversity through adaptation and speciation.

(left–right) Green Frog, Ruby-throated Hummingbirds, Crane Creek (Ohio), Snow Leopard

also provide valuable medicinals, including hormones, thyroid extracts, and estrogens. The slow-clotting blood of the endangered Florida manatee has aided researchers in the study and treatment of hemophilia, while the black bear, which loses no bone mass during its yearly five-month hibernation, may hold clues for the treatment of osteoporosis.

Industry — tourism, recreation, and manufacturing — realizes significant economic benefits from biological resources. Parks, reserves and wildlife refuges, in the United States and abroad, are becoming increasingly popular as places to view and enjoy wild species. In 2001, Americans spent approximately $108 billion on recreational activities associated with wildlife, including $35 billion on fishing, $21 billion on hunting, and $38 billion on wildlife watching activities like birding and nature photography. Visitors to national and state parks spend additional millions each year.

Manufacturing industries also owe a tremendous economic debt to biological resources. For example, jojoba, once considered a desert weed, yields a wax which retails to Japan for over $3,000 a barrel. The wax is used as an industrial lubricant in place of sperm whale oil.

Managing Biological Resources

Throughout history, people have amassed collections of rare and unusual plants and animals, both native and nonnative. In this section we'll look at current efforts to manage biological resources, with an emphasis on the pros and cons of various approaches. We also look at the legislation designed to preserve biological diversity, both in the United States and worldwide.

How Are Biological Resources Currently Managed?

For the most part, current efforts to manage biological resources concentrate on preserving plant and animal species or their genetic material (germ plasm). Both species and germ plasm may be preserved off site (ex situ) or on site (in situ). Although efforts to manage and protect biological resources by preserving habitats are increasing, management efforts have traditionally focused on off-site management (especially through aquariums and zoological and botanical gardens).

Off-Site Management of Plants

Off-site management and collection of plants has a long history. In about 2800 B.C., the Chinese Emperor Shen Nung established a garden of exotic medicinal plants. In the tenth and eleventh centuries, Islamic Moors of the Iberian peninsula created gardens of native and nonnative plants for the purposes of pleasure, inspiration, and study. Medieval Europeans viewed the collection and cultivation of plants as a way to recreate Eden. In the seventeenth and eighteenth centuries, the increased popularity of studying medicinal plants triggered the establishment of botanical gardens throughout Europe, and in the United States, privately owned gardens began to sprout up in the early eighteenth century.

Today, approximately 1,000 institutes worldwide maintain plant collections. They hold the primary responsibility for managing wild plant species outside their natural habitat. In the United States, important collections are housed at the Missouri Botanical Gardens, Boston's Arnold Arboretum, and the New York Botanical Gardens. The Royal Botanic Gardens at Kew, England, established in 1759, is probably the leading botanical garden in the world, housing over 50,000 species from almost every country.

(left–right) Zebra, Saddleback Caterpillar, Fall foliage (Ohio), Canada Geese

In addition to their recreational and educational functions, many contemporary botanical gardens sponsor research which adds to our store of botanical knowledge. They enhance conservation efforts by conducting basic biological research, including systematic biology, ecology and population genetics, on rare and/or little-understood species. They may also maintain collections of seeds or plant tissues. Their research and collections are often used by breeders in the quest to improve plant strains.

The Center for Plant Conservation, a network of 34 botanical gardens and research facilities in the United States, was established in 1984 to develop a "national collection" of all rare and endangered indigenous plants. It is an ambitious goal, considering that there are at least 3,000 such species, many located on private lands and many having only one or two known populations. By enlisting institutions in different biogeographic regions, the center has approximated the diversity of climates needed to compile and maintain a national collection. Each institution specializes in collecting and propagating species native to its region. For safety reasons, seeds are also stored at the Department of Agriculture's seed storage facilities. The cultivation of plants at participating institutions offers the species a secure environment while facilitating research into the plants' specific growing requirements and the general ecological niches they fill.

Despite these efforts, plant species usually receive far less attention from the public and government than do animal species, especially mammals. Perhaps it is the close kinship that we share with animals that makes it easier for us to appreciate them. Even so, from a pragmatic standpoint, plant conservation is far more important to human survival than is animal conservation. Plants form the base of all food chains; without them we would have neither food to eat nor oxygen to breathe. From a wider perspective, plants are also vital to other animal species. Botanists estimate that approximately 34,000 plant species are currently threatened with extinction; between 20 and 40 animal species depend, for their survival, upon each plant species. Thus, for every plant species faced with extinction, many more animal species may also disappear.

Off-Site Management of Animals

Much like plants, animal collections date back thousands of years. The Chinese Emperor Wen Wang built one of the earliest zoos, Ling Yu, the Garden of Intelligence, in the twelfth century B.C. When Cortés reached Montezuma's palace at Tenochtitlán in 1519, he found a vast menagerie of birds, mammals, and reptiles. The early European zoos of the late eighteenth and early nineteenth centuries, much like the earlier, private hunting preserves that inspired them, were mainly the province of royalty and the wealthy. Sadly, the early public zoos were primarily sources of entertainment, often leading to a "more is better" philosophy that neglected an animal's needs and comfort and led to an irresponsible attitude regarding the capture of wild animals; for example, obtaining a single baby chimpanzee might entail killing five or six adults.

Today, private collections, wildlife refuges, sperm banks, and zoological parks are all used to preserve animals outside their natural habitat. Since zoos are by far the best known of these means and since they incorporate elements of wildlife refuges and sperm banks, we will look more closely at how zoos work to preserve animal species.

Unlike old-style zoos, most modern institutions focus on caring well for small numbers of species. In this way, they are able to house more individuals of a species. Having adopted a philosophy in which the zoo is seen as an "ark" for the world's vanishing wildlife, they are dedicated to three interrelated objectives: education, recreation, conservation, and research.

The best zoos strive to meet the physical and psychological needs of their animals in a manner that is both educational and entertaining. Naturalistic exhibits

(left–right) Eastern forest stream, Sea Otter, Timber Rattlesnake, Golden Cheeked Warbler

approximate the animals' wild habitats, providing them with stimulation and allowing them to pursue their normal behaviors — a wolf scent-marking a post or an otter floating on its back in a pool, breaking open clams against a rock perched on its belly. Many zoos, for example, equip their chimpanzee enclosures with an artificial termite mound. In the wild, chimps use long sticks or straw to penetrate holes in termite mounds and draw out the tasty inhabitants. Because termites are extremely difficult to maintain, the artificial mounds usually contain honey, yet the chimps still enjoy fishing for the sweet treat. Visitors enjoy observing the animals' normal behaviors, and they learn much about the species and the intricate relationships between it, other species, and the environment.

Even the ways that zoos group animals can be educational. In the past, animals typically were grouped taxonomically. All the members of the cat family might be grouped in one general area, primates in another, and the horse family in yet another. Taxonomic groupings teach people to recognize genus types and the evolutionary relationships among species. While taxonomic grouping is common, more zoos are beginning to group their animals zoogeographically, on the basis of where they live. This arrangement mimics how various species occur and interact in the wild and emphasizes ecology and habitat. The Metropolitan Toronto Zoo was the first to be arranged zoogeographically. Covering 704 acres, it is divided into four pavilions: The Americas, Australasia, Africa, and Indo-Malaya. Numerous paddocks within each pavilion house species particular to different vegetation types within those broad geographic areas.

The latest concept in zoo design is in the works at the San Diego Zoo, one of the United States's best. In 1985, officials announced plans to reorganize the whole zoo around ten bioclimatic zones: tropical rain forest, tropical dry forest, savanna, desert, grassland, temperate forest, taiga, tundra, montane, and islands. The idea is to immerse visitors in the world of specific bioclimatic regions, including plant and animal life. When the project is finished, San Diego's will be the first zoo totally organized around the concept. Other zoos, including Chicago's Brookfield Zoo and New York's Bronx Zoo, have also developed bioclimatic exhibits.

Zoos can be a strong participant in global conservation efforts. They promise real hope as reservoirs of species and genetic diversity. However, to fulfill that potential, curators must maintain viable breeding stocks of species to supply their own and other zoos rather than adding to the pressure on wild stocks. In addition, captive breeding populations hold out the hope that zoos will someday be able to return animals which have disappeared in the wild to their native habitat — if and when there is any wild habitat to which to return them.

However, captive breeding is not simply a matter of producing as many young as possible. Rather, programs attempt to meet two objectives: to maximize the contribution of unrelated animals in order to reduce the effects of inbreeding and to attain a good age distribution in order to ensure a consistent number of individuals of breeding age. The International Species Inventory System (ISIS), a computer-based information system for wild animal species in captivity, was developed to overcome the problems caused by inbreeding, which include diminished fertility, reduced resistance to disease, and lessened competitive ability. ISIS serves nearly 600 zoological institutions in 72 countries, including about half of the world's zoos and aquariums. ISIS includes information on 2 million living specimens of 10,000 species, along with an additional 750,000 of their ancestors. It records basic biological information on each individual animal: age, sex, parentage, place of birth, and circumstance of death. This information is used to compile various reports and to analyze the status of captive populations. More importantly, it allows zoos to cooperate in the genetic and demographic management of their animals.

(left–right) Red Bat, Yellow-necked Caterpillars, Campion, Blue-banded Goby

FIGURE 15-30: Arabian oryx, Phoenix Zoo. A successful breeding program has increased numbers of the oryx, which has been reintroduced into its original range in Oman.

Due to breeding programs at zoos, many species have been brought back from the brink of extinction. For instance, by 1962, the Arabian oryx had been virtually exterminated in the wild, the victim of hunters in Land Rovers armed with automatic weapons. That year, the last few wild animals were captured and sent to the Phoenix Zoo, where a breeding program was instituted using the wild oryx as well as some already in captivity. By 1980, the "world herd" at the Phoenix Zoo numbered more than 320 individuals. In 1982, 14 oryx were released into the wild in southern Oman, where they have flourished under the official protection of the Sultan of Oman (Figure 15-30). Additional reintroduced populations now occur in Bahrain, Israel and Saudi Arabia, with approximately 886 individuals in the wild. Breeding programs have also saved the golden lion tamarin, Pere David's deer and Przewalski's horse, the original Mongolian wild horse, from extinction.

Despite the success of traditional captive breeding programs, they are plagued by several problems. Some animals, like the giant panda, simply do not breed well in captivity. Another problem is logistics. For example, in order for two black rhinos from different zoos to mate, one of them — an animal weighing thousands of pounds — must be moved to the other's institution. Doing so is costly and cumbersome and poses some risk to the animal. Breeding programs are also made more difficult by monomorphic species, species whose male and female look virtually identical. Many bird and reptile species are monomorphic. For a long time, the only reliable method zoos had to determine the sex of a monomorphic animal was to open up the animal and look at the gonads, or reproductive organs, directly. But surgery, even with today's improved surgical techniques and highly trained personnel, carries risk. Fecal steroid analysis, pioneered at the London and San Diego zoos, eliminates the need for surgery. The animal's feces are examined to determine the ratio of estrogen, the female hormone, to testosterone, the male hormone. Although each sex produces both male and female hormones, the female produces a greater amount of estrogen

Using ISIS data, the American Association of Zoological Parks and Aquariums has developed "species survival plans" for 106 threatened or endangered species. These plans indicate which animals should mate to preserve the maximum genetic diversity of the entire captive population across zoos. They also indicate how many animals should mate, and when, in order to create a good-sized population with an optimal age distribution.

(left–right) Hueston Woods State Park (Ohio), Crab Spider, Green Treefrog, Spotted Hyena

and the male, a greater amount of testosterone. Fecal steroid analysis can also be used to track an animal's reproductive cycle in order to tell when the animal is biologically ready to mate. Unfortunately, it requires sensitive and costly equipment, time and a technician, and many zoos do not have these resources.

Researchers have solved some of the problems that complicate breeding programs by adapting alternative assisted reproduction methods that have been used for some time in the domestic livestock industry. These methods include artificial insemination; embryo transplants; and cryopreservation, in which unfertilized eggs, sperm, and embryos arrested at an early stage of development are frozen in liquid nitrogen at -196°C and held for long-term storage.

On-Site Management of Plants and Animals

The move to preserve wild lands began in the United States over a century ago (see Chapter 16, *Public Lands*). In the past three decades, it has spread to many other countries, and by 2003, the number of protected areas worldwide had risen to 102,102 sites covering 4,656 million acres (18.8 million square km). North and Central America have set aside the highest percentage of land in protected areas (almost 12 percent); the countries of the former Soviet Union have set aside the least (just over one percent). Through their actions, some nations have shown an exemplary commitment to biodiversity protection. For example, almost 25 percent of Costa Rica's land has been given some sort of protected status; in addition to an extensive network of parks and nature reserves, the nation protects by law a certain portion of its remaining forests, many of which lie on private lands. Bhutan, bordered by Tibet to the north and India to the east, west and south, has set aside approximately 20 percent of its land area in wildlife sanctuaries, forest reserves, nature and wildlife reserves, and one national park. In 1986, the country rejected a World Bank project, the proposed Manas-Sankosh Dam on the Manas River in the south, that would have flooded one of Bhutan's most diverse and significant wildlife areas. The area lies in the center of what is now Royal Manas National Park.

Despite the successes of the past thirty years, several obstacles must be overcome before we can secure an effective global system of protected lands. First, while protected areas account for 11.5 percent of the Earth's land surface, existing parks and reserves do not represent the full complement of ecosystem types or species (Figure 15-31). In fact, a full six percent of this total is accounted for by the Greenland National Park. Because many protected areas are located in species-poor areas, such as mountains covered by rock and ice and semifrozen areas, they leave much of the world's biodivierstiy unprotected.

The 1992 Fourth World Congress on National Parks and Protected Areas addressed this and other biodiversity issues in what is known as the Caracas Action Plan (CAP). To encourage preservation of a wide range of ecosystems, CAP established a goal of having 10 percent of each of the world's major biomes protected by 2000. At the Fifth World Congress, held in 2003, scientists announced that the target had been met in nine of the 14 major biomes.

However, in many of the remaining areas it may be difficult or impossible to read CAP's stated goal. Less than 10 percent of the temperate, broadleaf forests that once cloaked Europe, the eastern United States, and parts of Asia is found in areas of low disturbance. Similarly, only 7.5 percent of the Mediterranean region's evergreen sclerophyllous forest remains in areas of minimal human disturbance, and only seven percent of the Atlantic forest of eastern Brazil remains undestroyed, much of it fragmented.

Fragmentation has led to another obstacle: size. Many protected areas are too small to hold viable populations of their largest carnivores and herbivores. Few of the world's protected areas are at least 2.5 million acres in size, a conservative estimate of the minimum amount of habitat needed to support viable populations of the largest mammals. Without consideration for

(left–right) Corn and squash, Domesticated horses, Florida Scrub Jay, Garter Snake

FIGURE 15-31: Lowland semi-evergreen rain forest, Thailand. Like rain forests around the world, this one is threatened by migrants seeking to settle and farm the land. Unless this endangered ecosystem receives adequate protection soon, it is likely to disappear.

the greater landscape, protected areas become isolated islands of life surrounded by developed or agricultural lands, and genetic erosion may eventually take its toll on populations. Moreover, relatively few protected areas embrace the full range of habitats exploited by species (such as migratory birds) during seasonal movements. It's unlikely that many countries would be able or will-

ing to set aside areas of 2.5 million acres. One approach would be to establish, within a broad area, a series of smaller protected parks connected by corridors, which would allow individuals to disperse and interbreed, thus reducing the risk of genetic erosion. This solution would also protect a greater range of habitat types. Additionally, even small areas (25 acres or so) are large enough to protect viable populations of plants, which account for much of the Earth's biodiversity.

The failure to provide for human residents has been a third obstacle hindering the development of an effective global system of protected lands. Many parks and reserves are subject to intensive agriculture, grazing, mining, poaching, and other activities; they are "protected" on paper only. The goal of integrated conservation-development projects is to reduce human impact on protected areas by providing local populations with sustainable, income-generating opportunities. In 1976, the United Nations Educational, Social, and Cultural Organization instituted the Man and the Biosphere (MAB) program to promote integrated ecological research and international cooperation in the field of environmental science. MAB has established a global system of Biosphere Reserves, which encompass concentric areas zoned for different uses. A core area is strictly managed for wildlife; human activities are prohibited, although existing settlements of indigenous peoples are allowed. In a buffer zone just outside the core area, research and some human settlements are permitted. Also permitted is light resource extraction, such as rubber tapping, collection of nuts, and selective logging. A second buffer zone, located outside the first, accommodates increased human settlements, tourism, and training.

A fourth obstacle to the establishment of protected areas is the debt burden of developing nations, many of which lie in the biologically rich tropical regions. Desperate for the short-term cash promised by the immediate exploitation of land, they may feel it is simply impossible to set aside large tracts. One solution to this problem is debt-for-nature swaps in which a conservation organization "buys" the debt of a nation at

(left–right) Scarab Beetle, African Rhinoceros, Hawkweed and Lupine, Crappie

a discount from the bank(s) which made the original loan, in exchange for a promise by the debtor nation to establish and protect a nature reserve.

Pros and Cons of Off-Site and On-Site Preservation

For many species, off-site preservation in a zoo or botanical garden offers the best or only hope for long-term survival. In some cases, species may be preserved off site until a time when they can be reintroduced to the wild. Off-site institutions also play a vital educational role. Studies conducted in the early 1980s at Yale University and elsewhere found that both adults and children learn about wildlife mainly through zoos.

The most obvious disadvantage of off-site preservation is that it is an alternative only for those species known to us. Thousands of unidentified species are not represented in zoos and botanical gardens, and if their wild populations are lost, the species will be lost forever. Off-site preservation is expensive; moreover, zoos, botanical gardens, and gene banks can maintain only limited numbers of species. U.S. zoos can probably preserve self-sustaining populations of only 100 of the over 4,000 known species of mammals.

In their natural habitat, species generally adapt to environmental changes. When we preserve species off site, however, they do not have the opportunity to change along with their natural environment. Further, captive species undergo evolutionary changes in response to their new environment in a matter of relatively few generations. For example, by the fifth to eighth generation in captivity, Przewalski's horse foals were being born outside of the species' traditional, sharply defined foaling season. In the wild, those foals would not survive. It seems probable that sustained preservation off site reduces the likelihood of a species' successful reintroduction to the wild.

In addition, we lack the basic biological knowledge essential for effective management of many exotic species, particularly in relation to captive breeding efforts. Finally, preserving species off site ignores the other components of their ecosystems. If we are to preserve organisms of all five kingdoms, and not just individual species of the plant and animal kingdoms, we must focus on ecosystem preservation.

On-site preservation offers the best hope for preserving maximum biological diversity. It is the only way in which the relationships between organisms can be preserved; many scientists contend that it is the relationships between organisms and between organisms and the nonliving environment that are of critical importance to the survival of the planet as we know it. On-site preservation also allows a species to evolve with its environment. It is generally less expensive than off-site preservation, although there are costs involved in law enforcement and protection of the designated preserve against poachers and development pressures. If on-site preservation is to be successful, however, effective legislation is clearly needed to protect the habitats upon which species depend.

No species better illustrates the importance — and potential — of habitat preservation than the ivory-billed woodpecker. Long believed extinct, the animal known as "The Lord God Bird" or "The Grail Bird" was rediscovered in 2004 in the Big Woods of eastern Arkansas, where private conservation groups and government officials have worked for decades to preserve a remnant of the South's once-plentiful bottomland swamp forest. For more information on the ivory-billed woodpecker, please go to www.EnvironmentalEducationOhio.org and click on "Biosphere Project."

What Is the Endangered Species Act?

Historically, most wildlife legislation in the United States dealt with the trade or transport of wild animals or animal products. Such legislation failed to protect

(left–right) Black-tailed Prairie Dog, Northern Gannets, Basidiomycete fungus, Blue-tailed Skink

living animals and their habitats and ignored threatened plant species. In recognition of this fact, conservationists began to work for the protection of species and habitats several decades ago. Their efforts resulted in the Endangered Species Act of 1973 (ESA).

The ESA mandated the establishment of a comprehensive list of all species faced with global extinction. Under the act, listed species are classified as either "endangered" or "threatened;" endangered species are in danger of extinction throughout all or a significant part of their range, while threatened species are those likely to become endangered in the foreseeable future (Figure 15-32). Terrestrial and freshwater aquatic species are listed by the Secretary of the Interior and regulated by the U.S. Fish and Wildlife Service (USFWS), while marine species are listed by the Secretary of Commerce and regulated by the National Marine Fisheries Service (NMFS).

Species listed under the ESA may not be harassed, harmed, hunted, trapped or killed, and may not be bought, sold, imported, or exported. The act also provides for the protection of the "critical habitat" essential to the survival and recovery of a listed species. All federal agencies must consult with the USFWS or NMFS before starting any project to ensure that no endangered or threatened species will be harmed by the project. Private citizens may also file legal lawsuits to stop a project on the grounds that it violates the ESA if that project is begun without provisions necessary to protect listed species. Despite these requirements, very few projects are halted because of a conflict with the act. Between 1979 and 1991, for example, over 120,000 projects were reviewed for conflicts with the ESA, but only 34 were canceled.

The Endangered Species Committee, composed of the seven Cabinet Secretaries, agency administrators and representatives from each of the states affected by the proposed project, can grant an exemption to the ESA under one of three conditions: (1) no reasonable and prudent alternatives to the project exist; (2) the project is of national or regional significance; or (3)

FIGURE 15-32: Kemp's Ridley sea turtle hatchlings. Like many other sea turtles, the Kemp's Ridley is an endangered species.

the benefits of the project clearly outweigh those of the alternatives. Informally known as the "God Committee" for its power to determine the survival or extinction of a species, this committee has been called upon only three times since its inception, a credit to the ability of the ESA to resolve conflicts and reach compromises while still protecting species.

The ESA is considered by many to be the most far-reaching and important environmental legislation ever written. Still, it has several weaknesses. First, simply listing a threatened or endangered species is expensive and time-consuming. The Secretary of the Interior or Commerce decides whether or not to list a species based on the best scientific and commercial data available. This decision must be made within a year, and if the

(left–right) American Crow, Velvet Ant, Deer Mice, Old Woman Creek National Freshwater Estuary (Ohio)

listing proposal is approved, the Secretary must publish an official notice in the Federal Register, inform the affected state and local governments along with any relevant scientific organizations, print a summary of the listing proposal in a local newspaper, and hold public hearings if requested to do so. Due to the chronic underfunding of the ESA, inadequate funds are available to finance the study and listing of all species which are probably in danger of extinction. By 2005, over 1,823 species had been listed under the ESA (988 endangered and 276 threatened in the United States; the remainder located in foreign countries), but thousands of other candidates for listing await further study and official consideration. Also, protection is more likely to be given to mammals and other highly visible animals than to plants or invertebrates, although the performance of the USFWS on this issue has improved in recent years.

Finally, politics can interfere with the protection of endangered species. (See Chapter 16, *Public Lands,* for a discussion of the current controversy surrounding the reintroduction of the endangered gray wolf to Yellowstone National Park). The first test of the ESA came in 1973, when a previously unknown fish species, a three-inch, tannish-colored perch dubbed the snail darter, was discovered in the Little Tennessee River. The river contained the darter's only known spawning habitat — the swifter portions of shoals (shallow areas) over clean gravel substrate in cool, low-turbidity waters. Unfortunately, that habitat was threatened by an impoundment under construction by the Tennessee Valley Authority (TVA). The TVA had begun working on the Tellico Dam in the late 1960s, but after the snail darter was listed as an endangered species in 1975, it seemed as if construction might be halted permanently. A court battle ensued, with the TVA and developers arguing that the ESA was never meant to protect insignificant species like the darter and environmentalists arguing that the act protected all rare species, not merely charismatic megafauna like the bald eagle. After a 1976 U.S. District Court ruling in favor of the TVA was reversed by an Appeals Court, the case was heard by the Supreme Court. Unswayed by arguments that $78 million had already been spent on the dam and that the darter was just a tiny fish anyway, the Supreme Court in 1978 came down on the side of endangered species, reasoning that the "plain intent" of the law is to save all species, whatever the cost.

The Supreme Court decision was not the last word on the Tellico Dam. Instead, Congress amended the ESA, creating the God Committee, which then took up the case of the snail darter. Much to the TVA's dismay, the God Committee refused to allow the dam project to proceed, citing both economic and ecological reasons. Nevertheless, in mid-1979, Senator Howard Baker and Congressman John Duncan, both from Tennessee, attached to an appropriations bill a rider exempting Tellico Dam from the ESA. The rider narrowly passed Congress, and after President Jimmy Carter refused to veto the bill, the project was on once again. The TVA completed Tellico Dam in late 1979, presumably extirpating the snail darter. Fortunately, as the controversy was unfolding throughout the 1970s, researchers had transplanted the fish to other nearby rivers, and some of those populations still survive. Several very small but naturally occurring populations have been found in other areas as well.

What Is CITES?

By the 1970s, the global community recognized the need for broad international action to save species. That recognition resulted in four conventions: the Convention on Migratory Species of Wild Animals, the World Heritage Convention, the Convention on Wetlands of International Importance Especially as Waterfowl Habitat, and the Convention on International Trade in Endangered Species of Wild Flora and Fauna (CITES). Of these, the most important and effective is CITES, which accords varying degrees of protection to more than 30,000 species of animals and plants, whether they are traded as live specimens or wildlife products, such

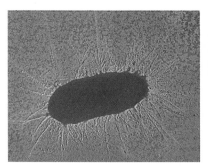

(left–right) *E. coli*, Raccoons, Woodland Damselfly, Peat bog (Finland)

as fur coats or dried herbs. As of 2003, 164 countries had signed CITES, including the United States. Member countries are required to designate a management authority to issue permits for trade in species listed in the CITES appendices. Member counties must also designate a scientific authority to provide scientific advice on imports and exports. Permits issued by CITES are the only legal, recognized permits for international trade in wild animals, plants, or wildlife products.

Though it can be effective, CITES does have weaknesses. Only nations that have ratified the convention are bound to its restrictions. Many countries that have not joined the convention, such as Taiwan, are significant participants in the wildlife trade. Moreover, a participating country can legally continue to trade in even the most highly endangered species simply by informing CITES of its intention to do so, an action known as taking out a reservation on a species. And as with all international conventions, enforcement is left to the countries involved. Consequently, there is wide variance in how well CITES is enforced in various nations. For example, Japan, although a participant in the convention, continues to import large quantities of prohibited items, including tortoise shell, rhino horn, and ivory. Exporting nations, which are usually the less-developed countries, need the foreign money brought in by the wildlife trade. Importing nations, on the other hand, are usually the more-developed countries. The primary enforcement problem faced by these nations is a lack of personnel qualified to detect and identify endangered species and wildlife products.

What Is the Convention on Biological Diversity?

Growing concern over biodiversity loss led to the Convention on Biological Diversity, first presented to the international community at the 1992 United Nations Conference on Environment and Development (com-monly called the Earth Summit). The Convention is an attempt to establish a comprehensive framework for stemming biodiversity loss. It recognizes the intrinsic value of biodiversity, its importance to human welfare, nations' sovereign right over their own biodiversity, and their responsibility for conserving it.

Conceptually, the Convention has four sections. The preamble and Articles 1-5 present objectives, definitions, principles, and jurisdictions. Articles 6-14 detail the commitments of each party to conserve biodiversity and promote its sustainable use. Articles 15-22 describe the relationships between parties with respect to implementing the Convention, including financial relationships, and between the Convention and other international legal instruments. The final section, Articles 23-42, concerns the administrative and procedural mechanisms by which the Convention will be implemented.

In 1993, the Clinton Administration signed the Convention but the Senate has so far failed to ratify it. Those who oppose the Convention cite concerns over its provisions on funding, intellectual property rights, and biotechnology. These issues find the world divided along a "North-South" split. With respect to funding, nations cannot agree on the amount of money that the industrialized North should provide for biodiversity protection in the South, nor on the mechanism that will be used to control and account for those funds. The developing countries of the South favor a mechanism in which they have an equal voice with donors in setting policies and priorities and allocating funds. The donor countries favor an institution over which they exercise effective control in policy and program decisions. The provisions on intellectual property rights and biotechnology are even more problematic. They concern different ways of regulating access to genetic resources (the majority of which are found in the South) and access to, and transfer of, technology and the products of biotechnology (which are generally under the control of the North).

(left–right) Caribou, Poison Dart Treefrog, Hermit Crab, Parnassian butterfly

Summary

All species are biological resources. Biological diversity, or biodiversity, refers to the variety of life forms which inhabit the Earth. Biodiversity is measured in terms of species diversity, the number of different species present in an area; ecosystem diversity, the variety of habitats, biotic communities, and ecological processes in the biosphere; and genetic diversity, the variation among the members of a single population of a species. Each member of a population has a unique genotype, the individual's complement of genes. The hereditary material of an organism is its germ plasm. The sum of all of the genes present in a population is a gene pool.

All ecosystems host unique associations of interdependent organisms. Endemic species are unique to their particular ecosystem and occur nowhere else on earth. Two ecosystems unequaled in terms of biodiversity are the tropical rain forests and the coral reefs.

Our cultural attitudes and beliefs affect how we value, use, and manage biological resources. A group of species which are highly valued by humans, such as endangered mammals or birds, have been called the charismatic megafauna. We label undesirable animals "vermin" and undesirable plants "weeds." Species hunted for sport are called game animals. Nongame animals are species not hunted for sport.

Cultural attitudes and beliefs give rise to many activities that pose a significant threat to biological resources and biodiversity. The single greatest threat facing wildlife is loss or degradation of habitat. Biological resources are also endangered by overharvesting and illegal trade, selective breeding, introduced species, and pollution.

Extinction is the elimination from the biosphere of all individuals of a group of organisms of any taxonomic rank (family, genus, or species). Whenever a species becomes extinct, biodiversity is diminished. Extinction is a natural process which has occurred throughout Earth's history. However, human activities have greatly accelerated the rate of extinction.

The disappearance of an individual species often has far-reaching effects. Diminishing biological diversity threatens ecosystem diversity. Decreasing genetic diversity threatens species diversity by making species more vulnerable to extinction. The loss of genetic variability is genetic erosion. It lessens a population's ability to adapt to changes in its environment. The minimum viable population size is the smallest population unit of a species that has a specified percent chance of surviving for a specified period of time, given foreseeable demographic, genetic and environmental occurrences, as well as natural catastrophes. This size has not been determined for most species.

The arguments for preserving biological resources and maintaining maximum biological diversity fall into five categories: ecosystem services; aesthetics; ethical considerations; evolutionary potential; and benefits to agriculture, medicine, and industry.

Wild plant collections and collections of exotic animals have a long history. For the most part, current efforts to manage biological resources concentrate on preserving plant and animal species or their genetic material. Both species and germ plasm may be preserved off site or on site. Private collections, wildlife refuges, sperm banks, and zoological parks are all means used to preserve animals outside their natural habitat. Captive breeding programs attempt to meet two objectives: to maximize the contribution

Discussion Questions

1. Explain the difference between genetic, species, and ecosystem diversity and explain why it is important to preserve each of these components of biological diversity.

2. What is extinction and how is it related to biological diversity? Research at least two species in the United States that are threatened with extinction (one plant and one animal) and prepare a status report telling its current population, the threats to its survival, and the role that the species plays in the community (its niche). What efforts, if any, are being made to preserve this species? In your opinion, are those efforts sufficient?

3. How do human activities threaten wildlife? How do they adversely affect domesticated species? Give at least two examples of each. Describe three things that could be done to lessen the human impact on wildlife and domesticated species.

4. What reasons are usually given in favor of preserving biological resources and maintaining maximum biological diversity? Briefly explain each.

5. Discuss how the role of zoos has changed over time. What are the implications for preserving genetic and species diversity?

6. In the United States, certain states and regions have suffered particularly high losses of biological diver-

(continued on next page)

(continued from previous page)

sity. One such region is the Southeast, which is the center of diversity worldwide for freshwater mussels. What factors contributed to this region's wealth of freshwater species? To answer, consider the region's topography and terrain and think in geologic terms; refer to Table 1-1 on page 12 for some clues. Conduct research to identify the factors that have contributed to the high rate of extinctions in the Southeast.

7. How are national parks, protected wilderness areas, and biosphere reserves different from private collections, zoos, and sperm banks? Explain the difference in terms of species, genetic, and ecosystem diversity. Which do you think is more effective, off-site or on-site preservation, and why?

8. Describe the provisions of both the Endangered Species Act and the Convention on Trade in Endangered Species of Wild Flora and Fauna. What are the strengths and weaknesses of each?

9. The Endangered Species Act has been due for reauthorization for several years. Contact your Congressional representative to determine his or her position on the Act and whether or not it is currently under consideration for reauthorization. Do you agree with your representative's view on the ESA? Why or why not?

10. Should the United States ratify the Convention on Biological Diversity? Why or why not?

of unrelated animals in order to reduce the effects of inbreeding and to attain a good age distribution in order to ensure a consistent number of individuals of breeding age.

National parks, protected wilderness areas, and biosphere reserves offer the best hope for preserving most species and for preserving maximum biological diversity. They are the only means by which the relationships between organisms can be preserved.

In the United States, most legislation and management efforts have been aimed at game species. However, as populations of individual wild species declined, laws were enacted to protect them. Historically, most wildlife legislation dealt with the trade or transport of wild animals or animal products. Such legislation failed to protect living animals and their habitats and ignored threatened plant species. In recognition of the need to protect a diversity of species (plant and animal) as well as habitats, the Endangered Species Act of 1973 was passed.

Throughout the twentieth century, countries have adopted various international treaties and conventions to protect migratory species; most of these are agreements that involve several countries in a specific region. Of the conventions developed to protect species on a global basis, the most significant is CITES, the Convention on Trade in Endangered Species of Wild Flora and Fauna. The Convention on Biological Diversity, introduced at the 1992 Earth Summit, is the first of its kind; its aim is to stem the loss of biodiversity through international cooperation.

KEY TERMS

biodiversity	endemic species	germ plasm
biological diversity	extinction	minimum viable population size
biological resources	game animal	
	gene	nongame animal
charismatic megafauna	gene pool	species diversity
charismatic minifauna	genetic diversity	vermin
	genetic erosion	weed
ecosystem diversity	genotype	

CHAPTER **16**

Public Lands

Harmony with land is like harmony with a friend; you cannot cherish his right hand and chop off his left. That is to say, you cannot love game and hate predators; you cannot conserve the waters and waste the ranges; you cannot build the forest and mine the farm. The land is an organism.

Aldo Leopold

What would the world be, once bereft
of wet and wildness?
Let them be left,
O let them be left.
wildness and wet:
Long live the weeds
and the wilderness yet.

Gerard Manley Hopkins, "Inversnaid"

Learning Objectives

When you finish reading this chapter, you will be able to:

1.   Identify and briefly describe the major types of federal lands and the agencies that manage each.

2.   Briefly explain the historical development of the system of federal public lands.

3.   Discuss the biological, physical, and social significance of the public lands.

4.   Identify some of the most important reasons to preserve the public lands.

5.   Describe the major threats that face public lands.

In 1872, the U.S. Congress set aside a vast stretch of land in Wyoming known for its natural beauty and scenic wonders as "a public park or pleasuring ground for the benefit and enjoyment of the people." In designating Yellowstone as a "nation's park," Congress established an American tradition of parks and preserves that would serve as a model for the world. In this chapter, we discuss those lands owned by all of us and managed by the federal government. We see just how vast and diverse these lands are, why they are biologically significant, what minerals and other resources they contain, and the diverse ways in which we use them. We also look at some of the threats that face our public lands (Figure 16-1).

FIGURE 16-1: Upper Falls, Yosemite National Park. Limiting the impact of people and cars is a particularly important issue in Yosemite, one of the most popular national parks.

Describing the Public Lands

The federal lands encompass an astounding diversity of ecosystems — from alpine meadows to prairies and coral reefs to mangrove swamps. They include the wind-swept Alaskan tundra and the sawgrass wetlands of the Everglades; the slick-rock canyons of Utah and the rocky shoreline of Maine; the bone-dry expanses of Death Valley and the tropical lushness of the Caribbean National Forest, Puerto Rico; the jagged, snow-covered peaks of the Rockies and the smooth, rounded slopes of the Great Smoky Mountains; the old-growth forests of the Pacific Northwest and the geologically young Great Lakes. The federal lands are a wealth of rich biological and physical resources. Because these resources are highly sought after for a wide variety of purposes, their presence on public lands can make management particularly tricky.

On the following pages, we will look at the different federal public lands and how they came to be, as well as consider why these areas are important to us biologically, physically, and socially.

What Are Public Lands?

The United States boasts one of the largest total acreages of publicly owned lands in the world — over 726 million acres (294 million hectares). That's approximately one-third of the entire country! Known as the **public domain**, these lands are so vast that fully one-half have never been surveyed. Owned jointly by U.S. citizens, the public domain is managed by various federal, regional, state, and local authorities, although the federal government oversees most of it. The vast majority of the public domain, a full ninety percent of it, lies in the western states and Alaska (Figure 16-2).

These lands have a rich and unique history. By 1800, the United States had begun to acquire territories outside of the original 13 states, either through treaties (usually broken) with native American tribes or through agreements with European powers. For example, with the Louisiana Purchase of 1803, the United States bought France's rights to much of the territory west of the Mississippi, significantly expanding the size of the young nation.

Many people, including Thomas Jefferson, believed that the best way to protect the expanding nation was to encourage settlement and development, particularly along the frontier. As a result, land disposal to private citizens, businesses, and state and local governments dominated federal policy for almost 150 years. The Homestead Act of 1862 granted 160 acres to anyone who settled the land and cultivated it for five years. Between 1860 and 1930, Congress gave more than 94 million acres (38 million hectares) to the railroads alone. Much of the clearcutting now occurring on private lands in the Pacific Northwest can be traced to the give-away of these railroad trust lands. Many other lands were given away to build schools and roadways.

In 1907, vacant public lands amounted to approximately 400 million acres (162 million hectares); by 1932, that figure had been cut in half. In the mid-1930s, President Franklin D. Roosevelt issued an executive order withdrawing all public lands within the lower 48 states from homesteading. He thereby effectively closed the public domain (outside Alaska) to major disposals, signaling a shift in policy from land disposal to reten-

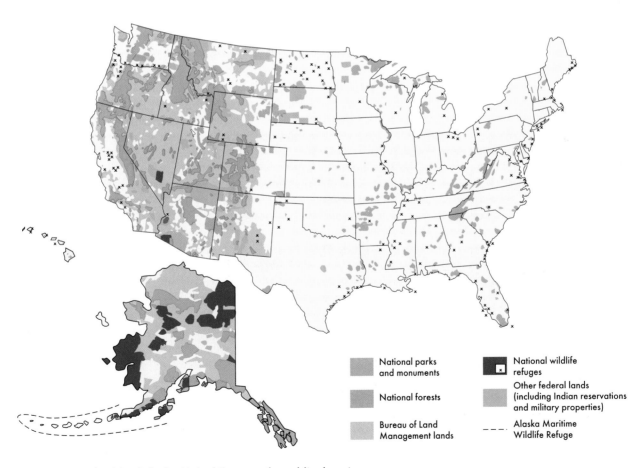

FIGURE 16-2: Federal lands in the United States — the public domain.

National parks and monuments

National forests

Bureau of Land Management lands

National wildlife refuges

Other federal lands (including Indian reservations and military properties)

Alaska Maritime Wildlife Refuge

tion of the remaining public lands. By this time, the government had either given away or sold 1.1 billion acres (445 million hectares).

The federal public lands are managed by numerous agencies, including the Department of Defense, Department of Energy, Bureau of Reclamation, Bureau of Indian Affairs, and the Tennessee Valley Authority. More than 90 percent of the public lands, however, fall into one of five classifications: the National Park System, National Wildlife Refuge System, National Forest System, National Resource Lands, and the National Wilderness Preservation System.

National Park System

The National Park System (NPS) is a network of 388 units totaling over 84 million acres (34 million hectares). In addition to 54 national parks, the system includes various monuments, historic sites, memorials, recreation areas, battlefields, seashores, wild and scenic rivers, parkways, scenic or historic trails, and other sites. Park units vary greatly in size. The largest, at 13.2 million acres (5.3 million hectares), is Alaska's Wrangell–St. Elias National Park; the smallest, at 0.02 acre, is Pennsylvania's Thaddeus Kosciuszko National Memorial.

The national parks are the country's best-known and most popular public lands. Many of the units within the system commemorate people and events that shaped our country, such as the Washington Monument, Lincoln Memorial, and Gettysburg National Battlefield. But the image of the park system is dominated by the large western parks — Yellowstone, Grand Canyon, Yosemite, and Glacier (Figure 16-3). Because of their natural grandeur and scenic wonders, they are called the "crown jewels" of the park system. They are the pride of our country, our counterpart to the cathedrals and castles of Europe, and a part of our shared heritage.

The NPS is managed by the National Park Service, Department of the Interior. In creating the service, Congress decreed that it should manage the parks in order to "conserve the scenery and the natural and historic objects and wildlife therein and provide for the enjoyment of the same in such manner and by such means as will leave them unimpaired for the enjoyment of future generations." Although many controversies arise over the struggle to make the parks accessible to citizens while preserving their natural character, national parks enjoy a greater degree of protection than other federal lands. Logging, for example, is prohibited; mining is also forbidden, except in those cases where a claim to the mineral rights had been filed prior to the area's designation as a park.

FIGURE 16-3: The Grand Canyon of the Colorado River. The first national parks were designated for their sublime beauty and grandeur.

While the concept of the "nation's parks" was immediately embraced by the American public, the early years of park management were plagued by problems. To begin with, there was no established park system, nor was there a standard method for adding units; parks were designated in a haphazard fashion. The Antiquities Act of 1906 gave the president the authority to preserve "national monuments," objects of cultural or historic significance on public lands. President Theodore Roosevelt, the first president to show a serious commitment to conservation, broadened the intent of the act in order to set aside spectacular natural areas such as Devil's Tower, Wyoming; the Petrified Forest and Montezuma's Castle, Arizona; and El Morro (Inscription Rock), New Mexico. However, since there was no unified, organized management, many of the parks suffered during their early years. Finally, the 1916 Organic Act established the Park Service to protect and manage the nation's parks.

The Park Service's first director, Stephen Mather, reasoned that the best way to protect the parks was to garner and solidify strong public support, and so he undertook several campaigns to encourage tourism. While Mather's strategy was successful, public enthusiasm for the growing park system had its drawbacks, and by the 1950s, many units suffered the effects of too many visitors and too little funding.

To meet the growing demand for outdoor recreation opportunities and ease pressure on existing parks and monuments, the Park Service began to establish national seashores and national recreation areas in the 1960s and 1970s. The service chose sites near urban areas in order to make it easy and convenient for people to use and enjoy the public lands. Battlefields, military sites, and other historic units were also added to the system, as were numerous monuments and open spaces in Washington, D.C. The seashores and recreation areas enabled people who had little hope or means of visiting the traditional parks to experience the public lands first-hand.

Throughout the 1980s, conservationists worked — with limited success — to expand and protect the park system. One focus of their efforts was the Mojave Desert, where heavy recreational use threatened the area's biological resources. In 1994, the California Desert Protection Act, hailed as the most significant piece of park and wilderness legislation in a decade, created Mojave National Park, expanded Death Valley and Joshua Tree national monuments and redesignated them as national parks, and designated 4.4 million acres of wilderness.

National Wildlife Refuge System

The National Wildlife Refuge System (NWRS) is administered by the U.S. Fish and Wildlife Service (USFWS), Department of the Interior. The NWRS consists of over 570 units, totaling over 97 million acres (39.2 million hectares) of land and water. Over 1,000 species of birds, mammals, reptiles, amphibians, fish, and plants — including some 60 endangered species — are found on the nation's wildlife refuges.

Although 96 percent of refuge units are in the lower 48 states, Hawaii and U.S. territories, 85 percent of the refuge system acreage is in Alaska. Wetlands account for about one-third of refuge lands (Figure 16-4). The smallest refuge, less than one acre, is at Mille Lacs, Minnesota; the largest is Alaska's 22-million-acre Yukon Delta National Wildlife Refuge.

The primary purpose of the refuge system is to provide habitat and haven for wildlife. Some units have been established to protect significant populations of a single species; others are located along the major north-south flyways, providing resting and feeding areas for migratory waterfowl and other birds. Still other units serve as sanctuaries for endangered species: Aransas National Wildlife Refuge in Texas, for example, has

FIGURE 16-4: Chincoteague National Wildlife Refuge, Virginia.

been critical to the recovery of the whooping crane population, while the Hawaiian Islands Refuge provides habitat for the Hawaiian monk seal and the green sea turtle. A number of refuges protect archeological artifacts and historically significant areas.

In addition to wildlife habitat, most refuges also permit secondary uses. Wildlife observation, photography, nature study, hiking, and boating are permitted in most refuges; some refuges also allow hunting, trapping, and fishing. The development of oil, gas, and mineral resources is permitted in certain refuges, at the discretion of the Secretary of the Interior. In refuges that have been established on land managed by another federal agency, such as the Army Corps of Engineers or the Department of Defense, the USFWS has only secondary authority. It is responsible for managing the wildlife in the area, but has no authority to prohibit various land uses, such as military aircraft overflights, even when those uses may pose a threat to resident animals or plants.

Much like the NPS, the NWRS had a rocky start. In 1903, President Theodore Roosevelt invoked the 1891 Forest Reserve Act to establish Pelican Island, Florida, as the first national wildlife refuge (Figure 16-5). Roosevelt contended that the island's mangrove thickets qualified as forest. Later, he used provisions of the 1906 American Antiquities Act to protect other areas as wildlife refuges, including the National Bison Range in Montana. By the time he left office in 1908, Roosevelt had designated 51 national wildlife refuges.

However, although Congress had approved the acquisition of land for new refuges under the Migratory Bird Conservation Act of 1929, it failed to provide funding. In 1934, President Franklin D. Roosevelt (FDR) named Aldo Leopold, J. N. Darling, and other prominent conservationists to a committee to seek funding for the refuges. The committee introduced the concept of the duck stamp, which hunters are required to purchase yearly; proceeds are used to establish new refuges. Since 1934, revenues from duck stamp purchases have been used to preserve over four million acres of wetland habitat.

In the 1940s, FDR merged the Bureau of Biological Survey, Department of Agriculture, with the Bureau of Fisheries, Department of Commerce, to form the USFWS within the Department of the Interior. The USFWS was charged with the mission to "provide, preserve, restore, and manage a natural network of lands and waters sufficient in size, diversity, and location to meet society's needs for areas where the widest spectrum of benefits associated with wildlife and wild lands is enhanced and made available." The number of refuges grew steadily over the next several decades, and in 1966, Congress formally established the National Wildlife Refuge System.

National Forest System

The National Forest System (NFS) consists of 155 national forests and 22 national grasslands. These units total approximately 192 million acres (77.7 million hectares), 8.5 percent of the United States — an area as large as Wisconsin, Michigan, Illinois, Indiana, Ohio, and Kentucky combined (Figure 16-6).

The national forests hold much of the public lands that have been designated wilderness areas in the lower 48 states, and are home to over 3,000 vertebrate species and 129 endangered or threatened species. The U.S. Forest Service, Department of Agriculture, manages the nation's forests and grasslands to accommodate a wide array of uses. These areas provide many recreational opportunities, including sight-seeing, driving, hiking, horseback riding, camping, canoeing, off-road vehicle use, picnicking, snowmobiling, and cross-country skiing. Lands are also leased to commercial enterprises such as ski resorts, youth camps, and cabin owners, as well as lumber and mining interests.

The first forest reserves were established in 1891 and placed under the control of the Department of the Interior. The Organic Act of 1897 gave the government the authority to protect and manage the forests. One year later, President Theodore Roosevelt named Gifford Pinchot the first chief forester of the Bureau of Forestry. In 1905, Pinchot managed to have the forest reserves

FIGURE 16-5: Pelican Island National Wildlife Refuge.

Figure 16-6: Olympic National Forest. This unit of the National Forest System blankets almost 650,000 acres in Washington.

transferred from the Department of the Interior to the Department of Agriculture. He then reorganized the Bureau of Forestry into the Forest Service.

Although Roosevelt's thinking was influenced by his friendship with both Gifford Pinchot and the preservationist John Muir, it was Pinchot's philosophy of **utilitarianism**, or conservation for economic reasons, that dominated official policy. Trained in Europe, Pinchot was committed to the scientific management of forests and the goal of sustained harvests. He set out the conservation principles, particularly the principle of multiple use, that would guide the Forest Service for decades. **Multiple use** is the management of land to accommodate various uses, both commercial and recreational. Pinchot also imposed grazing regulations and modest fees on ranchers, who had previously used the land free of charge.

Westerners were hostile to the conservation concerns of the Roosevelt administration, and they vented their hostility through their elected officials. In 1907, Congress passed legislation to prohibit new forest reserves in six western states without Congressional approval. Unwilling to veto the bill, Roosevelt directed Pinchot to quickly draw up plans for the addition of new western forests. The Forest Service responded with plans to add 16 million acres (six million hectares) in new reserves, enlarging the NFS to about 140 million acres (57 million hectares). Roosevelt approved the additions and

then signed the bill, which effectively prohibited any president from doing what he had just done!

Soon after this, easterners began to agitate for forests to be set aside in their region, largely to reverse deforestation and protect against flash floods. Although little public domain remained east of the Mississippi, the Weeks Act of 1911 granted the Forest Service the authority and the funds to acquire private timberlands. New Hampshire's White Mountain National Forest became the first national forest in the eastern United States. In the 1920s and 1930s, many more eastern forests and grasslands were added to the system, as landowners hurt by the Depression became anxious to sell land for much-needed cash.

The Forest Service's difficulties with western economic interests again came to the forefront after World War II, when in 1946, a handful of ranchers launched a serious attempt to dismantle the forest system. They met in Salt Lake City, Utah, and devised a plan to have the states take possession of the national forests as well as 142 million acres (58 million hectares) of grazing lands under the control of the Forest Service. According to the plan, supported by western politicians, a select group of ranchers would eventually buy the lands at nine cents per acre. When the historian and social critic Bernard DeVoto learned of the plan and described it in *Harper's Magazine*, the ensuing public outcry spoiled the ranchers' plan.

At about the same time, the management philosophy of the Forest Service underwent a dramatic change. Until World War II, the service was basically a caretaker agency. Government foresters acted as custodians of the national forests and taught wise management practices to private foresters. The postwar boom in the U.S. housing industry and the need to rebuild war-torn Europe created a skyrocketing demand for lumber. When private forests could no longer meet that demand, timber executives cast a covetous eye on the national forests. The Forest Service began to seek out the business of the timber industry, and harvesting in the nation's forests increased. However, contrary to popular perception, the national forests' contribution to domestic timber production is relatively small, approximately five percent.

National Resource Lands

Officially known as the National Resource Lands, the 261 million acres (107 million hectares) managed by the Bureau of Land Management (BLM), Department of the Interior, are more commonly called the

BLM lands. A 1987 BLM publication described them this way: "These lands in the lower 48 states are those which no one wanted; they were bypassed in favor of more promising areas during the course of settlement. The lands are the mountain tops, steep canyons, vacant valleys, and sandy deserts." They have been described as "stretches of picturesque poverty," "the lands nobody wanted," and "the lands no one knows" (Figure 16-7). Many units occur in a checkerboard fashion, interspersed with privately owned land, and many are unmarked. Compared to other federal lands, they host few visitors each year, in part because of a lack of public awareness about them. In fact, BLM lands are undoubtedly the most underappreciated of all federal lands. In addition to the land it manages, the BLM also leases mineral rights beneath another 200 million acres in national forests and private lands.

By the 1930s, the majority of the federal lands remained unreserved; they were not designated as parks, refuges, or forests. Most of these lands were severely degraded by overgrazing and other deleterious practices. To address concerns about overgrazing, the Taylor Grazing Act of 1934 authorized the Department of the Interior to set up grazing districts on 142 million acres (53 million hectares) of public range. It also established a federal grazing service to lease rangelands to established ranchers. In return for this authority, the Secretary of the Interior promised the livestock industry that grazing fees would be kept low and that bureaucracy within the administrative agency would be kept to a minimum. The result of this bargaining was a weak Grazing Service with district boards dominated by industry. In the late 1940s, the Grazing Service

was merged with the General Land Office to form the BLM, the nation's largest land-managing agency. Because of its origins, historically, cattle ranchers have had considerable influence on the BLM.

In 1976, the Federal Land Policy and Management Act (FLPMA) mandated that the BLM lands be managed to provide "the greatest good for the greatest number of people for as long as possible." Essentially a multiple-use planning mandate, FLPMA spelled out goals and procedures, but left BLM personnel with the authority to make resource decisions; it did not prioritize uses for BLM lands. Congress's reluctance to set policy priorities has resulted in tremendous pressure on the BLM (as well as the Forest Service and USFWS) by various commercial, recreational, and wilderness interest groups.

National Wilderness Preservation System

The National Wilderness Preservation System (NWPS) consists of 662 areas totaling 105 million acres (42 million hectares) that have been officially designated and set aside by Congress in order to preserve their wild state and the values associated with wilderness — scenic beauty; solitude; wildlife; geological features; archeological sites; and other features of scientific, educational, and historical value. Most wilderness areas lie within national parks, national forests, and national wildlife refuges, with a relatively small number found on BLM land. Wilderness areas are managed by the agency responsible for the public land unit in which they are found, and these agencies are charged with maintaining the pristine nature of wilderness areas. The largest wilderness areas, just under 9.1 million acres (3.7 million hectares), lies in Alaska's Wrangell-St. Elias National Park. The smallest, just six acres (two hectares), is in Florida's Pelican Island National Wildlife Refuge.

Most designated wilderness is in high alpine or tundra ecosystems. Well over half of all official wilderness acreage lies in Alaska alone, but two-thirds of all recreational use occurs on just 10 percent of wilderness areas in the lower 48 states, especially in California, North Carolina, and Minnesota. In the contiguous United States, where the demand for wilderness recreational use is the greatest, protected wilderness areas comprise less than five percent of the total land area. By conservative estimates, only another five percent qualifies for wilderness designation.

In the Wilderness Act of 1964, Congress instructed that wilderness areas be preserved in their essentially virgin state "for the use and enjoyment of the American

FIGURE 16-7: Cabezon Peak, near Albuquerque, New Mexico, a National Resource Land administered by the Bureau of Land Management.

people in such a manner as will leave them unimpaired for future use and enjoyment as wilderness." Accordingly, wilderness areas are open only for recreational activities such as hiking, sportfishing, camping, and nonmotorized boating (Figure 16-8). In some areas, hunting and horseback riding are also allowed. Roads and permanent structures are prohibited, as are logging and livestock grazing. Mining is also prohibited, unless the activity occurred before the area was designated as wilderness or if a claim was filed by December 31, 1983. Motorized vehicles and equipment are prohibited unless they are required in an emergency, such as fire control or a rescue attempt.

To minimize the impact of visitors on wilderness ecosystems and thus preserve the unspoiled nature of the areas, wilderness managers have adopted a variety of tactics. Officials may develop trails, issue permits for designated campsites, and limit the number of people hiking or camping. Additionally, wilderness managers may emphasize public education on wilderness values and use.

It was not until the twentieth century that scientists and naturalists began to realize the ecological importance of protecting wild lands and started to work for the establishment of a formal system of protected wilderness. In the 1920s, Arthur Carhart and Aldo Leopold, who were both working as Forest Service employees in the West, began to urge the Forest Service to preserve certain roadless areas in their natural, undisturbed state. In 1924, the Forest Service set aside a portion of Gila National Forest as the first "primitive area" within the national forests. Shortly thereafter, in the 1930s, Robert Marshall, Director of Forestry for the Bureau of Indian Affairs, created 16 new wilderness areas within Indian reservations. In essence, the wilderness preservation movement had begun.

From the beginning, the movement faced powerful enemies, particularly the logging, mining, and ranching industries and pro-development western politicians. With the increased pressures for greater development that followed World War II, conservationists realized that without comprehensive legislation protecting wilderness areas, it would become increasingly difficult to prevent development in pristine areas.

In 1957, Minnesota senator Hubert H. Humphrey introduced a bill into Congress that would give legal protection to wilderness. The opposition, consisting mainly of logging, mining, and ranching industries, was well-organized and well-financed, had significant political clout, and, ironically, was joined by the Park

FIGURE 16-8: The Absaroka-Beartooth Wilderness Area, located in the Gallatin, Shoshone, and Custer National Forests in Montana and Wyoming.

Service and the Forest Service, who felt that the bill stepped on their management toes. Despite the strong opposition, seven years and 66 different versions of the bill later the Wilderness Preservation Act of 1964 was signed into law by President Lyndon Johnson.

In addition to immediately designating 9.1 million acres (3.7 million hectares) of land in 54 sites, the Wilderness Preservation Act of 1964 directed the Park Service, Forest Service, and USFWS to review their holdings and recommend areas that qualified for protection (Figure 16-9). (Not until 1976 was the BLM also required to review its holdings.) Congress was given the responsibility to approve the recommendations and to designate areas for inclusion in the NWPS. The act also enabled Congress to move independently to designate new wilderness areas. This provision slowed the designation process somewhat, but it invited wider public participation, something preservationists felt necessary.

To secure congressional support for the wilderness bill, preservationists were forced to compromise on several points. Perhaps the most important compromise concerned the exploration and development of mining, prospecting, and oil and gas drilling in wilderness areas. These activities were allowed to continue until December 31, 1983; any claims filed by that date could be developed at any time in the future.

Despite the requirement that land management agencies periodically review holdings for wilderness preservation, these agencies resisted doing so because they viewed wilderness designation as being in conflict with agency responsibilities. The Forest Service in particular resisted designating wilderness areas, stating that any evidence of past alteration, even an abandoned mine shaft or overgrown track, eliminated an area from consideration for wilderness designation. To correct the situation that was fast developing, Congress passed the Wilderness Act of 1973. The act immediately added 16 areas, covering some 207,000 acres (83,806 hectares) in 13 states. By its action, Congress made plain its opinion concerning wilderness designation: If an area had sufficiently recovered from human abuses and was on its way back to a pristine condition, it could be included in the NWPS.

Why Should We Preserve Public Lands?

Public lands are culturally significant, rich in meaning and resonating in our collective memory. From the stark grandeur of Mount Denali to the other-worldly wonder of Mammoth Caves, from the now-quiet Civil War battlefields to the crowded monuments of Washington D.C., the federal lands encompass all that is best, most painful, and most important about our nation. Preserving these areas is preserving a piece of our history and our heritage. The natural wonder and vastness of untouched wilderness, for example, is a connection to our past and the struggles our ancestors faced as they carved our nation out of the North American continent. If we relegate wilderness to the history books, we will have obliterated a part of our national identity. If we preserve wilderness, we keep alive a fundamental part of these United States.

The personal benefits of exploring our national treasures are far-reaching. The beauty and expansive grandeur of public lands inspire awe and renew the

FIGURE 16-9: The Linville Gorge Wilderness in Pisgah National Forest, North Carolina, is one of the original components of the National Wilderness Preservation System. The wilderness area was enlarged in 1984 and now consists of 10,975 acres.

human spirit, thus enhancing mental and emotional health. Recreation on federal lands, such as hiking, can also enhance physical health; a visit to a national park or wildlife refuge can be a learning experience, a chance to gain knowledge about history, conservation, and science. Thus, federal lands enrich us: mind, body, and spirit.

In addition to these less tangible benefits, public lands contain many important resources. Some of these are described below.

Biological Resources

Public lands serve as reservoirs for genetic-, species-, and ecosystem-diversity. Preserving biological diversity is prudent: medicine, agriculture, and industry all benefit from the genetic characteristics possessed by wild species. Many people want to preserve biological diversity for aesthetic reasons; for them, sharing the planet with the greatest possible diversity of living organisms is a key ingredient in a high quality of life. Others argue that all species have a right to exist independent of their economic or aesthetic value to humans; for them, preserving biological diversity is an ethical or moral mandate. (Chapter 15, *Biological Resources*, includes a complete discussion of these and other arguments in favor of preserving biological diversity.) Regardless of motivation, to preserve biodiversity we must preserve the habitats upon which species depend; because public lands encompass a wide array of ecosystems, they protect a diverse complement of living organisms.

In addition to the diversity they encompass, the public domain is biologically significant for several other reasons. Public lands provide us ecosystem services free of charge. Forests and grasslands, for example, hold soil in place, minimizing erosion and helping to keep streams and rivers clean and clear. The filtering action of plants and trees in forests also improves air quality. Wetlands slow and absorb floodwaters, helping protect adjacent areas, and provide critical habitat for many species. Coral reefs buffer shorelines from the action of waves and tides and provide important habitat for many species.

Economic Resources

The federal lands contain substantial mineral wealth: one-quarter of the nation's coal, four-fifths of its huge oil shale deposits, one-half of its uranium deposits, and one-half of its estimated oil and gas reserves (including outer continental shelf deposits). Significant reserves of other strategic minerals needed for sophisticated technology and weaponry are also thought to lie on federal lands (Figure 16-10).

Public lands are also rich in renewable resources, particularly timber. Alaska and many of the western states have extensive forests that, if managed sustainably, can provide numerous benefits now and in the future. Properly managed and maintained, the grasslands of the public domain can support livestock grazing. Care must be taken, however, to ensure that herd sizes do not exceed the carrying capacity of an area. Semi-arid ranges, for example, can support fewer animals than grasslands that receive more abundant rainfall. The federal lands also hold one-half of the nation's naturally occurring steam and hot water pools, which can be used to generate electric power.

Scientific and Recreational Resources

Wilderness areas are valuable for a myriad of scientific research purposes. Using the wilderness ecosystem as a reference, scientists can gauge the effects of human actions and the resulting changes in similar but unprotected ecosystems. Further, wilderness areas may provide scientists with the clues and understanding required to restore the ecological health and integrity of altered ecosystems. Nature itself is likely the best teacher we have concerning how to heal ecosystems and restore stability. In addition, undisturbed areas are a natural laboratory in which evolutionary and ecological change can proceed without human influence or disruption.

Public lands also provide an arena for the growing segment of the population that enjoys and actively seeks outdoor experiences. As the popularity of camping, hiking, and nature study increases, the economic benefits derived from these pursuits will increase as well. Wilderness areas, for example, benefit the local tourist industry and businesses involved in the manufacture and sale of outdoor gear and equipment.

Managing the Public Lands

We use the federal lands for many purposes. Each use may incur a variety of environmental problems, from direct disruption of a natural ecosystem (as in logging) to indirect disruption (as in road building to

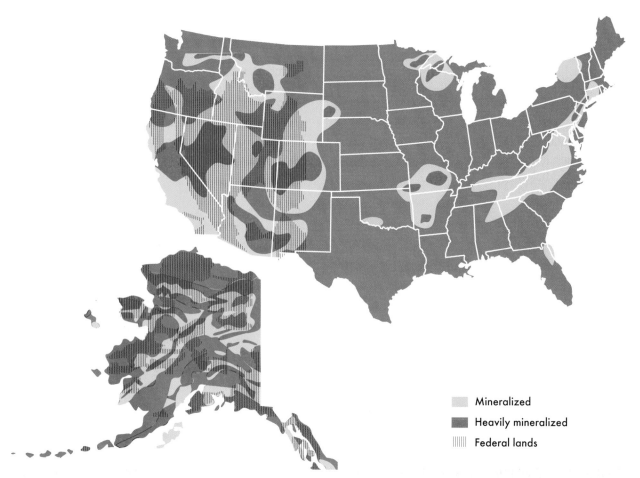

Mineralized

Heavily mineralized

Federal lands

Figure 16-10: Mineral deposits on federal public lands. Oil, oil shale, coal, uranium, copper, cobalt, nickel, zinc, titanium, platinum, and molybdenum are some of the minerals found on the public domain.

permit access to, and activity in, an area). The forests and BLM lands receive especially heavy commercial use, primarily logging, livestock grazing, and mining. These are **consumptive uses**: some portion of the resource is used up in the process. In contrast, the federal lands also offer opportunities for pursuits such as bird-watching, photography, and nature study, which are **nonconsumptive uses**, meaning that they do not use up or deplete the resource.

On all federal lands, conflicts arise among users. Recreational users vie for space and resources and argue over the correct use of the federal lands with ranchers, loggers, and miners. Conflicts even arise among recreational users, for example, hikers versus mountain bikers and off-road vehicle enthusiasts. At risk in all these conflicts is the integrity of the ecosystem. Once degraded, the "experience" and the benefits the ecosystem affords are altered and potentially lost.

What Threats Do Public Lands Face?

Although public lands "belong" to us, we hold them only for a short time before passing them on to our children and our children's children. The legacy we bequeath to future generations depends upon the actions we take today (see *What You Can Do: To Protect Public Lands*, page 373). We use federal lands for many purposes, both recreational and commercial. Because of this, public lands face a wide variety of problems that threaten the beauty and integrity of the lands and challenge the managing agencies. The difficulty of managing federal lands is compounded by the fact that, because a very diverse American public collectively owns public lands, managing agencies must try to meet the needs and interests of conflicting groups. In this section, we will look at some problems that are common to the public lands, including: overuse, insufficient funding, loss of wildlife, commercial use, activities on adjacent lands, and reluctance to designate public lands as wilderness areas.

Overuse

Many public lands are literally being "loved to death" by the crowds who visit them. The issue of overuse is particularly challenging because, in order to effectively deal with the problem, the agencies must, in some way, limit public use of lands that are in fact publicly owned. In 2003, recreational visits to the NPS exceeded 265 million, an increase of almost 30 percent since 1979. This figure is expected to soon reach 300 million and rise to 500 million by 2010. The impact of having so many visitors to an area can be great: Heavy traffic turns roads into parking lots, popular trails become eroded, vegetation is trampled, and wildlife is crowded into an ever-shrinking space. The visitors also suffer. Too often, crowded campsites, litter, noise, water pollution, and smog caused by automobile exhaust greet the visitor who comes to the park precisely to escape those conditions! The aesthetic and intangible qualities that once defined the nation's parks — serenity, solitude, tranquility, and sublime beauty — are significantly

diminished. Safety also has become a concern in recent years, as crime (especially vandalism) and drug use are becoming significant problems in some park units, particularly those near urban areas. To learn more about the problems causes by overuse, read *Environmental Science in Action: Yosemite National Park*, by going to www.EnvironmentalEducationOhio.org and clicking on "Biosphere Project."

Designated wilderness areas also face threats with increased visitation. As more people use wilderness areas, the very qualities that define the areas are degraded. This is especially true in the eastern states, specifically New England. Some areas may soon require a permit system to limit use and protect the resource, a step already taken in Minnesota's Boundary Waters Canoe Area Wilderness. To learn more, read *Environmental Science in Action: The Boundary Waters Canoe Area Wilderness*, by going to www.EnvironmentalEducationOhio.org and clicking on "Biosphere Project."

Insufficient Funding

Like many government entities, the agencies that manage public lands are plagued by insufficient funds to do the jobs that they are mandated to do. Without adequate funding, needed maintenance and repairs are delayed, the problems of overuse cannot be addressed, and overall agency management suffers. One way to address the lack of available funds for public lands is to charge higher fees for their use. Fees, however, are problematic because they can exclude lower-income Americans, and, since Americans already pay for public lands through taxes, the issue arises as to whether it is fair to make people essentially pay twice to use these lands.

In recent years, Congress has cut back on or eliminated many park projects deemed a priority by the Park Service. In their place, Congress has substituted other projects, many of which are either of low priority or have no priority with the service. As long as such Congressional "add-ons" are the norm, the national parks will continue to deteriorate.

In addition to park deterioration, insufficient funding has other less obvious adverse effects. As fewer park rangers handle increasing visitor numbers they may spend much of their time on crowd control and law enforcement. Many people fear that the old-time ranger — part naturalist, part teacher, part outdoorsperson — is giving way to a new breed of ranger, a park police officer. To its credit, the Park Service has recognized this problem and has developed a training program for employees. Twenty trainees per year from throughout the park system participate in the year-long program; they are exposed to a wide array of resource problems and issues, including air and water quality, integrated pest management, vegetation management, wildlife, minerals and mining, cultural resources management, natural resources law, and fire ecology. They also receive hands-on training in various parks and complete a research project on a specific management issue.

Insufficient funding also is a concern for the US-FWS. In addition to the care and maintenance of the NWRS, the USFWS is responsible for administering the Endangered Species Act, operating the nation's fish hatcheries, and other tasks geared to commercial interests. These functions all require significant funding, and the service departments that carry them out must compete for scarce money. Also, economic interests sometimes seem to take preference over managing the refuges as sanctuaries for all wildlife, plants as well as animals and nongame as well as game species.

The BLM also operates with too little funding. Its budget is just one-third that of the Forest Service, despite the fact that it is in charge of far more land than all the other services. Limited funds translate into too few personnel for the acreage being managed and too little money to develop and implement effective management plans. This occurs is in part because the

BLM is probably the least well-known land-managing agency. These factors have made the agency less effective than it must be in order to manage and protect the rich heritage of national resource lands.

Loss of Wildlife

Wildlife within the national parks is threatened by diminishing habitat, the elimination of predators, and poaching. Some of the parks are too small to maintain viable populations of large mammals. Off-road vehicles, legally and illegally operated, also create a hazard to wildlife by disrupting their normal activities and damaging soil and vegetation. Due to various factors, many western parks have lost populations of such species as grizzly bear, wolf, lynx, gray fox, bighorn sheep, jackrabbit, otter, mink, and pronghorn antelope. (Figure 16-11).

The elimination of predators in the public lands can wreak havoc within the ecosystem. Without their natural enemies, game such as elk, deer, bison, and moose become too plentiful and overbrowse the vegetation. As a result, many animals weaken and starve. By the early 1980s, the Park Service, at the urging of many wildlife biologists and conservation groups, drew up a plan for reintroducing gray wolves, a federally endangered species, to Yellowstone. Local ranchers opposed the plan, fearing that the wolves would prey on their livestock. Throughout the 1980s, the two sides remained at a stalemate. In 1991, Congress approved a bill supporting wolf reintroduction to several areas in the lower 48 states and provided funding for the USFWS to begin the reintroduction effort. Amid continued controversy, in 1995 and 1996, wolves were transported from Canada to Yellowstone; populations were also established at other suitable sites, including central Idaho. As part of a compromise to ranchers, these wolves are classified as "nonessential experimental populations," and do not enjoy the protections normally accorded endangered species. Additionally, an environmental group, Defenders of Wildlife, compensates ranchers for any livestock killed by wolves.

So far, the wolf reintroduction effort has been successful; in 2002, the USFWS announced that the goal of 30 breeding pairs with equal distribution throughout Idaho, Montana, and Wyoming for three successive years had been met, and by 2003, the Greater Yellowstone population numbered approximately 175 animals, including 13 breeding pairs. Once plans are approved for managing the wolf populations, the wolves will be removed from the endangered species list.

Figure 16-11: Ursus horribilis, grizzly bear. Grizzlies are omnivorous, feeding on berries as well as prey (especially elk and deer) and carrion.

Figure 16-12: Desert tortoise, Death Valley.

Poaching is also a serious threat to wildlife. Park rangers in Alaska consider trophy hunting to be the chief threat to wildlife in those units, and poaching is cited by managers at many parks nationwide as a serious and growing threat. The motive is often money. A mounted, record-size bighorn sheep head sells for about $50,000. Ground elk antler, sold as an aphrodisiac, can fetch $400 an ounce in Asian markets. An American black bear yields numerous "products": the gall bladder (ground for use in medicinal products) goes for about $18,000; the hide, $400 to $800; paws, $100 each; the skull, $20 to $50; and toe nails, $5 to $10 apiece.

It is impossible to determine how much poaching occurs on public lands, but officials believe that it is extensive, particularly in some parks and for some species. The species most vulnerable to poaching are those, like the grizzly bear, whose populations are already small because of habitat destruction and other causes. But heavy poaching can threaten even stable populations, since poachers take the best and fittest specimens (rather than the young, old, or infirm), leaving younger and weaker individuals to propagate the species.

The disruption of wildlife (including poaching and the theft of rare plants, especially cacti) is a significant problem on both BLM and National Forest lands. Barbaric human behavior is responsible for declines in the populations of some species. For instance, researchers documented a 30 to 60 percent decline in the tortoise population of the western Mojave Desert in California during the 1980s (Figure 16-12). According to the Audubon Society, 20 percent of the animals found dead were shot, ostensibly for target practice.

Commercial Use

In many cases, commercial uses post the greatest threat to public lands. These commercial uses include mineral and energy extraction, logging, ranching, concessionaires, and the commercial recreation business.

Mineral and Energy Extraction. Mineral and energy extraction can prove to be harmful to the environment, often displacing wildlife and polluting soil, water, and air. According to the Environmental Protection Agency's Toxic Releases Inventory, mining is the top toxic polluter of all U.S. industries. The 1872 Mining Law, which governs extraction of minerals on public lands, does not require mining companies to restore mined land, meaning that the clean-up costs are often borne by the public.

In many cases, federal law prohibits the development of mineral and energy resources on public lands, but there are some instances in which such activities can, and do, take place. For example, although the Mining in the Parks Act, passed in 1976, closed most national parks and monuments to new mineral claims, it upheld claims that were filed prior to that time. The rights to subsurface oil and mineral deposits within approximately 100 units are owned or claimed by parties other than the Park Service. For example, approximately 247 mining claims exist within Great Basin National Park, Nevada. Although none are currently being developed, due to low market prices, that could change

Mineral and energy exploration often undermine the primary purpose of the NWRS — to provide habitat and a haven for wildlife. According to the Refuge Recreation Act of 1962, any secondary or recreational use proposed for a refuge is supposed to be compatible with the system's primary purpose. The Secretary of the Interior has the authority to decide whether or not a use is compatible. In the 1980s, conflicts between preservationists and developers arose on many units of the refuge system. Many conservationists maintain that the Department of the Interior is too willing to permit harmful secondary activities. For example, in D'Arbonne Nation Wildlife Refuge, Louisiana, the development of a natural gas field is contaminating soil and water resources and may destroy foraging habitat for the endangered red-cockaded woodpecker.

Oil exploration on the coastal plain of the Arctic National Wildlife Refuge (ANWR), arguably the wildest area in the United States, has become one of the most hotly debated political issues in Congress. The coastal plain, on the Beaufort Sea, is home to many species and is the summer calving grounds of the 200,000-member porcupine caribou herd. Other sections of ANWR are designated wilderness areas and off-limits to development, but the coastal plain, specifically the "1002 section" (a thin slice of coastal plain 30 miles wide and 100 miles long), is not. Developing the area's oil resources would entail a work force of 6,000, four airfields (two large and two small), 100 miles of pipeline, two desalinization plants, seven large production facilities, 50 to 60 drilling pads, 10 to 15 gravel pits, one power plant, one seaport, and 300 miles of roads. The sights, smells, and sounds of airplanes and machinery would disturb the plain and other portions of the refuge, as well as Canada's Northern Yukon National Park to the east.

According to an Interior Department study, oil exploration and drilling in ANWR would cause "widespread, long-term changes in the wilderness character of the region." A draft report projected that the population of the porcupine herd would decline 20 to 40 percent, a loss of up to 80,000 caribou. That projection, along with all mention of the "unique and irreplaceable core calving grounds," was deleted from the final report. In fact, the Department of the Interior forbade USFWS employees in Fairbanks to use the term "irreplaceable." Wildlife biologists unaffiliated with the USFWS and conservationists in general consider the report to be so mangled with political ploys and word games as to be meaningless.

Alaska Senator Ted Stevens, a proponent of development, labels the Coastal Plain "the Saudi Arabia of North America." But the facts clearly show that this label is a misnomer. Interior's own report estimates that drilling in the area is expected to yield only 5.2 billion barrels of oil, a six-month supply at current U.S. consumption rates. Some experts believe even

those figures are high. According to an estimate by the Alaska Department of Natural Resources, there is a 95 percent chance that the plain contains no more than a five-day supply of oil.

So, despite its own admission of "widespread, long-term changes in the wilderness character" and the slim chances of finding economically valuable oil deposits, the Department of the Interior touted oil development as "ultimately ... in the best interest of preserving the environmental values" of the coastal plain. This assessment ignores the fact that every other portion of the United States's 1,100-mile Arctic coastline, onshore and offshore, is open to oil development. If ANWR is opened to exploration, it might well fall prey to the same problems oil development has wrought elsewhere in Alaska, with spills, stream siltation and contamination, and road construction disturbing pristine wildlife habitat. On the other hand, if the refuge is permanently closed to development, it may be the most important achievement in U.S. conservation efforts and signal a national willingness to forgo short-term economic incentives for preservation. As of 2004, the U.S. House of Representatives passed a bill that would open the Coastal Plain to exploration and drilling; the bill, however, is unlikely to be passed into law due to strong opposition in the Senate.

Energy exploration is also threatening the integrity of BLM lands. In Wyoming's Upper Green River Valley, lands managed by the BLM are being considered for extensive oil and gas projects which could disrupt the migration of nearly 100,000 antelope and mule deer. In 2003, the Department of the Interior ordered the BLM to stop managing specific areas in the Rocky Mountains as potential wilderness and begin opening these areas for oil and gas development.

Logging and Road Building. Especially within the Forest Service, an emphasis on logging has proven to be one of the most serious threats to the national forests' ecological integrity.

Despite the 1960 Multiple Use and Sustained Yield Act, which designated five different land uses (recreation, range, timber, watershed, and wildlife and fish habitat) as being equally important, many environmentalists maintain that the Forest Service is too closely allied with the timber industry, and that the agency's first priority has been to make timber available to private companies, primarily by building access roads. The road system in the national forests, at approximately some 400,000 miles, is eight times longer than the nation's interstate highway system! Critics argue that the agency's emphasis on logging over all other uses is economically and environmentally indefensible. After all, national forests currently meet less than five percent of the U.S. demand for timber, and they do so at a significant cost to taxpayers. Moreover, logging roads fragment habitat and accelerate erosion, particularly on steep slopes where water carves gullies into the hillside and may cause landslides (Figure 16-13).

In 1998, the Forest Service finally responded to its critics. Forest Chief Mike Dombeck announced that the agency would shift its emphasis to providing recreation and protecting watersheds. (Nationwide, 900 municipal watersheds that provide drinking water for millions of Americans originate in national forests.) He also instituted an 18-month moratorium on road-building in unprotected roadless areas of some of the national forests, providing the agency with the chance to develop a long-term road policy.

In January 2001, the Forest Service issued the Roadless Area Conservation Rule, which placed nearly 59 million acres (24 hectares) of pristine national forestland off-limits to virtually all road-building and logging. (More than one-half of acreage managed by the Forest Service is already open to these activities.) In protecting these areas, the Forest Service preserved current opportunities for public access and recreation. The roadless rule enjoyed broad, bipartisan public and Congressional support; the Forest Service received over four million comments in favor of the move to protect roadless areas. Even so, the Bush Administration repealed the roadless rule in May 2005, a move applauded by pro-industry western governors and the American Forest and Paper Association.

Another controversial issue with respect to logging in national forests is **clearcutting**, the practice of harvesting everything in an area, regardless of size, age, or species. Clearcutting is useful in some cases. Certain species, such as lodgepole pine, grow in even-aged stands with minimal species diversity; they depend on fires or clearcutting to level the forest and provide conditions (such as full sunlight) that favor the growth of seedlings. Moreover, clearcutting — like periodic forest fires — results in the creation of new habitat (shrubs and grasses) favored by some species of wildlife. If the clearcut is not too large, no more than 10 to 15 acres, the area can be reseeded naturally by the surrounding forest. Clearcutting also requires a less extensive road system than do other techniques. It is faster and cheaper than selective cutting, and it makes it easier to contain damage in stands that have been infested with insects. Finally, clearcutting promotes the growth

FIGURE 16-13: Heavy rains in 1996 resulted in extensive erosion from logging roads, Clearwater National Forest.

of trees of commercial value, such as the Douglas fir, which require broad, open light as seedlings.

Unfortunately, in many of the national forests, clearcutting has been used inappropriately: to harvest entire valleys, hundreds of acres of forest at a time, or in areas where it causes considerable damage, such as on steep slopes and close to streams or rivers (Figure 16-14). Under these conditions, clearcutting accelerates erosion and water runoff, which saps nutrients from the soil. Without the shading provided by trees growing on the banks, streamwater temperatures rise. Salmon, trout, and other temperature-sensitive species are adversely affected by the warmer temperatures, as is the sportfishing industry, which depends on healthy fish populations. Streams suffer siltation, particularly after heavy rains or during spring thaws, as soil runs unimpeded off the land. Silt in the streams makes it more difficult for downstream municipal water treatment plants to purify drinking water; silt in reservoirs means they must be dredged more frequently. Because

FIGURE 16-14: Clearcut in Olympic National Forest, Washington. Some of the old-growth forest of the Pacific Northwest is protected in national parks and wilderness areas, but much of it is found in 12 national forests where it is subject to logging.

clearcutting diminishes an area's aesthetic appeal and potential recreational use, it also can adversely affect local tourism.

With the 1976 National Forest Management Act, Congress limited the application of clearcutting and instructed the Forest Service to maintain species diversity, rather than simply to maximize the growth of commercially valuable species. It also reiterated the need for sustained-yield harvests, mandating that the sale of timber from each forest had to be limited to a quantity equal to or less than that which the forest could replace on a sustained-yield basis, provided that all multiple-use objectives were met.

In the late 1980s, amid growing concern by environmentalists and recreational users, the Forest Service adopted a policy aimed at reducing the amount of clearcutting in the national forests. Still, the technique remains a target of conservation groups, especially when it is applied in the old-growth forests of the Pacific Northwest. Consisting of cedar, Douglas fir, Sitka spruce, and hemlocks over 200 hundred years old, they are one of our country's last substantial stretches of virgin forest (Figure 16-15) and harbor a unique diversity of wildlife.

Below-cost timber sales, sales in which the price paid by loggers is lower than the costs incurred by the service to make the timber available, also have damaged the national forests. The Forest Service has always lost money on timber sales, but until the middle of the century the amount was minor, due to several reasons. First, the service didn't sell much timber; just about two percent of the annual harvest came from the national forests. Second, the service simply didn't offer timber sales that were economically or ecologically unsound. Third, because the service was not trying to push timber and to boost sales, it did not spend much money on road construction and improvement; in other words, its costs were relatively low. But after World War II, when the service began to actively seek the business of timber companies, it went to Congress to appropriate money for road construction. Congress agreed on the condition that the service's budget would be linked to an annual sales quantity; if a certain amount of lumber was not sold, the service would be in violation of the law, and its budget could be cut.

Although below-cost sales are sometimes appropriate, such as when an operation is already being undertaken to restore forest health, such sales have too often benefited neither the forest nor the federal treasury. No forest better illustrates the disastrous effects and costliness of ill-advised below-cost timber sales than the Tongass National Forest in Alaska (Figure 16-16). The 17-million-acre (seven-million-hectare) Tongass, our largest national forest, blankets most of Alaska's southeastern panhandle. Imagine that the Tongass was a ribbon of forest one mile wide. It would stretch over 26,562 miles, enough to completely encircle Earth at its equator! Old-growth forest (including some trees 300 to 800 years old and 150 to 200 feet tall), the largest concentration of grizzly bears and bald eagles on the planet and five species of Pacific salmon, to name just a few, all find refuge in the Tongass. Beginning in 1980, however, their refuge was threatened.

When the 1980 Alaska National Interest Lands Conservation Act (ANILCA) was passed, a provision of the bill required the government to spend at least $40 million annually to maintain a timber supply on the Tongass of 450 million board feet per year, regardless

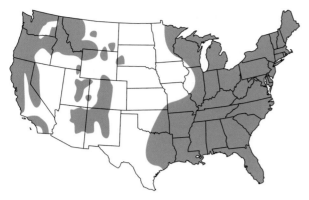

Virgin forests, 1620

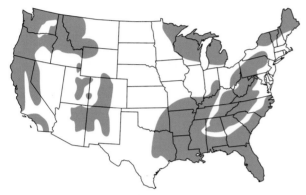

Virgin forests, 1850

Virgin forests, 1990

FIGURE 16-15: Virgin forests in the contiguous United States, 1620, 1850, 1990. Approximately 95 percent of the virgin forests have been cleared since 1620. According to the Wilderness Society, there may be as little as two million acres of old-growth left in Oregon and Washington, just 10 percent of the virgin forests that once covered these two states. The Forest Service and the logging companies argue that opponents of clearcutting old-growth substantially underestimate the area of virgin forest left standing.

of demand. The intent of the provision was to prevent the loss of jobs that would occur when commercial forest lands were designated for wilderness. Subsequently, the Forest Service used the funds to build new roads and to prepare timber sales for two southeastern Alaska pulp mills, the Louisiana Pacific Ketchikan and the Japanese-owned Alaska Pulp Corporation. Most of the timber taken from the Tongass was processed into dissolving pulp, a wood-fiber product, then shipped to Pacific rim countries, where it was used to produce cellophane and rayon.

Unlike funding requests for all other national forests, Tongass funds were not subject to Congressional review. To make matters worse, the two pulp mills enjoyed unprecedented 50-year contracts with the Forest Service. Under the terms of those contracts, the service had to sell two-thirds of the Tongass's annual timber supply to the mills at bargain-basement prices. According to the Southeast Alaska Conservation Council, Tongass timber management has cost U.S. taxpayers approximately one billion dollars.

Throughout the 1980s, the Forest Service's management of the Tongass was opposed by the region's major commercial fishing groups; subsistence users; hunting groups; tour operators; 16 of the 25 communities in southeast Alaska; and local and national conservation groups, including the Audubon Society, Wilderness Society, National Wildlife Federation, Sierra Club, American Forestry Association, and the National Taxpayers Union. They appeared to win a victory when, in 1990, the Tongass Timber Reform Act was passed. The act eliminated the $40 million annual appropriation and the mandated 450 million board feet supply and required that the contracts with the two pulp mills be renegotiated (those contracts have since been terminated). It also expanded an existing Tongass wilderness area and created five new wilderness areas which protect key fish and wildlife habitat. Unfortunately, the act has not halted below-cost timber sales on the Tongass.

With the repeal of the Clinton Administration's ban on logging and road building in roadless areas, the Tongass faces increased pressure from the logging and mining industries. The Forest Service defends the sales, maintaining that road building and logging benefits local communities by providing the basic infrastructure (such as roads and docks) that serve as access for recreation, hunting, fishing, and wildlife viewing; providing the basic transportation system between communities and ferry terminals; and creating high-paying job opportunities for local residents.

FIGURE 16-16: Tongass National Forest, Alaska. Receiving over eight feet of rain annually, the Tongass contains the last large tracts of North America's temperate coastal rain forest. Icy island peaks and frozen tundra rise above valleys darkened by ancient stands of Sitka spruce and hemlock. The area harbors a spectacular and diverse array of life.

Ranching. The Bureau of Land Management has often seemed to accommodate commercial interests at the expense of the land and wildlife; its policies have consistently favored grazing and mining over recreation, wilderness, wildlife habitat, and ecological functions. Those who oppose BLM policies sometimes refer to it as the Bureau of Livestock and Mining or the Bureau of Livestock Management.

Overgrazing, in particular, has severely damaged the vegetation in many areas (Figure 16-17). Too many livestock also mean that waterholes are depleted, streams are degraded by animal wastes, and riparian ecosystems are damaged or destroyed. As biologist Paul Ehrlich describes the problem:

> Most of the West is arid, and lush vegetation is largely concentrated within a few yards of waterways. Naturally, cattle congregate there, trampling the vegetation, beating down stream banks, and defecating and urinating into the streams. The resultant load of eroded soils and wastes, combined with increased stream tempera-

tures because of lack of shading, reduces or exterminates trout populations and makes water less pure for thirsty townships. The water-retaining capacity of riparian (stream-bank) ecosystems is reduced, making both droughts and floods more likely. Aesthetic and recreational values are also compromised, as camp sites become wall-to-wall cow pads and fish and game populations are reduced.

Not surprisingly, the cattle industry favors low grazing fees and a "hands-off" approach by the BLM. The Bureau charges much lower grazing fees than do private landowners, about $1.35 per animal unit month compared to $10–15. Low fees encourage overstocking and poor rotation plans, both of which lead to overgrazing. Some people argue that the public ranges are being damaged to subsidize a minority of ranchers who contribute less than four percent to the beef market: Only about 23,000 livestock producers have permits to graze federal rangelands; over 880,000 do not. Others, however, point out that the lack of parity with private

FIGURE 16-17: San Pedro Riparian Area, Arizona. Healthy riparian areas are vital to the ecological integrity of the West's often arid lands.

grazing fees is due to the lack of corresponding amenities. Private grazing fees typically include, for example, guaranteed weight gains, total supervision of livestock, and maintenance and improvement of pastures. Essentially, ranchers drop off their stock and pick them up later, fat and healthy. In contrast, ranchers who use the public lands are responsible for controlling and monitoring their herds and are not reimbursed or credited for low weight gains or the loss of animals.

In many potential wilderness study areas, the BLM has undertaken "range improvement" projects that involve such drastic changes in the landscape that they effectively eliminate the areas from inclusion in the wilderness system. A prime example is the BLM's use of **chaining**, a controversial management procedure used to reclaim rangelands; two bulldozers with a chain strung between move slowly along, tearing down trees and shrubs, which expand their range in the absence of natural fires or prescribed burning. After this, seed is applied to convert the denuded land to pasture. The purpose, according to the BLM, is to increase the lands' carrying capacity for sheep and cattle. Sixty percent of

the cost for "reclaiming" the land, as the agency calls its work, is borne by taxpayers; the ranchers' share is nine percent, with the state contributing the balance.

Efforts begun by the Clinton administration to raise grazing fees and reform grazing practices have so far been defeated by western Congresspersons. Fees will probably not be raised by any significant amount (if at all) unless there is sufficient public demand to offset the political clout of the cattle industry.

Concessionaires and Commercial Recreation. In recent years, concessionaires, businesses licensed to sell goods and services such as food and hotel space within the NPS, have come under fire. Park visitors often complain about the expense of housing, food, gift items, and recreational opportunities. Another common complaint is the overcommercialization prevalent in many parks; the circus atmosphere and tacky goods offered by some concessionaires diminish the aesthetic quality of the area. Finally, concessionaires return an average of just 2.5 percent of their gross receipts to the government. This money is put into a general fund for use in other programs; it is not earmarked specifically for park maintenance and restoration. Concessions contracts sometimes run for as long as 30 years and allow little room for renegotiation or competition. The result, critics argue, is that concessionaires enjoy a virtual monopoly in the national parks, realizing tremendous profits but failing to share in the burden of maintaining and preserving these areas.

In some park units, commercial recreational operations are a serious threat to the area's natural and cultural features and the quality of the visitor experience. For many visitors, "flight-seeing" — a segment of the tourist trade that enables people to view the country's natural wonders from small planes and helicopters — is a noisy intrusion on the serenity and solitude parks are supposed to offer. Low-flying aircraft can also disturb wildlife, especially at critical feeding, nesting, or mating periods. Arizona's Grand Canyon (the nation's busiest flight-seeing destination) and Hawaii's Haleakala are two of the most affected national parks. Currently, regulations restrict some air routes and ban low overflights in Grand Canyon National Park, but many people feel that these restrictions are not tight enough. They call for a complete ban on all flights in the Canyon and other selected parks, flight-free zones in all parks, and the use of quieter aircraft.

Activities on Adjacent Lands

When commercial activities occur in areas adjacent to public lands, they impact the public domain. For example, mining, logging, and other activities conducted on federal and private adjacent lands can adversely affect the ecological integrity of parks. Exploration and drilling for oil and gas outside Yellowstone threatens to disrupt the thermal activity of the geysers and hot springs; offshore drilling and clearcutting in adjacent national forests threatens Washington's Olympic National Park. Road building in the national forests, which surround many parks, disrupts wildlife that cross the park and forest borders. Residential and commercial development is encroaching on historic areas like the Antietam National Battlefield in Maryland and the Manassas National Battlefield in Virginia, sites of significant Civil War battles.

Another assault on the integrity of public lands is air pollution; winds carry pollutants from urban areas, smelters, factories, and coal-fired power plants to such remote parks as Capitol Reef and Canyonlands. Monitoring stations in Shenandoah and Great Smoky Mountains national parks record the highest levels of

sulfur and nitrogen of all units in the system. Sulfur dioxide is a contributor to acid precipitation, believed responsible for massive diebacks of red spruce at the higher elevations of the southern Appalachians. The highest emissions of sulfur dioxide have been traced to coal-burning power plants in the Ohio and Tennessee valleys. Nitrogen oxide, a product of combustion from all sources (including power plants and motor vehicles), is a component of ground-level ozone, or smog. In Great Smoky Mountains National Park, about a hundred species of plants and trees show signs of leaf damage or growth suppression from ozone. Smog has another, more obvious impact for the park visitor: it obscures spectacular views, such as the ones that built Shenandoah's and the Great Smoky Mountains' fame. Ozone is also a problem in most parks close to urban areas, the mountain parks of California (including Yosemite), and Acadia in Maine (Figure 16-18).

Lack of Wilderness Designation

Only lands included in the National Wilderness Preservation System are officially managed as protected wilderness; other wild lands remain undesignated and thus unprotected. Both those areas that enjoy official designation, "uppercase Wilderness," as the writer John G. Mitchell has called it, and those that do not, de facto (or, "in fact," if not in law) wilderness, face numerous threats to their ecological integrity. However, de facto wilderness is under greater threat, for unless it receives official protection, there is little chance that it will remain wild for long. Too many competing interests are eagerly waiting to mine the rock; graze the ranges;

FIGURE 16-18: Acadia National Park, Maine.

plumb the soil; dam the waters; cut the timber; or roar across the hills, sand dunes and dried salt beds in all-terrain vehicles, shattering the silence of wilderness as they unravel the delicate fabric of life it harbors.

It's important to keep in mind that the wilderness designation process is subjective, engendering controversy more often than preservation. The agencies charged with recommending areas for inclusion in the wilderness system must balance wilderness preservation against the myriad other uses of federal lands. At best, wilderness review is a difficult and challenging process. At worst, it can become a farce, a review in name only, flawed by bureaucratic antipathy and intense lobbying by powerful development interests.

The BLM's wilderness review and recommendation procedures in Utah illustrate some of the flaws in the designation process. Utah, with just less than one million acres of designated wilderness, contains the lowest amount of protected acreage among the western states. Many are among the wildest, most beautiful, and most interesting canyon lands our nation has to offer. Yet they remain unprotected. In the 1970s and 80s, millions of acres of de facto wilderness disappeared from the BLM study process after district managers arbitrarily decided that the land did not offer "outstanding opportunities" for solitude or "a primitive and unconfined type of recreation." (According to the Wilderness Act, a roadless area must have one of these two characteristics in order to qualify for inclusion in the NWPS.) Most of these lands were never surveyed; many were omitted on the basis of maps. Conservationists point out that the BLM was not given the authority to make such discretionary judgments. Moreover, they claim that the BLM, under the influence of economic interests, reserved the land for exploitation by ranchers, miners, and developers (especially hydropower developers). A coalition of 255 citizens groups, known as the Utah Wilderness Coalition, have identified and more carefully surveyed more than nine million acres of BLM lands that meet the criteria for wilderness and which deserve protection. Their work is embodied in the America's Red Rock Wilderness Act, which is now before Congress.

Utah is not alone, though; in almost every western state, there are significant conflicts over designation. In Colorado, for instance, the issue of federal reserved water rights in designated wilderness areas is an ongoing source of conflict. **Reserved water rights** ensure that upstream projects, such as dams and diversions for irrigation, will not deplete the water supply in an area. Under the Reagan administration, the federal government failed to press for reserved water rights. Arguing that water was not essential to the integrity of wilderness, the government effectively allowed special interest groups the right to drain rivers and streams before they reached protected wilderness areas. Economic self-interest compels some ranchers, farmers, and others to oppose reserved water rights for wilderness areas, with the result that wilderness designation for many areas in Colorado has been delayed.

Discussion Questions

1. Identify the major federal land-managing agencies and describe their missions. How do their missions differ and how do those differences affect the way the land is managed?

2. What are the most serious problems facing public lands? What are ways that these threats could be alleviated?

3. Briefly discuss the history of land acquisition and disposal that ultimately gave rise to the public domain.

4. Do you think that wilderness areas should be preserved or opened up for development? Give arguments to support your answer.

5. Identify and describe the national park, national forest, national wildlife refuge, or BLM land nearest you. (If there is no federal land close to you, choose the site that you would most like to visit.) Conduct library research or visit in person to find out what ecosystems it harbors. What rare species are found there? What environmental problems are of greatest concern? What management strategies are of greatest concern?

Summary

The history of the federal lands is largely one of land disposal to private interests. This tradition dates back to the early part of the nineteenth century when land disposal was seen as a way of encouraging western expansion and settlement and thus securing the nation's frontier. By the mid-1930s, the government had either given away or sold 1.1 billion acres.

Despite its tradition of land disposal, the government had begun to set aside land units for special purposes as early as 1832. Forty years later, Yellowstone was designated a "nation's park." The 1891 Forest Reserve Act facilitated the establishment of the national forests; this law was also invoked to establish the first national wildlife refuge, Pelican Island, Florida, in 1903. By the 1930s, the majority of the federal lands remained unreserved, not belonging to the National Park System, the National Wildlife Refuge System, or the National Forest System. The public domain, as it was called, was severely degraded due to overgrazing and other deleterious practices. Finally, in the late 1940s, the Bureau of Land Management was formed to oversee the remaining public domain.

The National Park System is a network of 388 diverse units totaling over 84 million acres. The Park Service, Department of the Interior, manages the system. According to Congressional mandate, the Park Service is to manage the parks in order to conserve their scenery, natural and historic objects, and wildlife for the enjoyment of and use by the public. The most serious threats to national parks are crowding and overuse, insufficient funding, threats to wildlife (diminishing habitat, elimination of predators, and poaching), commercial recreational opportunities, energy and minerals development, atmospheric pollution, and activities on adjacent lands.

The National Wildlife Refuge System, consisting of over 570 units, totals over 97 million acres of land and water. The refuges are managed by the Fish and Wildlife Service, Department of the Interior. Their primary purpose is to provide habitat and haven for wildlife. Threats to the refuges include the internal management structure of the Fish and Wildlife Service, which is plagued by too many competing responsibilities, harmful or incompatible secondary uses on many refuges, activities on adjacent lands, and political pressures.

The National Forest System consists of 155 forests and 22 grasslands encompassing almost 192 million acres. It is managed by the Forest Service, Department of Agriculture. The nation's forests and grasslands are managed to accommodate a wide array of commercial and recreational uses. The major threats to the system are an emphasis on logging over other uses, reliance on clearcutting, below-cost timber sales, and lack of wilderness designation.

The National Resource Lands are also managed to accommodate multiple uses. Administered by the Bureau of Land Management, Department of the Interior, the National Resource Lands, or BLM lands, are located primarily in the arid and semiarid western states and Alaska. BLM lands equal approximately 261 million acres. Threats include illegal harvesting and disruptive recreational use, commercial exploitation (especially grazing), and lack of wilderness designation.

After World War II, economic uses of land were emphasized over recreation and wilderness preservation. To counteract this emphasis, the Wilderness Preservation Act, introduced in 1957 and passed in 1964, directed the Park, Forest, and Fish and Wildlife services to recommend areas for protection and also enabled Congress to designate wilderness.

The Wilderness Preservation Act of 1964 designated 9.1 million acres (3.7 million hectares) of federal land as protected wilderness. These were the original constituents of the National Wilderness Preservation System, which now consists of 662 areas covering more than 150 million acres. The major threat facing the National Wilderness Preservation System is reluctance to designate public lands as wilderness areas.

The federal lands are biologically significant for several reasons. First, they encompass an astounding diversity of living ecosystems. Second, these diverse ecosystems perform a myriad of ecological functions, such as watershed protection. Finally, because the public lands encompass a wide array of ecosystems, they protect a diverse complement of living organisms. In addition to their biological wealth, the federal lands contain one-quarter of the nation's coal deposits, four-fifths of its huge oil-shale deposits, one-half of its uranium deposits, one-half of its naturally occurring steam and hot water pools, one-half of its estimated oil and gas reserves, and significant reserves of strategic minerals.

The federal lands are used for many purposes. Consumptive uses result in the depletion of a resource. Commercial consumptive uses include logging, mining, and oil development. The federal lands also offer a variety of recreational, scientific, and wilderness opportunities, many of which are nonconsumptive. However, even nonconsumptive uses can degrade an area if there are many users or if they are not careful about how they use the land.

Because the public domain is used for many purposes, a wide variety of problems threaten their beauty and integrity. The difficulty of managing federal lands is compounded by the fact that managing agencies must try to meet the interests of conflicting groups.

KEY TERMS

below-cost timber sales	consumptive use	public domain
chaining	multiple use	reserved water rights
clearcutting	non-consumptive use	utilitarianism

CHAPTER 17

Cultural Resources

Learning Objectives

When you finish reading this chapter, you will be able to:

1. Define cultural resources and differentiate between material and nonmaterial culture.

2. Discuss the six broad categories of why cultural resources are significant.

3. Provide examples of how environmental prob-lems and human actions hinder the preservation of cultural resources.

4. Describe at least three national efforts to pre-serve cultural resources.

5. Discuss the conventions enacted by the United Nations Educational, Scientific, and Cultural Organization (UNESCO).

Wilderness was never a homogeneous raw material. It was very diverse, and the resulting artifacts are very diverse. These differences in the end product are known as cultures. The rich diversity of the world's cultures reflects a corresponding diversity in the wilds that gave them birth.

Aldo Leopold

Culture is perhaps the only means by which individuals and nations can communicate: no common language is needed to feel the same fear of death, the same emotion at the sight of beauty, or the same anxiety at an uncertain future.

United Nations Educational, Scientific, and Cultural Organization

Imagine for a moment that representatives from every nation and people on Earth come together to create an immense tapestry. Some sit close to their neighbors, speaking and exchanging ideas with one another. Others are farther removed, and so they speak only among themselves. Each group uses materials brought from its homeland in order to fashion and decorate the cloth. The symbols and designs that the groups weave are based upon their cultures' unique beliefs, values, and accumulated knowledge. Imagine further that each succeeding generation carries on the work of the tapestry making. Although there may be subtle changes in the materials used or the designs wrought, all are based on materials and designs passed to the groups by their ancestors. The resulting tapestry is thus a kaleidoscope of color, texture and design, a richly varied creation. What a different tapes-try would result if all the groups had been given the same materials and instructed to weave the same design!

It is simplistic, perhaps, to equate cultural diversity to that varied tap-estry. Yet, our global cultural heritage abounds with diverse and unique resources: Independence Hall, where the Declaration of Independence was signed in Philadelphia; the special knowledge and traditions of tribal peoples in Australia, Africa, and South America; the temples and buildings of the Nepalese in the Kathmandu Valley; the creation beliefs of native Americans; and the pyramids of ancient Egypt. Indeed, the world's cultural resources bring pleasure, comfort, and wonder to the human existence. Only by recognizing and preserving these cultural resources can we bequeath to our children the splendid cultural tapestry we enjoy. Recognition depends on our willingness to expand our definition of what constitutes a resource. Preservation depends on our willingness and ability to protect the environments that have given rise to cultural resources. In this chapter, we describe cultural resources and examine efforts to manage them.

Describing Cultural Resources

In this section, we define what cultural resources are and discuss where they are found. We also examine why cultural resources are significant, as well as what factors threaten them.

What Are Cultural Resources?

A **cultural resource** is anything that represents a part of the culture (such as history, art, architecture, or archeology) of a specific people; all cultural resources arise from human thought or action. **Material culture** is defined as all tangible objects, such as tools, furnishings, buildings, sculptures and paintings, that humans create to make living in the physical world easier or more enjoyable. **Nonmaterial culture** is defined as

FIGURE 17-1: A boy participates in a school festival, Honiara City, Solomon Islands. Nonmaterial culture — intangible resources such as customs and traditions — provides a group of people with a cultural identity.

FIGURE 17-2: The Parthenon, Athens, Greece.

intangible resources, such as language, customs, traditions, folklore and mythology, that humans create to communicate with one another and to pass on their knowledge and experience to their descendants (Figure 17-1). Nonmaterial culture provides a group of people with a cultural identity, a sense of belonging.

Cultural resources reflect or are part of the environment. Indigenous or tribal peoples typically possess a broad and deep knowledge of the natural system in which they live. For instance, they are usually familiar with the rhythms of the seasons and with the native plants and animals. Because this knowledge arises from their environment, it is specific to that place. In other instances, cultural resources actually may be part of the environment. Temples, statues, and monuments are part of the cultural or built environment, and as such, they are important features of the landscape (Figure 17-2).

Historic preservation is the field concerned with preserving material cultural resources, which traditionally have been the focus of preservation organizations and programs. For example, the National Trust for Historic Preservation, the United States' largest preservation organization, defines a cultural resource as "a building, structure, site, object, or document that is of significance in [American] history, architecture, archeology, or culture."

Culture, however, includes all the behavior patterns, arts, beliefs, institutions, and other products of human work and thought. Thus, the traditional bias toward equating cultural resources with material culture ignores many nonphysical aspects of culture. Since the 1970s, the definition of cultural resources has been expanded to include living peoples and their differing cultural heritages. Slowly, we are realizing that the folklore, traditions, creation beliefs, and skills of a people are also resources. In 1984, the United Nations

Educational, Scientific, and Cultural Organization (UNESCO) instituted a formal program to preserve these and other components of our nonmaterial or living culture. Unlike historic preservation, **cultural resources management** is the discipline concerned with preserving both material and nonmaterial culture.

Where Are Cultural Resources Found?

Cultural resources can be found everywhere. Museums specializing in art, archeology, or natural history are filled with paintings, statues, and other cultural resources of many peoples and ages. Other artifacts, such as early American Indian pictographs, can be viewed at their original sites (Figure 17-3). Monuments, churches, and other buildings provide a dramatic record of the beliefs, values, and history of those who built them. Written records of different cultures can be found in literature, historical accounts, diaries, and official documents preserved in libraries and archives. Further, living history parks and exhibits bring the nonmaterial aspects of cultures alive through re-enactments and interactive programs for visitors.

Why Are Cultural Resources Significant?

The reasons cultural resources are significant can be grouped into six broad categories. Cultural resources provide us with a historical record of societies and their environments, are symbols of our heritage, are integral to cultural identity, are a valuable storehouse of environmental knowledge, provide economic benefits, and are essential for cultural diversity.

Historical Record

About A.D. 550, a group of native Americans settled in the high plateau country of what is now southwestern Colorado. These ancestral Pueblo people lived in the area for 700 years and built elaborate cities of sandstone perched in the sides of cliffs. The dwellings are a tangible record of a segment of human history on the North American continent. By studying the ancient dwellings and their surroundings, anthropologists have increased our knowledge about this land, the lives of the ancestral Pueblos, and how they modified their environment. This information may shed light on why

Figure 17-3: Pictographs in Panther Cave, Seminole Canyon State Historic Park, Texas. Contemporary Coahuiltecan Indians claim affiliation with these pictographs, which were painted over 4,000 years ago.

the ancestral Pueblo culture disappeared. Studying the environmental effects of past cultures also can enable us to better understand how our own societies modify the environment.

Cultural resources often link us with the past in a personal way; knowing the history of our environment enables us to see ourselves as part of a continuing, shared experience. History can be found in your own backyard or neighborhood, and you need not be a historian or an archeologist to study and enjoy the resources around you. For instance, if you find an arrowhead, food grinder, hide scraper, or pottery shard in a creekbed, you might wonder, how old is it? Who made it? Did its maker leave it here, or did it belong to someone else? You might go to the library and look for answers in a book about the native Americans who lived in your area. Or, your search might take you to a specialist in that subject. In any case, finding the artifact has caused you to wonder — and learn — about the past of the area you call "home."

By preserving cultural resources, we maintain a record of cultural and environmental history. As long as a cultural resource is preserved, it continues to evoke both wonder and curiosity and remains an object of study and enjoyment for succeeding generations.

Symbols of Our Heritage

A visitor to the museum and displays located at the Statue of Liberty and Ellis Island might expect to learn about the throngs of immigrants who entered the United States during the first half of the twentieth century. But visitors often come away with more than facts and figures. Their visit gives them a sense of the immigrant experience and a respect for the courage and determination of the immigrants. This heightened awareness expands their understanding and appreciation of the diversity that characterizes the United States. Because cultural resources enable us to personally experience our heritage, we often come to a better understanding of that legacy. Even those who never visit a particular resource can nevertheless cherish it as a symbol of their culture.

The Statue of Liberty, one of the most famous cultural resources in the world, is a potent symbol of the values important to the people of the United States. Lee Iacocca, former head of the Chrysler Corporation, served as Chairman of the Statue of Liberty–Ellis Island Centennial Commission, which advised the government on the restoration of the Statue of Liberty. *Liberty Enlightening the World*, as it was originally named, had suffered significant damage over the years due to its marine environment and other factors. It was restored in the early 1980s, in preparation for the Statue's 1986 centennial celebration. In speaking to groups about the restoration effort, which was funded through public donations, Iacocca stressed the Statue's symbolic importance:

> We didn't spend $230 million just so the statue won't fall in the harbor and become a hazard to navigation. We aren't fixing up Ellis Island so people will have a nice place to go on Sunday afternoon. We're doing it because we want to remember, and to honor, and to save the basic values that made America great. Values like hard work, dignified by decent pay. Like the courage to risk everything and start over. Like the wisdom to adapt to change. And maybe most of all — self-confidence. To believe in ourselves. Nothing is more important than that.

> We're not preserving a statue here; we're preserving all that she stands for. And if that's not worth remembering, and honoring, and saving…if that's not worth passing on to our kids…then let me ask you — what the hell is?

Cultural Identity

Cultural resources are a critical part of the psychological and social structure of the people to whom they belong. For instance, many immigrants to the United States retain a strong sense of cultural identity by living near and worshipping with others from the same background. By keeping alive aspects of their original culture such as language, food or customs, they retain a sense of their unique heritage, including attitudes and practices that can help them succeed in their new environment.

The process by which one culture adapts or is modified through contact with another is called **acculturation**. Acculturation can be unsettling to the people whose lifestyle is altered, especially if they do not choose to come in contact with the new culture, as did the European immigrants to North America, but instead have a new lifestyle forced upon them by a dominant culture, as did the native Americans. As their way of life changes, including how and where they live and what occupations they pursue, acculturated people may give up many of the practices of their native culture. But the outward features of a culture (such as housing and employment choices) can be abandoned more easily than can the knowledge and values that gave rise to those features. Thus, a people who have always lived in close association with the land may

find urban life unsettling and disruptive. Problems with acculturation are believed to contribute to a high incidence of alcoholism and suicide among some native American groups.

Storehouse of Environmental Knowledge

The accumulated wisdom of a specific people cannot be replaced. Consider, for example, the peoples of the rain forests. They hold the secrets of living and flourishing in what to us is a mysterious and hostile environment; for example, they know which plants to use for food and medicines and which to avoid (Figure 17-4). This highly developed awareness of and knowledge about the environment is characteristic of many indigenous groups (Figure 17-5). The more we learn about these groups, especially those who live outside the industrialized world, the more we realize we can learn from them. We would be wise to encourage preservation of their knowledge so that it can be used by both native and nonnative inhabitants.

The beliefs and technologies of a culture also might be adapted or applied in another culture. The *Foxfire* books, for instance, are compilations of skills and techniques from the Appalachian region. Some of these techniques, such as traditional agricultural methods, are becoming more attractive as we become increasingly aware of the environmental consequences and high costs of high-input, energy-intensive methods. In the first book of the series, editor Eliot Wigginton

FIGURE 17-4: A traditional doctor in Belize uses medicinal herbs to cure sore throats and infections.

dedicates the work "to the people of these mountains in the hope that, through it, some portion of their wisdom, ingenuity, and individuality will remain long after them to touch us all."

Economic Benefits

Maintaining, rehabilitating, or restoring material cultural resources is economically prudent. **Rehabilitation** is the act or process of returning a property to a state of utility through repair or alteration that makes possible an efficient contemporary use while preserving those features or portions of the property that are historically, architecturally, or culturally significant. **Restoration** is the process of accurately recovering the form and details of a property and its setting as it appeared at a specific period of time by means of the removal of later work or the replacement of missing earlier work.

In the long run, it costs less to maintain buildings than it does to repair damage caused by neglect. By identifying the causes of property damage, we can take steps to eliminate or control them and avoid expensive repairs. Moreover, it is sometimes cheaper to repair a building than to tear it down and replace it with another. Unlike new construction, restoration is labor-intensive and thus is not as influenced by rising materials costs.

Even more importantly, rehabilitation benefits an area economically by creating new jobs, both during construction and later (through new offices, shops, restaurants, and tourism activities); increasing property values in revitalized areas; increasing property tax revenues as rehabilitated buildings are returned to the tax rolls; and attracting new businesses, tourists, and visitors.

Cultural Diversity

Cultural diversity broadens our view. As we open ourselves to other cultures, we benefit from the mental and emotional stimulation other beliefs, customs, and practices offer us.

Preserving cultural diversity increases our chances for long-term survival. Our species certainly will face changing environmental conditions in the future — changes that may be dramatic. Cultural diversity can provide new options and new solutions, enabling us to tackle perplexing problems and to adapt to environmental change.

What Are the Threats to Cultural Resources?

Threats to our cultural heritage may be intrinsic or extrinsic. Intrinsic factors pertain only to material resources, such as buildings and monuments. A structure may deteriorate more rapidly than expected because of its location, the nature of the ground on which it stands, the soil type, faulty materials, and building defects.

Extrinsic factors affect both material and nonmaterial cultural resources. Objects of material culture are subject to long-term natural deterioration as a result of interacting environmental factors. These can be physical, chemical, or biological. Physical factors include temperature fluctuations, frost, heat, rain, humidity, and winds. Chemical factors usually act through the atmosphere and water; for example, oxidation and salt spray can deteriorate structures. Biological factors include microorganisms, termites, rodents (particularly rats), bird droppings, and vining vegetation.

Human activities also pose a threat to material culture. The use of wells and the construction of tunnels can alter water tables, possibly damaging structures. Mechanical vibrations from heavy road traffic, railways, subways, industrial plants, and supersonic flight also can cause structural damage.

Air pollution, particularly in the form of acid precipitation, can severely damage many materials used in monuments and buildings (Figure 17-6). Acids weaken limestone and marble and can corrode exposed

FIGURE 17-5: The agave plant is central to the culture of the Mescalero Apache.

Figure 17-6: Renaissance sculpture on Cloth Hall, Krakow, Poland. The sculpture's porous limestone has been damaged by airborne sulfuric acid from the smokestacks of nearby Nowa Huta Steel Mill. Poland has some of the worst air pollution in the world.

materials such as metal culverts, roofs, bridges and expressway support beams, unless they are painted and maintained. Acid precipitation also damages brick by dissolving the glassy fabric that holds the silica grains of brick together. The brick gradually weakens until it is little more than a "silica sponge."

Ultraviolet light combined with acid precipitation destroys wood. The acids weaken the cellulose fibers in wood, and the ultraviolet light decomposes lignin, the substance that holds the cellulose fibers together. Paint offers some protection.

Changing fashions are also a potential threat to material resources. When artistic and architectural styles become unfashionable, works in those styles may be neglected or deliberately altered or destroyed. In the 1850s, workers cleaning the walls of the Bardi Chapel in the Church of Santa Croce in Florence, Italy, uncovered frescoes painted by the fourteenth-century master Giotto di Bondone; the frescoes had been whitewashed a hundred years before. The frescoes were restored and have since added much to our knowledge of the work and influence of the artist.

War takes a terrible toll on material culture. In World War II, many cathedrals and public buildings throughout Europe suffered significant damage from bombing raids. More recently, during Croatia's struggle for independence from Yugoslavia in 1991, many historic structures within the beautiful, centuries-old city of Dubrovnik suffered irreparable damage.

Development is another human activity that threatens material culture. For example, as we learned in Chapter 13, *Water Resources,* Turkey is constructing a series of dams along the Tigris and Euphrates rivers in a major development effort known as the Southeastern Anatolia Project. The modern needs that the project will serve are in conflict with the need to preserve cultural resources. The dam system will provide electricity and irrigation water to farms and villages, but the reservoirs created by the dams will inundate an untold number of archeologically rich and virtually unexplored sites. Records indicate that one site, the ancient city of Hasankeyf, dates from the first millennium B.C. and served into the seventeenth century as a major center on an international trade route. Although full excavation of the Hasankeyf site would take about a hundred years, archeologists had only until 1999, when construction of the Ilisu Dam began. When completed in 2006, this dam will provide only electricity (it has no irrigation systems), and engineers estimate that silt build-up will put an end to its use in a mere 70 years.

Vandalism and pillaging degrade and destroy cultural resources. In 1991, vandals struck Naj Tunich, a cave in Guatemala containing 90 drawings which represent the only known large body of Mayan inscriptions and sketches; 23 of the drawings were smeared with mud, scratched, or struck with hard objects. Pillaging for profit is perhaps an even more serious problem. Artifact hunters equipped with metal detectors have pillaged Civil War battlefields in Virginia, Pennsylvania and Maryland, searching for shell fragments, uniform buttons, and other historic treasures. The most serious cases of looting in the United States occur at native American sites. Since the early 1970s, the value of native American artifacts has skyrocketed. By one estimate, 80 to 90 percent of the sites in Utah held by the Bureau of Land Management (BLM) have been destroyed by treasure hunters. This problem is not confined to the United States; sites across the world suffer the ravages of collectors, professional treasure hunters, and black-market traders.

The pillaging of a site is especially detrimental in several ways. First, important cultural resources — which are national property — are stolen. Second, pillaging disrupts a site to such an extent that the archeological record is ruined. The soil matrix, containing stores of information about how the inhabitants lived, what they ate and what their environment was like, becomes mixed and randomized. Finally, the desecration of a site that had religious significance for its society degrades the spiritual quality of the site.

Vandalism and pillaging are difficult to control. Although the National Park Service (NPS) and BLM have attempted to protect cultural resources in the United States, these activities continue. The federal lands are simply too vast to patrol adequately, and existing laws are not strong enough to halt the trade in illegally gotten artifacts. The Archeological Resources Protection Act, passed by Congress in 1979, levies stiff fines on anyone who removes archeological resources from public land or participates in their sale, purchase, transport, or receipt. However, the stipulation that law enforcement officials must prove that the artifact was taken from public property makes it difficult to prosecute cases, since defendants simply can assert that they obtained artifacts from private land.

Perhaps the greatest threat to nonmaterial cultural resources is acculturation. Whenever a group comes in contact with a foreign culture, it faces the difficult task of balancing the need to function in the new culture with the desire to retain its own distinct heritage. The dominant society can make that task easier — or far more difficult. Most often, minority cultures feel substantial pressure to conform to the practices of the dominant culture.

Managing Cultural Resources

In this section, we examine how cultural resources have been preserved, both in the United States and throughout the world.

How Have Cultural Resources Been Preserved in the United States?

Until fairly recently, the private sector was the force behind the preservation movement in the United States. In the late nineteenth and early twentieth centuries, preservation was at best a partnership of private groups and government; government action often came only at the insistence and prompting of citizens' groups. For instance, neither the Virginia legislature nor the federal government acted to preserve George Washington's Mount Vernon home. In 1853, private citizens rallied to the cause. The Mount Vernon Ladies Association of the Union, chartered in 1856, succeeded in raising the funds needed to save the estate, and in 1859, restoration began.

Despite its initial lack of involvement, the government became increasingly aware of the importance of cultural resources during the late 1800s and took on increasing responsibility throughout the twentieth century.

Early Government Involvement

The federal government did not formally acquire a historic property until 1864, when it purchased General Robert E. Lee's Virginia home, Arlington House. But the purchase was motivated by political considerations, not by the desire to preserve a national historic and architectural treasure. The mansion and its grounds were considered spoils of war and were thus the responsibility of the federal government. In 1883, the Lees were properly reimbursed, after the Supreme Court ruled that they were the rightful owners.

The government's concern for natural resources set a precedent for the preservation of cultural resources as well. The breathtaking natural beauty of the Yellowstone area led Congress to make it the first "nation's park" in 1872. Seventeen years later, Congress used its power to protect the cultural resources of Casa Grande in Arizona, making it the first park tract to be designated a national monument solely on the basis of its historic value. The area's native American ruins were threatened by continuing vandalism, and concerned private organizations and government groups prompted the 1889 designation.

The Antiquities Act of 1906 was the country's first major piece of legislation designed to safeguard cultural resources. Intended to preserve prehistoric sites, it gave the president the power to designate national monuments, establish regulations to protect archeological sites on public lands, and stipulate that research findings must be made accessible in public museums (Figure 17-7). Although an important first step, the Antiquities Act failed to consolidate jurisdiction for protected sites in one agency. Instead, the responsibility for each site was left to the department that had first dealt with it, resulting in duplicated efforts at some sites and no effort at others.

Figure 17-7: Three Rivers Petroglyph Site, Las Cruces, New Mexico. These designs, made by the Jornada Mogollon people (A.D. 900-1400), are administered as a federal unit by the Bureau of Land Management and thus are protected under the Antiquities Act of 1906.

World War II and Postwar Economic Growth

In August 1935, Congress passed the National Historic Sites Act. This act authorized the NPS to acquire national historic sites and to designate national historic landmarks. The NPS was directed to identify, maintain, manage, and interpret historic sites for the public benefit and to establish public education programs. Many of the nation's most important cultural resources are sited within units of the National Park System and so are managed by the NPS.

The government's commitment to historic preservation waned with the start of World War II. Of necessity, the private sector once again shouldered the responsibility of preserving the national heritage. These private efforts became increasingly important during the postwar period, when a national building boom threatened to destroy countless significant structures.

Realizing the danger to historic resources posed by the tremendous growth of the period, representatives from interested organizations nationwide met in 1947 to form the National Council for Historic Sites and Buildings. The council was to enlist members and generate interest in historic preservation nationally. The council also sought a Congressional charter to establish a national trust, a legal entity through which it could acquire and operate historic properties. The charter was granted in October 1949, and the National Trust for Historic Preservation became a reality. Five years later, the National Council for Historic Sites and Buildings merged into the National Trust for Historic Preservation. Today, this private, nonprofit organization continues to encourage public participation in the preservation of buildings, sites, and objects significant in American history and culture.

A new threat to material culture emerged during the late 1950s and the 1960s with the widespread popularity of urban renewal projects. **Urban renewal** consists of demolishing older buildings — often entire sections of a city neighborhood — to accommodate new and generally homogeneous structures. Because much of the demolition throughout the 1960s was indiscriminate, urban renewal actually fueled public sentiment in favor of historic preservation.

Preservation Efforts Since 1960

By the end of the 1950s, preservationists began to enjoy support at all levels of government. In 1960, the NPS initiated a listing of national historic landmarks which had "national significance in the historic development of the United States." In 1966, Congress passed the National Historic Preservation Act, which directed the Secretary of the Interior to develop and maintain the National Register of Historic Places, including districts, sites, buildings, structures, and objects important in American history, architecture, archeology, engineering, and culture.

The National Historic Preservation Act also provided grants to the states and the National Trust. These grants were to be used to conduct statewide surveys of historic resources, to formulate state historic preservation plans, and to assist in preserving the individual properties listed in the National Register. Finally, the act also created the Advisory Council on Historic Preservation, an independent agency to mediate between federal agencies and coordinate construction and preservation interests.

The National Environmental Policy Act of 1969 (NEPA), a wide-ranging piece of legislation that affected many segments of the environment, also affected the historic preservation movement. NEPA stressed the government's responsibility toward preservation and established the Council on Environmental Quality within the Office of the President. Among other duties, the council was charged with seeing that all federal activities took environmental considerations into account. NEPA, which required each federal agency to prepare a detailed environmental impact statement before pursuing any activity that would significantly affect environmental quality, also required a similar social impact statement for any activity affecting historic and archeological sites.

Two years after the passage of NEPA, President Nixon issued Executive Order 11593, the Protection and Enhancement of the Cultural Environment. The order directed each federal agency to inventory and nominate to the National Register all federal properties of historic value which it administered. While the National Historic Preservation Act of 1966 had granted protection from federal projects to those structures and sites listed in the National Register, Executive Order 11593 granted similar protection to federal properties that were not yet listed in the register. In doing so, the order formally recognized historic preservation as a national priority.

The popularity of historic preservation increased with the Economic Recovery Tax Act of 1981, which provided significant new investment tax credits for the rehabilitation of older or historic buildings. This economic incentive resulted in **urban revival**, the revitalization of neighborhoods as individuals, professional developers, and corporations rehabilitate older buildings, thus preserving or restoring their significant cultural features for private and commercial use. Urban revival spurred economic development in neighborhoods and historic districts throughout the country, resulting in increased tax revenues at the federal, state, and local levels.

In 1991, the NPS began an **applied ethnography** program to increase ethnic community involvement in the operation and preservation of several park units. A subdiscipline of anthropology, applied ethnography is concerned with recording an accurate account of a given people's culture. The NPS contends that because many park units are important historical sites for various ethnic communities, members of these communities can offer insight into how the parks can best be managed as cultural resources. For a fuller discussion of the applied ethnography program, see *Focus On: Enhancing the National Parks Through Applied Ethnography*.

How Have Cultural Resources Been Preserved Worldwide?

The chief international agency working to protect our global cultural heritage is the United Nations Educational, Scientific, and Cultural Organization (UNESCO), organized in 1945. A 1985 UNESCO publication, *Culture and the Future*, describes the organization's mission in this way:

> Only forty years ago, a fair part of our planet was emerging from a war that had cost humanity fifty million lives. A few enlightened and cultivated spirits decided to take up the old Socratic notion of eliminating ignorance and incomprehension, so as to succeed, at last, in preventing men from killing one another at regular intervals. Thus was UNESCO born, and one of its aims was and still is to offer culture to all so that they will love life. This has not yet been achieved, but it is not for want of trying.

UNESCO member nations agree to a binding legal framework that obliges them to protect the global

Enhancing the National Parks through Applied Ethnography

George Esber and Adolph Greenberg

Traditionally, the National Park Service (NPS) has operated as an insular federal bureaucracy dedicated to conservation and education while accommodating tourism. Authorized in 1916 by the Organic Act, the NPS has taken on the stewardship responsibilities for some 350 national park units, many of which are significant because they preserve cultural resources. Cultural resource management has been an integral part of the U.S. National Park System for many years. Normally, it involves interpreting and preserving the cultural materials and physical remains "abandoned" by the human communities that were removed from an area prior to its being earmarked for protected status. However, many of these cultural resources constitute critically important ancestral sites of various ethnic peoples, and while "abandoned" in a physical sense, these locales were never abandoned culturally by the affected communities. Because management decisions by the NPS have, until recently, typically followed a top-down administrative model, ethnic communities having long-term and deep-seated connections to certain park units were not consulted for cultural resource interpretation and preservation efforts.

In 1991, the NPS began hiring regional ethnographers whose responsibilities included research, liaison, and consultation with communities having traditional associations with each of the national park units. Mandated by a collection of laws, the NPS began the applied ethnography program to include ethnic communities, many of them American Indian, in park planning, management, and operation decisions. The early days of the program found the NPS bureaucracy sometimes willing and sometimes resistant to engage in ethnographic outreach. Target communities, on the other hand, were eager and waiting to be involved in what was, for some, the management of their own heritage at ancestral places.

While still in its infancy, the applied ethnography program has successfully developed community involvement in a number of parks. Associated communities have regained at least a partial sense of owning their heritage, and some park personnel have begun to manage in a more responsive fashion. Through the program, cultural resources have become better understood and therefore more appropriately protected, meaning that the interpretative story presented to the public at a national park more accurately reflects the history and cultural heritage of a people (Figure 17-8).

FIGURE 17-8: Cliff Palace, Mesa Verde National Park, Colorado. Through the applied ethnography program, many national parks offer a more legitimate representation of cultural resources.

cultural heritage and establishes international conventions concerning the protection of cultural treasures. Members are free not to ratify a convention, but if they do ratify it, they are bound to comply with its stipulations. UNESCO assists members in adapting their national legislation and administration to comply with convention requirements.

To date, UNESCO has established three conventions. The first concerns the protection of cultural property in the event of armed conflict. The second outlines means to prohibit and prevent the illegal import, export, and transfer of ownership of cultural property. This convention is intended to help put an end to the ages-old practice of pillaging works of art and archeological treasures. Many countries have been deprived of important pieces of their cultural heritage through these practices. Take, for example, the Rosetta Stone, a tablet of text uncovered in the mud of the Nile River by a French officer in Napoleon's engineering

corps. The text on the tablet is repeated three times: in hieroglyphics; in demotic script (the common, idiomatic form of the Egyptian language at the time of the stone); and in ancient Greek. By translating the Greek portion, French scholar Jean FranVois Champollion deciphered the meaning of the hieroglyphics, thus enabling scholars to read the literature of ancient Egypt. The Rosetta Stone, which is the key to our knowledge of ancient Egyptian language and culture, was removed from Egypt, taken to Paris, and later removed to Great Britain, where it now resides in the British Museum.

In 1972, UNESCO established its third convention, a document aimed at safeguarding property of outstanding universal value. It recognizes that the "deterioration or disappearance of any item of the cultural or natural heritage constitutes a harmful impoverishment of the heritage of all the nations of the world." Each nation conducts an inventory of the cultural and natural property within its borders and

Figure 17-9: World heritage sites. (a) A series of lakes, caves and waterfalls, Croatia's Plitvice Lakes National Park teems with wildlife, including bears, wolves, and many rare birds. (b) The Great Smoky Mountains National Park in Tennessee and North Carolina is home to more than 3,500 plant species and boasts what is likely the greatest variety of salamanders in the world.

composes a list of properties it deems to be of universal value. The lists are studied by the World Heritage Committee, a committee of representatives from 21 member nations. The committee decides which properties are to be included on the World Heritage List and assists nations in protecting these properties. As of July 2004, the World Heritage List included 788 properties in 134 countries (Figure 17-9).

Inclusion on the World Heritage List alone cannot guarantee that these treasures will be protected. Because many nations do not have the financial means to preserve their resources, the convention established a World Heritage Fund to aid members. The fund provides technical assistance, equipment, and supplies; finances studies to determine or prevent the causes of deterioration; and finances the planning of preservation measures and the training of local specialists in preservation and renovation techniques.

Perhaps our greatest opportunity to protect cultural resources worldwide is to preserve the natural environments that gave rise to them. This, in turn, depends upon the widespread adoption of the stewardship ethic that values all life, human and non-human, and respects natural and cultural diversity. By recognizing that we are a part of nature, we can view our cultures as integral parts of the living systems that sustain us. If we value and protect our environment, we can grow in our appreciation of cultural resources and act to preserve them. For some suggestions on personal action, see *What You Can Do: To Appreciate Cultural Resources.*

Summary

A cultural resource is anything that embodies or represents a part of the culture of a specific people. All cultural resources arise from human thought or action. Material culture includes all tangible objects humans create to make living in the physical world easier or more enjoyable. Nonmaterial culture is defined as intangible resources, such as language, customs, traditions, folklore and mythology, that humans create to communicate with one another and to pass on their knowledge and experience to their descendants. Cultural resources often reflect or are part of the environment. Historic preservation is concerned with preserving material cultural resources. Cultural resources management is concerned with preserving both material and nonmaterial culture.

Cultural resources are significant because they provide us with a record of societies and their environments, are symbols of our heritage, are integral to cultural identity, are a valuable storehouse of environmental knowledge, provide economic benefits, and are essential for cultural diversity.

Intrinsic threats to our cultural heritage pertain only to material resources and include the position of the structure, building materials, and design. Material and nonmaterial resources are threatened by extrinsic factors, including long-term natural causes and the action of humans, particularly war, development, and acculturation.

Until fairly recently, the private sector was the force behind the preservation movement in the United States. The Antiquities Act of 1906 was the country's first major piece of legislation to safeguard cultural resources. Intended to preserve prehistoric sites, it gave the president the power to designate national monuments and establish regulations to protect archeological sites on public lands. In 1935, Congress passed the National Historic Sites Act, which authorized the National Park Service (NPS) to acquire national historic sites and to designate national historic landmarks. In 1947, representatives from private organizations nationwide formed the National Council for Historic Sites and Buildings.

In 1960, the Park Service initiated a listing of national historic landmarks. In 1966, Congress passed the National Historic Preservation Act, which directed the Secretary of the Interior to develop and maintain a National Register of Historic Places. The National Environmental Policy Act of 1969 (NEPA) also affected the historic preservation movement because it required a social impact statement for any activity receiving federal funds that might affect historic and archeological sites. In 1971, President Nixon issued Executive Order 11593, which directed each federal agency to inventory and nominate to the National Register all federal properties of historic value that it administered. The Economic Recovery Tax Act of 1981 provided investment tax credits for rehabilitation of older or historic buildings. In 1991, the NPS began an applied ethnography program to encourage increased ethnic community involvement in the administration of several park units.

The chief international agency working to protect the global cultural heritage is the United Nations Educational, Scientific, and Cultural Organization (UNESCO), organized in 1945. To date, UNESCO has established three conventions. The first concerns the protection of cultural property in the event of armed conflict. The second outlines means to prohibit and

Discussion Questions

1. Why is it important to preserve nonmaterial cultural resources?

2. How would you convince your town council or board of commissioners to preserve the birthplace and early homestead of Rachel Carson? (Note: Rachel Carson was born in Springdale, Pennsylvania.)

3. What are the physical, chemical, and biological conditions that do the most damage to material culture?

4. What is the relationship between the National Environmental Policy Act (NEPA) and cultural preservation?

5. What can we learn by studying environmental history that would help us to solve current environmental problems?

6. Describe the cultural heritage of your area. Which native American tribes originally lived there? What lifestyle did they lead? When did nonnative settlers arrive in your area? From what countries did the majority of settlers arrive? What traditions, customs, celebrations, and holidays observed by your community are a reminder of those peoples?

7. Discuss the relationship between the natural environment of an area and the cultures that are associated with it.

prevent the illegal import, export, and transfer of ownership of cultural property. The third convention is aimed at protecting the world cultural and natural heritage by safeguarding properties of outstanding universal value. This convention established the World Heritage List, which as of July 2004, included 788 properties in 134 countries.

KEY TERMS

acculturation

applied ethnography

cultural resource

cultural resources management

historic preservation

material culture

nonmaterial culture

rehabilitation

restoration

urban renewal

urban revival

UNIT **V**

An Environmental Pandora's Box

Managing the Materials and Products of Human Societies

CHAPTER 18

Mineral Resources

Learning Objectives

When you finish reading this chapter, you will be able to:

1. Define and give examples of minerals.

2. Explain how mineral deposits formed in the Earth's crust.

3. Distinguish between proven and ultimately recoverable mineral resources.

4. Describe the steps and environmental consequences of the mining process.

5. Compare mineral availability and use in more-developed and less-developed countries.

6. Briefly explain how the United States manages its domestic and imported mineral supplies.

The entire lithosphere is an intricately interwoven fabric of many crystals: rocks and sand, gold and tin, diamonds and ice. Crystallization is perhaps the most important single process that creates the world we know. It is an expression of the supremely logical structure underlying all things — an orderliness usually hidden from our sight beneath the ever-changing masses of clouds and soil and sea.

Louise B. Young

At the end of the minerals- and energy-intensive development path taken by today's industrial nations lies ecological ruin. Mining enough to supply a world that has twice as many people, all using minerals at rates that now prevail only in rich countries, would have staggering environmental consequences.

John E. Young

Matter is neither created nor destroyed, but its form can be changed and it can be moved from place to place. Over the millennia, geological processes have changed — and continue to change — matter. Humans, too, transform materials. We extract minerals from the Earth and fashion them into a multitude of goods that make our lives more comfortable and enjoyable. Consider your automobile, the brick walls of your school, the concrete roadway you drive on, and the silicon chips that power your computer. The matter that comprises these objects was present at the birth of the planet, and it will be present one hundred years, five hundred years, five million years from now. In this chapter, we describe mineral resources and examine efforts to manage them.

Describing Mineral Resources

We tend to think of inorganic matter as immutable, unchanging. A rock is a rock, so to speak. But nothing could be further from the truth. A witness to the birth and development of the Earth would tell a story of change and transformation, a tale of process and becoming (Figure 18-1). In this section, we first look at how both the formation of the Earth and ongoing geological processes affect the planet's crust, that portion of the planet that holds the minerals so integral to human societies. We then describe mineral resources — their nature, formation, location, use (including patterns of production and consumption worldwide), and classification of deposits. We close this section with a look at the mining process and how it affects the environment and human health.

How Did the Earth Form and How Does It Change?

Most scientists theorize that when our solar system developed from a cosmic cloud about 4.6 billion years ago, small concentrations of matter began to condense at different distances from the newly formed sun, giving rise to the planets. The composition of the planets depended upon the matter that could condense at the different distances. The lighter, more volatile elements, unable to condense at high temperatures, were driven outward by the sun's heat. Eventually these elements and compounds cooled and became part of the planets in the far reaches of the solar system. The inner planets — Mercury, Venus, Earth, and Mars — were formed of denser, less volatile particles of stardust. Mercury and Venus formed in particularly hot portions of the cosmic cloud where water was unable to condense. Earth formed in a portion of the cloud where water

Figure 18-1: Mount Etna in Sicily. The eruption of volcanoes reminds us that the transformation of the Earth is an ongoing process.

was able to condense into a liquid form and remain on the planet's surface in that state.

Throughout its first few million years, the Earth underwent rapid changes. Gravity drew the planet's matter toward the center. The tremendous heat released by decaying radioactive elements within the Earth's mass caused the solid materials to melt, and the most dense minerals, such as iron, flowed inward to form the planet's metallic core. **Magma**, or melted rock, which contained less dense minerals, floated to the top, where it cooled and solidified, or crystallized, into **igneous rock**. These rocks formed the Earth's solid crust (Figure 18-2). Materials that cooled below the Earth's surface solidified more slowly.

Even today, in some places, such intense pressure builds up within the Earth that magma breaks through the surface. We call these sites volcanoes, and the magma that erupts from them is known as lava. Igneous rock is continually being formed by the cooling and crystallization of magma from deep within the Earth's crust or within the upper reaches of the mantle, the thick band that lies between the planet's core and crust.

Igneous rock is one of two predominant rock types on Earth; the other is sedimentary rock. **Sedimentary rock** is formed by the deposit of small bits and pieces of matter, or sediments, that are carried by wind or rain and then compacted and cemented. Sediments are continually eroded from the continents by atmospheric forces (water, wind, and ice) and chemical action.

Most sediments are deposited in the seas along the continental margins. As the piles of sediments grow larger, increasing pressure and rising temperatures produce physical and chemical changes in the underlying sedimentary rock. If the sedimentary pile is thick enough, material near the bottom may melt and form magma. Since the magma is less dense than the parent (sedimentary) rock, it tends to rise and is forced through the rock from which it originated. As the magma rises, it cools, forming new igneous rocks.

A third, less common type of rock is **metamorphic rock**. Metamorphic rock forms when rocks lying deep below the Earth's surface are heated to such a degree that their original crystal structure is lost. Different crystals form as the minerals that compose the rocks cool. For example, limestone that has been heated and recrystallized forms marble.

The Earth's crust is broken into about a dozen large, rigid plates, which slide by, collide with, or separate from each other as they move slowly across the Earth's surface. The boundaries between these plates are ar-

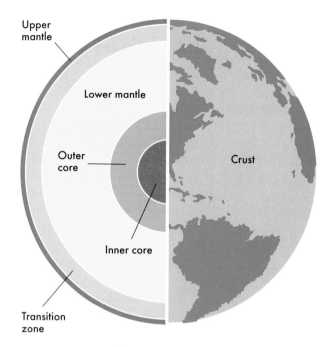

FIGURE 18-2: Internal structure of the Earth. The planet's mantle and crust surround its dense, metallic core. The mineral deposits exploited by humans lie within the Earth's crust.

eas of intense geologic activity, including mountain building, volcanoes, and earthquakes. Where plates collide, old crust is being destroyed and where plates separate, new crust is being formed. Thus, the surface of the Earth — composed of rocks and minerals — is continuously being destroyed, renewed, and reshaped. The movement of the Earth's plates, known as plate tectonic activity, helps us to understand the continual cycle of renewal that occurs on our planet.

Rock-forming processes demonstrate that over hundreds of millions of years the Earth refigures itself, transforms itself anew, so that the composition of rock changes over the ages. Even now, the Earth is changing — mountains are forming, the pressure that gives rise to earthquakes and volcanoes is building, sediments are eroding and being transported, and crust is being created and destroyed at plate boundaries. Because rocks are made of minerals, through this change, the mineral composition and distribution of the crust is also continually changing.

What Are Minerals?

Broadly defined, a **mineral** is a nonliving, naturally occurring substance with a limited range in chemical composition and with an orderly atomic arrangement. Minerals include materials such as granite and sand, nutrients such as phosphorus and nitrogen, precious metals such as gold and silver, and energy resources such as petroleum and coal.

A mineral occurs naturally in one of three forms: as a single element, such as gold; a compound of elements, such as calcite (the principal constituent of limestone); or an aggregate of elements and compounds, such as asbestos. Each mineral's unique chemical composition determines its physical properties. It is for their particular physical properties — strength, insulating or sealing capacity, electrical conductivity or beauty, for example — that minerals are valued.

Mineral resources are broadly classified as fuels or nonfuels. Nonfuel minerals are further classified as metallic or nonmetallic. Metals are a group of chemical elements characterized by malleability, ductility, thermal conductivity, and electrical conductivity. Malleability is the property of metals that allows them to be shaped or worked with a hammer or roller. It is the malleability of metals that allows us to produce wrought iron gates and copper pots and pans. Ductility is the capability of being drawn out and fashioned into a thin wire. Thermal and electrical conductivities are the metals' capacity to transfer heat and electricity.

Metals are classified in several ways. **Ferrous metals** contain iron or elements alloyed with iron to make steel. **Nonferrous metals** contain metallic minerals not commonly alloyed with iron. **Abundant metals** account for more than 0.1 percent of the Earth's crust, by weight. They include aluminum, iron, magnesium, and manganese. In contrast, copper, lead, zinc, tin, tungsten, chromium, gold, silver, platinum, uranium, and mercury comprise less than 0.1 percent of the Earth's crust and are known as **scarce metals**.

Nonmetals are simply those elements that do not possess the characteristics of metals. Nonmetallic mineral deposits provide a wide variety of industrial and structural resources. Industrial materials include sulfur, salts, fertilizers (composed of nitrogen, phosphorus, and potassium), abrasives, asbestos, and industrial diamonds. Structural materials include stone, cement, sand, and gravel.

How Do Mineral Deposits Form?

Geologic processes are responsible for the separation and formation of mineral deposits. As magma rises from the mantle or from deep within the Earth's crust, it begins to cool. The various minerals constituting magma crystallize at different temperatures; materials with a high melting point crystallize first. The heavier crystals sink and gradually form solid mineral deposits. As the lighter liquid magma continues to rise, it continues to cool, and other minerals begin to crystallize. In this way, varied deposits of minerals form throughout the mantle and crust.

Mineral deposits are not found in the crust in as orderly an arrangement as the preceding description might indicate. Deposits of one mineral are often mixed with deposits of others. There are several reasons for the dispersion of mineral deposits, including plate tectonic activity. As newly formed igneous rock rises to the crust's surface at major fracture zones beneath the oceans located at plate boundaries, it pushes the plates apart. The side of the plate farthest away from the spreading zone eventually bends and slides beneath the plate it is being pushed against. Although scientists do not completely understand the details of this process, many believe that the distribution of mineral deposits is related to the past and present fractures that are the boundaries of the moving plates.

Weathering and erosion also disperse mineral deposits, transporting some particles far from their parent rock. Chemical separation is responsible for forming many mineral deposits as substances are chemically dissolved and removed in solution. Gravity also plays a part. Heavier minerals do not remain suspended in solution as long as lighter minerals and particles. Instead, they tend to sink to the bottom of the fluid in which they are trapped and form deposits.

Where Are Minerals Found?

Minerals can be found everywhere — the oceans, the highest mountain peaks, the air — but typically they are bound up in rock within the Earth's crust, that portion of the planet readily accessible to us for mining and drilling. Just eight elements make up more than 99 percent, by weight, of the Earth's crust. Oxygen, found in combination with other elements to form minerals, accounts for about 46 percent of the Earth's crust by weight. The next seven most common elements are silicon, 28 percent; aluminum, 8.3 percent; iron, 5.6 percent; calcium, 4.2 percent; sodium, 2.4 percent; magnesium, 2.3 percent; and potassium, 2.1 percent.

Minerals are not evenly distributed; they differ in kind and amount from place to place. To be useful, a mineral must be profitable to extract. In most places, minerals are present in concentrations far too low to make their extraction profitable. Where a concentration is high enough to make mining economically feasible, the mineral deposit is known as an **ore**. Bauxite, for instance, is an aluminum-containing ore. If the amount of mineral per given volume is high, it is a high-grade ore; if the amount of mineral per given volume is low, it is a low-grade ore. Naturally, high-grade ores are more profitable to mine than low-grade ores.

Some nations are mineral-rich and others are mineral-poor. The United States, Russia, Canada, Australia, China, and South Africa are the world's major producers of many nonfuel minerals (Table 18-1). Less-developed countries (LDCs) in the Andean region and west coast of South America, Central Africa (especially a belt running through eastern Democratic Republic of Congo and central Zambia), and Southeast Asia are the leading producers of tin, manganese, cobalt, and chromium. Mineral production is not limited to these regions (Jamaica and Morocco, for example, are major producers of bauxite and phosphates, respectively), but they do have large concentrations of important minerals.

How Are Minerals Used?

Minerals play vital roles in technological processes: iron and coal in steel production; aluminum in the manufacture of such diverse products as cans and automotive parts; nitrogen and phosphorus as ingredients in fertilizers; titanium, manganese, cobalt, magnesium, platinum, and chromium in industrial processes and as components in aircraft, automobile engines, and other high-tech applications; limestone, gravel, sand, and crushed rock in construction and transportation (Table 18-2).

The more-developed countries (MDCs) consume the greatest share of mineral resources. The United States, Canada, Australia, Japan, and Western Europe, which together account for 15 percent of the world's population, use 61 percent of the aluminum produced each year, 60 percent of lead, 59 percent of copper, and 49 percent of steel. On average, a person in the developed world uses 50 times the amount of minerals per year as a person in the developing world. For example, the average American uses 49 pounds (22 kilograms) of aluminum per year, while the average Indian uses 4.4 pounds (2 kilograms) and the average African uses less than 2.2 pounds (1 kilogram). Iron is one exception; iron consumption is more evenly spread among nations, in part because several developing countries, including Brazil and India, are major iron and steel producers.

TABLE 18–1: Major Mineral-Producing Countries, 2002

Mineral	Country	Production (thousand metric tons)	Share of World Production (percent)
Bauxite	Australia	54,000	38
	Guinea	15,700	11
Chromium	South Africa	6,400	48
	Kazkhstan	2,370	18
Cobalt	Congo	12.5	26
	Zambia	10.0	21
Copper	Chile	4,580	34
	Indonesia	1,160	8
Gold	South Africa	0.4	16
	United States	0.3	12
Lead	Australia	683	23
	China	600	21
Manganese	South Africa	1,504	19
	Brazil	1,300	16
Molybdenum	United States	32.6	27
	Chile	29.5	24
Nickel	Russia	310	23
	Australia	211	16
Phosphate rock	United States	36,100	27
	China	23,000	17
Silver	Mexico	3	14
	Peru	3	13
Tin	China	80	32
	Peru	65	26
Tungsten	China	50	84
	Russia	3	6

Source: U.S. Geological Survey, Minerals Information, 2003.

Two important factors will influence the rate of increase of global demand: the growing human population and its reliance on mineral resources to support a rising standard of living worldwide.

The MDCs rely on about 80 minerals. Three-quarters of these either exist in abundant supply to meet anticipated needs or can be replaced by existing substitutes. **Critical minerals** are those considered essential to a nation's economic activity; **strategic minerals** are those considered essential to a nation's

TABLE 18–2: Primary Industrial Uses, Selected Minerals

Mineral	Industrial Uses
Aluminum	Transportation (32%), packaging (26%), building (16%), electrical (8%), consumer durables (8%)
Beryllium	Electronic and electronic components, aerospace, defense applications (80%)
Chromium	Stainless and heat resisting steel (68%), full-alloy steel (8%), superalloys (3%)
Cobalt	Superalloys (46%), cemented carbides (10%), paint driers (10%), magnetic alloys (10%), catalysts (9%)
Copper	Building construction (43%), electric and electronic products (24%), industrial machinery and equipment (12%), transportation equipment (12%), consumer and general products (9%)
Diamonds (industrial)	Machine manufacturing, mineral services, stone and ceramic production, abrasive industries, construction, and transportation equipment manufacturing
Gold	Jewelry manufacturing, industry
Lead	Transportation industry (71%), ammunition, television glass, construction (23%)
Molybdenum	Iron and steel production (75%); machinery (9%), electrical (4%), transportation (4%)
Nickel	Stainless and alloy steel production (49%), nonferrous alloys and superalloys (29%), electroplating (15%)
Platinum group metals	Automotive industry, catalysts, jewelry
Silver	Fabrication for arts and industry (90%), small companies and artisans (10%)
Tin	Cans and containers (30%), electrical (20%), construction (10%), transportation (10%)
Tungsten	Metalworking, drilling, mining, and construction (80%), lamp filaments, electrodes and electric components (7%), tool steels (6%), other steels and super alloys (6%)
Zinc	Galvanizing (54%), zinc based alloys (19%),

Source: *U.S. Geological Survey, Mineral Commodity Summary, 1998.*

defense. At present, there are no suitable alternatives for critical or strategic minerals. Cobalt, for example, is needed to produce high-strength, high-temperature alloys used in the aerospace industry, and platinum is unrivaled as a catalyst.

Scarcity of minerals is largely due to technical, social, and economic constraints rather than to the absolute constraint of finite resources. For example, the United States relies on imports of four critical and strategic minerals (chromium, cobalt, manganese, and the platinum group metals) from countries in politically volatile regions such as central and southern Africa. To learn more about these and other imports, read *Environmental Science in Action: Critical and Strategic Minerals* by going to www.EnvironmentalEducationOhio.org and clicking on "Biosphere Project."

How Are Deposits Classified?

Mineral deposits are classified primarily according to economic factors. An estimate of the total sum of a mineral found in the Earth is called the **resource base**. Because of the Earth's tremendous mass, the total tonnage of any one mineral, even those defined as scarce, is great. But most of these minerals are inaccessible, for they lie within the core and mantle of the planet. Hence, the resource base of a given mineral is a highly theoretical figure, and for most minerals, it does not indicate how much is ever likely to be available for use.

A mineral deposit that can be extracted profitably with current technology is called a **proven reserve** or **economic resource**. Proven reserves have been explored, measured, and inventoried. We know where to find proven reserves, we know how to extract and exploit them, we know how much it will cost to recover them, and we know we can make a profit on the recovery. **Subeconomic resources** are reserves that have been discovered but cannot yet be extracted for profit at current prices or with current technologies.

While proven reserves are an estimate of what is available and profitable now, an estimate of the total amount of a given mineral that is likely to be available for future use is called an **ultimately recoverable resource**. However, because it is based on assumptions about discovery rates, future costs, market factors and advances in extraction and processing technologies, such an estimate is difficult to determine.

It's also difficult to estimate how long mineral reserves and resources will last because of the variables involved. For salt, magnesium metal, lime and silicon, for instance, the estimated life expectancy is at least in the thousands of years. But other mineral resources, including aluminum, copper, cobalt, molybdenum, nickel and platinum group metals, may shrink significantly by 2025 if the projected world population of eight billion consumes resources at current U.S. rates.

Estimating the life expectancy for mineral reserves begs the question, "Will we run out of the minerals upon which industrial societies depend?" No one can answer that question with absolute certainty, but we can make some educated observations. Because minerals are nonrenewable (in human terms), deposits are finite. The higher our rates of consumption and the larger the size of the human population, the faster reserves will be depleted. As proven reserves become depleted, the cost is likely to rise. High prices and high demand encourage more exploration and may lead to the discovery of additional deposits, thus increasing the size of proven reserves. Those deposits, however, are likely to be of lower-grade ores. Four hundred years ago, for example, copper ores commonly contained about eight percent metal. Today, the typical copper ore contains less than one percent metal. Moreover, the exploitation of lower grade ores and less accessible deposits requires more energy and water and causes greater environmental damage. On the positive side, high prices and high demand also encourage conservation measures (which can extend the life of the supplies), recycling, the development of technological advances to mine previously unprofitable deposits, and the search for suitable substitutes. These activities ease pressure on dwindling mineral supplies.

What Are the Steps in the Mining Process?

It's a long way from a raw mineral to a finished product. The mineral ore must be located, extracted, and processed before it is ready for use. Each of these three steps is summarized below.

Location

The image of the old-time prospector, armed with a pick and trusty mule, is part of our frontier folklore, but the prospectors of yesteryear certainly would not recognize contemporary exploration techniques. Modern exploration relies on a knowledge of geology, particularly of crustal movements and the formation of mineral deposits, and sophisticated instruments. Some instruments, such as remote sensing devices carried aboard satellites, can identify rock formations or mounds indicative of the presence of mineral deposits. Others can detect changes in the Earth's magnetic field or the field of gravity that may be caused by concentrated mineral deposits. After a potential site is identified, rock samples are taken and analyzed for their mineral content.

Most high-grade ores, particularly those located in industrial nations, already have been identified and exploited. The ocean floor and open waters are two potential sources of minerals, but the latter is not a likely candidate for profitable mining soon. With few exceptions, such as sea salt, minerals are not present in high enough concentrations in seawater to make their extraction profitable or practical. In contrast, the ocean floor holds real potential for profitable mining. Manganese nodules approximately the size of potatoes have been found in abundance on the ocean floor in the deep waters of the Pacific. The nodules also contain lesser quantities of iron, cobalt, and copper. Concentrations of nickel also have been identified in the seabed.

Antarctica is another potentially significant source of minerals (Figure 18-3). Substantial deposits of petroleum and iron ore have been located; methane, copper, silver, and nickel deposits also may be locked below the continent's surface, which lies beneath ice sheets up to a mile thick. Exploiting the continent's resources would be difficult, however. Antarctica has the harshest climate in the world; the average temperature in the interior during the coldest months of the year is -96°F (-71°C). It may be too difficult or too expensive to conduct mining operations in such an inhospitable environment. Even if the economics proved favorable, the environmental consequences of mining in Antarctica may be devastating. Extracting and transporting oil in the Antarctic seas, the world's roughest, pose the danger of oil spills, which would threaten most of the marine environment's animal life (krill, seals and penguins, for example). At such cold temperatures, oil takes far longer to decompose. Also, a spill on ice would increase the ice's heat absorption, possibly causing it to melt. In 1991, environmental concerns prompted the countries that jointly govern the continent to continue a moratorium on mineral development, but many people fear that rising prices and minerals scarcity could lead to pressure to develop Antarctica's mineral wealth in the future.

Extraction

Extracting mineral ores is the step we commonly think of when we hear the word mining. **Mineral extraction** is the process of separating the mineral ore from the surrounding rock in which it is embedded. An ore may be extracted by subsurface or surface mining techniques.

Figure 18-3: Antarctic landscape. Scientific research and tourism are the only activities currently permitted on the Antarctic continent. Many people are hoping to make Antarctica a world park off limits to war and commerce.

Subsurface mining techniques are used to retrieve mineral deposits that lie deep underground. For coal and metallic minerals, a vertical shaft is dug to reach the level of the deposit, tunnels and rooms are blasted in the rock, and the ore is extracted and hauled to the surface. For oil and natural gas, a well is drilled into a reservoir deep underground, and the gas or oil, which is under pressure, rises to the surface. As an oil reservoir becomes depleted, heated water may be injected into the well to force the remaining oil to the surface.

Surface mining techniques are used to extract deposits located relatively near the Earth's surface. About 90 percent of the ores mined in the United States, and over two-thirds of ores worldwide, are extracted by surface mining. The overlying vegetation, soil and rock layers, collectively known as the **overburden**, are removed to expose the ore deposit. Because such large quantities of materials are displaced, surface mining results in more waste than does subsurface mining.

There are three types of surface mining: open pit surface mining, area strip mining, and contour strip mining.

As its name implies, **open pit surface mining** consists of digging a large pit and removing the exposed ore. This technique is commonly used to extract sand, stone, gravel, copper, and iron. At 3,960 feet (1,207 meters) deep, and visible from outer space, Utah's Bingham Canyon copper mine is the largest human excavation in the world; more than 6 billion tons (5.4 billion metric tons) of materials have been removed from the mine to yield nearly 17 million tons (15.4 million metric tons) of copper metal. The extensive pits created in the process of extracting limestone, granite, and marble are known as quarries.

Area strip mining is typically used in flat or rolling terrain. It is used most often to mine coal in the West and Midwest and phosphate rock in Florida, North Carolina, and Idaho. Bulldozers dig a trench and power shovels remove the ore. A second trench is then dug parallel to the first, and the overburden is removed and deposited in the first trench. Once the ore has been removed, a third trench, parallel to the second, is dug, and the process is repeated. If the land is not reclaimed, the result is a series of rolling hills, known as spoil banks, formed of rubble. The spoil banks, which resemble ocean waves or swells, erode easily.

Contour strip mining is used to extract deposits in hilly or mountainous terrain. In the United States, contour strip mining is used chiefly by coal miners in the Appalachian region. A power shovel cuts a series of terraces into a hillside, and the ore is removed at each terrace. The overburden from each new terrace is dumped onto the terrace below. If the land is not reclaimed, the result is a steep and highly erodible bank of rock and soil fronted by a wall of dirt. If the overburden is too thick to remove economically, huge drills called augers are used to burrow horizontally into the mountainside.

In much of Appalachia, coal companies have shifted from contour strip mining to a technique termed steep slope mining. Commonly called mountaintop removal, steep slope mining is the practice of clear-cutting a forested peak, scraping away the topsoil and vegetation, and then using explosives to remove layer after layer of rock until the mountaintop is flattened. The explosions used to blast the rock may pack 10 to 100 times the force of the Oklahoma City bombing. Once the coal seam is exposed, huge machines known as draglines scour the coal from the Earth.

Mountaintop removal is inarguably destructive. The rocky debris is dumped into nearby valleys, burying streams and forested hollows, destroying wildlife and habitat, contaminating water supplies, and — sometimes — obliterating entire towns. Residents who resist buy-out offers from mining companies live with constant noise (mining operations go on round-the-clock), dust, truck traffic, the pollution of local water supplies, and the ever-present threat of flash floods that occur more frequently when rains run off denuded land. As the mountain is eliminated layer by layer, the force of the explosions can damage homes, cars and wells, and flying or rolling debris poses a danger to children and adults playing or working outside. Ultimately, mountaintop removal irreversibly alters the environment, reducing the height of a summit anywhere from 600 to 1000 feet and transforming scenic landscapes of hardwood-forested mountains into a sprawling chain of sparsely vegetated, flat-topped fields.

The 1977 Surface Mining Control and Reclamation Act allowed mountaintop removal provided that operators meet certain standards to safeguard public health and the environment and reclaim the mined land to its approximate original contour. Unfortunately, few mining companies comply with health and environmental standards; and even federal and state regulators who are sympathetic to residents' plights find little support

for enforcing regulations in a region where coal is a powerful political ally — or enemy. Moreover, companies can request an exemption of the land-reclamation guideline if they show that there will be an industrial or commercial development on the flattened area after the land is mined. However, exemptions are routinely given even to companies whose permit applications state that the site will be used for "grazing" or "wildlife habitat" rather than the required industrial or commercial post-mine land use.

Until recently, opponents could sue mining companies to halt mountaintop removal based on environmental stipulations contained in the Clean Water Act. Specifically, the Clean Water Act prohibits the dumping of wastes into streams, and, in 1999, a federal judge ruled that burying streams under millions of tons of dirt was indeed a violation of federal law. In May 2002, however, at the urging of the coal industry, the Bush administration reclassified mining debris from objectionable "waste" to acceptable "fill." Because "fill" can be dumped legally into a waterway, mining companies can continue to fill valleys and streams with the debris from mountaintop removal.

Processing

Mineral processing consists of separating the mineral from the ore in which it is held and concentrating and refining the separated mineral. Processing plants, called smelters, are usually located near mining sites to reduce the cost of transporting raw ores.

The ore is physically or chemically treated to yield the desired mineral. For example, after a metal ore is mined, it is crushed, run through a concentrator to physically remove impurities, and reduced to a crude metal (that is, melted) at high temperatures. Refining removes any remaining impurities. The mineral is then ready for use.

How Does Mining Affect the Environment and Human Health?

Mining has significant adverse effects on both the environment and human health, and countries with a long history of mining bear inevitable scars. In the United States, for example, 83 of the 1,245 sites on the Superfund cleanup list are former mining sites. The largest Superfund site, the Clark Fork Basin site in Montana, stretches 136 miles (220 kilometers) along Silver Bow Creek and the Clark Fork River. For more than one hundred years, copper mining and smelting were conducted at the site. Even with regulations in place to control its environmental impact, mining continues to adversely affect the environment and human health. In countries that do little to regulate mining operations, particularly LDCs, mining creates environmental disaster areas (Figure 18-4). In the paragraphs below, we summarize the environmental and human health impacts associated with each step in the mining process.

Exploration is a relatively benign activity. However, energy is needed to locate mineral deposits; the more remote or inaccessible an area, the more energy required for exploration and study. Further, excavating and drilling for test samples, especially if done to a large extent in a limited area, can adversely affect sensitive ecosystems and the local water table.

Extraction degrades the environment in many ways. Surface mining, in particular, affects large areas of land. It disturbs the overlying soil and vegetation, with resulting erosion. Soil carried into nearby ponds or streams causes siltation, which poses a threat to fish and invertebrate populations. In addition, because surface mining rearranges the top layers of soil, it can lower the productivity of the land even if attempts are made to reclaim the site.

The wastes generated during extraction also adversely affect the environment and human health. As miners remove an ore deposit from the ground, they dump the **tailings** — the rock in which the ore is embedded — in large piles or in nearby ponds. Often, tailings contain hazardous materials (such as asbestos, arsenic, lead, and radioactive substances) that pose a health threat to

humans and wildlife. Mine drainage (from both surface and subsurface mines) may contain hazardous substances (such as sulfuric acid or cyanide) that are used to extract the ores; these substances can contaminate nearby surface waters and groundwater supplies, posing a health threat to living organisms (Figure 18-5).

Extraction adversely affects human health in other ways as well. Miners are at risk for developing diseases — such as black lung disease (coal miners) and lung disease and bronchial cancer (asbestos workers) — caused by the substances with which they come into contact. Further, excavation is a hazardous process; there is always the risk of mine collapses, underground explosions, fires, and accidents associated with heavy equipment. Modern techniques and safety precautions have reduced, but not eliminated, these hazards. In countries such as China, where health and safety regulations are lax or rarely enforced, the hazards are even greater. An estimated 7,000 to 20,000 workers die each year in China's coal mines.

Like extraction, processing takes a heavy toll on the environment. Tailings removed during the refining process are usually fine-grained. When piled on the ground, they can become airborne, contributing to air pollution, or they can leach into the soil (via rainwater), contaminating groundwater supplies. And because these tailings are so finely ground, contaminants that were bound up in solid rock — arsenic, cadmium, copper, lead and zinc, for example — are released to air and water. Tailings produced during refining also may contain traces of the toxic organic chemicals used

FIGURE 18-4: Gold mining in the Amazon. Hundreds of thousands of miners have entered the area in search of gold, clogging rivers with sediment and sludge.

FIGURE 18-5: Using cyanide to mine for gold outside of Lead, South Dakota. Netting covers cyanide-laced ponds to prevent birds from drinking the poisoned water and dying from the deadly poison.

to concentrate the ore. In addition, the huge amount of water needed to process ores becomes contaminated. The liquid and solid hazardous wastes that are by-products of ore processing must be safely disposed of or converted into less harmful materials. Finally, smelters emit substantial amounts of air pollutants, particularly soot, sulfur oxides and toxic particulates, such as arsenic, lead, and cadmium. In the western United States, pollutants released from copper smelters contribute to acid precipitation as well as to reduced visibility in some national parks.

Processing also adversely affects human health. Workers in some smelting industries are at increased risk for certain kinds of cancer. Arsenic smelter workers, for instance, have a lung cancer rate three times greater than average. Workers in cadmium smelters have twice the average lung cancer rate, and workers in lead smelters have higher than average rates for both lung and stomach cancer.

Managing Mineral Resources

Humans have used minerals since the day they discovered that they could fashion tools out of stone. In fact, many eras in human prehistory are named after the minerals, especially metals, that were first developed and widely used during that period — the Stone Age, the Bronze Age, the Chalcolithic (Copper) Age, and the Iron Age. Beginning around 1750, the Industrial Revolution ushered in the Energy Age, a period — still ongoing — of intense use of both fuel and nonfuel minerals. These minerals are essential to industrial societies and play a critical role in the economies of many nations; consequently, how they are managed is a topic of immense importance. In this section, we'll look at the international minerals industry and the factors that affect the industry. We'll then examine how the United States manages its minerals supplies, both domestic and imported. This section closes with a brief look at the Law of the Sea, the international agreement that provides for the management of mineral resources in the open ocean.

What Is the International Minerals Industry?

The geologic dispersal of minerals ties together the economies of producer nations and consumer nations.

The international minerals industry refers to the global trade in these resources.

There are three distinct groups of mineral-producing nations. The United States, Canada, Australia, and South Africa are all part of the industrialized world; they are major producers of a wide range of nonfuel minerals. Many industrialized nations import most or all of many important mineral resources. For instance, the United States, Japan, and the nations of the European Union (EU) rely on imports to meet more than 50 percent of their needs for copper, nickel, lead, tin, zinc, cobalt, iron ore, manganese, and chromium.

The second group of mineral-producing nations consists of the former eastern bloc countries. The former Soviet Union and allied member states of the Council for Mutual Economic Assistance (CMEA) were nearly self-sufficient in minerals. CMEA countries were economically interdependent; central planning, government subsidies, and barter agreements allowed member states to function with little regard for productivity and efficiency. Because the economies of CMEA nations were state-run, production levels were not tied to market demand; instead, they were based on the need for military planning and technology competition with the market economy world. The Soviet Union provided the inexpensive oil and technology assistance that enabled industries in CMEA nations to operate and produce the minerals and materials needed to support the central plan. With the dissolution of the Soviet Union and the formal break-up of the Council, mineral output in most former CMEA countries has dropped markedly.

The third group of mineral producers are the LDCs. They generally have one or two minerals as their primary source of income, accounting for the bulk of their export earnings. For example, copper and cobalt account for over half of Zambia's export earnings. Chile, Peru, and the Democratic Republic of Congo also rely heavily on copper exports. Indonesia, Bolivia, and Malaysia depend heavily on tin exports. Because their economies are so closely tied to the world minerals industry, falling demand and declining prices pose a threat to the economic, political, and social stability of these nations (Figure 18-6).

Some mineral-producing LDCs, including India, Thailand and Brazil, do have diversified economies. Brazil, for example, is among the world's largest producers of iron ore; the country also produces substantial quantities of petroleum, coal, bauxite, and other minerals. Much of its mineral production, however, is targeted for domestic trade. Minerals account for less than one percent of Brazil's export earnings.

Figure 18-6: Digging and sifting for gold, Burkina Faso.

What Factors Affect the International Minerals Industry?

Myriad factors affect the international minerals industry. Market supply and demand determine the price of minerals. Supply and demand, in turn, are influenced by the needs and uses of society, technological constraints, and economic forces. For example, after 1974, the demand for most metals slowed. Producers of metals did not anticipate this slowed demand; as a result, supply exceeded demand, and the price of metals fell.

While metal consumption is still growing, it is growing at a slower rate than before 1974. Two factors generally explain the downturn in global demand for metals. First, slow world economic growth has reduced the demand for metal products such as heavy machinery and equipment. Second, technological changes have reduced the amount of metal used in goods; the composition of many products is changing as the use of plastics and fiberglass increases. Additionally, improved technologies use less metal to develop products of equal or superior quality; the use of metals is thus becoming

more efficient. While efficiency will help to conserve metal supplies, producers must readjust their production strategies to remain competitive.

Supply and demand also are influenced by resource scarcity. Typically, political, economic, and technical factors — rather than actual reserves — determine scarcity. The oil embargo of 1973 by the Organization of Petroleum Exporting Countries (OPEC) raised fears of similar crises involving other "minerals cartels," but they did not materialize. Most experts agree that producer nations are unlikely to form cartels since they often depend on their mineral exports to secure foreign exchange. Botswana, Zambia, Liberia, Jamaica, Togo, Mauritania, Bolivia, the Democratic Republic of Congo, Chile, Peru, and Papua New Guinea derive a substantial portion of their foreign exchange from mineral exports. Such nations are unlikely to take any action that could disrupt exports, since a decline in export trades (caused by decreased demand due to rising prices) would almost certainly damage their economies.

Mineral prices tend to fluctuate wildly; the price of tin illustrates this fluctuation. In January 2000, the price of tin was about $6,000 per metric ton, a figure that fell to approximately $4,000 per metric ton by the end of 2001. By 2005, the market had rebounded, and the price of tin had more than doubled, to about $9,000 per metric ton. However, in general, there does seem to be a long-term downward trend in mineral prices. The economic value attached to minerals and other commodity exports (such as agricultural products) has fallen relative to the value of manufactured goods; in other words, the price of manufactured goods has risen more than the price of minerals. Consequently, mineral-exporting nations, particularly LDCs, must sell increasing amounts of minerals to "pay for" items such as tractors and fertilizers, usually exported by industrial nations. Strategic and critical minerals are exceptions to this long-term trend; because there are no substitutes for them at present, their price is likely to rise.

How Does the United States Manage Mineral Supplies?

The laws, regulations, and agreements a nation establishes in order to govern production, use, and commerce in minerals are known as a **minerals policy**. Unfortunately, the United States has no unified policy to guide mineral production, consumption, and conservation or the recycling of minerals and mineral-based products. In the following pages, we look more closely at how both domestic and imported mineral supplies are managed and at how Americans might conserve mineral resources.

Domestic Mineral Supplies

In the nineteenth century, mining was seen as a primary use for most public lands in the West, a belief that persists in some quarters today. The Mining Law of 1872 granted title to certain public lands as long as the claimant proved that the minerals were "valuable," conducted minimal mining activities, and paid the government $2.50 per acre. This law is still in effect for gold, iron, copper and others, and valid claims remain on some 28 million acres (11 million hectares) of public land (see Chapter 16). Those who develop claims under the 1872 law pay no royalties on the minerals they extract. Coal, oil, gas, potash and other sedimentary deposits, originally covered by the Mining Law of 1872, are currently managed under the Mineral Leasing Act of 1920, which allowed private individuals and companies to lease the rights to develop these resources but discontinued the

practice of selling the land outright. This act gave the Department of the Interior the discretion to approve or deny leasing requests, and it required current leasers to pay royalties to the government.

The Mining Law of 1872 contains no environmental provisions, meaning that taxpayers are often left to pay for the cleanup and reclamation of mining sites. In an effort to protect public lands, conservationists have pushed Congress to repeal the Mining Law of 1872 and replace it with a leasing system. Supporters of the law argue that mining creates jobs and helps local economies. Moreover, they point out that mining is a very expensive undertaking, which requires a significant financial investment; without the stipulations outlined by the Mining Law of 1872, many mining companies would be unable to operate profitably. Finally, those who wish to retain the law argue that every other nation subsidizes its minerals industry. They warn that if the United States revises the law and requires mining companies to pay royalties on minerals taken from public lands, the U.S. minerals industry will be less competitive in the global marketplace.

For the past several decades, opponents of the Mining Law of 1872 have sought to repeal or reform the legislation. One of the more recent efforts took place in January of 2003, when Representatives Nick Rahall (D-WV), Christopher Shays (R-CT), and Jay Inslee (D-WA) introduced the Mineral Exploration and Development Act of 2003 (H.R. 2141). The bill includes fiscal reform, environmental protection provisions, and a program to clean up abandoned mines. It also recognizes the value of other land uses besides mining. H.R. 2141 was sent to the Department of Interior for comment in May 2003, and has not seen any action since.

Increasingly, management of the nation's domestic mineral resources is tied to the larger debate over how to best use and protect public lands. This debate pits private economic interests (especially resource extraction industries like mining) against those who argue that the public welfare is best served by protecting and preserving our nation's parks, forests, and other commonly-owned lands. The focus of this debate is shifting to Alaska, where sizable reserves of many important nonfuel minerals, such as copper and nickel, have been identified and developed. The state also has several large deposits of critical and strategic minerals, including cobalt, the platinum group metals, chromite, tantalum, and columbium. In the past, high energy costs, the difficulty of mining in the Alaskan environment, and the lack of suitable infrastructures discouraged mineral development. But numerous factors — a disruption in foreign supplies, a substantial price increase on the metals market, or a push to diversify the Alaskan economy, which is highly dependent on oil — could increase pressure to develop these mineral deposits.

Imported Mineral Supplies

The United States relies on imports for supplies of many critical and strategic minerals. For some, production from domestic reserves cannot meet demand; for others, it is cheaper for the United States to import supplies from nations with high-grade ores than to develop its own lower-grade domestic reserves. In an effort to guard against shortages, the United States — the world's largest consumer and importer of critical and strategic minerals — began to stockpile important minerals soon after World War II. For most minerals, stockpile targets are set at a three- to four-year supply. In addition to providing a reserve in case of a sudden cutoff or embargo, stockpiles help to even out sharp fluctuations in mineral prices. In 1980, the stockpiles of many strategic minerals were short of target goals. Because of the Reagan administration's emphasis on defense, an effort was made to increase stockpiles to the target level, and in most cases this objective was accomplished. Currently, the United States has stockpiled enough reserves of critical and strategic minerals to last for several years.

Conservation of Mineral Resources

There are three basic ways to conserve mineral resources: find substitutes for minerals that are in short supply, reuse products and materials, and recycle.

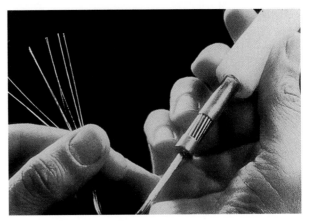

FIGURE 18-7: Cable using optical transmission fibers as a replacement for copper telephone wires. Advanced materials can replace traditional materials that may become scarce or expensive. Many have qualities that make them superior to the materials they replace.

In the area of substitutes, the emphasis is on **advanced materials**, materials that can replace traditional materials and have more of a desired property (such as strength, hardness, or thermal, electrical, or chemical properties) than traditional materials. Though advanced materials do use metals and other minerals, they generally use these substances in lesser amounts than traditional materials, or they use minerals that have not been widely used. Advanced materials offer the promise of energy conservation, better performance at lower prices, and reduced dependence on imports of critical and strategic minerals. Advanced materials have numerous important applications, particularly in the aerospace, automotive, packaging, and communications industries (Figure 18-7).

While advanced materials hold potential for reducing consumption of traditional materials, reusing and recy-

---

WHAT YOU CAN DO ➤ **To Conserve Mineral Resources**

➡ Purchase durable items that last, not disposable ones, and items that will not become obsolete. Repair items or replace worn parts in order to extend the useful life of objects.

➡ To dispose of a car battery, take it to a dealer who will recycle it for its lead content.

➡ Reuse objects whenever possible.

➡ Recycle cans, bottles, paper, and other materials.

➡ Write your Congressperson and ask him or her to work for the repeal of the Mining Law of 1872 in favor of a leasing system.

➡ Support corporate efforts to conserve minerals and develop safe substitutes. One way to do this is to patronize companies that are implementing conservation programs or that are developing alternatives to minerals.

cling materials are the two factors over which individuals have the greatest control. Reusing materials whenever possible — glass containers, for example — saves energy as well as mineral resources. Many materials that cannot be reused, such as aluminum beverage cans, can be recycled. Recycling conserves material resources and energy and reduces solid waste. Additional suggestions to promote mineral conservation are offered in *What You Can Do: To Conserve Mineral Resources.*

What Is the U.N. Law of the Sea?

As mineral exploration shifts from sovereign nations to global commons such as Antarctica and the oceans, the question of who will gain control of potential deposits becomes more important. In 1980, under the auspices of the United Nations, the nations of the world began to draft a treaty specifying who controls the ocean and its resources. The resulting U.N. Convention on the Law of the Sea (1982) established a 12-mile territorial sea for coastal nations, guaranteed passage of all ships through international straits, prohibited dumping of industrial wastes on the continental shelf, and addressed other issues related to military passage and ocean pollution. Importantly, the Law of the Sea established a 200-mile (322-kilometer) exclusive economic zone (EEZ) within which coastal nations control all fishing, marine life, and mineral rights. The treaty's most controversial provisions, however, were those concerning seabed mining in the open oceans, those areas outside each coastal nation's EEZ. The Law of the Sea created the International Seabed Authority (ISA), which grants license to companies to mine open-ocean seabed deposits and to collect taxes on the minerals produced.

U.N. Conventions become international law only when a predetermined number of countries both sign and ratify them. Moreover, they are binding *only* on those countries that actually ratify them. In 1982, President Ronald Reagan refused to sign the Law of the Sea, citing concerns over the process of granting mining rights, mandatory transfers of mining technologies from MDCs to LDCs, and the stipulation that the U.S. be responsible for 25 percent of the ISA's budget. Other industrialized nations, including the Soviet Union and many European countries, followed suit. Consequently, over the next 12 years, the treaty was revised to address the MDCs' concerns. Those revisions satisfied most MDCs, and the U.N. Law of the Sea entered into force in November 1994 after having been ratified by 60 countries, including Russia and the European Union. (Mexico ratified the original treaty in 1983, while Canada ratified it in 2003.)

Under President Bill Clinton, the United States signed the Law of the Sea in 1994, but after the Republicans gained control of Congress in that year's election, he did not send the treaty to the Senate for ratification. That action came a decade later, in February 2004, when President George W. Bush urged the Senate to ratify the Law of the Sea. Despite President Bush's support (as well as the support of the U.S. Navy), Senate approval is not a given. Critics, including Doug Bandow, a senior fellow at the Cato Institute, argue that the treaty is a "flawed document…designed to transfer wealth and technology from the industrialized states to the Third World." They maintain that the ISA is an inefficient and complicated bureaucracy likely to discriminate against U.S. mining interests. Supporters of the Law of the Sea contend that it is an extraordinary achievement in international law, protecting coastal nations' marine resources from exploitation and pollution, ensuring free navigation in straits and international waters, and establishing a framework within which nations can settle disputes.

Summary

Plate tectonic activity and rock-forming processes show that over hundreds of millions of years the Earth refigures itself, so that the composition of rock changes over the ages. Rocks are made of minerals, and thus the composition and distribution of minerals in the Earth's crust — that portion of the planet readily accessible to us — is also continually changing.

Minerals are nonliving, naturally occurring substances with a limited range in chemical composition and with an orderly atomic arrangement. A mineral may occur as a single element, a compound of elements, or an aggregate of elements and compounds. A mineral's unique chemical composition determines its physical properties. Minerals are broadly classified as fuels or nonfuels; nonfuels are further classified as metallic or nonmetallic. A metal is any of a group of chemical elements characterized by malleability, ductility, and conductivity of heat or electricity; nonmetals do not have these characteristics. Ferrous metals contain iron or elements alloyed with iron to make steel; nonferrous metals contain metallic minerals not commonly alloyed with iron.

Minerals are not evenly distributed throughout the Earth's crust. Additionally, deposits of one mineral are often mixed with deposits of others. The dispersion of mineral deposits is caused by fracturing of the Earth's crust, weathering and erosion, chemical separation, and gravity. A useful mineral is one that occurs in a concentration high enough to make it profitable to extract; a deposit of such concentration is known as an ore.

The greatest share of mineral resources is consumed by the MDCs. Global demand for most major minerals is expected to rise significantly as the human population continues to grow rapidly and societies continue to rely on mineral resources to support rising standards of living. Industrial nations rely on about 80 minerals, 75 percent of which exist in abundant supply to meet anticipated needs or can be replaced by existing substitutes. Strategic minerals are those vital to national defense; critical minerals are those considered essential to industry.

The resource base is an estimate of the total sum of a mineral in the Earth's crust; it includes deposits that we will probably never be able to extract and thus is not a very useful figure. A deposit that can be extracted profitably with current technology is called a proven reserve or economic resource. Subeconomic resources are known reserves that cannot be extracted at a profit at current prices or with current technologies. An estimate of the total amount of a given mineral that is likely to be available for future use is called the ultimately recoverable resource. It is based on assumptions about discovery rates, future costs, market factors, and future advances in extraction and processing technologies.

As mineral reserves become depleted, the cost is likely to rise. High prices and high demand encourage more exploration and lead to the discovery of more ores, increasing proven reserves. Other factors likely to ease the pressure on dwindling supplies include conservation measures, decreasing demand, technological advances that make it feasible to mine previously unprofitable deposits, and the development of substitutes.

Mining includes three major steps: location, extraction, and processing. Location is the most benign step; extraction is the most harmful. Acid mine drainage, water and air contamination from tailings and emissions,

Discussion Questions

1. What is the significance of understanding how minerals formed in the Earth's crust?

2. Should mineral exploration and development in less accessible areas, such as Alaska and Antarctica, be encouraged? Why or why not?

3. Suggest several ways that the availability of minerals might be extended.

4. What would you include in a unified minerals policy developed by the United States to ensure both an adequate supply of domestic and imported minerals and the environmental integrity of public lands?

5. How might the more-developed countries help less-developed countries become less dependent on minerals exports? What programs or actions could MDCs take in order to share, more equitably, the profits gained by exploiting the mineral wealth of LDCs?

disturbed vegetation and soil, and erosion are some of the most serious environmental consequences. Threats to human health include exposure to harmful substances, dangerous physical conditions associated with subsurface mining, and the heavy equipment of surface mining.

Mineral prices tend to fluctuate wildly in the short term, but over time their economic value has fallen relative to the value of manufactured goods. Mineral-exporting nations must sell increasing amounts of minerals to pay for imports such as tractors and fertilizers. Critical and strategic minerals are exceptions to this trend; because there are no substitutes for them at present, the price of these minerals is likely to rise.

The Mining Law of 1872 governs the development of mineral deposits on U.S. public lands. It grants title to certain public lands as long as the claimant proves that the minerals are "valuable," conducts minimal mining activities, and pays the government $2.50 per acre. The United States imports most critical and strategic minerals. Stockpiling prevents shortages, ensures a steady supply of important minerals, and evens out sharp price fluctuations.

There are three basic ways to conserve mineral resources: find substitutes for minerals that are in short supply, reuse products and materials, and recycle. Advanced materials have more of a desired property, such as strength or hardness, than traditional materials and are being used as replacements in several industries. Individuals have the greatest control over reusing and recycling materials, which saves energy as well as mineral resources.

The U.N. Law of the Sea took effect in 1994. It established a 12-mile territorial sea for coastal nations, recognized that those nations also have a 200-mile exclusive economic zone within which they control all resources, guaranteed passage of all ships through international straits, prohibited dumping of industrial wastes on the continental shelf, addressed issues related to military passage and ocean pollution, and created an international authority to oversee seabed mining. Its mining provisions are controversial in the United States, and as of 2005, the U.S. Senate had not signed the treaty.

KEY TERMS

abundant metals	mineral	resource base
advanced materials	mineral extraction	scarce metals
area strip mining	mineral processing	sedimentary rock
contour strip mining	minerals policy	strategic mineral
critical mineral	mountaintop removal	subeconomic resource
economic resource	nonferrous metals	subsurface mining
ferrous metals	open pit surface mining	surface mining
igneous rock	ore	tailings
magma	overburden	ultimately recoverable resource
metamorphic rock	proven reserve	

CHAPTER 19

Nuclear Resources

You're always balancing risks against possible benefits.

Dr. Joseph R. Castro

Even if no more nuclear waste were created, addressing that which already exists will require attention and investments for a period that defies our usual notion of time. The challenge before human societies is to keep nuclear wastes in isolation for the millennia that make up the hazardous life of these materials. In this light, no matter what becomes of nuclear power, the nuclear age will continue for a long, long time.

Nicholas Lenssen

Learning Objectives

When you finish reading this chapter, you will be able to:

1. Describe the different applications of nuclear technology.

2. Explain two types of exposure to radiation and the possible health effects.

3. Identify four types of waste produced by a nuclear power plant.

4. Identify the issues involved in long-term storage of nuclear waste.

A cusp is a point at which two curves meet. For the poetic among us, to approach a cusp is to come to a place where the path suddenly, drastically changes, where life is forever after altered. Individuals can reach and cross a cusp; so, too, can the human race. The use of fire, the development of the wheel and the first use of tools, all signaled dramatic changes in human history. Another such cusp occurred in 1945.

During World War II, the United States, in a race with Nazi Germany to develop atomic weapons, sponsored the Manhattan Project, a top-secret effort to create the world's first nuclear bomb. In 1945, Manhattan Project scientists succeeded in developing the bomb before the Germans, but the United States never used its new weapon against Hitler's regime. Instead, in the early morning hours of August 6, 1945, a U.S. warplane named the Enola Gay released the first atomic bomb on the city of Hiroshima, Japan. The ensuing destruction clearly demonstrated that humankind was capable of destroying life on a scale previously unimagined.

The attack on Hiroshima and a similar attack on the Japanese city of Nagasaki several days later bore witness to the awesome power of the atom. In the years that followed, researchers began developing other applications for nuclear power. Nuclear resources yielded significant benefits for medicine, agriculture, industry, and science (Figure 19-1). Nuclear power was used to generate electricity, avoiding the release of greenhouse gases and sulfur oxides that accompany the combustion of coal and oil. Ironically, the same resource that killed and maimed thousands of Japanese in an instant on that day in 1945 also saved the lives of untold numbers of cancer patients, improved the overall quality of life for many in the industrialized world, allowed researchers to push forward the frontiers of science, and provided electricity for millions.

The gains made through the use of nuclear resources have not come without cost: nuclear wastes continue to accumulate; water, land, and air have been contaminated by accidental and/or planned radioactive emissions from weapons facilities and power plants; and the threat of accidents or nuclear war darken the collective heart of humanity. In this chapter, we describe nuclear resources and examine efforts to manage them.

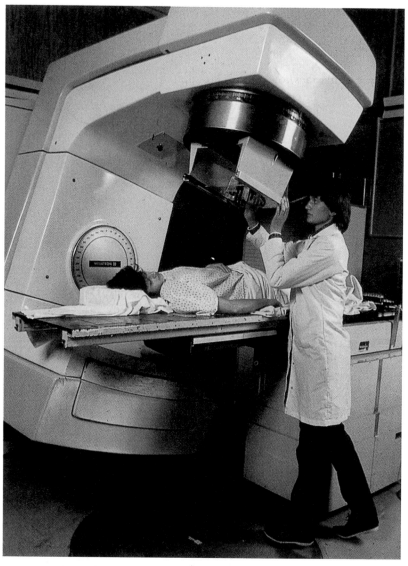

FIGURE 19-1: A cancer patient is treated using radiation therapy. Among the most important beneficial uses of nuclear resources are their medical applications.

Describing Nuclear Resources

In this section, we look closely at what nuclear resources are, how nuclear energy is released, how radiation affects human health and the environment, and how we use nuclear resources.

What Are Nuclear Resources?

Nuclear resources are derived from atoms, their energy, and the particles they emit. **Nuclear energy** is the energy released, or radiated, from an atom; this energy is called **radiation**. Radiation takes two basic forms: ionizing and nonionizing. **Ionizing radiation** travels in waves (x-rays, gamma rays) or as particles (alpha particles, beta particles). The energy level of ionizing radiation is high enough to remove electrons from atoms, creating charged particles called **ions**. These ions then can react with and damage living tissue. **Nonionizing radiation** (heat, lasers, light, radio waves) also can affect atoms, but its energy level is not high enough to create ions.

Ionizing radiation is released into the environment from both cultural and natural sources. Cultural sources such as nuclear testing, nuclear power plants, medical x-rays, luminous watch dials, color television sets, smoke detectors, and metals production account for only an estimated 18 to 32 percent of the radiation in the environment. The remaining radiation, about 68 to 82 percent, originates from natural sources, including the sun, outer space, soil, air, water, food, and rocks (Table 19-1). As we learned in Chapter 12, radon is a radioactive gas produced during the decay of uranium. Found in small quantities in water, soil, rocks and various building materials, it can enter buildings through cracks in the foundation or basement. Once inhaled, it releases particles that can cause lung cancer. Studies have shown radon gas to be the single largest source of natural radiation exposure. Because the gas is odorless, invisible and tasteless, only special tests can detect its presence. The discovery of radon in many homes in the late 1980s precipitated nationwide concern about the dangers of this natural source of radiation.

How Is Nuclear Energy Released?

Nuclear energy is released through three types of reactions: spontaneous radioactivity, fission, and fusion.

Spontaneous Radioactivity

Spontaneous radioactivity occurs when unstable atoms release mass in the form of particles (particulate radiation), energy in the form of waves (electromagnetic radiation), or both. The unstable nucleus emits particles and energy during a process called **radioactive decay**. For example, uranium 238 decays into thorium 234, then into protactinium 234, and eventually into the stable isotope lead 206 (Figure 19-2). Each radioactive element goes through its own unique sequence to reach stability.

Atoms of the same element that have different numbers of neutrons are known as **isotopes** of the element. Most elements have more than one naturally occurring isotope. Many more isotopes have been produced in reactors and scientific laboratories. The usual isotope of carbon is carbon 12 (six neutrons, six protons); a carbon atom with eight neutrons is the isotope carbon 14 and is radioactive. In addition to its usual form, hydrogen

---

TABLE 19–1: Sources of Radiation

Source	Dose (rem)[a] per year
Natural sources (67.6% to 82% of our radiation exposure per year)	
Concrete homes	.07–.10
Brick homes	.05–.10
Wooden homes	.03–.05
Cosmic rays	.045
Air, food, water	.025
Soil	.015
Cultural sources (18% to 32.4% of our radiation exposure per year)	
Diagnostic x-rays (per x-ray)	.02
Watches, color television sets, smoke detectors	.004
Air travel (round trip to London)	.004
Nuclear plant vicinity	.001

[a] Roentgen equivalent, man; the dose of an ionizing radiation that will have the same effect on living tissue as one roentgen of gamma ray or x-ray.

Type of radiation	Nuclide	Half-life
α	uranium 238	4.47 billion years
β	thorium 234	24.1 days
β	protactinium 234	1.17 minutes
α	uranium 234	245,000 years
α	thorium 230	80,000 years
α	radium 226	1,600 years
α	radon 222	3.823 days
α	polonium 218	3.05 minutes
β	lead 214	26.8 minutes
β	bismuth 214	19.7 minutes
α	polonium 214	0.000164 seconds
β	lead 210	22.3 years
β	bismuth 210	5.01 days
α	polonium 210	138.4 days
	lead 206	stable

FIGURE 19-2: The decay of uranium 238.

has two isotopes: deuterium, which has one neutron, and tritium, which has two neutrons.

Isotopes that release particles or high-level energy are called **radioisotopes**. Radiation emitted by radioisotopes is an example of ionizing radiation. The most common types of ionizing radiation are alpha, beta, and gamma radiation. **Alpha particles** consist of two protons and two neutrons and carry a positive charge. They are emitted at high speeds but travel only short distances before losing energy. Alpha particles can be stopped by a sheet of paper or by the skin, but they are dangerous when inhaled or when ingested through food or water. **Beta particles** are negatively charged particles that are emitted from nuclei. They are equivalent to electrons, but contain more energy. Beta particles arise when neutrons in the nucleus are converted into protons. The small amount of energy that is lost during the conversion process is the energetic beta particle. Beta particles can be stopped by a piece of wood an inch thick or a thin sheet of aluminum, but they can penetrate the skin. They are harmful when emitted inside the tissues of a living organism.

Gamma radiation is a powerful, high-energy wave; gamma rays are the most common form of ionizing electromagnetic radiation released from radioisotopes. Even a thick piece of lead or concrete will not stop all gamma rays. They can pass through the human body, causing damage as they do so. **X-rays**, a form of cosmic radiation, also can be produced by firing electrons at a target made of tungsten metal. When the electrons hit the metal, they give up their energy in the form of x-rays. Like gamma rays, x-rays can ionize atoms in living tissue, but they have considerably less energy and so are less penetrating (Figure 19-3).

Each stage in a decay sequence exists for a specific amount of time, measured in half-life. **Half-life** is the length of time it takes for one half of any given number of unstable atoms to decay. The half-life of a substance may vary from a split second to billions of years (Table 19-2). The half-life of a radioactive substance must not be confused with the amount of time before that substance can be released safely into the environment. Even after one half-life has gone by, 50 percent of a radioactive substance's atoms remain unchanged and hazardous.

Fission

A **fission reaction** occurs when an atom is split into two or more new atoms. When neutrons strike a large atom, such as uranium 235, electromagnetic radiation, alpha particles, beta particles, energy, and one or more neutrons from the atom's nucleus are released. The atoms formed when uranium is split are called **fission products**. Fission products are almost always radioactive. The neutron(s) released by each fissioned atom then can split other uranium 235 nuclei. **Critical mass** is the minimum amount of material that can sustain a chain reaction. A **chain reaction** is a self-perpetuating series of events that occurs when a neutron splits a heavy atom, releasing additional neutrons that cause other atoms to split in a like manner, thus emitting tremendous amounts of energy. This is the energy that is harnessed in commercial nuclear power reactors.

Fusion

A **fusion reaction** is the opposite of a fission reaction; nuclei are forced to combine. The sun is essentially a nuclear fusion reactor: two hydrogen nuclei fuse into a helium nucleus. The nuclei tend to repel one another because they are positively charged; therefore, a great amount of energy is required to fuse the nuclei.

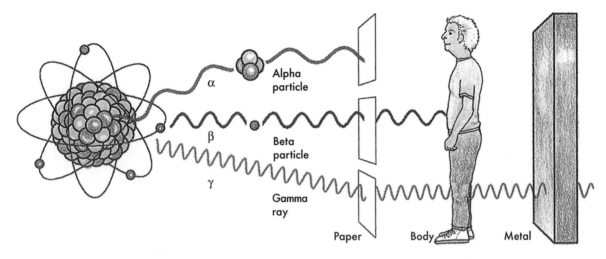

FIGURE 19-3: Penetrating power of alpha, beta, and gamma radiation. The three types of radiation possess different amounts of energy that determine their ability to penetrate different substances.

Once the repulsive force is overcome and the nuclei are joined, a huge quantity of high-energy radiation is released.

Scientists have tried to fuse various combinations of deuterium and tritium in the laboratory, but so far they have failed to get as much energy out of the reaction as they have put into it. One problem impeding fusion research is the difficulty in finding a material for the containment vessel. The material must be highly resistant to radiation and the extremely high temperatures (100 million °C or higher) that are required to form the plasma (an electrically neutral mixture of nuclei and electrons) in which the reaction takes place. Another difficulty is how to keep the plasma from striking the containment vessel's walls and losing the plasma particles' energy. One of the most promising containment technologies is magnetic containment, in which the plasma is suspended in a magnetic field. In 1991, scientists at the Joint European Torus (JET) laboratory in Culham, England, achieved the fusion of deuterium and tritium nuclei in a magnetically confined plasma. The reaction generated about 1.7 million watts of power in a burst lasting nearly one second. Although the experiment consumed 10 times more energy than it produced, it demonstrated that fusion is scientifically feasible.

For many, fusion holds the promise of "clean" nuclear energy, and some scientists predict that construction of a demonstrator reactor could begin around 2010, costing approximately $6 billion. Others speculate that we will have commercial fusion reactors producing electricity within the next 50 years. However, experts agree that building a commercial fusion reactor will be the world's most challenging feat of engineering. In fact, the technological problems that must be solved before we realize the practical applications of fusion energy prompt some detractors to say that day will never come. Furthermore, they caution that fusion energy, like fission, easily could be accompanied by problems we cannot even imagine today.

How Does Radiation Affect Human Health?

As ionizing radiation penetrates living tissue, it can destroy cells or alter their genetic structure. For example, when alpha particles come in contact with cells, which

TABLE 19–2: Half-Life of Selected Radioactive Materials

Radioactive Material	Half-Life (years)
Cesium 137	30
Neptunium 237	2,000,000
Plutonium 239	24,110
Strontium 90	30
Hydrogen 3 (Tritium)	12
Uranium 233	160,000
Uranium 235	714,000,000

Numbers following elements represent isotopes of those elements.

are largely composed of water, the sudden introduction of intense energy excites the water molecules. The positive charge of the alpha particles strips electrons from the water molecules, destabilizing their relationships with neighboring water molecules. Eventually, the destabilization may become so great that the cells can no longer function.

Ionizing radiation also can break or scramble chromosomes in the cell's nucleus. If the radiation dosage is high enough and the resulting damage too severe, the cell will die, often within hours. Those cells receiving nonlethal doses can continue to exist with altered DNA, reproducing abnormally for years, and may spawn cancerous cells and tumors.

The cumulative effect of cellular change caused by ionizing radiation can lead to a wide range of health problems, including cancers such as leukemia; degenerative diseases such as cataracts; mental retardation; chromosome aberrations; genetic disorders such as neural tube defects; and a weakened immune system. Various factors determine how radiation affects human beings; among the most significant factors are the amount of radiation (or dose) and the age and gender of the individual.

Amount of Ionizing Radiation

The amount of an individual's exposure to ionizing radiation is typically measured in two dosage units. The amount of radiation absorbed per gram of tissue is expressed by the *rad* (radiation absorbed dose). The international unit, which is replacing the rad, is called the *gray*; it is equivalent to 100 rads. The damage potential caused by a dose of radiation is expressed by the *rem* (Roentgen Equivalent, Man) and *millirem* (one-thousandth of a rem). The *sievert* is the international unit used to express damage potential. One sievert is equivalent to 100 rem.

Doses of radiation can be classified as low- or high-level. Low levels, which range from one to five rem per year, can still be health hazards because even small amounts of radiation can cause cell damage that accumulates over time. Some scientists believe that any amount of radiation, no matter how minute, affects biological systems in some way. Others believe there is a threshold below which radiation exposures are harmless (Table 19-3).

The government has established guidelines for the amount of radiation a person can safely withstand. The Nuclear Regulatory Commission (NRC) has set 0.17 rem per year (not including the 0.2 rem of background

TABLE 19–3: Health Effects, Exposure to Radiation	
Sudden, whole body exposure to radiation can cause the following general effects:	
1–100 rem	Nausea, vomiting
100–200 rem	Moderately depressed white blood cell count, not immediately fatal but long-term cancer risk
200–600 rem	Heavily depressed white blood cell count, blotched skin in four to six weeks, 80 to 100 percent probability of death
600–1,000 rem	Diarrhea, fever, blood-chemical imbalance in one to 14 days, almost 100 percent probability of death
Radiation exposure also can affect specific organs and systems:	
1–10 rem	Organs and tissues of high sensitivity: bone marrow, colon, stomach, breast, lung, thyroid
	Organs and tissues of moderate sensitivity: ovaries, intestines, pancreas, liver, esophagus, lymph, brain
	Organs and tissues of low sensitivity: bone, kidney, spleen, gallbladder, skin
50–500 rem	Brain and central nervous system: delirium, convulsions, death within hours or days
	Eye: death of cells in lens leading to opaqueness and cataracts that impair sight
	Gastrointestinal tract: nausea and vomiting within a few hours, bleeding of the gums, mouth ulcers, intestinal wall infection leading to death
	Ovaries and testes: acute damage can affect the victim's fertility or offspring
	Bone marrow: retards the body's ability to fight infections and hemorrhaging; may lead to leukemia

radiation from natural sources) as the maximum safe exposure for the general public. However, NRC limits for those working in the nuclear industry are almost 30 times higher — three rem within a 13-week period or a total of five rem annually. Radiation sickness is caused by doses ranging from 50 to 250 rem. Symptoms include nausea, vomiting, and a decrease in white blood cells. Doses of radiation over 500 rem are almost always fatal.

When a reactor at the Soviet Union's Chernobyl nuclear power plant exploded in 1986, it emitted tons of radioactive material into the environment. Victims living within a three- to four-mile radius of the plant stood only a 50 percent chance of surviving; those who survived suffered bone marrow and gastrointestinal damage. Even those living 60 miles away had an increased risk of leukemia and other forms of cancer for the next 30 years. International groups concerned with nuclear issues estimate that 12,000 people will die by mid-century as a result of the accident.

Age and Gender

Many studies have shown that people of varying ages and genders respond differently to similar amounts of radiation. For example, women are more likely to develop thyroid cancers than men. Infants and fetuses are more vulnerable to all types of cancers and disabilities than adults because of their faster rate of cell division.

Fetuses are especially likely to suffer severe effects from radiation exposure. If doses are high enough, a fertilized egg can develop lesions that will prevent its implantation in the uterus, thus causing a spontaneous abortion. Lower doses of radiation allow the fertilized egg to implant but can damage the fetus's developing organ systems. Such fertilized eggs can develop into live-born children with mild to severe defects.

How Does Radiation Enter and Affect the Environment?

Direct radiation exposure is exposure to the original radioactive source (such as a power plant or nuclear explosion). **Radioactive fallout** refers to dirt and debris contaminated with radiation; it can be produced by nuclear testing, the explosion of a bomb, or an accident at a nuclear power plant. Wind and rain can spread radioactive fallout far from its original source.

Indirect radiation exposure is exposure to radiation removed from its original source. Such exposure also can be dangerous, especially since it is often less obvious (Figure 19-4). For example, radiation contamination may travel from its original source through food chains. Grain grown in contaminated areas can pass on radioactivity to cows that consume it; it is then possible that the milk produced by the cows also will become radioactive. One major study documented indirect exposure through food chains resulting from atmospheric nuclear testing between 1945 and 1980, when the United States, the Soviet Union, France, the United Kingdom, and China detonated a total of 423 aboveground nuclear devices. These explosions released massive amounts of radioactive debris into the upper atmosphere, where air currents distributed it across over

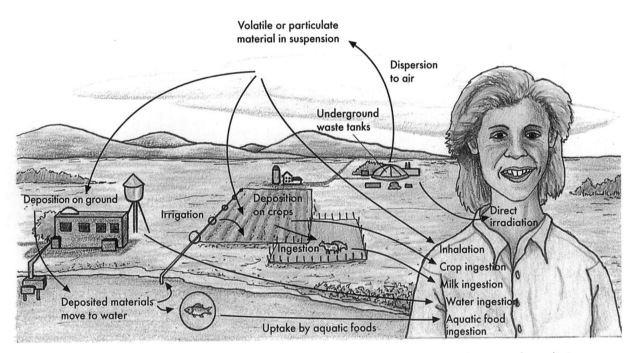

FIGURE 19-4: Major pathways of the environmental dispersion of radionuclides. Humans may be exposed to radiation directly, through contact with radioactive materials in the air or on the land, or indirectly, through food chains.

the world. As the debris returned to Earth, increasing levels of radiation began to enter various food chains (Figure 19-5).

In the mid-1980s, scientists investigating the potential consequences of a nuclear war raised the probability of what some call the ultimate environmental disaster: **nuclear winter**, the theoretical result of explosions and fires following a nuclear war that would pump vast quantities of smoke, soot, and debris into the atmosphere, effectively preventing sunlight from reaching huge areas of the globe. This sun block could result in temperature drops of 36° to 72°F (20° to 40°C). Given that drops of only a few degrees can drastically reduce photosynthetic rates, such a huge change would reduce agricultural yields and result in the mass extinction of countless plant and animal species, possibly including humans. Although scientists disagree over how severe and long-lasting the effects would be from any one nuclear war scenario, they generally agree that some sort of nuclear winter would take place.

How Are Nuclear Resources Used?

Nuclear resources are used in many ways. For example, small amounts of the radioactive element thorium added to other metals can produce stronger, lighter weight, and more heat-resistant metal alloys for use in aircraft engines and airframe construction. Similarly, some forms of uranium have been used in armor plating for tanks. The National Aeronautics and Space Administration (NASA) uses small plutonium reactors to power satellites such as Voyager II that travel outside our solar system. Radioactive materials also are used as power sources for robots that are designed to operate in environments hostile to humans, including nuclear power plants, burning buildings, outer space, and the ocean floor. The most widely recognized uses of nuclear resources, however, are in medical applications, electric generation, food preservation, and military applications.

Medical Applications

Nuclear resources have many medical uses, including the diagnosis and treatment of illness and the sterilization of supplies. Indeed, medical applications of nuclear resources represent the most significant human-made source of radiation exposure.

Nuclear technology has allowed diagnostic medicine to progress further in the last few decades than in the entire previous history of medicine. Radioisotopes have

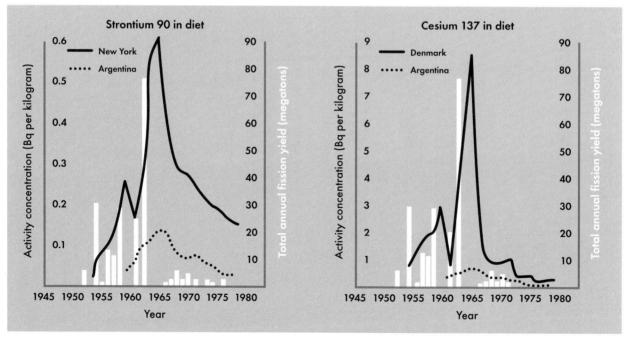

FIGURE 19-5: Relationship between nuclear testing and food-chain contamination by strontium 90 and cesium 137 produced during nuclear blasts. Concentrations of strontium 90 and cesium 137 (measured in becquerels per kilogram) in the total diet (graphed line, in black) are compared with the annual yield from atmospheric nuclear tests (white bars) measured in megatons. Because of worldwide weather patterns, exposure to the radioactive substances was much higher in the Northern Hemisphere (New York and Denmark) than in the Southern Hemisphere (Argentina).

played an important role in developing safe, accurate methods of diagnosing medical conditions.

The most common form of diagnostic nuclear medicine is scanning or imaging. During these procedures, the patient swallows, inhales, or is injected with radioactive compounds that travel through the body, continuously giving off gamma rays. Specially designed equipment detects and records the gamma rays, creating images of the body. Computers then interpret the test results.

These scans enable physicians to see inside the body without having to conduct traumatic exploratory surgery. They are used to detect conditions such as blood clots, tumors, coronary artery disease, bone fractures, and infections (Figure 19-6). Most of the radioactive compounds used in nuclear medicine are quickly excreted from the body, usually within hours or at most a day or two, so their radioactive effects are limited.

The most familiar use of radiation in therapeutic medicine is in the treatment of cancers. Radiation therapy involves focusing beams of radiation at tumors, injecting patients with radioactive materials chosen for their ability to combat specific types of cancer, or implanting radioactive materials directly into tumors. For example, solid cancerous tumors respond best to beta-emitting radioactive materials, while more diffuse tumors respond better to alpha emitters. Although radiation therapy is not a foolproof treatment for cancer, studies confirm that, in some cases, it can destroy or inhibit the further growth of a cancerous tumor. Radioactive materials also can be used as tracers to monitor the success of treatments for some disorders,

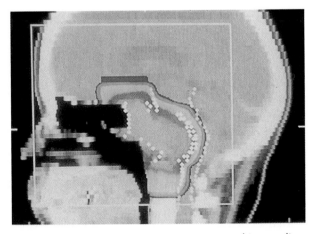

Figure 19-6: Brain scan. A computer-generated image displays focused, positively charged atomic particles. Proton beam therapy uses protons to destroy cancer cells rather than electrons, which are used in traditional radiation therapy.

usually to tell if specific dosages of medicines have been effective. Radiopharmaceuticals such as iodine 131 correct some types of thyroid conditions, such as overactivity and tumors. Continued refinement of radiation therapies promises to improve cancer treatment in the future.

Using radiation to sterilize medical instruments and supplies, such as surgical dressings, sutures, catheters, syringes and donor organs and tissues, is a rapidly growing practice. Instead of the traditional method of heating objects to kill contaminating bacteria and spores, medical supply companies use gamma rays to penetrate hermetically sealed packages, thereby sterilizing their contents without risking contamination during the packaging process. Sterilization by radiation, often referred to as the "cold process," offers the only method of sterilizing such heat-sensitive materials as plastic heart valves and medicines.

Electric Generation

The generation of electricity is one of the most widely recognized uses of nuclear resources. Conventional fission reactors work by sustaining the fissioning process. Typically, the fuel used is uranium 235 (U-235), a readily fissionable isotope of the element uranium. The concentration of U-235 in uranium ore is only about 0.07 percent, usually too low to sustain a chain reaction; uranium 238, a nonfissionable isotope, is the principal constituent of uranium ore. Therefore, before uranium can be used economically in a reactor, impurities must be removed and the concentration of U-235 raised to increase the incidence of fissioning. These tasks are accomplished by milling the ore to a granular form and using an acid to dissolve out the uranium. After it is converted to a gas, the fuel is purified. Its concentration of U-235 is enriched to approximately three percent, and the enriched gaseous form of uranium is converted into powdered uranium dioxide. The uranium dioxide powder is subjected to high pressure to form fuel pellets about 5/8-inch long and 3/8-inch in diameter. The U-235 in just three of these reactor pellets generates about the same amount of electricity as 3.5 tons of coal or 12 barrels of oil. The pellets are then placed into a 12-foot-long metal tube (about five pounds of pellets per tube), or fuel rod, which is made of a zirconium alloy that is highly resistant to heat, radiation, and corrosion.

The actual nuclear reaction occurs within the reactor core of a power plant, which is housed in a thick-walled structure called a **containment vessel**. Within the core are hundreds of fuel assemblies, each composed of groups of fuel rods. A material called a moderator is circulated between the rods and assemblies to slow down the neutrons emitted by the fission reaction, increasing the likelihood of fission and sustaining the chain reaction. Most commercial reactors worldwide use ordinary water, called light water, as a moderator. Others, including about half of the reactors in the former Soviet republics, use solid graphite, a form of carbon. Some reactors use heavy water as a moderator. Heavy water is a compound in which the hydrogen atoms are replaced by deuterium, the hydrogen isotope containing one neutron.

A coolant circulated throughout the reactor core absorbs and removes heat to prevent the fuel rods and other materials from melting. Water is used as a coolant in most water-moderated and graphite-moderated reactors. In some reactors, a gas or liquid metal is used as a coolant. If the coolant is water, it may be heated directly to steam, which then can be used to drive the generator turbines that produce electricity. In other nuclear reactors, coolants are kept under pressure and passed through heat exchangers to heat another body of water to its boiling point, creating steam to drive the turbines. Operators can control the rate of fission inside the reactor core by raising or lowering metal control rods that absorb neutrons into vacant holes within each fuel assembly.

Food Preservation

The use of low-level gamma radiation to kill pathogenic bacteria in foods is gaining acceptance with the Food and Drug Administration (FDA). In the irradiation process, gamma rays emitted by radioactive cobalt 60 kill bacteria, insects, and fungi on produce and extend shelf life by inhibiting enzyme activity. The food itself does not become radioactive. The FDA approved irradiation for wheat flour in 1963, white potatoes in 1964, and fruits and vegetables in 1986. In 1985, irradiation in pork processing was approved to destroy *Trichinella spiralis*, a microscopic roundworm found in the muscle tissue of hogs.

Those who support irradiation argue that it ensures a safe, wholesome food supply. Critics contend that more study and research are needed to determine if treating food with low-level radiation is safe for humans who consume the food. Other critics worry that irradiation will mask serious bacterial contamination resulting from inadequate plant sanitation or product mishandling.

Military Applications

Nuclear bombs and warheads are probably the most widely recognized military use of nuclear resources. An atomic bomb releases a tremendous amount of energy in a split second through fission and an uncontrolled chain reaction. These weapons use fissioned plutonium, which the military produces at several reactors in the United States. Nuclear reactors (using controlled chain reactions) also are used to power submarines and ships, eliminating the need to refuel frequently but posing a danger of contamination if the vessel is destroyed. Because military applications are intricately tied to the historical development of nuclear resources, they are discussed in further detail below.

Managing Nuclear Resources

In this section, we examine how nuclear resources have been managed historically. We also discuss some of the problems associated with their use.

How Has the Use of Nuclear Resources Changed?

It has been little more than 60 years since humankind first learned to tap the potential of nuclear resources. But in a period shorter than the average human life span, the use of nuclear resources — and society's reaction to them — has changed dramatically. For much of the past six decades, the dominant use of nuclear resources was in military applications. The later development of nuclear resources focused on peaceful uses. In the following paragraphs, we'll examine the history of both the military and nonmilitary uses of nuclear resources. In each case, public opposition to the use of nuclear resources has grown as our understanding of their complexity, benefits, and dangers has increased.

Military Uses of Nuclear Resources

Initially, nuclear resources were developed for use by the military. Before the United States dropped the first atomic bomb, scientists could only speculate on how such a huge concentration of radiation would affect living organisms, including humans. Of course, those near the bombs were killed instantly; the number of people killed at Hiroshima and Nagasaki soared to over 200,000 almost immediately. However, only after years of study and experimentation have we begun to realize the true range of radiation's devastating effects. One of the most sobering lessons concerns the effects of radiation on fetal development. For example, by studying children conceived in the weeks and months prior to the Hiroshima and Nagasaki bombings, researchers have established the window in pregnancy of 8 to 15 weeks during which the developing fetus is extremely susceptible to radiation exposure (Figure 19-7).

The Cold War and the Arms Race.

The attacks on Hiroshima and Nagasaki ended the war in the Pacific, but they signaled the dawn of the Cold War, an era marked by fear of a nuclear confrontation between the United States and the Soviet Union. In 1947, the United States established the Atomic Energy Commission (AEC) to control the use and disclosure of information on atomic power. When the Soviet Union detonated its first atomic bomb in 1949, the nuclear race between the two superpowers began, and strict control of the secrets of nuclear technology was deemed even more important to national security.

Ignorance about the effects of radiation led both the United States and the Soviet Union to conduct studies that today would be considered criminally careless. For example, on several occasions, U.S. troops were stationed within the blast zones of aboveground nuclear explosions to test their ability to negotiate radiation-contaminated terrain. In tests conducted under the auspices of the AEC, subjects were injected with uranium, fed fallout from atomic bomb tests, and made to breathe radioactive air.

In 1963, concern over radioactive fallout spurred 105 nations to sign the Nuclear Test Ban Treaty, an agreement to halt nuclear testing in the atmosphere, outer space, underwater, and anywhere else that fallout might spread beyond the borders of the country testing the weapon. As of 2004, 172 countries had signed the Nuclear Test Ban Treaty, meaning that they intend to comply with its provisions, and 116 had ratified the treaty, meaning that its provisions are enforced. The United States has signed, but not ratified, the Nuclear Test Ban Treaty.

One of the first attempts to halt the spread of nuclear weapons technology was the Nuclear Non-Proliferation Treaty, signed by 17 nations in 1968. This treaty stipulated that the United States and the Soviet Union would not provide nuclear weapons to other countries, nor would they assist nations in developing this technology.

Furthermore, the signatories agreed to facilitate the development of peaceful uses of nuclear energy. However, despite global concern, arsenals of nuclear weapons continued to grow, since both the United States and the Soviet Union argued that the only way to prevent a nuclear war was to be well prepared against one.

Attempts at Arms Reductions.

The history of negotiations for arms control suggests that reductions in nuclear weapon stockpiles will continue to develop slowly. Such negotiations are hampered by each country's concern that other parties might break an agreement and rebuild a nuclear arsenal.

For four decades, from the 1950s through the 1980s, the possibility of a nuclear confrontation between the world's superpowers, the United States and the Soviet Union, held center stage in world affairs. With the two countries' nuclear missiles targeted at one another, the specter of nuclear war was a real and immediate threat. But with the dissolution of the Soviet Union and the thawing of the Cold War, the superpowers seized the opportunity to reduce their nuclear arsenals. The Strategic Arms Reduction Treaty, or START I, led to a reduction in global arms, chiefly in delivery systems such as missiles. When the two nations agreed to detargeting — essentially, to no longer point their nuclear missiles at one another — many people assumed that the nuclear threat had all but disappeared. Yet, since

FIGURE 19-7: Kunizo Hanataka and his daughter, Yuriko. Yuriko was conceived three months before the bombing of Hiroshima. Her mother was about 2,400 feet from the hypocenter of the bomb. The radiation caused severe mental retardation in Yuriko. Kunizo operates a small barber shop from his home so that he can care for Yuriko, who requires full-time attention.

that period, negotiations have proceeded slowly. START II, which calls for the elimination of ballistic missiles and a two-thirds reduction in the number of strategic nuclear warheads, was drafted in 1992, but was not signed by the United States until 1996 and by Russia until 2001. START II entered into force on December 31, 2004. In 2002, the United States and Russia also signed the Strategic Offensive Reductions Treaty, which calls for a dramatic reduction in the total number of warheads — to between 1,700 and 2,200 — held by each county.

Clearly, the potential for nuclear confrontation remains; despite the arms reduction achieved through previous negotiations, the United States and Russia still maintain large stockpiles of nuclear warheads. And although the two nations no longer have their nuclear missiles targeted at one another, both sides retain their nuclear forces on full-alert status. At the threat of an imminent attack, they quickly could retarget launch missiles carrying nuclear warheads. Moreover, at least five other nations — Britain, China, France, India, and Pakistan — are known to have active nuclear programs.

Nonmilitary Uses of Nuclear Energy

Worried that the secrecy and public concern surrounding the atomic bomb would give nuclear energy a negative image, the federal government decided to encourage more constructive uses for nuclear resources. In 1953, the AEC made the development of economically competitive nuclear power a goal of national importance. That same year, President Dwight D. Eisenhower, in a speech entitled "Atoms for Peace," emphasized the need for productive uses of nuclear power.

The government's commitment to developing civilian nuclear capabilities led to the 1954 Atomic Energy Act, which opened the nuclear industry to the private sector. The first major developments were in the areas of medicine, biological research and agriculture, but by the end of the 1950s, interest began to turn toward using nuclear energy to produce electricity. This interest was spurred by lowered estimates of fossil fuel reserves, a growing demand for electrical power, and the 1956 war over the Suez Canal, a major route for transporting oil. In 1957, the world's first commercial nuclear reactor, the Shippingport nuclear power plant, came online near Pittsburgh, Pennsylvania.

Nuclear production of electricity increased slowly but steadily until the early 1970s, when an oil embargo by the Organization of Petroleum Exporting Countries (OPEC) compelled the United States and other industrialized nations to turn to alternative fuel sources. Before the 1973 OPEC embargo, nuclear energy produced only about five percent of electricity in the United States (compared to 17 percent produced by oil). However, during the 1970s and 1980s, the industry experienced rapid growth. By 1990, nuclear energy accounted for about 22 percent of U.S. electric generation (second only to coal) and nearly 20 percent worldwide. In more recent years, however, growth in the industry has slowed. In 2004, nuclear power accounted for approximately 20 percent of electricity in the United States and 23 percent worldwide. The only region poised to increase its reliance on nuclear power is Asia, where both China and South Korea have six reactors under construction. (Reliance on nuclear power in the United States and worldwide is discussed more thoroughly in Chapter 11, *Energy: Alternative Sources*.)

The stagnation in the nuclear energy industry can be attributed to several factors, chiefly high costs and technical problems. In the United States, costs have risen because of longer lead times for licensing and construction of plants and higher financing expenses. The global nuclear industry also has been hampered by accidents, such at those at the Three Mile Island nuclear facility near Harrisburg, Pennsylvania, and the Soviet Union's Chernobyl nuclear power plant. Such accidents have fueled public concern about the long-term consequences of nuclear resources. Many people advocate reducing our reliance on nuclear resources, even in nonmilitary applications. (See *What You Can Do: To Reduce Reliance on Nuclear Resources*.)

What Problems Accompany the Use of Nuclear Resources?

The use of nuclear resources involves serious problems, among them radioactive leaks from military and commercial facilities, accidents at nuclear power plants, disposal of nuclear wastes, decommissioning of old plants, and secrecy surrounding nuclear activities.

Radioactive Leaks

Although military and commercial nuclear facilities have accidentally leaked radioactive substances over the years, many of these incidents have gone unreported to the general public. Only recently have we begun to realize the scope of environmental contamination that has taken place from military and commercial nuclear facilities. For example, recent reports have disclosed that between 1952 and 1954, the Hanford Federal Nuclear Facility in the State of Washington released large quantities of radioactive ruthenium into the atmosphere on nine different occasions. (To learn more about these incidents, read *Environmental Science in Action: The Hanford Federal Nuclear Facility* by going to www. EnvironmentalEducationOhio.org and clicking on "Biosphere Project.") Other government reports have documented releases of over 395,000 pounds (177,750 kilograms) of uranium into the atmosphere from the Feed Materials Production Center (now called the Fernald Environmental Management Project) at Fernald, Ohio, just northwest of Cincinnati. One study commissioned by residents in the Fernald area estimated that those living near the plant have inhaled air containing almost 70 times the levels of uranium prescribed by the U.S. Department of Energy's (DOE) safety guidelines. Yet another frightening discovery occurred at the Rocky Flats munitions plant, only 50 miles upwind from Denver, where smokestack monitors registered plutonium emissions 16,000 times greater than the standard levels. Some environmentalists have referred to this situation as the "creeping Chernobyl."

Accidents at Nuclear Power Plants

Even more frightening than radioactive leaks is the possibility of a major nuclear accident — a **meltdown** in which the reactor core becomes so hot that the fuel rods melt, possibly burning through the containment vessel and the underlying concrete slab and boring into the Earth.

Researchers have investigated the probability of major nuclear catastrophes. A 1982 study conducted by the Sandia National Laboratory estimated that the chance that a catastrophic accident will occur in a one-year-old reactor is 20 million to one. However, other studies, which consider factors such as human error, estimate that a catastrophic accident will occur every 8 to 20 years.

As nuclear power plants age, the probability of accidents increases. In 2004, 30 percent of operating nuclear reactors in the United States were over 30 years old, with the oldest reactors still in operation licensed in 1969. Of reactors completed since 1976, only one has been permanently closed and no U.S. reactor (of any age) has closed since 1998. Aging plants often are plagued by the corrosion of steam generators and the embrittlement of steel pressure vessels through neutron bombardment. These conditions are difficult to remedy

and greatly increase the possibility of disaster due to overheated reactors. The Yankee Rowe power plant in western Massachusetts, one of the nation's oldest nuclear plants, was shut down in 1991 through the actions of the Union of Concerned Scientists, which charged that the reactor vessel had become so embrittled that any sudden temperature change could cause it to rupture, possibly leading to a meltdown.

Catastrophic events have occurred at nuclear plants worldwide (Table 19-4). For example, on april 26, 1986, one of the most devastating nuclear accidents ever rocked the Soviet Union. Operators at the Chernobyl nuclear plant turned off the plant's safety systems in order to conduct some unauthorized tests; while the safety systems were down, the plant's reactor number four overheated and exploded, spewing tons of cesium, iodine, uranium fuel, and other radioactive contaminants about three miles (five kilometers) into the atmosphere. The graphite moderator caught fire, further spreading contamination through smoke and flames. Thirty-one people died as a direct result of the accident, and 237 suffered severe radiation injuries.

TABLE 19–4: Selected Nuclear Facility Accidents

December 12, 1952 A partial meltdown occurred at Chalk River, Canada. This accident represents the first known malfunction of a nuclear power plant. No injuries were reported.

October 7, 1957 A plant similar in design to Chernobyl caught fire north of Liverpool, England. Two hundred square miles (518 square kilometers) of countryside were contaminated, and at least 33 cancer-related deaths were reported.

January 3, 1961 A steam explosion occurred at a military experimental reactor near Idaho Falls, Idaho. Three servicemen died. Their deaths were the first casualties in the history of U.S. nuclear reactor production.

March 28, 1979 A loss of coolant at the Three Mile Island plant near Harrisburg, Pennsylvania, caused the radioactive fuel to overheat. The buildup of heat led to a partial meltdown and the release of radioactive material into the atmosphere. The hazards posed to local inhabitants are still being debated.

March 8, 1981 Radioactive waste leaked from the storage tanks of a plant in Tsuruga, Japan. The leak was not disclosed until radiation was detected in the city's bay six weeks later.

January 6, 1986 One worker died and 100 were hospitalized when an improperly heated, overfilled container of nuclear material burst at a uranium-processing plant in Gore, Oklahoma.

April 26, 1986 The Chernobyl number four reactor exploded, spewing tons of radioactive materials three miles (five kilometers) into the atmosphere. Thirty-one workers died as a direct result of the explosion, and 237 received severe radiation burns. Surrounding communities were completely contaminated with radiation, which was also detected around the world.

Whole communities within an 18-mile radius had to be evacuated, with no hope of returning for several generations. In all, 135,000 people from 179 villages were evacuated, and topsoil and trees were stripped away and moved to an unspecified site. Within days, radiation was detected by countries around the world (Figure 19-8).

In 1975, the NRC published projected estimates of the worst damage a reactor accident could cause: 3,300 immediate human deaths; 45,000 human deaths from cancer; 45,000 radiation sickness victims requiring hospitalization; 240,000 people suffering from thyroid tumors; and 5,000 children born with genetic defects in the first generation following the accident. Accidents at Chernobyl, Three Mile Island, and other plants prove that such a disaster could occur; some experts say the worst case scenario has been avoided so far only by luck.

Disposal of Nuclear Wastes

Almost all waste produced by the nuclear industry is radioactive to some degree and therefore cannot be disposed of in the same manner as solid or even other hazardous wastes. Four classes of nuclear waste are produced — uranium mill tailings and low-level, transuranic, and high-level wastes — each of which poses a unique threat to the environment.

Uranium mill tailings are created when uranium ore is milled, or processed, into an enriched form for use in a reactor. Extracting just one pound of finished reactor fuel generates approximately 500 to 1,000 pounds (225 to 450 kilograms) of tailings. Because this amount is so large, the tailings frequently are piled in the open. There, wind and rain can disperse them in the environment. Over 200 million tons (180 million metric tons) have accumulated throughout the western United States alone. Tailings also have found their way

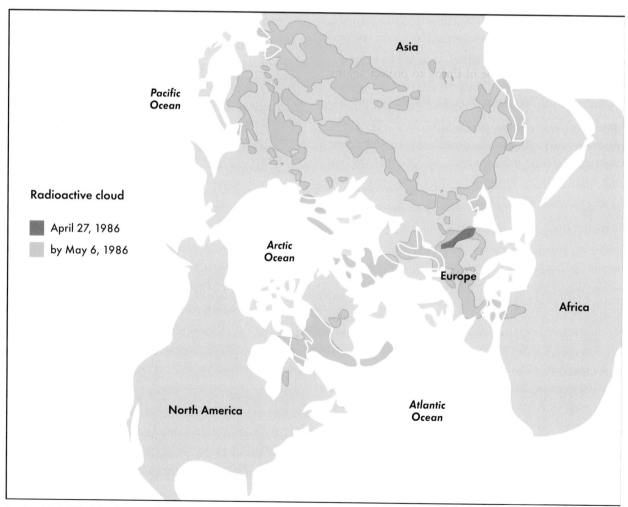

FIGURE 19-8: Worldwide spread of radiation from the accident at the Chernobyl power plant, 1986.
Source: *Lawrence Livermore National Laboratory.*

into such everyday objects as road pavement, fill material for construction sites, and sand for golf courses and children's sandboxes.

Low-level radioactive wastes are liquids and solids that will remain dangerous for a few hundred years or less. They include air filters, rags, protective clothing, and tools used in routine plant maintenance and fuel fabrication, as well as liquids used to cool reactors.

For several decades, much of the low-level waste produced in the United States was mixed with concrete, encased in steel drums, and dumped into the ocean. Although the U.S. Environmental Protection Agency (EPA) originally condoned this practice as safe, it reversed its opinion in the mid-1970s. Scientists found that many of the barrels had begun leaking, possibly eaten away by the corrosive salt water, and that radioactive leakage did not safely dilute and disperse; rather, it left long-lived hot spots of radioactivity. When the United States halted ocean dumping in 1983, over 90,000 barrels of radioactive materials already had been sunk in the Atlantic and Pacific oceans, some near commercial fishing zones.

In the past, U.S. nuclear facilities also disposed of solid low-level waste by burying it in long trenches, usually without any type of lining to protect against leaching. Materials often were packed in cardboard, wooden, and metal containers which were sometimes damaged in the dumping process, allowing the radioactive material to spill directly into the trench. As groundwater moved through the trenches, radioactive leachate would slowly migrate from the site.

The Low-Level Radioactive Waste Act was passed in 1980 to control the growing number of cases of groundwater contaminated by low-level waste. This act required each state to either establish a low-level waste site or gain access to another state's facility by 1992. Although researchers have investigated diverse disposal methods, including shallow-land burial, aboveground vaults, modular concrete canisters, earth mound bunkers, mined cavities and augured holes, they have yet to determine that any one method is most effective.

Transuranic radioactive waste contains human-made radioactive elements or any element with an atomic number higher than that of uranium. These elements, which include plutonium (half-life, 24,000 years), americium (half-life, 430 years) and neptunium (half-life, two million years), must be handled and stored with extreme care. Inhaling or absorbing even a tiny speck of plutonium can cause cancer or death. Unfortunately, the nearly 12 million cubic feet (339,000 cubic meters) of transuranic waste that have

been generated over the last 30 years often have not been properly isolated from the environment.

High-level radioactive wastes remain radioactive for tens of thousands of years or more. They consist primarily of spent, or used, reactor fuel and fission products created when spent fuel rods are reprocessed. The uranium 235 in each fuel rod lasts about three to four years, after which it can no longer sustain the fissioning process and must be removed and replaced. Most commercial plants store used rods on site. Studies predict that the United States will accumulate roughly 65,000 tons (59,000 metric tons) of spent fuel by 2010. By the end of 2010, it is estimated that 78 reactors will have no storage space left for spent fuel. Many military reactors reprocess the rods by separating the remaining uranium 235 and plutonium 239 (which is produced during the fission reaction and can be used as fuel for atomic weapons) from the other fission products.

Spent fuel rods and the fission products produced during reprocessing pose major challenges for permanent, safe disposal. Presently, there is no satisfactory procedure for isolating these highly radioactive wastes from the environment for the tens of thousands of years necessary to guarantee the safety of future generations. Much of this waste is stored in concrete and steel silos and tanks or in metal drums, all of which are exposed to various climatic conditions. The combination of weathering and internal corrosion has led to alarming amounts of leakage from these containers.

As the problems involved in storing radioactive wastes mount, scientists are searching for a permanent method of disposal. At this time, land-based repositories are the option of choice. Although past plans have been abandoned due to uncertainty about how well waste can be contained under specific geologic conditions, DOE scientists feel that they are capable of constructing safe, permanent waste disposal sites underground. The Waste Isolation Pilot Plant (WIPP) near Carlsbad, New Mexico, is a research and development facility designed to demonstrate the safe disposal of defense-related transuranic radioactive waste. Located 2,150 feet below the surface in a 2,000-foot-thick salt bed, its tunnels extend over 10 linear miles. In 1999, the first waste shipments were sent to WIPP. Some experts worry that water can enter the repository and carry highly radioactive plutonium 239 to the Pecos River, but others maintain that the site is safe.

One of the more difficult questions facing the planners of such a site — for reasons of public concern as well as geologic criteria — is where it should be located. Fear of radioactive waste leads to a "not in my backyard" (NIMBY) mentality. Most individuals see the need for safe, permanent storage facilities for radioactive waste, but no one wants a facility in his or her neighborhood. Others go even further to a NIABY position, "not in anybody's backyard."

In 1982, Congress attempted to address the problem of location by passing the Nuclear Waste Policy Act (NWPA). The NWPA directed the DOE to select two land-based repositories (one in the eastern half of the country and one in the western) by investigating and comparing potential sites. The first site was proposed to be operational by the mid-1990s. Once a site was established, machinery would lower canisters of waste packed in lined cells 2,000 to 4,000 feet below ground. Special vehicles would transport the cells through tunnels to a permanent storage spot. Once filled, the entire maze of holes, tunnels and shafts would be backfilled, sealed, and protected by law from drilling and mining.

From three potential western sites, the DOE chose an area at Yucca Mountain, Nevada, characterized by volcanic rock, a dry location, and a deep water table (Figure 19-9). However, critics point out that the department did not use the comprehensive selection process outlined by the NWPA, focusing instead on sites it had thought of all along. Despite the prospect of more jobs and a government incentive of $1 million per year, Nevada residents were outraged at being designated the country's nuclear waste dump with little choice in the matter. The indefinite postponement of a search for the eastern site only made matters worse.

Although the Nevada state government fought to prevent the creation of the repository, the site eventually will be installed as planned: In 2002, President George W. Bush signed a Congressional resolution calling for the construction of the Yucca Mountain facility. However, delays and complications have multiplied original cost estimates by 1,500 percent, and

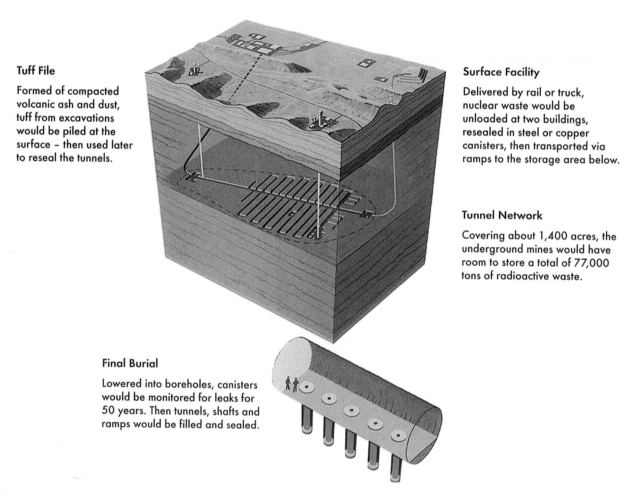

Tuff File

Formed of compacted volcanic ash and dust, tuff from excavations would be piled at the surface – then used later to reseal the tunnels.

Surface Facility

Delivered by rail or truck, nuclear waste would be unloaded at two buildings, resealed in steel or copper canisters, then transported via ramps to the storage area below.

Tunnel Network

Covering about 1,400 acres, the underground mines would have room to store a total of 77,000 tons of radioactive waste.

Final Burial

Lowered into boreholes, canisters would be monitored for leaks for 50 years. Then tunnels, shafts and ramps would be filled and sealed.

Figure 19-9: Land-based repository, Yucca Mountain, Nevada. The federal government proposes to place sealed canisters of radioactive waste in bore holes 1,000 feet below the mountain.

burial will not begin until 2010 at the earliest. The feasibility studies alone have already cost more than $1.7 billion. While land-based repositories may be a satisfactory solution to the radioactive waste problem from a technical viewpoint, the political ramifications may impede their widespread use in the future.

Decommissioning of Old Plants

Perhaps the most overlooked cost associated with nuclear power is the disposal of plants once they are too old to function. In 1989, the Shippingport plant, a commercial prototype capable of producing only seven percent of the power of an average reactor, was dismantled, and the reactor was buried at the Hanford facility in the State of Washington (Figure 19-10). This small operation cost $98 million.

What is the best way to deal with the host of much larger commercial reactors that will be retired in the next few decades? Plant owners can entomb the entire reactor building in concrete and bury it, although they cannot be certain that the concrete will be able to contain the contamination over the centuries. The preferred method is to decommission the plant by decontaminating and dismantling the reactor and then disposing of the resulting radioactive parts and wastes at an appropriate facility. Estimates for the cost of decommissioning an average-size plant range from

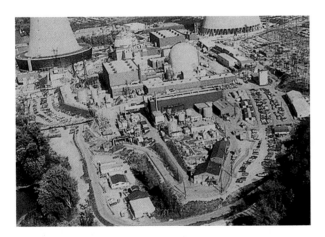

FIGURE 19-10: The Shippingport Nuclear Facility, before and after decommissioning.

$50 million to $3 billion — sometimes more than the cost of building a new plant. Experts predict that a dismantled plant will produce approximately 653,176 cubic feet (18,000 cubic meters) of low-level waste, enough to cover a football field 13 feet deep.

Secrecy Surrounding Nuclear Activities

Secrecy has played a prominent role in the development of nuclear resources. When the U.S. government began developing nuclear power in the early 1940s, it stressed the importance of withholding information for reasons of national security. Scientists working on the Manhattan Project were not permitted to share information with their peers in the scientific community. All discoveries and developments became the secret property of the federal government. In 1946, Congress even passed legislation establishing the death penalty as the punishment for anyone revealing atomic secrets.

The nature of the AEC has contributed significantly to government secrecy regarding nuclear issues. This government agency was established to promote, regulate, and license the nuclear industry and to control the flow of information to those outside the agency. Because its military operations were deemed vital to national security, the AEC also was granted the unprecedented power to virtually regulate itself. Self-regulation exempted the AEC from outside reviews of its spending and from external regulation of defense-related impacts on the environment. By 1974, the AEC had evolved into the NRC, responsible for licensing and regulating commercial nuclear facilities, and the Energy Research and Development Administration (ERDA), responsible for energy research and the construction of nuclear weapons. In 1977, ERDA became the DOE, which inherited the privilege of self-regulation originally granted to the AEC.

Critics point out that the government has used its privilege of secrecy to cover up incompetence and carelessness that puts the lives of countless citizens at risk. Indeed, over the last 45 years, more than 600,000 people have worked at U.S. weapons facilities without knowing whether or not they were being exposed to harmful emissions of radiation. When cover-ups do come to light, it is usually because of public pressure to release information pertaining to nuclear facilities.

The problem of secrecy is not confined to the United States. The Soviet government attempted to hide the 1986 Chernobyl disaster but was unsuccessful because of the extent of radioactive contamination. Thirty-six hours after the accident, technicians at a Swedish nuclear power plant noted disturbing signs of high radiation, the first indication to the rest of the world that a nuclear accident had taken place. The Soviet Union's reluctance to reveal information about the disaster prevented other countries from sending immediate aid and put the lives of people in nearby countries at risk without their knowledge.

Discussion Questions

1. What is nuclear energy and how is it produced?

2. What are some of the immediate and long-term health effects of exposure to radiation?

3. Explain the difference between direct and indirect exposure to radiation, using the accident at Chernobyl as an illustration.

(continued on following page)

Summary

Nuclear resources are derived from atoms, their energy, and the particles they emit. Nuclear energy is the energy released, or radiated, from an atom; this energy is called radiation. Radiation takes two basic forms: ionizing and nonionizing.

Nuclear energy is released through three types of reactions: spontaneous radioactivity, fission, and fusion. Spontaneous radioactivity occurs when unstable atoms release mass in the form of particles, energy in the form of waves, or both during radioactive decay. Isotopes that release particles, or high-level energy, are called radioisotopes. The most common types of ionizing radiation are alpha, beta, and gamma radiation. Alpha particles consist of two protons and two neutrons and carry a positive charge. Beta particles are negatively charged particles emitted from nuclei. Gamma radiation is a powerful electromagnetic wave. X-rays, a form of cosmic radiation, also can be produced by firing electrons at tungsten metal. Half-life is the length of time it takes for any radioactive substance to lose one-half of its radioactivity.

A fission reaction occurs when an atom is split into two or more new atoms. The atoms formed when uranium in a nuclear reactor is split are called fission products. Critical mass is the smallest amount of fuel necessary to sustain a chain reaction. A chain reaction is a self-perpetuating series of events that occurs when a neutron splits a heavy atom, releasing additional neutrons that cause other atoms to split. In a fusion reaction, nuclei are forced to combine, or fuse. For many people, fusion holds the promise of "clean" nuclear energy. However, fusion is not yet feasible for commercial electric generation.

As ionizing radiation penetrates living tissue, it can destroy cells or alter their genetic structure. The effects of radiation on living organisms depend on several factors, including dose, age, and gender.

Direct radiation exposure is exposure to the original radioactive source. Indirect radiation exposure is exposure to radioactive substances through sources such as food chains. Radioactive fallout is dirt and debris contaminated with radiation.

Nuclear resources are used in a variety of ways, including medical applications, electric generation, food preservation, and military applications. Early on, nuclear resources were developed primarily for military use. Later, nonmilitary uses of nuclear resources, especially the generation of electricity, took center stage. In each case, however, public opposition grew as the dangers inherent in the use of nuclear energy became apparent.

The use of nuclear resources has been accompanied by many environmental and social problems. Among the most serious are radioactive leaks, accidents at nuclear power plants, disposal of radioactive wastes, decommissioning of old plants, and secrecy surrounding nuclear activities.

(continued from previous page)

4. Describe three applications of nuclear technology other than electric generation.

5. How has the end of the Cold War changed the nature of the debate over nuclear weapons?

6. What are some of the social and environmental problems involved with the use of nuclear resources?

KEY TERMS

alpha particle	high-level radioactive waste	nuclear resources
beta particle		nuclear winter
chain reaction	indirect radiation exposure	radiation
containment vessel	ionizing radiation	radioactive decay
critical mass	ions	radioactive fallout
direct radiation exposure	isotope	radioisotope
fission products	low-level radioactive waste	spontaneous radioactivity
fission reaction	meltdown	transuranic radioactive waste
fusion reaction	nonionizing radiation	uranium mill tailings
gamma radiation	nuclear energy	x-ray
half-life		

CHAPTER 20

Unrealized Resources: Solid and Hazardous Wastes

Learning Objectives

When you finish reading this chapter, you will be able to:

1. Explain what is meant by "unrealized resources" and give some examples of them.

2. Differentiate among solid waste, municipal solid waste, and hazardous waste.

3. Describe how the generation and management of solid wastes vary worldwide.

4. Identify major waste disposal methods in the United States and describe each of them.

5. Define resource recovery and pollution prevention and explain why they are preferable to waste disposal.

6. Identify three major pieces of U.S. legislation that regulate hazardous wastes.

The biggest challenge we will face is to recognize that the conventional wisdom about garbage is often wrong.

William Rathje

In an era when a chemical spill can kill thousands, when wastes buried 20 years earlier can contaminate a community's water supplies, when our environmental ignorance can plant the seeds for later diseases, environmentalists and industrialists cannot continue to draw false battle lines.

Jay Hair

Lyndhurst, New Jersey, is home to a museum unlike any other. Paintings do not grace its walls, nor do models of dinosaurs prowl its galleries. Instead, the museum, located at the Hackensack Meadowlands Environment Center, displays common artifacts of contemporary American life — trash. Visitors view a first-rate display of old tins, empty bottles, rusting bicycles, crumpled chicken wire, squashed milk cartons, and long-emptied cereal boxes. As they do, they come face to face with one of our society's most visible gifts to future generations: our refuse. It is an unusual but appropriate museum for our modern culture of consumption. But is our society really so different from the many civilizations that have preceded it? At base, the answer is no. Humans have been throwing away their leftovers and unwanted artifacts for tens of thousands of years. What distinguishes contemporary society from previous cultures is the amount and type of our throwaways. Because of a growing population, the volume of wastes is much greater. Our technological, consumer-based culture provides more products in more elaborate packaging, both of which are made of more synthetic materials and much of which we use once and throw away. In addition, many of these products contain hazardous substances. In this chapter, we learn about "unrealized resources" — materials that are usually thrown away but could be used to benefit both human and natural systems.

Describing Unrealized Resources

Many solid and hazardous wastes are misplaced or **unrealized resources** — materials and substances termed "wastes" only because people do not have the foresight to recognize their value or the political power to stop their misuse. In this section, we define solid and hazardous wastes and examine their effects on humans and the environment.

What Is Solid Waste?

Waste comes in many shapes and forms. **Solid waste** (also called **refuse** or **trash**) refers to any variety of materials that are rejected or discarded as being spent, useless, worthless, or in excess. **Garbage**, while sometimes used as a synonym for solid waste, actually refers strictly to animal or vegetable wastes resulting from the handling, storage, preparation, or consumption of food.

The **solid waste stream** refers to the collective and continual production of all refuse; it is the sum of all solid wastes from all sources. An estimated five billion tons (4.55 billion metric tons) of solid waste are produced each year in the United States alone (Figure 20-1). The two largest sources of solid wastes are agriculture (producing animal manure, crop residues, and other agricultural by-products) and mining (producing waste rock, dirt, sand and slag, the material separated from metals during the smelting process). About 10 percent of the total solid waste stream is generated by industrial activities (producing scrap metal, slag, plastics, paper, and sludge from treatment plants).

Common solid waste, materials generated by households, businesses and institutions, is referred to as **municipal solid waste** (MSW). About three percent of the United States' solid waste stream is made up of MSW. Paper and paperboard account for the largest portion (52 percent) of MSW by weight. Yard wastes are the next most abundant material, accounting for just over 12 percent, while food wastes contribute nearly 12 percent. Plastics account for 11.3 percent; metals, 8 percent; glass, 5.3 percent; and wood, 5.8 percent. According to the U.S. Environmental Protection Agency (EPA), 236 million tons (214 million metric tons) of MSW were generated in the United States in 2003, equivalent to about 4.5 pounds (2 kilograms) per person, per day. Taking into account recycling and composting efforts, that figure drops to 164 million tons (149 million metric tons), equivalent to 3.1 pounds (1.4 kilograms) per person, per day.

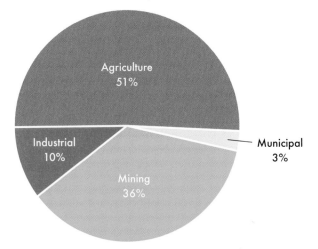

Sources of Solid Waste, United States

Agriculture 51%
Industrial 10%
Mining 36%
Municipal 3%

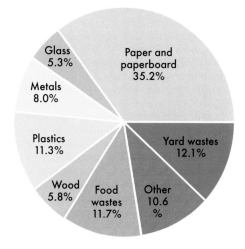

Sources of Municipal Solid Waste, United States

Glass 5.3%
Metals 8.0%
Plastics 11.3%
Wood 5.8%
Food wastes 11.7%
Other 10.6%
Yard wastes 12.1%
Paper and paperboard 35.2%

Figure 20-1: Sources of solid waste and composition of municipal solid waste. Agriculture and mining are the largest sources of solid waste, with industry and municipalities accounting for much smaller proportions. Municipal solid wastes are dominated by paper and cardboard.

What Are Hazardous Wastes?

Hazardous wastes are solid or liquid wastes that can adversely affect human health and the environment. This broad definition includes elements such as lead, compounds such as polychlorinated biphenyls (PCBs), and the products of infectious agents like bacteria and protozoa.

When hazardous wastes are discussed on television or in newspapers, the phrase "toxic and hazardous substances" often is used. Although the terms "toxic" and "hazardous" may seem to be interchangeable, they are not always synonymous. For our purposes, "toxic" implies the potential to cause injury to living organisms. Almost any substance can be toxic if the concentration is high enough or if an organism is exposed to the substance long enough. "Hazardous," however, implies that there is some chance that an organism will be exposed to a toxic substance and that the exposure will result in harm.

The term "hazardous waste" may conjure images of leaking, corroded barrels and abandoned, desolate buildings (Figure 20-2), but these images represent only one part of the range of hazardous substances

in the environment. Toxic and hazardous substances can be found virtually everywhere, and many occur naturally. Some are as old as the Earth itself; lead and radium 222 (contained in uranium rock), for instance, are part of the Earth's crust. Others are found in the planet's biota; many plants and animals, for example, manufacture chemicals that are poisonous in varying degrees to humans and other animals. Toxic and hazardous substances also may be synthetic, manufactured by humans through physical or chemical processes. These substances are used in industrial processes to manufacture products such as plastics, inks, and pharmaceuticals. Pesticides, often used to ensure cosmetically attractive fruits and vegetables, contain many potentially dangerous chemicals. Countless other common household products — including cleaners, shoe polish, motor oil, antifreeze, and insecticides — fill the American home with an amazing array of hazardous wastes.

What Materials Can Be Recovered from the Solid Waste Stream?

Every day in the United States, people routinely throw away potentially useful resources. Some of the most promising materials are aluminum, paper, cardboard, glass, plastics, iron and steel, tires, and used oil.

Aluminum

Aluminum, the most abundant metal on Earth, is never found in a free state; to be useful, it must be separated from aluminum ore. In 2001, 3.2 million tons (2.9 million metric tons) of aluminum entered the solid waste stream, representing 1.4 percent of total MSW generated. Nearly two million tons (1.8 million metric tons) was used for containers and packaging. The remainder was used for durable and nondurable goods, such as house siding, gutters, window and door frames, and lawn furniture. Reusing aluminum rather than mining for new metal results in tremendous energy savings and reduces water and air pollution. By far, the largest source of aluminum for reuse is cans; more than 66 percent of all aluminum cans are recycled.

Paper

One of the most frequently recycled materials, paper can be made into a variety of products, such as newsprint, paper bags, and insulation. To make paper, fibers

Figure 20-2: In the 1970s, Love Canal, an excavated but never-used canal near Niagara Falls, New York, embodied the threat of hazardous substances. From 1942-1953, Hooker Chemicals and Plastics Corporation dumped its industrial wastes into the canal. In the ensuing decades, the wastes contaminated the soil and water, and residents began to report unusually high incidences of health problems, ranging from burning eyes to migraine headaches to various forms of cancer. This photo shows an abandoned house in the Love Canal community shortly after residents were evacuated by the State in 1978. Two years later, President Jimmy Carter declared Love Canal a federal disaster area, leaving what was once an average working-class neighborhood nothing more than a ghost town.

from various sources (such as wood, corn stalks, and other organic matter) are bonded together to form a continuous sheet. However, making new paper products from recycled paper saves both trees and energy and is thus more environmentally and commercially cost-effective. Because 100-percent-recylced paper is not of sufficient quality for many purposes, recycled paper typically is combined with virgin paper to create products.

Of the paper recycled, most fits into three basic categories: newsprint, mixed office waste, and waste paper from paper-converting plants, which make products such as envelopes and business forms. Sixty-three percent of all newsprint in the United States is recovered for recycling. Recycling newspapers is important since this item constitutes the largest single component by volume of most landfills (15 to 20 percent). The automotive industry, a major market for recycled newsprint, uses recycled newspaper in the insides of automobiles; the average car contains about 60 pounds of recycled newspaper. There is also a growing trend among U.S. papermakers to use old newsprint to produce new newsprint.

Mixed office waste consists of computer paper, stationery, and other white paper. If sorted, mixed office waste can be recycled into high-grade paper. If unsorted, it can be made into roofing felt or filler ply in paperboard. In either case, the recycled product will contain a percentage of **post-consumer waste** — paper that was used before being recycled.

High-grade waste paper from paper-converting plants consists of machine trimmings and clippings from envelopes, forms, books, and catalogues. If sorted, the waste paper can go back into the same product line. If unsorted, it can be used in filler ply in paperboard. In either case, the recycled product will contain a percentage of **pre-consumer waste** — paper that was not used before being recycled.

Cardboard

Almost 70 percent of corrugated cardboard containers, which are generated in bulk from retail stores and factories, are recycled. Large handlers of corrugated cardboard, such as supermarkets, typically bale the cardboard and sell it directly to recycling mills. Factories and retail outlets, which handle smaller amounts of the material, often depend upon individuals who collect the cardboard and sell it to waste paper processors. Individuals in China and other less-developed nations earn their living by recycling cardboard (Figure 20-3).

Glass

In the United States in 2001, 21 percent of glass containers were recycled. Nearly 90 percent of recycled bottles are made into new bottles. Recycled glass is also used to manufacture fiberglass and to produce glasphalt (a material used in street paving), bricks, tiles, and reflective paint on road signs.

According to the Glass Packaging Institute, about 25 percent of any new glass container is made with recycled glass. The recycled glass (bottles and jars) is crushed into small pieces, which are then passed over a magnetic belt to remove any ferrous metals. The end product, known as cullet, is mixed with sand, soda ash, and limestone and melted to produce molten glass. Using cullet conserves the raw materials (sand, soda ash, and limestone) of glass and saves energy; since cullet melts at a lower temperature than the other materials, the furnace temperature can be reduced 10°C for every 10 percent of cullet used in the mixture. Using cullet also extends the life of the furnace and reduces particulate emissions to the atmosphere. Recycled glass can account for as much as 83 percent of molten glass without adversely affecting its quality. However, the quality of the final product depends on using cullet that is free of contaminants.

Plastics

In 2003, plastics made up 11.3 percent, by weight, of the municipal solid waste stream. This volume and the durability of plastics make them one of the most potentially valuable of all unrealized resources. While plastic recycling has risen rapidly since the late 1980s, less than 10 percent of all plastics are recycled. At present, only a few of the many types of plastics (each of which has a different chemical composition) are being recycled in significant quantities. These are polyethylene terephthalate (PET), high-density polyethylene (HDPE), and polystyrene.

The PET soft drink container is the most commonly recycled plastic container. About 35 percent of all PET containers are recycled into various products such as fiber filling for sleeping bags and outerwear, bathtubs, swimming pools, paint, automotive parts, and containers for nonfood products.

HDPE is the plastic used in milk jugs. Recycled HDPE is used to make toys, trash cans, flowerpots, piping, traffic cones, and other milk jugs. HDPE is also being used to create "plastic lumber" for railroad ties, decking for boat piers and docks, and fencing. Unlike

Figure 20-3: A child enjoys a ride alongside a cardboard recycler in China. Cardboard recyclers are a common sight in Chinese communities.

wood, plastic lumber does not rot, splinter, chip, or need painting.

Polystyrene is used in items such as fast food carryout containers ("clam shells"), cups, and plates. It can be cleaned, converted into pellets, and used to make building insulation, packaging material, and plastic lumber for walkways, benches, and other items.

Iron and Steel

Scrap iron and steel products, such as discarded automobiles and large appliances, can be melted and reformed into new products. Most "tin" cans — for such food items as soups, canned vegetables, and pork and beans — are actually steel cans. Until recently, just a small portion of steel (primarily "tin" and bimetal cans) was recycled, but this situation is changing. In 2003, 58 percent of steel containers, including food, paint and aerosol cans, were recycled. In addition, 14.2 million tons of scrap iron were recycled from automobiles, as well as 2.6 million tons from old appliances.

Tires

About 270 million tires are thrown away every year. Because many service stations now charge a fee to take them back, landfilling tires poses several problems. Besides taking up large amounts of space, intact tires hold pockets of air, which makes them tend to rise to the surface, releasing gases and shifting other waste around. In most places, old tires cannot be landfilled unless they are shredded, and so they may be disposed of illegally — one reason tires are often found dumped in ravines and creekbeds or left on beaches. Tires, however, are a potentially valuable resource. About 11 percent of discarded tires are recycled annually by various methods. Many can be retreaded, prolonging the life of the tire and conserving rubber. Retreading uses only about one-third of the energy needed to produce new tires, and a retreaded tire provides about three-quarters of the mileage of a new tire. Tires that cannot be retreaded can be used in a number of ways. Strung together, they act as effective controls against landslides and beach erosion, and they can be used as artificial reefs, providing habitat for marine life. Some power plants (primarily in Oregon, Washington, and

California) use old tires for fuel; combustion of the tires produces enough energy to generate electricity. Considering the number of discarded tires, the benefits of recycling can be significant. A word of caution, however: burning tires generates a significant amount of air pollution unless the plant is equipped with pollution control devices.

Used Oil

The EPA estimates that over 1.4 billion gallons (4.9 billion liters) of used oil are generated annually, a figure equivalent to about 84,000 barrels per day. Motor oil accounts for about 60 percent of used oils; the remaining 40 percent consists of industrial substances such as hydraulic, metal working, and cooling oils. About 67 percent of the used oil generated in the United States each year is recycled, usually as a fuel. The remaining 33 percent, some 429 million gallons (about two billion liters), is disposed of in landfills or is simply dumped on the ground or down storm sewers, where it poses a threat to local surface water and groundwater.

How Does Waste Production Vary Worldwide?

The United States generates more MSW, per capita, than any other more-developed country (MDC) — about 4.5 pounds (2 kilograms) per day. Canada and the Netherlands come in second and third, at 3.75 and 3.0 pounds per day, respectively. At less than two pounds per person per day, Germany and Sweden generate the least amount of MSW per capita.

The waste stream of less-developed countries (LDCs) is typically much smaller because items are reused and recycled far more often. People in Mexico, China, the Philippines, and other LDCs recycle out of economic necessity. It is not unusual to find communities built around open dumps in large urban areas in the developing world (Figure 20-4). The poor who live in these communities scavenge in dumps for food and any items that can be recycled and sold. Government officials are reluctant to close the dumps and landfill the refuse because it would be an expense few LDCs can afford.

FIGURE 20-4: The city garbage dump in Recife, Brazil, 1995. Each day, men, women, and children search for items that can be eaten or sold.

Most of the world's hazardous wastes are generated and disposed of by the chemical, primary metal, and petroleum-refining industries in MDCs, although LDCs use significant quantities of pesticides (many of them imported from the United States and Europe, where their use may be banned). According to the EPA, in 2001, Americans generated 1.6 million tons (1.4 million metric tons) of household hazardous waste per year; 56 percent of that waste was buried in landfills, another 29 percent was recycled; and the remaining 15 percent was burned (usually to generate energy). Every day, hazardous substances such as pesticides, used motor oil, mercury-containing batteries and light bulbs, and lead paint are thrown out with yesterday's newspaper and potato peels.

What Are the Health Effects of Hazardous Wastes?

The health effects of hazardous wastes are almost as numerous and diverse as the substances themselves. Much is known about the effects of acute exposure to high concentrations of some chemicals, but little is known about the effects of chronic exposure to low concentrations of most chemicals.

Acute toxicity is the occurrence of serious symptoms immediately after a single exposure to a substance. Methyl isocyanate, for example, is acutely toxic; a single exposure can maim or kill. **Chronic toxicity** is the delayed appearance of symptoms until a substance accumulates to a threshold level in the body after repeated exposures to the substance. This may occur weeks, months, or years after the initial exposure. Lead is chronically toxic; the longer the exposure, the greater the health risks (especially to children, who are particularly susceptible to lead poisoning). These risks include behavioral disorders, hearing problems, brain damage, and death.

There are four general ways in which hazardous substances affect long-term human health. **Carcinogenic** substances cause cancer in humans and animals. **Infectious** substances contain disease-causing organisms. **Teratogenic** substances affect the fetus; they may cause birth defects or spontaneous abortions or otherwise damage the fetus. **Mutagenic** substances cause genetic changes or mutations, which then appear in future generations. Most, but not all, mutagens turn out to be carcinogenic; so far, all known carcinogens have proven to be mutagens as well.

We have seen that hazardous substances can affect human health in serious ways. Moreover, the production, use, and disposal of these materials have important social consequences. Often, those at greatest risk from toxic contamination are the poor and people with little political clout: migrant laborers who work in fields sprayed with pesticides, factory workers who are exposed to low levels of toxic heavy metals over a prolonged period of time, and residents of inner cities who live in old apartments containing lead-based paint and asbestos insulation (Figure 20-5). Indeed, hazardous substances historically have precipitated numerous incidents of large-scale chemical contamination. The names of communities in such far-flung places as Minamata, Japan; Times Beach, Missouri; and Bhopal, India have become synonymous with chemical disasters. In the early 1950s, people in Minamata, Japan, mysteriously began dying or giving birth to severely deformed children. It was determined that the mercury sludge being dumped into the sea by a local factory was contaminating the fish and that people who ate these fish were suffering from what became known as Minamata Disease. The nightmare of severe contamination by hazardous substances perhaps is best illustrated by the case of Bhopal, India, site of the deadliest chemical accident ever (see *Focus On: Bhopal*).

Much concern over hazardous wastes has focused on how these compounds affect human health, but they also pose a substantial threat to other species and to the natural systems upon which all life depends. Spills of toxic substances, such as the release of crude oil into Prince William Sound, Alaska, by the tanker *Exxon Valdez*, can contaminate surface water and groundwater, killing fish and other aquatic life and reducing species diversity in the contaminated area. Toxins can also contaminate the soil and air, sometimes dramatically upsetting the dynamic balance within the ecosystem.

From an environmental standpoint — which takes into account the health of both humans and wildlife — the contaminants of greatest concern are those that are able to resist the natural processes of degradation. **Persistent organic pollutants** (POPs) are synthetic compounds that are highly toxic, very stable, and soluble in fats and oils; they are also easily dispersed. Because of these characteristics, POPs can have devastating, long-lasting, and far-spread environmental impacts even if released to the environment at extremely low levels. A short-term exposure to a high level of a POP can be fatal or result in serious injury or illness, including damage to the liver or nervous system. However, chronic exposure to lower doses can also be

Figure 20-5: Pesticide being sprayed on fruit trees, Po River delta, Italy. Workers exposed to pesticides are at increased risk for many diseases.

very harmful. A wide array of health problems — in both humans and animals — is believed to be linked to long-term exposure to low doses of POPs.

In recognition of the grave dangers posed by these substances, nearly 100 nations began negotiations in 1998 to craft a global treaty to eliminate 12 of the most widespread and insidious POPs (Table 20-1, page 446). The treaty, known as the Stockholm Convention, entered into force in May 2004. One of the highlights of the treaty is a move to embrace the "precautionary principle," which would prevent and eliminate POPs at their source and allow action to be taken before there is conclusive scientific proof of damaging effects. This provision shifts the burden of proof to those whose activities or products threaten harm to human health and the environment.

Managing Unrealized Resources

As we continue to produce an increasing amount of waste, we face a growing challenge to manage it in ways that protect human health and the environment. In this section, we discuss various methods of solid and hazardous waste disposal. We also examine legislation guiding the management of these unrealized resources.

How Do Living Systems Manage Waste Products?

The decomposition of organic materials into their constituent compounds, which makes them available for reuse by other organisms, is one of the most fundamental biological processes. Soil microbes such as bacteria and fungi secrete enzymes that break down organic materials; the decomposed materials provide the organisms with energy. **Aerobic decomposition** occurs in the presence of oxygen by oxygen-requiring organisms. Some organisms do not require oxygen; when these organisms are present, **anaerobic decomposition** occurs.

A material that can be decomposed by natural systems is said to be **biodegradable**. Organic compounds, those containing carbon, can be readily degraded by decomposers when sufficient oxygen is available. Unfortunately, much of our waste is synthetic and not biodegradable. Plastics are perhaps the best example of a durable, nonbiodegradable substance. Because of their

TABLE 20–1: Twelve POPs Under an International Ban

Pollutant	Category	Chief Use	Secondary Use
Hexachlorobenzene (HCB)	Pesticide	Fungicide for seed treatment of wheat, onions, sorghum	Industrial by-product
Mirex	Pesticide	Insecticide used to fight fire ants, harvester ants, leaf cutter ants, mealy bugs, and harvester termites	Fire retardant in plastics, rubber, and electrical goods
Chlordane	Pesticide	Broad-spectrum crop insecticide used on vegetables, small grains, maize, other oilseeds, potatoes, sugarcane, sugar beets, fruits, nuts, citrus, cotton, and jute	Home lawns and gardens
DDT	Pesticide	Insecticide used chiefly to control disease vectors such as mosquitoes	Crop insecticide (especially cotton)
Endrin	Pesticide	Insecticide used mainly on field crops like cotton and grains	Rodenticide to control mice and voles; also used to control birds
Toxaphene	Pesticide	Insecticide used primarily to control pests on crops such as cotton	Insecticide used to control ticks and mites on livestock and to kill fish in lakes
Heptachlor	Pesticide	Insecticide used mainly to control soil insects and termites	Insecticide used to combat cotton insects, other crop pests, and grasshoppers; also used to control malaria
Aldrin and dieldrin	Pesticides	Insecticides used for crops like corn, potatoes, and cotton	Insecticide used for termite control
Polychlorinated biphenyls (PCBs)	Industrial chemicals	Used in electrical transformers and large capacitors, as heat exchange fluids, as paint additives, in carbonless copy paper, and in plastics	
Hexachlorobenzene (HCB)	Industrial chemical	Used to make fireworks, ammunition, and synthetic rubber	By-product of the manufacture of industrial chemicals including carbon tetrachloride, perchlorethylene, trichloroethylene, and pentachlorobenzene
Dioxins	Unintentional by-product of the manufacture of other chemicals, like pesticides, polyvinyl chloride, and other chlorinated solvents		
Furans	Unintentional by-product; major contaminant from PCBs; by-product often bonded to dioxin; furans have the same biological effect as dioxins but are less potent		

Pesticides Note: Contact with pesticides occurs when humans or wildlife breathe contaminated air, eat contaminated food, or drink or wash in contaminated water. Exposure to the unborn occurs when the fetus absorbs chemicals that the mother has accumulated.

Industrial Chemicals Note: PCBs are known to cause adverse effects in wildlife and acutely exposed human populations. There are some documented estrogenic effects in wildlife; human fetal exposures have been associated with neural and development changes and long-term effects on intellectual function. HCB is toxic when inhaled, ingested, or when it contacts the skin; a known animal carcinogen and a "possible" human carcinogen, it is classified by the World Health Organization as an "extremely hazardous" product. HCB has been shown to be harmful to stomach, intestines, liver, and kidneys; can affect nervous system, and cause reproductive and developmental defects. It can also cross the mammalian placenta to affect the unborn.

Unintentional By-product Note: The confinement of hospital waste, municipal waste, hazardous waste and car emissions, as well as the combustion of coal, peat and wood, can result in emissions that contain dioxins and furans. Dioxins are created when chlorine is incinerated in the presence of certain precursors to dioxin. Polyvinyl chloride (PVC), vinylidene chloride (plastic wrap), chlorinated solvents, paint strippers, and pesticides are sources of chlorine in incinerators. Also responsible for the formation of dioxins are processes used by metal smelters, refineries, and cement kilns. Chlorinated dioxins are believed to interfere with fundamental biochemical messenger systems, resulting in reproductive disturbances, diminished intellectual capacity, and cross-generational toxic effects.

Source: Adapted from Issue Brief, January 1999, World Wildlife Fund.

chemical structure, plastics cannot be decomposed by bacteria and other organisms. Marketing claims that certain plastic bags and other items are biodegradable are misleading. Biodegradable substances, such as cornstarch, are integrated into the matrix of these items, and it is the cornstarch, not the plastic, that decomposes. As the cornstarch is digested by microbes, the bag breaks up into smaller and smaller pieces of plastic, but the total volume of plastic remains the same. In fact, such a "biodegradable" item typically contains more plastic than its "nonbiodegradable" counterparts; the cornstarch tends to lower a material's tensile strength, and the manufacturer must increase the total amount of plastic to make the item.

What Is Incineration?

The burning, or **incineration**, of wastes — purification by fire — dates back to biblical Jerusalem. Crude incinerators can still be found in the backyards of many American homes, in the form of rusty 55-gallon drums. Used to burn household paper and plastic wastes, these "burn barrels," emit a substantial amount of particulates, greenhouse gases, and, depending on what is being burned, toxic chemicals.

Modern large-scale incineration, which was first used to dispose of industrial hazardous wastes in the 1960s, is a complex and highly refined process. During incineration, organic materials are burned at high temperatures — typically ranging from 1800°F (980°C) to 2500°F (1400°C) — to break them down into their constituent elements, chiefly hydrogen, carbon, sulfur, nitrogen, and chlorine. These elements then combine with oxygen to form inorganic gases like water vapor, carbon dioxide, and nitrogen oxides. After combustion, the gases pass through a pollution control system to remove acidic gases and particulate matter prior to being released to the environment.

As space in landfills decreases and disposal costs rise, interest in incineration increases. In 2001, in the United States, there were 97 waste-to-energy incinerators with the capacity to burn up to 95,000 tons of MSW per day. **Waste-to-energy incinerators** burn garbage to produce heat and steam, which can then be used to generate electricity. These plants can be profitable; in the late 1990s, 111 waste-to-energy plants incinerated 15 percent of the United States' solid waste, providing 2,650 megawatts of electricity — enough to power 1.2 million homes and businesses.

One major advantage of incineration is that it permanently reduces or eliminates the hazardous character of the waste. In order to receive an operating permit, an incineration facility must demonstrate a 99.99 percent destruction and removal efficiency for each principal organic hazardous constituent in the feed material. This means that for every 10,000 molecules of an organic compound that enter the incinerator, one molecule is released to the air. Another advantage of incineration is that it substantially reduces the volume of the waste being disposed. Virtually all of the organic content of the waste is combusted, leaving only the inorganic fraction behind as incinerator ash.

Problems Associated with Incineration

Ideally, incinerators allow only carbon dioxide and water vapor to escape into the atmosphere. The combustion process, however, is very complex, involving thousands of physical and chemical reactions. Depending on the composition of the waste and the particular combustion conditions, an infinite number of combustion products may be released — some of them hazardous. If some of the waste does not reach a high enough temperature to be completely destroyed, it can form organic compounds known as dioxins and furans, which have been linked to severe acne, liver damage, digestive disorders, and cancer. If they escape the combustion chamber, these and other organic compounds are released to the atmosphere.

Another serious drawback of incineration is that it cannot destroy toxic heavy metals present in the waste. Most metals will remain in the incinerator ash; however, some may be released as particulates or vapors that have escaped the pollution control system. Heavy metals such as arsenic, barium, beryllium, cadmium, chromium, lead, mercury, nickel, and zinc are frequently found in hazardous waste and, if released in emissions, may have adverse effects on humans and the environment.

What Is Landfilling?

Until the mid-1970s, landfills were often little more than open dumps. There were no guidelines for their location, construction and operation, and there were no restrictions against dumping of hazardous substances. In 1976, the Resource Conservation and Recovery Act (RCRA) established classifications for various types of

Bhopal, India

The twentieth-century nightmare of severe contamination by toxic substances is best illustrated by the case of Bhopal, India, site of the deadliest chemical accident ever. On the night of December 2-3, 1984, water leaked into underground storage tank 610 at Union Carbide India's pesticide production facility in Bhopal, India. The tank contained 50,000 gallons of methyl isocyanate (MIC), which is used to make the pesticides Temik and Sevin. The MIC reacted violently with the water and quickly began to convert from a liquid to a gas. The tank's pressure raced from two pounds per square inch at 10:20 P.M. to 30 pounds per square inch at 12:15 A.M. By 1:00 A.M., the rapidly expanding MIC vapor ruptured the tank's safety valve and escaped into the night air. A temperature inversion trapped the dense cloud of MIC and other unidentified gases and aerosols near the ground, preventing it from dissipating.

The poisonous cloud drifted through the shantytowns just outside the plant's gates, instantly killing hundreds in their sleep. The path of the gas took it through homes, barns, a railway station and a hospital, spreading death and sickness across a 25-square-mile (65-square-kilometer) area of Bhopal. As people awoke gasping for breath, they flooded into the streets. Thousands attempted to flee the invisible, odorless vapor. In the chaos, some ran toward the plant. Others, blinded by the gas, stumbled into one another and were trampled by the crush. By morning, 2,500 people were dead. An estimated 200,000 were exposed to the gas (Figure 20-6).

We will never have an accurate count of the human toll. In 1989, the Indian government listed 3,828 dead. Ten years after the catastrophe, in 1994, the count was over 6,000. The government lists almost 23,000 victims as permanently injured with another 181,000 who suffered temporary injury or disability. Many of the dead were children and elderly people. Their lungs, wracked by the MIC, filled with fluid, causing a death analogous to drowning. Many survivors suffered permanent blindness, paralysis, or severely diminished lung capacity. Pregnant women miscarried or delivered premature, deformed babies.

The protracted and bitter negotiation between Union Carbide and India over compensation for the victims of the disaster lasted more than four years. Carbide claimed that the water was added to the tank by a disgruntled worker bent on sabotage. Unexpectedly, a settlement was announced on February 14, 1989, calling for Union Carbide to pay $470 million in damages (India had been seeking $600 million). Based on the Indian government's official count, this works out to $2,238 for each man, woman, and child killed or injured. The Indian Supreme Court unanimously upheld the settlement in 1991, but rejected the part of the settlement granting Union Carbide immunity from criminal prosecution. The criminal trial was still slowly proceeding in the state court of Bhopal as the world marked the tenth anniversary of the tragedy in December 1994. In November of that year, Union Carbide sold all of its holdings in India, and pledged to spend $20 million of the sale's proceeds to build and maintain a hospital for the victims.

FIGURE 20-6: A Bhopal survivor, 1987. An aging woman, her eyes damaged by the Bhopal gas disaster, protests against a proposed out-of-court settlement between the Indian government and Union Carbide.

landfills and developed operating guidelines for each. Today's technology has made considerable progress since the time of the open dump.

Municipal refuse and nonhazardous industrial wastes are disposed of in **sanitary landfills** that must meet EPA standards (Figure 20-7). In these landfills, the bottom and sides are lined with thick layers of clay or plastic so that precipitation which percolates through the landfill is prevented from entering underlying groundwater. Leachate moves through a network of drains throughout the landfill to recovery points where it is collected for treatment. The local groundwater is monitored for signs of contamination; monitoring continues for up to 30 years after the landfill is closed.

EPA regulations also mandate that operators control the production of methane gas within the landfill. As garbage decomposes, gases are emitted; methane accounts for about 55 percent of these gases; carbon dioxide, 45 percent; water vapor, 4 percent; and trace gases, the remainder. Because methane is explosive, its buildup is a potential danger. Moreover, unlike natural gas, methane does not lie in a reservoir; rather, it is emitted continuously from the surface of the landfill and migrates outward underground, where it may seep into basements of nearby homes and buildings, posing a safety threat.

Recovery systems can help to eliminate the potential danger caused by the production of methane. Collection pipes sunk vertically into the landfill tap the decomposition gas and bring it to the surface. There, it is dissipated to the air, burned, or collected through a vacuum system. The gas may be cleaned to extract the methane, which can then be sold or used to generate electricity.

Sanitary landfills are usually divided into a series of individual cells. The portion of the site being filled is called the working face. Each day, the collected garbage is spread out in thin layers, bulldozed to compact the volume, and covered with a layer of dirt about six inches deep or a layer of plastic or both to discourage animals from foraging. (The dirt or plastic cover also prevents oxygen from reaching the garbage, thereby inhibiting decomposition.) It is then capped with a layer of clay and earth, and native grasses are planted to hold the soil in place, minimizing erosion. When the entire site is filled, it is capped with a final layer of clay and earth and landscaped.

Hazardous wastes must be disposed of in **secure landfills**, which are specially engineered to prevent the

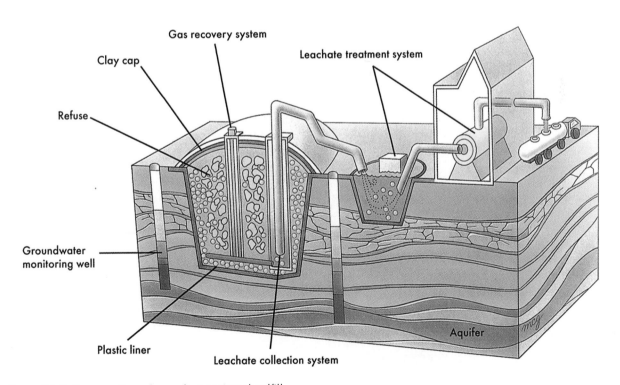

FIGURE 20-7: Cross-section of a modern sanitary landfill.

escape of leachate into the surrounding environment (Figure 20-8). Secure sites are chosen on the basis of their geology and relation to surface waters and aquifers. The best sites have a thick layer of dense clay or chalk between the landfill and any underlying aquifers. This layer resists the downward flow of leachate, which helps protect the groundwater. The bottom of the landfill is compacted and slightly graded toward a lower, central area. A thick waterproof plastic liner along the sides and bottom prevents leachate from flowing out of the landfill. A layer of sand or gravel covers the liner, along with a network of perforated pipes, which are part of the leachate collection system. Gravity draws leachate through the sloping pipes and down to a sump pump. Once the leachate is pumped from the landfill, it is treated before being disposed of. A leachate detection and collection system located beneath the plastic liner is separated from the clay bottom by a second plastic liner, which further minimizes the possibility of leachate escaping from the landfill.

Before being placed in a secure landfill, wastes are sorted into groups in order to prevent chemical reactions. The landfill is divided into compartments; each compartment accepts only one group of materials.

Landfill operators are required to maintain records of exactly what materials are in what part of the landfill. These records are important in the event that problems with the landfill develop in the future.

After the landfill is completely filled, it is covered with a thick clay cap and possibly a layer of waterproof plastic (much like the bottom liners). The entire landfill is then covered with earth and planted with vegetation to prevent erosion. The cap is sloped to cause rainwater to drain away from the landfill, thereby reducing the amount of leachate. Clay caps are subject to many stresses that may lead to their failure: uneven settling of the landfill contents, animal and human activity, and natural processes such as freezing and thawing and goelogic moevment. Consequently, the landfill operator must monitor the leachate collection system indefinitely for any increase in leachate production that would signal that the landfill cap had failed.

Problems Associated with Landfills

The EPA estimates that as many as 80 percent of the United States' landfills — which have a lifespan of about 10 years — may close by 2009. Most old landfills

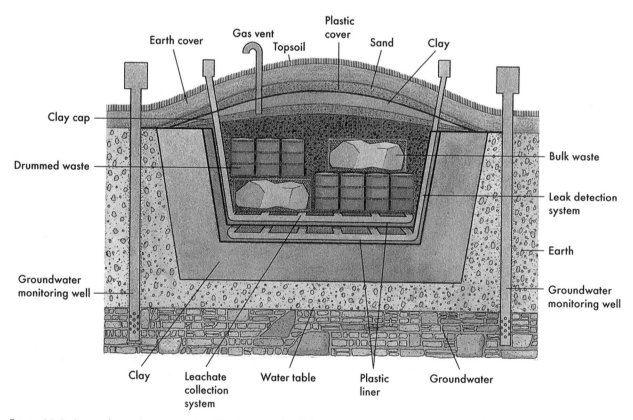

FIGURE 20-8: Secure hazardous waste landfill. A secure landfill is designed to contain hazardous wastes, thereby preventing hazardous leachate from contaminating groundwater, nearby surface waters, and soils.

are not replaced by new ones, usually due to a lack of appropriate sites.

A clay bed is considered the best substrate for a landfill, but clay sites are not evenly distributed throughout the country. High water tables or permeable underlying geologic strata (which allow water to percolate, leaching pollutants from the fill and carrying them into groundwater) prohibit the siting of new fills. In much of Florida, for example, the water table is too high to facilitate landfilling. Certain areas, particularly in the Northeast, are simply too congested.

As landfill space becomes increasingly scarce, the cost of landfilling rises. Between 1982 and 2002, landfill disposal fees rose by more than 75 percent. Although the average cost to dump wastes remained fairly consistent between 1993 and 2002 (approximately $34 per ton), start-up costs for new landfills increased and were often passed on to the consumer.

In those areas where landfill sites are not readily available, shipping or transporting wastes to out-of-state landfills significantly increases the cost of waste disposal. Even so, in many areas, this option is currently less expensive than incineration or recycling. Private haulers, who contract with communities in northeastern states, routinely ship wastes to states such as West Virginia, Ohio and Michigan where disposal costs, called **tipping fees**, are lower. Tipping fees at Ohio landfills, for example, are just one-fourth the cost of disposal on the East Coast. New York and New Jersey ship their wastes the farthest (New York's refuse is shipped as far away as Louisiana, Missouri, and New Mexico) and to the most locations (both send refuse to 11 other states). Mounting public concern over limited landfill space and health and environmental effects is causing many trash-importing states to reconsider the practice.

Once a landfill has been sited, operators still face numerous problems, several of which concern leachate contamination. The EPA found that even sanitary landfills can contain more than 200 chemicals, including such carcinogens as arsenic, methylene chloride, and carbon tetrachloride. Rainwater percolating through the fill can leach chemicals from the refuse and carry them with it as it continues to seep downward. If leachate reaches an underground aquifer, it may contaminate the water. Because about half of the U.S. population relies at least in part on groundwater for its drinking water, the problem is potentially serious.

Many older landfills, including hundreds that have already closed, may leave a toxic environmental legacy. Built during a time when refuse was less toxic or when protection standards were nonexistent, many of these fills are located in areas once considered "undesirable," such as wetlands or along riverfronts. Unfortunately, wetlands and river frontage are two areas where leachate contamination is most likely to occur, since the underlying geologic strata make it easy for water to filter downward and migrate to the nearest aquifer, stream, or river.

What Is Deep-Well Injection?

Used to dispose of hazardous wastes, **deep-well injection** involves pumping liquid wastes deep underground into permeable rock formations that geologists believe will contain the wastes permanently (Figure 20-9). A

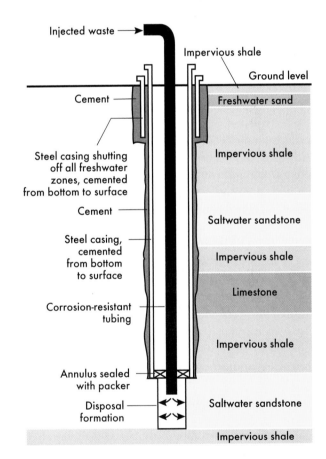

Figure 20-9: Deep-well injection. Liquid hazardous waste is injected in corrosion-resistant tubing within a steel casing. Cement around the casing further isolates the injection system from the environment. The waste is pumped under high pressure deep underground into permeable rock formations, such as sandstone, which permit the movement of liquids. As long as the sandstone disposal layer lies far below the lowest level of nearby aquifers and is separated from them by impervious layers of rock, such as shale, the wastes should be properly contained.

high-pressure pump forces the liquids into small spaces or pores in the underground rock, where they displace the liquids (water and oil) and gases originally present. Sedimentary rock formations like sandstone are used because they are porous and allow the movement of liquids. The disposal layer must be well below the lowest freshwater aquifer and must be separated from the aquifer by a thick layer of nonpermeable rock, such as shale, which is impervious to the liquid wastes.

Since the 1880s, the petroleum industry has used deep-well injection to dispose of salt water produced when drilling for oil. Disposing of hazardous liquids by deep-well injection, however, is a relatively recent development. The EPA estimates that about eight percent of all hazardous waste produced in the United States is injected deep into the ground. Most hazardous waste injection wells are located in the Great Lakes region and along the Gulf Coast.

Problems Associated with Deep-Well Injection

Theoretically, deep-well injection systems, when properly constructed, operated and monitored, may be the most environmentally sound disposal method for toxic and hazardous materials. If the substances are injected into a stable and receptive layer that is below the lowest drinking water aquifer, that aquifer is less likely to become contaminated than if the substances are placed in a landfill above it. However, injection wells can fail and pose direct threats to groundwater.

The history of hazardous waste injection wells illustrates the potential dangers of this disposal method. In 1968, an over-pressurized well in Erie, Pennsylvania, erupted. Roughly four million gallons (15 million liters) of injected hazardous material escaped over a period of three weeks. Probably the worst known case of environmental damage due to the failure of a hazardous-waste-injection well took place in Vickery, Ohio, in 1983. Inadequate design, operation, and maintenance procedures on the part of the operator, Ohio Liquid Disposal, led to the massive migration of about 20 million gallons (76 million liters) of hazardous substances from six injection wells. The Ohio EPA ordered the owner, Chemical Waste Management, to pay $12 million in federal and state fines and to spend an additional $10 million to upgrade the facility.

Many people feel that even with improved management of deep wells and stricter regulatory oversight, the lack of precise knowledge about the exact fate of hazardous substances after injection prevents this management option from being a final solution to the disposal of liquid hazardous wastes.

What Is Ocean Dumping?

Historically, coastal communities worldwide have dumped sewage and industrial wastes into the ocean. Because the ocean seems so vast, little thought was given to the effects of ocean dumping on the marine ecosystem. But, as we learned in Chapter 6, *Ecosystem Degradation*, both the volume and toxicity of wastes adversely affect marine life. In the United States, increased awareness of these effects resulted in the Ocean Dumping Ban Act of 1988, which prohibited all ocean dumping of municipal sewage sludge and industrial wastes after June 30, 1992. Unfortunately, the dumping of wastes from private cruise vessels, pleasure craft, and fishing boats continues, despite its ban under an international treaty.

Problems Associated with Ocean Dumping

Ocean dumping poses a serious threat to the marine environment. Because plastics do not disintegrate, they are particularly dangerous to seals, sea turtles, sea birds, and other marine animals. Six-pack yokes and fishing line can entangle an animal, causing it to drown or strangle. These plastic nooses can also cause deep wounds and fatal infections. Even apparently innocuous items, like plastic foam cups, balloons and sandwich bags, can prove lethal in the marine environment. Turtles may mistake plastic bags for jellyfish, a favored morsel, and seabirds may mistake spherical resin pellets, the raw material used to make plastic, for fish eggs. If plastics are ingested, they can become lodged in an animal's intestinal system, resulting in intestinal blockage and ulceration.

Many communities and states have enacted laws to combat the problems caused by ocean dumping, both intentional and unintentional. Baltimore, Maryland; Louisville, Kentucky; and the state of Florida prohibit the intentional release of balloons into the atmosphere. Several states have passed laws mandating that six-pack plastic yokes be manufactured from biodegradable plastic (a measure that does not prevent the ingestion of small particles of plastic), and in 1989, Maine and Massachusetts became the first states to ban the use of six-pack plastic yokes altogether.

Because the oceans are a global common, efforts to eliminate dumping must be international in scope. The

Convention on the Prevention of Marine Pollution by Dumping of Wastes and Other Matter, an international treaty established in 1972, prohibits the dumping of certain materials and regulates ocean disposal of others, fosters regional agreements, and attempts to develop a mechanism for assessing liability and resolving disputes. As of October 2001, 78 parties had entered into the Convention, including the United States, Canada, and Mexico. The International Convention for the Prevention of Pollution from Ships and its annexes, or amendments, regulate ocean dumping and the activities of sea-faring vessels. Annex I attempts to eliminate international pollution by oil and other harmful substances and to minimize their accidental discharge. Annex V prohibits the disposal of plastics at sea and restricts the ocean dumping of other types of refuse. By December 2001, the United States, Canada, Mexico, and 158 other countries had ratified this treaty.

How Are Hazardous Wastes Treated to Reduce Environmental Risks?

Most hazardous wastes can be treated to reduce the threat that they will contaminate the environment if released. Often, treatment of a hazardous waste will significantly reduce the volume and toxicity of the material that is ultimately disposed. Methods of treatment can be classified as chemical, physical, or biological.

Chemical treatment processes alter the chemical structure of the constituents to produce a waste residue that is less hazardous than the original waste (Table 20-2). The altered constituents may also be easier to remove from the waste stream. Treatment is often accomplished by the addition of other chemicals to the waste.

Physical treatment, which does not affect the chemical structure of the wastes, can take many forms. **Phase separation processes**, such as filtration, centrifugation, sedimentation and flotation, separate a waste into a liquid phase and a solid phase, both of which can then be more easily treated. **Component separation processes** remove certain hazardous substances from a waste stream without the use of chemicals. Some have been developed to remove inorganic components; others remove organics. A third type of physical treatment, solidification, is becoming increasingly popular because it reduces the mobility of the hazardous constituents, minimizing the chance that hazardous leachates will be formed. **Solidification** transforms a liquid, semi-solid, or solid hazardous waste into a more stable solid product. Solidification does not destroy the hazardous components in the waste nor does it make them available for recovery and recycling. The technique simply entombs the hazardous constituents in a concrete or asphalt block, thus isolating them from the environment for an extended period of time.

Biological treatment, or **bioremediation**, is fast becoming a preferred option for hazardous waste treatment. Bioremediation uses both naturally occurring and genetically engineered microorganisms to decompose organic chemical wastes into harmless by-products. For example, bacteria can be used to convert polychlorinated biphenyls into water, carbon dioxide, and simple salts. Petroleum refineries often use landfarms, a type of bioremediation, to successfully to treat their oily wastes. In this process, operators spread wastes on a parcel of land and till them into the soil. Aerobic microorganisms in the upper 12 inches of the soil decompose the organic chemicals in the waste. Nutrients like nitrogen and phosphates may be added to bolster microorganism populations.

Bioremediation appears promising because it can be used to treat both soil and water at lower costs, with less disruption to the site, and often without generating toxic by-products. Biological treatments often cost far less than disposal options, and in some cases, they can be so successful that no disposal is needed. Unlike incineration, however, which takes only minutes to destroy a hazardous material, biological treatment of a hazardous waste site takes months or years to complete.

What Is Resource Recovery?

Resource recovery is the taking of useful materials or energy out of the waste stream at any stage before ultimate disposal. Recycling and composting are two major types of resource recovery.

Recycling

Recycling is the collection, processing, and marketing of waste material for use in new products. One of the problems with the way we currently manage our wastes is that all materials are simply mixed together. Mixing wastes greatly reduces the value of recyclable materials. Paper, for instance, can be easily soiled if mixed with food or other organic wastes. Not surprisingly, then, efficient and effective collection and sorting are critical to a successful recycling effort.

TABLE 20–2: Chemical Treatment Methods

Electrolysis is the reaction of either oxidation (loss of electrons) or reduction (gain of electrons) taking place at the surface of electrodes. Electrolytic processes can be used for reclaiming heavy metals from contaminated aqueous wastes.

Neutralization is a reaction that adjusts the pH of either an acid (low pH) or a base (high pH) by the addition of the other to achieve a neutral substance (pH 7). This commonly used treatment method may not always make a substance less hazardous, but it can aid in the removal of heavy metals, prevent metal corrosion, or provide a more manageable material to be recycled.

Oxidation is used to detoxify wastes by causing the transfer of electrons from the waste chemical being oxidized to an oxidizing agent. Some common oxidizing agents and their uses are:

- **Ozone** is used in the oxidation of cyanide to carbon dioxide and nitrogen and the oxidation of phenols to less toxic compounds. Ozone is an unstable molecule, which makes it a highly reactive oxidizing agent. In the presence of ultraviolet light, ozone can oxidize halogenated organic compounds that are otherwise resistant to ozone oxidation.

- **Hydrogen peroxide,** also a powerful oxidizing agent, can be used to treat wastes that contain phenols, cyanides, sulfur compounds, and metals.

- **Potassium permanganate** can oxidize aldehydes, mercaptans, phenols, and unsaturated acids. It has been used to destroy organics in wastewater and drinking water.

Precipitation is a process that converts soluble metal ions into insoluble precipitates, which can be easily removed from the waste stream by filtration. It involves the addition of substances that change the chemical equilibrium of the waste stream and induce precipitation. Some chemical precipitators are:

- **Alkalines,** such as lime or caustic soda, which, when added to the waste stream, raise its pH, causing a decrease in the solubility of certain metal ions.

- **Sulfides** are used to precipitate heavy metals ions as insoluble metal sulfides. Hydrogen sulfide, sodium sulfide, and ferrous sulfide are most commonly used for this purpose. Sulfide precipitation is very pH sensitive.

- **Sulfates** of zinc and iron (II) can be used to remove cyanide from an aqueous waste stream.

- **Carbonate** precipitation of metal ions is common because carbonates can easily be filtered out of aqueous waste.

Although precipitation with chemicals is an effective and reliable treatment process, the resulting sludge contains high levels of metals and other toxic materials. After the recoverable metals are removed, the sludge requires ultimate disposal. Industries that use precipitation include inorganic chemicals, metals forming and finishing, pharmaceutical manufacturing, and textile mills.

Reduction is the chemical opposite of oxidation; electrons are transformed from a reducing agent to the waste being reduced. This process may reduce toxicity or encourage precipitation. There are many common reducing agents. A major application of chemical reduction uses sulfur dioxide, sodium bisulfites, or ferrous salts to reduce the extremely toxic chromium (VI) to the much less toxic chromium (III), which can then be removed by precipitation as hydroxide.

Source separation is the act of separating recyclable items from other wastes at the point of generation, prior to collection. Separating recyclables at their source keeps the items clean and thus preserves their value. Once they are separated, recyclables can be taken to a drop-off or buy-back center.

Unless sufficiently motivated, whether by economic or moral reasons, many people find drop-off and buy-back centers inconvenient and simply do not bother to make the effort to recycle. Consequently, many municipalities are looking to **curbside collection**: recyclable materials are set out at the curb and picked up weekly, often with the regular refuse. Some communities require residents to sort recyclables and place each type (paper, glass, aluminum, plastic) in a separate container. Others allow **co-mingled recycling**, the collection of different types of recyclables in a single container. Co-mingling is more convenient for the homeowner and can make collection simpler and faster, but it increases the cost of the recycling program, since workers must be hired to separate incoming wastes.

The chief advantage of curbside collection programs is that they usually increase the number of residents participating in a recycling effort. Curbside programs can be expensive in that they require major capital expenditures for collection vehicles and involve significant operating costs (for labor, maintenance, administrative, and transportation expenses), particularly if separate trips are made to collect regular refuse and recyclables. A high participation rate can offset costs, since revenue generated by the fees imposed for collecting the recyclables increases with each participant. Other revenue is generated by the sale of curbside materials. In some communities, theft has become a problem; recyclables left at the curb are easy pickings for thieves, resulting in lost revenues that could have been used to fund the program's operating expenses.

Recent years have witnessed the emergence of a number of economic factors that encourage recycling. As landfill capacity dwindles, fees have become prohibitive for many communities. Every ton of recycled material saves a municipality the cost of its disposal. According to the Worldwatch Institute, these savings may be as high as $100 per ton in parts of New York, New Jersey, and Pennsylvania. These powerful incentives are slowly bringing about an increase in the rate of recycling.

A number of states, including Pennsylvania and New Jersey, have enacted mandatory recycling laws. By requiring consumers to recycle goods and materials, the laws ensure a supply of — but not necessarily a demand for — recyclables. As long as demand is low, recyclable materials will simply pile up — collected from homes and businesses but not given new, useful life.

To increase market demand for recyclables, many state governments are taking innovative steps. Eleven states, for example, have enacted laws requiring newspaper publishers to increase the amount of recycled paper they purchase. Unfortunately, the U.S. federal government has been slow to enforce legal provisions that call for reducing waste at the source and encouraging recycling. The government has not established nationwide standards for products with recycled content have not been established, creating difficulty for suppliers of recycled and recyclable materials, whose markets often have varying requirements for what constitutes a recycled product.

Nations throughout the world are developing commercial and consumer programs to maintain a recycling market. Belgium has instituted an "eco-tax" on virgin paperboard used to package food; the tax is waived if the paper is made of recycled material. Japan has an extensive recycling program that extends to nearly all of the country's 3,255 municipalities. Neighborhood drop-off points receive four types of discards: glass, metal, and paper; hazardous materials and batteries; burnables such as kitchen waste, soiled paper, and some plastics; and non-recyclables such as construction debris and other plastics.

Municipal Solid Waste Composting

Refuse composting operations convert organic wastes into useful material. Some municipalities pick up leaves and yard waste and compost them in rows on fields. The material is turned periodically to encourage biodegradation. When the waste has composted, it is sold or given back to members of the community for use as a soil conditioner. A pilot effort in Guelph, Ontario, begun in 1989, combined composting and recycling efforts into a "wet-dry" program. Residents place all organic materials (food scraps, wet paper towels, leaves, and grass) in a container labeled "wet." Traditional recyclables (bottles, cans, newspapers, and some plastics) are put into a second container labeled "dry." The program achieved a combination recycling-composting rate of 60 percent, leading officials to make it permanent and to expand it to all city residents in 1995. Facilities have been built to handle both the recyclable and compostable materials.

In compostable refuse operations, waste arriving at a composting facility is sorted to remove recyclable materials. The remaining waste is reduced in volume by a procedure such as shredding and passed through a metals separator. A second screening step removes other noncompostable materials such as textiles and leather. The compostable wastes are moved to a chamber specially designed to accelerate the composting process; temperature, moisture, and oxygen content are closely monitored in order to maximize the rate of decomposition. After decomposition, a final screening removes any remaining traces of noncompostable materials. The compost is set out to air, or cure. This compost is usually given away or sold for only a nominal fee because it is difficult to guarantee a consistent product. Composting can also be done on a smaller scale (Figure 20-10). Homeowners can compost yard waste to produce a soil conditioner for gardens and lawns.

FIGURE 20-10: Backyard composting in Cincinnati, Ohio. Decomposition occurs more rapidly when the composted materials are turned periodically. The steam rising from this compost results from heat generated by decomposition.

In **co-composting** operations, yard waste or compostable refuse is mixed with sewage sludge. In one method, the refuse is mixed with sludge in digesters, vats about six feet (1.8 meters) deep and 100 feet (30 meters) in diameter. The digesters stir the mixture for five to seven days as the bacteria decompose the refuse and sludge. After the mixture has been sufficiently decomposed, it is moved from the vats and set in mounds to cure. The entire co-composting process can take anywhere from 30 to 180 days.

What Is Pollution Prevention?

Soaring disposal costs, shrinking disposal capacity, the increasing threat of financial liability for unsound disposal methods, and the rising cost of raw materials have led industries to turn to pollution prevention. **Pollution prevention** means reducing or eliminating the creation of pollutants through increased efficiency in the use of raw materials, energy, water, and other resources. The term includes equipment modifications, process modifications, product redesign, substitution of raw materials, and worker training. Pollution prevention views hazardous raw materials and wastes as resources that can and should be managed in environmentally sound ways. Prevention strategies can be grouped into two broad categories: **Source reduction** strategies reduce the amount of hazardous substances used and the volume of wastes produced; **waste minimization** strategies reduce the volume of hazardous wastes that must be disposed of.

Source reduction keeps machinery and pipe systems in good order and working properly to minimize the volume of hazardous material that must be used in a given process. In process distillation, systems capture and reuse solvent vapors that would otherwise have escaped to the atmosphere, thus increasing the usefulness of a given quantity of solvent and reducing a source of pollution. Redesigning manufacturing processes can also reduce the volume of hazardous resources used. Modifications include substituting water for organic solvents, altering rinsing procedures, and using better housekeeping measures in general. AT&T, for instance, now uses a cleaning solution based on citrus oils rather than chloroflourocarbons, which deplete the ozone layer when emitted into the atmosphere. Source reduction was declared to be the national policy of the United States in the Pollution Prevention Act of 1990.

Waste minimization strategies recover resources, either materials or energy, from the waste stream. Examples include the reclamation of titanium from the residues of utility company pollution control systems and the recovery of precious metals from plating operation liquids. While these are positive developments, it is important that proposed hazardous resource recovery operations be carefully scrutinized and regulated to prevent sham recycling operations.

One of the corporate leaders in pollution prevention is the 3M Company. The goal of the company's aggressive waste-minimization program, called Pollution Prevention Pays (3P), is to reduce the amount of pollution and wastes generated at its many operations around the world. Started in 1975, 3P included 5,265 different projects at its operations worldwide by 2003. Between 1975 and 2003, the 3P program prevented 1.1 million tons (0.99 million metric tons) of pollutants from entering the environment and realized savings of more than $950 million. The environmental goals for 3M for 2005 and beyond include improving energy efficiency by 20 percent, reducing waste by 25 percent, and reducing volatile air emissions by 25 percent. Ultimately, 3M wants to reduce waste and environmental releases to as close to zero as possible.

What Legislation Affects the Management of Hazardous Wastes?

Of the federal laws drafted to address the management of hazardous wastes, three are of primary importance. Both the Comprehensive Environmental Response, Compensation, and Liability Act (CERCLA), more commonly known as "Superfund," and the Resource Conservation and Recovery Act (RCRA) directly address issues concerning hazardous wastes. The Toxic Substances Control Act differs from Superfund and RCRA in that it regulates the production and use of toxic and hazardous substances.

Superfund

Superfund was created in response to several environmental crises in the late 1970s. Enacted in 1980, it empowered the federal government to clean up hazardous waste sites and to respond to uncontrolled releases of hazardous substances into the environment.

The Superfund act has several key objectives: to develop a comprehensive program for cleaning up the worst hazardous waste sites; to force responsible parties to pay for site cleanup; to create a multi-billion dollar Hazardous Substance Response Trust Fund to pay for the remediation of contaminated sites where no responsible parties can be held accountable and to respond to emergency situations involving hazardous substance releases from these sites; and to advance scientific and technological capabilities for the environmentally sound management, treatment, and disposal of hazardous resources.

The EPA manages Superfund, which is financed by taxes on the manufacture or import of crude oil and commercial chemical feedstocks. The agency identifies and prioritizes the most serious sites for long-term cleanup using a mathematical scoring system called the Hazardous Ranking System (HRS), which assesses the relative risks posed by various sites. The HRS score for any site is determined by taking into account three factors: the likelihood that a hazardous substance will be released from the site, the amount and toxicity of hazardous substances at the site, and the location of nearby populations that might be affected by contamination at the site. Those sites that receive a high ranking are placed on the agency's National Priorities List (NPL) and slated for a complex, often lengthy, cleanup. If needed, short-term cleanups can be scheduled for sites that do not receive a high ranking — known as non-NPL sites — in order to address any health and environmental risks.

If the EPA decides that a hazardous waste site poses significant potential harm to human or environmental health, the agency can use Superfund money to respond to the crisis (Figure 20-11). Whenever possible, those responsible for the release of hazardous wastes are forced to reimburse the fund for such expenses. (To read more about Superfund remediations, refer to *Environmental Science in Action: Woburn Industri-Plex Superfund Site* by going to www.EnvironmentalEducationOhio.org and clicking on the "Biosphere Project.") By 2004, there were 1,244 sites on the NPL; the EPA estimates that it will cost $31 billion to clean them up. An additional 39,783 non-NPL sites have been assessed; of these, 9,245 remain active with some short-term cleanup taking place. No further remedial action is planned for the remaining 30,438 non-NPL sites.

Superfund was envisioned as a one-time, short-term effort at cleaning up the country's most polluted hazardous waste disposal sites. However, it has been criticized by many who claim the program is a bureaucratic nightmare, an ineffective and wasteful use of taxpayers' monies in which billions of dollars are spent on non-cleanup activities (lawyers' fees, administration, etc.).

Indeed, nearly 20 years after Superfund's inception, only a few dozen sites have been declared clean and stable enough to be removed from the NPL. Moreover, some critics question whether even these sites have been adequately remediated for the long term.

Resource Conservation and Recovery Act

Whereas Superfund cleans up sites polluted by the unwise disposal practices of the past, the Resource Conservation and Recovery Act (RCRA), enacted in 1976, establishes a legal definition for hazardous wastes and sets guidelines for managing, storing, and disposing of these materials in an environmentally sound manner. Congress granted the EPA the authority to determine which wastes are to be regulated as legally hazardous and which are not. RCRA is intended to prevent the creation of new Superfund sites by severely limiting the land disposal of hazardous wastes and by regulating the generation of these materials.

FIGURE 20-11: Workers in protective gear help to clean up a Superfund site.

Regulating Hazardous Wastes. According to RCRA, if a waste has been found to be fatal to humans at low doses or has been shown through animal studies to be dangerous to humans, the EPA can place it on its official list of hazardous wastes.

RCRA also specified four hazardous waste characteristics: ignitability, corrosivity, reactivity, and toxicity. If a waste exhibits one or more of these four characteristics, it is legally hazardous. Ignitable wastes are easily combustible or flammable during routine handling, such as gasoline, paint, degreasers, and solvents. Corrosive wastes dissolve steel or burn skin and can corrode containers and escape into the environment, and are found in rust remover, cleaning solutions, and battery acid. Reactive wastes are unstable under normal conditions and undergo rapid or violent chemical reactions with water or other materials. Examples are cyanide plating wastes, bleaches, and other oxidizers. Toxic wastes are identified by a standard test, the Toxicity Characteristic Leaching Procedure (TCLP), designed to simulate the release of toxic contaminants by leaching. If the waste contains any of 39 heavy metals, pesticides, or other organic compounds above their maximum concentration levels, the waste is legally toxic.

It is relatively easy to determine if industrial wastes exhibit any of the four characteristics; however, the characteristics are not adequate to identify wastes that are hazardous due to biological properties — carcinogenicity, mutagenicity, teratogenicity, or infectivity — which are much more difficult to detect. To be considered hazardous and therefore subject to regulation, substances with these properties must be specifically identified as hazardous by the EPA. Testing a substance for such biological characteristics is time consuming and expensive. Consequently, the disposal of many potentially dangerous substances is not regulated because the EPA has not yet performed sufficient testing to determine whether to add them to its hazardous waste lists.

Cradle-to-Grave Management of Hazardous Wastes. Prior to RCRA, there was no regulatory system that established accountability for hazardous waste management practices. Consequently, those who disposed of hazardous waste often used unsound methods and kept no records of what the wastes were, where they came from, and where they were disposed of. RCRA established a system of "cradle-to-grave" regulation of hazardous wastes. From the time a hazardous waste is generated, a record must be kept of what the waste is, who generated it, who transported it, who treated it, and who disposed it. This record, called a manifest, provides the basis for assuring that wastes are handled, treated, and disposed in compliance with RCRA regulations. Facilities that treat, store, or dispose hazardous waste must obtain an operating permit from the EPA or the state.

Toxic Substances Control Act

The Toxic Substances Control Act (TSCA) concerns toxic commercial products rather than hazardous wastes. TSCA regulates the production, distribution, and use of chemical substances that may present an unreasonable risk of injury to human health or the environment. The EPA determines whether new chemical products present unreasonable risks by weighing their expected environmental, economic, and social benefits.

Manufacturers of chemicals are required to notify the EPA before producing new chemical substances or processing chemicals for significant new uses. This allows the EPA to review and evaluate information about the potential health and environmental effects of new products before they can cause environmental damage. TSCA gives the EPA has the authority to take regulatory action to control new, potentially risky chemicals.

Summary

Waste comes in many shapes and forms. A solid waste is any material that is rejected or discarded as being spent, useless, worthless, or in excess. Some solid wastes are hazardous and must be treated differently than nontoxic refuse. Garbage is animal or vegetable waste resulting from the handling, storage, preparation, or consumption of food. The solid waste stream is the collective and continual production of all refuse. The two largest sources of solid wastes are agriculture and mining. Municipal solid waste is refuse generated by households, businesses, and institutions. Solid and hazardous wastes are misplaced or unrealized resources — materials and substances termed "wastes" only because people do not have the foresight to recognize their value or the political power to stop their misuse.

Hazardous wastes are chemicals that can adversely affect human health and the environment. Although the terms "toxic" and "hazardous" often are used interchangeably, they are not always synonymous. "Toxic" implies the potential to cause injury to living organisms. Almost any substance can be toxic if the concentration is high enough or if an organism is exposed to the substance long enough. "Hazardous" implies that there is some chance that an organism will be exposed to a toxic substance and that the exposure will result in harm.

Many toxic and hazardous substances occur naturally in the Earth's crust and biota; others are manufactured by industrial processes. Hazardous wastes are released into the environment as by-products, as end products, or through the use and disposal of manufactured products.

Materials such as aluminum, paper, cardboard, glass, plastics, iron and steel, tires, and used oil can be recovered from the solid waste stream for reuse. The recovery rate of these materials, however, is typically much lower than what could be achieved to maximize the use of unrealized resources.

Both the consumption of goods and the management of solid and hazardous wastes vary from country to country. Among industrialized nations, the United States and Canada generate about twice as much refuse per person as do Germany, and Sweden. The waste stream of less-developed countries is typically much smaller than that of more-developed countries.

Acute toxicity occurs when a substance triggers a reaction after a single exposure. Chronic toxicity is the delayed appearance of symptoms until a substance accumulates in the body to threshold levels after repeated exposures. Carcinogenic substances cause cancer in humans and animals. Infectious substances contain disease-causing organisms. Teratogenic substances affect the fetus. Mutagenic substances cause genetic changes or mutations, which then appear in later generations.

Taking into account the health of both humans and wildlife, the contaminants of greatest concern are those that are able to resist the natural processes of degradation. Persistent organic pollutants (POPs) are synthetic compounds that are highly toxic, very stable, and soluble in fats and oils; they also disperse easily. Because of these characteristics, POPs can have devastating, long-lasting, and far-spread environmental impacts even if released to the environment at extremely low levels.

Aerobic decomposition occurs in the presence of oxygen by oxygen-requiring organisms. Anaerobic decomposition occurs in the presence of

Discussion Questions

1. What kinds of unrealized resources do you use and discard in your home and your school or office? Which of these resources could be reused or re-cycled? Identify resource recovery centers in your area and find out what kinds of materials they accept.

2. Compare the waste generation patterns of the United States and Canada with those of other nations. What factors might be responsible for the differences?

3. In what ways is hazardous waste a national social problem? In what ways is it an international social problem?

4. Choose one of the 12 persistent organic pollutants listed in Table 20-1 and research its uses and health and environmental effects. Is it banned in the United States or elsewhere, and if so, why? After completing your research, determine whether or not you think the global community should outlaw this substance. Defend your answer.

5. Most waste-to-energy incinerators in the United States are located in the Northeast, which has three times as many such plants as the West. What factors do you think account for the Northeast's greater reliance on incineration?

(continued on next page)

(continued from previous page)

6. How does a sanitary landfill differ from a secure landfill? Can leachate leak from a secure landfill? If so, how?

7. What is bioremediation? How is it different from other methods of treating hazardous waste?

8. Discuss various methods of resource recovery and pollution prevention. What are the advantages and disadvantages of each?

9. Imagine that a contaminated hazardous waste site has been identified in your community. Over the years, many companies have disposed of their waste there or have paid someone else to dispose it for them. What law(s) would apply to the situation, and what types of organizations would be involved in the cleanup?

organisms that do not require oxygen. A biodegradable material can be decomposed by natural systems. Organic compounds, those containing carbon, can be readily degraded by decomposers when sufficient oxygen is available. Unfortunately, much of our waste is synthetic and not biodegradable.

Incineration is a primary method for hazardous waste disposal. Incinerators greatly reduce the volume of waste; they also permanently reduce or eliminate the hazardous character of the waste. Waste-to-energy incinerators burn waste to produce electricity.

Sanitary landfills are designed to receive nonhazardous wastes. Secure landfills hold hazardous wastes. One key problem is that old landfills often are not replaced by new ones, usually due to a lack of appropriate sites. In those areas where landfill sites are not readily available, wastes are transported to out-of-state landfills. Disposal costs are called tipping fees. Once a new landfill has been sited, operators still face numerous problems, including leachate contamination.

Deep-well injection involves pumping liquid wastes deep underground into rock formations. Theoretically, deep-well injection systems, when properly constructed, operated and monitored, may be the most environmentally sound disposal method for hazardous materials. However, injection wells can fail and pose direct threats to groundwater.

Historically, coastal communities have dumped sewage and industrial wastes into the ocean. However, ocean dumping poses a serious threat to the marine environment. Many communities, states, and nations have enacted laws to combat this threat.

Prior to disposal, most hazardous wastes can be treated to reduce the threat that they will contaminate the environment. Often, treatment of a hazardous waste will significantly reduce the volume and toxicity of the material that is ultimately disposed. Methods of treatment can be classified as chemical, physical, or biological. Chemical treatment processes alter the chemical structure of the constituents to produce a residue that is less hazardous than the original waste. Physical treatment, which does not affect the chemical makeup of the wastes, can take many forms, including phase separation, component separation, and solidification. Biological treatment, or bioremediation, uses microorganisms to decompose organic chemical wastes into harmless by-products.

Resource recovery is the taking of useful materials or energy out of the waste stream at any stage before ultimate disposal. Recycling and composting are two major types of resource recovery.

Pollution prevention means reducing or eliminating the creation of pollutants through increased efficiency in the use of raw materials, energy, water, and other resources. Pollution prevention views hazardous raw materials and wastes as resources that can and should be managed in environmentally sound ways. Pollution prevention strategies can be grouped into two broad categories. Source reduction strategies reduce the amount of hazardous substances used and the volume of wastes produced; waste minimization strategies reduce the volume of hazardous wastes that must be disposed of.

In the United States, toxic or hazardous substances are managed according to the provisions of three major laws. The Comprehensive Environmental Response, Compensation, and Liability Act of 1980 (CERCLA), or Superfund, authorizes the federal government to respond directly to uncontrolled releases or threatened releases of hazardous substances into the environment. The Resource Conservation and Recovery Act of 1976 (RCRA) defines hazardous wastes; sets guidelines for managing, storing, and disposing of hazardous wastes; and establishes a cradle-to-grave regulation system, in which manifests must be kept to track wastes from generation to disposal. The Toxic Substances Control Act (TSCA) regulates the production, distribution, and use of toxic commercial products that may present an unreasonable risk of injury to human health or the environment.

KEY TERMS

acute toxicity

aerobic decomposition

anaerobic decomposition

biodegradable

bioremediation

carcinogenic

chronic toxicity

co-composting

co-mingled recycling

component separation processes

curbside collection

deep-well injection

garbage

hazardous wastes

incineration

infectious

municipal solid waste

mutagenic

persistent organic pollutant

phase separation processes

pollution prevention

post-consumer waste

pre-consumer waste

recycling

refuse

resource recovery

sanitary landfill

secure landfill

solid waste

solid waste stream

solidification

source reduction

source separation

teratogenic

tipping fee

trash

unrealized resources

waste minimization

waste-to-energy incinerator

Glossary

Abiotic component (abiota)   Non-living part of the physical environment.

Abortion   An elective surgical procedure in which the lining of the uterus is removed along with the developing embryo.

Abundant metal   Any metal that constitutes greater than 0.1 percent, by weight, of the Earth's continental crust. Examples include iron, aluminum, manganese, titanium, and magnesium. Compare *Scarce metal*.

Abyssal zone   The bottom waters of the ocean beyond the continental shelf, usually below a depth of 1,000 meters.

Acculturation   The process by which one culture is modified through contact with another.

Acid drainage   The seepage of sulfuric acid solutions from mines and mining wastes dumped at the surface.

Acid precipitation   Rain, snow, or fog that contains higher than normal levels of sulfuric or nitric acid, which may damage forests, aquatic ecosystems, and cultural landmarks.

Acid rain   See *Acid precipitation*.

Acid surge   A period of short, intense acid deposition in lakes and streams as a result of the release (by rainfall or spring snowmelts) of acids stored in soil or snow.

Active solar system   System for temperature control that utilize devices such as fans and pumps to enhance the distribution of collected and stored solar heat. See *Passive solar system*.

Actual rate of increase   The true growth rate of a country, the actual rate is based upon the natural rate of increase but includes the effects of migration. See *Natural rate of population increase*.

Acute pollution effects   These effects occur immediately upon or shortly after the introduction of a pollutant, and they are readily detected.

Acute toxicity   The ability of a substance to trigger an adverse reaction immediately after an organism is exposed to it.

Adaptation, genetic   A genetically determined characteristic that improves an organism's ability to survive and transmit its genes to the next generation.

Adaptation, physiological   A change in an organism's physiological response to a stimulus or substance that occurs after prolonged exposure to that stimulus or substance.

Adsorption   A tertiary treatment whereby organic compounds bind, or stick, to carbon particles but do not chemically combine with them.

Advanced materials   In the area of substitutes, materials that exhibit greater strength, greater hardness, or better thermal, electrical, or chemical properties than a traditional material.

Aerobic   Requiring oxygen.

Aerobic decomposition   The degradation of organic material by living organisms in the presence of oxygen.

Aerobic respiration   A process requiring oxygen to release energy.

Age distribution   Percentage of the population or number of people of each sex at each age level in a population.

Age-specific fertility rate   The num-ber of live births per 1,000 women of a specific age group per year.

Agrichemicals   Fertilizers, herbicides and pesticides, many of which are derived from petroleum, that are applied to plant crops to increase yield or reduce pest damage.

Agriculture   The intentional tending of a particular plant species for human use.

Agroecology   The study of agroeco-systems and the long-term management of agricultural land. Its aim is to match agricultural sustain-ability with ecological principles.

Agroecosystem   A cropfield ecosystem.

Airborne toxins   Hazardous chemical pollutants that have been released into the atmosphere and are carried by air currents.

Air pollutant   Any substance present in or released to the atmosphere that adversely affects human health or the environment.

Air pollution alert   Level of air pollution dangerous to humans defined as an air quality index reading of 100 or greater.

Air quality index   Index showing the frequency and degree of air pollution in a particular region, usually determined by the pollutant with the highest measured level.

Albedo   Reflectivity, or the fraction of incident light that is reflected by a surface.

Algae   A group of relatively simple aquatic organisms (one-celled, multi-celled, or colonial) containing chlorophyll that occur in fresh and salt water, and also in damp terrestrial areas.

Algal bloom   An extensive growth of algae in a body of water, sometimes as a result of an increase in the phosphate and nitrate content of the water.

Alkaline   Having the properties of or containing a soluble base. See *pH*.

Alpha particle   Slow-moving particle consisting of two protons and two neutrons which cannot penetrate paper or skin but is dangerous when inhaled or ingested.

Alternative energy resource   Any energy resource other than fossil fuels that is available but not widely used.

Amino acids   Small units that combine in various ways to form larger protein molecules; these units are essential for growth and development.

Ammonia   A colorless gas composed of nitrogen and hydrogen (NH_3); the main form in which nitrogen is available to living cells.

Anaerobic   Not requiring oxygen.

Anaerobic decomposition   The degradation of organic material by living organisms in the absence of oxygen.

Anaerobic respiration   A process that does not require oxygen to release energy.

Anemia   A condition in which the number of red blood cells, amount of hemoglobin, or total volume of blood is deficient that may be caused by various diseases, poisoning, or malnutrition; causes weakness and fatigue.

Anthracite   It is a hard coal and has the highest carbon content and the lowest moisture content of all types of coal. It is a rare coal with low sulfur content and high heating value.

Anthropocentric worldview   The belief that humans are not part of nature and that nature is that part of the world devoid of human influence. Humans are viewed as having dominion over all living things.

Anthropocentrism   See *Anthropocentric worldview.*

Anthropogenic   An event or consequence of human action.

Antinatalist policy   A policy that is designed to prevent increases in fertility or to lower existing fertility and birth rates. Compare *Pronatalist policy.*

Applied ecology   A scientific discipline that measures and attempts to predict the ecological consequences of human activities and recommend ways to limit damage to, and to restore, ecosystems.

Applied ethnography   A subdiscipline of anthropology concerned with recording an accurate account of a given people's culture.

Aquaculture   The production of aquatic plants or animals in a controlled environment.

Aquatic   Pertaining to marine and freshwater environments. See *Terrestrial.*

Aquifer   A water-bearing geologic formation composed of layers of sedimentary material.

Arable land   Region that is able to sustain agriculture on an annual basis.

Area strip mining   A method of mineral extraction in which the entire deposit is removed from the surface down. This technique is used in flat or rolling terrain.

Asbestos   A silicate of calcium and magnesium, asbestos occurs in long, threadlike fibers. Because it does not burn and does not conduct electricity. it has been widely used in insulation, vinyl floor tiles, cement and other building products.

Asbestos has been found to cause lung cancer and asbestosis.

Associated gas   Natural gas that is found with petroleum and can be extracted only when the petroleum is brought to the Earth's surface and refined. See *Nonassociated gas.*

Atom   The smallest unit of an element that retains the unique characteristics of that element composed of subatomic parts called protons, neutrons, and electrons.

Atmosphere   A 500-kilometer thick layer of colorless, odorless gases known as air that surrounds the Earth and is composed of nitrogen, oxygen, argon, carbon dioxide, and other gases in trace amounts.

Autotrophs   See *Producers.*

Bacteria   Unicellular, or rarely multicellular, prokaryote organisms. Bacteria have various shapes, occurring as cocci (spherical)), bacilli (rod-shaped), and spirilla (helical). They are present in soil, water, and air and as free living symbionts, parasites, and pathogens.

Barrier contraception   A type of birth control that relies on a physical device, such as a condom or diaphragm, to prevent sperm from reaching the egg.

Below-cost timber sales   The sale of timber to lumber companies at prices lower than the actual cost incurred by the Forest Service to make that timber available.

Benthic community   Association of organisms, primarily invertebrate animals, which lives at the bottom of a body of water.

Beta particle   Fast-moving particle, similar to an electron but containing more energy, which can penetrate the outer layers of skin and cause damage to the body.

Big bang theory   The theory that the universe — all matter, energy, and space — arose from an infinitely dense, infinitely hot point called a singularity which exploded and began to expand 13 to 20 billion years ago.

Binding force   The attractive force that holds together the constituents of an atom, atomic nucleus, or molecule.

Bioaccumulation   The storage of chemicals in an organism in higher concentrations than are normally found in the environment.

Biocentric worldview   The belief that humans are a part of nature and that humans are subject to all natural laws. This worldview recognizes an inherent worth in all life and maintains that humans are no more or less valuable than all other parts of creation.

Biocentrism   See *Biocentric worldview.*

Biodegradable   Materials that can be broken down and rendered harmless by living systems.

Biodiversity   The variety of life forms that inhabit the Earth; biodiversity includes the genetic diversity among members of a population or species as well as the diversity of species and ecosystems. See *Biological diversity.*

Biogas   The methane-rich gas produced from the fermentation of animal dung, human excreta, or crop residues.

Biogeochemical cycle   A series of biological, chemical, and geological processes by which materials cycle through ecosystems.

Biointensive agriculture   A form of organic agriculture that relies on the use of biological controls to maintain croplands and strives to put all available land to use.

Biological diversity   The variety of life, including the genetic diversity among members of a population or species, the species themselves, and the range of communities and ecosystems present on Earth.

Biological oxygen demand (BOD)   A measure of the amount of oxygen needed to decompose the organic matter present in water.

Biological resources   All plant and animal species, domesticated and wild.

Biomagnification   The accumulation of chemicals in organisms in increasingly higher concentrations at successive trophic levels.

Biomass   The sum of all living material in a given environment.

Biomass energy   Organic matter such as woods and crop wastes used as a fuel. One of the oldest and most versatile energy sources capable of providing high-quality gaseous, liquid, and solid fuels.

Biome   A major ecological community of organisms, both plant and animal, usually characterized by the dominant vegetation type, for example, a tundra biome and a tropical rain forest biome.

Bioremediation   The use of living organisms (bacteria) to clean up hazardous substances in the environment.

Biosphere   The thin layer of air (atmosphere), water (hydrosphere), and soil and rock (lithosphere) that surrounds the planet and contains the conditions to support life.

Biotic component (biota)   That part of the environment comprised by living organisms.

Biotechnology   The industrial use of living microorganisms, such as bacteria and other biological agents, to perform chemical processing or to produce materials such as animal food. See *Genetic engineering*.

Biotic potential   The maximum growth rate that a population could achieve, given unlimited resources and ideal environmental conditions.

Bitumen   A thick, high-sulfur and tarlike liquid, contained in tar sands. High-pressure steam is used to force the bitumen from the sandstone; it is then purified and upgraded to produce synthetic crude oil.

Bituminous   An abundant coal with medium to high sulfur content and high heating value.

Bottle bill   Legislation to encourage recycling by requiring consumers to pay a deposit on bottle or can beverage containers, which is refunded when the containers are returned to a food market or recycling center.

Breeder reactor   A type of fission reactor which eventually produces more fuel (plutonium 239) than it consumes.

Cancer   A general term for a disease process in which the growth control mechanism of cells becomes altered and the cells reproduce with abnormal rapidity.

Capitalism   In this economic system, the buyers and sellers in the marketplace determine demand, supply and price. See *Pure market economy*.

Carbon dioxide   A colorless gas (CO_2) formed as a product of animal respiration and decay and combustion of animal and plant matter. An unregulated primary pollutant originating chiefly from fossil fuel combustion; it is the chief greenhouse gas.

Carbon monoxide   A colorless, odorless toxic gas (CO) formed as a product of incomplete combustion of carbon. A regulated primary pollu-tant originating chiefly from imperfect combustion in fuel engines.

Carcinogenic   Cancer-causing.

Carnivores   Organisms that eat only animals. They feed indirectly on plants by eating herbivores.

Carrying capacity   The population size that can best be supported by the environment over time.

Carryover stocks   The total amount of grain in storage when the new harvest begins. An important indicator of food security.

Cash crops   Crops grown for export rather than domestic consumption. Examples include coffee, tobacco, and cotton.

Cellulolytic bacteria   Bacteria that feed on the cellulose in plants and plant products such as wood and paper.

Centers of diversity   Where the genetic diversity for a particular species was the greatest the earliest domestication of that plant occurred.

Centrally planned economy   See *Pure command economy*.

CFC   See *Chlorofluorocarbon*.

Chaining   A forest management technique in which a chain is strung between two bulldozers and then pulled along, tearing down any trees in its path.

Chain reaction   A series of reactions in which each reaction causes additional and increasingly larger reactions.

Charismatic megafauna   A group of species, typically large mammals, which are highly valued by humans.

Charismatic minifauna   Small and inconspicuous species valued for their intrinsic worth.

Chemical sensitizer   A substance that causes an individual to become extremely sensitive to low levels of that particular substance or to experience reactions to other substances.

Chemosynthesis   The synthesis of organic compounds within an organism, with chemical reactions providing the energy source.

Chemotrophs   Producer organisms that use the energy found in inorganic chemical compounds (rather than light energy from the sun) for their energy needs.

Childhood mortality rate   The number of children between the ages of 1 and 5 who die per 1,000 births per year.

Chlorofluorocarbon   A chemical used as a coolant, solvent, or propellant which is a catalyst in the destruction of atmospheric ozone.

Chronic pollution effects   These effects act in the long term; they are not noticed until several years or decades after the introduction of the pollutant.

Chronic toxicity   The ability of a substance to cause an adverse reaction in an organism after it accumulates in the organism through repeated exposures.

Chronic undernutrition   The consumption of too few calories and too little protein over an extended period of time.

Clear-cutting   A forest management technique in which an entire stand of trees is felled and removed.

Climate   The long-term weather pattern of a particular region.

Climate change   A shift in the Earth's long-term weather patterns caused by an anthropogenically enhanced greenhouse effect.

Climax community   Ecosystem development that results in the association of organisms best adapted to the physical conditions of a defined geographic area.

Clockwork universe   The view of the nature of the universe, proposed by Francis Bacon in the seventeenth century, that God established the laws of nature and that science could discover God's patterns and arrive at truth.

Cloning   See *Tissue culture*.

Coal   A solid composed primarily of carbon with small amounts of hydrogen, nitrogen, and sulfur compounds and is derived from deposits of plant matter.

Coal gasification   The conversion of solid coal to a gas that can be burned as fuel.

Co-composting   A form of composting in which organic wastes are mixed with sewage sludge to yield a high-quality, nutrient-rich material.

Cogeneration   The use of waste heat from one process to power a second process, for example, the use of steam generated in the production of electricity to heat buildings.

Coliform   A bacteria normally inhabiting the colon, for example, *Escherichia coli* and *Enterobacteria aerogenes*, high levels of which in drinking water or swimming water may indicate the presence of disease-causing organisms in the water.

Combined sewer system   Sewer system in which pipes that carry sewage are coupled with pipes that remove rainwater from streets after storms. Compare *Separate sewer system*.

Co-mingled recycling   The collection of different types of recyclable items in a single container.

Commensalism   An association of two species in which one benefits and the other neither benefits nor is harmed.

Common law   A system of law, originating in England, based on custom and past court decisions rather than on government or ecclesiastical statutes.

Community   All of the populations of organisms that interact in a given area at a given time.

Compensation   Monetary award for damages.

Competition   An interaction between two or more organisms that are striving to attain the same limited resource, such as food, water, space, or mates.

Competitive exclusion principle   The principle that different species with similar requirements sometimes compete to the elimination of one of them.

Component separation   A waste-treatment method in which particular hazardous materials are removed from a waste stream without the use of chemicals.

Composite   An advanced material that is a matrix of one material reinforced with fibers or dispersions of another.

Composting   A waste-disposal method in which bacteria decompose biodegradable wastes such as paper, food scraps, and yard wastes to form compost or humus.

Compound   Molecules composed of two or more different elements.

Computer model   In environmental science, a sophisticated mathematical equation that helps applied ecologists understand how ecosystems respond to stress and to predict the effects of management strategies on a particular environment.

Confined aquifer   A groundwater deposit in which the overlying material is impermeable, restricting the downward flow of water to the water table.

Conservation   A management philosophy defined as wise use, or careful and planned management, of an area or resource in order to provide for its continued use.

Conservation biology   A scientific discipline dedicated to protecting, maintaining, and restoring the Earth's biological diversity.

Conservation revolution   Environmental movement in the 1970s in the United States which increased awareness of the country's energy problems and led to environmental legislation.

Conservation tillage   Numerous farming practices that disturb the soil as little as possible in order to reduce soil erosion, lower labor costs, and save energy. See also *Low-till planting, No-till planting.*

Consumers   Organisms that cannot produce their own food and eat by engulfing or predigesting the fluids, cells, tissues, or waste products of other organisms.

Consumerism   Wasteful consumption of resources to satisfy wants rather than needs.

Consumptive use   An activity that depletes some portion of a resource.

Containment vessel   An enclosure surrounding a nuclear reactor designed to prevent the accidental release of radiation.

Continental shelf   The gently sloping sea floor around the Earth's land masses.

Contour plowing   The farming practice of tilling the soil along the natural contour of the land in order to protect against erosion.

Contour strip mining   A method of mineral extraction in which a series of shelves or terraces are cut on the side of a hill or mountain to remove a mineral from a deposit near the Earth's surface.

Contour terracing   The farming practice of constructing broad "steps" or level plateaus into the hillside and locating swales, or trenches, at the edge of the terraces to act a catch basins in order to reduce soil erosion from the unimpeded runoff of rainwater.

Conventional agriculture   The practice of maintaining croplands through the use of synthetic fertilizers, pesticides and herbicides, and heavy machinery.

Conventional fuel   An energy source, primarily fossil fuels, that is currently widely used.

Cooperation   An interaction between members of the same species or an association of two dissimilar species that aids in the survival of one or both parties.

Coral reefs   A limestone ridge near the surface of the sea formed from the calcareous skeletons of reef-building coral and other marine organisms.

Cornucopian view of population growth   The belief that population growth is a positive influence on future economic growth and that any problems arising from excess population will be resolved by human inventiveness.

Cost-benefit analysis   A technique used to compare the estimated costs (losses) of a proposed project with the estimated benefits to be gained.

Cover crop   A crop planted to cover the ground after a main crop is harvested in order to reduce erosion and maintain soil fertility.

Creationism   The belief that living organisms were created by an all-powerful being (God), according to the record set forth in the Bible.

Critical mass   The amount of a fissionable material necessary to sustain a chain reaction.

Critical mineral   A mineral considered essential to a country's economic activity.

Crop rotation   The farming practice of changing the crop planted in a particular area in successive years to avoid soil depletion and maintain soil fertility.

Cross-media pollutant   A pollutant that moves through various mediums (soil, water, air) from one environment to another.

Crude birth rate   The number of live births per 1,000 people, per year. Compare *Crude death rate.*

Crude death rate   The number of deaths per 1,000 people, per year. Compare *Crude birth rate.*

Cryopreservation   A method of preserving living material by freezing and storing in liquid nitrogen at very low temperatures.

Cultural carrying capacity   The optimal size that the environment can sustain in perpetuity, with a given technology, standard of living, and associated patterns of resource use.

Cultural diversity   The variety of values, attitudes, beliefs, and actions within a population and in different populations.

Cultural eutrophication   The accelerated aging of a body of water caused by the addition of nutrients originating from human activities.

Culture of consumption   A culture or society based on the premise that *one can never have enough,* a belief fueled by advertising. Consumption is an end in itself; people are encouraged to buy more in order to keep the economy expanding.

Culture of maintenance   A culture or society based on the premise that one's quality of life should be judged on intangibles (such as satisfaction, personal happiness, and fulfillment) rather than material possessions.

Cultural resource   Anything that embodies, symbolizes, or represents a part of the culture of a specific people.

Cultural resources management   A field of interest concerned with preserving both material culture and living culture.

Culture   The beliefs, attitudes, values, behaviors, and practices of a group or society.

Curbside collection   Method of source separation in which households remove recyclables from other garbage and place them in separate containers to be picked up with the garbage.

Cyanobacteria   Algae-like prokaryotes that live in marine or fresh water. Also known as blue-green algae.

DDT   Dichlorodiphenyltrichloro-ethane, a chlorinated hydrocarbon that was at one time widely used in the United States as a pesticide.

Debt-for-nature swap   An agreement between a debtor nation and a bank in which the bank agrees to absolve a loan if significant amounts of undisturbed land are set aside for conservation.

Decommissioning   The process of dismantling and decontaminating a nuclear reactor once it has reached the end of its useful life.

Decomposers   Heterotrophs that digest, outside of their cells and bodies, the tissues of dead organisms or waste products. See also *Microconsumers.*

Deep ecology   A belief in the unity of humans, plants, animals, and the Earth.

Deep-well injection   A waste disposal method in which liquid wastes are pumped deep underground into rock formations that geologists believe will contain them permanently.

De facto wilderness   A wilderness area not protected by official designation.

Deforestation   The cutting down and clearing away of forests.

Demand side management   A promising approach to energy production and consumption that encourages energy efficiency by electric utilities, electricity users and electric appliance manufacturers.

Demographic fatigue   A condition brought on by sustained rapid population growth in which a government lacks the financial resources to deal effectively with threats such as natural catastrophes and disease outbreaks.

Demographic transition   The movement of a nation from high population growth to low growth as it moves through stages of economic development.

Demographic trap   The inability of a country to pass the second phase of demographic transition, which is characterized by high birth rates and low death rates.

Demography   The statistical study of characteristics of human populations, such as size, distribution, age, income.

Density   How closely people are grouped, especially with respect to the degree of resource use within defined geophysical or political boundaries.

Density-dependent factor   A biotic factor in an environment that limits population growth and is more pronounced when population density is high.

Density-independent factor   An abiotic factor in an environment that sets upper limits on the population, and is unrelated to population density.

Dependency load   The number of dependents (those under age 15 or over 65) in a population.

Desalination   The process by which salt and suspended minerals in seawater are removed to make drinkable water.

Desert   A land area in which evaporation exceeds precipitation, usually where annual precipitation is less than 250 millimeters per year.

Desertification   Land degradation in arid, semi-arid, and dry, subhumid regions resulting mainly from adverse human impact.

Detritus food web   A food web that includes several levels of consumers that derive energy from decomposing plant and animal material or animal waste products.

Detritivores   Organisms that consume dead or decaying tissues or organic wastes. Also known as detritus feeders.

Dicotyledon   The most numerous of the two subclasses of flowering plants in which the embryo is typified by two seed leaves. The leaves show a wide variety of shapes, although the veins within the leaves are arranged in the form of an irregular net. These species contain most of the hardwood trees as well as shrubs and herbs.

Direct radiation exposure   Exposure to a source of radioactivity. Compare *Indirect radiation exposure.*

Dispute resolution   The process of negotiation and compromise in which two disputing parties meet face to face and reach a mutually acceptable solution to the dispute.

Disruption   An acute disturbance in a natural system that can be traced directly to a specific human activity. Compare *Disturbance.*

Disturbance   A variation in some factor in an ecosystem beyond the normal range of variation resulting in a change in the ecosystem.

Disturbance ecology   A branch of ecology concerned with assessing the impact of stress on organisms, populations, and ecosystems.

Diversity-stability hypothesis   School of thought which suggests that biodiversity promotes resistance to environmental disturbances.

DNA   Deoxyribonucleic acid, a complex molecule found in the cells and chromosomes of almost all organisms that acts as the primary genetic material.

Domestication   An evolutionary process in which genes useful for survival in captivity prevail over genes necessary for survival in the wild.

Doubling effect   The hypothesis that a doubling in atmospheric carbon dioxide will increase world-wide temperature by 9°F (3°C), causing a number of diverse climatic changes.

Doubling time   The number of years it takes for a population to double assuming current growth rates remain the same.

Dry steam reservoir   A rare and highly preferred geothermal energy source in which underground water deposits have already been vaporized.

Dynamic   A condition of continual change.

Dynamic equilibrium   An ecosystem's ability to react to constant changes thereby maintaining relative stability.

Ecological economics   A transdisciplinary field of study that addresses the relationships between ecosystems and economic systems in the broadest sense.

Ecological succession   The process by which an ecosystem matures; it is the gradual, sequential, and somewhat predictable change in the composition of the community. See also *Ecosystem development.*

Ecological toxicology   Subdiscipline of ecology concerned with the effect of toxins on population dynamics, community structure, and ecosystems.

Ecology   The study of the structure, function, and behavior of the natural systems that comprise the biosphere.

Economic goods   Products such as clothing or appliances. Also known as manufactured capital.

Economic growth   An increase in the capacity of the economy to produce goods and services. It is generally viewed as the best way to modernize.

Economics   The study of the production, consumption, and distribution of goods and services.

Economic reserve   See *Proven reserve.*

Economy   A system of production, distribution, and consumption of economic goods or services.

Ecosystem   A community of self-sustaining, self-regulating organisms interacting with the physical environment within a defined geographic area.

Ecosystem damage   An adverse alteration of a natural system's integrity, diversity, or productivity.

Ecosystem destruction   The conversion of a natural system to a less complex human system.

Ecosystem development   Process by which an ecosystem matures, taking into account the accompanying modifications in the physical environment brought about by the actions of living organisms. See also *Ecological succession.*

Ecosystem disruption   A rapid change in the species composition of a community that can be traced directly to a specific human activity.

Ecosystem diversity   The variety of habitats, communities, and ecological processes in an ecosystem.

Ecosystem services   Functions or processes of a natural ecosystem that provide benefits to human societies.

Ecotone   The transitional zone of intense competition for resources and space between two communities.

Ecotoxicology   The study of the effect of toxins on population dynamics, community structure, and, ultimately, ecosystems.

Ectoparasite   A parasite that lives on the outside of the host's body. Compare *Endoparasite.*

Education   The processes through which students learn.

Effluent   Any substance, usually a liquid, that enters the environment through a point source such as an industrial site or a sewage plant outlet.

Elements   Substances that cannot be changed to simpler substances by chemical means. Each one has a given name and symbol.

Emigration   Migration out of a country or region.

Emissions standards   The maximum amount of a specific pollutant permitted to be legally discharged from a particular source in a given environment.

Emissivity   The relative power of a surface to reradiate solar radiation back into space in the form of heat, or long-wave infrared radiation.

Endangered species   A plant or animal species that no longer can be relied on to reproduce in numbers sufficient to ensure its survival.

Endemic species   A plant or animal species confined to or exclusive to a specific area.

Endocrine disrupters   Synthetic compounds that interfere with the endocrine systems of living organisms, causing problems in growth, development, and reproduction.

Endoparasite   A parasite that lives inside of the host's body cavity, organs, or blood. Compare *Ectoparasite.*

End product substitution   A method of minimizing hazardous substances in the environment in which consumers choose products that will generate a minimum volume of hazardous resources.

Energy   The ability to do work, to move matter from place to place, or to change matter from one form to another.

Energy efficiency   A measure of the percentage of the total energy input that does useful work and that is not converted into low temperature, diffuse, excess heat.

Entropy   A phenomenon where energy constantly flows from a high-quality, concentrated, and organized form to a low-quality, randomly dispersed and disorganized form.

Environment   The system of interdependent living and nonliving components in a given area over a given period of time, including all physical, chemical, and biological interactions.

Environmental activism   An umbrella term that encompasses both efforts to stop damage to natural systems caused by human activity and efforts to proactively manage resources and natural systems in an environmentally sound manner.

Environmental degradation   A change in the Earth's life support systems that interferes with the Earth's capacity to maintain a maximum range of tolerances for life.

Environmental education   Information and training to awaken and explore a person's intuitive value of the natural world.

Environmental ethics   Principles of respect and care for the natural world.

Environmental impact statement   An analysis of a proposed project to predict its likely repercussions on the social and physical environment of the surrounding area.

Environmental law   A law system organized to use all of the laws in a nation's legal system to minimize, prevent, punish, or remedy the consequences of actions that damage or threaten the environment.

Environmentally sound management   A method of managing any resource, natural or cultural, which minimizes or prevents environmental degradation. A stewardship ethic is key to environmentally sound management.

Environmental problem solving method   A five-step interdisciplinary and goal-directed model to develop workable solutions, which can be applied to almost any situation.

Environmental Protection Agency   The U.S. government agency respon-sible for federal efforts to control pollution of air, land, and water by human activities and to sponsor re-search and education on the impact of humans on the environment.

Environmental racism   Racial discrimination in environmental policymaking, enforcement of regulations and laws, and targeting of communities of color for toxic waste disposal and siting of polluting industries.

Environmental refugees   People forced to abandon their homes because the land can no longer support them.

Environmental resistance   Limiting factors that exert a controlling influence on population size.

Environmental science   Narrowly defined as the study of the human impact on the physical and biological environment of an organism. In its broadest sense, it also encompasses the social and cultural aspects of the environment. Compare *Environmental studies.*

Environmental studies   An interdisciplinary field that attempts to understand and to solve the problems caused by the interaction of natural and cultural systems. Compare *Environmental science.*

Epidemiology   The study of the distribution and causes of health disorders.

Epilimnion   Upper layer of a lake, which is heated by the sun and thus lighter and less dense than the underlying water.

Equilibrium   A condition of balance or stability.

Escherichia coli   See *Coliform.*

Estuary   A shallow, nutrient-rich, semienclosed coastal body of water usually composed of fresh and salt water.

Ethic   A system or code of morals that governs or shapes attitudes and behavior.

Ethics   A branch of philosophy concerned with standards of conduct and moral judgment.

Euphotic zone   The area in a body of water where light penetrates enough to support photosynthesis and populations of plankton.

Eutrophication   A process of nutrient enrichment of a lake, stream, or estuary by natural or human activities that set in motion a mix of physical, chemical, and biological changes that lead to the natural aging of a body of water.

Eutrophic lake   A lake that is warm and shallow, with a low oxygen content and a relatively high amount of nutrients, phytoplankton, and other organisms.

Evaporation   The process of energy absorption that enables a liquid to change state and become a gas.

Even-aged harvesting   A timber harvesting technique that replaces the forest with a stand in which all the trees are about the same age. Compare *Uneven-aged harvesting.*

Evolution   The process of change with continuity in successive generations of organisms. Compare *Creationism;* See *Natural selection.*

Evolutionary potential   The potential for a new species to arise from an ancestor through adaptation and specialization over millions of years.

Exclusive economic zone   Waters extending 200 miles from a country's shoreline within which it has exclusive rights to exploit mineral and fish resources established by the United Nations Law of the Sea Convention in 1977.

Exponential growth   A rate of growth in which a quantity increases by a fixed percentage of the whole in a given time period. When plotted on a graph, exponential growth creates a J-curve. See *Geometric growth.* Compare *Linear growth.*

External costs   The harmful social or environmental effects of the production and consumption of an economic good that are not included in the market price of the good.

Extinction   The complete disappearance of a species from the biota.

Extreme poverty   Subsisting on less than $1 a day.

Facilitation model of succession   A theory that proposes that one community prepares the ecosystem for a subsequent community.

Family planning   Measures that enable parents to control the number of children they have (if they so desire) and the spacing of their children's births.

Famine   The widespread scarcity of food, with subsequent suffering and starvation in the population.

FAO   The Food and Agriculture Organization, a specialized agency of the United Nations, founded in 1945 to coordinate programs of food, agriculture, forestry, and fisheries development in order to improve the standards of living of rural populations and combat malnutrition and hunger.

Fauna   The total animal population that inhabits an area.

Feedback   Any factor which influences the same trend that produced it.

Ferrous metals   Metallic minerals containing iron or elements alloyed with iron to make steel.

Fertility   The actual bearing of offspring. A more accurate indicator of the potential for future population growth.

Fertilizer   Material added to soil to supply essential nutrients for crop growth.

Field gene bank   Storage facility that preserves genetic plant material through annual germinations.

First Law of Thermodynamics   Law that states that during a physical or chemical change energy is neither created nor destroyed, but it may be changed in form and moved from place to place. Also known as first law of energy. See also *Second law of thermodynamics.*

Fishery   Concentrations of particular aquatic species in a given aquatic area suitable for commercial exploitation.

Fission products   The radioactive by-products of a fission reaction.

Fission reaction   Reaction that occurs when neutrons hit the unstable nucleus of a radioactive atom, splitting it into two smaller fragments while discharging more neutrons and large amounts of energy.

Flora   The total plant population that inhabits an area.

Fluidized bed combustion   Process for burning coal more efficiently, cleanly, and cheaply, in which a mixture of powdered coal and limestone is suspended by a stream of hot air during combustion.

Fly ash   Extremely fine particles of ash produced when coal is burned in modern forced-draft furnaces, particularly those associated with electric power plants.

Food chain   Successive steps within an ecosystem illustrating the energy transfers between organisms as a result of consumption.

Food security   The ability of a nation to feed itself on an ongoing basis.

Food web   Interlocking food chains, woven into complex associations that describe the feeding relationships among organisms in a community and the movements of energy and materials.

Forestry   The practice of planting, tending, and managing forests primarily for the exploitation of timber for commercial or local subsistence needs.

Formaldehyde   A substance used as a preservative for biological specimens and numerous other products such as foam insulation, carpets, and pressed wood products.

Fossil fuel   Any naturally occurring carbon or hydrocarbon fuel derived from anaerobic decomposition of organic material in the Earth's crust, including natural gas, oil, coal, oil shale, and tar sands.

Fossil groundwater   Aquifers formed thousands of years ago when the Earth's rainfall pattern was much greater.

Fragmentation   The process by which a frontier forest becomes a patchwork of cropland, logging roads, and smaller, discrete forest areas that do not function as intact ecosystems.

Frontier ethic   Belief by American settlers, which is still popular today, that natural resources are inexhaustible or will regenerate and that exploration will discover new, untapped resources.

Frontier forests   Expansive tracts of contiguous forest, largely untouched by human activities, that function as intact ecosystems.

Frontier mentality   See *Frontier ethic.*

Fungicide   A chemical used to kill fungi.

Fungus   A saprotrophic or parasitic organism that may be unicellular or made up of tubular filaments and which lacks chlorophyll.

Fusion reaction   Reaction that occurs under extremely high temperatures in which individual nuclei of a particular element fuse to form another element.

Gaia hypothesis   James Lovelock's worldview that maintains that the Earth is a self-regulating, living organism.

Game animal   Any animal hunted chiefly for sport.

Gamma radiation   Short, intense burst of electromagnetic energy given off by a radioactive substance, capable of penetrating lead or concrete one meter thick and of causing tissue damage in living organisms.

Garbage Animal or vegetable waste resulting from the handling, storage, preparation or consumption of food.

Gaseous cycle The circulation of a gas through the environment, primarily in the atmosphere, as materials circulate from the air, through land and water, and back again.

Gene The fundamental physical unit of heredity that transmits information from one cell to another and thus from one generation to another.

Gene bank Storage facility to preserve genetic material through various freezing and drying methods.

Gene pool The sum of all the genes present in a population of organisms.

General fertility rate The number of live births per 1,000 women of childbearing age per year.

Generalist A species that can survive in many different habitats. Compare *Specialist*.

Genetic diversity Variation among the members of a single population of species.

Genetic erosion The reduction of genetic diversity in a gene pool due to decreasing population size.

Genetic modification The human manipulation and transfer of genes from one organism to another to improve the productivity or survivability of economically important organisms.

Genotype The genetic makeup of an organism or group of organisms.

Geopressurized reservoir A geothermal energy source in which underground water and methane gas are subjected to extremely high temperatures and pressures.

Geometric growth A rate of growth in which a quantity increases by a fixed percentage of the whole in a given time period. When plotted on a graph, geometric growth creates a J-curve. See *Exponential growth*. Compare *Linear growth*.

Geothermal energy Energy generated by the natural heat and pressure occurring beneath the Earth's surface.

Germ plasm The genetic material of an organism.

Giardiasis The most common waterborne disease in the United States today; it is caused by the ingestion of the protozoan *Giardia lamblia*.

Global environment The sum of all living organisms, the relationships among organisms, and the relationships between organisms and their physical surroundings.

Global warming The increase in global temperature predicted to arise from increased levels of carbon dioxide, methane, and other gases in the atmosphere. See *Climate change*.

Government An established system of administration through which a nation, state, or district is ruled.

Grassed waterway A trench that is planted in grass to slow the course of runoff and absorb excess water and eroding soil.

Grassroots loan Small-scale loans to individuals in a community rather than to capital-intensive development projects.

Great hunger belt Equatorial region in which most of the world's hungry live, spanning parts of Southeast Asia, the Indian subcontinent, the Middle East, Africa, and Latin America.

Greenhouse effect The prevention of the reradiation of heat waves to space by carbon dioxide, methane, and other gases in the atmosphere. The greenhouse effect makes possible the conditions that enable life to exist on Earth.

Greenhouse gas A gas that contributes to the greenhouse effect, such as carbon dioxide, chlorofluorocarbons, ozone, methane, and nitrous oxide.

Green manure A cover crop, such as alfalfa, that is planted and then plowed under to improve soil structure and fertility.

Green marketing The promotion of products based on claims that they are benign to the environment.

Green revolution A group of measures to improve agricultural productivity in less-developed countries including the development of high-yield cereal varieties, their adoption and diffusion in the third world, the increased use of fertilizers, pesticides, and irrigation.

Gross national product (GNP) Total market value in current dollars of all goods and services produced by an economy during a year. Compare *Net national product*.

Gross primary productivity (GPP) The total amount of energy produced by autotrophs over a given period of time.

Groundwater Water found underground in water-bearing porous rock or sand or gravel formations recharged by rainfall and infiltration from surface waters and wetlands.

Growth rate See *Natural rate of increase*.

Habitat The specific environment or geographic region in which a species is found.

Half-life The period of time it takes for half of the atoms of a radioactive material to decay into the next element of the decay process, ranging from minutes for some low-level radioactive materials to billions of years for some highly radioactive materials.

Hazardous air pollutants A varied group of chemical substances, numbering in the hundreds, known to cause or suspected of causing cancer or other serious human health effects or ecosystem damage.

Hazardous substance A toxic substance that has the potential to threaten human health and the environment if poorly handled or stored.

Hazardous waste A substance defined as any solid, liquid, or gaseous waste which, due to its quantity, concentration, or physical, chemical, or infectious characteristic, may cause or significantly contribute to an increase in mortality or serious illness; or pose a substantial present or potential hazard to human health or the environment when improperly stored, transported, disposed of, or otherwise managed.

Herbicide Any chemical substance, usually synthetic, used to kill plant life.

Herbivores Organisms that eat green plants directly.

Heterotrophs See *Consumers*.

High-density polyethylene (HDPE) A modern plastic, used in packaging, that can be recycled.

High-input agriculture See *Conventional agriculture*.

High-level radioactive waste Nuclear reactor fuel or fission products that will remain radioactive for tens of thousands of years and must be handled or stored with extreme care.

High-pressure cell Atmospheric condition in which air sinks toward the ground.

High-yield variety (HYV) Special genetic strain of a hybrid crop developed by scientists to permit substantial gains in crop yield.

Historic preservation The preservation of material culture representative of events in a society's history.

Homeostasis The maintenance by an organism of a constant internal environment, such as regulation of blood sugar levels by insulin; the process involves self-adjusting mechanisms.

Hormonal contraception A type of birth control that alters a female's hormone levels to inhibit production of the egg.

Hot dry rock reservoir A geothermal energy source in which subsurface rock is heated to high temperatures as a result of the intrusion of molten rock from the Earth's center into the Earth's crust.

Human capital Skill and labor used to manufacture goods and products.

Human development index For a given country, a measure of the average quality of life based on three indicators: life expectancy at birth, literacy rates, and real GNP per capita. Proposed by the United Nations.

Humus Partially decomposed plant and animal matter that is the organic component of soil and gives it the ability to retain water and maintain a high nutrient content.

Hunger The inability to acquire or consume an adequate quality or sufficient quantity of food through non-emergency food channels.

Hybridization The crossing of one or more varieties of a species to produce an offspring with particular desired qualities.

Hydrocarbon An organic compound containing hydrogen and carbon and often occurring in fossil fuels. A primary pollutant originating chiefly from fossil fuel combustion.

Hydropower A means of producing electricity which exploits the energy present in falling water. Water at the top of a fall or dam is in a high state of gravitational potential energy; as the water drops, its potential energy is converted to kinetic energy. As the water strikes the blades of a turbine, spinning the turbine shaft, the kinetic energy is converted to mechanical energy, which can be used to drive a generator to produce an electric current.

Hydrological cycle The circulation of water through bodies of water, the atmosphere, and land.

Hydroponics The growing of plants in a nutrient-rich solution.

Hydrosphere The sum of free water in solid or liquid state on the Earth's surface.

Hypolimnion Lower layer of a lake, which receives little heat from the sun and tends to be colder than overlying water. Compare *Epilimnion.*

Igneous rock Rock formed by the cooling and crystallization of magma.

Illegal harvesting The taking of plants or animals prohibited by law in terms of season, age, size, or species.

Immanence The belief that the spiritual (God) is embodied in the living world and in all its components as well as in the interrelationships among them.

Immigration Migration into a country or region.

Incineration The burning of wastes.

Index of sustainable economic welfare (ISEW) A measure of economic well-being that adjusts the per capital GNP according to inequalities in income distribution, resource depletion, loss of wetlands, loss of farmland and the cost of air and water pollution.

Indicated reserves Mineral or fossil fuel deposits that are thought to exist and are likely to be discovered and available for use in the future. Also known as inferred reserves.

Indicator species A species that indicates, by either its presence or absence, certain environmental conditions.

Indirect radiation exposure Exposure to radiation removed from its original source, for example, exposure to radioactive substances transmitted through the food chain. Compare *Direct radiation exposure.*

Indoor air pollution Airborne contaminants present inside homes, offices and other buildings that pose a health threat.

Industrial smog Air pollution consisting chiefly of sulfur oxides and particulates that is emitted to the air from industrial and manufacturing facilities.

Inertia The ability of a ecosystem to resist change.

Infant mortality rate The number of infants who die before age 1 per 1,000 births per year. It is widely considered the single best indicator of a society's quality of life.

Infectious Disease-causing.

Inferred reserves See *indicated reserves.*

Inhibition model of succession The theory that certain species are able to inhibit or prevent succession within an ecosystem.

Injunction A court order to do or refrain from doing a specified act.

Insecticide Any chemical substance, usually synthetic, used to deter, repel, or kill insects.

Integrated pest management (IPM) The combined use of several methods, biological and chemical, to control insect pests.

Integrated solid waste management The combined use of various methods of waste disposal, including recycling, composting, incineration, and landfills, to minimize the negative impact of the waste stream on the environment.

Interconnection A concept that recognizes the relatedness of all things. Humans are seen as part of, not separate from, nature.

Internal costs The direct costs of production, for example in growing corn, the seed, fertilizer and labor.

Interspecific competition Competition between members of different species for limited resources such as food, water, or space.

Intraspecific competition Competition between members of the same species for limited resources such as food, water, or space.

Intrauterine device (IUD) A birth control device implanted in a woman's uterus that prevents the attachment of the zygote to the wall of the uterus.

In vitro preservation A method of preserving plant tissue for future propagation by storing it under regulated temperature and light conditions.

Ion An atom or atom group electrically charged by the loss or gain of electrons.

Ionizing radiation Energy released from a fission reaction which is powerful enough to pull electrons away from the atoms of materials that it strikes, producing positively charged ions; can be extremely damaging to living tissue.

Isotope One of two or more forms of the atom of an element that differs from the other forms in the number of neutrons.

J-curve Pattern formed by exponential growth.

Kerogen A waxy, combustible organic matter found in oil shales.

Keystone species A species that has a significant role in community organization due to its impact on other species in the community.

Kinetic energy Energy associated with the motion of matter.

Kwashiorkor Childhood disease arising from protein insufficiency.

Lake A body of standing water that occupies a depression on the Earth's surface and is completely surrounded by land.

Land ethic The principle developed by ecologist Aldo Leopold of cooperation between humans and other biospheric components.

Landfilling The disposal of waste in excavated sites or in sites such as former quarries, abandoned mine workings, gravel pits, and clay pits.

Land race A variety of a species adapted to specific local conditions such as climate and soil type.

Landscape Many ecosystems taken together.

Landscape ecology The holistic study of a geographic area: the distri-bution of ecosystems and the move-ment of plants, animals, nutrients, and energy among those ecosystems. Of primary importance are the relations between human society and its living space, both the natural and managed components of the landscape.

Land subsidence The gradual sinking of the ground surface to a lower level.

Land use The way in which a particular parcel of land is used.

Land use planning The practice of studying information on a specific parcel of land and its possible uses before choos-ing a course of action.

Law of the Conservation of Matter Law that states that during a physical or chemical change matter is neither created nor destroyed, but it may be changed in form and moved from place to place.

Law of the minimum Law that states that survivability is primarily deter-mined by the minimum amounts of limiting fac-tors in an ecosystem.

Law of tolerances Range of abiotic factors within which an organism can survive from the minimum amount of a limiting factor that the organism requires to the maximum amount that it can withstand.

LD50 test Test to determine the lethal dose of a substance, defined as the dose after which 50 percent of test subjects die.

Leachate The liquid formed when rainwater percolates downward through landfilled wastes, picking up contaminants on its way into the surrounding environment.

Lead pollution The accumulation of lead in organic tissue, primarily in plants, animals, and humans, which may produce behavioral changes, blindness, and ulti-mately death.

Less-developed country (LDC) A country that has low to moderate indus-trialization and low to moderate average GNP per person. See *Third world*. Com-pare *More-developed country*.

Lifeboat ethics A guide to action based on the belief that each nation is like a lifeboat, with limited capa-city (resources) and that people in the lifeboats with enough resources to support their passengers easily cannot rescue everyone from the overcrowded lifeboats without jeopardizing their own safety.

Life-cycle cost The sum of the initial purchase cost plus the operating costs incurred over the lifetime of the product.

Life expectancy The average number of years a newborn can be expected to live. This is a good indicator of a nation's standard of living.

Lignite An abundant coal with low sulfur content and low heating values. See *Subbituminous coal*. Compare *Anthracite coal, Bituminous coal*.

Limiting factors Abiotic and biotic regulators that determine the distribution and success of living organisms.

Limnetic zone Area of a body of fresh water where light penetrates enough to support photosynthesis and populations of plankton.

Linear growth A rate of growth that occurs when a quantity increases by a fixed amount in a given time period.

Lithosphere The rigid outer layers of the Earth's crust and mantle.

Litigation The act of initiating a lawsuit; it is sometimes used to address environmental problems.

Littoral zone Shallow area of a near-shore body of water where light penetrates to the bottom and rooted vegetation dominates.

Low-input sustainable agriculture The practice of maintaining croplands through limited use of synthetic fertilizers, pesticides, and herbicides.

Low-level radioactive waste Nuclear reactor fuel or fission products that will remain dangerous for a few hundred years or less.

Low-till planting The practice of till-ing the soil just once in the fall or spring, leaving half or more of previous crop residue on the ground's surface.

Macroclimate The average weather pattern, including temperature and pre-cipitation, of a region.

Macroconsumers Organisms that feed by ingesting or engulfing part or entire bodies of other organisms, living or dead. They include herbivores, carnivores, om-nivores, scavengers, and detritivores.

Macronutrients Chemicals needed by all living organisms in large quantities for the construction of proteins, fats, and carbohydrates.

Magma Melted rock found deep within the Earth's crust where it is exposed to tre-mendous pressure and high temperatures.

Malabsorptive hunger The body loses its ability to absorb nutrients from food consumed. Can be caused by parasites in the intestinal tract or by a severe protein deficiency.

Malnuourishment The consumption of inadequate levels of specific nutrients essential for good health.

Malthusian view of population growth Belief, first proposed by eigh-teenth-century English economist Thomas Malthus, that population increases geo-metrically while food increases arithmeti-cally, thus causing population to exceed sustainable levels.

Manufactured capital See *economic goods*.

Marasmus Childhood disease arising from insufficient amounts of protein and calories.

Marker-assisted selection Using genetic information to speed up and improve conventional plant and animal breeding.

Marsh An area of spongy, waterlogged ground dominated by grasses with large numbers of surface water pools.

Mass burn incinerator Plant that burns all incoming waste, regardless of its composition, at temperatures as high as 2400°F.

Material culture Tangible objects, such as tools, furnishings, buildings, and art works, that humans create.

Maternal mortality ratio In any given year, the number of women's deaths due to pregnancy and childbirth complications per 100,000 live births.

Matter That which constitutes the substance of physical forms, has mass, oc-cupies space, and can be quantified.

Mediator A neutral third party called on to facilitate negotiations: to help dis-puting parties see their common interests, define the problem, and find a solution.

Meltdown The overheating of a nuclear reactor core so that fuel rods melt, burn through the containment vessel, and bore toward the center of the Earth releasing large amounts of radiation to the environment.

Mesosphere An atmospheric layer that extends from the top of the stratosphere to about 56 miles above the Earth.

Metamorphic rock A type of rock that forms when rocks lying deep below the Earth's surface are heated to such a degree that their original crystal structure is lost. As the rock cools, a new crystalline structure is formed.

Microclimate The weather conditions just above the surface of the round.

Microconsumers Organisms that feed on waste products of living organisms or the tissue of dead organisms. They digest materials outside of their cells and bodies, through the external activities of enzymes, and then absorb the predigested materials into their cells. They are often referred to as decomposers.

Micronutrients Substances needed in trace amounts for the construction of proteins, fats, and carbohydrates.

Migration Movement from one geographic area to another for the purpose of establishing a new residence.

Mineral Any nonliving, naturally occurring substance with a limited range in chemical composition and with an orderly atomic arrangement.

Mineral extraction The process of separating the mineral ore from the surrounding rock in which it is embedded.

Mineral processing A process that consists of separating the mineral from the ore in which it is held and concentrating and refining the separated mineral.

Mineral ore A naturally occurring deposit which is economically valuable.

Minerals policy Laws, regulations, and agreements established by a nation to govern mineral production, use, and commerce.

Minimum critical diet The minimum amount of food needed to remain healthy and maintain body weight, assuming little physical activity.

Minimum viable population The minimum number of individuals required to maintain a viable breeding group without suffering short-term loss of genetic variability.

Mixed economic system An economic system that combines elements of market, command, and traditional systems.

Moderate poverty Living on $1 to $2 a day.

Molecule The smallest physical unit of an element or compound, consisting of one or more like atoms in an element and two or more different atoms in a compound.

Monocotyledon One of the two classes of flowering plants, characterized by having embryos with one seed leaf, narrow leaves with a parallel vein system, scattered vas-cular bundles with no cambium, and flower parts arranged in multiples of three. Members of this class include the economically important grasses and grain groups.

Monoculture The extensive cultivation of one or two profitable crops.

Monomorphic species A species in which the male and female look almost identical.

Monotheism The belief in a single God who created the universe but is separate from and outside of His creation.

Morals Standards of right and wrong behavior.

More-developed country (MDC) A country that has a high degree of industrialization and moderate to high average GNP per person. Also known as an industrialized country, developed country, or first world. Compare *Less-developed country*.

Multiple chemical sensitivity An illness characterized by an intolerance of one or more classes of chemicals.

Multiple use The management of an area for several different purposes, for example forestry, recreation, and wildlife habitat.

Municipal solid waste Waste generated by households and small businesses.

Mutagenic Tending to increase the frequency or extent of genetic mutations.

Mutation A random change within the genetic material of an individual that can be passed on to that individual's offspring.

Mutualism A mutually beneficial interaction between two species.

National Environmental Policy Act (NEPA) Legislation passed in 1969 that charges federal agencies with restoring and maintaining environmental quality throughout the country and requires federal agencies to prepare environmental impact statements for any major project.

National ambient air quality standards Regulations for ambient concentrations of particular substances, that is, concentrations in the outside air that people breathe. Primary air quality standards are designed to protect human health and welfare; secondary standards protect the environment.

Natural capital Natural resources, such as energy and minerals, used to manufacture products.

Natural gas A fossil fuel composed of several different types of gases of which methane, ethane, and propane are most prevalent.

Natural rate of increase Difference between the crude death rate and the crude birth rate expressed as a percentage.

Natural selection The process which enables individuals with traits that better adapt them to a specific environment to survive and outnumber other, less well-suited individuals.

Nature The sum of all living organisms interacting with the Earth's physical and chemical components as a complete system.

Negative feedback Any output that interferes with the trend which produced it.

Negative population growth A situation in which deaths outnumber births, resulting in a decline in absolute numbers.

Negligence A failure to exercise the care that "a prudent person" usually takes resulting in an action or inaction that causes personal or property damage.

Nekton Free-swimming aquatic organisms. Compare *Plankton*.

Neritic zone The part of the euphotic zone that overlies the continental shelf or surrounds islands and supports most of the ocean fisheries.

Net energy efficiency Found by determining the efficiency of each energy conversion in a process or system that includes two or more energy conversions.

Net national product (NNP) An economic indicator based on the gross national product but which factors in the depletion or destruction of natural resources. Compare *Gross national product*.

Net primary productivity (NPP) The total amount of energy produced each year at the producer level minus what producers need for their own life processes, and thus available to other organisms in the community.

Niche Defines the functional role of an organism within its community. It is the complete ecological description of an individual species.

Nitrogen fixation The conversion of atmospheric nitrogen into a form of nitrogen usable by plants.

Nitrogen oxide A regulated primary pollutant originating chiefly from the high-temperature combustion of fossil fuels.

Nonassociated gas Free-flowing natural gas that is found apart from petroleum reserves. See *Associated gas*.

Nonbiodegradeable A substance that cannot be broken down by natural systems, and which is thus unusable to organisms.

Nonconsumptive use An activity that does not deplete a resource.

Nonferrous metals Minerals that contain metallic minerals not commonly alloyed with iron.

Nongame animals Species that are not hunted for sport.

Non-ionizing radiation Energy such as radio waves, heat, and light released from a fusion reaction which is not powerful enough to create ions and is not generally considered a health risk.

Nonmaterial culture Intangible resources, such as language, customs, traditions, folklore, and mythology.

Nonpoint-source pollution Pollution that cannot be traced to a specific source but rather comes from multiple generalized sources.

Nonrenewable resource A resource, such as fossil fuels, that exists in finite supply or is consumed at a rate faster than the rate at which it can be renewed.

No-till planting The practice of sowing seeds without turning over the soil, thus allowing plants to grow amid the stubble of the previous year's crop. The stubble acts as a mulch to fertilize the soil and prevents it from drying out.

Not in my backyard syndrome (NIMBY) Opposition to having waste and pollutants processed near one's home or community.

Nuclear energy The energy contained within the nucleus of the atom, this energy can be utilized through nuclear fission to create electricity.

Nuclear resources Resources derived from atoms, their energy, and the particles they emit.

Nuclear winter Theory that a global nuclear war would pump such large quantities of smoke, soot, and debris into the atmosphere that sunlight reaching the Earth would be severely reduced causing the Earth's temperature to drop substantially.

Nuisance The most common cause of action in the field of environmental law. It is a class of wrongs that arise from the unreasonable, unwarrantable, or unlawful use of a person's own property that produces annoyance, inconvenience, or material injury to another.

Null hypothesis School of thought which promotes the view that ecosystem functions or processes are insensitive to the addition or deletion of species.

Ocean power Energy that can be derived from the seas through means such as harnessing tides and thermal currents.

Ocean thermal energy conversion A relatively new technology that involves harnessing the temperature difference between warm surface water and colder, deeper layers by alternately vaporizing and condensing a working fluid.

Oil See *Petroleum*.

Oil shales A minor fossil fuel source composed of fine-grained, compacted sedimentary rocks that contain varying amounts of a solid, combustible, organic matter called kerogen.

Old-growth forest Uncut, virgin forest which may contain massive trees hundreds of years old.

Oligotrophic lake A lake that is cold and deep, with a high oxygen content and a relatively low amount of dissolved solids, nutrients, and phytoplankton.

Omnivores Organisms that consume both plants and animals.

Open pit surface mining A method of mineral extraction in which a large pit is dug and the exposed ore removed.

Opportunistic species A species that takes advantage of the weaknesses of other species or its own ability to exploit temporary habitats or conditions.

Optimum population size The number of individuals an environment or habitat can best support.

Ore A mineral deposit that is concentrated enough to make mining it economically feasible.

Organic agriculture The practice of maintaining cropland without the use synthetic fertilizers, pesticides, or herbicides.

Organic compound A compound that contains carbon. Its carbon atoms may be combined with other carbon atoms or with atoms of one or more other elements.

Overburden Overlying vegetation, soil, and rock layers that are removed to expose ore deposits for surface mining.

Overdraft The withdrawal of water from an aquifer faster than the aquifer can naturally replenish itself.

Overgrazing Consumption of vegetation on rangeland by grazing animals to the point that the vegetation cannot be renewed or is renewed at a rate slower than it is consumed.

Overpopulation A situation in which the number of people in an area cannot be supported adequately by the available resources, leading to declining standards of living and failure to fully realize human potential.

Oxygen cycle The natural circulation of oxygen through the environment.

Ozone An atmospheric gas (O_3) that when present in the stratosphere helps protect the Earth from ultraviolet rays, but when present near the Earth's surface is a primary component of urban smog and has detrimental effects on both vegetation and human respiratory systems.

Ozone layer A layer of concentrated ozone in the stratosphere about 20 to 30 miles (32 to 48 kilometers) above the Earth's surface.

Pantheism The belief that multiple gods are responsible for the various forces and workings of nature.

Parasitism An association of two species in which one benefits and the other is harmed.

Parent material The raw mineral material from which soil is eventually formed.

Particulates Tiny particles of solids or liquid aerosols that are light enough to be transported in the air and may cause respiratory problems in humans and atmospheric haze in the environment.

Passive solar system Use of the natural forces of conduction, convection, and radiation to collect, store, and distribute solar heat.

Pasture Grassland used for livestock production.

PCBs Polychlorinated biphenyls, a group of at least 50 widely used compounds which accumulate in food chains and may produce harmful effects in organisms.

Pelagic zone The area in an ocean so deep that light cannot penetrate and organisms must rely on nutrients that filter down from above.

Pellagra A disease associated with a niacin-deficient diet that can cause weakness, spinal pain, convulsions, and idiocy.

Perennial polyculture A mixture of self-sustaining, or perennial, crops.

Permafrost A layer of permanently frozen ground beneath the Earth's surface found in frigid regions.

Perpetual resources Resources that originate from a source that is virtually inexhaustible, at least in time as measured by humans. Examples include the sun, tides, falling water, and winds.

Persistent organic pollutants (POPs) Synthetic compounds that share four common characteristics: toxicity, stability, solubility, and mobility.

Persistent pollutant A pollutant, including those that are nonbiodegradable and those that are only slowly biodegradable, which accumulates in natural systems over time.

Petroleum A liquid fossil fuel composed primarily of hydrocarbon compounds with small amounts of oxygen, sulfur, and nitrogen compounds. Also called crude oil.

pH Numeric value that indicates the relative acidity or alkalinity of a substance on a scale of 0 to 14, with acid solutions below 7, neutral solutions at 7, and basic solutions above 7.

Phase separation process A physical treatment, such as filtration, centrifugation, sedimentation, and flotation, which separates wastes into liquid and solid components.

Photochemical reaction A reaction induced by the presence of light.

Photochemical smog An atmospheric haze that occurs above industrial sites and urban areas resulting from reactions between pollutants produced in high temperature and pressurized combustion processes, such as the combustion of fuel in a motor vehicle. The primary component is ozone.

Photosynthesis The process of using the sun's light energy by chlorophyll-containing plants to convert carbon dioxide (CO_2) and water (H_2O) into complex chemical bonds forming simple carbohydrates such as glucose and fructose.

Phototoxicity The increase in the harmful effects of a pollutant that occurs due to the presence of sunlight.

Phototrophs Organisms containing chlorophyll that produce complex chemicals bonds through photosynthesis. See *Producers*.

Photovoltaic cell A promising new solar technology that relies on a process that occurs when light hits certain light-sensitive materials, called semiconductors. A direct electrical current is funneled into a wire leading from the cell.

Photovoltaic solar systems Systems that transform sunlight directly into electricity utilizing light-sensitive material to absorb the solar energy.

Phytoplankton Microscopic floating plants and algae that function as the major producer in aquatic systems.

Pioneer species The initial species that occur in primary succession, organisms that are capable of breaking down weathered rock and minerals to form soil, or the initial species in a habitat after a major disturbance.

Plankton Microscopic plants (phytoplankton) or animals (zooplankton) that float passively or swim weakly in a body of water. Compare *Nekton*.

Poaching Trespassing on a preserve or private property to hunt; also, the killing of animals that are legally protected.

Point-source pollution Pollution that can be traced to an identifiable source.

Politics The principles, politics, and programs of government.

Pollutant A substance that adversely alters the physical, chemical, or biological quality of the Earth's living systems or that accumulates in the cells or tissues of living organisms in amounts that threaten the health or survival of that organism.

Pollution prevention Reducing or eliminating the creation of pollutants through increased efficiency in the use of raw materials, energy, water, and other resources.

Polyculture The cultivation of a variety of crops suited to the parti-cular climate and soil of the area.

Polyethylene terephthalate (PET) A recyclable type of plastic.

Pool An area of relatively low oxygen content in running-water habitats that houses the consumers and decomposers of biomass. Compare *Riffle*.

Population The size, density, demographic distribution, and growth rate of any definable nation, region, or continent. Individuals of a particular species with definable group characteristics.

Population density The number of individuals per unit of space.

Population momentum Tendency of a population to continue to increase in absolute numbers despite declines in fertility rate due to a large base of childbearing women.

Population policy A government's planned course of action designed to influence and regulate its constituents' choices or decisions on fertility or migration.

Population profile A graphical representation of the age distribution of a population.

Positive feedback An output that promotes a trend.

Post-consumer waste Paper that has been previously used in a consumer product and can be used again in a second product. Examples include computer paper, mimeograph paper, office stationary, and other white paper.

Potential energy Any energy that can be released to do work.

Power The rate at which work is performed or the rate at which energy is expended to do work.

Precedent A legal decision or case that may serve as an example, reason, or justification for a later one.

Pre-consumer waste Paper destined for use in a consumer product that has not been used previously in a consumer product. Examples include the machine trimmings and clippings that result when envelopes, business forms, books, and catalogues are manufactured.

Precycling The conscious effort to purchase merchandise that has a minimal adverse effect on the environment.

Predators Organisms that obtain their food by eating other living organisms.

Predator–prey interaction Relationship between two organisms of different species in which one organism, the predator, feeds on another organism, the prey.

Prescribed burn Deliberately set fire in a forest to prevent more destructive fires or to kill off unwanted plants that compete with a desirable species.

Preservation A strict form of conservation in which use of a resource or area is limited to nonconsumptive activities. Also, the process of maintaining the existing condition of a historical site.

Pressurized hot water reservoir A geothermal energy source in which underground water, heated under intense pressure, produces a mixture of steam, scalding water, and dissolved materials.

Prey Living organisms that serve as food for other organisms.

Primary consumers In a food chain, organisms that consume producers (green plants).

Primary pollutants Pollutants that are emitted directly into the atmosphere where they exert an adverse influence on human health or the environment. The six primary pollutants are carbon dioxide, carbon monoxide, sulfur oxides, nitrogen oxides, hydrocarbons, and particulates. All but carbon dioxide are regulated in the United States.

Primary productivity   The rate at which autotrophs store energy over a given period of time. Compare *Gross primary productivity, Net primary productivity*.

Primary succession   The development of a new ecosystem in an area previously devoid of organisms.

Producers   Self-nourishing organisms. Given water, nutrients and a source of energy, they can produce the compounds necessary for their survival. They form the basis of the food web. Also known as autotrophs.

Profundal zone   Deep area of a body of fresh water where light cannot penetrate and organisms must rely on nutrients that filter down from above.

Pronatalist policy   A policy that encourages increased fertility and a higher birth rate. Compare *Antinatalist policy*.

Proven reserves   Deposits of minerals or fossil fuels that have been located, measured, and inventoried and can be or are currently being extracted at a profit. Also known as an economic reserves.

Public domain   Land owned by U.S. citizens and managed by various federal, regional, state, and local authorities.

Pure command economy   An economic system in which the government makes all the decisions about what and how much to produce and how to distribute goods and services.   Also known as a centrally planned economy. Compare *Pure market economy*.

Pure market economy   (also known as pure capitalism) An economic system in which buyers and sellers in the marketplace determine demand, supply and price. The government does not interfere with the market. See *Capitalism*. Compare *Pure command economy*.

Pyramid of biomass   Conceptual tool used to illustrate that total biomass tends to decrease at each subsequent trophic level and the size of each individual organism tends to increase.

Pyramid of energy   Conceptual tool used to illustrate the inefficiency of energy transfers from one trophic level to another.

Pyramid of numbers   Conceptual tool used to illustrate the tendency toward large population on lower trophic levels and small population on higher trophic levels.

Quality of life   A complex set of indicators that provides a definition of the general condition of a human population in a given area.

Rad   Unit used to measure the amount of radiation absorbed per gram of tissue.

Radiation   Energy released by an atom; radiation takes two basic forms, ionizing and nonionizing radiation.

Radiational cooling   The reradiation of heat from the ground after sunset faster than from the air masses above it, keeping the cooler, denser air near the Earth's surface.

Radioactive decay   The process by which a radioactive atom seeks stability by emitting particles and energy so that the number of neutrons will eventually equal the number of protons in the nucleus.

Radioactive fallout   Dirt and debris contaminated with radiation which is produced by atomic tests and spread throughout the environment by winds and rain.

Radioisotope   Isotope that releases particles or high-level energy.

Radon 222   A naturally occurring radioactive gas arising from the decay of uranium 238 which may be harmful to human health in high concentrations.

Rain shadow effect   The phenomenon that occurs as a result of the movement of air masses over a mountain range. As an air mass rises to clear a mountain, the air cools and precipitation forms. Often, both the precipitation and the pollutant load carried by the air mass will be dropped on the windward side of the mountain. The air mass is then devoid of most of its moisture; consequently, the lee side of the mountain receives little or no precipitation and is said to lie in the rain shadow of the mountain range.

Range   The extent or distribution of a species.

Rare species   A species at risk of local or general (worldwide) extinction because of its small total population.

Reactor core   The hundreds of fuel assemblies comprised of fuel rods that make up the nuclear reactor.

Real gross national product   The gross national product adjusted for any rise in the average price of final goods and services.

Real gross national product per capita   The real gross national product divided by the total population; it gives some idea of how the average citizen is faring economically.

Recycling   The recovery and reuse of materials from wastes.

Redundancy hypothesis   School of thought which contends that some species in an ecosystem can be lost with no adverse effect on the ecosystem as long as others which play the same role persist. According to this view, only minimum diversity is needed for an ecosystem to function properly.

Reforestation   The planting of trees on land previously covered by forest that was removed by natural or human agency.

Refuse   Solid wastes.

Refuse-derived fuel incinerator   Plant that separates incoming waste before burning to remove noncombustible items and recyclable materials.

Rehabilitation   The process of repairing and altering a historical site while maintaining features of the site with significant historical, architectural, and cultural value.

Relative poverty   Having a household income level below a given proportion for the national average.

Religion   The expression of human belief in and reverence for a superhuman power.

Rem   Measure of the damage potential caused by a specific dose of radiation. The Nuclear Regulatory Commission has set 0.17 rem annually as the maximum safe exposure for the general public.

Renewable resources   Resources that are resupplied at rates faster than or consistent with use; consequently, supplies are not depleted.

Replacement fertility   The fertility rate needed to ensure that the population is just "replaced" by its offspring, ranging from 2.1 to 2.5 depending on the mortality rate of the population.

Reserved water rights   The policy insures that upstream projects, such as dams and diversions for irrigation, will not deplete the water supply of an area.

Resiliency   The capacity of an ecosystem to return to a state of dynamic equilibrium after a stress.

Resource   Something that serves a need, is useful, and is available at a particular cost.

Resource base   An estimate of the total amount of a fossil fuel or mineral contained in the Earth's crust based on proven, subeconomic, and indicated reserves.

Resource recovery   An umbrella term referring to the taking of useful materials or energy out of the solid waste stream at any stage before ultimate disposal, for example, recycling, composting.

Respiration The process by which organisms produce energy by capturing the chemical energy stored in food.

Restoration The process of returning a historical site to its original condition by making such changes as removing later alterations and adding missing features.

Restoration ecology A branch of applied ecology that has three major goals: repairing, restoring, or replacing native biotic communities; maintaining the present diversity of species and ecosystems by finding ways to preserve biotic communities or to protect them from human disturbances so they can evolve naturally; and increasing knowledge of biotic communities.

Restricted land use Using land for only one or two purposes. Also known as exclusionary land use. Compare *Multiple land use.*

Rhizobium A rod-shaped bacterium in the genus *Rhizobium*, capable of fixing nitrogen in the root nodules of beans, clover, and other legumes.

Rhythm method Form of birth control based on the natural cycle of ovulation in the female.

Ridge tilling The practice of planting a crop on top of raised ridges to prevent soil erosion.

Riffle An area of high oxygen content in running-water habitats that houses the producers of biomass. Compare *Pool.*

Riparian Relating to the banks of a natural waterway, such as a river.

Rivet hypothesis School of thought which likens species in an environment to the rivets that hold together an airplane. Removing too many species (rivets) may cause the ecosystem (airplane) to collapse.

Rule of 70 Rule for finding the amount of time required for a population to double in numbers: doubling time is equal to 70 divided by the annual growth rate.

Running water habitats Aquatic ecosystems, such as streams and rivers, that have continuously moving currents of water. The speed of the current influences both the composition of the channel and the oxygen content.

Runoff Precipitation that flows along land contours from high elevations to low elevations.

Salinity A measure of the concentration of dissolved salts in water.

Salinization The deposition on farmland of salty minerals during the evaporation of irrigation water.

Sanitary landfill Landfill designed to receive only nonhazardous waste.

Saprotroph An organism that uses enzymes to feed on waste products of living organisms or tissues of dead organisms. Compare *Autotroph, Heterotroph.*

Scarce metal Any metal that constitutes less than 0.1 percent, by weight, of the Earth's continental crust. Compare *Abundant metal.*

Scavengers Heterotrophs that consume entire dead organisms.

Science Body of systematized knowledge about nature and the physical world that has been derived from observation, study, and experimentation.

Scientific method Process of observation, hypothesis development, and experimentation.

Seasonal hunger The period of time between the new harvest and the point at which reserves from the previous harvest run out.

Secondary consumers In a food chain, organisms (usually animal) that consume primary consumers.

Secondary pollutants Pollutants formed from the interaction of primary pollutants with other primary pollutants or with atmospheric compounds such as water vapor.

Secondary succession The regrowth that occurs after an ecosystem has been disturbed, often by human activity. Although some organisms are still present, the ecosystem is set back to an earlier successional stage.

Second Law of Thermodynamics Law that states that with each change in form some energy is degraded to a less useful form and given off to the surroundings, usually as low-quality heat. See also *First Law of Thermodynamics.*

Secure landfill A landfill designed to receive hazardous and toxic wastes with specially engineered systems to prevent the escape of these substances into the environment and reduce the production and release of leachate.

Sedimentary cycle Involves materials that move primarily from land to oceans and back to the land again.

Sedimentary rock Rock formed by the deposit of small bits and pieces of sediments that are carried by wind or rain and then compacted and cemented.

Selective cutting The cutting of intermediate-aged, mature, or diseased trees in an uneven-aged forest stand either singly or in small groups.

Self-sufficiency A way of life in which basic needs of food, clothing, and shelter are met and sustained by the efforts of the immediate community without recourse to goods and services produced by others.

Sense of the Earth An intimate knowledge of the local environment — location and use of plant species, habits and movements of animals, and seasonal weather patterns. Often best exhibited by early hunter-gatherer societies. Also known as sense of place.

Separate sewer system Consists of sanitary sewers which carry only wastewater, and storm sewers which carry only rainwater and melting snow. Compare *Combined sewer system.*

Septic tank An underground concrete tank large enough to accommodate the wastewater flow from a building in which bacteria liquefy solids and the liquid then passes out to a drainage field.

Shelterwood technique A timber harvesting technique that consists of two cuts. In the first cut only a portion of trees are harvested to open up the forest and allow some light in; the majority of the trees are left to naturally re-seed the area. After the young seedlings are established, the second cut removes the oldest and best trees.

Siltation The obstruction of streams and lakes by soil deposited as a result of erosion.

Singularity The infinitely hot and dense dot from which, according to the big bang theory, the entire universe arose.

Slash-and-burn A method of land clearing in which vegetation is cut, allowed to dry, and then set on fire prior to the cultivation of the soil and the planting of crops.

Sludge A viscous, semisolid mixture of bacteria- and virus-laden organic matter, toxic metals, synthetic organic chemicals, and settled solids removed from domestic and industrial wastewater at sewage treatment plants.

Smog A dense, discolored haze containing large quantities of soot, ash, and gaseous pollutants such as sulfur dioxide and carbon dioxide.

Soil A naturally occurring mixture of inorganic chemicals, air, water, decaying organic material, and living organisms.

Soil degradation A deterioration of the quality and capacity of a soil's life-supporting processes.

Soil erosion The accelerated removal of soil through various processes at a greater rate than soil is formed.

Soil fertility   Measure of the soil's mineral and organic content; the ability of a soil to supply the required type and amount of nutrients for optimum growth of a particular crop when all the other growing factors are favorable.

Soil horizon   A horizontal layer formed as a soil develops and distinct in color, texture, structure, and composition.

Soil loss tolerance level   The amount of soil that can be lost through erosion without a subsequent decline in fertility. Also known as a soil's T-value or replacement level.

Soil productivity   The ability of the soil to sustain life, especially vegetation.

Soil profile   A vertical profile of soil horizons.

Soil structure   The tendency of the particles of a soil to form larger aggregates (crumbs, chunks, or lumps) largely depending on the amount of clay and organic material it contains. Also known as soil tilth.

Soil texture   The coarseness of a soil, depending largely on the proportion of sand, silt, and clay it contains.

Solar energy   Radiant energy originating from the sun.

Solar pond   A method of generating electricity using salt water and fresh-water heated by the sun.

Sole source aquifer   An aquifer that serves as the principal drinking water source for a community.

Solidification   A waste disposal method in which liquid, semisolid, or solid hazardous waste is transformed into a more benign solid product by entombing the waste in a concrete block.

Solid waste   Material that is rejected or discarded as being spent, useless, worthless, or in excess. Also known as refuse or garbage.

Solid waste stream   The sum of all waste from all sources — industrial, agricultural, residential.

Source reduction   Reducing the amount of waste initially entering the waste stream by product changes such as minimizing excessive packaging, extending the useful life of products or reducing the amount of hazardous substances used and the volume of wastes produced.

Source segregation   Keeping used hazardous resources apart from nonhazardous resources in order to reduce the overall volume of material that must be treated as hazardous.

Source separation   The removal of recyclables from the waste stream and the sorting of recyclables to maintain the purity and value of the recyclable material.

Spaceship Earth   Term coined by environmentalists of the 1960s which likens the closed system of a spaceship to the planet Earth, with humans as the pilots of the spaceship, responsible for its well-being.

Specialist   A species that requires one particular type of habitat for survival. Compare *Generalist*.

Speciation   The separation of populations of plants and animals, originally able to interbreed, into independent evolutionary units (or species) which can no longer interbreed because of accumulated genetic differences.

Species   A group of individuals or populations potentially able to interbreed and unable to produce fertile offspring by breeding with other sorts of animals and plants.

Species diversity   The variety of plants and animals in a given area.

Split estate   Situation in which the federal government owns a piece of land, but private citizens or companies own the minerals beneath the surface.

Spontaneous radioactivity   The release of mass in the form of particles, energy in the form of rays, or both, by unstable atoms of a particular element, such as uranium.

Standard of living   Quality of life for the majority of people in a given population.

Standing water habitats   Relatively closed ecosystems with well-defined boundaries that generally contain both inlet and outlet streams.

Staples   Principal edible plants essential to people's diets.

Starvation   The consumption of insufficient calories to sustain life.

Statutory law   The body of acts passed by a local or state legislature or Congress.

Steady-state economy (SSE)   An economy characterized by a constant level of human population and a constant level of artifacts.

Sterilization   Nonreversible method of birth control in which the sperm tubes or the oviducts are altered to prevent the occurrence of fertilization.

Stewardship ethic   A guide for behavior based on the belief that humans should act as caretakers and nurturers of the natural world.

Stewardship worldview   A way of percieving reality that humans have a responsibility to care for the Earth.

Still bottom   The concentrated, highly toxic mixture of metals and solvents that results from resource recovery distillation techniques.

Strategic mineral   A mineral that is essential to national defense but which exists in relatively short supply.

Stratosphere   An atmospheric layer extending from 6 or 7 miles to 30 miles above the Earth's surface.

Stratospheric ozone depletion   The thinning of the ozone layer in the stratosphere; occurs when certain chemicals (such as chlorofluorocarbons) capable of destroying ozone accumulate in the upper atmosphere.

Strict liability   In reference to products, means that if harm results from a product, the maker of that product is liable for the harm done. The same principle applies to activities.

Strip cropping   The practice of alternating rows of grain with low-growing leaf crops or sod.

Stunting   An indicator of chronic malnutrition, it is calculated by comparing the height-for-age of a child with a reference population of well-nourished children.

Subbituminous   An abundant coal with low sulfur content and low heating values. See *Lignite*.

Subeconomic resources   Mineral concentrations or fossil fuel deposits that have been discovered, but from which the resource cannot be extracted at a profit at current prices or with current technologies.

Subsistence farming   The production of food and other necessities to satisfy the needs of the farm household.

Subsoil   Relatively infertile lower layer of soil in which dissolved minerals accumulate.

Subsurface mining   A method of mineral extraction in which a shaft is dug down to the level of the deposit and the ore is then extracted and hauled to the surface.

Succession   The gradual, sequential, and somewhat predictable changes in the composition of an ecosystem's communities from an initial colonization of an area by pioneer organisms to the eventual development of the climax community.

Sulfur oxide A primary pollutant originating chiefly from the combustion of high-sulfur coals; it is regulated in the United States.

Superfund A fund originally established by the U.S. government in 1980 to clean up hazardous waste sites and to respond to uncontrolled releases of hazardous substances.

Surface mining A method of mineral extraction in which the overlying vegetation, soil, and rock layers are removed to expose an ore deposit.

Surface waters Bodies of water such as lakes and rivers recharged by precipitation that flows along land contours.

Sustainability The goal of ecological economics; it is a relationship between dynamic human systems and larger dynamic, but normally slower-changing, ecological systems, in which the effects of human activities remain within limits, so as not to destroy the diversity, complexity, and function of the ecological support system.

Sustainable In harmony with natural systems and acting to maintain the health and integrity of the environment.

Sustainable agriculture Farming methods, such as crop rotation and the use of organic fertilizers, that protect the soil and restore its fertility.

Sustainable development Managing the economy and renewable resources of an area for the common good of the entire community and the environment.

Sustainable resource use The use of renewable resources at rates that do not exceed their capacity for renewal.

Sustained yield The amount of harvestable material (of a renewable resource) that can be removed from an ecosystem over a long period of time with no apparent deleterious effects on the system.

Symbiosis Any intimate association of two dissimilar species regardless of the benefits or harm derived from it.

Synergistic effect An interaction between two substances that pro-duces a greater effect than the effect of either one alone. An interaction between two relatively harmless components in the environment to form a more potent pollutant.

Tailings The rock surrounding a valuable mineral in an ore deposit which is discarded.

Takings Stipulation in the Fifth Amendment which allows the federal government to take private property for legitimate public use but requires the

government to provide the landowner(s) with just compensation.

Tar sands A minor fossil fuel deposit composed of sandstones which contain bitumen, a thick, high-sulfur, tarlike liquid that may be purified and upgraded to synthetic crude oil.

Ten percent rule The concept that, in general, 90 percent of available energy is lost as low-quality heat when members of one trophic level are consumed by members of another.

Teratogenic Causing malformations in fetuses.

Terrestrial Of or pertaining to land environments.

Territorial waters Any area of water over which an adjacent country claims jurisdiction; an area within which a country has the sole right to exploit mineral and fish resources.

Tertiary consumers In a food chain, organisms such as carnivores and omnivores that eat secondary consumers.

Thermal pollution An undesired rise in temperature in an environment above that which occurs through natural solar radiation; caused by the release of warm substances into the environment.

Thermal stratification The seasonal process that large lakes in temperate zones undergo in which an epilimnion, hypolimnion, and thermocline develop within the lake. See *Epilimnion, Hypolimnion, and Thermocline.*

Thermocline The area of sharp temperature gradient that exists between the epilimnion and the hypolimnion.

Thermosphere An atmospheric layer that extends from 56 miles to outer space.

Third world Those countries located mainly in Africa, Asia, and Latin America that are neither industrial market economies (first world) nor centrally planned economies (second world).

Threatened species A species that is severely exploited at present or inhabits an area of major environmental disturbance, is unlikely to adapt to those changes, and will most likely become endangered.

Throughput The natural capital used to produce economic goods.

Throwaway society A society in which objects are manufactured to be short-lived, disposable, and nonrepairable and people habitually discard objects rather than repairing, reusing, or recycling them.

Tidal energy Energy originating from fluctuations in the ocean's water level.

Tilth The tendency of the particles of a soil to form larger aggregates (crumbs, chunks, or lumps) largely depending on the amount of clay and organic material it contains. Also known as soil structure.

Tipping fees Disposal cost for dumping garbage in a landfill.

Tissue culture The production of plants from individual plant cells rather than plant seeds. Also known as cloning.

Tolerance model of succession Theory that subsequent communities in an ecosystem are not determined by present communities, but by an increased level of tolerance to environmental changes.

Topography The shape and contour of land formations.

Topsoil Highly fertile layer of soil found immediately below the surface litter and composed primarily of humus, living organisms, and minerals.

Tort A claim for which a civil suit can be brought by an injured plaintiff.

Total fertility rate The average number of children a woman will bear during her life, based on the current age-specific fertility rate.

Toxicant A chemical that can cause serious illness or death. Also called toxin.

Toxicology The study of poisonous materials, their effects, and their antidotes.

Toxic substance A chemical substance that adversely affects human health and the environment.

Traditional economy A self-sufficient economic system where people grow their own food and make the goods.

Transpiration The loss of water vapor through the pores of a plant.

Transuranic radioactive waste Human-made radioactive elements such as plutonium, americium, and neptunium, that have atomic numbers higher than that of uranium.

Trash Solid wastes.

Trash conversion The burning of municipal and industrial refuse and waste to produce electrical energy.

Trespass The unwarranted or uninvited entry upon another's property by a person, a person's agent, or an object that he or she caused to be deposited there. The property can be land, material possessions, or even one's own body.

Trickle drip irrigation An irrigation technique in which water is delivered slowly to the base of plants through perforated or permeable pipes or tubes.

Trophic level   A group of organisms with the same relative position in the food chain.

Trophic level   The various levels of producers and successive steps removed from the producers.

Trophy animal   An animal valued for its antlers, horn(s), hide, beauty, or symbolism.

Troposphere   The atmospheric layer that extends from the Earth's surface to 6 or 7 miles above the surface.

Turnover   The mixing of the upper layer and lower layer of a lake, which most often occurs in the spring and fall, due to dramatic changes in surface water temperature.

T-value   See *Soil loss tolerance level.*

Ultimately recoverable resource   An estimate of the total amount of fossil fuel that will eventually be recovered based on discovery rates, future costs, demand and market values, and future technological developments.

Unconfined aquifer   Groundwater deposit in which the overlying material is permeable, allowing water which permeates the soil to move freely downward to the water table. Unconfined aquifers are susceptible to contamination.

Undernourishment   The consumption of inadequate levels of protein and calories which over an extended period of time gradually weakens an individual's capacity to function properly and to ward off disease.

Uneven-aged harvesting technique   A timber harvesting technique that usually involves three or more cuts spaced over the average lifetime of trees in a particular forest. Compare *Even-aged harvesting.*

Underweight   A measure of a child's weight-for-age compared to a reference population of well-nourished children.

Unrealized resource   Material that is usually thrown away, but could be used to benefit human and natural systems.

Upwelling   An upward movement of ocean water masses that brings nutrients to the surface and creates a region of high productivity.

Uranium mill tailings   Wastes created when uranium ore is milled or processed into an enriched form which can be used in a reactor.

Urbanization   An increase in the number and size of cities.

Urban renewal   The practice of demolishing older buildings in order to accommodate new and generally homogeneous structures.

Urban revival   Revitalization of neighborhoods for private or commercial use.

Urban sprawl   The spread of urban areas; in general, it substantially reduces farmland.

Utilitarianism   Conservation for economic reasons.

Vermin   Animal species that are regarded as undesirable.

Vital statistics   Population statistics such as births (natality), deaths (mortality), and immigration.

Waste exchange   A process which brings together companies that have waste and companies that want to recover and reuse those resources.

Waste minimization   An umbrella term that refers to industrial practices that minimize the volume of products, minimize packaging, extend the useful life of products, and minimize the amount of toxic substance in products.

Waste-to-energy incinerator   Special plant engineered to burn garbage to produce heat and steam that is then used to produce electricity.

Wasting   Acute malnutrition that causes a recent and substantial weight loss.

Waterlogged   Air spaces in the soil fill with water for too long a period of time. Caused by irrigating farmland continuously without allowing it to lie idle.

Water mining   Severe overdrafts that can lead to depletion, often lowering the water table so drastically that further extraction is no longer economically feasible.

Water scarcity   The chronic lack of renewable freshwater.

Water stress   The episodic lack of renewable freshwater.

Water table   The depth at which an aquifer begins.

Watershed   The entire runoff area of a particular body of water.

Wave power   The stored kinetic energy contained within ocean waves, which can be used to generate electricity.

Weather   The day-to-day pattern of precipitation, temperature, wind, barometric pressure, and humidity.

Weed   A plant species that is regarded as undesirable.

Wetland   Vegetated land area that is occasionally or permanently covered by water and acts as a natural boundary between land areas and bodies of water.

Wilderness   The domain of nature, a region undisturbed by human artifacts and activity. An area in which both the biotic and abiotic communities are minimally disturbed by humans.

Wilderness ecosystem   An ecosystem in which both the biotic and abiotic components are minimally disturbed by humans.

Wildlife management   Field of study that seeks to sustain populations of wildlife species.

Windbreak   A row or groups of trees and shrubs planted along the wind-ward side of fields in order to reduce wind erosion. Also known as a shelterbelt.

Wind farm   A large utility project consisting of many wind turbines that produce electricity for the surrounding population.

Wind power   Energy that originates from air currents and can be collected using windmills or wind turbines.

Wind turbine   A device for produ-cing electricity in which large blades or rotors are driven by wind.

Wise use movement (WUM)   Both an anti-environmental and anti-conservation ethic and an umbrella name for a coalition of some 3,000 groups, many of them funded by developers and extractive industries (such as timber and mining).

Work   The product of the distance that an object is moved times the force used to move it.

X ray   An ionizing radiation that is produced by bombarding a metallic target with fast electrons in a vacuum, is capable of penetrating various thicknesses of solids, and is a powerful mutagen.

Zero population growth   The growth rate at which births are equal to deaths.

Zone of leaching   The layer of soil between topsoil and subsoil through which water and dissolved minerals pass.

Bibliography

NWF: National Wildlife Federation

PRB: Population Reference Bureau

UN: United Nations

USDA: United States Department of Agriculture

USDOE: United States Department of Energy

USEPA: United States Environmental Protection Agency

USFWS: United States Fish & Wildlife Service

USGS: U.S. Geological Survey

Chapter 1

Abbey, Edward. *Desert Solitaire*. New York: Ballantine Books, 1968.

American Society of Zoologists. *Science as a Way of Knowing: Human Ecology*. Thousand Oaks, Calif.: American Society of Zoologists, 1985.

Berry, Wendell. *A Continuous Harmony: Cultural and Agricultural Essays*. New York: Harcourt Brace Jovanovich, 1972.

Botkin, Daniel B. *Discordant Harmonies: A New Ecology for the Twenty-First Century*. New York: Oxford University Press, 1990.

Callicott, J. Baird. *Companion to a Sand County Almanac: Interpretive and Critical Essays*. Madison: University of Wisconsin Press, 1987.

Carson, Rachel. *The Sense of Wonder*. New York: Harper & Row, 1956.

Carson, Rachel. *Silent Spring*. Boston: Houghton Mifflin, 1962.

Chown, Marcus. "The Big Bang." *New Scientist* 22 October 1987.

Commoner, Barry. *The Closing Circle: Nature, Man, and Technology*. New York: Knopf, 1972.

Commoner, Barry. *Making Peace with the Planet*. New York: Pantheon Books, 1990.

Devall, Bill, and George Sessions. *Deep Ecology: Living As If Nature Mattered*. Salt Lake City, Utah: Smith, 1985.

"Eating Your House." *CNN Money*. Available: http://money.cnn.com/2003/10/01/commentary/everyday/sahadi/index.htm. 27 May 2005.

Ehrlich, Anne H., and Paul R. Ehrlich. *Earth*. New York: Franklin and Watts, 1987.

Fladers, Susan. *Thinking Like a Mountain: Aldo Leopold and the Evolu-tion of an Ecological Attitude Toward Deer, Wolves, and Forests*. Columbia: University of Wisconsin Press, 1974.

Hardin, Garrett. "The Tragedy of the Commons." *Science* 162 (1968): 1243–1248.

Hodgson, Bryan. "Buffalo Back Home on the Range." *National Geographic* 18.5 (1994).

"Inside the New American Home." *Time* October 14, 2002: 64-75.

Joseph, Lawrence E. *Gaia: The Growth of an Idea*. New York: St. Martin's Press, 1990.

Kaplan, Robert D. "The Coming Anarchy." *The Atlantic Monthly* February 1994.

Kerr, Richard A. "No Longer Willful, Gaia Becomes Respectable." *Science* 240 (1988): 293–296.

Leopold, Aldo. *A Sand County Almanac and Sketches Here and There*. New York: Oxford University Press, 1949.

Lewin, Roger. "Case Studies in Ecology." *Science* 232 (1986): 25.

Lovelock, James. *Gaia: A New Look at Life on Earth*. New York: Oxford University Press, 1979.

_____. *The Ages of Gaia*. New York: W.W. Norton & Company, 1988.

Lubchenco, Jane. "Entering the Century of the Environment: A New Social Contract for Science." *Science* (23 January 1998): 491-497.

Margulis, Lynn, and Lorraine Olendzenski. *Environmental Evolution*. Cambridge, Mass.: MIT Press, 1992.

Marsh, George Perkins. *Man and Nature*. New York: Scribner, 1864.

_____. *The Earth as Modified by Humans*. New York: Arno, 1874.

McLuhan, T. C. *Touch the Earth. A Self-Portrait of Indian Existence*. New York: Dutton, 1971.

Miller, Alan. "The Influence of Personal Biases on Environmental Problem Solving." *Journal of Environmental Problem Solving* 17.2 (1983): 133–143.

Mills, Stephanie. *In Praise of Nature*. Washington, D.C.: Island Press, 1990.

Nash, Roderick. *The American Environment: Readings in the History of Conservation*. Reading, Mass.: Addison-Wesley, 1968.

_____. *The Rights of Nature: A History of Environmental Ethics*. Madison: University of Wisconsin Press, 1988.

_____. *Wilderness and the American Mind*. New Haven, Conn.: Yale University Press, 1967.

Natural Resources Defense Council. "Summary and Analysis." *Damage Report: Environment and the 104th Congress*. Available: http://www.nrdc.org/nrdc/nrdcpro/dmg/chap1.html.

_____. "Legislative Assaults Turned Back." *Damage Report: Environment and the 104th Congress*. Available: http://www.nrdc.org/nrdc/nrdcpro/dmg/chap2.html.

_____. "The Damage Done — Environmentally Damaging Legislation Enacted by the 104th Congress." *Damage Report: Environment and the 104th Congress*. Available: http://www.nrdc.org/nrdc/nrdcpro/dmg/chap3.html.

Orians, Gordon H. "The Place of Science in Environmental Problem Solving." *Environment* 28 (1986): 12–21.

Parker, Tom. *In One Day*. Boston: Houghton Mifflin, 1984.

"Planet of the Year: Endangered Earth." *Time* (2 January 1989): 24–63.

Rifkin, Jeremy. *Declaration of a Heretic*. Boston: Routledge & Kegan Paul, 1985.

Rothkrug, Paul, and Robert L. Olson. *Mending the Earth: A World for Our Grandchildren.* Berkeley, Calif.: North Atlantic Books, 1991.

Sagan, Dorion. *Biospheres: Metamorphosis of Planet Earth.* New York: McGraw-Hill, 1990.

Schumacher, Ernest. *Small Is Beautiful: Economics As If People Mattered.* New York: Harper & Row, 1973.

Science 277 (25 July 1997). Special issue on human-dominated ecosystems.

Silk, Joseph. *The Big Bang. The Creation and Evolution of the Universe.* San Francisco: Freeman, 1980.

Speth, Timothy C. "The Ecological Lessons of the Past: An Anthropology of Environmental Decline." *The Ecologist* 19.3 (1989).

State of the World 2004: A Worldwatch Institute report on progress toward a sustainable society. New York: W.W. Norton & Company, 2004.

State of the World 2003: A Worldwatch Institute report on progress toward a sustainable society. New York: W.W. Norton & Company, 2003.

"Teens, boomers find common ground on environment." *CNN Earth Story Page* (21 April 1998). Available: http://www.cnn.com/EARTH/9804/21/environ.survey/index.html. 22 April 1998.

Tuxill, John, and Chris Bright. "Losing Strands in the Web of Life." *State of the World 1998: A Worldwatch Institute report on progress toward a sustainable society.* New York: W.W. Norton & Company, 1998.

Udall, Stewart. *The Quiet Crisis.* New York: Holt, Rinehart and Winston, 1963.

U.S. Department of Interior, National Park Service. *Natural Resource Year in Review.* Denver: NPS, 1997.

Vital Signs 2003. Worldwatch Institute. New York: W.W. Norton & Company, 2003.

Wattenberg, Ben J. *The Good News Is the Bad News Is Wrong.* New York: Simon & Schuster, 1984.

Westbroek, Peter. *Life As A Geologic Force: Dynamics of the Earth.* New York: W.W. Norton & Company, 1991.

Wilson, Edward O. *Biodiversity.* Washington, D.C.: National Academy Press, 1988.

Chapter 2

"Aid Flows." *United Nations Development Programme Website.* Available: http://www.undp.org/hdro/iaid.htm. 24 February 1999.

"The Antarctic Treaty System." *The Antarctica Project Website.* Available: http://www.asoc.org/general/ats.htm. 5 March 1999.

"Antarctic World Park Campaign." *The Antarctica Project Website.* Available: http://www.asoc.org/campaign.htm.

Baer, Richard A. Jr. *The Ethical Quality of Life: Ecology, Religion, and the American Dream* : 7-14.

Baumol, William J., and Wallace Oates. *The Theory of Environmental Policy.* New York: Cambridge University Press, 1988.

Berry, Thomas. *The Dream of the Earth.* San Francisco: Sierra Club Books, 1988.

Berry, Thomas, and Thomas Clark, S. J. *Befriending the Earth.* Mystic [Conn.]: Twenty-Third Publications, 1991.

Berry, Wendell. "Getting Along with Nature" *Country Journal* (1983): 46, 50-55.

BNA Editorial Staff, eds. *U.S. Environmental Laws.* Washington, D.C.: Bureau of National Affairs, 1988.

Braus, Judy A. and David Wood. *Environmental Education in the Schools: Creating a Program that Works!"* Peace Corps, Information Collection and Exchange, 1993.

Brittan, Samuel. *Capitalism With a Human Face.* Brookfield [VT]: Aldershot, 1995.

Brown, Lester. *Building a Sustainable Society.* New York: W.W. Norton & Company, 1981.

Brown, Lester R., Christopher Flavin, and Sandra Postel. "Banking on the Environment." In *Saving the Planet: How to Shape an Environmentally Sustainable Global Economy.* New York: W.W. Norton & Company, 1991.

_____. "Green Taxes." In *Saving the Planet: How to Shape an Environmentally Sustainable Global Economy.* New York: W.W. Norton & Company, 1991.

_____. "Reshaping Government Incentives." In *Saving the Planet: How to Shape an Environmentally Sustainable Global Economy.* New York: W.W. Norton & Company, 1991.

The Coalition on the Environment and Jewish Life. *A Jewish Response to the Environmental Crisis.* Available: http://www.jtsa.edu/org/coejl/Pages/Pages/whatis.htm. 1 March 1999.

Cohen, Michael J. *Across the Running Tide.* Freeport, Me.: Cobblestone, 1979.

_____. *Prejudice Against Nature: A Guidebook for the Liberation of Self and Planet.* Freeport, Me.: Cobblestone, 1984.

Constanza, Robert. *Ecological Economics: The Science of Management and Sustainability* . Columbia [NY]: Columbia University Press, 1991.

Dalai Lama of Tibet. "A Universal Task." *EPA Journal* 17.4 (1991).

Daly, Herman E. and Kenneth N. Townsend. *Valuing the Earth.* Cambridge [MA]: MIT Press, 1993.

"Debt-For-Nature Swaps." *Conservation International Website.* Available: http://www.conservation.org/web/aboutci/strategy/dfn-swap.htm. 5 March 1999.

Decker, Daniel J., and Gary R. Goff, eds. *Valuing Wildlife: Economic and Social Perspectives.* Boulder, Colo.: Westview Press, 1987.

Durning, Alan Thein. *How Much Is Enough? The Consumer Society and the Future of the Earth.* New York: W.W. Norton & Company, 1992

Environmental Law Institute. *Environmental Law Deskbook.* Washington, D.C.: Environmental Law Institute, 1989.

Evangelical Environmental Network. *About Us.* Available: http://www.esa-online.org/about. 27 May 2005.

Floresta: Healing the Land and Its People. Available: http://floresta.org/program.htm. 2 March 1999.

Grameen Foundation USA Website. Available: http://www.grameenfoundation.org. 27 May 2005.

Gray, David B. *Ecological Beliefs and Behaviors: Assessment and Change.* Westport, Conn.: Greenwood Press, 1985.

Hawken, Paul. "The Ecology of Commerce." *INC* 14.4 (April 1992): 93-100.

Hardin, Garrett. "The Tragedy of the Commons." *Science* 162 (1968): 1243–1248.

Hargrove, Eugene C. *Foundations of Environmental Ethics.* Englewood Cliffs, N.J.: Prentice-Hall, 1989.

International Court of Justice Website. Available: http://www.icj-cij.org/. 27 May 2005.

Iozzi, Louis A. "What Research Says to the Educator — Part Two: Environmental Education and the Affective Domain." *Journal of Environmental Education* 20.4 (1989): 6–14.

Knickerbocker, Brad. "Bully Pulpit: Clergy Preach Conservation," *The Christian Science Monitor* 3 February 1997.

Leopold, Aldo. *A Sand County Almanac, and Sketches Here and There.* New York: Oxford University Press, 1949.

"Loans by sector." *Asian Development Bank Website.* Available: http://adb.org/About/Statisitics/lending.asp. 24 February 1999.

Mardenfeld, Sandra. "Children Create a Club for a Cleaner Environment." *The Christian Science Monitor* (30 September 1997): 15.

Meeker, Joseph W. "The Assisi Connection." *Wilderness* 51 (1988): 61–64.

"Mexico Sets International Precedent with First Ever 'Debt for Environmental Education' Swap." *Environmental Communicator* (May/June 1997): 15.

Monroe, Martha C., and Stephen Kaplan. "When Words Speak Louder Than Actions: Environmental Problem Solving in the Classroom." *Journal of Environmental Education* 19.3 (1988): 38–42.

Narveson, Jan and Tony Smith. "Free Market Environmentalism." *Ag Bioethics Forum* (Spring 1995): 2-7.

Nash, James A. "Frugality: Outdated Virtue or the Foundation of Economic Justice and Sustainability? (lecture at Miami University) 19 October 1994.

Nash, Roderick. *The Rights of Nature.* Madison: University of Wisconsin Press, 1989.

National Religious Partnership for the Environment. *History.* Available: http://www.nrpe.org. 27 May 2005.

Neihardt, John. *Black Elk Speaks: Being the Life Story of a Holy Man of the Oglala Sioux.* New York: W. Morrow & Co., 1932.

Newsome, Melba. "To have dominion in the earth." *The Amicus Journal* (Winter 1999): 15-17.

Nixon, Will. "In a land of malls, some educators are trying to help children find their way back into the woods." *The Amicus Journal* (Fall 1997): 31.

_____. "Personality, play, and a sense of place." *The Amicus Journal.* (Summer 1997): 31.

The North American Coalition for Christianity and Ecology (NACCE) Website. Available: http://www.nacce.org. 27 May 2005.

"On Eve of Clinton Visit, World Bank Grants China $330 Million for High Polluting Coal Project." *SEEN.* Available: http://www.seen.org/chinaprelease.html. 27 May 2005

Orr, David. *"Ecological Literacy: Education and Transition to a Postmodern World.* Albany: State of New York Press, 1992.

Planning Group of the National Forum for Partnerships Supporting Education about the Environment. *Education for Sustainability: An Agenda for Action.* Washington, D.C.: GPO.

Plant, Judith, ed. *Healing the Wounds.* Philadelphia: New Society Publishers, 1989.

"The Protocol on Environmental Protection to the Antarctic Treaty." *The Antarctica Project Website* (24 July 1998). Available: http://www.asoc.org/currentpress/protocol.htm. 4 March 1999.

Raphael, Ray. *Tree Talk: The People and Politics of Timber.* Washington, D.C.: Island Press, 1981.

Rees, William E. "The Ecology of Sustainable Development." *The Ecologist* 20.1 (1990): 18–24.

Repetto, Robert, et al. *Wasting Assets: Natural Resources in the National Income Accounts.* Washington, D.C.: World Resources Institute, 1989.

Ridker, Ronald G., and William D. Watson. *To Choose a Future: Resource and Environmental Consequences of Alternative Growth Paths.* Baltimore: Johns Hopkins University Press, 1980.

Rifkin, Jeremy. *Declaration of a Heretic.* Boston: Routledge & Kegan Paul, 1985.

Roberts, Elizabeth, and Elias Amidon. *Earth Prayers from Around the World: 365 Prayers, Poems, and Invocations for Honoring the Earth.* New York: Harper-Collins, 1991.

Rolston, Holmes. *Environmental Ethics, Duties to and Values in the Natural World.* Philadelphia: Temple University Press, 1988.

_____. *Philosophy Gone Wild: Essays in Environmental Ethics.* Buffalo, N.Y.: Prometheus Books, 1986.

Rossi, Vincent. "The Eleventh Commandment: Toward an Ethic of Ecology." *Epiphany Journal* (1981): 1-11.

Simons, Marlise. "Modern Creed for Ancient Church." *New York Times* 19 October 1997.

"Hazardous Waste Sites Defined." *EPA Superfund Website.* Available: http://epa.gov/superfund/accomp/ei/state.htm. 27 May 2005.

"Statistical Updates: August 1998." *Grameen Bank Website.* Available: http://www.grameen.com/bank/supdates.html. 3 March 1999.

Stone, Christopher D. *Should Trees Have Standing? Toward Legal Rights for Natural Objects.* Los Altos, Calif.: Kaufman, 1973.

Swimme, Brian. *The Universe is a Green Dragon.* Santa Fe, N.M.: Bear and Company, 1984.

"Trends in Advertising Volume." *Television Bureau of Advertising.* Available: http://www.tvb.org/nav/build_framest.asp?url=/rcentral/MediaTrends/Track/Trends_In_Advertising_Volumer.asp. 27 May 2005.

Turner, Tom. *Wild by Law: The Sierra Club Legal Defense Fund and the Places It Has Saved.* San Francisco: Sierra Club Books, 1990.

Warner, Keith. "SWEEP: a voice for the poor and God 's Creation...with a Franciscan accent." *GreenCross* (1997): 3, 16-17.

"Wise Use: What Do We Believe?" *The Center for the Defense of Free Enterprise.* Available: http://www.eskimo.com/~rarnold/wiseuse.html. 27 May 2005.

White, Lynn., Jr. "The Historical Roots of Our Ecological Crisis." *Science*: 51-59.

World Conservation Union. *Caring for the Earth: A Strategy for Sustainable Living.* Gland, Switzerland; 1991.

Ziffer, Karen A., J. Martin Goebel, and Susanna Mudge. "Ecotourism." *Orion* 9.2 (1990): 42–45.

Chapter 3

Carpenter, Steven R., and James F. Kitchell. "Consumer Control of Lake Productivity." *Bioscience* 38 (1988): 764–769.

Evans, F. C. "Ecosystem and the Basic Unit in Ecology." *Science* 123 (1956): 1127–1128.

Gittleman, John L. *Carnivore Behavior, Ecology, and Evolution.* Ithaca, N.Y.: Comstock, 1989.

Naiman, Robert J., Jerry M. Meillo, and John E. Hobbie. "Ecosystem Alteration of Boreal Forest Streams by Beaver." *Ecology* 67 (1986): 1254–1269.

"Integrated Assessment of Hypoxia in the Northern Gulf of Mexico." *National Science and Technology Council, Committee on Environment and Natural Resources Website.* Available: http://www.nos.noaa.gov/products/hypox_finalfront.pdf. 2000.

Odum, Eugene. "The Emergence of Ecology as a New Integrative Discipline." *Science* 195 (1977): 1289–1292.

Perry, John S. "Managing the World Environment." *Environment* 28. 1 (1986): 10–40.

Royce-Malgren, Carl H., and Winsor H. Watson III. "Modification of Olfactory Related Behavior in Juvenile Atlantic Salmon by Changes in pH." *Journal of Chemical Ecology* 13.3 (1987): 533–546.

Simon, Herbert A. *The Colonization of Complex Systems in Hierarchy Theory.* New York: Patee, 1973.

Smith, Robert Leo. *Ecology and Field Biology.* New York: Harper & Row, 1980.

Whittaker, Robert H. *Communities and Ecosystems*, 2nd ed. New York: MacMillan, 1975.

Worster, David. *Nature's Economy: The Roots of Ecology.* San Francisco: Sierra Club Books, 1977.

Chapter 4

Alliance for the Chesapeake Bay. *Bay Journal* 3.10; 4.1-8, 10 (1994-1995).

Bolin, Bert. "The Carbon Cycle." *Scientific American* 223.3 (1970): 124–132.

Borman, F. H., and Gene E. Likens. "The Nutrient Cycles of an Ecosystem." *Scientific American* 223.3 (1970): 92–101.

Cook, E. K., and R. A. Berner. *The Global Water Cycle.* Englewood Cliffs, N.J.: Prentice-Hall, 1987.

Costanza, Robert, et al. "The value of the world's ecosystem services and natural capital." *Nature* 387 (15 May 1997).

Daily, Gretchen, et al. "Ecosystem Services: Benefits Supplied to Human Societies by Natural Ecosystems," *Issues in Ecology* 1997.

Delwiche, C. C. "The Nitrogen Cycle." *Scientific American* 223.3 (1970): 136–146.

Devey, Edward S. "Mineral Cycles." *Scientific American* 223.3 (1970): 148–158.

Ehrlich, Paul R., and Harold A. Mooney. "Extinction, Substitution and Ecosystem Services." *Bioscience* 33 (1983): 248–254.

Hobbie, John, Jon Cole, Jennifer Dugan, R.A. Houghton, and Bruce Peterson. "Role of Biota in Global CO2 Balance: The Controversy." *Bioscience* 34 (1984): 492-498.

"Nutrient Pollution of Coastal Rivers, Bays, and Seas." *Issues in Ecology Website.* Available: http://www.esa.org/science/Issues/FileEnglish/issue7.pdf. 2000.

Jordan, Carl F. *Nutrient Cycling in Tropical Forest Ecosystems: Principles and Their Application in Management and Conservation.* New York: Wiley, 1982.

"Ecosystems & Human Well-being: A Report of the Conceptual Framework Working Group of the Millennium Ecosystem Assessment." *Millennium Ecosystem Assessment Website.* Available: http://pdf.wri.org/ecosystems_human_wellbeing.pdf. 2003.

Kellog, W.W., R. D. Cadle, E. R. Allen, A. L. Lazarus, and E. A. Martell. "The Sulfur Cycle." *Science* 175 (1972): 587–596.

National Academy of Science. *Productivity of World Ecosystems.* Washington, D.C.: National Academy of Science, 1975.

Phillipson, J. *Ecological Energetics.* New York: St. Martin's Press, 1966.

Pimm, Stuart L. *Food Webs.* New York: Chapman and Hall, 1982.

Prigogine, Ilya, Gregoire Nicoles, and Agnes Babloyantz. "Thermodynamics and Evolution." *Physics Today* 25.11 (1972): 23–28; 25.12 (1972): 138–141.

Schindler, David W. "Evolution of Phosphorus Limitation in Lakes." *Science* 195 (1977): 260–262.

Svensson, Bo H., and R. Soderlund, eds. "Nitrogen, Phosphorus and Sulfur, Global Cycles." *Ecological Bulletins* (No. 22). Stockholm: Royal Swedish Academy of Sciences, 1976.

Vitousek, Peter et al. "Human Alteration of the Global Nitrogen Cycle: Causes and Consequences," *Issues in Ecology* 1997.

Whittaker, Robert H. *Communities and Ecosystems*, 2nd ed. New York: MacMillan, 1975.

Chapter 5

Allee, W. C. "Cooperation Between Species." *American Scientist* 60 (1951): 348–357.

Boucher, D. H., S. James, and K. H. Keeler. "The Ecology of Mutualism." *Annual Review of Ecological Systems* 13 (1982): 315–347.

Cheng, T. E. *Aspects of the Biology of Symbiosis.* Baltimore: University Park Press, 1971.

Clements, Frederic E. "Plant Succession: An Analysis of the Development of Vegetation." Publication No. 242. Carnegie Institution of Washington, 1916. (Reprints in book form. New York: Wilson, 1928.)

Cornell, Joseph H., and Ralph G. Slayter. "Mechanisms of Succession in Natural Communities and Their Role in Community Stability and Organization." *American Naturalist* 111 (1977): 1119–1114.

den Boer, P. J. "The Present Status of the Competition Exclusion Principle." *Trends in Ecological Evolution* 1 (1986): 25–28.

Egerton, Frank N. "Changing Concepts of the Balance of Nature." *Quarterly Review of Biology* 48 (1973): 322–350.

Ehrlich, Paul R., and Peter H. Raven. "Butterflies and Plants: A Study of Coevolution." *Evolution* 18 (1965): 586–608.

Ewald, P. W. "Host-Parasite Relations, Vectors, and the Evolution of Disease Severity." *Annual Review of Ecology And Systematics* 14 (1984): 365–485.

Gause, G. F. "Ecology of Populations." *Quarterly Review of Biology* 7 (1932): 27–46.

Hardin, Garrett. "The Competitive Exclusion Principle." *Science* 131(1960): 1292–1297.

Head, Suzanne, and Robert Heinzman, eds. *Lessons of the Rainforest.* San Francisco: Sierra Club Books, 1991.

Johnson, Kris H., et al. "Biodiversity and the productivity and stability of ecosystems." *Tree* (9 September 1996): 372-377.

May, Robert M. "Thresholds and breakpoints in ecosystems with a multiplicity of stable states." *Nature* (6 October 1997): 471-477.

McGrady-Steed, Jill, Patricia M. Harris and Peter J. Morin. "Biodiversity regulates ecosystem predictability." *Nature* (13 November 1997): 162-165.

McIntosh, Robert P. *The Background of Ecology: Concept and Theory.* New York: Cambridge University Press, 1985.

Moffett, Mark W. "Life in a Nutshell." *National Geographic* 175.6 (1989): 782–796.

Naeem, Shahid, et al. "Declining biodiversity can alter the performance of ecosystems." *Nature* (21 April 1994): 734-737.

Naeem, Shahid, et al. "Empirical evidence that declining species diversity may alter the performance of terrestrial ecosystems." *Phil. Trans. R. Soc. Lond. B.* (1995): 249-262.

Odum, Eugene P. "Population Regulation and Genetic Feedback." *Science* 159 (1969): 1432–1437.

Patten, Bernard C., and Eugene P. Odum. "The Cybernetic Nature of Ecosystems." *American Naturalist* 118 (1981): 886–895.

Pimm, Stuart L. "The complexity and stability of ecosystems." *Nature* (26 January 1984): 321-326.

Pontin, A. J. *Competition and Coexistence of Species.* Boston: Pitman Advanced Publishing Program, 1982.

Shugart, Herman H. *A Theory of Forest Dynamics: The Ecological Implications of Forest Succession Models.* New York: Springer-Verlag, 1984.

Tilman, David, David Wedin, and Johannes Knops. "Productivity and sustainability influenced by biodiversity in grassland ecosystems." *Nature* (22 February 1996): 718-720.

Tilman, David and John A. Downing. "Biodiversity and stability in grasslands." *Nature* (27 January 1994): 363-365.

Whitmore, Timothy C. *An Introduction to Tropical Rain Forests.* New York: Oxford University Press, 1990.

Chapter 6

"100,000 Gallons of Coal Slurry Spill into Streams." *The Herald-Dispatch Website.* Available: http://www.herald-dispatch.com/2002/October/09/LNtop2.htm. 9 October 2002.

Arbaugh, M., Bytenerowicz, A., Grulke, N., Fenn, M., Poth, M., Temple, P., and Miller, P. 2003. Photochemical smog effects in mixed conifer forests along natural gradient of ozone and nitrogen deposition in the San Bernardino Mountains. Environment International 29: 401-406.

Blaikie, Piers M., and Harold Brookfield. *Land Degradation and Society.* New York: Methuen, 1987.

Bluewater Network Website. Available: http://www.earthisland.org. 27 May 2005.

"Reefs at Risk: A map-based Indicator of Potential Threats to the World's Coral Reefs." *World Resources Institute Website.* Available: http://pubs.wri.org/pubs_description.cfm?PubID=2901. 1998.

Collins, Mark. *The Last Rain Forests: A World Conservation Atlas.* New York: Oxford University Press, 1990.

"Deforestation: Causes and Implications." *Global Futures Foundation.* Available: http://future500.org/articles/3. 27 May 2005.

DeLong, Jeff. "Preliminary Study: Engine Emissions Damaging to Lake Tahoe Aquatic Life." *Reno Gazette-Journal* 9 April 1998.

Diem, J. 2003. Potential Impact of ozone on coniferous forests of the interior southwestern United States. Annals of the Association of American Geographers 93(2): 265-280.

"Fact Sheet: President Announces Wetlands Initiative on Earth Day." *White House Website.* Available: http://www.whitehouse.gov/news/releases/2004/04/20040422-1.html. 22 April 2004.

Freedman, Bill. *Environmental Ecology: Impacts of Pollution and Other Stresses on Ecosystem Structure and Function.* San Diego, Calif.: Academic Press, 1989.

"Global Coral Reef Monitoring Network. Status of Coral Reefs of the World: 2002 Report." Available: http://www.aims.gov. au/pages/research/coral-bleaching/scr2002/scr-00.html. 2002.

Holmstrom, David. "Water 'Motorcycles' Catch Negative Spray." *The Christian Science Monitor* 20 May 1998.

Kannan, Kurunthachalam, et al. "Butyltin Residues in Southern Sea Otters (Enhydra lustris nereis) Found Dead Along California Coastal Waters." *Environmental Science and Technology* 32.9 (1998).

Laurance, William F. "Gaia's Lungs" *Natural History* March 1999.

Mattingly, David. "Earth's Coral Reefs in Decline, Researchers Say." *CNN Interactive* (30 December 1997). Available: http://cnn.com/EARTH/9712/30/year.of.reef/index.html. 8 January 1998.

"National Resources Inventory. 2002 Annual Report." *USDA Website.* Available: http://www.nrcs.usda.gov/technical/land/nri02/nri02wetlands.html.

"National Survey of MTBE and Other VOCs in Community Drinking-Water Sources." *USGS Website.* Available: http://sd.water.usgs.gov/nawqa/vocns/nat_survey.html. 1998.

"Nowhere Near No-Net-Loss." *NWF Report.* Available: http://www.nwf.org/nwfwebadmin/binaryVault/Nowhere_Near_No-Net-Loss.pdf.

People for Puget Sound Website. Available: http://www.pugetsound.org/releases/2004_05_13.html

"President Focuses on Wetlands Conservation." *USFWS Website.* Available: http://www.fws.gov/feature/wetlandsprwebversion.html. 22 April 2004.

Safina, Carl. *Song for the Blue Ocean.* New York: Henry Holt and Company, Inc., 1997.

Science 277 (25 July 1997). Special issue on human-dominated ecosystems.

Sears, Paul B. *Deserts on the March.* Washington, D.C.: Island Press, 1988.

"Special Report: Fragments of the Forest." *Natural History* 107.6 (July/August 1998).

"Spreading Deserts Threaten Africa." *FAO Hypermedia Collections on Desertification: Factfiles* (1990). Available: http://www.fao.org/LIBRARY/Desert/fact-e.htm. 16 June 1998.

Sugal, Cheri. "Forest Loss Continues." *Vital Signs.* Worldwatch Institute. New York: W.W. Norton & Company, 1997.

Taylor, George E., Jr., et al. "Air Pollution and Forest Ecosystems: A Regional to Global Perspective." *Ecological Applications* 4.4 (November 1994).

Thompson, Jon. "East Europe's Dark Dawn." *National Geographic* 179.6 (1991): 36–69.

UN Convention to Combat Desertification Website. Available: http://www.unccd.int/main.php.

U.S. Government Accounting Office. "MTBE Contamination From Underground Storage Tanks". Report GAO-02-753T. Available: http://www.gao.gov/new.items/d02753t.pdf.

Vesilind, P. Aarne. *Environmental Pollution and Control.* Stoneham, Mass.: Butterworth, 1983.

Vital Signs 2003. Worldwatch Institute. New York: W.W. Norton & Company, 2003.

Waid, John S. *PCB's and the Environment.* Boca Raton, Fla.: CRC Press, 1987.

"West Virginia Settles With Coal Company Over Slurry Spill into River." *Associated Press Website.* Available: http://www.enn.com/news/2003-10-08/s_9222.asp. October 8, 2003.

Chapter 7

"2004 World Population Data Sheet." *Population Reference Bureua,* 2004.

"About UNFPA: Frequently Asked Qustions." *UNFPA Website.* Available: http://www.unfpa.org/about/faqs.htm. 27 May 2005.

Banister, Judith. *China's Changing Population.* Stanford, Calif.: Stanford University Press, 1987.

Bianchini, Bob. "Different America Predicted for 2050." *The Billings Gazette* [Billings, Montana] No. 153 (29 September 1993).

Brown, Becky J., Mark G. Hanson, Diana M. Liverman, and Robert W. Meredith, Jr. "Global Sustainability: Toward A Definition." *Environmental Management* 11.6 (1987): 713–719.

Coale, Ansley J., and Edgar M. Hoover. *Population Growth and Economic Development in Low-Income Countries.* Princeton, N.J.: Princeton University Press, 1958.

Conly, Shanti R and Nada Chaya. "Educating Girls: Gender Gaps and Gains." *Population Action International.* Available: http://www.populationaction.org/resources/publications/educating_girls/index.htm. 27 May 2005.

Cook, James. "More People Are a Good Thing: An Interview with Julian Simon." *Forbes* (21 December 1981): 70–72.

Cornelius, Diana. Research Demographer, Population Reference Bureau. Personal communication.

Cutler, M. Rupert. "Human Population: The Ultimate Wildlife Threat." *Vital Speeches* 53 (1987): 691–697.

Daily, Gretchen C. and Paul R. Erlich. "Population, Sustainability, and Earth's Carrying Capacity." *Bioscience* 42.10 (1992): 761-770.

"A Demographic Portrait of Asian Americans," *PRB Website.* Available: http://www.prb.org. 19 November 2004.

Eekelaar, John M. *An Aging World: Dilemmas and Challenges for Law and Social Policy.* Oxford: Clarendon Press, 1989.

Ehrlich, Paul, and Anne Ehrlich. *The Population Explosion.* New York: Simon & Schuster, 1990.

_____. "The Population Explosion: Why Isn't Everyone as Scared As We Are?" *The Amicus Journal* 12.1 (1990): 22–29.

Feldman, Linda. "Education for Girls Credited for Drop in Teen Birth Rates in Third World," *The Christian Science Monitor,* February 13, 1997.

Gelbard, Alene and Carl Haub. "Population Explosion Not Over for Half the World." *Population Today* 26.3 (March 1998).

Grier, Peter. "Advocacy Group Cites Family Planning Efforts." *The Christian Science Monitor* 9 March 1993.

"How To Attain Population Sustainability." *World Population Awareness.* Available: http://www.overpopulation.org/pop-sustainability.html. 27 May 2005.

"Largest Metropolitan Areas in the World." *About Website.* Available: http://geography.about.com/library/weekly/aa072897.htm. 22 November 2005.

"Latinos and the Changing Face of America." *PRB Website.* Available: http://www.prb.org. 19 November 2004.

Lee, Sharon M. "Asian Americans: Diverse and Growing." *Population Bulletin* 53.2 (June 1998).

Livernash, Robert and Eric Rodenburg. "Population Change, Resources, and the Environment." *Population Bulletin* 53.1 (March 1998).

"Maternal Mortality." *UNICEF Website.* Available: http://www.childinfo.org/eddb/mat_mortal/. 27 May 2005.

"Maternal Mortality in 2000: Estimates Developed by WHO, UNICEF and UNFPA," *ReliefWeb Website*. Available: http://wwwnotes. reliefweb.int. 19 February 2005.

Mazur, Laurie Ann, ed. *Beyond the Numbers: A Reader on Population, Consumption, and the Environment*. Washington, D.C.: Island Press, 1994.

McFalls, Joseph A., Jr. "Population: A Lively Introduction." *Population Bulletin*, 4th ed. 58.4 (December 2003).

Meadows, Donella H., Dennis Meadows, Jorgen Randers, and William Behrens. *The Limits to Growth*. New York: The New American Library, Inc., 1974.

"Millennium in Maps: Population." National Geographic Society. Washington, D.C., 1998.

Moffet, George. *Critical Masses: The Global Population Challenge*. New York: Viking Penguin Ltd., 1994.

"New Population Policies: Advancing Women's Health and Rights," Population Reference Bureau, 2001.

"Transitions in World Population." *Population Bulletin* 59, no 1. Washington, D.C: Population Reference Bureau, 2004.

"Principal Agglomerations of the World." *City Population Website*. Available: http://www. citypopulation.de/World.html. 22 November 2004.

"Special Issue on Rebuilding Cities." *The Amicus Journal* Summer 1992.

"State of the World Population 2004," *United Nations Population Fund Website*. Available: http://www.unfpa.org/swp/2004/english/ch1/ page7.htm#1. 23 February 2005.

State of the World 2004: A Worldwatch Institute report on progress toward a sustainable society. New York: W.W. Norton & Company, 2004.

State of the World 2004: A Worldwatch Institute report on progress toward a sustainable society. New York: W.W. Norton & Company, 2005.

Thurow, Lester R. "Why the Ultimate Size of the World's Population Doesn't Matter." *Technology Review* 89 (1986): 22–29.

United Nations Population Division, "World Population Prospects: The 2002 Revision (Highlights)," 26 February 2003.

Vital Signs 2003. Worldwatch Institute. New York: W.W. Norton & Company, 2003.

"Wage Gap Increases Between Women and Men, US Census Report." *Feminist Majority Foundation*. Avaialble: http://www.feminist. org/news/newsbyte/uswirestory.asp?id=8622. 27 May 2005.

"Where Are We Now? Report Card Finds Mixed Results Worldwide 10 Years After 179 Governments Pledged to Improve Health and Women's Status." *Population Action International*. Available: http://www.populationaction.org/news/press/news_083104_GRT.htm. 27 May 2005.

"Why Population Matters." Population and Habitat Campaign, National Audubon Society.

"Why Population Matters to Natural Resources." *Population Action International Website*. Available: http://www.populationaction. org/resources/factsheets/factsheet_13.htm. 27 May 2005.

"World Population Awareness News Digest." *World Population Awareness Website*. Available: http://www.overpopulation.org/newsScan. html. 27 May 2005.

"World Population Highlights," Population Reference Bureau, 2004.

Zero Population Growth. *ZPG's Urban Stress Test*. Washington, D.C.: Zero Population Growth, 1988.

Chapter 8

Atwood, J. Brian. "Aid for Family Planning," *The Christian Science Monitor* 13 February 1997.

"Advancing Women's Health and Rights." *Population Reference Bureau Website*. Available: http://www.prb.org/content/navigationmenu/ PRB/aboutPRB/Population_Bulletin2/New_ Population_Policies_Advancing_Womens_

Health_and_Rights.htm. 19 February 2005.

The Alan Guttmacher Institute, "U.S. Teenage Pregnancy Statistics." New York. 19 February 2004.

Balter, Michael. "On World AIDS Day, a Shadow Looms Over Southern Africa." *Science* 282 (4 December 1998).

Cordell, Dennis D., and Joel W. Gregory. *African Population and Capitalism: Historical Perspectives*. Boulder, Colo.: Westview Press, 1987.

Ecological Society of America. "ESA Passes Resolution on Human Population." *Bulletin of the Ecological Society of America* (December 1994): 203.

Feldman, Linda. "Debate Revives Over Funds for Global Family Planning," *The Christian Science Monitor* 13 January 1997.

"Frequently Asked Questions: Mifepristone/ Mifeprex (RU-486)." *Family.org Website*. Available: http://www.family.org/cforum/fosi/ bioethics/faqs/a0027731.cfm. 26 February 2005.

"General Facts and Statistics." *Tennpregnancy. org Website*. Available: http://http://www. teenpregnancy.org/resources/data/genlfact.asp. 26 February 2005.

Harris, Marvin. *Cannibals and Kings: The Origins of Culture*. New York: Random House, 1977.

"The Impact of Illegal Abortion." *The Abortion Access Project Website*. Available: http://www.abortionaccess.org/AAP/publica_ resources/fact_sheets/illegalabortion.htm. 26 February 2005.

"Joint Report Details Escalating Global Orphans Crisis Due to AIDS." *United States Agency for International Development Website*. Available: http://www.usaid.gov/press/releas- es/2002/pro020710.html. 28 February 2005.

Kasun, Jacqueline R. *The War Against Population: The Economics and Ideology of World Population Control*. San Francisco: Ignatius Press, 1988.

Kirkby, R. J. R. *Urbanization in China: Town and Country in a Developing Economy, 1949–2000*. New York: Columbia University Press, 1985.

Lancaster, Henry O. *Expectations of Life: A Study in the Demography, Statistics, and History of World Mortality*. New York: Springer-Verlag, 1990.

Mazur, Laurie Ann. *Beyond the Numbers*. Washington, D.C.: Island Press, 1994.

"Medical Abortion: An Alternative for Women." *Center for Reproductive Rights Website*. Available: http://www.reproductiver- ights.org/pub_fac_medabor.html. 26 February 2005.

Ornstein, Robert E., and Paul Ehrlich. *New World, New Mind: Moving Toward Conscious Evolution*. Garden City, N.Y.: Doubleday, 1989.

Russell, Thornton. *American Indian Holocaust and Survival: Population History Since 1492*. Norman: University of Oklahoma Press, 1987.

Simon, Julian Lincoln. *Population Matters: People, Resources, Environment and Immigration*. New Brunswick, N.J.: Transaction, 1990.

State of the World 2004: A Worldwatch Institute report on progress toward a sustainable society. New York: W.W. Norton & Company, 2004.

State of the World 2003: A Worldwatch Institute report on progress toward a sustainable society. New York: W.W. Norton & Company, 2003.

Vital Signs 2003. Worldwatch Institute. New York: W.W. Norton & Company, 2003.

Wattenberg, Ben J. *The Birth Dearth*. New York: Pharos Books, 1989.

"Where Are We Now?" *Countdown 2015 Website*. Available: http://www.countdown2015. org.

Chapter 9

"Biotechnology and Food Security." *United Nations Food and Agriculture Website*. Available: http://www.fao.org.

_____. "Outgrowing the Earth: Food Security Challenge in an Age of Falling Water Tables and Rising Temperatures." New York: W.W. Norton & Company, 2005.

"Breaking the Cycle of Poverty." *United Nations World Food Programme Website.* Available: http://www.wfp.org/aboutwfp/introduction/hunger_stop.asp?section=1&sub_section=1. 29 March 2005.

Brown, Lester R., Gary Gardner and Brian Halweil. "Beyond Malthus: Sixteen Dimensions of the Population Problem." *Worldwatch Paper 143.* Washington, D.C.: Worldwatch Institute, 1998.

Byron, William. *The Causes of World Hunger.* New York: Paulist Press, 1982.

"Commercial Fishing in a State of Collapse." *Animal Protection Institute Website.* Available: http://www.api4animals.org/71.htm. 4 April 2005.

"Commodity Markets: Global Trends, Local Impacts." *United Nations Food and Agriculture Website.* Available: http://www.fao.org/newsroom/en/focus/2005/89746/index.html. 17 March 2005.

"Contributions to Global Food Security: the CGIAR Helps the Poorest Farmers." *United Nations Food and Agriculture Website.* Available: http://www.fao.org.

"Counting the Hungry." *United Nations World Food Programme Website.* Available: http://www.wfp.org/aboutwfp/introduction/counting_the_hungry/infodiag_popup/infodiag_popup.html. 30 March 2005.

Crosson, Pierre R., and Norman J. Rosenberg. "Strategies for Agriculture." In *Managing Planet Earth: Readings from Scientific America.* New York: W. H. Freeman and Company, 1990.

Danhoff, Clarence H. *Change in Agriculture: The Northern United States, 1820–1870.* Cambridge, Mass.: Harvard University Press, 1969.

_____. *How Much Is Enough? The Consumer Society and the Future of the Earth.* New York: W.W. Norton & Company, 1995.

Ehrlich, Paul R., and Anne H. Ehrlich. "Population, Plenty, and Poverty." *National Geographic* 174 (1988): 914–946.

"Faces of the Hungry." *United Nations World Food Programme Website.* Available: http://www.wfp.org/aboutwfp/introduction/hunger_who.asp?section=1&sub_section=1. 29 March 2005.

"Fast Food: WFP's Emergency Response." *United Nations World Food Programme Website.* Available: http://www.wfp.org/aboutwfp/introduction/hunger_fight.asp?section=1&sub_section=1. 29 March 2005.

"Feeding the Cities." *United Nations Food and Agriculture Website.* Available: http://www.fao.org.

"Fish Catch Leveling Off." *Earth Policy Institute Website.* Available: http://earth-policy.org/Indicators/indicator3.htm. 28 March 2005.

"Food Aid to Save and Improve Lives." *United Nations Food and Agriculture Website.* Available: http://www.fao.org.

"Food Outlook." *United Nations Food and Agriculture Website.* Available: http://www.fao.org/docrep/007/j3877e/j3877e02.htm. 22 March 2005.

"Food Security and the Environment," *United Nations Food and Agriculture Website.* Available: http://www.fao.org.

"Food Quality and Safety." *United Nations Food and Agriculture Website.* Available: http://www.fao.org.

Galbraith, John Kenneth. *The Nature of Mass Poverty.* Cambridge, Mass.: Harvard University Press, 1979.

Gupte, Pranay. *The Crowded Earth: People, Politics and the Population.* New York: W.W. Norton & Company, 1984.

"How Much Do We Consume?" *World Resources Institute Website.* Available: http://earthtrends.wri.org/features/view_feature.cfm?theme=6&fid=7. 22 March 2005.

"Hunger, Humanity's Oldest Enemy." *United Nations World Food Programme Website.* Available: http://www.wfp.org/aboutwfp/intoduction/hunger_what.asp?section=1&sub_section=1. 29 March 2005.

"India Likely to Export Wheat to Algeria." *Trading Charts Website.* Available: http://news.tradingcharts.com/futures/6/6/64854366.html. 30 March 2005.

"Investing in Development." *Millinneum Project Website.* Availble: http://www.unmillenniumproject.org/reports/index.htm. 13 April 2005.

"Is World Food Security Deteriorating?" *The Globalist Website.* Available: http://www.theglobalist.com/DBWeb/StoryID.aspx?StoryID=4030. 22 March 2005.

Jackson, Wes. *New Roots for Agriculture.* San Francisco: Friends of the Earth, 1980.

Kendall, Henry W., and David Pimentel. "Constraints on the Expansion of the Global Food Supply." *Ambio* 23.3: 198-205.

Knickerbocker, Brad. "World Dawdles as Its Fisheries Decline." *The Christian Science Monitor* 23 September 1998.

"Land of Plenty, Lives of Desperation." *Global Policy Forum Website.* Available: http://globalpolicy.org/socecon/hunger/2003/1102congoplenty.htm. 4 April 2005.

Lenihan, John, and William W. Fletcher. *Food, Agriculture and the Environment.* San Diego, Calif.: Academic Press, 1976.

Levi, Carolyn. "Growing Fish Salad: An Experiment in Integrated Aquaculture." *Nor'Easter* 3.1 (1991): 15–17.

Mann, Charles C. "Crop Scientists Seek a New Revolution." *Science* 283 (15 January 1999): 310-316.

Mellor, John. "The Entwining of Environmental Problems and Poverty." *Environment* (November 1988): 8.

Moffett, George. "'Super Rice' May Ease World Food Crisis." *The Christian Science Monitor* 26 October 1994.

"Our Threatened Oceans." *Biper USA Website.* Available: http://biperusa.biz/SeaBounty.htm. 24 March 2005.

"Preventing Child Deaths." *World Future Society Website.* Available: http://www.wfs.org/trendnd03.htm. 22 November 2005.

"Projection of World Fishery Production in 2010." *United Nations Food and Agriculture Organization Website.* Available: http://www.fao.org/fi/highligh/2010.asp. March 28, 2005.

"The Right to Food," *United Nations Food and Agriculture Website.* Available: http://www.fao.org.

Sachs, Jeffery D. "The End of Poverty." *Time Magazine.* Vol.165, No. 11. 42-54.

Sachs, Jeffery D and Pedro A. Sanchez. "We Can End World Hunger." *World Ark.* November/December 2004. 6-13.

Shepard, Jack. *The Politics of Starvation.* New York: Carnegie Endowment for International Peace, 1975.

"Social Safety Nets." *United Nations Food and Agriculture Website.* Available: http://www.fao.org.

"Spectrum of Malnutrition," *United Nations Food and Agriculture Website.* Available: http://www.fao.org.

"State of Food Insecurity in the World 2004." *United Nations Food and Agriculture Website.* Available: http://www.fao.org.

"State of the Oceans, Part 1: Eating Away at a Global Food Source." *National Institute of Health Website.* Available: http://ehp.niehs.nih.gov/members/2004/112-5/focus.html. 24 March 2005.

State of the World Fisheries and Aquaculture. Rome, 2004.

State of the World 2004: A Worldwatch Institute report on progress toward a sustainable society. New York: W.W. Norton & Company, 2004.

State of the World 2003: A Worldwatch Institute report on progress toward a sustainable society. New York: W.W. Norton & Company, 2003.

"Top Ten Fish Missing from the Ocean and Missing from Fish Markets." Marine Fish Conservation Network News Release. 30 September 2004.

Tufts University Center on Hunger, Poverty & Nutrition Policy. "Summary of U.S. Hunger Estimates: 1984 to the Present," March 1993.

"Underpinning the Foundations of Food Production: the SGRP and Conservation of Genetic Resources." *United Nations Food and Agriculture Website.* Available: http://www.fao.org.

Vital Signs 2003. Worldwatch Institute. New York: W.W. Norton & Company, 2003.

"Water and Food Security." *United Nations Food and Agriculture Website.* Available: http://www.fao.org.

"Why Does Hunger Exist?" *United Nations World Food Programme Website.* Available: http://www.wfp.org/aboutwfp/introduction/hunger_causes.asp?section=1&sub_section=1. March 29, 2005.

"Women, Agriculture and Food Security." *United Nations Food and Agriculture Website.* Available: http://www.fao.org.

"World Grain Consumption and Stocks, 1960-2004." *Earth Policy Institute Website.* Available: http://www.earth-policy.org.

"World Health Report 2002." *World Health Organization Website.* Available: http://www.who.int/whr/2002/overview/en/. March 29, 2005.

Chapter 10

"About Natural Gas Vehicles." *The Natural Gas Vehicle Coalition Website.* Available: http://www.ngvc/ngvc.nsf/bytitle/fastfacts.htm. March 9, 2005.

"Analysis of Oil and Gas Production in the Arctic National Wildlife Refuge." *Energy Information Administration of the USDOE,* Washington D.C., March 2004.

"Annual Energy Outlook 2005." *Energy Information Administration of the USDOE Website.* Available: http://www.eia.doe.gov/oiaf/aeo/forecast.html. March 10, 2005.

Appenzeller, Tim. "The End of Cheap Oil." *National Geographic* 205 (June 2004).

Baldauf, Scott. "World's Oil May Soon Run Low." *The Christian Science Monitor* 23 September 1998.

Commoner, Barry. *The Poverty of Power: Energy and the Economic Crisis.* New York: Bantam Books, 1977.

"Crude Oil and Total Petroleum Imports Top 15 Countries." *Energy Information Administration of the USDOE Website.* Available: http://www.eia.doe.gov/pub/oil_gas/petroleum/data_publications

Deudeny, Daniel, and Christopher Flavin. *Renewable Energy: The Power to Choose.* New York: W.W. Norton & Company, 1983.

"Energy in Focus," *British Petroleum Website.* Available: http://www.bp.com/statisticalreview2004.

Energy Information Administration. *World Energy Use and Carbon Dioxide Emissions, 1980-2001.* May 2004.

_____. *International Energy Outlook.* May 2004.

"Energy Policy Act of 2003 Will Create Nearly 1 Million Jobs, Further Stimulate Economy." *US Senate Committee on Energy and Natural Resources Website.* Available: http://energy.senate.gov/news/rep_release.cfm?id=215172. March 17, 2005.

"Global Nuclear Capacity Trend Will be Downwards, Says Green Report." *Greens-European Free Alliance Group Website.* Available: http://www.greens-efa.org. December 8, 2004.

"International Energy Outlook 2004." *Energy Information Administration of the USDOE Website.* Available: http://www.eia.doe.gov/oiaf/ieo/world.html. March 5, 2005.

Knickerbocker, Brad. "The Big Spill." *The Christian Science Monitor* (22 March 1999): 1, 10.

Natural Gas Vehicle Coalition. "Questions and Answers About Natural Gas Vehicles." *NGVC Website.* Available: http://www.ngvc.org./ngv/ngvc.nsf/bytitle/fastfacts.htm. 27 May 2005.

Nuclear Energy Agency Annual Report 2003.

"Oil and Gasoline Prices Hit New Record Today." *US Senate Committee on Energy and Natural Resources Website.* Available: http://energy.senate.gov/news/rep_release.cfm?id=233962. March 17, 2005.

"Oil Shale Activities." *Office of Fossil Energy of the USDOE Website.* Available: http://www.fossil.energy.gov/programs/reserves/npr/NPR_Oil_Shale.html. March 9, 2005.

Resources, *Resources for the Future,* Issue 156 (Winter 2005).

Schneider, M. and A. Froggatt. "The World Nuclear Industry Status Report 2004." Commissioned by the Greens-EFA Group in the European Parliament. December 2004.

Shogan, Colleen. "Presidential Campaigns and the Congressional Agenda: Regan, Clinton, and Beyond." Presented on November 22, 2004 at the Woodrow Wilson International Center for Scholars, The Congress Project, Washington, DC.

State of the World 2004: A Worldwatch Institute report on progress toward a sustainable society. New York: W.W. Norton & Company, 2004.

State of the World 2003: A Worldwatch Institute report on progress toward a sustainable society. New York: W.W. Norton & Company, 2003.

United Nations Development Program. *World Energy Assessment 2004.* New York. 2004.

"U.S. Coal Supply and Demand: 2004 Review." *Energy Information Administration of the USDOE Website.* Available: http://www.eia.doe.gov/cneaf/coal/page/special/feature.html. 5 March 2005.

U.S. Department of Energy. *Comprehensive National Energy Strategy.* Washington, D.C.: April 1998.

_____. *Fossil Energy Strategic Plan: Meeting 21st Century Challenges* March 1998.

_____. Energy Information Administration. "Energy Consumption by Source, 1949-1997." *EIA Website.* Available: http://www.eia.doe.gov/pub/energy.overview/aer/aer0103.txt. January 1999.

_____. "Table 1.2: Energy Overview." *Annual Energy Review 1997* (July, 1998). *EIA Website.* Available: http://www.eia.doe.gov/pub/energy.overview/monthly.energy/mer1-2. 7 January 1999.

_____. "World Oil Demand." *EIA Website.* Available: http://www.eia.doe.gov/emeu/ipsr/t24.txt. January 1999.

_____. Office of Fossil Energy. *Clean Coal Technology. Office of Fossil Energy Website.* Available: http://www.fe.doe.gov/coal_power/cct.html. 8 January 1999.

Vital Signs 2003. Worldwatch Institute. New York: W.W. Norton & Company, 2003.

"World Coal Institute Institute Fact Cars." *World Coal Institute Website.* Available: http://wci.rmid.co.uk/web/content.php?menu_id=5.9.2

Chapter 11

Beggs, Sandy L. "Fulfilling Small Hydro's Worldwide Potential." *World Directory of Renewable Energy Suppliers and Services,* 1996.

"Energy Actions You Can Take." *EnviroLink.* Available: http://www.envirolink.org/topics.html?topic=Energy&topicsku=2002109191036&topictype-topic&do=catsearch&catid=8. 27 May 2005.

Energy Information Administration. "Renewable Energy Trends 2003, With Preliminary Data for 2003." Washington, DC. July 2004.

Geo-Heat Center. "Where are Geothermal Resources Located?" *Geo-Heat Center Website.* Available: http://geoheat.oit.edu. 27 May 2005.

"Geothermal Power: FAQs." *Renewable Energy Policy Project Website.* Available: http://www.crest.org/articles/static/995653330_5.html. 30 March 2005.

Hagerman, George. "OSPREY and LIMPET." *Islander Magazine* 1 (1996).

International Nuclear Safety Center. *INSC Homepage.* Available: http://www.insc.anl.gov/. 27 May 2005.

"Landfill Gas (Methane) Recovery and Utilization: Reading and Resource List," *USDOE Website*. Available: http://www.eere.energy.gov/concimerinfo/reading_resources/vg7.html. 8 April 2005.

McCully. Patrick. *Silenced Rivers*. London: Zed Books, 1996.

"Nuclear Energy Generation In Oecd/Nea Countries Will Continue To Grow At 0.9 Per Cent Per Year Until 2010." *OECD Nuclear Energy Agency* (2 June 1998). Available: http://www.nea.fr/html/general/press/98-5.html. 27 May 2005.

Nuclear Energy Institute. "Nuclear Energy Basics." *Nuclear Energy Institute Website*. Available: http://www.nei.org/basics/pp.html. January 1999.

"Nuclear Power Reactors." *Uranium Information Centre Website*. Available: http://www.uic.com.au/nip64.htm. 10 March 2005.

Pilkington Solar International. *Status Report on Solar Thermal PowerPlants*. GmbH: Cologne, Germany, 1996.

"Renewable Energy 2000: Issues and Trends." *Energy Information Administration Website*. Available: http://www.eia.doe.gov/cneaf/solar.renewables/rea_issues/rea_issues_sum.html. February 2001.

Quillen, Ed. "At Home in a High-Altitude Think Tank." *Country Journal* (July 1986): 42–51.

State of the World 2004: A Worldwatch Institute report on progress toward a sustainable society. New York: W.W. Norton & Company, 2004.

State of the World 2003: A Worldwatch Institute report on progress toward a sustainable society. New York: W.W. Norton & Company, 2003.

U.S. Department of Energy, National Renewable Energy Laboratory. *National Wind Technology Center*. Available: http://www.nrel.gov/wind/. 27 May 2005.

_____. *Ocean Thermal Energy Conversion*. Available: http://www.nrel.gov/otec/what.html. 27 May 2005.

Vital Signs 2003. Worldwatch Institute. New York: W.W. Norton & Company, 2003.

Zhnag, Dan. "China Eases the way for Steady Progress." *Windpower Monthly* November 1997.

Chapter 12

"Acidification." *Swedish Environmental Protection Agency Website*. Available: http://www.internat.naturvardsverket.se/documents/pollutants/kalka/forsure.html. 5 February 2005.

"Acid News: Report No. 4, Decmber 2004." *Swedish NGO Secretariat on Acid Rain Website*. Available: http.www.acidrain.org/AN4-04.htm#CAFÉ. 7 February 2005.

"Acid Rain." *U.S. Environmental Protection Agency Website*. Available: http://www.epa.gov/airmarkets/acidrain/index.html. 26 January 2005.

"Acid Rain Program 2003 Progress Report." *U.S. Environmental Protection Agency Website*. Available: http://www.epa.gov/airmarkets/cmprpt/arp03/summary.html. 2 February 2005.

"Acid Rain Program: 2002 Progress Report." U.S. Environmental Protection Agency. 2003.

"Acid Rain, Sleet, and Snow Sour Soils and Streams." *National Parks Conservation Association Website*. Available: http://www.npca.org/across_the_nation/visitor_experience/code_red/acid_precipitation.asp. 5 February 2005.

"Air Pollution Causes and Effects." *Health and Energy Website*. Available: http://healthandenergy.com/air_pollution_causes.htm. 9 February 2005.

"Air Quality Planning & Standards." *USEPA Website*. Available: http://www.epa.gov/oar/oaqps/takingtoxics/p1.html#9. 26 January 2005.

"Alaskan Glaciers Melting Faster." *British Broadcasting Channel (BBC) News Website*. Available: http://news.bbc.co.uk/2/hi/science/nature/2137205. 5 February 2005.

"Alaska's Melting Glaciers Evidence of Global Warming." *KTUU.com Website*. Available: http://www.ktuu.com/CMS/templates/master.asp?articleid=7688&zoneid=4. 5 February 2005.

"Asthma Facts and Figures." *Asthma Foundation of America Website*. Available: http://www.aafa.org/display.cfm?id=8&sub=42. 10 February 2005.

"Asthma Facts May 2004, Indoor Environments Division. EPA 402-F-04-019." *USEPA Website*. Available: http://www.epa.gov/asthma/images/asthma_fact_sheet_en.pdf.

Ayers, Harvard, Jenny Hager, and Charles E. Little, ed. *An Appalachian Tragedy*. Hong Kong: Sierra Club Books, 1998.

"China–On the Fast Track to Lowering Emissions." *Corning Emissions Control Technology Magazine Online*. Available: http:www.corning.com/environmentaltechnologies/emissions_control_technology_magazine/archived_issues/2004-1/article1.asp. 7 February 2005.

"Clean Air Act–Clear Skies." *The League of Conservation Voters Website*. Available: http://www.lcv.org/fedfocus/fedfocus.cfm?ID=3525&c=11. 31 January 2005.

"Climate." *USEPA Website*. Available: http://yosemite.epa.gov/oar/globalwarming.nsf/content/Climate.html. 26 January 2005.

"Climate of 2003 Annual Review." *National Oceanic and Atmospheric Administration Website*. Available: http://www.ncdc.noaa.gov/oa/climate/research/2003/ann/global.html. 15 January 2004.

"Climate Change 2001: Synthesis Report." *An Assessment of the Intergovernmental Panel on Climate Change*. United Kingdom, 2001.

Using the Climate Vulnerability Index to Assess Vulnerability to Climate Variations. Centre for Ecology & Hydrology, Wallingford, Oxfordshire OX10 8BB, UK

Clow, David W., Sickman, James O., Striegl, Robert G., Krabbenhoft, David P.,

Elliott, John G., Dornblaser, Mark, Roth, David A., and Campbell Donald H. Changes in the chemistry of lakes and precipitation in high-elevation national parks in the western United States, 1985–1999. Water Resources Research, v. 39, no. 6, 1171, doi:10.1029/2002WR001533, 2003.

"Code Red: America's Five Most Polluted National Parks." *National Parks Conservation Association Website*. Available: http://www.npca.org/across_the_nation/visitor_experience/code_red/default.asp. 14 February 2005.

Coughenour, Michael B. and De-Xing Chen. "Assessment of Grassland Ecosystem Responses To Atmospheric Change Using Linked Plant-Soil Process Models." *Ecological Applications* 7.3 (1997): 802-827.

"Current Evidence of Climate Change." *UNFCCC Website*. Available: http://unfccc.int/essential_background/Feeling_the_heat/items/2904.php. 27 May 2005.

Daily, Gretchen C., Ed. *Nature's Services: Societal Dependence on Natural Ecosystems*. Washington, D.C.: Island Press, 1997.

Dale, Virginia. "The Relationship Between Land-Use Change and Climate Change." *Ecological Applications* (August 1997): 753-769.

"Dirty Skies: The Bush Administration's Air Pollution Plan." *Natural Resources Defence Council Website*. Available: http://wwwnrdc.org/air/pollution/qbushplan.asp. 12 February 2005.

Driscoll, C.T., G.B. Lawrence, A.J. Bulger, T.J. Butler, C.S. Cronan, C. Eagar, K.F. Lambert, G.E. Likens, J.L. Stoddard, K.C. Weathers. 2001. Acid Rain Revisited: advances in scientific understanding since the passage of the 1970 and 1990 Clean Air Act Amendments. Hubbard Brook Research Foundation. Science Links™ Publication. Vol. 1, no.1. Available: http://www.hbrook.sr.unh.edu/hbfound/report.pdf. July 2004.

Eizenstat, Stuart E. "Combating Global Warming." *The Christian Science Monitor*. (10 November 1998): 19.

_____. "Curbing Global Warming, Not Prosperity." *The Christina Science Monitor* (18 February 1998): 18.

"Emissions Scorecard for units affected by the Acid Rain Program, including SO2, NOx, CO2, and Heat Input. 2001. (Updated in April 2003)." *USEPA Website*. Available: http://www.epa.gov/airmarkets/emissions/score01/index.html.

"Enforcing the Clean Air Act to Protect Parks and People," *National Parks Conservation Association Website*. Available: http://www.npca.org/across_the_nation/visitor_experience/code_red/clean_air_act.asp. 14 February 2005.

"EPA's Air Quality Standard for Particulate Matter, June 1998." *Wisconsin Department of Natural Resources Website*. Available: http://www.dnr.state.wi.us/org/aw/air/hot/newPM-stand.htm. 15 February 2005.

"EPA To States: Clean Up Your Act." *The Associated Press Website*. Available: http://www.cbsnews.com/stories/2004/04/09/tech/main611121.shtml. 15 April 2004.

"EPA Issues First Draft Ozone Criteria Document, January 31, 2005," *American Lung Association Clean Air Standards Website*. Available: http://www.cleanairstandards.org/article/articleview/377/1/41/. 10 February 2005.

"EPA's Methylmercury Guideline is Scientifically Justifiable for Protecting Most Americans, But Some May Be at Risk." *The National Academies Website*. Available: http://www4.nationalacademies.org/news.nsf/isbn/0309071402?OpenDocument. 9 February 2005.

"EPA Releases 2002 Toxic Release Inventory: Right-to-Know Compromised." *Office of Management and Budget (OMB) Watch Website*. Available: http.www.ombwatch.org/article/articleview/2259/1/1?TopicID=1. 12 February 2005.

"EPA Scientists Recommend Tighter Health and Visibility Standards for PM, January 31, 2005." *American Lung Association Website*. http://wwwcleanairstandards.org/article/articleview/378/1/41/. 15 February 2005.

"EPA Acid Rain Program 2002 Progress Report. Fall, 2003, Clean Air Markets Division. EPA 430-R-03-011." *USEPA Website*. Available: http://www.epa.gov/airmarkets/cmprpt/arp02/2002report.pdf.

"Europe's smokers feel heat." *The Christian Science Monitor Website*. Available: http://www.csmonitor.com/2003/1024/p01s04-woeu.html. 24 October 2003.

"Executive Summary–The Clear Skies Initiative," *The White House Website*. Available: http://www.whitehouse.gov/news/releases/2002/02/clearskies.html. 12 February 2005.

"Frequently Asked Questions about Air Pollution." *The Mercury News Website*. Available: http://weather.mercurynews.com/auto/mercurynews/health/pollutionfaq.asp. 7 February 2005.

"Glaciers in Retreat a Recurring Theme from Around the Globe." *Nichols College Website*. Available: http:///www.nichols.edu/departments/Glacier/glacier_retreat.htm. 26 January 2005.

"The Global Commons: Climate Brief: Early Spring, Late Winter." *World Resources Institute Website*. Available: http://pubs.wri.org/pubs_content_text.cfm?ContentID=1540. 5 February 2005.

"Global Temperature Trends: 2004 Summation." *National Atmospheric and Space Administration Website*. Available: http://www.giss.nasa.gov/data/update/gistemp/2004/. 26 January 2005.

Grace, John, et al. "Carbon Dioxide Uptake by an Undisturbed Tropical Rain Forest in Southwest Amazonia." *Science* (3 November 1995): 778-780.

Graedel, Thomas E., and Paul J. Crutzen. "The Changing Atmosphere." *Scientific American* 261.3: 58–68.

"Greenhouse Gases, Climate Change, And Energy." *Energy Information Administration Website*. Available: http://www.eia.doe.gov/oiaf/1605/ggccebro/chapter1.html. 26 January 2005.

"Growing Scientific Evidence of Global Warming Becoming an 'Urgent Priority' for Business Leaders, September 7, 2004." *The Conference Board Website*. Available: http://www.conference-board.org/utilities/pressDetail.cfm?press_ID=2465. 31 January 2005.

Halpin, P.N. "Global Climate Change and Natural-Area Protection: Management Responses and Research Directions." *Ecological Applications* 7.3 (1997): 828-843.

Hamilton, Martha M. "Linking Business's Climate and Earth's." *The Washington Post* 12 January 1997.

"Health Effects of Ozone and Particle Pollution." *American Lung Association Website* Available: http://lungaction.org/reports/sota04_heffects.html#pp. 27 May 2005.

"Illegal Trade in Ozone Depleting Substances." *Europa Website*. Available: http://europa.eu.int/comm/environment/ozone/illegal_trade.htm. 14 February 2005.

Intergovernmental Panel on Climate Change. Third Assessment Report. 2001. J. H. McCarthy et al., editors, Cambridge University Press, Cambridge UK, 1032 pp. Available: http://www.grida.no/climate/ipcc_tar/

Intergovernmental Panel on Climate Change. "Climate Change and Biodiversity." April 2002 Technical Paper V. Available: http://www.ipcc.ch/pub/tpbiodiv.pdf

"Ireland Stubs Out Smoking in Pubs." *BBCnews.com Website*. Available: http://news.bbc.co.uk/1/hi/world/europe/3577001.stm. 29 March 2004.

"Is Leaded Gasoline Still Around?" *NRTC Communications Website*. Available: http://www.nrtco.net/~lead/leadgas.htm. 5 February 2005.

Keeling, C.D. and T.P. Whorf. 2004. Atmospheric CO2 records from sites in the SIO air sampling network. In Trends: A Compendium of Data on Global Change. Carbon Dioxide Information Analysis Center, Oak Ridge National Laboratory, U.S. Department of Energy, Oak Ridge, Tenn., U.S.A. Available: http://cdiac.esd.ornl.gov/trends/co2/sio-mlo.htm.

"Kilimanjaro's Ice Archive." *British Broadcasting Channel News Website*. Available: http://news.bbc.co.uk/2/hi/science/nature/2337023.stm. 5 February 2005.

"Kyoto Protocol." *Wikipedia, the Free Encyclopedia Website*. Available: http://en.wikipedia.org/wiki/Kyoto_Protocol. 31 January 2005.

"The Kyoto Protocol–A Brief Summary." *Europa Website*. Available: http://europa.eu.int/comm/environment/climat/kyoto.htm. 13 January 2005.

Lynch, Colum F. "Warm up to idea: Global warming is here." *The Amicus Journal* (Spring 1996): 20-25.

"Massachusetts Indoor Smoking Ban Begins." *The Associated Press Website*. Available: http://www.cnn.com/2004/US/Northeast/07/05/smoking.ban.ap/. 5 July 2004.

"Metropolitan Areas Most Polluted by Short-term Particle Pollution." *American Lung Association Website*. http://www.lungusa.org/site/pp.asp?c=dvLUK9O0E&b=50752. 14 February 2005.

Michener, William K., Elizabeth R. Blood, Keith L. Bildstein, Mark M. Brinson, and Leonard R. Gardner. "Climate Change, Hurricanes and Tropical Storms, and Rising Sea Level In Coastal Wetlands." *Ecological Applications* 7.3 (1997): 770-801.

Moore, Curtis A. "Warming Up To Hot New Evidence," *International Wildlife* (January/February 1997).

"Mercury." *USEPA Website*. Available: http://www.epa.gov/mercury/exposure.htm. 26 January 2005.

"Mercury." *Clear the Air Website*. Available: http://www.cleartheair.org/mercury/. 8 February 2005.

"Mercury and Health." *North Carolina Department of Environment and Natural Resources Website*. Available: http:www.p2pays.org/mercury/health.asp. 8 February 2005.

"Most Americans Face Smog Danger." *CBSnews.com Website*. Available: http://www.cbsnews.com/stories/2004/04/29/health/main614671.shtml. 29 April 2004.

National Research Council. *Biologic Markers of Air-Pollution Stress and Damage in Forests*. Washington, D.C.: National Academy Press, 1989.

National Research Council, 2001a: *Climate Change Science: An Analysis of Some Key Questions, Committee on the Science of Climate Change*, National Academy Press, Washington DC, 29 pp.

Natural Resources Defense Council. "Breathtaking: Premature Mortality Due to Particulate Air Pollution in 239 American Cities," 1996.

"New Report Finds 20 States Increased Warnings for Mercury in Fish in 2003." *Clear the Air Website*. Available: http://www.cleartheair.org/proactive/newsroom/release.vtml?id=25801. 8 February 2005.

North Cascades Glacier Climate Project Website. Available: http://www.nichols.edu/departments/glacier/what.htm.

"Overview of Leaded Gasoline and Sulfur Levels in Gasoline and Diesel, November 14, 2002." *International Fuel Quality Center Website*. Available: http://www.ifqc.org. 2005.

"Overview of the Human Health and Environmental Effects of Power Generation: Focus on Sulfur Dioxide (SO2), Nitrogen Oxides (NOx) and Mercury (Hg)." *USEPA Website*. Available: http://www.epa.gov/air/clearskies/pdfs/overview.pdf. July 2004.

Parmesan, Camille. "Climate and species' range." *Nature* 382 (29 August 1996).

"Particulate Matter (PM) Research." *USEPA Website*. Available: http://www.epa.gov/pmresearch/. 26 January 2005.

Patz, J.A. A Human Disease Indicator for the Effects of Recent Global Climate Change PNAS, Oct 2002; 99: 12506 - 12508.

"The Plain English Guide to the Clean Air Act." *USEPA Website*. Available: http://www.epa.gov/oar/oaqps/peg_caa/pegcaa03.html#topic3f. 10 February 2005.

"Pollution Locator/Smog and Particulates/Rank Counties by Exposures." *Scorecard Website*. Available: http://www.scorecard.org/env-releases. 10 February 2005.

Pounds, Alan, Michael P.L. Fogden and John H. Campbell. "Biological response to climate change on a tropical mountain." *Nature* (15 April 1999): 611-615.

"Recent Articles Highlighting the North Cascade Glacier Climate Project." *Nichols College Website*. Available: http://www.nichols.edu/departments/Glacier/glacier_articles.htm. 26 January 2005.

"Regional Haze Shrouds Scenic Vistas." *National Parks Conservation Association Website*. Available: http://www.npca.org/across_the_nation/visitor_experience/code_red/visibility_loss.asp. 5 February 2005.

"Releases: Chemical Report (Data Source: 2002 Data Update as of August 2, 2004)." *USEPA Website*. Available:http.www.epa.gov. 12 February 2005.

"Report Ranks Cities and Counties Threatened by Dirty Air." *American Lung Association Website*. Available: http://www.lungusa.org/site pp.asp?c=dvLUK9O0E&b=50986. 14 February 2005.

"Response of Surface Water Chemistry to the Clean Air Act Amendments of 1990. January 2003, National Health and Environmental Effects Research Laboratory. EPA 620/R-03/001." *USEPA Website*. Available: http://www.epa.gov/ord/htm/CAAA-2002-report-2001-rev-4.pdf.

Rifkin, Jeremy. *Entropy: Into the Greenhouse World*. New York: Bantam Books, 1989.

Rodó, X., Pascual, M., Fuchs, G. and Faruque, A. S. G. ENSO and Cholera: A Nonstationary Link Related to Climate Change? PNAS, Oct 2002; 99: 12901 - 12906.

Root, Terry L. and Stephen H. Schneider. "Ecology and Climate: Research Strategies and Implications." *Science* (21 July 1995): 334-341.

Schneider, Stephen H. "The Changing Climate." *Scientific American* 261.3 (September): 70–79.

_____. "Detecting Climatic Change Signals: Are There Any "Fingerprints?" *Science* (21 January 1994): 341-347.

Slaper, Harry, et al. "Estimate of ozone depletion and skin cancer incidence to examine the Vienna Convention achievements," *Nature* 384 (21 November 1996).

"South American Glacier's Big Melt." *British Broadcasting Channel News Website*. Available: http://news.bbc.co.uk/2/hi/science/nature/3200450.stm. 5 February 2005.

"State of the Air: 2004." *American Lung Association Website*. Available: http://lungaction.org/reports/sota04intro.html. 14 February 2005.

"State of Climate Science: October 2003." *Union of Concerned Scientists Website*. Available: http://www.ucsusa.org/global_environment/global_warming/page.cfm?pageID=1264.

State of the World 2004: A Worldwatch Institute report on progress toward a sustainable society. New York: W.W. Norton & Company, 2004.

State of the World 2003: A Worldwatch Institute report on progress toward a sustainable society. New York: W.W. Norton & Company, 2003.

"Study Blasts Growing Use of Coal-fired Power Plants." *Health and Energy Website*. Available: http://healthandenergy.com/coal.htm. 8 February 2005.

Suplee, Curt. "Unlocking the Climate Puzzle." *National Geographic* (May 1998): 38-72.

"Sweden's National GHG Inventory Reports, 1990-2002." *Swedish Environmental Protection Agency Website*. Available: http://www.internat.naturvardsverket.se/index.php3?main=/documents/pollutants/kalka/acidrain/acidrain.htm.

"TED Case Studies." *American University Website*. Available:http://www.american.edu/projects/mandala/TED/russcfc.htm. 14 February 2005.

"Toxic Air Pollutants." *USEPA Website*. Available: http://www.epa.gov/air/toxicair/newtoxics.html. 26 January 2005.

"Toxics and Health." *Envirohealthaction Website*. Available: http://envirohealthaction.org/. 8 February 2005.

"Toxic Release Increased in 2002, Study Says." *The New York Times*. 23 June 2004.

"2002 Toxics Release Inventory (TRI): Public Data Release Report." *USEPA*. June 2004.

Treshow, Michael, and Franklin K. Anderson. *Plant Stress from Air Pollution*. New York: Wiley, 1990.

"Two New Studies Link Smog to Premature Death, November 14, 2004." *American Lung Association Website*. Available: http://wwwcleanairstandards.org/article/articleview/352/1/41/. 15 February 2005.

"U.S. Consumer Product Safety Commission. An Update on Formaldehyde: 1997 Revision (CPSC document #725)." *USEPA Website*. Available: http://www.epa.gov/iaq/pubs/images/formaldehyde_cpsc.pdf.

USEPA Air – Indoor Air Quality (IAQ) Website. Available: http://www.epa.gov/iaq/index.html.

U.S. Environmental Protection Agency. "EPA Assessment of Risks from Radon in Homes." June 2003, Office of Radiation and Indoor Air. EPA 402-R-03-003. Available: http://www.epa.gov/rpdweb00/docs/assessment/402-r-03-003.pdf. 27 May 2005.

U.S. National Park Service. Air Quality in the National Parks. 2nd Edition. 2002. National Park Service Air Resources Division. Available: http://www2.nature.nps.gov/air/pubs/aqnps.htm#download.

"U.S. PIRG Reports." *U.S. Public Interest Research Group Website*. Available: http://uspirg.org/spirg.asp?id2=8822&id3=USPIRG&#1. 12 February 2005.

Vital Signs 2003. Worldwatch Institute. New York: W.W. Norton & Company, 2003.

Wellburn, Alan. *Air Pollution and Acid Rain: The Biological Impact*. New York: Wiley, 1988.

"Scientific Assessment of Ozone Depletion: 2002, Global Ozone Research and Monitoring Project—Report No. 47, 498 pp., Geneva, 2003." Available: http://www.unep.org/ozone/pdf/scientific-assessment2002.pdf.

Chapter 13

Carson, Rachel. *The Edge of the Sea*. Boston: Houghton Mifflin, 1955.

Cousteau, Jacques-Yves. *The Cousteau Almanac: An Inventory of Life on Our Water Planet*. Garden City, N.Y.: Doubleday, 1981.

Cummins, Joseph E. "The PCB Threat to Marine Mammals." *The Ecologist* 18.6 (1988): 193–195.

Dobrzynski, T., et al., *Oceans at Risk*, Oceana, 2002.

Dziegielewski, Benedykt, and Duane Baumann. "The Benefits of Managing Urban Water Demands." *Environment* 34.9 (November 1992).

"Earth Trends," *World Resources Institute Website*. Available: http://earthtrends.wri.org. February 2005.

Edgerton, Lynne T. *The Rising Tide*. Washington, D.C.: Island Press, 1990.

Environmental Protection Agency, *Small Systems Guide to Safe Drinking Water Regulations: the Simple Tools for Effective Performance (STEP) Guide Series*, 2004.

Gleick, Peter. "Basic Water Requirements for Human Activities: Meeting Basic Needs." *International Water* 21.2 (1996).

Global International Waters Assessment, Newsletter, March 2004.

Horan, N. J. *Biological Wastewater Treatment Systems: Theory and Operation*. New York: Wiley, 1990.

"H.R. 2762 Wetlands and Watershed Management Act of 1997." *Wetlands Regulation Center Website*. Available: http://www.wetlands.com/legis/hr2762.htm. 27 May 2005.

McClurg, Sue, *A Colorado River Compromise*, Western Water, November/December 2000.

"National Water Summary on Wetland Resources," *USGS Website*. Available: http://water.usgs.gov/nwsum/WSP2425/. March 1, 2005.

Platt, Anne McGinn. "Safeguarding the Health of Oceans." *Worldwatch Paper 145*. Washington, D.C.: Worldwatch Institute, 1999.

Reisner, Marc. *Cadillac Desert: The American West and Its Disappearing Water*. New York: Penguin Books, 1986.

Reisner, Marc, and Sarah F. Bates. *Overtapped Oasis: Reform or Revolution for Western Water*. Washington, D.C.: Island Press, 1989.

State of the World 2004: A Worldwatch Institute report on progress toward a sustainable society. New York: W.W. Norton & Company, 2004.

State of the World 2003: A Worldwatch Institute report on progress toward a sustainable society. New York: W.W. Norton & Company, 2003.

Tarlock, Dan, et al. *Water in the West: Challenge for the Next Century*. The Western Water Policy Review Advisory Commission, June 1998.

"The U.N. World Water Development Report: Water for People, Water for Life." *The U.N. World Water Assesment Programme Website*. Available: http://www.unesco.org/water/wwap/wwdr/index.shtml. February 24, 2005.

Vital Signs 2003. Worldwatch Institute. New York: W.W. Norton & Company, 2003.

"Vital Water Graphics," *United Nations Environment Programme Website*. Available: http://www.unep.org/vitalwater/. November 2004.

Western Water Policy Review Advisory Commission. *Water in the West: Challenge for the Next Century*. National Technical Information Service, 1998.

World Wildlife Fund, *Living Planet Report*, 2004.

Chapter 14

American Farmland Trust website. Available: http://www.farmland.org/

Berry, Wendell. *A Continuous Harmony*. San Diego: Harcourt Brace Jovanovich, 1970.

_____. *The Unsettling of America: Culture & Agriculture*. San Francisco: Sierra Club Books, 1977.

Bouwman, A. F., ed. *Soils and the Greenhouse Effect*. New York: Wiley, 1990.

Dunkle, Richard L. "Big or Small, All Farmers Gain From Research." *Agricultural Research Magazine*. Beltsville, MD: USDA/Agricultural Research Service, 2 November 1997.

Ebeling, Walter. *The Fruited Plain: The Story of American Agriculture*. Berkeley: University of California Press, 1979.

"EPA: Pesticides - 2000/2001 Pesticide Sales and Useage Report," *USEPA Website*. Available: http://www.epa.gov/oppbead1/pestsales/01pestsales/useage2001.html. 3 February 2005.

Faulkner, Edward H. *Plowman's Folly and a Second Look*. Washington, D.C.: Island Press, 1987.

"World Food Summit Five Years Later." *Food and Agriculture Organization of the United Nations*. 2002. Available: http://www.fao.org/worldfoodsummit/

"Government Study Finds Widespread Herbicide Contamination in Mississippi River Basin." *The Land Stewardship Letter* 10.1 (Winter 1992).

"How Safe Is Our Produce?" *Consumer Reports* 64.3 (March 1999).

International Irrigation Management Institute. "Water Management in the Twenty-First Century." *Annual Report 1997*. Colombo [Sri Lanka], 1998.

Little, Charles E. *Green Fields Forever: The Conservation Tillage Revolution in America*. Washington, D.C.: Island Press, 1987.

Logan, William Bryant, ed. "Living Soil." *Orion* 11.2 (1992).

Mollison, Bill. *Permaculture: A Practical Guide for a Sustainable Future*. Washington, D.C.: Island Press, 1990.

Pesticide Action Network North America (PANNA). "Global 'No Pesticides Day' Launched." *Panups* 18 December 1998.

Sorensen, A. Ann, Richard P. Greene, and Karen Russ (American Farmland Trust). *Farming on the Edge*. DeKalb, Illinois: Northern Illinois University, March 1997.

"State of the Land." *National Resource Conservation Service*. Available: http://www.nrcs.usda.gov/technical/land. 27 May 2005.

State of the World 2004: A Worldwatch Institute report on progress toward a sustainable society. New York: W.W. Norton & Company, 2004.

State of the World 2003: A Worldwatch Institute report on progress toward a sustainable society. New York: W.W. Norton & Company, 2003.

Steiner, Frederick. *The Living Landscape: An Ecological Approach to Landscape Planning*. New York: McGraw-Hill, 1991.

Tarlock, Dan, et al. *Water in the West: Challenge for the Next Century*. The Western Water Policy Review Advisory Commission, June 1998.

Tyson, Ann Scott. "Urban Sprawl's Appetite for Rich Farmland," *The Christian Science Monitor* 21 March 1997.

U. S. Organic Farming in 2000-2001: Adoption of Certified System. Available: http://ers.usda.gov/publications/aib780/aib780a.pdf

Vital Signs 2003. Worldwatch Institute. New York: W.W. Norton & Company, 2003.

Chapter 15

"2001 National Survey of Fishing, Hunting, and Wildlife-Associated Recreation." *USFWS Website*. Available: http://www.census.gov/prod/2002pubs/FHW01.pdf

"The 2004 IUCN red list of threatened species." *World Conservation Union Website*. Available: http://www.redlist.org

"Arabian Oryx." *Animal Info Website*. Available: http//www.animalinfo.org/species/artiperi/oryxleuc.htm.

"Atlantic Forest." *Global 200 Ecoregion*. Available: http://www.wwfus.org/action/global200/new/focal.htm.

Bertness, Mark D. "The Ecology of a New England Salt Marsh." *American Scientist* 80 (1992): 260–268.

"Box Score, Endangered Species." *U.S. Fish and Wildlife Service Division of Endangered Species* (28 February 1999). Available: http://www.fws.gov/boxtbl.html. 27 May 2005.

Brown, Lester, Christopher Flavin, and Sandra Postel. "Protecting the Biological Base." *Saving the Planet: How to Shape an Environmentally Sustainable Global Economy.* New York: W.W. Norton & Company, 1991.

Chadwick, Douglas. "Dead or Alive – The Endangered Species Act." *National Geographic* 187.3 (1995): 2–41.

"Cites at a Glance." *TRAFFIC Network.* Available: http://www.traffic.org/factfile/factfile_cites.html. 27 May 2005.

"Convention on Biological Diversity." *UNEP Website* (28 May 1998). Available: http://.www.biodiv.org/convention/default.shtml. 27 May 2005.

The Coral Reef Alliance Website. Available: http://www.coralreefalliance.org/. 27 May 2005.

Defenders of Wildlife. *Saving America's Wildlife: Renewing the ESA.* July 1995.

Durrell, Lee. *State of the Ark.* Garden City, N.Y.: Doubleday, 1986.

Ehrlich, Paul R., and Anne H. Ehrlich. *Extinction: The Causes and Consequences of the Disappearance of Species.* New York: Ballantine Books, 1983.

Ehrlich, Paul R., and E. O. Wilson. "Biodiversity Studies: Science and Policy." *Science* 253 (1991): 758–762.

"Endangered Species." *National Wildlife Federation.* Available: http://www.nwf.org/pubs/positions/faendang.html. 5 February 1999.

"Endangered Species Information." *USFWS Website.* Available: http://www.endangered.fws.gov/wildlife.html

Erwin, Terry L. "Tropical Forests: Their Richness in Coleoptera and Other Arthropod Species." *Coleopterists Bulletin* 36.1 (1982): 74-75.

Hallé, Francis. "A Raft Atop the Rain Forest." *National Geographic* 178.4 (1990): 129–138.

"Illegal Trade Endangering Many Species." The Washington Times. 2003. Available: http://www.washtimes.com/world/20031102-111354-3357r.htm

International Species Information System Website. Available: http://www.isis.org/CMSHOME/ 27 May 2005.

Lowther, Jason et al. 2002. "Crime and Punishment in the Wildlife Trade." World Wildlife Fund (WWF) publication.

Mann, Charles C., and Mark L. Plummer. "The Butterfly Problem." *Atlantic* 269.1 (1992): 47–73.

Matthiessen, Peter. *Wildlife in America.* New York: Viking Press, 1987.

McNeely, Jeffrey, et. al. *Conserving the World's Biological Diversity.* Gland, Switzerland: International Union for Conservation of Nature and Natural Resources, 1990.

"Millions of Americans Enjoy Wildlife-Related Recreation, Pumping Billions Into National Economy, Survey Shows." *U.S. Fish and Wildlife Service News Release* (8 July 1997). Available: http://news.fws.gov/historic/1997/19970708.odf. 27 May 2005/

Mitsch, William J., and James G. Gosselink. *Wetlands* 2nd ed. New York: Van Nostrand Reinhold Co., 1993.

Norton, Bryan G. *The Preservation of Species: The Value of Biological Diversity.* Princeton, N.J.: Princeton University Press, 1986.

Pounds, Alan, Michael P.L. Fogden and John H. Campbell. "Biological response to climate change on a tropical mountain." *Nature* (15 April 1999): 611-615.

"Probe of International Reptile Trade Ends With Key Arrests." *U.S. Fish and Wildlife Service News Release* (15 September 1998). Available: http://www.fws.gov/r9extaff/pr9851.html. 3 February 1999.

"Putting a Price Tag on Nature's Bounty." *Science* 276 (16 May 1997).

"Rates of Rainforest Loss." *Rainforest Action Network.* Available: http://www.ran.org/info_center/factsheets/04b.html. 27 May 2005.

Reisner, Marc. "The Sting." *The Amicus Journal* 10.2 (1988): 40–47.

Ritter, Michael. 2005. Online textbook "The Physical Environment: An Introduction to Physical Geography." Available:

http://www.physicalgeography.net

Roush, Wade. "Putting a Price Tag on Nature's Bounty." *Science* 16 May 1997.

"Snail Darter." *U.S. Fish and Wildlife Service Division of Endangered Species.* Available: http://www.fws.gov/i/e/sae15.html. 27 May 2005.

"Southwestern Amazon Rain Forests." *Global 200 Ecoregion.* Available: http://www.wwfus.org/action/global200/new/25_rainf.htm. 25 January 1999.

Spotts, Peter N. "Plant Triage: Which Imperiled Species Should Be Saved?" *The Christian Science Monitor* (15 April 1998): 12.

State of the World 2004: A Worldwatch Institute report on progress toward a sustainable society. New York: W.W. Norton & Company, 2004.

State of the World 2003: A Worldwatch Institute report on progress toward a sustainable society. New York: W.W. Norton & Company, 2003.

"State of the World's Forests 2003." *Food and Agriculture Organization Website.* Available: http://www.fao.org/DOCREP/005/Y7581E/y7581e05.htm. 22 February 2005.

Thorne-Miller, Bruce, and John Catena. *The Living Ocean: Understanding and Protecting Marine Biodiversity.* Washington, D.C.: Island Press, 1990.

"Tropical Forests." *WWF Canada Factsheets.* Available: http://www.wwfcanada.org/facts/trofor.html. 22 January 1999.

UNEP World Conservation Monitoring Centre website. Available: http://www.unep-wcmc.org/right.htm

"What are Coral Reefs." *World Wildlife Fund.* Available: http://www.panda.org/about_wwf/what_we_do/marine/what_we_do/coral_reefs/about/index.cfm. 27 May 2005.

"What is CITES?" *Convention on International Trade in Endangered Species (CITES) Website.* Available: http://www.cites.org. 25 February 2005.

Wilson, Edward O. *Biodiversity.* Washington, D.C.: National Academy Press, 1988.

_____. *Biophilia.* Cambridge, MA.: Harvard University Press, 1984.

_____. "Rain Forest Canopy: The High Frontier." *National Geographic* 179.12 (1991): 78–107.

_____. "Threats to Biodiversity." *Scientific American* 261.3 (1989): 108–16.

Worldwatch Institute. *Vital Signs.* New York: W. W. Norton & Company, 2003.

Yost, Nicholas, et al. "Wetlands: Through Murky Waters." *NAEP Newsletter* 17.2 (1992).

Chapter 16

"The 1999 Grazing Fee, Surcharge Rates, and Penalty for Unauthorized Grazing Use." *Bureau of Land Management Website* (4 February 1999). Available: http://www.blm.gov/nhp/efoia/wo/fy99/im99-062.html. 2 March 1999.

Adams, Ansel. *The American Wilderness.* Boston: Little, Brown, 1990.

Agee, James K., and Darryll R. Johnson. *Ecosystem Management for Parks and Wilderness.* Seattle: University of Washington Press, 1988.

Boyle, Robert H. "The Killing Fields." *Sports Illustrated* 78.11 (22 March 1993): 62-69.

Bryne, John Barry. "Clinton Roadless Plan: Bold But Incomplete." *Sierra Club Website* (March 1998). Available: http://www.sierraclub.org/planet/199803/road.asp. 27 May 2005.

"The Citizen's Proposal for Wilderness in Utah: Frequently Asked Questions." *The Utah Wilderness Coalition Website*. Available: http://www.uwcoalition.org/faq/proposal.html. 25 March 2005.

Drabelle, Dennis. "Obey the Law and Tell the Truth." *Wilderness* (Fall 1994): 29-31.

Findley, Rowe. "Along the Santa Fe Trail." *National Geographic* 179.3 (1991): 98–123.

_____. "Will We Save Our Own?" *National Geographic* 178.3 (1990): 106–136.

Fox, Stephen. *The American Conservation Movement*. Madison: The University of Wisconsin Press, 1981.

"Gray Wolf Recovery Status Report." *U.S. Fish and Wildlife Service Website*. Available: http://fire.mountain-prairie.fws.gov/wolf/wk05202005.htm. 27 May 2005.

"Grazing Facts." *Natural Resources Defense Council Website*. Available: http://www.nrdc.org/land/use/fgrazef.asp. 27 May 2005.

Hammond, Herb. *Seeing the Forest Among the Trees*. Vancouver: Polestar Press, 1992.

Harrison, Robert Pogue. *Forests: The Shadow of Civilization*. Chicago: University of Chicago Press, 1992.

Howard, Malcolm. "'Flight-seeing' Puts Peace of National Parks Up in the Air." *The Christian Science Monitor* (2 April 1996): 10-11.

Kluger, Jeffrey. "Deep divide," *Time* 10 February 1997.

Leopold, Aldo. *Aldo Leopold's Wilderness: Selected Early Writings by the Author of A Sand County Almanac*. Harrisburg, Penn.: Stackpole, 1990.

Lopez, Barry. *Arctic Dreams: Imagination and Desire in a Northern Landscape*. New York: Scribner, 1986.

_____. *Crossing Open Ground*. New York: Scribner, 1988.

_____. *Desert Notes: Reflections in the Eye of a Raven*. New York: Avon Books, 1990.

_____. *Of Wolves and Men*. New York: Scribner, 1978.

_____. "Understanding the Role That Humans Have Played in Shaping America's Forest and Grassland Landscapes." *USDA Forest Service* (7 September 1994): 1-9.

_____. "What on Earth Have We Done to Our Forests?" *USDA Forest Service* (24 January 1994): 1-22.

_____. *River Notes: The Dance of the Herons*. Kansas City [KS]: Andrews and McMeel, 1979.

McPhee, John. *Encounters with the Archdruid*. New York: Farrar, Straus & Giroux, 1971.

Meadows, William H. "National Forests: Turn the Bulldozers Around." *The Christian Science Monitor* (24 February 1998): 19.

"Meet the Forest Service." *U.S. Forest Service Website*. Available: http://www.srs.fs.usda.gov/careers/meet_the_forest_service.htm. 27 May 2005.

Muir, John. *The Yosemite*. New York: Century, 1912.

Naar, Jon and Alex J. Naar. *This Land is Your Land*. New York: Harper Collins, 1993.

Nash, Roderick. *Wilderness and the American Mind*. New Haven [CT] Yale University Press, 1967.

"The National Forest Roadless Area Rule." *Natural Resources Defense Council Website*. Available: http://www.nrdc.org/land/forests/qroadless.asp. 25 March 2005.

National Wilderness Preservation System Website. Available: http://www.wilderness.net/nwps/. 27 May 2005.

"The National Park System Caring for the American Legacy." *National Park Service Website*. Available: http://www.nps/gov/legacy/mission.html. 27 May 2005.

Nixon, Will. "Paradise Lost?" *The Amicus Journal* (Spring 1998): 16-22.

Norse, Elliot. "What Good Are Ancient Forests?" *The Amicus Journal* 12.1 (1990): 42–45.

Phillips, Michael K. and Douglas W. Smith. *The Wolves of Yellowstone*. Voyageur Press, Stillwater MN. 1996.

"A Red Light for Subsidized Logging Roads." *The Wilderness Society's Quarterly Newsletter* (Winter 1998-99): 3, 7.

Reisner, Marc. "The Sting." *The Amicus Journal* 10.2 (1988).

"Tongass Forest Facts." *Southeast Alaska Conservation Council Website*. Available: http://www.seacc.org/ForestFacts.htm. 25 March 2005.

"Tongass National Forest: Forest Management." *USDA Forest Service Website*. Available: http://www.fs.fed.us/r10/tongass/forest_facts/faqs/forestmgmt.shtml. 25 March 2005.

The Wilderness Society. "Of dreams and courage," *Wilderness America* December 1996.

Wilderness.net Website. Available: http://www.wilderness.net. 27 May 2005.

Wood, Daniel B. "National Parks Scale Peak of Popularity." *The Christian Science Monitor* (10 July 1998): 1, 7-9.

Zaslowsky, Dyan, and the Wilderness Society. *These American Lands*. New York: Holt, Rinehart and Winston, 1986.

Zuckerman, Seth, and The Wilderness Society. *Saving Our Ancient Forests*. Venice, Calif.: Living Planet Press, 1991.

Chapter 17

"About World Heritage." *UNESCO World Heritage Center Website*. Available: http://whc.unesco.org/en/convention.20 April 2005.

Cultural Survival, Inc. "Deforestation: The Human Costs." *Quarterly* 6.2 (1982).

Daniel, John. "Stealing Time." *Wilderness* 53.188 (1990): 18–38.

Denslow, Julie S., and Christie Padoch. *People of the Tropical Rainforest*. Berkeley: University of California Press, 1988.

Errahmani, Abdelkader Brahim. "The World Heritage Convention: A New Idea Takes Shape." *Courier*. Paris, France: United Nations Educational, Scientific and Cultural Organization, 1990.

Ki-Zerbo, Joseph. "Oral Tradition as a Historical Source." *Courier*. Paris, France: United Nations Educational, Scientific and Cultural Organization, 1990.

National Trust for Historic Preservation. *The Brown Book*. Washington, D.C.: The Preservation Press, 1983.

Robbins, Jim. "Violating History." *National Parks* (July/August 1987): 26–31.

Spirn, Anne Whiston. "From Uluru to Cooper's Place: Patterns in Cultural Landscape." *Orion* 9.2 (1990): 32–39.

Sugarman, Aaron. "The Treasures of America ... Looted!" Condé Nast *Traveler* July 1992.

Ward, Diane Raines. "In Anatolia, a Massive Dam Project Drowns Traces of an Ancient Past." *Smithsonian* 21.5 (1990): 29–40.

Wigginton, Eliot, ed. *The Foxfire Book*. Garden City, N.Y.: Doubleday, 1972.

"The World Heritage List." *UNESCO Website*. Available: http://www.unesco.org/whc/heritage.htm#debut. 27 May 2005.

Chapter 18

"Appalachia is Paying Price for White House Rule Change." *Washington Post Website*. Available: http://www.washingtonpost.com/wp-dyn/articles/A6462-2004Aug16.html. 16 March 2005.

Bates, Robert L., and Julia A. Jackson. *Our Modern Stone Age*. Los Altos, Calif.:Kaufmann, 1982.

"Bush Administration Rule Change Would Allow Burying Waterways," *Environmental Media Services Website*. Available: http.www.ems.org/mountaintop_removal/rule_change.html. 16 March 2005.

"China Coal Mine Blast Death Toll Rises to 66," *The All I Need Website*. Available: http://theallineed.com/news/0410/225920.htm. 16 March 2005.

Door, Ann. *Minerals Foundations of Society*. Alexandria, Va.: American Geological Institute, 1987.

Earthworks 1872 Mining Law Website. Available: http://www.earthworksaction. org/ewa/1872.cfm.

Humphries, M. "Mining on Federal Lands." Congressional Research Service Brief IB89130. September 15, 2003. Available: http://www. ncseonline.org/nle/crsreports/03Sep/IB89130. pdf.

Kennecott Utah Copper Website. Available: http://www.kennecott.com/

Lessons and Activities of Discovery and Appreciation. Mineral Information Institute (1992): 27-50.

"Mineral Commodities Summaries 2004." USGS Website. Available: http://minerals. usgs.gov/minerals/pubs/mcs/2004/mcs2004. pdf.

Mikesell, Raymond F. Non-Fuel Minerals: Foreign Dependence and National Security. Ann Arbor: University of Michigan Press, 1987.

Organisation for Economic Cooperation and Development, OECD Core Set. *Environmental Indicators*. Paris: OECD, 1994.

Library of Congress Website. Available: http:// thomas.loc.gov.

United Nations. *Division for Ocean Affairs and the Law of the Sea*. Available: http://www. un.org/Depts/los. 27 May 2005.

USEPA Superfund Website. Available: http:// www.epa.gov/superfund/index.htm.

Young, John E. and Aaron Sachs. "The Next Efficiency Revolution: Creating a Sustainable Materials Economy." *Worldwatch Paper 121*, September 1994.

Chapter 19

Arquilla, John and Andrew Goodpaster. "Nuclear Weapons: Time to Phase Them Out?" *The Christian Science Monitor* 18 December 1996.

Beckman, Peter R., et al. *The Nuclear Predicament: An Introduction*. Englewood Cliffs, N.J.: Prentice-Hall, 1989.

Center for Defense Information. 2003. Total suspected nuclear weapons. Available: http:// www.cdi.org

Energy Information Agency Website. Available: http://www.eia.doe.gov.

Flavin, Christopher. *Reassessing Nuclear Power: The Fallout from Chernobyl*. Washington, D.C.: Worldwatch Institute, 1987.

Gale, Robert Peter, and Thomas Hauser. *Final Warning: The Legacy of Chernobyl*. New York: Warner Books, 1988.

Gershey, Edward L., et al. *Low-Level Radioactive Waste: From Cradle to Grave*. New York: Van Nostrand Reinhold, 1990.

"India's Nuclear Forces, 2002," *Bulletin of the Atomic Scientists Website*. Available: http:// www.thebulletin.org/article_nn.php?art_ ofn=ma02norris_38. 16 March 2005.

Megaw, James. *How Safe? Three Mile Island, Chernobyl and Beyond*. Toronto: Stoddart, 1987.

Murray, Raymond L., and Judith A. Powell. *Understanding Radioactive Waste*. Columbus, Ohio: Battelle Press, 1988.

U.S. Spent Fuel Update: Year 2000 in Review. Available: http://www.nacworldwide.com/pdf/ SR_SF2000.pdf

Nader, Ralph, and John Abbots. *The Menace of Atomic Energy*. New York: W.W. Norton & Company, 1977.

National Academy of Sciences. 2001. A Comprehensive Nuclear Arms Reduction Regime: Interim Report. Available: http://www.nas.org/

Nealy, Stanley M. *Nuclear Power Development: Prospects in the 1990s*. Columbus, Ohio: Battelle Press, 1989.

"Nuclear Fact 2004." *Nuclear Energy Institute Website*. Available: http://www.nei.org/doc. asp?catnum=2&catid=106.

Office of Civilian Radioactive Waste Management, Yucca Mountain Project Website. Avaiable: http://www.ocrwm.doe.gov/ymp/index.shtml.

"Pakistan's Nuclear Forces, 2001," *Bulletin of the Atomic Scientists Website*. Available: http:// www.thebulletin.org/article_nn.php?art_ ofn=jf02norris. 16 March 2005.

Probst, Katherine and Michael McGovern. "long-term stewardship and the nuclear weapons complex: The challenge ahead." *1998 Resources for the Future*, 1998.

Rifkin, Jeremy. *Declaration of a Heretic*. Boston: Routledge & Kegan Paul, 1985.

Sneider, Daniel. "New Allies in the Bid to Ban Nuclear Arms," *The Christian Science Monitor* (5 December 1996).

U.S. Department of Energy. *Environmental Restoration and Waste Management: An Introduction, Student Edition* DOE/EM-0104.

Uranium Institute. "Uranium" *UI Facts*, 1997.

Wagner, Henry N., Jr., and Linda E. Ketchum. *Living with Radiation: The Risk, the Promise*. Baltimore: Johns Hopkins University Press, 1989.

White House web site. 2000. Fact Sheet: START II Treaty Agreement. Available: http://www.fas.org/nuke/control/start2/docs/ start2fact00.htm. 27 May 2005.

"Text of Strategic Offensive Reductions Treaty." *White House Website*. Available: http://www.whitehouse.gov/news/releas- es/2002/05/20020524-3.html.

"Yucca Mountain Statement." *White House Website*. Available: http://www.whitehouse. gov/news/releases/2002/02/20020215- 11.html.

State of the World 2005: A Worldwatch Institute report on progress toward a sustainable society. New York: W.W. Norton & Company, 2005.

Chapter 20

"The 12 POP's Designated for International Action." *World Wildlife Fund Issue Brief* January 1999.

Bagcji, Amalendu. *Design, Construction, and Monitoring of a Sanitary Landfill*. New York: Wiley, 1990.

Barnett, Richard. "Federal Implementation of Permanent Paper Standards." *Abbey* Newsletter (April 1992). Available: http://palimpsest. stanford.edu/byorg/abbey/an/an16/an16-2/ an16-207.html. 28 January 1999.

Brown, Lester R., Christopher Flavin, and Sandra Postel. "Reusing and Recycling Materials." *Saving the Planet: How to Shape an Environmentally Sustainable Global Economy*. New York: W.W. Norton & Company, 1991.

Bullard, Robert D. "Grassroots Flowering." *The Amicus Journal* 16.1 (1994): 32.

Colborn, Dumanoski, Myers. *Our Stolen Future; Are we threatening our fertility, intelligence, and survival? — a scientific detective story*, New York: Dutton, 1996.

"Convention on the Prevention of Marine Pollution by Dumping of Wastes and Other Matter, 1972." *International Maritime Organization Website*. Available: http://www. imo.org/Conventions/contents.asp?topic_ id=258&doc_id=681.

Fiske, John. "Americans Recycle More Paper Despite Low Demand." *The Christian Science Monitor* 20 October 1997: 13.

"Frequently Asked Questions about recycling and Waste Management." *USEPA Website*. Available: http://permanent.access.gpo.gov/ websites/epagov/www.epa.gov/epaoswer/non- hw/muncpl/faq.htm#7. 24 May 2005.

"Garbage Studies." *The University of Arizona Website*. Available: http://bara.arizone.edu/gs/ htm. 27 May 2005.

The Glass Packaging Institute, Recycling Website. Available: http://www.gpi.org/Recycling.html.

Hahn-Baker, David. "Rocky Roads to Consensus." *The Amicus Journal* 16.1 (1994): 41.

"Hazardous Waste." *Annenberg/CPB Learner. Org Website*. Available: http:www.learner. org/exhibits/garbage/hazardous.html. 24 May 2005.

The Aluminum Association, Inc Website. Available: http://www.aluminum.org. 27 May 2005.

Institute for Local Self-Reliance. *Beyond 40 Percent: Record-Setting Recycling and Composting Programs*. Washington, D.C.: Island Press, 1990.

"International Convention for the Prevention of Pollution from Ships, 1973, as Modified by the Protocol of 1978 Relating Thereto (MARPOL 73/78)." *International Maritime Organization Website*. Available:

http://www.imo.org/Conventions/contents. asp?doc_id=678&topic_id=258.

Kharbanda, O. P., and Ernest A. Stallworthy. *Waste Management: Toward a Sustainable Society*. New York: Auburn House, 1990.

Kirschner, Elisabeth. "Love Canal Settlement." *Chemical & Engineering News* 27 June 1994: 4–5.

Landay, Jonathan. "Iran May Pose First Test of Chemical-Arms Ban," *The Christian Science Monitor* 28 April 1997.

"London Convention." *USEPA Website*. Available: http://www.epa.gov/owow/OCPD/ icnetbri.html.

"MARPOL 73/78" *USEPA Website*. Available: http://www.epa.gov/OWOW/OCPD/marpol. html.

McCrady, Ellen. "Clinton Mandates Recycled Paper in Executive Branch, Circumvents Permanence Standards." *Abbey Newsletter* (October 1993). Available: http://palimpsest. stanford.edu/byorg/abbey/an/an17/an17-5/ an17-502.html. 27 May 2005.

"Municipal Solid Waste: Basic Facts." *USEPA Website*. Available: http://www.epa.gov/ep-aoswer/non-hw/muncpl/facts.htm. 24 May 2005.

"Municipal Solid Wastes - Basic Facts." *U.S. Environmental Protection Agency Website*. Available: http://www.epa.gov/epaoswer/non-hw/muncpl/facts. 27 May 2005.

"Municipal Solid Waste in the United States: 2001 Final Report (2001 data)." *USEPA Website*. Available: http://www.epa.gov/epaoswer/non-hw/muncpl/pubs/msw2001.pdf

"NPL Site Status Information." *EPA Superfund National Priority List*. Available: http://www.epa.gov/superfund/sites/npl/stuatus.htm. 27 May 2005.

Organisation for Economic Co-operation and Development, OECD Core Set. *Environmental Indicators*. Paris: OECD, 1994.

Piasecki, Bruce. *Beyond Dumping: New Strategies for Controlling Toxic Contamination*. Westport, Conn.: Quorum Books, 1984.

Rathje, William L. "Once and Future Landfills." *National Geographic* 179.5 (1991): 116–134.

Rathje, William L., and Cullen Murphy. *Rubbish! The Archaeology of Garbage*. New York: Harper Collins, 1992.

Reed, Sherwood C., E. Joe Middlebrooks, and Ronald W. Cities. *Natural Systems for Waste Management and Treatment*. New York: McGraw-Hill, 1988.

Reisch Mark and David Michael Bearden. "Superfund Fact Book." *Congressional Research Service* (3 March 1997). Available: http://www.ncseonline.org/NLE?CRSreports/Waste/waste-1.cfm?&CFID=229799&CFTOKEN=12014330. 27 May 2005.

"Scientists Find New Uses for Old Tires." *Earth Vision Reports* (4 December 1998). Available: http://www.gnet.org/ColdFusion/News_Page1. cfm?NewsID=5816. 7 December 1998.

Selke, Susan E. *Packaging and the Environment: Alternatives, Trends, and Solutions*. Lancaster, Penn: Technomic, 1990.

Senior, Eric. *Microbiology of Landfill Sites*. Boca Raton, Fla.: CRC Press, 1990.

Small, Gail. "War Stories." *The Amicus Journal* 16.1 (1994): 38.

"Status of International Conventions." *The International Tanker Owners Pollution Federation Limited Website*. Available: http://www.itopf. com/convstat.html. 17 February 1999.

"Steel continues to be the backbone of recycling in America." Press release, 2004. *Steel Recycling Institute Website*. Available: http://www. recycle-steel.org/2003rates.pdf.

Stockholm Convention on Persistent Organic Pollutants (POPs) Website. Available: http:// www.pops.int/

Trinh, Judy. "Energy from trash tires." *E: The Environmental Magazine Website*. Available: http://www.findarticles.com/p/articles/mi_ m1594/is_4_13/ai_90191331. 2002.

Warmer Bulletin. Journal of the World Resource Foundation. January 1998.

Wentz, Charles A., Jr. *Hazardous Waste Management*. New York: McGraw-Hill, 1989.

Wolf, Nancy, and Ellen Feldman. *Plastics: America's Packaging Dilemma*. Washington, D.C.: Island Press, 1990.

Credits

Photographs

Prologue

Figure P-1, NASA; Figure P-2b, Michael Holford Photographs; Figure P-2d, Michael Wright/Miami University; Figure P-3a, The Granger Collection, New York; Figure P-3b, Michael Holford Photographs; Figure P-3c, Adolph Gottleib, reprinted by permission of the Hakone Open-Air Museum

Chapter 1

Figure 1-1a, David Osborne/Miami University; Figure 1-1b, Hardy Eshbaugh/Miami University; Figure 1-1c, John Vankeat/Miami University; Figure 1-1d, Hardy Eshbaugh/Miami University; Figure 1-1e, David Osborne/Miami University; Figure 1-1g, Hardy Eshbaugh/Miami University; Figure 1-1h, James Foley/Miami University; Figure 1-3, Mark Edwards/Still Pictures; Figure 1-5, Patricia Kaufman; Figure 1-6, Michael Vanni/Miami University

Chapter 2

Figure 2-1, Carlos Guarita/Still Pictures; Figure 2-2, Courtesy of the Burton Historical Collection, Detroit Public Library; Figure 2-3, The Wilderness Society; Figure 2-4, Patricia Kaufman; Figure 2-5, Patricia Kaufman; Figure 2-6, Nigel Dickinson/Still Pictures; Figure 2-7, Nigel Dickinson/Still Pictures; Figure 2-8, Mary Nemeth; Figure 2-9, Hefner Zoology Museum/Miami University

Chapter 3

Figure 3-5, David Gorchov/Miami University; Figure 3-6, David Gorchov/Miami University; Figure 3-7, Hardy Eshbaugh/Miami University; Figure 3-8, Dolph Greenberg/Miami University; Figure 3-9, John Vankat/Miami University; Figure 3-10, Tischer, Gary/US Fish and Wildlife Services; Figure 3-11, Dolph Greenberg/Miami University; Figure 3-12, John Vankat/Miami University; Figure 3-13, Carole Katz and Jonathan Levy; Figure3-15, Brad Hinckley; Figure 3-16a, Gerald D. Tang; Figure 3-16b, D. Foster, WHO/Visuals Unlimited

Chapter 4

Figure 4-1, Dolph Greenberg/Miami University

Chapter 5

Figure 5-1, Assistant Regional Director-External Affairs/US Fish and Wildlife Services; Figure 5-2a, Dolph Greenberg/Miami University; Figure 5-2b, P. Frenzen/USDA Forest Service, 1999; Figure 5-7a, Jay Mager/Miami University; Figure 5-7b, Mr. Mohammed Al Momany/Agaba, Tordan of NOAA Photo Library; Figure 5-8, Greg Koch/US Fish and Wildlife Service, Red Wolf Recovery Program; Figure 5-9, Hardy Eshbaugh/Miami University; Figure 5-10, Thomas Collins; Figure 5-13a, Gerald and Buff Corsi/California Academy of Sciences; Figure 5-13b, Hardy Eshbaugh/Miami University; Figure 5-14 a and b, David Osborne/Miami University

Chapter 6

Figure 6-1, Paul C. Baumann/U.S. Fish and Wildlife Services; Figure 6-3, Division of Public Affairs/US Fish and Wildlife Service; Figure 6-4, John and Karen Hollingsworth/US Fish and Wildlife Service; Figure 6-6, James Foley/Miami University; Figure 6-7, Mark Edwards/Still Pictures; Figure 6-9, Steven Norris/Miami University

Chapter 7

Figure 7-1a, Matrix International, Inc./Karen Kasmauski; Figure 7-1b, Paul Berg; Figure 7-4, UN/DPI Photo; Figure 7-7, UN/DPI Photo; Figure 7-8, Jeff Widener/AP/Wide World Photos; Figure 7-9, Mark Edwards/Still Pictures; Figure 7-10, FAO Photo; Figure 7-11, FAO Photo; Figure 7-12, Jorgen Schytte/Still Pictures; Figure 7-13, UN/DPI Photo 1

Chapter 8

Figure 8-1, © UN/DPI Photo; Figure 8-2, Ralph Morse/Life Magazine © Time Inc.; Figure 8-5, © Secure the Future/Bristol-Myers Squibb Company; Figure 8-7, James Foley/Miami University; Figure 8-8, Friedrich Stark/Peter Arnold, Inc.; Figure 8-9, Patricia Kaufman

Chapter 9

Figure 9-1, Hardy Eshbaugh/Miami University; Figure 9-4, © Wolcott Henry; Figure 9-5, © GRACE; Figure 9-6, Mark Edwards/Still Pictures; Figure 9-7, © UN/DPI Photo; Figure 9-9, L. Groseclose/Courtesy The United Nations; Figure 9-10, Hardy Eshbaugh/Miami University; Figure 9-11, Susan Hoffman/Miami University; Figure 9-12, © Matt Bradley/Heifer International; Figure 9-13, © Darcy Kiefel/Hefner International; Figure 9-14, WFP/Crispin Hughes

Chapter 10

Figure 10-1, © Sarah Leen; Figure 10-4, Pennsylvania Historical & Museum Commission; Figure 10-7, Heldur Netocny/Still Pictures; Figure 10-8, Mark Edwards/Still Pictures; Figure 10-10a, Robert Harbison/The Christian Science Monitor; Figure 10-10b, © Michael Wright/Miami University

Chapter 11

Figure 10-1, © Sarah Leen; Figure 10-4, Pennsylvania Historical & Museum Commission; Figure 10-7, Heldur Netocny/Still Pictures; Figure 10-8, Mark Edwards/Still Pictures; Figure 10-10a, Robert Harbison/The Christian Science Monitor; Figure 10-10b, © Michael Wright/Miami University

Chapter 12

Figure 12-4, AP/MAX/Wide World Photos; Figure 12-6a, Austrian Alpine Club; Figure12-6b, Heinz Slupetzky; Figure 12-7, NASA: Goddard Space Flight Center; Figure 12-9, Mark Edwards/Still Pictures; Figure 12-10, Andre Maslennikov/Still Pictures; Figure 12-11a and b, South coast Air Quality Management District, California; Figure 12-15, Allegheny Conference on Community Development; Figure 12-17, Doug Bryant/D. Donne Bryant Stock;

Chapter 13

Figure 13-1a, Steven Norris/Miami University: Figure 13-1b, Dolph Greenberg/Miami University; Figure 13-1c, Steven Norris/Miami University; Figure 13-1e, Mr. Mohammed Al Momany, Agaba, Jordan/NOAA: The Coral Kingdom Collection; Figure 13-6, Hartmut Schwarzbach/Still Pictures; Figure 13-7, Aceh Besar (Lhok Nga)/FAO: Figure 13-8, David Gorchov/Miami University; Figure 13-9, Daniel W. Gotshall/Visuals Unlimited; Figure13-11a, Ron Nichols/U.S. Department of Agriculture; Figure 13-11b, Mark Edwards/Still Pictures; Figure 13-12, Dolph Greenberg/Miami University; Figure 13-15, The Orlando Sentinel; Figure 13-16, David Osborne/Miami University

Chapter 14

Figure 14-4a, Cecilia Franz/Miami University; Figure 14-4b, Kenneth Hanf; Figure 14-5, George Risen/NGS Image Collection; Figure 14-6, Cecilia Franz/Miami University; Figure 14-8, U.S. Department of Agriculture; Figure 14.9, Lynn Betts/U.S. Department of Agriculture; Figure 14-10, U.S. Department of Agriculture; Figure 14-12, The Land Institute

Chapter 15

Figure 15-1a, Ed Reschke; Figure 15-1b, Dennis Kunkel/PHOTOTAKE; Figure 15-3a, Dolph Greenberg/Miami University; Figure 15-3b, Wendell R. Haag/USDA Forest Service, Southern Research Station; Figure 15-5, Michael Wright/Miami University; Figure 15-6a, David Gorchov/Miami University; Figure 15-6b, Paul D. Heideman; Figure 15-7, Mark Edwards/Still Pictures; Figure 15-8, Michael Wright/Miami University; Figure 15-9, Hardy Eshbaugh/Miami University; Figure 15-10, Viesti Associates; Figure 15-11, Andy Jones; Figure 15-12, David Osborne/Miami University; Figure 15-13, David Osborne/Miami University; Figure 15-14, Nicholas de Vore III/Bruce Coleman, Inc.; Figure 15-15, David Osborne/Miami University; Figure 15-16, Jeff Davis/Northwest School District, Cincinnati; Figure 15-17, Dolph Greenberg/Miami University; Figure 15-18, Andre Bartschi/Still Picture; Figure 15-19, Robert Maier/ANIMALS ANIMALS; Figure 15-20, David Gorchov/Miami University; Figure 15-21, Bates Littlehales/Earth Scenes; Figure 15-22, Randall B. Henne/Dembinsky Photo Associates; Figure 15-23, Environmental Investigation Agency; Figure 15-25, Michael Wright/Miami University; Figure 15-27, Michael Nichols Photography; Figure 15-28, Hardy Eshbaugh/Miami University; Figure 15-29, Louise Van Vliet/Miami University; Figure 15-30, The Phoenix Zoo; Figure 15-31, Mark Edwards/Still Pictures; Figure 15-32, Susan Rice/U.S. Fish and Wildlife Service

Chapter 15 – Gallery

Page 334 (left–right), Hardy Eshbaugh/Miami University; Dolph Greenberg/Miami University; Dolph Greenberg/Miami University; Hardy Eshbaugh/Miami University. Page 335 (left–right), Robert Scholl; Louise Van Vliet/Miami University; Hardy Eshbaugh/Miami University; Dolph Greenberg/Miami University. Page 336 (left–right), Dolph Greenberg/Miami University; Donald Kaufman and Cecilia Franz/Miami University; Hardy Eshbaugh/Miami University; Lisa Breidenstein. Page 337 (left–right), Kenneth Hanf; Steven Norris/Miami University; Curtis Carley/U.S. Fish and Wildlife Service; Dolph Greenberg/Miami University. Page 338 (left–right), Alan Cady/Miami University; Steven Norris/Miami University; Hardy Eshbaugh/Miami University; Dolph Greenberg/Miami University. Page 339 (left–right), Hardy Eshbaugh/Miami University; Dolph Greenberg/Miami University; Dolph Greenberg/Miami University; David Cavagnaro. Page 354 (left–right), Dolph Greenberg/Miami University; Dolph Greenberg/Miami University; Steven Norris/Miami University; Hardy Eshbaugh/Miami University. Page 355 (left–right), Kenneth Hanf; Dolph Greenberg/Miami University; Alice Kahn/Miami University; Dolph Greenberg/Miami University. Page 356 (left–right), Dolph Greenberg/Miami University; Dolph Greenberg/Miami University; Dolph Greenberg/Miami University; Dolph Greenberg/Miami University. Page 357 (left–right), David Cavagnaro; Alan Cady/Miami University. Page 358 (left–right), Division of Wildlife/Ohio Department of Natural Resources; Donald Kaufman and Cecilia Franz/Miami University; Dolph Greenberg/Miami University; Hardy Eshbaugh/Miami University. Page 359 (left–right), Dolph Greenberg/Miami University; Susan Hoffman/Miami University; Lisa Breidenstein; Hardy Eshbaugh/Miami University. Page 360 (left–right), Dolph Greenberg/Miami University; Dolph Greenberg/Miami University; David Cavagnaro; Dolph Greenberg/Miami University. Page 362 (left–right), Donald Kaufman and Cecilia Franz/Miami University; Dolph Greenberg/Miami University; Alice Kahn/Miami University; David Osborne/Miami University. Page 364 (left–right), Dolph Greenberg/Miami University; David Osborne/Miami University; Donald Kaufman and Cecilia Franz/Miami University; Bruce Bunting/World Wildlife Fund. Page 365 (left–right), Dolph Greenberg/Miami University; David Russell/Miami University; Dolph Greenberg/Miami University; Dolph Greenberg/Miami University. Page 366 (left–right), Donald Kaufman and Cecilia Franz/Miami University; Mike Boylan/U.S. Fish and Wildlife Service; Division of Wildlife/Ohio Department of Natural Resources; Steve Maslowski/U.S. Fish and Wildlife Service. Page 367 (left–right), Michael Wright/Miami University; David Cavagnaro; Dolph Greenberg/Miami University; Edward S. Ross. Page 368 (left–right), Michael Wright/Miami University; Division of Wildlife/Ohio Department of Natural Resources; Edward S. Ross; Hardy Eshbaugh/Miami University. Page 369 (left–right), Donald Kaufman and Cecilia Franz/Miami University; Donald Kaufman and Cecilia Franz/Miami University; Archbold Biological Station/U.S. Fish and Wildlife Service; Division of Wildlife/Ohio Department of Natural Resources. Page 370 (left–right), David Russell/Miami University; John and Karen Hollingsworth/U.S. Fish and Wildlife Service; Dolph Greenberg/Miami University; Division of Wildlife/Ohio Department of Natural Resources. Page 371 (left–right), Dolph Greenberg/Miami University; Dolph Greenberg/Miami University; Michael Wright/Miami University; Edward S. Ross. Page 372 (left–right), Michael Wright/Miami University; Edward S. Ross; Edward S. Ross; Donald Kaufman and Cecilia Franz/Miami University. Page 373 (left–right), Dennis Kunkel/PHOTOTAKE; Joseph Jacquot/Miami University; David Cavagnaro; Orie Loucks/Miami University. Page 374 (left–right), Hardy Eshbaugh/Miami University; Michael Wright/Miami University; Edward S. Ross; Evi Buckner. Page 375, (left–right), U.S. Fish and Wildlife Service; David Cavagnaro; Donald Kaufman and Cecilia Franz/Miami University; Edward S. Ross. Page 376 (left–right), David Cavagnaro; Division of Wildlife/Ohio Department of Natural Resources; David Cavagnaro; Division of Wildlife/Ohio Department of Natural Resources. Page 377 (left–right), D. Diwurst/U.S. Fish and Wildlife Service; Hardy Eshbaugh/Miami University; Thomas Wissing/Miami University; Louise Van Vliet/Miami University.

Chapter 16

Figure 16-1, Steven Norris/Miami University; Figure 16-2, U.S. Fish and Wildlife Service; Figure 16-3, Dolph Greenberg/Miami University; Figure 16-4, Debbie McCrensky/U.S. Fish and Wildlife Service; Figure 16-5, Rob Schallenberger/U.S. Fish and Wildlife Service; Figure 16-6, Dolph Greenberg/Miami University; Figure 16-7, U.S. Department of the Interior, Bureau of Land Management; Figure 16-8, Dolph Greenberg/Miami University; Figure 16-9, Cecilia Franz/Miami University; Figure 16-11, Bob Stevens/U.S. Fish and Wildlife Services; Figure 16-12, U.S. Department of the Interior, Bureau of Land Management; Figure 16-13, Bill Haskins/Wildlands Center for Preventing Roads; Figure 16-14, Daniel Dancer/Still Pictures; Figure 16-16, John Vankat/Miami University; Figure 16-17, U.S. Department of the Interior, Figure 16-18, Dolph Greenberg/Miami University;

Chapter 17

Figure 17-1, Heldur Netocny/Still Pictures; Figure 17-2, Ruth McCleod; Figure 17-3, Dolph Greenberg/Miami University; Figure17-4, Nigel Dickinson/Still Pictures; Figure 17-5, Dolph Greenberg/Miami University; Figure 17-6, James Nachtwey/Magnum Photos; Figure 17-7, Dolph Greenberg/Miami University; Figure 17-8, Dolph Greenberg/Miami University; Figure 17-9a, Louise Van Vliet/Miami University; Figure 17-9b, Dolph Greenberg/Miami University

Chapter 18

Figure 18-1, D. Decobecq/Still Pictures; Figure 18-3, Hardy Eshbaugh/Miami University; Figure 18-4, Julio Etchart/Still Pictures; Figure 18-5, Patricia Kaufman; Figure 18-6, Mark Edwards/Still Pictures; Figure 18-7, Courtesy AT&T, Bell Laboratories

Chapter 19

Figure 19-1, Michael Goitein/Massachusetts General Hospital/National Cancer Institute; Figure 19-6, Michael Goitein/Massachusetts General Hospital/National Cancer Institute; Figure 19-7, Karen Kasmauski/Woodfin Camp & Associates

Chapter 20

Figure 20-2, Institue of Environmental Sciences/Miami University; Figure 20-3, Patricia Kaufman; Figure 20-4, FAO Photo; Figure 20-5, Mark Edwards/Still Pictures; Figure 20-6, Rueters/UPI/Bettmann; Figure 20-10, Hamilton County Solid Waste Management District; Figure 20-11, Clean Sites, Inc.

Illustrations, Tables, and Music Lyrics

All unit openers, Susan Friedmann

Grateful acknowledgment is hereby made to HarperCollins Publishers to reprint illustrations from *Biosphere 2000*, first edition.

Prologue

"Lyin' Eyes," by Don Henley and Glen Frey. Copyright © 1975 by Cass County Music/ Red Cloud Music. All rights reserved

Chapter 1

Table 1-1, Geology by Stanley Chernicoff. Copyright © 1999 by Houghton Mifflin Company. Used with permission

Chapter 3

Figure 3-1, Michael Wright/Miami University; Figure 3-2 a and b, Michael Wright/Miami University; Figure 3-4, Michael Wright/Miami University, Figure 3-17, Michael Wright/ Miami University

Chapter 4

Figure 4-2, Michael Wright/Miami University; Figure 4-3, Michael Wright/Miami University; Figure 4-4, Michael Wright/Miami University; Figure 4-5, Michael Wright/Miami University; Figure 4-6, Michael Wright/Miami University; Figure 4-7, Michael Wright/Miami University; Figure 4-8, Michael Wright/Miami University; Figure 4-9, Michael Wright/Miami University; Figure 4-10, Michael Wright/Miami University; Figure 4-11, Michael Wright/ Miami University; Figure 4-12, Michael Wright/Miami University; Figure 4-13, Michael Wright/Miami University

Chapter 5

Figure 5-3, Michael Wright/Miami University; Figure 5-4, Michael Wright/Miami University; Figure 5-5, Michael Wright/Miami University; Figure 5-6, Michael Wright/Miami University; Figure 5-11, Figure from Ecology and Our Endangered Life-Support Systems by Eugene P. Odum. Copyright © 1989 by Sinauer Associates, Inc., Publishers. Reprinted by permission of the publisher; Figure 5-12 a and b, Figure from Ecology and Our Endangered Life-Support Systems by Eugene P. Odum. Copyright © 1989 by Sinauer Associates, Inc., Publishers. Reprinted by permission of the publisher

Chapter 6

Figure 6-2 a and b, Michael Wright/Miami University;

Chapter 7

Figure 7-6, Adapted from Population Reference Bureau

Chapter 8

Figure 8-3, Adapted from Population Reference Bureau

Chapter 9

Figure 9-8, Michael Wright/Miami University

Chapter 10

Figure 10-3, Courtesy of Cincinnati Gas & Electric Company. Reprinted by permission.

Chapter 12

Figure 12-2, Michael Wright/Miami University; Figure 12-5, Adapted from State of the World 1999; Figure 12-12, "A Brief History of Lung Cancer" from Biology: The Science of Life, third edition, by Robert A. Wallace et al. Copyright © 1991 by HarperCollins Publishers, Inc.; Figure 12-14, Michael Wright/Miami University; Figure 12-16, Adapted from National Air Quality and Emissions Trends Report, 1197;

Chapter 13

Figure 13-4, Ohio Environmental Protection Agency; What You Can Do, From "Tips to Stem the Flow of Water Waste," Chemology, October 1991. Based on "50 Ways to Save a Drop," published by the Southwest Florida Water management District. Reprinted by permission.

Chapter 14

Figure 14-1, Michael Wright/Miami University; Figure 14-3, Susan Friedmann; Figure 14-11, "Ten-Year Crop Rotation" by Alexandra Schulz, as appeared in Harrowsmith Magazine, September/October 1989; Table 14-1, Adapted from Farming on the Edge, American Farmland Trust/Center for Agriculture in the Environment

Chapter 15

Figure 15-2, Susan Friedmann; Figure 15-26, Adapted from , Figure 15-4, *World Resources 1994-95*. Copyright © 1994 by the World Resources Institute. Reprinted by permission; Figure 15-24, Adapted from *Gaia: An Atlas of Planet Management* by Norman Myers

Chapter 16

Figure 16-15, Adapted from *Saving Our Forests* by Seth Zuckerman

Chapter 19

Figure 19-2, "Radiation" from Radiation: Doses, Effects, Risks by United Nations Environment Programme. Copyright © 1985 by United Nations Environment Programme. Reprinted with permission; Figure 19-3, Michael Wright/Miami University; Figure 19-4, Michael Wright/Miami University; Figure 19-9, United Nations Environment Programme

Chapter 20

Figure 20-1, U.S. Environmental Protection Agency

Index

A

Abiota, 8
Abiotic components, of ecosystem,
  52-53, 63
    abiotic factors, 63
    energy, 52-53
    limiting factors, 65-67
    matter, 53
    regulators, 65
Abortion, 156, 166-67
    legal vs. illegal, 166
Absaroka-Beartooth Wilderness Area, 369
Abundant metals, 405
Abyssal zone, of ocean, 277
Acadia National Park, 249, 380
Accidents, at power plants, 431-32
Acculturation, defined, 390
Acid drainage, 207
Acid fogs, 250
Acid precipitation, 248-52
    acid rain program, 264
    air pollution and, 248-52
    effects of, 251-52
    formation of, 249-50
    locations of effects of, 250
    management of, 264
    resistance to, 251
    transport of, 249-50
Acid rain. *See* Acid precipitation
Acid surges, 252
Acidified lake, 251
Acorn, as food, 97
Acorn moths, 97
Active system, solar energy, 224
Actual rate of increase, population, 135
Acullia angustipennis, 92
Acute pollution effects, 108
Acute toxicity, 444
Adirondack mountain range, 248, 251
Adjacent lands, and public lands
  management, 382
Admiral butterfly, 98
Adsorption, 289
Advanced materials, as mineral
  substitutes, 415
Advantages/disadvantages
    of biomass energy, 232
    of geothermal energy, 230
    of hydropower, 228
    of nuclear energy,223-24
    of ocean power, 231

of solar energy, 226
of solid waste 233
of wind power, 227
Aerobic decomposition, 445
Aerobic respiration, 71
Aesthetic quality, of water, 272
Aesthetics, of biological diversity, 346
Affluenza (PBS), 19
Afghanistan, IMRs in, 145
Africa
    central, 54-55
    east, 55, 97
    hunger in, 181
    leaded gasoline in, 243
    poaching in, 338
    southern, 58
    sub-Saharan, 133
    west, 54, 131, 139
African-American children, IMRs in, 146
African Green Revolution, 193-94
African rhinoceros, 354
African Sahel, 283
Africanized bees, 112
Agassiz National Wildlife Refuge, 59
Agave plant, 57, 391
Age, of nuclear resources, 425
Age distribution, of human population,
  133-35
Age-specific fertility rate, 132
Agrichemical crops, list, 308
Agricultural areas in peril, 303
Agricultural land management, history of
  U.S., 304
Agricultural Stabilization and Conservation
  Service (ASCS), 306
Agriculture
    biological diversity and, 348
    defined, 186-87
    early, 153
    effect on environment, 306-9
    history of, 186-88
AIDS, 142, 146, 156
    farming and, 174
    prostitution and, 149
    in sub-Saharan Africa, 157
Air emissions, locations of, 264
Air pollutant, defined, 240
Air pollution, 240-266
    acid precipitation, 248-52
    air toxics, 254
    climate change, 243-47
    controls, 261

factors affecting levels of, 258-60
human health and, 255
indoor, 256-58
management of, 260-66
primary pollutants, 241-43
smog, 252-54
stratospheric ozone depletion, 247-48
temperature inversions, 259-60
topography and, 259
weather and, 258
Air pollution, management of, 260-66
    acid precipitation, 264
    air toxics, 263-64
    Clean Air Act, 260
    Clear Skies Initiative, 264-65
    criteria pollutants, 260-63
    Kyoto Protocol, 265-66
    Montreal Protocol, 266
    Toxic Releases Inventory, 263-64
Air quality, 262-63
    economy and, 35
    improvements, 262
    standards, 262
    trends, 263
Air resources, 238-69
    air pollution, 240-66
    atmosphere, defined, 239
    atmosphere and life, 239-40
    discussion questions, 267-68
    key terms, 269
Air toxics, 254, 263-64
    effects of, 254
    hazardous air pollutants, 254
    management of, 263-64
Alaska, 29, 38, 61, 62, 87
    forests of, 61
    mineral revenues and, 414
    tundra, 362
Alaska National Interest Lands
  Conservation Act (ANILCA), 378
Alaska Peninsula, 119
Alaska Pipeline, 204, 217
Alaska Pulp Corporation, 379
Albedo, 240
Alcohol, as fuel source, 231
Alerce, 120
Algae, 280
Algal blooms, 66
Allegheny River, coal and, 210
Alpacas, 59
Alpha particles, 422
Alps, 246

Altamont Pass, California, 226
Alternative resources, of energy, 200
Aluminum, 406
    recycling of, 217
    solid waste stream, 440
Amazon
    forest, 117
    mining in, 411
    river basin, 54-55
    See also Tropical rain forest
Amazonia, Brazil, 326
American Antiquities Act, 366
American Association of Zoological Parks
    and Aquariums, 352
American crow, 356
American Farmland Trust (AFT), 302-3
American Forest and Paper Association, 377
American Forestry Association, 379
American Indian culture, 388
American Lung Association (ALA), 262-63
American Lung Association State of the Air
    (report), 262
American Minor Breeds Conservancy, 339
American Southwest, water in, 284-85
Amino acids, 171
Amu Dar'ya tributary, 284
Anaerobic decomposition, 445
Anaerobic respiration, 71, 287
Andes Mountains, 188-89, 246
Anemone, 330
Angola, 29
Animal unit month (AUM), 380
Animalia, 319
Animals, as food, 172
Ankara, Turkey, smog in, 252
Ant, 97, 356
Antarctic landscape, 409
Antelope, 59, 343
Anthropocentric worldview, 28
Anthropogenic degradation, 107
Anthropogenic pollutants, list, 247
Anthropology, of early life styles, 154
Antietam National Battlefield, 380
Antinatalist policy, 158
Antipater, 233
Antiquities Act, 393-94
ANWR (Arctic National Wildlife Refuge),
    29, 209, 214-15, 376
Apache, 391
Apalachicola River, 276
Appalachian mountains, 61, 248, 380
    ozone and, 254
Appalachian region, 390
    strip mining in, 410
    Foxfire (book series), 390
Applied ethnography, 395-96
Aquaculture, 191
Aquariums, 352
Aquifer, 275
    depletion of, 309
    types of, 275
    water classification, 275-76
Arabian oryx, 352
Aracama Desert, 57
Arachnids, 92
Aral Sea, 284
Aransas National Wildlife Refuge, 365

Araviapa Canyon, 322
Arbor vitae, 60
Arboreal mammals, of Costa Rica, 55
Archeological Resources Protection Act, 393
Arctic, warming of, 17
Arctic National Wildlife Refuge (ANWR),
    29, 209, 214-15, 376
Area strip mining, 409-10
Argiope, 92
Aridity zones, 115
Arkansas, 139
Arkansas River, 340
Arlington House, 393
Arms race, 429
Arms reductions, 429-430
Army Corps of Engineers, 366
Arnold Arboretum, 349
Arsenic, 207
Arthropods, 57
Aryan race, 159
Asbestos, 256, 258
Ash tree, 91, 95
Asian continent, water resources
    management on, 284
Associated gas, defined, 205
Aswan High Dam, 228
AT&T, 456
Ataturk Dam, 284
Athens, Greece, 388
Atmosphere, 238-40
    composition of, 8
    defined, 239
    of other planets, 240
    primitive, 10
Atolls, 330. *See also* Coral reefs
Atomic Energy Act, 233
Atomic Energy Comission (AEC), 429, 436
Atomic weapons, 419
"Atoms for Peace" (Eisenhower), 430
ATP (phosphorus cycle), 80
Attenborough, David, 87
Attractive force, of water, 272
Audobon Society, 375, 379
Australia, 55
    desert, 57
    endemic species in, 321
    forests of, 60
    grasslands, 59
    Great Barrier Reef in, 328
    growth rates, 131
    mutton in, 177
    southern, 58
    trees of, 55
Autotrophs, 63
Availability
    of coal, 203-4
    of natural gas, 206
    of petroleum, 204-5
Average global temperature, 243-45

B

Baby Boom, 134
Bacon, Francis, 32
Bacteria, indoor air pollution and, 256
Baer, Richard A., Jr., 29
Bagasse, 231

Bahamas, 99
Bahrain, 352
Baker, Howard, 357
Banana slug, 322
Bandow, Doug, 416
Bangladesh, 139
    hillsides in, 185
    hunger and, 181
    mangroves in, 335
    sea levels and, 246
    shrimp ponds in, 191
    starvation in, 175
Bardi Chapel, 392
Bare rock, and primary succession, 90
Barrel, defined, 200
Barrier methods, of birth control, 162
Barro Colorado Island, 55
Bascuit Bay, 331
Basidiomycete fungus, 342, 346, 355
Basketweaving, 328
Bass, 67, 342
Bat, 80, 341, 351
Bauxite, 406
Bean, winged, 348
Bear, 375, 398
Bear berry, 95
Beaver dams, 67
Beech-maple forest, 91
Beef, 177
Bees, social structure of, 98
Beetle, 92, 324, 327, 354
Beijing, air pollution in, 242
Bench terracing, 115
Bernarnos, George, 6
Best available technology (BAT), 286
Best conventional technology (BCT), 286
Beta particles, 422
Beymer, Betsy A., 143-45
Bhopal, India, 17, 444, 448
Bibliography, B1-B16
Big Bang theory, 10
Bilge water, in Great Lakes, 340
Binding force, 221
Bingham Canyon copper mine, 409
Bioaccumulation, 109
Biocentric worldview, 29
Bioclimatic exhibits, in zoos, 351
Biodegradable materials, 111, 445-47
Biodiversity, 319, 344-49
Biogas, 232
Biogeochemical cycles, 77
Biological diversity, 319, 344-49
    aesthetics, 346
    agriculture, 348
    ecosystem services, 344-46
    ethics, 346
    evolutionary potential, 346
    industry, 349
    medicine, 348-49
    reasons to preserve, 344-49
    See also Diversity
Biological oxygen demand (BOD),
    279-80
Biological resources, 318-60
    beliefs/attitudes about, 321-23
    biodiversity, 319, 344-49
    charismatic megafauna, 321

Funding, of public lands management 373-74
Fungi, 319
Fungicides, 308
Fungus gnat larvae, 97
Fusion reaction, 422-23
Future use
    of biomass energy, 231-32
    of geothermal energy, 230
    of hydropower, 227-28
    of nuclear energy, 222-23
    of ocean power, 230-31
    of solid waste, 232
    of wind power, 226

G

Gaia, 297
Gaia hypothesis, 240
Gaia Peace Atlas, 198
Gallatin National Forest, 369
Gallium arsenide, 226
Gall-making fly, 92-93
Game animals, 322
Gamma radiation, 422
Gandhi, Indira, 159
Ganges-Brahmaputra River basin, 186
Gannets, 355
Garbage
    burning of, 447
    decomposition of, 449
    defined, 439
    solid waste, 439
Garden of Intelligence, 350
Garter snake, 353
Gas (natural)
    industries using, 205
    reserves in U.S., 206
    use of, 205
    world reserves, 206
Gas prices, 214
Gaseous cycles, 78-80
Gasoline, leaded, 243
Geese, Canada, 349
Genangan area of Indonesia, 185
Gene, defined, 321
Gene bank, 189, 191, 321
General fertility rate, 132
General Land Office, 368
General Motors, 20
Genesis, Book of, 28
Genetic diversity, 321
    biological resources, 321
    of food crops and livestock, 189-91
    and soil resources management, 309-10
Genetic engineering, 339
Genetic erosion, biological resources, 344
Genetic library, 339
Genetic material, 80
Genetic modification (GM), 192
Genotype, 321
Geothermal energy, 229-30
    advantages/disadvantages, 230
    drawbacks to, 230
    future use of, 230
    present use of, 230
Geothermal power plant, 229
Germ plasm, 321

Germany
    acid precipitation and, 250
    population profile, 134
Gettysburg National Battlefield, 364
Giant clam, 339
Giardia lambia, 279
Giardiasis, 279
Gienaga Grande, 332
Gila National Forest, 369
Giotto di Bondone, 392
Glacial till, 90
Glacier(s), 246
    acid precipitation and, 249
    receding of, 247
Glacier Bay, 342
Glacier Bay National Park, 61
Glacier National Park, 339, 364
Glass, in solid waste stream,441
Glass Packaging Institute, 441
Global Coral Reef Monitoring
    Network, 108
Global environment, 8
Global Footprint Network, 16
Global hunger, failure to respond to, 182
Glossary, G1-G18
God Committee, 357
Goddard Institute for Space Studies, 245
Gold mining, 411, 413
Golden-cheeked warbler, 350
Golden eagles, wind farms and, 227
Golden rice, 192
Goldenrod fields, 92-93
Goldfinches, 93
Goldfish, 67
Goldilocks effect, 240
Gonservation International (CI), 323
Goodall, Jane, 345
Government, as social factor, 37-38
Government involvement, in cultural
    resources management, 393
Grail Bird, 355
Grain
    food resources,173-75
    imports/exports, 175
    prices, 175
    production, locations of, 174
    radiation and, 425
    sorghum, 307
    stocks, measurement of, 174
Grand Canyon, 364, 380
Grand fir, 60
Grand Tetons National Park, 344
Grande Carajas program, Brazil, 325
Grass, 334
Grasses, of tundra, 62
Grasshoppers, 64, 73
Grassland(s), 58
    Australia, 59
    temperate, 58-59
Grass-pink, 99
Gray whale, 113
Grazing, on public land, 367-68
Grazing Service, 368
Great Barrier Reef, 328
Great Basin National Park, 375
Great egret, 321
Great Lakes, 66, 110, 112, 286, 362
    bilge water in, 340

development of coastal areas at, 290
Great Lakes Legacy Act, 287
Great Lakes region, injection wells in, 452
Great Plains, 31, 175, 304
Great Smoky Mountains, 343, 380
    ozone and, 254
Great Smoky Mountains National Park, 60,
    249, 380, 398
Great spangled fritillary, 338
Greece, 89
Green frog, 348
Green manure, 311
Green Revolution, 187-88, 193-94
    in Africa, 193-94
Green sea turtle, 366
Green taxes, 37
Green treefrog, 352
Green vine snake, 102
Greenberg, Adolph, 396
Greenhouse effect, 240
Greenhouse gases, 207, 243-44
    China and. 266
    coal and, 207
    trends in, 243-44
Greenland iris, 319
Greenland National Park, 353
Grey parrot, 340
"Grim Payback of Greed" (Durning), 127
Gristmill, water-powered, 233
Grizzly bear, 375
Grosbeak, 337
Gross national income in purchasing power
    parity (GNI PPP), 139
Gross national product (GNP), 34, 194
Gross primary productivity (GPP), 71
Groundfish, 176
Groundwater, 275
    classification of, 275
    defined, 272
    depletion, 309
    diagram of, 276
    management of, 290
    management, 290-91
Growth rates, of human population, 128-36
    age distribution, 133-35
    factors in human population, 132-36
    fertility, 132-33
    migration, 135-36
    population measurement, 128-31
    world population, 130
Guadalupe Mountains National Park,
    57, 318
Guano, 80
Guatemala, 37, 228
Guinea, hydropower in, 228
Guinea-Bissau, fertility rate, 133
Gulf Coast, injection wells on, 452
Gulf of Mexico, 66, 290
Gulf states, dead zone and, 66
Gunnison National Park, 228
Gunnison River, 292

H

Habitat, 321
    coral reef, 330
    defined, 49
    loss of, 323, 337

India, 20
    hunger and, 181
    population in, 139
    wheat exports from, 184
Indian Ocean, 331
Indicated reserves, 202
Indicator species, defined, 66
Indifference, to species variety, 322
Indirect radiation exposure, 425
Individuals, and fossil fuel efficiency, 217
Indonesia, 117
    fertility rate, 133
    hunger and, 181
    soil erosion in, 185
    tsunamis in, 182
Indonesian archipelago, 54, 327
Indoor air pollution, 256-58
    asbestos, 258
    defined, 256
    effects of, 257-58
    formaldehyde, 257
    radon 222, 257
    sources of, 256
    tobacco smoke, 258
Indus River basin, 186
Industrial Revolution, 7, 32, 154, 211
    extinctions since, 341
    greenhouse gases and, 244
Industry, 216
    biological diversity and, 349
    energy conservation and, 216
    and fossil fuels 216
    industrial smog, 252
    industrial uses of mineral
        resources, 407
Inertia, 97
Inez, Kentucky, 108
Infant mortality rate (IMR), 144-46
Infant/childhood mortality, 144-46
Infanticide, 154
Infectious substances, 444
Inferred reserves, 202
Initiation ceremony, 29
Inland wetlands, 331
Inscription Rock, 365
Insecticides, 308
Insects
    aquatic, list, 66
    goldenrods and, 92
    See also individual species
Inslee, Jay, 414
Integrated pest management, 313
Internal costs, 35
International Conference on Population
    and Development (ICPD), 147, 156,
    159-61
International Convention for the
    Prevention of Pollution from Ships, 453
International Food Policy Research
    Institute, 280
International Irrigation Management
    Institute, 309
International minerals industry, 412-14
International Seabed Authority (ISA), 416
International Species Inventory System
    (ISIS), 351-52
Interspecific competition, 98-100

Intraspecific competition, 98
Intraspecific cooperation, 99
Intrauterine device (IUD), 166
Introduced species, biological resources,
    339-40
Inversions, temperature, 259-60
Investing in Development (report), 193
Iodine deficiency, 181
Ionizing radiation, 421, 424
Iraq, IMRs in, 145
Iron Bridge Sewage Treatment Plant, 292
Iron deficiency, 181
Iron/steel, in solid waste stream, 442
Irrigation, 114, 309
ISIS, 351-52
Isotopes, 421-22
Israel, 57, 352
Ivory-billed woodpecker, 355

J

Jackson, Wes, 315
Jambias, 338
Jane Goodall Institute, 345
Japanese beetle, 92, 112
Java, hillside erosion in, 185
Jay, Florida scrub, 353
Jefferson, Thomas, 363
Jellyfish, 318
    plastic and, 281
John Paul II, 26
Johns Hopkins School of Hygiene and
    Public Health, 246
Johns Hopkins University, 166
Johnson, Lyndon, 370
Joint European Torus (JET), 423
Jornada Mogollon people, 394
Joshua Tree, 57, 365
Judaism, 28
Judeo-Christian belief, 28

K

Kangaroo, 59
Kathmandu, Nepal, 278
Kayaop Indians, 189
Kazakstan, 161, 177
Kazi Ya Mwanamke, 143-45
Kelp, 340
Kemp's Ridley sea turtle, 356
Kendall, Henry, 152
Kennedy, John F., 26
Kenya, 116, 135
    growth rate in, 130
    ICPD and, 160
Kerogen, 206
Key terms
    air resources, 269
    biological resources, 360
    cultural resources, 400
    ecosystem degradation, 122
    ecosystem development, 106
    ecosystem function, 86
    ecosystem structure, 69
    energy, alternative sources of, 236
    food resources, 197
    fossil fuels, 219

    human population dynamics 151, 169
    mineral resources, 418
    nuclear resources, 437
    public lands, 385
    social factors, of biosphere, 46
    soil resources, 317
    state of biosphere, 25
    unrealized resources, 461
    water resources, 296
Keystone species, 67
"Killer" bees, 112
Kilowatt (kw), defined, 199
Kilowatt-hour (KWH), defined, 199
King crab, Alaskan, 177
King penguins, 322
Kingdoms, of living organisms, 319-20
Koyukuk National Wildlife Refuge, 87
Krygyz Republic, 161
Kubla Khan, palace of, 19
Kunming, China, 35
Kwashiorkor, 180
Kyoto Protocol, 38, 265-66
Kyrgyzstan, mutton in, 177

L

Lake(s). *See also individual lakes*
    ecosystem development, 95-96
    parts of, 273
    thermal stratification in, 274
Lake Erie, 66, 81, 95
Lake Superior, 284
Lake Tahoe, 108
Lake Victoria, 284, 339
Lake Washington, 81
Lampsilis cardium, 322
Land classification system, 302
Land ethic, 32
Land Institute, 315
Land management, in U.S., 304
Land races, defined, 186
    land use effect on soil, 302-3
Land uses, list, 376
    land-based repository, nuclear resources
        management, 434
Landfill, 447-51
    design of, 451
    secure, 449
Landscape, defined, 51
Language, 48
Lanzhou, China, air pollution in, 243
Laos, IMRs in, 145
Lappe, Frances Moore, 186
Largemouth bass, 342
Larval fish, 18
Lascaux, France, 153
Latin America, 112. *See also individual*
    countries
    hunger in, 181
    population and, 140
Latino population, in U.S., 136
Laudholm Beach, 17
Laurance, William F., 119
Law of conservation of matter, 53
Law of the Sea, 416
Law of tolerances, 66-67
LDCs. *See* Less-developed countries

N